T0140070

Advances in Intelligent Systems and Computing

Volume 916

Series Editor

Janusz Kacprzyk, Systems Research Institute, Polish Academy of Sciences, Warsaw, Poland

Advisory Editors

Nikhil R. Pal, Indian Statistical Institute, Kolkata, India
Rafael Bello Perez, Faculty of Mathematics, Physics and Computing, Universidad Central de Las Villas, Santa Clara, Cuba
Emilio S. Corchado, University of Salamanca, Salamanca, Spain
Hani Hagras, Electronic Engineering, University of Essex, Colchester, UK
László T. Kóczy, Department of Automation, Széchenyi István University, Gyor, Hungary
Vladik Kreinovich, Department of Computer Science, University of Texas at El Paso, EL PASO, TX, USA
Chin-Teng Lin, Department of Electrical Engineering, National Chiao Tung University, Hsinchu, Taiwan
Jie Lu, Faculty of Engineering and Information Technology, University of Technology Sydney, Sydney, NSW, Australia
Patricia Melin, Graduate Program of Computer Science, Tijuana Institute of Technology, Tijuana, Mexico
Nadia Nedjah, Department of Electronics Engineering, University of Rio de Janeiro, Rio de Janeiro, Brazil
Ngoc Thanh Nguyen, Faculty of Computer Science and Management, Wrocław University of Technology, Wrocław, Poland
Jun Wang, Department of Mechanical and Automation Engineering, The Chinese University of Hong Kong, Shatin, Hong Kong

The series "Advances in Intelligent Systems and Computing" contains publications on theory, applications, and design methods of Intelligent Systems and Intelligent Computing. Virtually all disciplines such as engineering, natural sciences, computer and information science, ICT, economics, business, e-commerce, environment, healthcare, life science are covered. The list of topics spans all the areas of modern intelligent systems and computing such as: computational intelligence, soft computing including neural networks, fuzzy systems, evolutionary computing and the fusion of these paradigms, social intelligence, ambient intelligence, computational neuroscience, artificial life, virtual worlds and society, cognitive science and systems, Perception and Vision, DNA and immune based systems, self-organizing and adaptive systems, e-Learning and teaching, human-centered and human-centric computing, recommender systems, intelligent control, robotics and mechatronics including human-machine teaming, knowledge-based paradigms, learning paradigms, machine ethics, intelligent data analysis, knowledge management, intelligent agents, intelligent decision making and support, intelligent network security, trust management, interactive entertainment, Web intelligence and multimedia.

The publications within "Advances in Intelligent Systems and Computing" are primarily proceedings of important conferences, symposia and congresses. They cover significant recent developments in the field, both of a foundational and applicable character. An important characteristic feature of the series is the short publication time and world-wide distribution. This permits a rapid and broad dissemination of research results.

** Indexing: The books of this series are submitted to ISI Proceedings, EI-Compendex, DBLP, SCOPUS, Google Scholar and Springerlink **

More information about this series at http://www.springer.com/series/11156

Michael E. Auer · Thrasyvoulos Tsiatsos
Editors

The Challenges of the Digital Transformation in Education

Proceedings of the 21st International Conference on Interactive Collaborative Learning (ICL2018) - Volume 1

 Springer

Editors
Michael E. Auer
Carinthia University of Applied Sciences
Villach, Kärnten, Austria

Thrasyvoulos Tsiatsos
Department of Informatics
Aristotle University of Thessaloniki
Thessaloniki, Greece

ISSN 2194-5357   ISSN 2194-5365   (electronic)
Advances in Intelligent Systems and Computing
ISBN 978-3-030-11931-7   ISBN 978-3-030-11932-4   (eBook)
https://doi.org/10.1007/978-3-030-11932-4

Library of Congress Control Number: 2018968529

© Springer Nature Switzerland AG 2020
This work is subject to copyright. All rights are reserved by the Publisher, whether the whole or part of the material is concerned, specifically the rights of translation, reprinting, reuse of illustrations, recitation, broadcasting, reproduction on microfilms or in any other physical way, and transmission or information storage and retrieval, electronic adaptation, computer software, or by similar or dissimilar methodology now known or hereafter developed.
The use of general descriptive names, registered names, trademarks, service marks, etc. in this publication does not imply, even in the absence of a specific statement, that such names are exempt from the relevant protective laws and regulations and therefore free for general use.
The publisher, the authors and the editors are safe to assume that the advice and information in this book are believed to be true and accurate at the date of publication. Neither the publisher nor the authors or the editors give a warranty, express or implied, with respect to the material contained herein or for any errors or omissions that may have been made. The publisher remains neutral with regard to jurisdictional claims in published maps and institutional affiliations.

This Springer imprint is published by the registered company Springer Nature Switzerland AG
The registered company address is: Gewerbestrasse 11, 6330 Cham, Switzerland

Preface

ICL2018 was the 21st edition of the International Conference on Interactive Collaborative Learning and the 47th edition of the IGIP International Conference on Engineering Pedagogy.

This interdisciplinary conference aims to focus on the exchange of relevant trends and research results as well as the presentation of practical experiences in Interactive Collaborative Learning and Engineering Pedagogy.

ICL2018 has been organized by Aristotle University of Thessaloniki, Greece, from 25 September to 28 September 2018, in Kos Island.

This year's theme of the conference was "Teaching and Learning in a Digital World".

Again outstanding scientists from around the world accepted the invitation for keynote speeches:

- Stephanie Farrell, Professor and Founding Chair of Experiential Engineering Education at Rowan University (USA)—2018–19 President of the American Society for Engineering Education. Speech title: Strategies for Building Inclusive Classrooms in Engineering.
- Demetrios Sampson. Ph.D. (ElectEng) (Essex), PgDip (Essex), B.Eng./M. Eng. (Elec) (DUTH), CEng—Golden Core Member, IEEE Computer Society—Professor, Digital Systems for Learning and Education, University of Piraeus, Greece. Speech title: Educational Data Analytics for Personalized Learning in Online Education.
- Rovani Sigamoney, UNESCO Engineering Programme. Speech title: UNESCO—Engineering the Sustainable Development Goals.

In addition, three invited speeches have been given by

- Hans J. Hoyer, IFEES, United States of America. Speech title: The work of IFEES and GEDC towards a new Quality in Engineering Education.
- David Guralnick, Kaleidoscope Learning, United States of America. Speech title: Creative Approaches to Online Learning Design.

Furthermore, five very interesting workshops and one tutorial have been organized:

- Tutorial titled "Improving Practical Communication Skills Through Participation in Collaborative English Workshops" by Edward Pearse Sarich (Shizuoka University of Art and Culture, Japan); Mark Daniel Sheehan (Hannan University, Japan), and Jack Ryan (Shizuoka University of Art and Culture, Japan).
- Workshop titled "Evaluation of Experimental Activities by Diana Urbano and Maria Teresa Restivo" (University of Porto, Portugal).
- Workshop titled "Machine Learning and Interactive Collaborative Learning" by Panayotis Tzinis and Irene Tsakiridou (Google Developer Experts).
- Workshop titled "Teaching and Learning Electrical Engineering and Computer Science in High School with a STEM Approach" by Arturo Javier Miguel-de-Priego (Academia de Ingeniería y Ciencia Escolar, Perú).
- Workshop titled "Introduction to BLE System Design Using PSoC® 6 MCUs" by Patrick Kane (Cypress).

Since its beginning, this conference is devoted to new approaches in learning with a focus to collaborative learning and engineering education. We are currently witnessing a significant transformation in the development of education. There are two essential and challenging elements of this transformation process that have to be tackled in education:

- the impact of globalization on all areas of human life and
- the exponential acceleration of the developments in technology as well as of the global markets and the necessity of flexibility and agility in education.

Therefore, the following main themes have been discussed in detail:

- Collaborative learning
- Lifelong learning
- Adaptive and intuitive environments
- Ubiquitous learning environments
- Semantic metadata for e-learning
- Mobile learning environments applications
- Computer-aided language learning (CALL)
- Platforms and authoring tools
- Educational mashups
- Knowledge management and learning
- Educational virtual environments
- Standards and style guides
- Remote and virtual laboratories
- Evaluation and outcomes assessment

- New learning models and applications
- Research in engineering pedagogy
- Engineering pedagogy education
- Learning culture and diversity
- Ethics and engineering education
- Technical teacher training
- Academic–industry partnerships
- Impact of globalization
- K-12 and pre-college programmes
- Role of public policy in engineering education
- Women in engineering careers
- Flipped classrooms
- Project-based learning
- New trends in graduate education
- Cost-effectiveness
- Real-world experiences
- Pilot projects/Products/Applications.

The following special sessions have been organized:

- Entrepreneurship in Engineering Education (EiEE 2018)
- Digital Technology in Sports (DiTeS)
- Talking about Teaching 2018 (TaT'18)
- Multicultural Diversity in Education and Science
- Tangible and Intangible Cultural Heritage digitization and preservation in modern era (TICHE-DiPre)
- Advancements in Engineering Education and Technology Research (AEETR).

Also, the "1st ICL International Student Competition on Learning Technologies" has been organized in the context of ICL2018.

The following submission types have been accepted:

- Full paper, short paper
- Work in progress, poster
- Special sessions
- Round-table discussions, workshops, tutorials, doctoral consortium, students' competition.

All contributions were subject to a double-blind review. The review process was very competitive. We had to review near 526 submissions. A team of about 375 reviewers did this terrific job. Our special thanks go to all of them.

Due to the time and conference schedule restrictions, we could finally accept only the best 186 submissions for presentation.

Our conference had again more than 225 participants from 46 countries from all continents.

ICL2019 will be held in Bangkok, Thailand.

Michael E. Auer
ICL Chair
Thrasyvoulos Tsiatsos
ICL2018 Chair

Committees

General Chair

Michael E. Auer   CTI, Villach, Austria

ICL2018 Conference Chair

Thrasyvoulos Tsiatsos   Aristotle University of Thessaloniki, Greece

International Chairs

Samir A. El-Seoud   The British University in Egypt, Africa
Neelakshi Chandrasena   University of Kelaniya, Sri Lanka, Asia
  Premawardhena
Alexander Kist   University of Southern Queensland,
                                  Australia/Oceania
Arthur Edwards   Universidad de Colima, Mexico, Latin America
Alaa Ashmawy   American University Dubai, Middle East
David Guralnick   Kaleidoscope Learning New York, USA,
                                  North America

Programme Co-chairs

Michael E. Auer   CTI, Villach, Austria
David Guralnick   Kaleidoscope Learning New York, USA
Hanno Hortsch   TU Dresden, Germany

Technical Programme Chairs

Stavros Demetriadis   Aristotle University of Thessaloniki, Greece
Sebastian Schreiter   IAOE, France
Ioannis Stamelos   Aristotle University of Thessaloniki, Greece

IEEE Liaison

Russ Meier   Milwaukee School of Engineering, USA

Workshop and Tutorial Chair

Barbara Kerr   Ottawa University, Canada

Special Session Chair

Andreas Pester   Carinthia University of Applied Sciences, Austria

Demonstration and Poster Chair

Teresa Restivo   University of Porto, Portugal

Awards Chair

Andreas Pester   Carinthia University of Applied Sciences, Austria

Publication Chair

Sebastian Schreiter   IAOE, France

Senior PC Members

Andreas Pester   Carinthia University of Applied Sciences, Austria
Axel Zafoschnig   Ministry of Education, Austria
Doru Ursutiu   University of Brasov, Romania
Eleonore Lickl   College for Chemical Industry, Vienna, Austria
George Ioannidis   University of Patras, Greece
Samir Abou El-Seoud   The British University in Egypt
Tatiana Polyakova   Moscow State Technical University, Russia

Program Committee

Agnes Toth	Hungary
Alexander Soloviev	Russia
Anastasios Mikropoulos	Greece
Armin Weinberger	Germany
Athanassios Jimoyiannis	Greece
Charalambos Christou	Cyprus
Charalampos Karagiannidis	Greece
Chris Panagiotakopoulos	Greece
Christian Guetl	Austria
Christos Bouras	Greece
Christos Douligeris	Greece
Chronis Kynigos	Greece
Cornel Samoila	Romania
Costas Tsolakis	Greece
Demetrios Sampson	Greece
Despo Ktoridou	Cyprus
Dimitrios Kalles	Greece
Elli Doukanari	Cyprus
Hanno Hortsch	Germany
Hants Kipper	Estonia
Herwig Rehatschek	Austria
Igor Verner	Israel
Imre Rudas	Hungary
Ioannis Kompatsiaris	Greece
Istvan Simonics	Hungary
Ivana Simonova	Czech Republic
James Uhomoibhi	UK
Jürgen Mottok	Germany
Martin Bilek	Czech Republic
Matthias Utesch	Germany
Michalis Xenos	Greece
Monica Divitini	Norway
Nael Barakat	USA
Pavel Andres	Czech Republic
Rauno Pirinen	Finland
Roman Hrmo	Slovakia
Santi Caballé	Spain
Stavros Demetriadis	Greece
Teresa Restivo	Portugal
Tiia Rüütmann	Estonia
Vassilis Komis	Greece
Viacheslav Prikhodko	Russia

Victor K. Schutz	USA
Yiannis Dimitriadis	Spain
Yu-Mei Wang	USA

Local Organization Chair

Stella Douka   Aristotle University of Thessaloniki, Greece

Local Organization Committee Members

Hippokratis Apostolidis	Aristotle University of Thessaloniki, Greece
Agisilaos Chaldogeridis	Aristotle University of Thessaloniki, Greece
Olympia Lilou	Aristotle University of Thessaloniki, Greece
Andreas Loukovitis	Aristotle University of Thessaloniki, Greece
Angeliki Mavropoulou	Aristotle University of Thessaloniki, Greece
Nikolaos Politopoulos	Aristotle University of Thessaloniki, Greece
Panagiotis Stylianidis	Aristotle University of Thessaloniki, Greece
Christos Temertzoglou	Aristotle University of Thessaloniki, Greece
Efthymios Ziagkas	Aristotle University of Thessaloniki, Greece
Vasiliki Zilidou	Aristotle University of Thessaloniki, Greece

Contents

Engineering Pedagogy Education

Collaborative Learning

Work-in-Progress: Development of a Framework to Foster Collaborative Learning Among Engineering Students Using Moodle Mobile App

A. Peramunugamage[1(✉)], H. U. W. Ratnayake[2(✉)],
and S. P. Karunanayaka[2(✉)]

[1] University of Moratuwa, Colombo, Sri Lanka
anuradhask@uom.lk
[2] Open University of Sri Lanka, Colombo, Sri Lanka
{udithaw, spkar}@ou.ac.lk

Abstract. Students' learning patterns vary from person to person. Some students prefer to learn by seeing, hearing, reflecting, acting or reasoning logically while others prefer to learn by visualizing, drawing analogies and building mathematical models. Traditionally, Engineering education has focused on content-based or design-based approaches that allowed students to develop problem solving skills. Use of technology for instructional purposes or communication has not been very effective in Sri Lankan universities in the past. This research is to find out how we can improve Sri Lankan Engineering students' engagement in studies via collaborative learning by using mobile technology. In addition, research will compare the effectiveness of different online collaborative learning methodologies that use mobile based and web-based learning and teaching platforms in Sri Lanka.

Keywords: Collaborative learning · Engineering education ·
Mobile application introduction

1 Introduction

Students' learning patterns vary from person to person. Some students prefer to learn by listening to lectures, some prefer to learn by reading textbooks and notes; these can be supplemented by reflecting and acting, reasoning logically and intuitively, memorizing, visualizing and drawing analogies, and building mathematical models. The same applies to teachers. Some teachers mainly lecture, others demonstrate or discuss; some focus on principles and others on applications; some emphasize memory and others understanding. Therefore, understanding a concept will depend on the students' learning style and the teacher's teaching style (Felder and Silverman, 1988). Therefore, the teacher has to take on this challenge and help the students to learn the most effective way.

Twenty-first century students view themselves as participants in the creation of information and new ideas (Leadbeater and Wong, 2010). Therefore, they are active learners rather than mere listeners (Scott, 2015). Such interactive learning leads to the

© Springer Nature Switzerland AG 2020
M. E. Auer and T. Tsiatsos (Eds.): ICL 2018, AISC 916, pp. 3–13, 2020.
https://doi.org/10.1007/978-3-030-11932-4_1

development of metacognition by improving student participation, which tends to shift teacher-centered education into the collaborative mode. The advent of digital tools for learning and education, though a drastic change, has provided an opportunity for the student to carry his own devices to the classroom or any other place that is conducive to learning (Institute for the Future, 2013). As discussed above, students have different learning styles. Therefore, the teacher is obliged to find out the most suitable technology that can be readily adapted to match the learner's individual learning style. This requires them to move to a twenty-first century education system as it can offer a more personalized learning environment (Scott, 2015). However, technology is not the sole solution but it is certainly an enabler that can enhance the collaboration among students and teachers. As suggested by the research of Redecker et al. (2011), a mix of different technologies will be able to transform the learning process into different modes, as needed. Open Educational Resources (OER), web-based multimedia production, and distribution tools incorporating text, audio, photos, and videos, and the integration of social media are some of the strategies that can be used to implement a technology-enhanced education (Mcloughlin and Lee, 2007).

Powerful relationships can enhance the quality of learning, collaboration, communication, and competencies until the student masters these skills (Rand Education, 2012). To enhance the collaboration among teachers and learners, institutions used Learning Management Systems (LMS) that provided online discussion forums, electronic delivery of readings and assignments, and the electronic posting of graded assignments (Scott, 2015). Traditionally, Engineering education focused on content-based or design-based curriculum, either of which allowed students to develop problem-solving skills. Recently, accreditation bodies have added team building and collaborative problem-based learning to the undergraduate Engineering curriculum. However, in previous studies there seems to be a gap in coverage regarding Collaborative Mobile Learning in respect of Engineering Education in which highly technical subjects and skills development subjects are included.

2   Purpose or Goal

Use of technology for instructional design or communication has not been very effective in Sri Lankan universities. Mainly, it has been teacher-centered, theater based education in which teachers distributed printed learning materials or sent course material via the web or used a VLE such as MOODLE. Materials mainly distributed are PowerPoint slides and video links. Further, there is minimal interaction between students and the teacher due to several reasons. The tight schedule imposed by their studies, heavy workload and large classes has led to this situation. Though the universities have now adopted learning management systems (LMS) like Moodle, teachers are not able to take full advantage of same due to various reasons. Therefore, students are unable to derive the maximum benefit from these developments. This research attempts to find out how we can improve Sri Lankan Engineering students' absorption of course related knowledge via collaborative learning by using mobile technology. Further, what is the current status of electronic interactions (in terms of frequency and diversity of channels[1]) that occur among Engineering students and their peers and instructors? Also, how can Moodle Mobile App based instructional materials be designed and developed as tools to enhance student-student and student-instructor interactions in Engineering education?

3 Background

Higher educational institutions adopt technologies to enhance the operations of the institution or to make quantitative or qualitative changes to the learning environment or to achieve all three (Kirkwood and Price, 2016). Hence, use of technology may differ as it may be utilized to achieve blended learning, e-learning or hybrid courses, audio or podcasts, video resources or lectures or games, multimedia tools, virtual laboratories or fieldwork, blogs, collaborative tools or wikis, online discussion boards or conferences or forums, e-Portfolios, online course resources, electronic voting or personal response systems and assistive technologies (Price and Kirkwood, 2014). Universities and Colleges Information Systems Association (UCISA) have done several longitudinal analyses based on surveys that were conducted from 2001 to 2016. The top three barriers to Technology Enhanced Learning (TEL) developments are lack of time, money and the requisite knowledge by academic staff. They have found that Moodle is the most popular Virtual Learning Environment (VLE) platform while commercial VLEs (Blackboard Classic, WebCT and Version 9) and enterprise or institutional VLEs are also commonly used in UK universities (Voce et al., 2016). Likewise, Moodle is the most widely used VLE in Sri Lanka. It is a free and open source educational platform that embeds common types of learning resources and various educational activities; it can also be upgraded by various modules and plugins and is able to collect various logging data for all student interactions. These are the reasons for its popularity.

Traditionally, engineering education has been content-centered, design-oriented, and permeated by an emphasis on problem-solving skills. More recently, team building concepts and collaborative problem-solving techniques have been added (Bourne et al., 2005). The fact is that the learning styles of most engineering students and the teaching styles of most engineering professors are incompatible in several respects. Many or most engineering students are inclined towards visual, sensory, inductive, and active approaches, while some of the most creative students adopt a global approach. In contrast, most engineering education is based on auditory, abstract (intuitive), deductive, passive, and sequential approach. Therefore, there is a huge mismatch between engineering teachers and students (Felder et al., 2000). These incompatibilities lead to poor student performance, professorial frustration, and a loss to society of many potentially excellent engineers (Felder and Silverman, 1988). Twenty-first century engineering students need to have strong communication and teamwork skills, which they lack (Bourne et al., 2005).

To address these two issues, the accreditation bodies revised their guidelines and advocated a mixed-mode approach by using both traditionally taught courses, and

---

[1] SMS, phone call, email, Moodle, WhatsApp, Viber, Skype, etc.

adding project-based components to satisfy industry needs (Felder and Silverman, 1988). Nevertheless, the adoption of quality online learning has been slow to take hold in the engineering education field (Bourne et al., 2005). Undergraduate engineering programs lag behind in adopting online methodologies because these cannot be readily applied to laboratory work and mathematical courses, or used with design tools. However, there is much potential for the use of technology in online engineering education in future. These barriers can be overcome because technology has already contributed much to support online engineering education by making possible virtual laboratories, computer simulations and augmented reality applications (Bourne et al., 2005). The free and open source educational platform called Moodle will be used in this study to support collaborative learning as it is now commonly used by teachers and students worldwide (Kasimatis et al., 2010). According to the preliminary study done in 2017, it was found that practically all the engineering faculties in Sri Lanka used the Moodle web-based LMS for course material sharing, module registration, assignment submission, etc. Since Moodle is independent of operating systems, the end users can access it through most web browsers (e.g. Firefox, Internet Explorer, Chrome, etc.). Moreover it embeds many common types of learning resources and various educational programs that can be used to create well-structured learning scenarios. It contains many activities and communication modules, collects various logging data for all student interactions, contains advanced grading methods like rubrics and enriched rubrics and enables teachers to provide a variety of activities such as quizzes, electronic journals etc. online, both with and without time limits (Al-dous and Samaka, 2015).

3.1   Mobile Collaborative Learning

As described by Dyckhoff et al. (2012), mobile learning becomes interactive when used in a collaborative environment (Lee et al., 2011; Dyckhoff et al., 2012). It provides both formal and informal learning associated with the interaction and exchange of information. Spikol (2008) describes that collaborative learning through mobile devices has been investigated mainly because of the universal availability, versatility and mobility offered by these devices. According to Jain et al. (2011), collaborative mobile learning is an activity that allows transparent collaboration by empowering the social negotiation space of group members, by enabling coordination between their activity states, encouraging members' mobility, facilitating mediation in interactivity, allowing organization of the managed material and enabling students to collaborate in groups through wireless networks supporting social face-to-face communication.

By integrating a variety of media like video clips, instant messages, photos, music, simulations and animations, collaboration can even be entertaining, making it more appealing to users (Ayodele and Olalekan, 2017). The rich media mix possible in mobile learning instructional content can create meaningful engagement with the learner through stimulating his intellect by getting him involved in a collaborative learning setting. In addition to using the mobile device as a tool or platform for Computer Supported Collaborative Learning, researchers have claimed that mCSCL can increase a learner's active participation level in activities by providing more opportunities for the learner to instantly interact with his/her peers (Amara et al., 2016; Sung et al., 2017). Therefore, in the mobile learning environment, features such as

Lecture Video, Lecture Note, Audio, Quiz and Test, Assignment, Discussion and Grade components are generally included (Lee et al., 2011).

4 Approach

In this study, a mixed method approach is used within a Design Based Research (DBR) Framework. DBR does not evaluate an innovative product or intervention. It helps to guide similar research and future development by performing a cycle of research. Figure 1 shows the design based research approach that will be used in this research as described by Amiel and Reeves (2008) in education technology research (Amiel and Reeves, 2008). The primary objective of this research is to enhance the engineering students' interaction by implementing collaborative learning through the Moodle Mobile application.

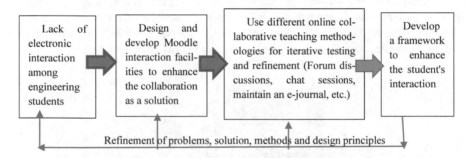

Fig. 1. Design-based research approach for the research (Amiel and Reeves, 2008)

A literature survey and a desk study will be done to identify the current habits of interaction among Engineering students and their peers and teachers. The frequency of the interactions and the use of technology for teaching and learning will also be determined by conducting a questionnaire survey among teachers and students with open and closed-end questions to collect quantitative and qualitative data. Random sampling will be used to select the sample size for questionnaire distribution and structured interviews. To summarize the collected data so that the patterns and relationships can be easily and clearly understood, shared variance Exploratory Factor Analysis (EFA) will be conducted to measure such factors as motivation to use ICT for teaching, barriers against the use of ICT, and perception of the need to use TEL. Further, one-way and two-way Analysis of Variance (ANOVA) will be used to find the significance of the correlation between mobile usage and demographic factors. Based on the students' performance, those engineering courses that act as bottlenecks will be identified. After identifying those courses, some of them will be redesigned according to the mixed method teaching approach; this will include teacher-centered classes and online collaborative activities using the Moodle Mobile App. The data collected through the Moodle VLE will be used to evaluate the students' interactions with their peers and teachers. To monitor students' interactions with different online activities such as reading assessment, portfolios, journal submissions, quiz, online discussion forums, chat forums, lesson assignments etc., web links will be used. Based on the level of interaction the students' requirements in respect of mobile based instructional design will be carried out for the experiment. This will be a highly challenging task as already observed by other researchers in the field.

In this research the MOODLE Mobile App will be used to conduct all experiments. Moodle Mobile App enables a swift exchange of information so that the teacher can upload all the necessary materials and students will have instant access to same on their devices as shown in Fig. 2. When students have difficulties attending lectures in person they can view them later. Even if they happen to live in a distant place, the whole process of collaboration will be far easier. This will encourage them to start live-blogging, per-form peer evaluation and correction, even anonymously if necessary. They could monitor forums and deliver podcasts for other students, maintain learning diaries so that everyone involved can track their progress, engage in online discussions and collaboration, and perform revisions that will help them to keep track of important updates, assignments, forum posts, calendar updates and other class activities. Further, time-related calculations can be monitored through the Moodle LMS and after the peer evaluation and teacher evaluation activity the grades can be calculated.

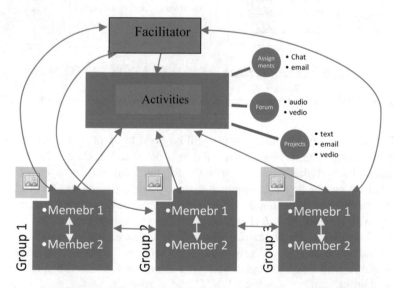

Fig. 2. Conceptual Design for interaction monitoring

Lesson by lesson interaction will be monitored. Furthermore, students' feedback will be used to improve the following lesson's instructional design. This process will be done iteratively to refine the intervention. Collected data will be analysed by using quantitative measurements. Mean and Standard deviation will be used to find the dispersion or the spread of student access levels and interaction levels. A series of structured interviews will be conducted on selected courses by an instructor or teacher and the students engaging in collaborative learning via Moodle. Teachers will be selected from a different discipline and a different age group. Focus group interviews will be conducted with students who have followed the experimental course. A small number of subject matter experts (key informants) will be interviewed to identify the barriers/challenges they faced and how they overcame those challenges and the initiatives they have taken to enhance the interaction between students and teacher when designing any online course material for engineering undergraduates. Thematic analysis will be conducted to analyse the interviews and focus group session data. After gathering all the data and analysis from study one, study two and study three, the findings will be used to develop a tool for Moodle to monitor interactions among fellow students and between students and teachers, both when learning and teaching. The tool will be validated through pilot courses.

5   Analysis

A total of around 850 first-year engineering undergraduate students were given access to the Google online questionnaire and informed of the purpose of the study. They were asked to fill in the questionnaire during their learning management system (LMS) introductory lab session. Therefore, each student had the opportunity to access the questionnaire link through his or her own Moodle account with a desktop computer via the Internet. Students were informed that their participation was voluntary and that they could opt out of the survey if they wanted. The data was collected anonymously.

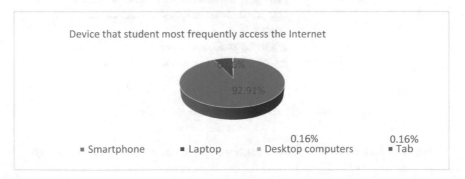

Fig. 3. Distribution of students according to the frequency with which they used the device to access the Internet

There were 650 respondents; however, 29 responses were rejected due to incompleteness of the data provided. Figure 3 shows the distribution of students according to the frequency with which they used the device to access the Internet. All the students used the Internet daily, either using a smartphone (92.91%), laptop (6.76%), tab or desktop computer.

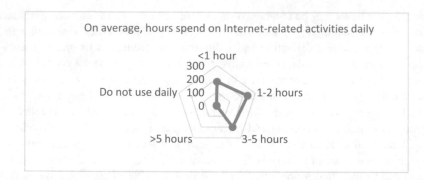

Fig. 4. Distribution of students who spend time daily on Internet related activities

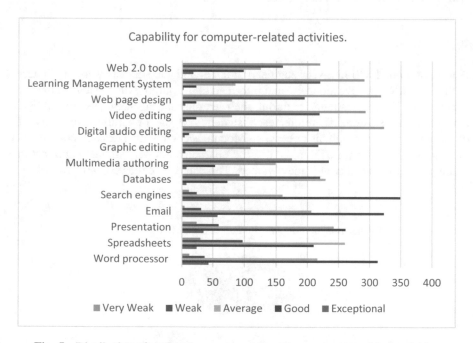

Fig. 5. Distribution of students' engagement in various computer-related activities

The findings show that out of the 621 participants 421 (68%) were male students and 200 (32%) were female students. Therefore, it was a male dominated sample, typical in the context of the Sri Lankan engineering industry. In this study, 80% of those sampled owned a laptop computer while the other 20% were planning to buy a computer within 12 months. 99% of the students had a social media account. Figure 4 shows how the students spent their time on Internet related activities daily. It is clear that a majority of them spent more than an hour daily accessing the Internet. Most of the students (90%) had used those hours to browse their Facebook account, Twitter account, connect with Viber or WhatsApp or engage in some other social media activity, rather than do something constructive. Students were asked to rank various computer-related activities according to how often they accessed same. This is shown in Fig. 5. It clearly shows that students were keen on search engines and email related activities. Moreover, more than 50% of students had used word processing, spreadsheet and presentation software. The majority of students did not use Learning Management Systems (LMS), video editing software, web page designing, graphic designing, digital audio editing or databases.

6 Anticipated Outcomes

This research will focus on the areas that are not covered by the research done in this area with respect to online collaborative learning through the use of mobile applications in Engineering education. In addition, this research will compare the use of different online collaborative learning methodologies in mobile based and web-based learning and teaching platforms in Sri Lanka. This research will explore results by using the design based research approach and will contribute to the extant knowledge on technology-enhanced learning environment by introducing the most suited learning aid for mobile based learning. Further, the Moodle mobile interface will be redesigned to enhance the flexibility and user-friendliness needed for collaborative learning. Finally, a mobile learning framework will be introduced to enhance collaborative learning among Engineering students.

7 Conclusions

Engineering students and teachers mainly communicate through face to face conversation. However, initial studies have shown that students are ready for mobile devices but tend to use it mostly for social activities rather than educational purposes. Students need some time to adapt to new strategies and once they get used to technology assisted interactions they will find it easier to work with their peers. Whereas there is some enthusiasm for the use of technology there are also some barriers that will affect its adoption in the context of Sri Lankan education. By considering those factors design solutions can be recommended to enhance interactions through the use of technology to foster collaborative learning among Engineering students, by means of Moodle Mobile App.

Acknowledgements. The authors acknowledge the support received from the LK Domain Registry in publishing this paper. The conclusions and/or recommendations expressed in this paper are those of the author and may not necessarily reflect the views of the LK Domain Registry.

References

Al-dous, K.K., Samaka, M.: The design and delivery of hybrid PBL sessions in moodle. Int. J. Educ. Inf. Technol. **9**, 105–114 (2015)

Amara, S., et al.: Group formation in mobile computer supported collaborative learning contexts: a systematic literature review. Educ. Technol. Soc. **19**(2), 258–273 (2016)

Amiel, T., Reeves, T.C.: Design-based research and educational technology: rethinking technology and the research agenda. Educ. Technol. Soc. **11**, 29–40 (2008)

Ayodele, M., Olalekan, H.: Learning techniques school effects of collaborative learning styles on performance of students in a ubiquitous collaborative mobile learning environment, **8**(3), 268–279 (2017)

Bourne, J., Harris, D., Mayadas, F.: Online engineering education: learning anywhere, anytime. J. Eng. Educ. **94**(1), 131–146 (2005)

Dyckhoff, A.L., et al.: Design and implementation of a learning analytics toolkit for teachers. Educ. Technol. Soc. **15**(3), 58–76 (2012)

Felder, R.M., et al.: The future of engineering education II. Teaching methods that work, Chem. Engr. Educ. **34**(1), 26–39 (2000). https://doi.org/10.1.1.34.1082

Felder, R., Silverman, L.: Learning and teaching styles in engineering education. Eng. Educ. **78**, 674–681 (1988). https://doi.org/10.1109/FIE.2008.4720326

Jain, M., Birnholtz, J., Cutrell, E., Balakrishnan, R.: Exploring display techniques for mobile collaborative learning in developing regions. In: Proceedings of the 13th International Conference on Human-Computer Interaction with Mobile Devices and Services. ACM Press, Stockholm (2011)

Kasimatis, A. et al. (2010) Using moodle and e-assessment methods during a collaborative inquiry learning, pp. 1–7

Kirkwood, A., Price, L.: Technology-enhanced learning and teaching in higher education: what is "enhanced" and how do we know? A critical literature review. Learn. Media Technol. **39**(1), 6–36 (2016). https://doi.org/10.1080/17439884.2013.770404

Leadbeater, C., Wong, A.: Learning from the Extremes. Cisco, 3–4 (2010)

Lee, K.B., Clark, R., Nosekabel, H.: Developing mobile collaborative learning applications for mobile users. iJIM, **5**(4), 42–48 (2011)

Mcloughlin, C., Lee, M.J.W.: Social software and participatory learning: pedagogical choices with technology affordances in the web 2.0 era introduction: social trends and challenges. Ascilite **2007**, 664–675 (2007)

Price, L., Kirkwood, A.: Using technology for teaching and learning in higher education: a critical review of the role of evidence in informing practice. High. Educ. Res. Dev. **33**(3), 549–564 (2014)

Rand Education: Teachers' Matter: Understanding Teachers Impact on Student Achievement, 1, p. 693 (2012)

Redecker, C., Ala-Mutka, K., Bacigalupo, M., Ferrari, A., Punie, Y.: Learning 2.0: The impact of Web 2.0 Innovations on Education and Training in Europe. Office for Official Publications of the European Communities, Luxembourg (2009)

Scott, C.L.: The Futures of Learning 3: what kind of pedagogies for the 21st century? Education Research and Foresight, pp. 1–21 (2015). https://doi.org/10.1016/j.pse.2015.08.005

Spikol, D.: Playing and Learning Across Locations: Identifying Factors for the Design of Collaborative Mobile Learning (2008)

Sung, Y.-T., Yang, J.-M., Lee, H.-Y.: The effects of mobile-computer-supported collaborative learning: meta-analysis and critical synthesis. Rev. Educ. Res. **87**(4), 768–805 (2017)

Voce, J., et al.: 2016 Survey of Technology Enhanced Learning For Higher Education in the UK, p. 233 (2016). http://immagic.com/eLibrary/ARCHIVES/GENERAL/UCISA_UK/U080902B.pdf

Collaborative Learning in Data Science Education: A Data Expedition as a Formative Assessment Tool

Olga Maksimenkova[1(✉)], Alexey Neznanov[1], and Irina Radchenko[2]

[1] National Research University Higher School of Economics, Moscow, Russia
{omaksimenkova, aneznanov}@hse.ru
[2] ITMO University, St. Petersburg, Russia
iradche@gmail.com

Abstract. The paper addresses the questions of data science education of current importance. It aims to introduce and justify the framework that allows flexibly evaluate the processes of a data expedition and a digital media created during it. For these purposes, the authors explore features of digital media artefacts which are specific to data expeditions and are essential to accurate evaluation. The rubrics as a power but hardly formalizable evaluation method in application to digital media artefacts are also discussed. Moreover, the paper documents the experience of rubrics creation according to the suggested framework. The rubrics were successfully adopted to two data-driven journalism courses. The authors also formulate recommendations on data expedition evaluation which should take into consideration structural features of a data expedition, distinctive features of digital media, etc.

Keywords: Data expedition · Data science · Collaborative technologies · Education

1 Introduction

It is well known that *Data Science* (DS) is relatively young and rapidly growing area. The discussions about DS and its understanding as a field are still heat topic of current interest [1]. To be clear in this paper we will follow the definition of DS from IBM Analytics [2]:

> Data Science is an interdisciplinary field that combines machine learning, statistics, advanced analysis, and programming. It is a new form of art that draws out hidden insights and puts data to work in the cognitive era.

Naturally, the youthfulness of DS is one the main reasons why *data science education* (DSE) is shaping today. However, several teaching and learning techniques and methods have been already introduced to the courses in this area [3, 4]. It is interesting that the actuality of DSE increases not only to data scientists but to the specialists of the other fields as well (e.g., education, medicine, journalism). This explains by that millions of different information systems produce petabytes of data every hour all over the world and specialists and experts need skills in DS. Thus, many communities

© Springer Nature Switzerland AG 2020
M. E. Auer and T. Tsiatsos (Eds.): ICL 2018, AISC 916, pp. 14–25, 2020.
https://doi.org/10.1007/978-3-030-11932-4_2

postulate that DS is a new foundation of Data Literacy (good example is [5]). The context of this paper lays in the field of DSE and it focuses on a data expedition DE as a collaborative formative assessment technique as it is follows from the definition by School of Data [6]:

> Data expedition (DE) is a learning by doing computer-supported collaborative learning technique, which is applied to DSE.

The purpose of any DE is to create a data-driven digital media artefact in 2–3 weeks by collaborating in small (2–3 persons) groups. A typical DE consists of several stages: data gathering/data collecting, data processing/data analyzing, data visualization and making a product based on data [7]. Naturally, this structure makes a DE an attractive and appropriate tool for complex skills in DS shaping. Moreover, the authors of this paper have great confident that a DE may successfully acts as a powerful collaborative assessment technique.

Surely, the idea of active learning and formative assessment implementation in DS courses is not new. Thus, these last few years practitioners of DSE began to document and reflex their experience in blended learning [8], flipped classroom [9], and project-based learning assessment [10] in implementation in DS courses. Moreover, several valuable examples of education DEs were introduced [7, 11]. However, the question of how to evaluate a digital media artefact obtained during a DE is still acute.

The paper explores a DE as a collaborative assessment technique in DS, focuses on its evaluation and introduces a framework for rubrics design.

2   Background

As the authors share the position that an assessment should be agreed with Merrill's principles of instruction [12] and be designed according to learning objectives (outcomes) [13]. They also strongly believe that the clear understanding of data scientists' competencies is mandatory to design evaluation tool suitable to DSE. This section aims to distinguish crucial features of assessment in DSE. By reviewing the literature on DS curriculum design, teaching and learning approaches we collect the learning outcomes to be evaluated and generalize current teaching practices of DS-oriented programs.

Convenient to our goals data scientist's job descriptions are given in [8, 14, 15]. We will use a short-list of requirements to an ideal data scientist [8]. He/she must:

- be of an analytical and exploratory mindset,
- have a good understanding of how to do data-based (quantitative) research,
- possess statistical skills and be comfortable handling diverse data,
- be clear, effective communicators with the ability to interact across multiple business levels,
- have a thorough understanding of their business context.

Anderson and colleagues in 2014 [16] claimed that difficulties with the curriculum design relate to multidisciplinary learning objectives of DS programs/courses. The valuable result of this work is the DS curriculum topic list. In 2016 Mills and colleagues [17] reported that the most of information systems programs begin offering

DS-related courses. So, it is not surprising that in 2017 De Veaux and colleagues [18] specified the list by Anderson and colleagues, and highlighted that DS is not the direct sum of skills in statistics, computer science and mathematics, and introduced key competencies for bachelor in DS: computational and statistical thinking, mathematical foundations, model building and software foundation, data curation, and knowledge transference – communication and responsibility. These outcomes give us several clear directions in evaluation tool construction, but all these initiatives did not suggest any recipes of assessments.

In 2016 Hicks and Irizarry [19] formulated several principles of teaching DS:

- organize the course around a set of diverse case studies;
- integrate computing into every aspect of the course;
- teach abstraction, but minimize reliance on mathematic notation;
- structure course activities to realistically mimic a data scientist's experience;
- demonstrate the importance of critical thinking/skepticism through examples.

This requirements, frameworks and principles shaped the basis of practical DSE. Thus, Brunner and Kim in [20] described the design of an introductory DS course. Because the course was delivered through MOOC platform, the assessments' types were partly predefined by the platform: quizzes, programming assignment and peer assessment. Asamoah and colleagues [21] also reported on introductory DS course which is based on interdisciplinary approach. Both papers are quite circumstanced, and their results may be directly reused by the educators, but they mostly focused at courses structure and their agreement with curriculum and recommendations that were mentioned above [16, 18]. So, there were no special approaches to DSE or learning outcomes evaluation introduced in these papers.

These last few years several publications that let us feel enthusiastic about the future of active learning and formative assessment techniques in DSE have been appeared. This year Ryan in [8] called blended model of education a powerful approach to DSE. The model engages as industry as academia to educational processes. Ryan also reviled the impetuous grow of number of DS programs in different universities. Moreover, innovative educational technologies spreading among DS educators all other the world. Thus, Eybers and Hattingh in [9] reported on flipped class room approach implementation to post-graduate DS students. Despite the paper contains the impressive results it does not mention any assessment type which was utilized during the instructions. More relevant to our topic research by Giabbanelli and Mago [10] explored the role of teaching computational modeling during DSE and suggested a relevant course. Unlike Brunner and Kim [20] Giabbanelli and Mago used a project-based learning assessment to evaluate complex DS skills. The results of each projects had to be a research proposal and using real data was obligatory. Unfortunately, the authors do not in detail developed the topic about features of projects, grading rules and students' reaction.

We can resume that a full-fledged assessment in DS courses should reflect complex nature of DS. So, taking into consideration growing interest to active learning techniques from DS educational society formative assessment seems to be suitable to DSE.

We may also note that the complexity of assessment procedures in DSE grows from the interdisciplinary of DS and consequently from composite structure of DS courses and their learning outcomes.

3   Educational Data Expedition from the Evaluation Point of View

A data-driven digital media artefact acts as a main objective result of a DE. From the significant properties of DEs we can deduce requirements on digital media artefacts. But for educational DE we should consider additional aspects. The short-list of what should be obligatory taken into consideration during evaluation is follows.

- **The processes of digital media artefact creation**. They should be fixed in DE diary by participants of a collaboration. Automatic software logs gathering is not enough!
- **The quality of data citation**. All the data should be cited according to data citation principles [11].
- **Correctness and reproducibility**. We need the measures that is taken to ensure correctness and reproducibility of data transformations – it is very hard task.
- **The complexity, interactivity, and design of final artefacts**. We give an advantage to good storytelling [22] with interactive data dashboards [23] and infographics.

3.1   Data Expedition Evaluation Criteria

As it is follows from [6] the evaluation of data expedition reflects interdisciplinary nature of DS and consequently is rather complicated. Wherefore, as far as educational data expedition is a prolonged in time procedure we formulated several criteria which responds as to process as to result. Because each criterion may be expressed on various levels, we developed rubrics. Most of the criteria have five levels with comprehensive description. The list of criteria is follows.

1. **Data expedition dairy** (experiment journal) for research data expeditions. The criterion reflects the reproducibility of the results of completed data expedition.
2. **Using scripts** written in one of the programming languages. The criterion allows to evaluate the laboriousness of data preparation/visualization/analysis processes.
3. **Using special software** for data preprocessing or analysis. The criterion stands for ability of implementation data analysis by means of professional tools.
4. **Programming code**. The criterion allows to validate the independency of scripts' development.
5. **Data tiding**. The criterion allows to estimate the accuracy of quantitative statistical results.
6. **Data analysis**. The criterion reflects the level of data analytics skills of a team such as settling a hypothesis, testing hypothesis, etc.

7. **Data visualization**. The criterion allows to estimate the accuracy of quantitative statistical results and skill of a team in preparing interpretable results of data analysis.

Given criteria are quite general and give us a frame to evaluate educational DE. Surely, every DE requires either extra criteria or detailing of specified above.

3.2 Supportive Software

Software using during each DE play a significant role in evaluation. At first, it supports collaboration and helps to identify and measure an author's contribution to processes and results. At second, special tools allow teams to collect, version, and share artifacts of their work as inside as outside of a team. So, a DE uses a lot of software for collaborative work, data sources access, data processing, and publishing data-driven products. Supportive software may be grouped as followed (with some examples in brackets).

1. Integration, communication, and management tools
   (a) Rich collaborative environments (Microsoft Office 365, Google G Suite).
   (b) Project management tools (Trello, Microsoft Teams, MeisterTask).
   (c) Online communication tools (Microsoft Skype, Telegram).
2. Search and external discussions tools
   (a) Universal search engines (Google Search, Microsoft Bing).
   (b) Specialized search engines (Wolfram Alpha).
   (c) Social Media Services (Facebook, Twitter, Instagram).
   (d) QA services (Data Science Stack Exchange, Public Lab, Quora).
3. Data processing tools
   (a) Universal cloud data storages (DropBox, Microsoft OneDrive, Google Drive, Yandex Disk).
   (b) Code repositories (GitHub, BitBucket, GitLab).
   (c) Data analysis and visualization tools (R Studio, Jupyter Notebook, Orange, Microsoft Excel, Microsoft Machine Learning Studio, Tableau, OpenRefine, Infogram, Plotly).
   (d) Open data platforms (CKAN, Zenodo).
   (e) Specialized data extraction tools (Import.IO, Kimono, Scrapy, ABBYY FineReader).
   (f) Data validation tools (CSVLint, Data Package Validator, Data Hub LOD Validator).
4. Media publishing platforms (Medium, Microsoft Sway, Tilda, GitBook, Wordpress, Blogger).

We should note that collaborative technologies utilization is one of the main trends in software development. It is amazingly corresponded with the ability to evaluate a collaborative work of a team.

4 Data Expedition Implementation in Universities

The DE was adopted to two educational activities in parallel. Both were implemented to data-driven journalism master-students of different universities. So, the digital media artefact was a data-driven article. The type of article was defined within a DE and the rubrics were expanded by extra requirements.

Note that the prerequisites for these groups, the rules of their shaping, and instructional design were different. So, the paper does not provide any comparison between these groups and their results.

4.1 Instructional Design

Data expeditions were implemented twice in the spring term of 2016/2017 academic year. Two groups of students from different universities took part in educational data expeditions. First group was first-year students of "Data Journalism" master program at National Research University Higher School of Economics, Moscow (HSE). Second group consisted mostly of students of European University, Saint-Petersburg (EU) and selected researchers who could join the event. After we will address these groups as HSE-students and EU-participants.

Both data expeditions were given as a summative assessment at the end of educational modules. HSE-students were engaged into data expedition after a part of Scientific Research Seminar which followed several blocks of trainings in DS. So, the students were familiar with the basics of Python-programming, mathematical statistics, data analysis and open data retrieving. EU-participants did not work with open data previously, and they had basic skills in statistics and data processing only.

4.2 The Data Expedition in HSE: Features and Schedule

The data expedition in HSE was conducting at the same time as in EU but was slightly different. At first, only HSE-students of the first-year "Data Journalism" master program was involved into this activity. At second, the participation was obligatory because the DE acted as an assessment after several modules of the Scientific Research Seminar, "Open data" and "Introductory programming" courses. This explains by that the DE was targeted to evaluate complex skills in data-driven journalism which should have been shaped by the end of the first year of education. At third, DE was preceded by short in-class session where students divided into working teams and took part in discussion on evaluation rubrics. At fourth, collaboration between participants and administrators was supported by Microsoft Office 365 Education (using OneNote Class Notebook [24]).

The module of the Scientific Research Seminar was delivered by the authors and had took 8 academic hours. First 4 h were spent on practice in cleaning data. The last 4 h were reserved for rubrics introduction and discussion, group shaping and data-expedition introduction.

The purposes of the data expedition were (1) to engage students in collaborative work with real open data; (2) to push students to generalize their knowledge and experience in programming, data analysis and data science tools; (3) to show students directions for future activities by giving formative feedback; (4) to give a feedback on learning outcomes achievements and troubles with the instructional design of the master program to academic and administrative supervisors.

For these purposes and according to these purposes evaluation criteria from Sect. 3.1 were specified by adding: (1) a journal of an experiment to evaluate processes of DE; (2) a structure and a specification for the content of a paper to evaluate the quality of result digital media; (3) the requirements to journalistic, popular-science or science text.

This year total number of students who was engaged to the data expedition was 28. They were divided into 9 groups without any randomization and with the only limitation on a group size. The group size is recommended [7] to be not more than 3 persons.

4.3 The Data Expedition in EU: Features and Schedule

Participants was invited to the DE through the announcement which was posted in Internet[1] and everyone who was interested in work with open data was asked to send his/her CV with an indication of his/her scientific interests and experience in scientific research. So, the participants had different levels of knowledge and skills.

The data expedition was running for two weeks. All the participants (14) were divided into small groups of two people. They were immediately provided with educational materials, posted on GitHub.[2] Participants were also invited to contribute their materials on GitHub. A digital media which was expected at the end of DE was a draft of a scientific paper based on data.

During the DE all the participants attended lessons which were targeted except others to fill the gaps in their knowledge and skills. Each lesson included the following sections: (1) lesson plan, (2) tasks, (3) assessments, (4) supportive information and additional material, (5) instructions, (6) useful links, (7) the list of tools and software, (8) summary, (9) the assignment(s) for the next lesson. Assignments to work online were given after every class. A link to rubrics allowed to improve students' comprehension on assessment criteria. It also contains variants of assessment for each criterion.[3]

The criteria introduced in Sect. 3.1 were modified and we presented 8 criteria in total: the data expedition participation's diary, data processing software, data analysis tools, data cleaning, data analysis, data visualization, paper's structure and paper's content. Every criterion had 4 discrete levels: 0, 0.25, 0.75, and 1 (where 0 – nothing has been done, 1 – a criterion is completely fulfillment).

The progress of the participants' team work was collected in a Google Spreadsheet.

---

[1] https://eu.spb.ru/forthcoming-events/17625-opencitydataworkshop.

[2] https://github.com/iradche/Data-Expedition-in-EU.

[3] https://docs.google.com/spreadsheets/d/1fDgEbYeI1P87ob_MaS-oWcbro_BV0oj-Ofr9PeUXgn4/pubhtml.

5 Results and Discussion

5.1 Data Collections

As it is easy to see from Sect. 3.1, in our study we have two general populations because data expeditions were started independently in two higher schools and had some differences. The first general population consists of the first year "Data journalism" master students of NRU HSE, Moscow. All the participants were asked to fill in an electronic post-survey. The volume of a research sample (HSE-students) is 9 observations. The sample differs from the population, because each group delegates a member to provide feedback via the post-survey.

The second general population included EU-students and several researchers who joined the data expedition. In this case, post-survey was taken place too and the volume of the sample (EU-participants) is 14 observations.

5.2 Tools and Methods

EU-participants were asked to pass a pre-questionnaire that was presented as a Google Form with eight closed-ended questions about students experience in working with open data. After the DE students were asked to participate in a post-survey. It contained four closed-ended questions and three open-ended questions about their impressions, behavior and intentions on further work with open data.

Before final grades publication all the HSE-students were invited to take part in a post-survey. The post-survey was implemented as a Windows Form hosted at the Microsoft Office 365 Education platform. The questionnaire contained 3 multiple choice questions and 2 open-ended questions. The questions were targeted to get the information about learners' behavior during the data expedition and satisfaction of assessment.

As far as it was a pilot implementation and both research samples are rather small we use simple learning analytics for data analysis and interpretation.

5.3 The Data Expedition in HSE

The results of the post-survey demonstrate students that found their participation in data expedition very useful. From the first row of Table 1 it is easy to see that none of them reported on useless of data expedition. This also evidently proved by open answers, for example, see Responses #1 and #3 in Table 2. The data from the second row in Table 1 allow us to conclude that prolongated nature of a data expedition is a problem to master students. Probably, we may reduce this problem if students are prepared to this type of assessment during the instructions. The last row demonstrates that the third part of students feel themselves inconvenient working in group.

Naturally, that a pilot reviled several problems in integration of data expeditions to daily educational practice. For example, Response #4 shows that a student was expected new data and did not understand that he/she had to demonstrate some skills in working with data. Similar picture is observed in Responses #5 and #6. Respondents did not recognize their experience in programming, data analysis and mathematical statistics as the preliminary to the data expedition. This clearly shows the power of the data expedition as a formative assessment tool. It punishes students to clue their previous learning experience altogether.

Table 1. HSE post-survey. Multiple choice answers distribution.

Question	33%	67%
Was your participation to the data expedition useful for you?	Yes	Probably Yes
How was your time allocated during the data expedition?	Little by little, but regularly	I worked not regularly. Sometimes for a long, sometimes there were no time at all
Give a characteristic of your interaction with the other participants?	I prefer working alone	I had enough interaction, I am satisfied

Table 2. HSE post-survey. Selected answers to open questions.

What were useful for you in data expedition?	Response #1: "I have created universal code which I am going to use in the other projects"
	Response #2: "Data expedition is a completely new for me type of a project. So, to accomplish the task the time had to be distributed in a separate way. But, each project goes to "Experience" storage and this is good"
	Response #3: "Practical experience in writing data-driven article"
What were not satisfactory for you in data expedition?	Response #4: "On the one hand, we were free in the selection of topic but on the other the result would have been limited stricter. I mean clear requirements to our work. Moreover, it would be better to work with new not previously publicized data (not only links to open data portals which are we familiar with and have no interest). But the freedom in topic selection should be kept"
	Response #5: "I did not like that we immediately got down to independent work. At the lessons before the data expeditions we only clean data. It would be better to pass through all the stages with teachers' control"
	Response #6: "I wish more practice before a data expedition"

5.4   The Data Expedition in EU

In this section, we present the main results of questionnaires and DEs for EU-participants. Table 3 shows the partial results of pre-questionnaire in EU.

The results of this questionnaire allow to make a conclusion that participants had not experience with open data and open data visualization. Most of them already have written scientific papers in their native language and nearly half of them had an experience with scientific papers in English. All participants noticed in post-questionnaire that this DE was quite useful for them and reported on their desire for taking part in further data expeditions. These results do not surprising because groups were gathered students with different background.

The pilot demonstrated that the developed framework for rubrics reduces laboriousness of grading rules design. Grading rules for the first DE were similar to the second one and contained all the criteria from Sect. 3.1. The main differences were in data analysis methods, application field knowledge for data analysis interpretation, experience in software tools.

Table 3. Results of pre-questionnaire in EU, percent.

Questions	Yes, %	No, %
Have you previously worked with open data?	28.6	71.7
Have you previously worked with online services for open data visualization?	7.1	92.9
Have you previously written scientific papers in your native language?	92.9	7.1
Have you previously written scientific papers in English?	35.7	64.3
Is it interesting for you to learn about Open Science and its implementation for research?	92.9	7.1

6 Recommendations

Based on results of the pilot we can formulate the recommendations to educational DEs evaluation:

1. Evaluation should take into consideration two aspects of a DE: orientation to the predefined goal and multi-stage nature of main DE processes.
   (a) Predefined goal allows one to check correspondence between task description and final digital media artefacts.
   (b) Multi-stage nature of DE processes allows one to assess involvement and impact of participants based on DE diary and logs.
2. The features of digital media artefact which is created during a DE should be reflected in grading rules.
3. Proposed rubrics should be discussed with DE's participants for common understanding of goals and better involvement in the methodology of DE.
4. Grading of data citation, reproducibility and provenance should follow available guidelines (see [24, 25] as an example).

Some responses on the pilot's post-survey allow us to carry out the directions of the future work on educational DEs' methodology:

1. Assessment in form of DE should be implemented after a short supportive block which aims to refresh relevant knowledge and skills.
2. Working groups should be completed according to rubrics. For example, if programming skills mentioned in grading rules, the groups should contain at least one programmer.

7   Conclusion

The context of the paper lays in the field of data science education. The paper contributed to the implementation of educational DEs and focused on their assessment role. Thus, the paper introduced the framework for rubrics design.

Using DE is highly beneficial for interdisciplinary educational programs where Data Science meets domain knowledge. We share the experience of using DE in data journalism programs. Evaluation criteria were developed according to the professional requirements to data scientists' competencies which are documented in recommendations to DS curriculum design and jobs' descriptions. Our practice proved that a DE is a good example of an interactive collaborative learning technique. Next academic year we plan to improve the assessment quality, to expand it into related domains, and to gather additional statistics and opinions of students with different background.

Acknowledgment. The article was prepared within the framework of the Basic Research Program at the National Research University Higher School of Economics (HSE) and supported within the framework of a subsidy by the Russian Academic Excellence Project "5-100".

References

1. What is Data Science? http://cds.nyu.edu/research/ (2013). Accessed 14 July 2017
2. Data Science. https://www.ibm.com/analytics/us/en/technology/data-science/ (2017). Accessed 14 July 2017
3. Anderson P., Nash T., McCauley R.: Faciliatating programming success in data science courses through gamified scafollding and Learn2Mine. In: Proceedings of the 2015 ACM Conference on Innovation and Technology in Computer Science Education (2015)
4. Jormanainen I., Sutinen E.: An open approach for learning educational data mining. Koli Calling'13, Koli (2013)
5. Bhargava R., Deahl E., Letouzé E., Noonan A., Sangokoya D., Shoup N.: Beyond data literacy: reinventing community engagement and empowerment in the age of data (2015). Accessed 26 June 2017
6. School of Data. Data expeditions. https://schoolofdata.org/data-expeditions/ (2013) Accessed 29 May 2017
7. Radchenko I., Sakoyan A.: On Some Russian Educational Projects in Open Data and Data Journalism. Open Data for Education. Linked, Shared, and Reusable Data for Teaching and Learning, pp. 153–165. Springer (2016)
8. Ryan L.: The Data Science Education and Leadership Landscape, pp. 85–106. Elsevier (2017)
9. Eybers S., Hattingh M.: Teaching data science to post graduate students: a preliminary study using a "F-L-I-P" class room approach. International Conferences ITS, ICEduTech and STE 2016 (2016)
10. Giabbanelli, P., Mago, V.: Teaching computational modeling in the data science era. Procedia Comput. Sci. **80**, 1968–1977 (2016)
11. Data Citation Synthesis Group: Joint Declaration of Data Citation Principles. In: Martone, M. (ed.) FORCE11. San Diego CA (2014)
12. Merrill, M.: First principles of instruction. Educ. Tech. Res. Dev. **50**(3), 43–59 (2002)

13. Fink L.: Creating Significant Learning Experiences: An Integrated Approach to Designing College Courses, p. 320. Wiley (2003)
14. MIT: Data Scientist Position Description. https://ist.mit.edu/sites/default/files/about/org/roles/Data_Scientist_Position_Description_v4.pdf (2015). Accessed 14 July 2017
15. Guidance: Data Scientist: Role Description. https://www.gov.uk/government/publications/data-scientist-role-description/data-scientist-role-description (2017). Accessed 14 July 2017
16. Anderson P., Bowring J., McCauley R., Pothering G., Starr C.: An undergraduate degree in data science: curriculum and a decade of implementation experience. In: Proceedings of the 45th ACM technical symposium on Computer science education (2014)
17. Mills, R., Chudoba, K., Olsen, D.: IS programs responding to industry demands for data scientists: a comparison between 2011–2016. J. Inf. Syst. Educ. **27**(2), 131–140 (2016)
18. De Veaux R., Agarwal M., Averett M., Baumer B., Bray A., Bressoud T., Bryant L., Cheng L., Francis A., Gould R., Kim A., Kretchmar M., Lu Q., Moskol A., Nolan D., Pelayo R., Raleigh S., Sethi R., Sondjaja M.: Curriculum guidlines for undergraduate programs in data science. Annu. Rev. Stat. Appl. **4**, 2.1–2.16 (2017)
19. Hicks S., Irizarry R.: A Guide to Teaching Data Science. https://arxiv.org/vc/arxiv/papers/1612/1612.07140v1.pdf (2016). Accessed 13 July 2017
20. Brunner, R., Kim, E.: Teaching data science. Procedia Comput. Sci. **80**, 1947–1956 (2016)
21. Asamoah D., Doran D., Schiller S.: Teaching the Foundations of Data Science: An Interdisciplinary Approach. Pre-ICIS Business Analytics Congress Conference (2015)
22. Kosara, R., Mackinlay, J.: Storytelling: the next step for visualization. IEEE Comput. **5**, 44–50 (2013)
23. Logi Analytics: The Definitive Guide to Dashboard Design. http://logianalytics.com/dashboarddesignguide/ (2017). Accessed 18 July 2017
24. Microsoft, OneNote Class Notebook. https://www.onenote.com/classnotebook (2017). Accessed 23 June 2017
25. Rauber A., Asmi A., van Uytvanck D., Pröll S.: Identification of Reproducible Subsets for Data Citation, Sharing and Re-Use. https://www.rd-alliance.org/system/files/documents/TCDL-RDA-Guidelines_160411.pdf (2016). Accessed 17 May 2017

Collaborative Learning: The Group is Greater than the Sum of its Parts

Mathew Docherty$^{(\boxtimes)}$

University of Applied Sciences Upper Austria School of Engineering,
Stelzhamerstraße 23, 4600 Wels, Austria
mathew.docherty@fh-wels.at

Abstract. How can our social brain aid our academic learning? Which role do our emotions play in learning? Can collaborative learning lead to improved academic achievement and, arguably more importantly, help knowledge transfer into the real world? This paper researches neuroscience, cognitive psychology, social interdependence and social learning examining the role of collaboration in learning. It researches empirical data, including meta-analyses and studies, into collaborative interactive learning, in order to support the argument that collaboration leads to increased academic achievement compared to students who learn via traditional methods. Finally, it identifies six factors crucial to effective collaborative learning: accountability, versatility, ambience, comprehensibility, amalgamation and recapitulation (AVACAR).

Keywords: Collaborative learning · Neuroscience · Cognitive psychology · Social learning · Learner motivation · Shared intentionality

1 Introduction

For the purpose of this paper, cooperative learning shall be understood as synonymous with collaborative learning. Collaborative learning is an "instruction method in which students at various performance levels work together in small groups toward a common goal" [1] and differs fundamentally from classic approaches where students are only responsible for their own progress. According to Barkley, Cross and Major, it is "two or more students laboring [sic] together and sharing the workload equitably as they progress towards the intended learning outcomes" [2]. Collaborative learning is linked to increased learner motivation as well as increased information retention levels, when compared to individual learning [3], and enhances critical thinking through peer discussion and shared responsibility [1]. Developmental psychologists have long purported that our levels of cognitive development are not only dependent upon stored cognitive knowledge but also on the ability to implement it through external guidance. Vygotsky, the Russian developmental psychologist, saw the role of education to give learners experiences through collaboration and stated that "human learning presupposes a specific social nature" [4], This viewpoint allows us to see the capabilities of our learners not only from a historical, 'what has already been learnt' stance, but from a potential 'what could be developed' viewpoint.

© Springer Nature Switzerland AG 2020
M. E. Auer and T. Tsiatsos (Eds.): ICL 2018, AISC 916, pp. 26–33, 2020.
https://doi.org/10.1007/978-3-030-11932-4_3

It is with cognitive integration and assimilation that knowledge transfer and relevance for real-world scenarios is obtained, as it is ultimately the job of education to prepare learners for the real world. Lieberman stated, in his paper on Education and the Social Brain; "For more than 75 years, studies have consistently found that only a small fraction of what is learned in the classroom is retained even a year after learning [5]", surely we must then pay head to neuroscience and social learning in order to improve long-term recall and therefore make learning for life more relevant. Immordino-Yang and Damasio presented evidence in studies into both the early- and late-acquired prefrontal damage patients, that knowledge and reasoning divorced from emotional implications and learning lack meaning and motivation and are of little use in the real world" [6].

Cognitive learning was first researched by Lev Vygotsky and Jean Piaget, both of whom studied children's development throughout childhood. Piaget took the standpoint that learning occurs from the self to the social aspect, whereas Vygotsky purported that it indeed goes from social to the individual [7]. However, both agreed that there is a social aspect to learning. Vygotsky went on to develop a concept, called the zone of proximal development, which he defined as "the distance between the actual developmental level as determined by independent problem solving and the level of potential development as determined through problem-solving under adult guidance or in collaboration with more capable peers [4], clearly identifying the need for guidance and cooperation in learning. Tomasello, known for his work on shared intentionality, which refers to collaborative interactions in which participants share psychological states with one another [8], says that human social learning is concerned with providing others with helpful information and forming shared intentions, a statement that is built upon the Vygotskian idea "that what makes human cognition different is not more individual brainpower, but rather the ability of humans to learn through other persons and their artefacts, and to collaborate with others in collective activities" [8]. With this, the notion of the brain as a social entity was born, enabling human capabilities to increase through shared intelligence [9].

In what follows, the science behind social-cognitive learning and interdependence will be investigated, decades of studies into the effects on learning researched and key factors crucial to its effectiveness explained.

2   Literature Review

For the purpose of this paper, collaborative learning has been sub-divided into the following categories: cognition, social interdependence and social learning.

2.1   Cognition

Cognitive psychology is the study of cognitive processes including memory, generally broken down into working (or short) term and long-term memory. The transition between these two domains is of particular interest to educational science and, more specifically, learning theorists. Cognitive theory suggests that new information is introduced by connecting it to, or building upon, previous knowledge and is, therefore, a highly individual process which should be reflected in our chosen instructional methods. However, proponents also confer that learner interaction increases achievement as the information processing is intensified [10].

Learning unites education and neuroscience: According to Gallistel and Matzel [11] there are plausible connections between synaptic transmission to learning and memory, yet in all the plethora of books covering the topic of neuroscience there are no references to education or pedagogy [12]. They go on to state that adaptive plasticity is the process by which the brain changes, at a neurophysiological level, as a response to its cognitive environment [12], could this characteristic be the key to enhancing learning? Hebb proposed that learning could be enhanced through the neural circuit, ultimately creating the Hebbian reinforcement rule, also known as synaptic plasticity; synchronised neural pathways become more efficient in response to repeated coincident stimulation of the synapses along the route [13]. This is referred to in neuroscience as long-term potentiation (LTP) and has relevance on associative learning and memory [11].

Additionally, neurobiologists have recently discovered the potential of emotions and neural activity in cognition [14] and have recognised that effective learning only happens when the learner is not under stress, something that Krashen [15] called the affective filter in his groundbreaking work on language acquisition. Krashen explains that "Low anxiety appears to be conducive to second language acquisition, whether measured as personal or classroom anxiety" [15], this is backed up by neuroscience as Goswami states "When a learner is stressed or fearful, connections with frontal cortex become impaired, with a negative impact on learning" [14]. Immordino-Yang and Damasio conclude that "When we educators fail to appreciate the importance of students' emotions, we fail to appreciate a critical force in students' learning [...] we fail to appreciate the very reason that students learn at all." [6]. Depicted in Fig. 1 is the influence of emotion on cognition.

2.2   Social Interdependence

Social interdependence originated as part of the gestalt psychology at the University of Berlin in the early 1900s, with Kurt Lewin first proposing the concept of group dynamics and interdependence (1943, 1948, 1951), but these theories were not implemented until the 1960s [16]. Since then social interdependence has become widely accepted in pedagogy and is one of the most preferred teaching methods [17]. Social interdependence is well described as "the outcomes of individuals [...] affected by their own and others' actions" [16], both positively as well as negatively. Johnson & Johnson [16] outline five aspects that contribute to effective cooperation: positive interdependence, individual accountability, promotive interaction, the appropriate use of social skills, and group processing.

Deutsch [18] first proposed that group member interdependence, both positive and negative, could have a collective effect on the group's success as a whole, and Johnson, Johnson and Smith further concluded that it also has an effect on the individual's long-term knowledge retention [17].

Individual accountability is the factor that any one person's performance has an influence on the group performance and uses group dynamics and peer pressure to motivate the individuals to contribute and try their best [19], and exists when "the

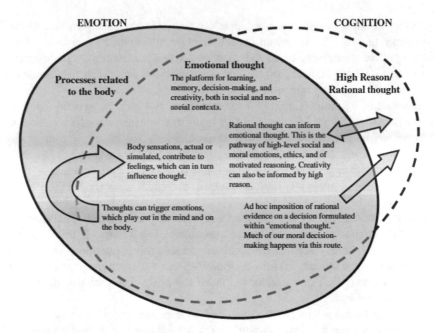

Fig. 1. The evolutionary shadow cast by emotion over cognition influences the modern mind.

performance of each individual member is assessed and the results are given back to the individual and the group to compare against a standard of performance" [16].

Promotive interaction occurs when learners support and aid one another when working towards a common goal and is characterized by learners exchanging information, trusting and supporting each other, working for mutual benefit, providing peer feedback whilst offering constructive criticism and therefore promoting higher quality decision making and greater creativity [16].

The appropriate use of social skills is required in order to cooperate effectively as a team on the tasks set. These include interpersonal and small group skills [16] and aim to further promote interaction.

Group processing allows learners to reflect upon actions and make collective decisions in order to improve the effectiveness of their actions. It allows learners to show appreciation for each other's contributions which in turn leads to improve cooperation group-esteem and effort [16].

2.3 Social Learning

The brain is wired to be social. According to Lieberman a human's high brain to body ratio, or encephalization, and the fact that we coexist in large harmonious groups, indicates that the human brain has evolved to be mentally aware of other's behaviours, motives, goals, thoughts, feelings and dispositions and, unique to humans, to compare them to one's own [5]. Lieberman goes on to explain that our brains are wired to explore and master the social world but that our education system engages learners in the working memory, which actually causes conflict and reduces the learning effect [5]. He proposes that instructors should in fact structure information in terms of plausible social cognitive narratives, or in the case of science, technology, engineering and mathematics (STEM) subjects, on the socially motivated aspects of the learning process, in order to improve retention.

Peer-assisted-learning; learning for social reasons, not learning to be tested. A study done by Bargh and Schul [20] showed that learners who learn in order to teach others had better recall rates than those who learned purely for memorization purposes, therefore indicating that social motivation for learning improves recall. Such peer-tutoring may require prior specific training and is probably best "guided by the provision of structured materials amongst which a degree of student choice may be available" [21].

Shared Cooperative Activity is the coordination of parties whilst performing tasks with mutual intentions and adhering to appropriate behaviours. It involves mutual responsiveness, whereby each participant is responsive to the intentions and actions of the other, and commitment, both to the completion of the activity and collaboration with counterparts [22]. The instructor has to set the goals clearly so that the intentions and goals are the same.

The social learning theory, as developed by Bandura, sees learning as a social process whereby learning occurs through either experiencing it or through observation and imitation and explains it as continuous reciprocal interaction between cognitive, behavioural and environmental interactions [23]. A view also shared by social and cognitive constructivists, for example Vygotsky in his zone of proximal development (ZPD) model, as he put it "an essential feature of learning is that it awakens a variety of internal developmental processes that are able to operate only when the [learner] is interacting with people in his environment and in cooperation with his peers" [4]. Perhaps the best example of social learning is language acquisition, a process of observation and imitation that starts early in life and which is unquestionably successful.

Student-centred social methods make learning an active process, which is not only preferred by learners [24] but has been proven to increase academic achievement and reduce failure rates [25]. Here, active involvement allows learners to engage in higher-order thinking [26], such as peer-instruction, content synthesis and critical analysis [27]. As they perform the given tasks students collaborate, communicate with and support each other with their specific skill-sets whilst working towards a common goal.

3   Current Situation

Many studies into collaborative interactive learning have, to date, seen increased academic achievement compared to students who learnt via traditional methods [28–30]. Johnson and Johnson found, in their meta-analysis of over 600 studies, that "students working together cooperatively learned much more [...] compared to competitive and individualistic learning" [31]. Springer et al. found, in their meta-analysis into the effects of small-group learning in STEM subjects, that small-group work led not only to greater academic achievement compared to traditionally taught counterparts, but also that learner's persistence and attitudes toward learning improved [10]. These findings were confirmed in another meta-analysis by Bowen [28]. Romero [32] performed a meta-analysis of 30 studies into cooperative learning published between 1995 and 2007 and concluded that they had a medium improvement effect (mean of .308) on student achievement in science, which was directly relatable to the amount of structuring that the interventions had. Hake [29] reported, by way of pre and post-test scores, that collaboration between peers or with the instructor significantly improved conceptual understanding. In a meta-analysis of 225 active learning studies of STEM courses, Freeman et al. [25] found that academic achievement improved by about 6% and that students in classes with traditional lecturing were 1.5 times more likely to fail than were students in classes with active learning. Chickering and Gamson [24] also found that active participation, whereby students are engaged in solving problems through discussion and collaboration, improves knowledge understanding, therefore, recall and that active participation is also preferred by students.

Decades of research into long-term recall has found that the vast majority of classroom learned knowledge is lost within the first year [5], this is supported by the so-called 'forgetting curve hypothesis' [33]. It has also been found that the knowledge loss-rate is greatly determined by the type of practice involved, with retrieval practice proving to produce the least significant losses [34] and that knowledge cannot simply be transferred from instructors to the learner's minds but must be assimilated by students in order to be remembered [2].

Nevertheless, some studies have found contrary academic effects to collaborative interactive learning (Borresen (1990), O'Brien and Peters (1994), Shearn and Davidson (1989), and Smith (1984) in [10]). It is the author's belief that these shortcomings can be attributed to insufficient or incorrect task planning, design and implementation and, in order to overcome these by no means minimal challenges, the following conclusions and recommendations have been drawn.

4 Conclusions and Recommendations

There is no single panacean method that can be prescribed in order to gain optimal collaborative learning, it will, as always, depend upon subject matter, resources, levels and aims, there are, however, guidelines that can be recommended. Below are the author's recommendations based on the research performed in this paper:

Task Design – Tasks should not have predetermined outcomes. They should be versatile, with several possible trajectories, leading to multiple acceptable solutions and should allow for peer feedback, negotiation of meaning [10] and opportunities for constructive criticism. Selecting specific predesigned tasks for students to perform [2] whereby all members of the group actively collaborate on individual aspects of a greater whole. This means setting group goals that no one individual could accomplish [3]. However, the achievement of the group goal should be evaluated via individual performance of each group member [35].

Recapitulation – In order to increase long-term potentiation and therefore recall, repetition should be inherent in all tasks set. The author suggests methods such as the Postcard where new group constellations are formed and learners are required to instruct peers on their previously allocated task. This peer-tutoring supports learning for social reasons will also improve recall rates [20].

Roles – The role of the instructor is now as a facilitator for learning [1], as a guide [36], to scaffold, support and motivate the learners to realise their own personal potentials. A crucial element of this to have the collaborative tasks so well documented and explained that the students can immediately work together towards a solution, removing the need for the instructor to lecture then on procedural matters at all. The instructor must create small, no more than four, homogenous groups [1] and intervene if any issues occur in order to enable a low anxiety working atmosphere by maintaining a low affective filter [15].

In short, collaborative tasks should be designed to inherently include each and every one of the following aspects:

A V A C A R:

Accountability	Use evaluation via individual performance
Versatility	Not one fixed, but multiple acceptable solutions
Ambience	Enable a low anxiety working atmosphere
Comprehensibility	Well documented self-explanatory tasks
Amalgamation	Tasks that no one individual could accomplish
Recapitulation	Repetition should be inherent in all tasks

References

1. Gokhale, A.A.: Collaborative learning enhances critical thinking (1995)
2. Barkley, E.F., Cross, K.P., Major, C.H.: Collaborative learning techniques: a handbook for college faculty. Wiley (2014)
3. Johnson, R.T., Johnson, D.W.: Cooperative learning in the science classroom. Sci. Child. **24**, 31–32 (1986)
4. Vygotsky, L.S.: Interaction between learning and development. Mind Soc. 79–91 (1978)
5. Lieberman, M.D.: Education and the social brain. Trends Neurosci. Educ. **1**, 3–9 (2012). https://doi.org/10.1016/j.tine.2012.07.003
6. Immordino-Yang, M.H., Damasio, A.: We feel, therefore we learn: the relevance of affective and social neuroscience to education. Mind Brain Educ. **1**, 3–10 (2007). https://doi.org/10.1111/j.1751-228X.2007.00004.x
7. Solso, R.L., MacLin, O.H., MacLin, M.K.: Cognitive Psychology: Pearson New International Edition. Pearson Higher Ed (2013)
8. Tomasello, M., Carpenter, M.: Shared intentionality. Dev. Sci. **10**, 121–125 (2007). https://doi.org/10.1111/j.1467-7687.2007.00573.x
9. Sloman, S., Fernbach, P.: The knowledge illusion: Why we never think alone. Penguin (2018)
10. Springer, L., Stanne, M.E., Donovan, S.S.: Effects of small-group learning on undergraduates in science, mathematics, engineering, and technology: a meta-analysis. Rev. Educ. Res. **69**, 21–51 (1999). https://doi.org/10.3102/00346543069001021
11. Gallistel, C.R., Matzel, L.D.: The neuroscience of learning: beyond the hebbian synapse. Annu. Rev. Psychol. **64**, 169–200 (2013). https://doi.org/10.1146/annurev-psych-113011-143807

12. Geake, J., Cooper, P.: Cognitive neuroscience: implications for education? Westminst. Stud. Educ. **26**, 7–20 (2003). https://doi.org/10.1080/0140672030260102
13. Hebb, D.O.: The Organization of Behavior (1949)
14. Goswami, U.: Annual review neuroscience and education. Most **74**, 1–14 (2004). https://doi.org/10.1348/000709904322848798
15. Krashen, S.: Principles and Practice in Second Language Acquisition (1982)
16. Johnson, D.W., Johnson, R.T.: An educational psychology success story: social interdependence theory and cooperative learning. Educ. Res. **38**, 365–379 (2009). https://doi.org/10.3102/0013189X09339057
17. Johnson, D.W., Johnson, R.T., Smith, K.: The state of cooperative learning in postsecondary and professional settings. Educ. Psychol. Rev. **19**, 15–29 (2007). https://doi.org/10.1007/s10648-006-9038-8
18. Deutsch, M.: A theory of co-operation and competition. Hum. relations. **2**, 129–152 (1949)
19. Wentzel, K.R.: Relations of social goal pursuit to social acceptance, classroom behavior, and perceived social support. J. Educ. Psychol. **86**, 173 (1994)
20. Bargh, J.A., Schul, Y.: On the cognitive benefits of teaching. J. Educ. Psychol. **72**, 593 (1980)
21. Topping, K.J.: The effectiveness of peer tutoring in further and higher education: a typology and review of the literature. High. Educ. **32**, 321–345 (1996)
22. Bratman, M.E.: Shared cooperative activity. Philos. Rev. **101**, 327–341 (1992)
23. Bandura, A., Walters, R.H.: Social Learning Theory. Prentice-hall Englewood Cliffs, NJ (1977)
24. Chickering, A.W., Gamson, Z.F.: Seven principles for good practice in undergraduate education. AAHE Bull. Mar. 3–7 (1987). https://doi.org/10.1016/0307-4412(89)90094-0
25. Freeman, S., Eddy, S.L., McDonough, M., et al.: Active learning increases student performance in science, engineering, and mathematics. Proc. Natl. Acad. Sci. **111**, 8410–8415 (2014). https://doi.org/10.1073/pnas.1319030111
26. Bloom, B.S., Englehard, M.D., Furst, E.J., et al: Taxonomy of educational objectives: the classification of educational goals. Handbook I cognitive domain. New York, 16, 207 (1956). https://doi.org/10.1300/j104v03n01_03
27. Schultz, R.B.: Active pedagogy leading to deeper learning: fostering metacognition and infusing active learning into the GIS&T classroom. Teach. Geogr. Inf. Sci. Technol. High. Educ. 133–143 (2012). https://doi.org/10.1002/9781119950592.ch9
28. Bowen, C.W.: A quantitative literature review of cooperative learning effects on high school and college chemistry achievement. J. Chem. Educ. **77**, 116 (2000)
29. Hake, R.R.: Interactive-engagement versus traditional methods: a six-thousand-student survey of mechanics test data for introductory physics courses (1998)
30. Prince, M., Bucknell, U.: Does active learning work? A review of the research. J. Eng. Educ. **93**, 223–231 (2004)
31. Johnson, D.W., Johnson, R.T.: Cooperation and Competition: Theory and Research. Interaction Book Company, Edina, MN, US (1989)
32. Romero, C.C.: Cooperative learning instruction and science achievement for secondary and early post-secondary students: a systematic review. Colorado State University (2009)
33. Ebbinghaus, H.: Über das gedächtnis: untersuchungen zur experimentellen psychologie. Duncker & Humblot (1885)
34. Karpicke, J.D., Roediger, H.L.: The critical importance of retrieval for learning. Science (80-) **319**, 966–968 (2008). https://doi.org/10.1126/science.1152408
35. Stump, G., Hilpert, J., Husman, J., et al.: Collaborative learning in engineering students: gender and achievement. J. Eng. Educ. **100**, 475–497 (2011)
36. King, A.: From sage on the stage to guide on the side. Coll. Teach. **41**, 30–35 (1993)

Student Activation in iOER Maker Spaces

Manfred Kaul[(✉)]

University of Applied Sciences Bonn-Rhein-Sieg, Sankt Augustin, Germany
Manfred.Kaul@h-brs.de

Abstract. Interactive Open Educational Resources (iOER) package both information and software to interact with its information in a single component. In Maker Spaces students are following their own ideas, building individual products in small projects, learning about production skills and appropriate machinery. This paper describes concepts and implementation of iOER Maker Spaces based on advanced web technologies, how they contribute to student activation and finally reports on our practical results. iOER Maker Spaces contribute to the societal task of enriching our digital common good.

Keywords: OER · iOER · Maker space · WWW · Web components ·
Digital common goods · Learning analytics · Learning data mining

1 Digital Common Goods

1.1 World-Wide Education

For several decades, so many ideas for digital transformation in education have been developed and digital gadgets implemented. However, when you return home to your own classroom, they often turn out to be too hard to be implemented in your own environment. This is caused by incompatibilities and missing standards: nice gadgets do not fit into your own system at home. They do not become digital common goods available for every classroom. For example, there are so many different learning management systems (LMSs) and dedicated plugins, but they do not fit in other LMSs. If you are lucky and have the same LMS, then different versions might be incompatible. Real complex didactic scenarios that ambitious educators strive for cannot be realized on these shaky grounds. Unfortunately, learning management systems are bad integration platforms [5].

Additionally, LMSs are closed environments, so that LMS boundaries become an obstacle to cross-university, let alone international classroom collaboration. A worldwide integration platform for digital common educational good enabling "world-wide education" is urgently needed.

1.2 WWW as Digital Common Good

Fortunately, there is a most successful platform to start with and to build upon, the World-Wide Web. When Tim Berners-Lee invented the WWW, he had not purely technical concepts in mind, but intended to establish a universal platform for digital common goods for all of us: *"There are a few principles which allowed the web, as a*

© Springer Nature Switzerland AG 2020
M. E. Auer and T. Tsiatsos (Eds.): ICL 2018, AISC 916, pp. 34–45, 2020.
https://doi.org/10.1007/978-3-030-11932-4_4

platform, to support such growth. By design, the Web is universal, royalty-free, open and decentralized. Thousands of people worked together to build the early Web in an amazing, non-national spirit of collaboration; tens of thousands more invented the applications and services that make it so useful to us today, and there is still room for each one of us to create new things on and through the Web. This is for everyone." [1]. Representing European values, the WWW as a digital common good is obviously *the* European platform for digital economy, in which all of us can participate with equal opportunity in generating value and wealth.

1.3   Open Educational Resources

Open Educational Resources (OER) are gaining momentum in perception and efficacy and are contributing to innovative pedagogy. Interactive OER (iOER) goes one step further, packaging both information and interaction logic in a single module [2]. Whereas OER is restricted to texts and images only, iOER comprises all kinds of interactivity such as quizzes, exercises and even interactive games. Being open, iOER contributes to digital common goods (Fig. 1):

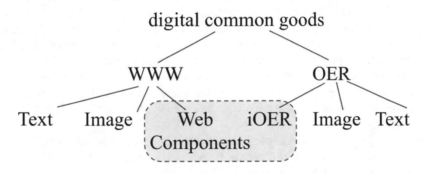

Fig. 1. Bringing interactivity to digital common goods

1.4   Web Components and iOER

Web components are a recent feature of the Web platform introducing modularization via Custom Elements and Shadow DOM [3] standardized by W3C [4]. The Web technologies HTML, CSS, and JavaScript code are packaged into coherent modules that can be imported into any Web page by everyone. This is a browser-native technology: you need nothing else other than a standard browser to run the software. No installation is needed. HTML provides the content and CSS the layout, while JavaScript code bundles the software for interactivity. This makes web components the ideal solution for implementing iOER that is intended for modular, platform independent world-wide distribution of educational resources. The emerging web component technology is a game changer for online learning [5].

   A major success factor of the WWW has been how its independent technologies HTTP, HTML, CSS3, and ECMAScript have evolved in parallel. This kind of parallel innovation process is now being continued on the next higher level with independent web components, providing new basic building blocks from which more complex solutions can be composed in new and creative ways by everyone.

Unfortunately, many systems and apps, even those built on top of the web platform, fail to harness these benefits and achievements: Despite the openness of the underlying technology, the systems and apps themselves are closed and centralized causing some kind of narrow-mindedness or *"filter bubble"* [6]. The original concept paper on the WWW of Tim Berners-Lee in 1989 [7] envisioned that people browsing the web should also find information for which they did *not* originally search for, that means, which opens their mind to the new and unforeseen. Today, *"living in the filter bubble"* has become a major negative effect on political behavior and our society. Many systems and apps, albeit based on the WWW, have had a bad effect on society due to ignoring the first and foremost WWW principles.

Since 2014 mobile phones have generated most of the traffic on the Internet. Interestingly, the portion of mobile web usage has been dropping constantly, reaching 6% in 2018. Most traffic thus no longer originates from the WWW, but rather from closed company-driven apps such as WhatsApp or Facebook. Apps are entry doors into a centralized kind of a digital platform economy and create potential risks: The private app owner has enormous power and can exert influence, manipulate or control the masses without any societal control mechanism. On the other hand, the *central* architecture makes apps vulnerable to oppression. In 2018, the Russian government tried to shut down the Telegram app that was popular amongst its opponents [8]. This could not have happened to the web; Decentralized services and web components have the distinctive advantage that they cannot be shut down. The WWW and its basic principles need to be reinforced again for the sake of a decentralized approach to a digital platform economy that is used by a free society, for free scientific research and free education.

Web components are part of the browser platform, not a new technology on top of it. You do not have to stick to a certain framework in order to use web components. You can even mix and match components built from different frameworks, or with no framework at all inside the same web page. The browser is the integration platform. This gives framework and component developers more freedom. In fact, there are very different approaches to supporting component building by frameworks: either data-driven like ccm.js [9], or HTML document- and ES6 module-driven like Polymer [10]. No matter which approach is taken, the various kinds of components can be used together in the same web page. The web browser becomes the integration platform: not only for technologies, but for applications, languages, memes and cultures as well.

Web components enhance the web platform as a whole: With custom elements [3], the basic language of the web, HTML, becomes extensible and customizable. Everyone can invent her or his own custom HTML tags introducing new capabilities into the web for everyone, available world-wide, and not dependent on a single server. Once a new custom element is published on the WWW, it can be distributed, used, serviced and managed in a fully decentralized manner. It can never be deleted or shut down by cutting connections to a single server. There is no central power that could exert influence, manipulate or control usage via web components.

Custom elements are the basic building blocks. For a restricted application domain an adequate subset of these blocks is selected to serve the special demands of the domain. By selecting a subset of custom elements, a Domain Specific Language (DSL) is developed. For the e-learning domain, for example, a dedicated DSL serves as a world-wide integration language for e-learning applications. The following DSL was used for specifying weekly exercise pages in our courses for the past two years. A typical HTML page using custom elements can be seen in Fig. 2:

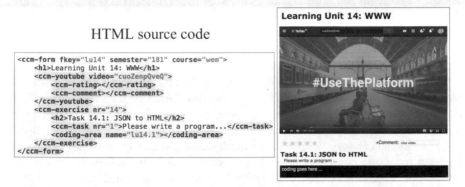

Fig. 2. HTML source code and rendered web page using an E-Learning Domain Specific Language

In Fig. 2, all HTML tags with a hyphen ("-") are custom elements encapsulating the web component with the same name, which renders its content into a separate section of the web page via JavaScript. All other tags are standard HTML tags. Both kinds of tags can be mixed freely: The DSL is an *embedded DSL*, HTML being the mother language. This is client-only technology and a browser standard [4].

The page given in Fig. 2 describes the learning unit no. 14 in the course "Web Engineering". Embedded is a video which can be commented and rated by the students. The page ends with a programming task with a coding area to fill in the program code. In summary, an e-learning HTML page is composed of web components as Lego building blocks. These building blocks are freely combinable to form new artifacts within this e-learning DSL.

There are two kinds of DSLs: (1) the outer DSL, a collection of custom elements from which more complex artifacts are composed, and (2), an inner DSL for every custom element: The space between opening tag and closing tag of a custom element is parsed, but not rendered into the page by the standard HTML parser. In web component terminology, the parsing result is called LightDOM. Every custom element may independently decide how to render itself and how to understand and interpret its attributes and its LightDOM. This means, any kind of language is allowed as inner DSL. No namespace for separation is necessary, as the enclosing tags of the custom elements separate the LightDOM code from the rest of the page. The different nested DSLs for the different custom elements may evolve independently in parallel as needed in their application areas. The WWW success factor of parallel evolution also applies on the next level of the WWW, the web components tier. The WWW is clearly evolving from a reader/writer space for hypertext documents to a Maker Space for web components, which is particularly useful in the e-learning domain. Web components and domain specific languages are very powerful tools and serve as an integration platform for a "World-Wide Educational Maker Space".

2 World-Wide Educational Maker Space

Physical Maker Spaces (also known as open fabrication laboratories) are well-known workshops democratizing the access to cutting-edge technology and high-tech machinery (3D printing etc.) fostering bottom-up and grassroots innovation. In Maker Spaces students are building individual products in small projects following their own ideas, learning about production skills and appropriate machinery.

Virtual Maker Spaces serve the same purpose, but belong to the virtual space. Web components in the World-Wide Web are the foundation of a "World-Wide Maker Space" available anytime, anywhere, anyhow, because web browsers are available on all major platforms today. Web components serve different roles in Maker Spaces: They are simultaneously the products of individual and collective manufacturing, the building tools, the construction, planning, composing, communicating, and collaboration tools, delivering cutting-edge technology and high-tech machinery through the same universal component technology. Technology is no obstacle to easy usage, because the complexity is completely abstracted away and hidden behind simple HTML custom tags.

Manufacturing new components is done by coding, configuration, adaptation, or generation. The JavaScript coder implements new web components natively in JavaScript, or by using a library like ccm.js [9]. Non-coders take a configuration and edit or adapt it according to their needs. Web component generators are web components generating other web components. Such kind of recursion can be entered into the building process as well:

A builder component is a web component that is paired with another web component in order to support the manufacturing process of building similar components. For example, a fill-in-the-blank builder component [11] allows for building new quizzes simply by filling out a web form. The new components are uploaded to a component gallery such as the Web Component Cloud (W2C) [12] or Digital Maker Space (DMS) [13]. W2C is a web portal. DMS is a web component itself fulfilling the recursion principle. From such a gallery, other educators can select, bookmark and download web components of their choice according to their individual needs.

Of course, educators want to monitor their students' performance. Web components are therefore also paired with analytics components [11] that allow students' activities and outcomes to be displayed in real time. Overview charts help educators to spot challenging tasks and to provide assistance when needed. All in all, the different types of educational web components start to become a well-formed e-learning ecosystem within the WWW.

3   iOER Maker Space Concepts

For getting students active, educators need methods and digital media for integrating more activity into students' learning habits. iOER brings interaction and collaboration to OER. In practice, however, a single educator is overburdened when producing all digital media on her or his own. Interactive media are even more demanding, because interaction logic has to be programmed as well. OER repositories help to spread the workload across several colleagues. By simplifying OER production and providing better production tools in a Maker Space environment, students themselves are engaged in iOER production as well. The purpose of our approach is to introduce student iOER production as an additional track of activation in the didactic design using our Digital Maker Spaces [13]. Multi-step workflows also mean that students engage in building up databases, from which quizzes and learning games are generated.

The Lego principle of web components leads to a nice bottom-up approach, "changing the language to suit the problem" [18]: First, there is a colorful chaos of Lego building blocks, with which everything can be built. You do not have to decide on the final outcome before you start. You do not need a big master plan upfront, executing it unthinkingly thereafter. With the bottom-up approach, you can just start building and keep a creative mindset along the way, carefully deciding every single step as you go. Student activation is thereby directed towards a creative mindset coming up with new ideas that even educators would never imagine. Rather than being restricted to the planning phase, creativity now comprises the implementation phase as well. This resembles the actual habits in today's technology-driven world in a far better way.

Maker spaces deviate from the Lego analogy in the following aspect: every time a new web component has been implemented, the Lego construction set is extended by a new building block, which is available any number of times, because "digital" means "copy at no cost". Every new web component brings new power to the WWW everywhere for everyone. The expressiveness of the web language evolves with every new component in two ways: (1) the new component can be built into any web page and as a subcomponent into other components, and (2) every new component may bring a new inner DSL to the domain.

Some repositories for iOER already exist: LearningApps.org [14], for instance, is a large collection of HTML widgets programmed in JavaScript, which can be embedded into any HTML page. But the underlying web technology is iFrame, so that no data exchange between the widget and the rest of the HTML page and web application is allowed at all. iFrames "break the one-document-per-URL paradigm, which is essential for the proper functioning of the web (think bookmarks, deep-links, search engines, …)" [15]. With iFrames, learning apps cannot be recombined to form more complex apps. Another restriction of LearningApps.org is that learning analytics cannot be built into these apps due to iFrame restrictions. The purpose of our approach is to overcome all these restrictions by basing iOER on more advanced browser technology, in particular on web components [3].

Digital Maker Spaces also offer a rich environment for multi-encoding learning objectives, methods and parallel paths using different iOER components thereby supporting a multitude of learner types. Tim Berners-Lee´s original vision of the WWW was to provide digital common goods for opening our minds. Our digital learning environments can only be "mind openers" if they provide a multitude of encodings of the learning objectives on different cognitive levels with different learning approaches and methods. This is clearly a task for society, not a single educator working alone. iOER Maker Spaces provide an environment and infrastructure for this societal endeavor.

4   Student Activation in iOER Maker Spaces

According to the Lego principle, web components can be configured and recombined cyclically: from the standard basic blocks, new larger blocks can be built, bundled as a new iOER component and stored into the repository again, so that it can be used as a basic block in the next step. The student workflow is cyclic, drawing from the iOER repository in the beginning, and pushing back new components at the end, thereby enriching the learning environment at every turn. Students are trained to work towards enriching the digital common good as a common ground. A balance between giving and receiving help in understanding the learning objectives is achieved. The role of the educator is to introduce students into the Digital Maker Space, to set the rules of the workflow, to observe the overall student activity and to guide and coach in the learning process. By combining learning apps with logging, monitoring and analyzing web components, learning analytics and learning data mining has been added and implemented in our Digital Maker Space [13].

For example, in our "work&study" project funded by our Federal Ministry of Education and Research [16], our mission was to conduct research on the student activity of non-traditional students in blended learning settings. We implemented blended learning courses by using our iOER approach, intensively equipped with activity logging. Using elaborate click statistics, the difference in learning habits between traditional students (TS) and non-traditional students (NTS) could be examined in detail. In fact, only 23% of our students undertaking full-time study courses are actually 100% full-time students. Most students have other commitments that take up their time (i.e. earning money for living in a part or full-time job, raising children, or caring for elderly parents or grandparents). For 61%, part-time work was the main time constraint. Over 36% of the students had relevant working experience. About half of our students are in fact non-traditional students, trying to survive in a traditional full-time study scheme. E-learning was greatly welcomed as it allowed them to plan and use time more flexibly. 56% of the NTS are satisfied with mobile learning compared to 36% of the TS. Only 7% of the TS had substantial experience with mobile learning before participating in the course, compared with 22% of the NTS. TS are more skeptical towards mobile learning and prefer to be in the classroom on campus with face-to-face contact with their instructor and fellow students. 72% of the NTS expressed a desire to have more mobile learning units, as against only 46% of the TS.

In our project "Digital Maker Space" [17], our iOER approach is extended to Maker Spaces offering better equipment and tools for educator and student production of iOER components. For example, "Build similar app" is the name of a button in the

Maker Space for starting a configuration editor. This again is a standard web component in itself. By uniformly basing iOER components as well as editors on web standards, universal editing and production tools could be implemented.

iOER brings interactivity to OER. There are several kinds of interactivity: navigating, answering questions, quizzes, fill-in-the-blanks, games, writing texts, problem solving, project planning and management, virtual lab experiments, peer communication, cooperation and collaboration. Without openness and integration on a single platform, however, students are lost in the multitude of apps and educators are excluded from student's activity with no options to log, monitor, analyze or guide it. With separate apps from app stores, educators are at the mercy of the app manufacturer, which rarely provides a separate instructor interface. Even if such monitoring interfaces for instructors are provided by some of the app manufacturers, educators have to face the even bigger problem of dealing with the multitude of those interfaces delivering heterogeneous data that would be virtually impossible to integrate into a meaningful overview. In our approach, therefore, all kinds of student activity are supported by web components based on the WWW platform, and are logged, monitored, and analyzed, so that corrective action can be taken as necessary. In particular, "logging" being a web component on its own, can be deliberately chosen to be included in all those interaction components, switched on or off just as needed. If switched on, mass logging data will be collected during virtual classroom operation. Educators monitor aggregated activity data in real time. Analyzing algorithms extract important information from log data and issue helpful alerts. For instance, the author used the "*logging*" component in a software engineering course, in which a digital clock with three buttons was given as a web component that was integrated into a weekly exercise web page. The exercise task was to extract a state machine as a UML diagram from student observations while experimenting with the buttons of the clock, for which no operating instructions were given. Imagine, you bought a digital clock and you have lost the leaflet of operating instructions. Students were confronted with this challenging situation leaving experimentation for homework. With logging every button click, students pressed the buttons about 100 times on average in order to solve the task. The TS needed 15% more clicks than the NTS. Alerts indicated statistical outliers performing no click at all or, on the other extreme, over 10,000 clicks.

Most exercise and training tasks in university courses produce irrelevant results that end up in a drawer or waste bin somewhere. But students want to engage in real-world projects that have an impact on the real world. This is also important for their intrinsic motivation. The WWW has a very low entry level for engagement: anybody can publish directly without asking for permission. This holds true for the reader/writer web of hypertext documents and the user/maker web of components, as well. In our course on Web Engineering we organized a "web component contest", in which every student had to program a new web component on her or his own. The contest results were published via a web component named "component marketplace", in which all components were listed with their name, description and preview functionality. All students could test and peer review their fellow students' solutions via commenting and rating [12]. Several students have deliberately chosen to publish their web component on GitHub under the MIT License [17], so that the code was freely available for everyone on the web even outside the classroom. Every single student implemented only a small component with manageable effort, but all components together had quite an impact, not only for getting grades in the course, but also for enhancing the capabilities of the WWW for themselves and for everyone forever. This kind of real-world impact fostered their intrinsic motivation.

As regards rating and ranking in student contests, the same artifacts in the real world occur in the classroom as well: groups of students conspire to down rate foreign solutions and uprate their own, not taking into account their actual quality, thereby distorting the overall ranking. Although annoying, this effect occurs in real-world rankings as well. The instructor has to be aware of this dysfunction and should not make grades dependent on these distorted student rankings.

In Web Engineering courses, students are active in iOER Maker Spaces by coding their own iOER components. In other courses several other activation pathways without JavaScript programming are available:

- Activation by Composition: students contribute new compositions of already existing components. They compose larger components from smaller ones by sequencing or nesting them. Because all components are embedded into HTML as the mother language, sequencing and nesting is achieved in the standard way of HTML tags (see Fig. 2).
- Activation by Configuration: students configure pre-existing components in new ways. The new configuration can be created using a configuration editor, a drag and drop interface, via web form or some appropriate inner DSL, either mark-down or mark-up, in JSON format or some other formal or semi-formal language [11, 12, 20].
- Activation by "Creating a similar app": starting from an existing component, students use the corresponding builder app in order to build a similar app. The builder app may be a configuration editor, a web form, the fill-in-the-blanks web component or a DSL grammar-driven tool offering a menu of appropriate choices at every step. From metadata about the app, a builder generator can generate a corresponding builder app [11, 12, 19].
- Activation via Content Creation: most web components are generic and can be used with any content. Students can thus contribute to OER creation by creating new content for existing web components (quiz, fill-in-the-blank, etc.) themselves. In a multi-step learning workflow (learnflow), students are (1) asked to read a scientific text in a first step, then (2) to enter a question and type the correct answer into a database in a second step. After that, the question-answer database generates quizzes or fill-in-the-blanks (3). The next student task is to answer the quizzes generated (4), to rate them and to comment on them (5). From the rating, the most popular questions are extracted (6). Every single student contributes only one single question and a correct answer, but all students together produce a complete new quiz or fill-in-the-blank exercise, thereby experiencing the power of teamwork.
- Activation via learning apps: last but not least, students are made active by appropriate interactive learning games such as quizzes, fill-in-the-blanks, team-building, problem-solving apps, commenting or rating apps as well as specific contests [11, 12, 20].

All in all, the activation workflow is cyclic: on the one hand, learning apps are used to train students in digital literacy, enabling them to build new learning apps in one way or the other as described above. On the other hand, this development in turn increases the amount of learning apps for all of us. Learning and teaching in iOER Maker Spaces means using and contributing to digital common goods.

Learning analytics sum up information about the evoked student activity and the effectiveness of the different learning apps and of the activation methods. In our approach, the anonymous logging of all interactions can be switched on in all learning apps easily. Interesting facts about our apps, methods and anonymous student behavior can be derived from collected mass data. For example, the NTS submitted more elaborate solutions than the TS: their programming code is 8% longer on average and their free text solutions and explanations 16% longer. In lab experiments, by contrast, the NTS are more focused and produced 13% shorter solutions. They needed 14% fewer steps in their experiments to arrive at a solution.

All students had to submit their own solutions to given problems on a weekly basis with a deadline of Sunday at 8:00 pm. 62.5% of 187 students submitted their solution at least once in the last hour before the deadline. 96% submitted their solution at least once during the weekend. However, the majority started on Monday and began submitting parts of their solution. By Wednesday 75% had delivered their solution for the first time. The data showed that they had been active all week, finalizing their solution on Saturday or Sunday in most cases. There are always some students that are active during different parts of the day, even at midnight or at 3 am. The most favorable time span was between noon and 9:00 pm.

Our mobile learning units have been equipped with web components for video, texts, images, PDF viewer, quiz and exercise tasks. For the NTS, the PDF viewer was the most popular mobile component (67%), quizzes the second (56%). The TS preferred quizzes (68%) over PDF viewer (64%). Video only received 44% of the votes from the NTS and 57% from the TS. Video seems to be overrated for mobile learning.

Mobile phones are the dominating Internet platform today. All our students have smartphones, providing universal access to mobile learning anytime, anywhere. But the screen is rather small and many learning activities in STEM subjects, for example programming tasks, are very difficult to perform on these devices. Therefore, laptops are the preferred device (74%) at least in coding-heavy courses.

Social activities are one of the most important types of activities in the physical as well as the virtual world. Therefore, in our courses we ask students to start small project teams in which problems are solved in teamwork. Team building is supported by our tiny web component called "teambuild" [11], in which only two buttons are available for every student, one for joining and one for leaving a team, in real time using web sockets. This means that teams can be established in just a few minutes, even on large courses with several hundred students. The method is highly scalable, as the time for team building remains constant even if the number of students is increased. On a course with 270 students, teams were joined 298 times and left 93 times. 113 teams were run in parallel, 52 of which kept completely stable the whole semester for 16 weeks. Only one person left in 39 teams, 2 people in 17 teams, 3 people in 2 teams and only more than 3 people left in 3 teams. All in all, team building turned out to be relatively stable and fast. The instructor had to intervene personally only twice for 5 min each, in order to resolve team conflicts that could not be resolved by the participants on their own. On the other hand, all teams were very active keeping their GitHub repositories, forums and wikis up to date. Student teams struggle to find their identity and to give it a name: The team name was changed 764 times in total.

STEM students should practice teamwork during their courses. Our Kanban component [11] supports the lean management method via the collaborative drag-and-drop of Kanban cards showing the result in real time. Project tasks are written down on Kanban cards together with responsibility, priority and workload. All Kanban cards are sorted in three lines, "*ToDo*", "*Doing*", "*Done*". Based on our statistics, the NTS turned out to be more focused and more effective in getting things done. They accepted responsibility for project tasks more often and ensured completion more successfully.

If practical exercises are based on weekly tasks, students often want to receive a "master" solution afterwards. But many tasks do not have a master solution. On the contrary, there are many possible ways, techniques, tools and methods. Our solution is our web component "*show_solutions*" [17], which provides the list of all student solutions after the deadline together with rating and commenting [12]. Ticking the "*show solutions*" checkbox displays the best rated top ten solutions. The "*more*" button allows students to scroll through an additional 10 solutions, adding another "*more*" button dynamically, until all solutions are shown. Statistical results on using the "*more*" button showed a 36% difference between student groups: the NTS tend to browse through more solutions more often. Usage of the initial "*show*" button was virtually equal. It is evident that NTS tend towards intense in-depth study more often.

5 Summary

By directing student activation towards iOER creation on the WWW platform, students make a public contribution to digital common goods. Their activity inside the virtual classroom has an impact on the power of the web everywhere. Apps such as Facebook, Uber and Airbnb implement a platform economy of "the winner takes it all", leaving many losers behind. The WWW platform has many winners, giving equal opportunities to many people, small and medium-sized enterprises and startups. Getting students active by focusing them on iOER creation on the open Web platform thus contributes to societal wealth in a far better way than the closed app economy. Student activation in iOER Maker Spaces also becomes civic activation with a societal impact.

Acknowledgment. The author would like to thank the work&study project team [16], in particular André and Tea Kless for their excellent work on *ccm* [9]. The author would like to express his gratitude to the funding agencies: the German Federal Ministry of Education and Research (BMBF) for funding the work&study research, ref. no. 16OH21056. Finally, the German Stifterverband and the Ministry of Culture and Science of the German State of North Rhine-Westphalia also awarded the author a digital fellowship [17] for developing the Digital Maker Space.

References

1. Berners-Lee, T.: Welcome to the Web's 25th Anniversary. Dated 12 Mar 2014 (2018). https://www.w3.org/webat25/news/tbl-web25-welcome
2. Shank, J.D.: Interactive Open Educational Resources: A Guide to Finding, Choosing, and Using What's Out There to Transform College Teaching, 1st edn. Jossey Bass (2014)
3. webcomponents.org (2018). https://www.webcomponents.org/introduction
4. W3C: Web Components specifications (2018). https://github.com/w3c/webcomponents
5. Kaul, M., Kless, A., et al.: Game changer for online learning driven by advances in web technology. In: Proceeding E-Learning, (MCCSIS). Lisbon (2017)
6. Pariser, E.: The Filter Bubble: What the Internet is Hiding from You. Penguin (2012)
7. Berners-Lee, T.: Information management: a proposal; hand conversion to HTML of the original MacWord (1989). https://www.w3.org/History/1989/proposal.html
8. MacFarquhar, N.: Russian court bans telegram app after 18-minute hearing. https://www.nytimes.com/2018/04/13/world/europe/russia-telegram-encryption.html. Accessed 13 April 2018
9. ccm: Client-side Component Model (ccm). https://github.com/ccmjs/ccm
10. Polymer (2018). https://www.polymer-project.org/blog/
11. Kless, A.: ccm-components (2018). https://github.com/ccmjs/akless-components/
12. Kless, T.: W2C (2018). https://tkless.github.io/w2c/components.html
13. Digital Maker Space (DMS) (2018). https://ccmjs.github.io/digital-maker-space/dms/
14. LearningApps.org (2018). https://learningapps.org/
15. Stackoverflow: Why developers hate iframes? (2018). https://stackoverflow.com/questions/1081315/why-developers-hate-iframes
16. work&study: Homepage of project "Work and Study - offene Hochschulen Rhein-Saar". German Federal Ministry of Education and Research, ref. no. 16OH21056 (2018). http://www.work-and-study.info
17. Kaul, M.: Digital Maker Space; Fellowship for Innovations in digital higher education (2017). https://www.stifterverband.org/digital-lehrfellows/2017/kaul
18. https://teropa.info/blog/2014/08/10/web-components-and-bottom-up-design.html (2014)
19. Eck, L.: Factory for *ccm* components. M.Sc. project and thesis, H-BRS. https://github.com/ccmjs/ccm-factory (2018)
20. Kaul, M.: ccm-components. https://github.com/ccmjs/mkaul-components (2018)

The Didactic-Technology Challenges for Design of the Computer Supported Collaborative Teaching

Stefan Svetsky$^{(\boxtimes)}$, Oliver Moravcik, Pavol Tanuska, and Zuzana Cervenanska

Slovak University of Technology, Bratislava, Slovakia
{stefan.svetsky, oliver.moravcik, pavol.tanuska, zuzana.cervenanska}@stuba.sk

Abstract. A design of the Computer Supported Collaborative Learning (CSCL) represents a big challenge from an informatics point of view. In general, any computer support of teaching requires existing didactic algorithms for design of the informatics algorithms and writing programming codes. In comparison with technical processes, which are well-standardized, due to the complexity of teaching processes, such algorithms are missing. Moreover, the computer support of collaborative activities requires using the virtual spaces of servers, networks and clouds. Additionally, the lifetime of software and hardware is too short for supporting long-life learning. Contextually, a complex, didactic-technology approach is not researched enough within the fields such as technology-enhanced learning (TEL) or educational technology. This paper presents the authors' didactic-informatics approach related to these challenges. It is based on developing the in-house education-specific software, knowledge representation, personalized IT infrastructure, and a specific way of solving the knowledge transmission between off-line and online environments to support collaborative activities of teachers and researchers. This complex personalized approach seems to be beyond the state-of-the-art (e.g., a utility model is used).

Keywords: Computer supported collaborative learning · Long-life learning · Technology-enhanced learning · Educational technology · Knowledge transmission

1 Introduction

The integration of ICT into education represents a very complex problem – despite the great progress in digital technology. Specific challenges related to digital technology, TEL or educational technology can be found in [1–4], for example. Key trends accelerating technology adoption and important developments in educational technology for higher education are discussed in [5]. In the HoTEL report [6], it was analyzed which technologies are being researched in TEL (where they were classified in three types: reviews of existing technologies, identification of new trends, and forecasting reports). The report contains a list of reviewed FP6 and FP7 projects. Especially realistic is the TEL monography [1], which is still useful, because many

© Springer Nature Switzerland AG 2020
M. E. Auer and T. Tsiatsos (Eds.): ICL 2018, AISC 916, pp. 46–57, 2020.
https://doi.org/10.1007/978-3-030-11932-4_5

actual problems were predicted there. For example, it is mentioned that professors "also need the support of their assistants, who in turn require institutional support to prepare courses, to model collaborative learning and perform evaluations jointly with the lead professor" or that "preparing a virtual course may require twice the time needed to prepare a traditional one and require a great deal of technological expertise." It is emphasized that "this can be a frustrating experience, at least in the beginning. That is why we find it necessary to have a design and technology team behind every professor, which assists with the design of Web pages, design of materials, design of visual aids, aesthetic enhancements, production (including pre and postproduction needs), and the technological platforms used for interaction." The issues mentioned above automatically relate to the issues of CSCL, which is a part of TEL. In comparison to the past, only collaborative activities are added, which technologically require the shared virtual spaces or clouds (Note, they did not exist at the time of publishing the TEL monography). The state-of-the-art in CSCL one can find e.g., in [7].

Relatively frequent is the critique of the lower level of the TEL, i.e., the computer support of teaching and learning (see e.g., [8]). For instance, it is stated in [9] that the "Higher Education-TEL" community has been just as ineffective as the school-level "ICT" community at delivering real improvements in education. As was presented on the Online EDUCA conference in Berlin, "the European Commission, which has funded a series of large academic research projects into TEL, explicitly recognized that there was a general 'absence of evidence' that the projects had achieved lasting impact" (at this time the London KnowledgeLab reported the premise that "evidence of digital technologies producing real transformation in learning and teaching remains elusive"). It is discussed in [9], as well, that the TEL acronym for Technology Enhanced Learning should be terminologically understood as (1) T: education-specific technology (2) E: by teaching students a technology-enhanced curriculum, and (3) L: Teaching (not learning) is what we do. Similarly, the attention was drawn to the tendency of a range of technologies to reduce pedagogical interactions to a series of data-field transactions of information [10]. Within the UK survey, which was focused on the TEL tools that are being used by institutions to support learning, teaching and assessment activities – mostly "global" VLEs (Virtual Learning Environments) – were evaluated as the key tools, (e.g., Moodle, Blackboard Learn) [11]. Surprisingly, only nine percent (9%) of the VLEs currently used were in the category "Other VLE developed in-house". The absence of education-specific personalized software for TEL was emphasized in the survey. The same is also stated in [3], i.e., that the suitable software in the field of TEL is still missing.

According to the authors, this criticism is related to the didactic - informational imbalance of existing approaches. The authors link these challenges to the absence of a knowledge-based approach. They emphasize the question of knowledge in their publications by arguing that teaching and learning are knowledge-based process. So, if one wants to computerize teaching processes, one must design the suitable knowledge representation and start from the didactic background. It is important because the automation of teaching is related to mental processes. From the existing learning theories, the constructivism or radical constructivism seems to be interesting because knowledge is considered as a mental representation. The learner is not a passive recipient of knowledge, but that knowledge is constructed by the learner. Such a constructivist approach to teaching – including linking cybernetics and the theory of knowledge – is presented, especially, by von Glasersfeld [12]. The authors' research on TEL is also linked to cybernetics regarding their model of the knowledge representation, which is discussed in this paper (see also [13]). In general, the design of the TEL or CSCL regarding the question of knowledge requires additional ontological and semantics approaches. Such issues with a focus on robotic education are discussed in [14, 15], or on cloud-based learning [16], soft computing [17], and teaching the blind students, respectively [18].

The authors' research started around ten years ago with a vision that knowledge workers (teachers, researchers) should be equipped with IT tools for the complex support of knowledge-based processes. So, the in-house education-specific database beta-software, BIKE(E), has been designed for performing the batch knowledge processing, which enables the teacher to automate his/her teaching based on the simulation of teachers' activities. The complex didactic-technology approach of the authors is briefly discussed in the paper, especially the principles of knowledge representation and processing; barriers of ICT integration into teaching; the IT support of STEM courses of study and enlarged IT support.

2   The Principle of Knowledge Representation and Processing

The authors' research approach is not designed to compare such didactic situations when one group uses computers and other one practices "traditional learning". Contrary to this, how the teacher develops his/her educational technology is up to him/her; this simulates cognitive functions of mental processes associated with the teacher's activities needs (the in-house beta-software BIKE(E)-Batch Information and Knowledge Editor and Environment). It provides the teacher with a personalized infrastructure (including VLE) within which the teacher's and students' knowledge are transmitted between online and off-line environments (e.g., between the teacher's personal computer and the cloud). The knowledge processing and transmission complement the traditional face-to-face teaching. The human-curriculum knowledge is processed by using a model of the virtual knowledge representation, which is both human-readable and machine-readable (where both the teacher and computer use the same knowledge). The BIKE(E) enables the teacher to process knowledge analogically to a calculating machine, which processes numbers. Because the virtual knowledge is only an empty data structure, it is crucial how users (teacher, researcher, students, and knowledge workers) design the inserting of their knowledge into this default structure. After that, ontology and semantics are automatically created, and the data becomes semi-structured. Then they can be better controlled by programming codes.

In general, engineering study requires more self-study activities than standard learning. Thus, the teacher does not have didactically enough time for each student, as it would be needed. Moreover, the amount of contextual information regarding each topic of courses of study is "huge", so one should navigate in this vast amount of information. Therefore, it is a guiding role of the teacher to prepare teaching material of high quality that generalizes the topics of teaching. It is common that a university teacher constructs teaching and learning materials for lectures and exercises.

In practice, these teachers only have a limited time for planning, e.g., minimally 50 min for the lecture or exercise. Moreover, within this limited time, the teacher must manage several topics. It is a paradox that only a minimal number of scientific papers deal with this time factor.

The authors' research started with the idea, which was presented on the e-Learning brokerage event in Warszawa (2006) – that a teacher as a knowledge worker should be equipped with technology analogically as are a contemporary soldier. This was a driving factor within their research under the umbrella of TEL. One should be aware that computers were not invented for teaching. Thus, teachers need a suitable knowledge representation and many specific software or hardware tools to cover automation of teaching processes. Additionally, because the processes are knowledge-based, he/she must identify curriculum knowledge, processes, and steps, and design didactic algorithms (methodology) as a background to be able to write informatics algorithms and programming codes. The in-house software BIKE(S)/WPad as an all-in-one teaching support tool was also tested and developed continually for IT support of collaborative activities which requires knowledge exchange and managing. In this context, these categories of collaboration were implemented until now: Student–Student | Student–Teacher | Teacher–Students | Teacher–Teachers | Teacher–Researchers.

The research on the automation of teaching processes showed that design of the didactic issues or a "direct" computer support for collaborative teaching is potentially solvable [19]. Contrary to this, in order to solve the adaptation of the in-house software to digital technology (e.g., to Windows, networks), or to overcome the incompatibility of software and hardware on the market, it is complicated and very time-consuming. From the automation point of view, it is the weakest point of the automation. Moreover, a short lifetime of digital technology was found as an important technological barrier for developing long-term education computer support. Despite this fact, several case studies resulted as outcomes from the implemented computer support of teaching (ca 2000 bachelors students participated during the long-term authors' research). The preliminary outcomes showed that the research on CSCL requires the teacher to solve both didactics solutions (to indicate and define didactic algorithms), and software and hardware solutions (– especially, to solve the knowledge transmission between off-line and online environments). The long-term suitability of the operatively-developed in-house software was achieved thanks to the approach on how to solve the unstructured data transmission (i.e., the knowledge transmission). The related utility model is registered on the Slovak patent office [20]. Consequently, another patented solution is worked out in relation to a vision of an educational robot (Note: As for the in-house software, it is not possible to patent software in the European Union).

In terms of learning outcomes, tens of contribution to the conferences and journals were continually published by authors, including the ICL Conferences (2012–2017). The publications content was targeted to: (1) the empirical participatory action research, when the teacher designs his/her own educational technology in parallel (as the technology-driven or educationally-driven computer support, and active learning); (2) building the personalized Virtual Learning Environment by using the in-house software and associated IT infrastructure; and (3) the computer-supported collaborative teaching (CSCL) for solving the collaborative activities of students and researchers. Presently, e.g., the collaborative activities are tested on the IBM BOX-cloud by the

V4 + AcaRD Consortium of six academic subjects that was initiated within the conference ICL 17. The authors' approach resulted in many added didactic and informatics values, which are briefly summarized in Table 1.

Table 1. Added didactics and informatics values within tens of case studies (basic, CSCL)

Added didactic values	Added informatics values
Identification of didactic algorithms for lectures, exercises, self-study	Developed informatic algorithms associated with didactic algorithms
Production of the extended teaching, study and self-study materials	Tailored personalized ICT infrastructure with communication channels (combination of off-line and online)
Shortening teacher's time for preparation of the teaching and constructing study materials (lectures, exercises, self-study)	The Utility model used for transmission of knowledge and files (feedback, shared activities)
Elimination of passivity of students and minimizing plagiarism when using WPad	Shared virtual spaces for collaboration and knowledge exchange on faculty's cloud and IBM cloud (international team)
Model of the elimination of passivity and absence of students within distance learning	Educational software WPad installed on each computer in classroom used for several courses of study
Enabling feedback (not possible without communication channels)	Educational software for cloud and virtual environments (client-server)
Knowledge exchange – the international collaboration of V4 teachers	Apps – Database applications for research, teaching, learning and self-study
Audio additions of English language into the course of study "Background of environmental protection"	Personalized research on technology-enhanced learning, Digital libraries (scientific heritage), Learning analytics
Models of collaborative activities (collaborative writing of semester works and study material for chemistry student- created content for students)	Visualization tutorials and computing support for chemistry, language support for self-study and academic writing
Extended assessment variations (summative and formative assessment), support of diploma and bachelor's works	Research areas within FP7/Horizon: Technology-enhanced learning, Learning Analytics, HLT, CSCL
Automation of associated teacher's activities (research, publishing, information and knowledge management, administrative activities)	Universal - personalized all-in-one multipurpose software/system that does not exist in the market
Creation of knowledge base of C ++ programming codes by students/teacher	Modelling visualization and natural marking for STEM support

3 Barriers of ICT Integration into Teaching

As for the integration of digital technology into education, in contrast to long-term teaching needs, the life-cycle of technological tools is shorter. In general, in the case of an ICT support of teaching, computers work reliably commonly for three to five years. After that period, they work slower and slower. This fact is rarely discussed in scientific literature. It is automatically supposed that any computer support of teaching functions perfectly. However, it is not true from the point of view of the teacher who needs sustainable, long-term personalized support. In [21], these issues are discussed by authors in more detail. There, the obstacles related to the barriers of ICT integration into teaching are divided into several categories:

- Internet Browsers
- Changes of Hardware and Software
- Servers and Networks
- Students and the Human Factor

It must be once more emphasized that computers were not invented for teaching. Therefore, this gap stemming from ICT state-of-the-art tools can be considered as a standard situation. The authors' research has shown that the best solution to the problem of shorter life time of technology is to develop both their own educational software and personalized ICT infrastructure for educational data transmission. In the case of the all-in-one educational software application BIKE(E)/WPad, the life cycle is not ended even after around ten years of integration of ICT into engineering teaching within the author's research on TEL. In addition, the sustainability for long-term teaching is automatically assured while operating system Windows continues to exist. Moreover, the principle of the approach is "evil" because of the independence of virtual knowledge representation from software and hardware. [The database application could be also written by using any other database platform.]

Another specific teaching and learning issue is the knowledge transfer. The issue of the educational data transmission (the content transfer) is less described in scientific literature than the content processing, despite the fact that it is very time consuming. It does not matter how many students have the teacher, the data transmission requires one to build a complex ICT infrastructure with paths for knowledge transfer (as the files in various formats) between off-line and online devices. In the case of CSCL, this issue can be crucial, i.e., how to transfer knowledge between tens of students or researchers. It is common as well that teachers use some computers and e-devices. Contextually, Fig. 1 illustrates the teacher's infrastructure used by the main author of this paper as a combination of home and work computers, clouds and online virtual spaces (– just the Windows 10 is already used presently).

It should be also emphasized that the computer time is very important since computer time is typically very important for the process of transmitting knowledge. In the case of online/off-line transmission of files or folders, it requires the user to take minutes – even tens of minutes (especially when using at-home WIFI with a weak signal or when copying giga-bytes of files between computers, networks and e-devices).

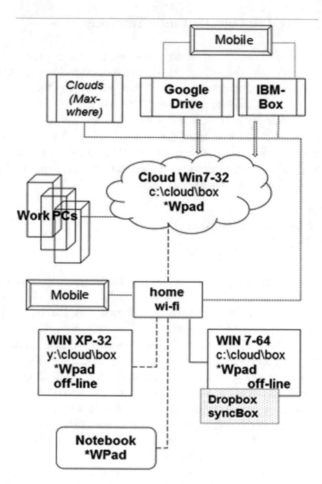

Fig. 1. Example of the personal teacher's off-line/online ICT infrastructure

4 The IT Support of STEM Courses of Study

Within around ten years of the authors' research on TEL-support of bachelor's-level teaching, several categories of courses of study were supported, e.g., Background of environmental protection, Programming languages, Occupational health and safety, Chemistry, Industrial management. The research showed that each of them required specific didactic approaches, so also different informatic approaches. This means that one course of study required e.g., more text formats than images, more working off-line than online, or more active or passive activities. For example, the IT support of STEM teaching required more visualization, which has, in general, a richer information content. A specific problem was how to solve chemical formulas and equations without using any specific software. The most effective method was using the paper and chalk and blackboard in the classroom. It could also be mentioned that PHP-programming codes were used for training in areas such as calculation of chemistry tasks, including writing a set of tutorials related to teaching photosynthesis.

4.1   Basic IT Support - the Design of Educational Packages

Because there was no educational technology which could cover such a large range of problems in teaching, the education-specific application WPad was developed on the principle of simulation of teacher's or student's activities. [It is a selected part of the in-house software BIKE(E).] This means, that the programming codes were written based on didactics steps primarily, i.e., up to the teacher needs. The complex didactic-technology approach can be explained in the case of designing the IT support of STEM when solving the modelling and simulation tasks which require a higher level of visualization. Principally, the approach must be in compliance with the didactic goals and algorithms required by the teacher.

In this case, the teacher wants to design the online application aimed at the self-study of distance learning students, and a set of exercises when creating the simulation projects by individual students.

Specification of the teacher's requirements requires that the curriculum content should be created by using texts, images, videos and files in the native format of the Witness software, which is used for exercises, as follows:

- images and texts as the navigation elements
- videos as the motivation and teaching elements
- files in the native format of the simulator (*.mod) with access to Witness software - it should enable students' direct launching and browsing of the files

As for the specifications, she would consider as the ideal case when enabling students to relink the problem areas with the offer of manuals and scripts, i.e., to support students to solve the individual problem tasks by using:

- hypertext links for instructions, including the enlarged help-function
- keyword search in scripts of the course of study
- search in the knowledge base consisting of simulation models and instructional files.

Figure 2 illustrates how to fulfill the teacher's requirements for IT support of the course of study. The in-house software BIKE(E)/WPad will work as a black box, which would join the inserted educational "big data" from the associated teaching resources. In the "black box", the educational packages will be constructed, which will be used by students and teachers within the exercises. The design of the educational packages is just being started with intention to also obtain model templates that could be useful for the planned design of the educational robot within the V4 + AcaRD consortium (a project of academic subjects of so-called Visegrad countries).

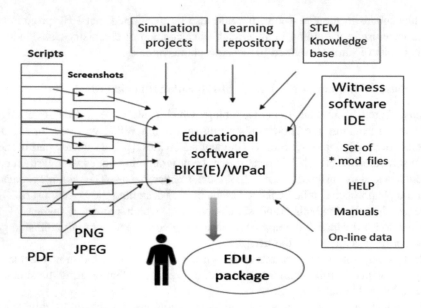

Fig. 2. Scheme of the design of educational packages

4.2   Enlarged IT Support – the Design of Collaborative Teaching and Learning

Commonly, a general or global software and hardware are mechanically used; this is not automatically suitable already for the standard teaching. Mostly, it is tested by users to see whether or not it works; therefore, didactic aspects are not primary. Additionally, in comparison with the standard teaching, any collaborative teaching and learning requires a specific methodology for collaborative activities and enlarged IT support. Namely, "collaborative" means practically that a group of users needs to share virtual spaces for the collaborative work or larger number of files must be processed in a unified manner.

As it is apparent from the previous description of the design of educational packages, a teacher always must play the key role (not technology). Thus, he/she should formulate didactics goals and procedures (algorithms) which would be a background for writing the informatics algorithms and the design of suitable IT infrastructure. Because the collaborative activities are not a part of standard teaching, the additional methodology must be prepared before a design of any IT support. This is very rarely discussed in scientific papers because it is automatically supposed that the didactic algorithms exist. However, each teacher knows that if one wants to teach whatever subject collaboratively, it is relatively a hard issue. Moreover, from the informatics point of view, it requires very specific software and hardware for the shared information and knowledge exchange. The "knowledge exchange" practically means that the crucial problem is how to solve both knowledge transmission between off-line and online environments and how to share knowledge among many users (it requires shared virtual spaces or cloud environments). Knowledge transmission means that data or files are off-line or transmitted online between computers and networks. Another specific issue is that if one wants to automate teaching, one must additionally also solve the adaptation of one's programming to the Windows operating system. In other words, the teacher also must solve the personalized file management, i.e., write the analogical programming codes as are performed by the Windows Explorer, Internet Explorer, or Internet browsers, in general. The authors' research showed that this is the most time-consuming programming activity. Table 2 illustrates it on the example of data transfer/transmission.

Table 2. Examples of data transmission/transfer cases when teaching STEM

Data transmission/transfer cases from needed for the complex IT support	
Adoption to Windows: file management and conversion of formats (pdf, images)	Data transfer via WIFI and e-mail communication
FTP is needed (teacher's computers - virtual space of the institute) Data compatibility from other software outputs (ASCII unification)	Data transfer from the EU, national, local, and institutional resources and information systems (CORDIS, AIS)
Data transfer from other database platforms, Excel, text/audio editors Data transfer from the internet services, databases and repositories	Teacher's computers - Faculty's cloud Teacher's computers - Global clouds (e.g., Google Drive)
Data transfer from the internet browsers Data transfer from the USB and external discs	Data transfer from the mobiles, digital camera, e-carriers International networking (V4 + AcaRDC): Researchers' computers – IBM Box cloud

5 Conclusions

One of the authors' finding is that any IT support of the knowledge-based teaching processes, i.e., the simulation of mental processes, requires synchronized solving both didactics and technology issues. It means that any computerization of every didactic activity requires solving the didactics and technology simultaneously. The didactic-technology challenge is that any suitable education-specific technology does not exist. The author's approach is primarily based on the selection of the didactic situations that need to be computerized. This may be e.g., collaborative writing of semester works or scientific articles; a methodical procedure to eliminate plagiarism, a support of STEM; or the mass production of study materials. When the didactics algorithms are defined, they are processed by the education-specific in-house software BIKE(E)/WPad within the personalized off-line/online IT infrastructure. In comparison with the IT support of standard teaching, the design of the CSCL required solving the data (knowledge) transmission between off-line and online environments (virtual spaces, clouds, networks). This brings absolutely a new situation, especially from personalized point of view. Programming codes of the in-house software must also cover the adaptation to existing technology, especially how to transfer the didactic content of knowledge between online and off-line environments. This requires solving the compatibility issues and elimination of the above-mentioned technology barriers. Additionally, the specific approach is needed for solving the computer-mediated communication. [Until now, these communication levels were tested: man-machine, machine-machine, man-machine-machine-man, people-machine-machine-people.] As it was explained on examples from the teaching practice, this all requires the specific enlarged IT support which is complex and time-consuming. In this context, the future research will be aimed at the didactic-technology design of the educational packages and computer solutions for the knowledge transfer, including automation of teaching processes based on the idea of the educational robot which is being worked out.

Acknowledgment. This research work was supported by the financial resources from the International Visegrad Fund's Strategic Grant No. 21810100 "V4 + Academic Research Consortium for integration of databases, robotics and languages technologies". The academic consortium started within the previous ICL Conference in Budapest (2017).

References

1. Goodman, S.P., et al.: Technology-Enhanced Learning: Opportunities for Change, p. 2002. Laurence Erlbaum Associates, Mahwah, NJ, USA (2002)
2. Balacheff, N., Ludvigsen, S., Jong, T, Lazonder, A., Barnes, S. (eds.): Technology - Enhanced Learning. Principles and Products. Springer, XXVI, 326 p (2009)
3. Martens, A.: Software engineering and modelling in TEL (2014). In: Huang, R., Kinshuk, N.-S.C. (eds.) Book: The New Development of Technology Enhanced Learning Concept, Research and Best Practices, pp. 27–40. Springer (2014)
4. Laurillard, D.: Digital technologies and their role in achieving our ambitions for education (2018). https://www.researchgate.net/publication/320194879_Digital_technologies_and_their_role_in_achieving_our_ambitions_for_education
5. Johnson, L., Adams Becker, S., Estrada, V., Freeman, A.: NMC Horizon Report: 2015 Higher Education Edition. The New Media Consortium, Austin, Texas (2015)
6. Nápoles, C.L.P., Montandon, L. (eds.): D 1.1.2 Emerging Technologies Landscape: Report on Field Research Results. Public report final 7-30/4/2013-M7 audience. HoTEL (2013). [cit. 2014-07-15]
7. Smith, B.K., Borge, M., Mercier, E., Lim, K.Y. (eds.).: Making a difference: prioritizing equity and access in CSCL. In: 12th International Conference on Computer Supported Collaborative Learning (CSCL) 2017, Vol. 2. Philadelphia (2017)
8. Kinchin, I.: Avoiding technology-enhanced non-learning. Br. J. Educ. Technol. **43**(2), 43–48 (2012)
9. Weston, C.: The problem with Technology Enhanced Learning. Ed Tech Now Blog (2012). http://edtechnow.net/2012/12/05/tel/
10. Lundie, D.: Authority, autonomy and automation: the irreducibility of pedagogy to information transactions. Stud. Philos. Educ. **35**(3), 279–291 (2016)
11. Walker, R., Voce, J., Swift, E., Ahmed, J., Jenkins, M., Vincent, P.: 2016 Survey of Technology Enhanced Learning for Higher Education in the UK. UCISA TEL Survey report 2016. University of Oxford (2016)
12. Von Glasersfeld, E.: Cybernetics and the Theory of Knowledge (2002). http://www.vonglasersfeld.com/255

13. Svetsky, S., Moravcik, O.: Some aspects of teaching processes computerization. In: Future Technologies Conference. FTC 2017: IEEE, 2017, Vancouver, Canada, pp. 1–5 (2017)
14. Haidegger, T.: Developing and maintaining sub-domain ontologies. In: Proceedings of Standardized Knowledge Representation and Ontologies for Robotics and Automation. Workshop at IEEE/RSJ IROS, 2014, Chicago, IL (2014)
15. Tolgyessy, M., Hubinský, P.: The Kinect sensor in robotics education. In: Proceedings of RiE 2011, 2nd International Conference on Robotics in Education. Vienna, Austria, 2011, pp. 143–146 (2011)
16. Shyshkina, M.: The general model of the cloud-based learning and research environment of educational personnel training. In: Auer, M., Guralnick, D., Simonics, I. (eds.), Teaching and Learning in a Digital World. ICL, 2017. Advances in Intelligent Systems and Computing, vol. 715. Springer, Cham (2017)
17. Volná, E.: Introduction to Soft Computing. Bookboon.com (2013). https://bookboon.com/en/introduction-to-soft-computing-ebook
18. Mikulowski. D., Pilski, M.: Ontological support for teaching the blind students spatial orientation using virtualsound reality. In: Advances in Intelligent Systems and Computing book series (AISC, volume 725) Interactive Mobile Communication Technologies and Learning, Proceedings of the 11th IMCL Conference, pp. 309–316. Springer (2018)
19. Svetsky, S., Moravcik, O.: The implementation of digital technology for automation of teaching processes. Presented at the Future Technologies Conference 2016, San Francis-co (2016). http://ieeexplore.ieee.org/stamp/stamp.jsp?tp=&arnumber=7821632
20. UV 7340: The connection of an unstructured data processing system using a specific data structure. Industrial Property Office of the Slovak Republic (2014). https://wbr.indprop.gov.sk/WebRegistre/UzitkovyVzor/Detail/45-2014
21. Svetsky S., Moravcik O.: Some barriers regarding the sustainability of digital technology for long-term teaching. In: Arai, K., Bhatia, R., Kapoor, S. (eds), Proceedings of the Future Technologies Conference (FTC) 2018. FTC, 2018. Advances in Intelligent Systems and Computing, vol. 880. Springer, Cham (2019)

Digital Transformation in Collaborative Content Development

András Benedek$^{(\boxtimes)}$

Budapest University of Technology and Economics, Budapest, Hungary
benedek.a@eik.bme.hu

Abstract. At the turn of the Millennium by the Digital Transformation (DT), we are moving fast from the appearance of human-machine interactive communication to the formation of spatial independence of human communicational possibilities. These two landmarks implying new pedagogic challenges figuratively also symbolize thresholds, one of which we have already crossed, and hesitating at the other of which we are faced with an ever complex transformation of the world of education. Recently in the Vocational Education and Training (VET) schools by the Information Technology (IT), we are moving fast from the appearance of human-machine interactive communication to the formation of spatial independence of human communicational possibilities. However, effective communication and work-oriented collaboration between man and man remained the most successful pedagogical tool. These landmarks implying new didactic challenges figuratively also symbolize thresholds, one of which we have already crossed, and now we are at the other which we are faced with a new complex transformation need. We were starting from our original construction of the collaborative open content development (OCD), which was based on the results of several learning content digitalizing projects, was built upon the recognition of the change in the teaching-learning paradigm. That was the main reason for analyzing and development of VET teachers' attitude to the OCD. Referring to the new collaborative developments implemented in the VET practice, besides the conclusions, we also formulate proposals on the new teaching competencies.

Keywords: VET content development · Digital transformation ·
Collaborative learning

1 Digitalization and Open-Up in the Content Development

Although the online learning resources and multi-media e-learning representations include more dynamic (flash, video) content, the "logic" of building up learning materials has changed little – in fact, visual content is only a completion to verbal communication. In the process of DT, the Collaborative Content Development based on interactivity seems an innovative method. Developing Open Educational Resources (OER) with students' participation means a potential for improving the content and the

© Springer Nature Switzerland AG 2020
M. E. Auer and T. Tsiatsos (Eds.): ICL 2018, AISC 916, pp. 58–67, 2020.
https://doi.org/10.1007/978-3-030-11932-4_6

methodology. The new pilot curriculum development and the applied IT solutions open source and commercial Learning Management System (LMS), and the flexible management of micro-content is capable of surpassing traditional VET teaching and learning.

In our new world, which, owing to the IT tools, differs from any of the previous ones, we also have to reconsider plenty of pedagogical questions. In the world of interactions [1], the channels and community-making elements of interpersonal communication have transformed (especially in the web 2.0 environment), and the more and more open access to learning contents and Digital Pedagogy [2] has made a considerable try to renew pedagogical thinking after the millenary and firstly [3] at the universities. Nowadays in the VET content development the rapidly changing elements determined by digital transformation aspects [4], and the dynamics of the changes are difficult to be forecasted. Even the Content Development is under a transformation, the old structures are no more able to respond to the new challenges [5]. We are less and less able to manage these changes within the rigid frameworks of the qualification systems and closed vocational structures. It has a significant impact on our research the "Trialogical Learning" approaches [6] that provide new impulses in the online learning environment. In the last decade the cognitivist learning theory [7] and. the significance of collaborative pedagogical effects has been demonstrated worldwide practice. This topic has strong VET didactical features [8], and is partly connected to the endeavours which strive to shape the alternatives of the traditional VET curricula in a learning environment determined by modern ICT (info-communication technology) in the learning process and also in the construction of the curriculum as well [9, 10].

This paper deals with a new practical aspect of general DT, using an approach to VET content development by the collaborative way. By introducing the OCD model and its description, our papers aim is to demonstrate a possibility of methodological innovation in the DT process. Subsequently, we describe the initiatives taken in the practice of the VET and the results achieved at the empirical stage of the project. In this sense, the natural way of communication between human beings relies on the application of images, icons or comics where film clips represent an interesting manifestation of unique, relatively short narrations (they quickly penetrated the world of teenagers in the past decade in the form of television advertisements and YouTube clips). So an important preliminary aspect of our research was the realization that both learning and the teaching process by the DT have become increasingly open systems. This openness is partly the result of a change of paradigm and partly the application of IT tools themselves that rendered learning a form of communication no more depending on either space or time constraints. Learning in this latter sense is about information that is no longer characterized by closed text blocks but images and other elements presented in particular selection algorithms.

2 Purpose of Our New Content Development Model

The "modern" learning materials in the frame of the Hungarian VET schools having developed by the end of the 20th century remained unchanged concerning their linear structure, the dominance of written texts and static image conveyance in terms of verbal and image communication. Interdisciplinary approaches are increasingly acknowledged, however, we should note here that accelerating technical development results in the continuous restructuring of technical culture and educational contents. Our purpose has been to implement DT for changing the process of vocational Content Development using collaborative elements in practice. These traditionally focus on information, so a strong competition may be foreseen between traditional curricula focusing on quantity and new ones representing a more complex approach. This contradiction may only be resolved by educational institutions if it is willing to modernize its knowledge transfer system that is traditionally rigid and divided into subjects and create the didactics for the new, integrated approach. According to the widely used DT, we made efforts to implement Collaborative Content Development.

The rapid penetration of IT tools since the second Millennium, in particular, presents a challenge regarding the practice and methods of education (just consider mobile communication, hybrid solutions in smartphones, multiple functions or apps). Hence, we selected the new possibilities of visual learning as a subject of our research, responding to the main trends of these changes. In addition to theoretical research, our work also included innovative didactic development aiming at creating a new system of teaching on higher complexity, regarding both the activation of students in the development of content.

3   The Open Content Development (OCD) Model

Concerning the new didactical model supported by DT became an important precondition of the online assessment of learning and innovation to be able to evaluate the methodological impacts of the system and to prove our preliminary working hypotheses on the wider professional base. All this can only be implemented in practice established with the help of which the rather diverse activities are organized into a system applying up-to-date technological solutions, as well Fig. 1. presents this input-outputs.

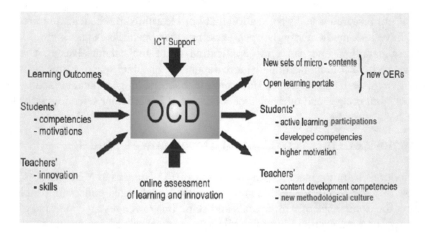

Fig. 1. The OCD model and its inputs-outputs

The OCD model, which was based on the results of several learning content digitalizing projects, was built upon the recognition of the change in the teaching-learning paradigm presented above. The new competencies connected to info-communication technologies include, at the most basic level, search for, evaluation, storing, creation, introduction, and transmission of information of multimedia technology, communication through the internet and the capability of participating in collaborative networks.

However, in the development of teaching methods and especially the renewed learning environment, there has been a specific "spatial" transformation going on in terms of the transmission of visual information. In image content services (see, for instance, the success of Facebook, Instagram or YouTube), an epochal development has taken place. Digital image creation, the storing, editing, and sharing of pictures and their distribution in the network communication systems have become essential learning activities during the latest decade but have been integrated into school teaching to a very small extent. The theories have already made it clear to us that the images bear considerable emotional and rational impacts on knowledge acquisition and our experiences. Modern curriculum theory drew the attention to the synergic effects of cognitive and affective functions already decades ago, which makes the pedagogical role of using images outstandingly important.

As Fig. 2 shows the main actors and activities in the OCD process, we primarily wished to refer, as an opportunity of the methodological renewal of vocational teacher training, to the practical possibilities of collaborative work and visual communication, i.e. imagination that has always existed but is outstandingly important today as innovation factors. In the present pedagogical practice, the limits detectable in the teachers' personal resistance are connected to special human or psychological features.

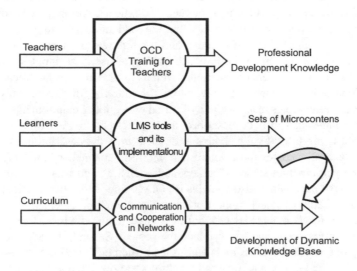

Fig. 2. Actors and activities in the OCD process

Concerning the input, the Learning Outcomes (LO) database is an input service for the users of the system that offers orientation as well as information on the already existing sources.

4   Approach and Intervention on the VET Practice

Our conception on the digitalization of the new kind learning units is aimed at involving students and willing teachers and undertaking active participation in building a learning material construction that is created with an open-access approach – providing the possibility of community content development for the concerned learners' groups or classes as well. It is also an innovative activity of ours to provide huge background storage and cloud services which this process needs. In the frame of technical university vocational teacher training programmes our OCD model, which was based on the results of several learning content digitalizing projects between 2015–2017, was built upon the recognition of the change in the teaching-learning paradigm presented above. In addition to the appearance of the new learning functions, and this digital transformation offering new opportunities through which we started disseminate last two semesters.

The OCD model main input factors are the descriptions of the Learning Outcomes (LO) in accordance with the current formal education; these descriptions connected to the requirements of the European Framework for Education and Training and it is easy to achieve their operationalization at the itemization level. The "input" competencies and motivations of the learners/students attending the given course are also of great importance and must be treated with the necessary diversification; these factors can have a considerable impact on the successfulness of education. Since our research are focused at secondary level VET and the methodological development of the teachers active in VET, we must refer to the fact that the students' motivation in this type of education is generally below the average, and for this reason, it makes a great pedagogical challenge. Concerning the input, the LO database is an input service for the users of the system that offers orientation as well as information on the already existing sources. This solution does not only allow the filtering of theoretically possible recurrences, but the topics making a part of actual educational requirements but lacking learning resources can also be described more precisely. In fact, this is a divided database that is produced by developers in order to help the description of innovation objectives, as a service for those interested. The implemented methodical system is open for the participating students/learners and educators and it is accessible through the research web page which also contains BYOD (Bring Your Own Device) proposals that serve the methodological support of the innovations implemented.

As the first result our model was built on three main activities: first, it offered a special training for the teachers who joined the project. Second, it made it possible for the participants to use the LMS tools and to learn methods necessary to handle these. This process started during the training by getting to know and applying the MOODLE system. Finally, in the evolving network, we initiated a communication focused on development collaboration between the teacher and students. Thus we considered the concluded personal development knowledge, the pile of the newly elaborated 120 micro contents, the possibility that a dynamic knowledge base can be created by the participants and the open contents created by them as output factors.

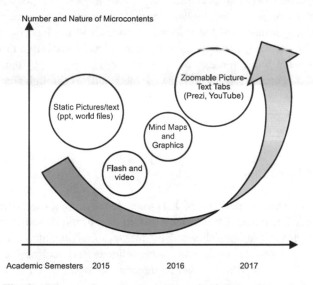

Fig. 3. Change of nature of micro-content last two semesters

Owing to the nature of action research, the expected results of the research can be perceived already in the first phase (2017–2018). The evolution of the VET schools' network is much more dynamic than expected before, the network of 12 schools that was planned to be created during three years has already been born, which indicates the interest in this process. 74 teachers joined the 30-hour blended training that was determined as the precondition of formal participation, which means that 5–7 vocational teachers joined the development program in each school. The number of the micro-content units related to the vocational learning units elaborated by the teachers is now over one hundred, and as a result of the teachers' work, the first results of the students' learning content development have also appeared in our model of open learning content development. The organization of the micro-contents, their labeling and the creation of the possibility to analyze them show two significant results. On one hand, the representational characteristics of the micro-contents created within the frames of open content development have become searchable in several dimensions.

Our research attributes special importance to the appearance of visual elements, the diversity of their functions and the growth in the role they play in enriching multi-modality (Fig. 3). On the other hand, the thematization of the micro-contents, the applied presentation methods (e.g. adopting dynamic pictorial presentation – flashes, videos) and the illustration of structures on mind maps clearly indicate characteristics that point perceivable differences regarding the age and educational parameters of the teachers' and the students' population. The examination in the field of VET schools offers well adoptable methodological potentials on the topic of Inclusion and Exclusion. It eases the activation of the disadvantaged students potentially dropping out – some of them with social disadvantages while others with learning difficulties having evolved in the previous school grade – as well as their joining in the development processes, and offers the opportunity to apply the cooperative pedagogical methods in the teaching process of vocational subjects. It also reduces the possible lack of educational materials or school books, and through the applied presentation methods, it also offers the opportunity for the students to connect to the learning materials (with examples or case studies) with new mobile communication tools and through their own views. On the other hand, through the constructive description of the personal tasks, this method allows creative work with the talented students through sharing the content elements created by them or the pedagogical elaboration of the experiences in order to create classroom communication processes in which each student and teacher, as it is typical in open systems, can actively participate.

5   Actual Outcomes

In the practical implementation of OCD model, the number of participating VET teachers was 87. The output factors can be arranged into three groups: we consider the content development results elaborated by the teachers/students, which can be presented in the forms of micro-contents, case studies or practical problem solutions, as direct outputs. These outputs can also be understood as single OER units and can be accessible in the system for others for further use. The information gained from measurement relating to student participation, the development of competencies and the levels of motivation is an important result and makes the direct or indirect effects of the OCD model possible to be analyzed. The improvement in teachers' competencies, which can be proved with measurement, appears here, as well; as a hypothesis, it is at the same time treated as the essential specialty of the OCD model and the core field of methodological development. This is how, during our research, it has become a new answer to old questions that online content units should be made more than an "experimental teaching method" by extending the frameworks of VET teaching and learning.

At the beginning our project we undertook to elaborate a new complex curriculum which on one hand offered a framework for a general course on the analysis of system theory and on the other hand it allowed students to prepare case studies by which they could participate in the development of the curriculum of the course. We followed the general rules of OER development, and we started teaching the "System theory" course to a major group of exactly the students questioned in our survey; an e-learning resource was also made for the course, and this material, together with the solutions of the exercises, was administered in the Moodle system.

The constantly evolving *Sysbook* platform (W1) connects to the OCD model; this platform introduces the open learning resource elaborated in relation to a certain comprehensive topic: About systems and management for everyone [11]. This platform (W2) presents the students' innovations, as well, and after an expert evaluation, the development results can be published here. In *Sysbook* several case studies [12] show the application of system view. In the different examples it is investigated, what is considered a system, how is the system connected to its environment, what are the input signals and what are the output signals? How can the model of the system be built? What are the requirements set for the system? Which balance and energy considerations have to be applied? Can we control the system? How to control the system?

Considering the open approach the students can contribute to the content of *Sysbook*. They can develop their own contents related to systems and controls they are familiar with. After a review, this area can be supplemented with new materials. Till now some examples are temperature control of a terrarium, speed control, model of the learning lab experiment, the model of motor, etc. (Fig. 4).

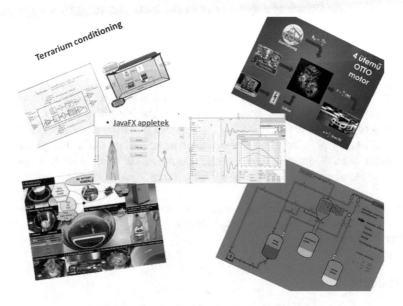

Fig. 4. Systems appearing in the *Sysbook*

6   Conclusions and Recommendations

Matching the peculiarities of Hungarian vocational education, we develop methodology training for complex school subjects to develop the educational competencies needed for designing and applying complex curriculum units in the training of vocational teachers and practical training. The basic idea is to provide a base for a new, medium-term research and development project that will aim at applying the developed methods and procedures in practice, designing and applying new e-learning curricula. The conservative approach of the teachers training to new initiatives in the field of open content development mostly roots in its general attitude. We should note here that "new" methods of VET are much more open-minded regarding interactive collaborative systems. A good example of this attitude is the adaptation of OCD model initiated in 2016 by a new project, organizing open online in-service training courses in a new formulated VET pilot school network. In the case of new experiments, the co-operation and the procedure are sufficient and results are encouraging for the future. Our research in the frame of the OCD model essentially focused on the Collaborative Content

Development in-class activities of vocational teachers and the application of new educational methods.

The activities of teachers are based both on conventional criteria such as knowledge, skills and attitudes and new teaching competencies as follows:

- to formulate vocational training topics and learning objects as a problem;
- to activate the student seeking solutions for the problems;
- to create a description as micro-content about the given problem;
- to analyze the problem as a system (inputs, outputs, the task of the control);
- supporting, organizing and controlling the cooperation among the students;
- recognition further links to the other systems/problems;
- commitment and responsibility for professional development by collaborative activity.

According to the current procedural approach, applying such solutions at the level of open content development requires decisions at the institutional level. Utilizing the advantages of digital transformation and the fact that it allows accessibility in space and time, comfort and personal time management in acquiring the learning material, we succeeded to open the process of content development and make it possible for teachers and students to share new knowledge by collaborative work.

Acknowledgment. This research project of the MTA-BME Open Content Development Research Group is funded by the Content Pedagogy Research Program of the Hungarian Academy of Sciences.

References

1. Paavola, S.: The knowledge creation metaphor. In: Seel, N.M. (ed.) Encyclopedia of the Sciences of Learning, pp. 1682–1684. Springer, New York (2012)
2. Benedek, A., Molnár, Gy.: ICT in education: a new paradigm and old obstacle. In: Leist, A., Pankowski, T. (eds.) ICCGI 2014: The Ninth International Multi-Conference on Computing in the Global Information Technology. Sevilla, Spain, 2014.06.22-IARIA, pp. 54–60 (2014)
3. Benedek, A., Molnár, Gy.: New approaches to the E-content and E-textbook in higher education. In: Gómez Chova, L., López Martínez, A., Candel Torres I. (eds.) INTED2015 Proceedings: 9th International Technology, Education and Development Conference. Madrid, Spain: International Academy of Technology, Education and Development (IATED) 2015, pp. 3646–3650 (2015)
4. Beetham, H., Sharpe, R. (ed.): Rethinking Pedagogy for a Digital Age: Designing for 21st-Century Learning (2nd edition). Routledge, Abingdon. See more (2013). http://www.dtransform.eu/
5. Nore, H.: Re-contextualizing vocational didactics in norwegian vocational education and training. Int. J. Res. Vocat. Educ. Train. (IJRVET) 2(3)(Special Issue), 182–194 (2015). https://doi.org/10.13152/ijrvet.2.3.4
6. Paavola, S., Hakkarainen, K., Sintonen, M.: Abduction with dialogical and trialogical means. Logic J. IGPL 14(2), 137–150 (2006)
7. Siemens, G.: Connectivism A learning theory for the digital age. Int. J. Instr. Technol. Distance Learn. 2(1) Available online at (W3), (ISSN 1550–6908) (2005)

8. Gessler, M., Herrera. M.L. (eds.): Special edition: vocational didactics. Int. J. Res. Vocat. Educ. Train. 2(3) (2015)
9. Colons, A., Halverson, R.: Rethinking Education in the Age of Technology. Teacher College Press, New York (2009)
10. Benedek, A., Molnár, Gy.: E-teaching and digitalization at BME. In: New Technologies and Innovation in Education for Global Business: 19th International Conference on Engineering Education - PROCEEDINGS. Zagreb, 2015.07.20 2015.07.24. (the University of Zagreb, Faculty of Economics & Business) Zagreb (2015)
11. Vámos, T., Bars, R., Sik, D.: Bird's eye view on systems and control – general view and case studies. In: 11th IFAC Symposium on Advances in Control Education, ACE 2016, Bratislava, Slovakia, IFAC-PapersOnline Vol. 49, Issue 6, pp. 274–279 (2016)
12. Benedek, A., Horváth, J. Cz.: Case Studies in teaching system's thinking. In: 11th IFAC Symp. on Advances in Control Education, ACE 2016, Bratislava, Slovakia, Preprints, (ed. M. Huba and A. Rossiter), IFAC-PapersOnline, Vol. 49 Issue 6, pp. 286–290. (W1) (2016). http://sysbook.sztaki.hu/ (W2) http://sysbook.sztaki.hu/studentarea.php (W3) http://www.elearnspace.org/Articles/connectivism.htm

Teaching Programming
and Design-by-Contract

Daniel de Carvalho, Rasheed Hussain, Adil Khan, Mansur Khazeev,
JooYong Lee, Sergey Masiagin, Manuel Mazzara$^{(\boxtimes)}$, Ruslan Mustafin,
Alexandr Naumchev, and Victor Rivera

Innopolis University, Innopolis, Russia
{d.carvalho,r.hussain,a.khan,m.khazeev,j.lee,s.masiagin,
m.mazzara,a.naumchev,v.rivera}@innopolis.ru

Abstract. This paper summarizes the experience of teaching an introductory course to programming by using a correctness by construction
approach at Innopolis University, Russian Federation. In this paper we
claim that division in beginner and advanced groups improves the learning outcomes, present the discussion and the data that support the claim.

1 Introduction

Formal methods are still struggling to get a broad acceptance in industry world-
wide, and Russia is not an exception. Innopolis is a new IT city [7], incorporating
a technopark and a university, aiming at prioritizing the development of IT and
software engineering in Tatarstan and in the Russian Federation. Innopolis University (IU) is pioneering several research and pedagogical projects and experiments with innovative teaching methods and curricula. One of the numerous
peculiarities of this innovation has been the decision to teach formal methods
and correctness by construction together with programming since the first year
of the bachelor program. In particular the Eiffel programming language [9] is
used as a programming instrument and Design by Contract as a methodological
and conceptual tool [8].

This paper summarizes the experience accumlated by followinng this pedagogical approach. It also reports on the course structure and answers questions related to setup, programming language and chosen paradigm. The work
is structured as following: Section 2 motivates the choice of adopting Eiffel as
first programming language to be studied. Section 3 describes the structure of
the course and the Design by Contract approach adopted. Section 4 reports
empirical results on our teaching effort: a poll was presented to students and
some data collected. Finally, the numerical data presented is then analyzed and
commented in Sect. 5.

© Springer Nature Switzerland AG 2020
M. E. Auer and T. Tsiatsos (Eds.): ICL 2018, AISC 916, pp. 68–76, 2020.
https://doi.org/10.1007/978-3-030-11932-4_7

2   Eiffel as First Language

The choice of Eiffel and Design by Contract as programming and methodological tools for first year bachelors has been long discussed inside the university. After four years we could not find any evidence suggesting the need for a change. Instead, we will consider some data supporting the idea that the course worked out succesfully, both in terms of content and organization. It is however worth motivating the decision more in detail.

Which programming language is better to start studying for the beginners? No single answer exists. In general, it is easier to answer to the opposite question: "What programming language is better not to start with?". Experience has shown that teaching a specific language from scratch in order to satisfy a specific and urgent needs may not bring individuals to develop into a skilled and versatile professionals. Professional experience has shown cases of individuals who improvised themselves as Visual Basic programmers from scratch, or moved from FORTRAN or COBOL to Java because of some local business need. Often the immediate emergency was patched, but the correct mindset and basic skills required by an experienced professional were not developed. Sometimes such an emergency is inevitable, but developing a quality curriculum for a top-level university requires more care and deeper analysis of what programming is and programmers need, with some initial and pedagogical detachment from raw business needs.

Worldwide, examples of good pedagogical approaches for programming are not missing. There are a few preliminary considerations to be done in order to follow these succesfull steps . First, what programming paradigm we want to use? There is a general tendency to prefer the Object Oriented Programming (OOP) Paradigm [17] as starting point since it helps students developing abstraction skills and design method. This approach, however, is not without its critics: some believe that Object Orientation may deal too much (and too early) with design and interface aspects and not enough with algorithmic details and imperative flow structure. According to this view, procedural programming would be better to start, while Object Orientation should be introduced in advanced courses. Of course, this depends on how the course is organized and taught, but the concern is serious. The school of thought privileging OOP usually concentrates on languages like Java [1] or C# [4] in order to take into account business demand. The school of procedural programming sometime concentrate on purely academic languages like Pascal [5], with the benefit of simplicity, or widespread languages like C [6], offering broader flexibility (and related complexity). There are other paradigms too, for example the Functional (Lisp [15], ML [11], Scala [12] and Haskell [14]), which has attracted renewed interest in recent years, and Logic (Prolog [2]).

The second general observation is that any computer scientist or software engineer will learn a number of programming languages over the course of his career. In particular, everyone will learn one or more of the dominant languages such as, today, Java, C#, C, C++ or Python. So the choice of the first programming language is not exclusive of others; rather, it is a preparation for others, and should emphasize development of the skills needed to learn programming.

(In fact, an increasing number of students have done some experience with Java or other languages before they even join the university program.)

When desiging an introduction to programming course is also important to reflect on how much emphasis (if any) shoud be put on formal reasoning, software quality and correctness by construction. The University is the ideal time of life for learning new concepts and, at the same time, build the foundations of one's knowledge and mindset. Establishing a broad and deep basis is also the best way to make sure that students not only receive sufficient initial training to obtain a first job, but acquire the extensive long-term intellectual skills to pursue a successive career over several decades: the technologies will change, particularly in such a quickly evolving field as Information Technology, but the principles acquired during university study, if thought without dogma and with an open mind, will remain useful.

As a result, a broad school of thought supports the idea that the introductory programming course and the first programming language should emphasize Computer Science foundations and formal reasoning in order to strengthen a mindset leading to development of quality software. Eiffel and Design-by-Contract are just one possible technological and methodological solutions to implement such philosophy, and it is the one collectively chosen for our *Introduction to Programming* in order to provide the adequate mindset to future professionals. This path is not free of controversy. The experience inherited from ETH Zurich[1] is positive [13], and the course was well received by students. We aim at repeating the success in different contexts, though an adaptation phase is necessary and benefits of the approach may not appear as immediate.

3   Course Structure and Approach

In this section we will discuss how the course *Introduction to Programming I* is structured at Innopolis.

The *Introduction to Programming I* course (was called *Object Oriented Programming* the first year it was delivered) is a 6 ETCS course delivered to first year bachelor students at IU over 15 weeks with 2 academic hours of frontal lectures and 4 hours of laboratory exercises every week. The team is composed by a Principal Instructor (PI) in charge of delivering lectures (PI has changed twice in the past years – although they both work in the same team) and Teaching Assistants (TA) in charge of delivering laboratory exercises. PIs are formal method experts, and TAs are researchers or PhD students in the area. The foundational ideas on which the course is based are:

– The foundation for programming lies on mathematical and logical bases
– Identifying and fixing bugs early is cost effective, hence the emphasis on correctness by construction (in synergy with testing)
– Explain and delivering these points to the students so that this knowledge is passed to their future job environments

---

[1] https://www.ethz.ch/en.html.

Frontal lectures are given in English to all students and there is no differentiation between the level of English proficiency of students. Lab sessions, on the other hand, are split into 4 categories: by the level of experience in programming, *Beginners* or *Advanced*, and by the natural language, *Russian* (native language for the majority of students) and *English*. Students had the ability to select which level they belong to based on their perception. The decision is done at the beginning of the course and students are not allowed to change once they have decided. In order to successfully pass the course, all students need to pass all evaluations. Evaluations do not make any assumptions about the level and are the same for all students. As overall structure, the course cover the foundations of imperative programming, from the notion of variable to control flow structure, but keeps tightly an object orientation introducing very early concepts as classes, objects and methods. Soon enough inheritance and polymorphism are also introduced. There is nothing new in exposing millenials to an OOP language as first programming language. These kind of experiments appeared as early as in the 80s and become very common in the 90s. The peculiarity of our approach is exposing millenials to the notion of *Software Contract* using the metaphor of business contract. Design by Contract (DbC) [10] is an approach to achieve the so-called correctness by construction [3]. Correctness by Construction makes use of foundations of logic, concepts that are taught by a *Discrete Math* course which our students need to attend either prior or in parallel to our course. *Introduction to Programming I* is not a course project (somehow along the Russian academic tradition), the evaluation is based on a set of smaller assignments, a mid-term exam and a final exam. It has to be admitted that this represents somehow a limitation, since it is difficult to relate the course with the activity of the companies already based in Innopolis. There are also quizzes that are not graded and their purpose is to provide feedback to students.

The notion of contract is introduced in the very first weeks of the course as an instrument to embed specification into the code and being sure that such a specification is checked for violation at run time. Tools for static verification also exist, for example Autoproof [16], however these are only introduced towards the end of the course. The natural perception for students, at least to those who have been exposed to programming languages before, is that specification and code do not go together. It has been observed that students do not initially understand the reason of using math and logic concepts to specify the behavior of code. The perception changes when they can actually see the importance of specifying the 'what' so to properly implement the 'how': specification and code are not two separate artifacts, they go together and proceed together and we can automatically trace and verify their consistency. This is the very idea of formal methods, and it is something our students have been never exposed to before in the totality of cases.

Students are introduced the idea of DbC without making any neat distinctions (syntactical or semantical) between the code itself and the contract, in fact a contract is presented for what it is: part of the code integrating with the imperative aspects and the modularization and reuse of code peculiar of OOP (natively supported by Eiffel). It has been noticed that students with no programming experience absorb this fact without any problem or objections, and indeed in a completely natural way. This is not true for students with a bias due to occasional and superficial previous programming experience. The bias is even stronger in students that consider themselves fluent programmers in some other language. This observation reinforces the fact that formal methods can more easily thought when there is no a priori bias. The course also introduces the concept of testing, its taxonomies and different approaches for testing measurements. Students quickly grasp the idea of using contracts as unit tests for features. It was noticed that advanced students, after having understood the main idea behind DbC, often ask about the use of contracts in the automatic generation of input values for test cases for features. They are also curious about the necessity of testing in the presence of contracts since contracts can be used for the formal verification of correctness. This observation reinforces the fact that students grasp the concepts and master them in a natural way.

In order to understand how the course, and in particular the notion of DbC, helps students to better grasp programming concepts, a questionnaire was given to those who passed Introduction to Programming in Fall 2016. We asked a single question:

– Did Design by Contract help you to grasp better software concepts presented in the continuation of your study? (definitely not/ not /neutral / yes/ definitively yes)

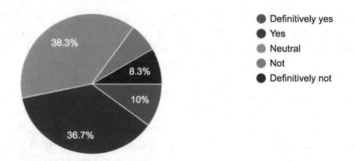

Fig. 1. Results of the question: Did DbC help you to better grasp software concepts presented in the continuation of your study?

Figure 1 shows that students found DbC useful to better grasp programming concepts.

4   Results

This Section is devoted to present the data and analysis of the students' performance for the *Introduction to Programming* course after implementing the choices and structure described in previous sections. We analyzed the data of students' grades who attended the course at Innopolis University and asked ourself whether the separation in beginners and advanced is useful to the pedagogical process. In this paper, we do not analyze the effect of the language division: English vs. Russian. This aspect will be explored in the future.

In Fig. 2 we report the distribution of the overall final grade of the course for Fall 2016. It is noticeable that higher grades go to advanced and lower to beginners. This suggests that the self assessment is informative for the continuation of the course and labs at different levels can be treated separately. Instructors for example can assume a better understanding of programming for advanced students. Table 1 provides some further confirmation of this fact and summarizes the results of the major grading milestones for Fall 2016. This data suggests that the self assessment is sound, i.e. students are able effectively to capture their programming skills. In particular, in the final exam advanced performed about 10% better on average. It is worth noticing however, that the best grade was obtained instead by an outlier, i.e. a beginner student who performed better than anyone else. The presence of outliers does not invalidate the general scheme. To the contrary, it is expected than some students decide to attend a group different from the self perceived level for different reasons, for example to study with a friend or to have a more comfortable environment. The presence of successful beginners can also be explained in terms of attitude. As discusses in Sect. 3, it has been observed how students that consider themselves good programmers have a bias against learning a new language and a new methodology, while beginners naturally absorb new ideas. Clearly, this bias may end up in being an inhibitor of success.

The data collected in Fall 2016 and reported here seemed to suggest that the division Beginner/Advanced was informative. To find more evidence of that, in Fall 2017 we compared the self assessment with the results of an actual entry test that students undertook at the beginning of the semester. The test was about computer science in general, and in particular on programming and data structures. Students were never informed about the results of such a test, so that their self assessment was not biased. Figure 3 reports on how accurate the students' self-assessment (x-axis) compares with the actual test assessment (y-axis), for instance, 68 students who assessed themselves as *advanced* were in fact advanced students based on their grades of the course. Results show with a higher level of confidence the usefulness of the division beginner/advanced.

5 Discussion

Introduction to Programming divides students in four groups on the basis of a technical self assessment and a choice of preferred teaching language (Russian or English). In this paper we analyzed only the division based on programming skills while the language choice will be investigated in future. The data collected and analyzed supports the idea that such an organization is useful, i.e. the students self assessment is a good predictor of technical skills, with the exception of

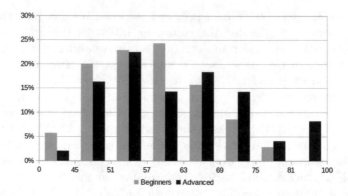

Fig. 2. Total grade distribution in *beginners* and *advanced* groups

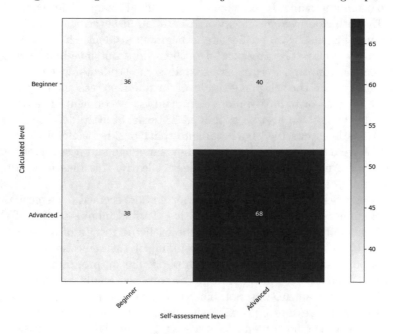

Fig. 3. Self-assessment vs. entry test

Table 1. Cumulative results

	Average	Max	Min	
Mid-term exam	21.11	32.4	7.6	**Beginners**
	23.80	37.2	11.2	**Advanced**
	12.70%	14.81%	47.37%	**Difference**
Final exam	20.28	31.8	3	**Beginners**
	22.21	30	12	**Advanced**
	9.55%	−5.66%	300.00%	**Difference**

some outliers that may be explained by the bias of advanced against learning a new language. The results of this paper are however not fully conclusive, and further studies are required. On top of this conclusion, we report the result of a questionnaire that shows how the majority of students who passed the course believe that DbC is useful to better grasp software concepts presented in the continuation of Innopolis curriculum. All this supports the idea that the course is effective, both in terms of content and methodological approach, and for what concerns the actual organization. After further confirmation of the results we plan to introduce changes to the course. In particular, we plan to inform students of their grades in the technical test. This will allow them to make an informed choice when they self assess their technical skills. Participation to group may remain up to students' decision, but this decision will be supported by objective information.

References

1. Arnold, K., Gosling, J., Holmes, D.: The Java Programming Language, 3rd edn. Addison-Wesley Longman Publishing Co., Inc, Boston, MA, USA (2000)
2. Bratko, I.: Prolog Programming for Artificial Intelligence. Addison-Wesley Longman Publishing Co. Inc, Boston, MA, USA (1986)
3. Chapman, R.: Correctness by construction: A manifesto for high integrity software. In: Proceedings of the 10th Australian Workshop on Safety Critical Systems and Software - Volume 55. pp. 43–46. SCS 2005, Australian Computer Society, Inc., Darlinghurst, Australia, Australia (2006). http://dl.acm.org/citation.cfm?id=1151816.1151820
4. Hejlsberg, A., Torgersen, M., Wiltamuth, S., Golde, P.: C# Programming Language. Addison-Wesley Professional, 4th edn. (2010)
5. Jensen, K., Wirth, N.: PASCAL User Manual and Report. Springer, New York, NY, USA (1974)
6. Kernighan, B.W.: The C Programming Language. Prentice Hall Professional Technical Reference, 2nd edn. (1988)
7. Kondratyev, D., Tormasov, A., Stanko, T., Jones, R.C., Taran, G.: Innopolis university-a new it resource for Russia. In: 2013 International Conference on Interactive Collaborative Learning (ICL). pp. 841–848 (Sept 2013)
8. Meyer, B.: Applying "design by contract". Computer 25(10), 40–51 (Oct 1992). https://doi.org/10.1109/2.161279
9. Meyer, B.: Eiffel: The Language. Prentice-Hall Inc., Upper Saddle River, NJ, USA (1992)
10. Meyer, B.: Object-oriented Software Construction, 2nd edn. Prentice-Hall Inc., Upper Saddle River, NJ, USA (1997)
11. Milner, R., Tofte, M., Macqueen, D.: The Definition of Standard ML. MIT Press, Cambridge, MA, USA (1997)
12. Odersky, M., Micheloud, S., Mihaylov, N., Schinz, M., Stenman, E., Zenger, M., et al.: An overview of the scala programming language. Technical report (2004)
13. Pedroni, M., Meyer, B.: The inverted curriculum in practice. SIGCSE Bull. 38(1), 481–485 (Mar 2006). https://doi.org/10.1145/1124706.1121493
14. Peyton Jones, S.: Haskell 98 Language and Libraries: the Revised Report. Cambridge University Press (2003)

15. Steele Jr., G.L.: Common LISP: The Language, 2nd edn. Digital Press, Newton, MA, USA (1990)
16. Tschannen, J., Furia, C.A., Nordio, M., Polikarpova, N.: Autoproof: Auto-active functional verification of object-oriented programs. In: 21st International Conference on Tools and Algorithms for the Construction and Analysis of Systems. Lecture Notes in Computer Science, Springer (2015)
17. Wegner, P.: Concepts and paradigms of object-oriented programming. SIGPLAN OOPS Mess. 1(1), 7–87 (Aug 1990). https://doi.org/10.1145/382192.383004

A Collaborative Approach for Practical Applications in Higher Education

Dorian Cojocaru[(✉)] and Anca Tanasie

University of Craiova, Craiova, Romania
cojocaru@robotics.ucv.ro, ancatanasie@gmx.de

Abstract. As part of the University of Craiova, a comprehensive university, the Faculty of Automation, Computers Electronics offers a number of bachelor programs, such as: Control System, Multimedia Systems, Computers, Applied Electronics, Mechatronics and Robotics. Other study programs are offered at master and doctoral level. Following the tradition and specific local perception, both influenced by the social, the technical and economic context, study programs have been perceived differently by stakeholders, professors and students. The System Engineering study programs are considered to be more theoretical oriented, given with their curricula and students' skills. The programs within the Computer and Information Technology field are ranked higher if considering the demand of the labour market and the opportunity to find better-paid jobs. Mechatronics and Robotics programs are the newer ones and by consequence not well known from the point of view of curricula and skills offered to diploma holders, and not so demanded in relation to other programs. In this paper we shall be presenting the methods used to develop new practical activities in order to support the students' ability to discover new things, and to solve practical problems even if they do not yet have a higher theoretical background related to the imposed task.

Keywords: Collaborative learning · Adaptive and intuitive environments paper publishing · Online journals · Styles · How-to

1 Introduction

1.1 Context

As part of the University of Craiova, a comprehensive university, the Faculty of Automation, Computers Electronics offers a number of bachelor programs, such as: Control System, Multimedia Systems, Computers, Applied Electronics, Mechatronics and Robotics. Other study programs are offered on a master and doctoral level.

Following the tradition and specific local perception, both influenced by the social, technical and economic context, the study programs are being perceived differently by stakeholders, professors and students.

The System Engineering study programs are considered, together with their curricula and the skills of their students, more theoretically oriented. The programs within the Computer and Information Technology field are ranked higher if considering the demand of the labour market and the opportunity to find better-paid jobs. Mechatronics

© Springer Nature Switzerland AG 2020
M. E. Auer and T. Tsiatsos (Eds.): ICL 2018, AISC 916, pp. 77–86, 2020.
https://doi.org/10.1007/978-3-030-11932-4_8

and Robotics programs are the newer ones and by consequence, not well known from the point of view of curricula and skills offered to diploma holders, and not so demanded in relation to other programs.

In Romania, by tradition, these two programs are developed within the Mechanic and Industrial Engineering faculties. The University of Craiova is the only higher education institution in Romania where these two programs have been started and developed within the frame of a Control and Computer faculty. Leaving from the placement of Mechatronics as an integrating specialization for Mechanics, Electronics, Computers and Automation technical fields, and as Robotics maybe the most representative application, this approach seems to be justified.

The main activities are research and teaching in the fields of robotics, mechatronics, computer vision and smart materials. Supporting these activities, our facilities include 7 labs, well equipped with hardware and software. The group consists of more or less 20 researchers including PhD students.

The group obtained a useful experience (national and international grants and contracts) in the following fields of research: hyper redundant robots; intelligent robot control; educational systems; distributed parameter systems; discrete optimization problems; robot vision; mechanical impact study; mobile robot control trajectory; robotics for prosthesis; smart materials and structures for medical applications; biomimetic/bionics. The group assures the management of Robotics Society of Romania.

The Mechatronics and Robotics program's promoters try to support their students by training their skills in relation to the ability to self-discover solution for practical problems, not always using higher theoretical base, and to work in teams, cooperating in order to fulfil the task. Leaving from this target a new collaborative educational technology was designed in the frame an EU specific program. An alternative way to achieve the intended purpose is to involve the students in an important number of technical contests among teams.

1.2 Goal

The effects of the Bologna process implementation for the Romanian higher education system have already been studied in the framework of few EU funded programs in which the authors of this contribution were involved. Regarding this implementation of the Bologna process in EU, we have identified different approaches. In some countries' cases the Bologna model has been implemented without eliminating the traditional forms of higher education organization. If there are also EU countries where the Bologna model has been minimally implemented, Romania is part of the latest EU accession countries' group where important efforts have been put into following the EU directives, as adequately as possible.

From the Bologna process perspective, the most important change in the Romanian engineering higher education has been the reduction in the number of years of study from five to four. In fact, in many cases, the universities tried to keep the volume of theoretical courses just as before, and no important changes has taken place in practical applications, labs and internships.

In this paper we shall be presenting the methods used to develop new practical activities in order to support the ability of students to discover new things, and to solve practical problems even if they do not yet have a higher theoretical background related to the imposed task. By doing so, Mechatronics and Robotics students will have greater confidence in their skills and they will contribute to a better image of this type of diploma among employers.

2 Collaborative Education Technology

2.1 Laboratory Activities

In the framework of a FP7 program called Practice-based Experiential Learning Analytics Research and Support – PELARS a new collaborative educational technology has been designed and tested [1]. The main idea was to put students to work together, in teams, once they received a practical task to solve.

In the Romanian higher education system for engineering, lab activities are standardized according to rules imposed by ARACIS (The Romanian Agency for Quality Assurance in Higher Education). By consequence, there are scarce differences in the development of lab activities between different engineering programs.

One rule is that the ratio between the total number of course hours and the total number of non-course activities hours (seminar, lab, project, internship), for an entire four years bachelor's program in engineering, must be between 0.8 and 1.2. Taking into account that such a program includes at least 240 h of internship, the most frequent situation is that a disciple has 2 course hours and 1 or 2 lab hours each week. A study year consists of 2 semesters, each of them with a 14 weeks' duration. Following the already mentioned example, both Digital Electronics and Computer Vision has 14 lab sessions - 2 h each.

There are professors considering that this old fashion way to organize the lab activities only in sessions of two hours is not the best solution. If you need to propose a more complex task you must divide it in two or three sessions, in two or three consecutive weeks. Considering our two examples, we can appreciate that two hours lab is more appropriate for Digital Electronics (second year of study), but for Computer Vision (four year of study), where more complex applications could appear, dividing them into two or three parts is somehow difficult.

According to ARACIS rules, each lab in line with each discipline's objectives should be supported by the adequate equipment [8, 10]. At least 50% of the lab activities must be experimental ones by opposition to software simulations. For informatics labs, at least every two students should have one computer.

Given the surface of the lab and the number of available devices and equipment, at the same time in a lab, about 15 students will be working. Comparing the already mentioned two disciplines we can say that for Digital Electronics the students will work together in groups of 2–3 students. Because the number of available technical installations (computer vision systems, machine vision systems) is smaller in the case of Computer Vision, in this case, the number of groups of students will be higher and more students will be working together.

It is compulsory to produce laboratory guides, printed or in electronic format, and to make them available for students before each lab season. In these guides the development of the lab work is described including:

- The purpose of the work,
- A short theoretical background with necessary links for bibliography,
- The available equipment with their technical characteristics,
- The plans/algorithms of the experiments,
- The necessary data acquisition and processing,
- The content of the individual reports.

For bachelor engineering level, lab activities are more drive-on instead based on self-discover. The reason could be that during bachelor studies the students learn to integrate existing technical parts into a functional system in order to solve common problems. Self-discovering, finding new solutions, designing and research activities are more targeted during master studies.

The weak points are at least the following: the need to have/learn a lot of theoretical knowledge, a fixed plan for everyone, a weak cooperation, the difficulty in measuring the degree of cooperation, not enough support for self-discovering neither for solving errors.

It is compulsory to grade each student, during each lab session and finally for the entire activity. Considering the number of students and the way they work together for the same job, it is sometime difficult to provide an accurate differentiation between students and their work.

The professors try to introduce more self-discovering and cooperative activities during project activities. Following the ARACIS rules, it is compulsory to have at least four projects during a bachelor program. The project could be integrated in the evaluation of the discipline or could be considered a distinct discipline with his distinct evaluation.

It is important to identify the subjects and activities that could be suitable for the new collaborative educational technology based by the concept of learning by doing. A survey has been implemented and the main subjects of interest where [3]:

- Designing hands-on, physical learning activities (How do to decide which activities to make more hands-on, physical for students, and which to make more lecture-driven or demonstrative? What resources to use when designing hands-on, physical activities? What (if any) challenges there are in designing hands-on, physical projects for students?).
- Examples of effective activities (What are the most successful learning activities in design for the classes; What specific elements of these learning activities are expected to perform best? What elements are expected not to work well? Why? How could these be improved? How to introduce the activity to students? What it is expected that students would like about this activity? What don't they like? How to motivate students to engage in this activity?)
- Materials for Hands-on, Physical Learning Activities (Are there necessary physical implements and materials always available in the classroom? What are these and why are they important? What practical constraints there are in developing physical

activities for students (e.g.: scale of materials, manual dexterity of students, economic or budget constraints, etc.)
- Assessment for hands-on, physical projects (What formative assessment methods are used to assess student learning in hands-on, physical projects? What skills and understanding is it useful to measure by these means? How effective do you feel these methods are? What summative assessment methods do you use to assess student learning, in hands-on, physical projects? What skills and understanding it is useful to measure with these? How effective these methods are expected to be? It is useful to use formal or informal peer assessment? How is this structured and how is it managed? How to monitor student progress during the activity? It is useful to manage student progress within the activity? Is the activity structured into different stages? Is it useful to support students' reflection on what they've learnt from the activity? When and how?)

As the main objective is to propose a new method for learning by practice, it is necessary to implement both a technical support based on adequate hardware equipment and software resources, and a set of didactical procedures.

The research methods must be adapted to these different purposes. From didactical point of view, it is necessary to know [5]:

- How practical works (labs) were organized until now, what are the strong points and the weak ones,
- How can we introduce the new methods, what do we need from this point of view in terms of material and human resources,
- How to fulfil the accreditation/evaluation demands and/or to propose new criteria for an adapted accreditation,
- How to evaluate student's activity during the new organized practical works (learning analytics),
- How to evaluate the overall obtained results.

Based on upper considerations, we identified as research methods the "ethnographic methods" [2]. Participant observation is a very important tool. Data collection using interviewing, observation, and document analysis are useful qualitative methods. The participant observation should have an increased role. In this context, we used the definition according to which observation is "the systematic description of events, behaviours, and artefacts in the social setting chosen for study". Following the results obtained and identified in specialised literature, the research activities will involve: active looking, improving memory, informal interviewing, and writing detailed field notes. From similar sources we know that using observations we will be able to determine: interactions, participants' communication the amount of time spent on various activities. Though very useful, is more difficult to find a relation between the nonverbal expressions of feelings and the participation of the students to the lab activities. It was already underlined that it is useful to use the participants' observation, a qualitative method, together with additional strategies such as interviewing, document analysis, surveys, questionnaires, or other more quantitative methods. Observation is for this type of activities, a data collection method, and for this reason it is necessary to associate it with coherent methods for recording and analysing the data. In analysing the collected data, as part of research it would be useful to employ the skills of the team, such as teaching experience, research experience and the experiences in evaluation of the quality in higher education.

Having as main purpose to support two main skills - team cooperation and self-discover the problem solution, the new education technology proposes to students to exchange information and ideas from the preparation phase (see Fig. 1) [4, 7].

Fig. 1. Collaboration during the preparation phase

Students have access to information, not necessary of a high theoretic level, using tablets or smartphones.

During the practical work team members cooperate between them by exchanging tools and parts, and by solving individual parts of the complex task and finally integrating the results (see Fig. 2a).

Fig. 2. a) Collaboration during practical work. b) Visual programming.

Students could use software tools offering visual support and could solve the problem by connecting functional blocks from libraries (see Fig. 2b). It is not necessary to know in detail the structure and the functioning of every block. The inputs and the outputs of every block and the possibility to interconnect them must be known.

A structured working area, including tables and supports, has been designed. During the work, the necessary parts and devices as well as the hands of students are identified using computer vision techniques (see Fig. 3a). Thus monitor all activities becomes possible: how students exchange information and parts, who gives and who receives.

Fig. 3. a) Parts to be recognized by computer vision techniques. b) Example of specific reports.

During the final stage, students are asked to test the application and to register a report concerning the entire work done [9]. The supervisor of the practical activity is supported in the assessment by specific reports correlating the activities with the performer and the moment in time (see Fig. 3b).

2.2 Collaboration During Competitions

Alternatively, the involvement of students in a variety of national and international competitions should contribute to improving their ability to cooperate by working in teams, to solve new unknown practical problems. The field of Mechatronics and Robotics offers a wide range of such competitions involving mobile robots, PLC programming for different types of machines, computer vision, robot programming and other [6].

An example for this type of competition has been organized using the PELARS education technology, and it started off with 8 groups of student on 2 to 4 stations in round 1 of the activity. The students have 75 min to solve task 1 in groups of 3. The PELARS team evaluated the solutions, and the quarter-finalists proceed to round 2 and the other groups proceed to the b-group.

Round 2 took place during the next day where the finalists (2 to 4 groups) competed in solving task 2 during 90 min sessions. Again, judges chose the semi-finalists for the final round, but all teams proceeded for user testing purposes.

The final round took place on the third day and the two final teams competed but the next two teams competed as well. The task time is 120 min. The judges awarded first, second, third, and other prizes for the competitors.

The competition follows the standard PELARS process as listed below: informed consent by all parties (students and staff); pre-Survey with coded linkage to individuals and group formation; workshop introduction and training sessions; tasks 1 to 3; end of task session debriefing and brief semi-structure interview for usability; post-Survey for each task.

The proposed subject has been linked to the prototype of an automobile interface of the future. Between the near future of driverless automobiles and today's current cars there is the opportunity of hybrid control for these vehicles. For the PELARS competition, groups of three students will compete in three heats to ultimately design the hybrid interface the car of the future.

The teams first compete to design an interactive toy in session 1, learning the PELARS programming system basics. Session 2 provided a more complex task of designing a color sorting machine, so students can go into depth with the toolkit. Session 3 is the big challenge of designing hybrid interface for the automobile of the future, where accelerometers, color and light sensors, motor controllers, buttons, come into play with the group's concept and prototyping skills to design and build a working prototype in 120 min.

Task 1 has as purpose to introduce the students to the PELARS visual programming system with basic input/output that includes button, light sensor, piezoelectric components, RGB LED, potentiometer along with logic blogs of swap, map, interval, and fade. The task is to build an interactive toy that includes these and other blocks (Fig. 4).

Fig. 4. Student teams participating to a competition.

The learning goal for this task is to get students used to prototyping an open-ended task with a systematic engineering design approach.

Task 2 was a more complex challenge that included the use of the motor controller, color sensor, and the if-then logic block. The counter and the trigger logic blocks will be introduced if necessary. Task 2's learning goals are more engineering focused on solving the classic case of the colored candy sorter figuring out how to move the objects of the sensor and mechatronic control of sorting. Task 3 was in fact the real competition, where the teams needed to conceptualize the future hybrid human and autonomous car interface for near future driving.

In order for the students to build their prototypes for the competition, other educational components where needed: blocks (Technic from Mindstorms), cardboard, small wood blocks, dowels, tape, and other materials so the students can build their prototypes.

The research objectives where linked with the tentative of finding answers to the following questions:

- How do groups leverage knowledge from the PELARS system to improve their performance over the three sessions?
- How does the data collected by system compare with the learner's perception of their performance?
- Do the groups' phases of planning, building, and reflecting change as they become more experienced?

3 Conclusion

The paper deals with a specific problem: the change in performing practical works for engineering students by proposing a new collaborative education technology. The main idea is to support students to learn to work in teams and to cooperate in solving practical problems. They are encouraged to discover new solutions even they do not have the highest theoretical background regarding the field of application. We have used the experience acquired during a FP7 research program and the information obtained after the end of this program throughout didactical activities and, respectively evaluation of the quality assurance.

Acknowledgment. "Practice-based Experiential Learning Analytics Research And Support (PELARS)", STREP, FP7-ICT-2013-11.

References

1. http://www.pelars.eu/
2. Hammersley, M., Atkinson, P.: Ethnography: Principles in Practice. Routledge, London (1995)
3. Mavrikis, M., Gutierrez-Santos, S., Geraniou, E., Noss, R.: Design requirements, student perception indicators and validation metrics for intelligent exploratory learning environments. Pers. Ubiquit. Comput. 17(8), 1605–1620 (2012)
4. Martinez-Maldonado, R., et al.: Capturing and analyzing verbal and physical collaborative learning interactions at an enriched interactive tabletop. Int. J. Comput. Support. Collab. Learn. 8(4), 455–485 (2013)
5. Friesel, A., Cojocaru, D., Avramides, K.: Identifying how PELARS-project can support the development of new curriculum structures in engineering education. In: The 3rd Experiment@International Conference (exp.at'15), 2–4 June 2015, University of Azores (Ponta Delgada, São Miguel Island, Azores, Portugal) (2015). ISBN 978-989-20-5753-8, paper #92
6. Popescu, D., Cojocaru, D., Popescu, L.C., Petrisor, A., Popescu, R.: To learn engineering through competitions. In: The 26th Annual Conference on European Association for Education in Electrical and Information Engineering EAEEIE, 1–2 July 2015, Copenhagen, Denmark (2015)
7. Cojocaru, D., Friesel, A., Spikol, D.: Supporting STEM knowledge and skills in engineering education - the PELARS project. In: ASEE International Forum, Concurrent Paper Tracks Session II Courses, 25 June 2016, New Orleans, Louisiana, USA (2016)
8. Cojocaru, D., Friesel, A., Tanasie, R.T.: Adapting the accreditation procedures to a new educational technology. In: 3rd International Conference on Education and Training Technologies (ICETT 2016), 24–26 Aug 2016, Turku, Finland (2016)
9. Cojocaru, D., Spikol, D., Friesel, A., Cukurova, M., Valkanova, N., Rovida, R., Tanasie, R. T.: Prototyping feedback for technology enhanced learning. Int. J. Educ. Inform. Technol. 10, 144–151 (2016). ISSN 2074-1316
10. Cojocaru, D., Tanasie, R.T., Friesel, A.: Adapting the accreditation procedures to a new educational technology. Int. J. Inf. Educ. Technol. 7(11), 851–857 (2017)

Peer Teaching in Tertiary STEM Education: A Case Study

Niels Heller[✉] and François Bry

Institute for Informatics, Ludwig-Maximilian University of Munich,
Munich, Germany
niels.heller@ifi.lmu.de

Abstract. This article reports on a novel higher-education course format exploiting choreographed peer reviews and self corrections so as to reduce to a minimum the teachers' involvement. The novel course format was motivated by the necessity to run examinations for all courses during all terms, even though almost all courses are offered only every second term. As a consequence and because of a very high students to teacher ratio, many students have to prepare for examinations without sufficient assistance. This article describes the novel course format and reports on its evaluation in a case study. The evaluation indicates that most students benefit from the novel course format but that it is less efficient than traditional formats based on a much higher teachers' involvement. The major weakness of the novel format is an insufficient dedication of some students to their reviewing. The article suggests and discusses possible measures to address that weakness.

Keywords: Peer review · Collaborative learning ·
Learning environments

1 Introduction

This article reports on a novel format for higher-education introductory courses exploiting peer reviews and self corrections so as to reduce to a minimum the teachers' involvement. The novel course format relies on a well-thought choreography of peer and self correction so as first to provide fast feedback through peer reviews to all students and to ensure a good learning through self-correction. The novel course format was motivated by the necessity to run examinations for all courses during all terms in a bachelor course of studies in computing and related fields such am bio-informatics and media informatics, even though almost all courses are offered only every second term. As a consequence and because of students to teacher ratios of over 800 for professors and over 70 for teaching assistants, many students have to prepare for examinations without sufficient assistance.

The positive impact of peer reviews of students' homework on the learning of both reviewers and reviewees has been demonstrated in former studies [7,8,24].

© Springer Nature Switzerland AG 2020
M. E. Auer and T. Tsiatsos (Eds.): ICL 2018, AISC 916, pp. 87–98, 2020.
https://doi.org/10.1007/978-3-030-11932-4_9

Even though exploiting peer review in higher sciences, technology, engineering, and mathematics (STEM) education is promising, this has been so far rarely undertaken and therefore rarely studied. To the best of the authors' knowledge, this article is the first proposal of a course format exploiting choreographed peer reviews and self corrections. An appropriate choreography is important for several reasons: It gives students precise tasks to perform, it provides common time periods for these tasks keeping the students' learning "in phase", a pre-condition of peer review, and ensuring a collective experience turning a group of students into a learning community sharing common goals and therefore motivating to help each other.

The novel course format proposed in this article has been tested and evaluated in an introductory computer science course for Bachelor students, an introduction to functional programming with the programming language Haskell. To this aim, the web-based learning platform Backstage supporting the sophisticated multi-phased choreography of peer reviews and self-correction of the proposed novel course format has been conceived and implemented. The learning platform in addition to peer review and self-correction services also provides learning material and communication tools which the students can use to perform their learning assignments, discuss the learning material among themselves, and perform their homework. Backstage also provides coding services: Students can write, compile, and test code without leaving the learning platform. The platform's coding services are used in the novel course format both for one student's own learning, for her review of her peers' code and for her self-correction. A description of the platform, along with its user interface, is given on the projects homepage.[1]

The evaluation of the novel course format reported about in this article is both quantitative and qualitative. The evaluation's focus was the course format's learning effectiveness. To this aim, the quality of both, the homework and the peer reviews delivered by the students on the learning platform has been assessed by human experts. Furthermore, the course attendance, that is, the participation to the assigned homework and peer review and self-correction task has been tracked. The evaluation was guided by the following research questions:

1. What is the peer reviews' quality?
2. Does the quality of the reviews delivered, respectively received, by students correlate with their examination performances?
3. Does solving homework assignments correlate with examination performances as it is the case with traditional course formats?
4. What is the students' attitude towards the novel course format?

To answer the first research question, a simple categorizing assessment scheme of the peer reviews' quality has been worked out. Using that scheme, two teaching staff members categorized independently of each other all of submitted reviews (Kohen's $\kappa = 0.85$). This evaluation revealed that 28% of the reviews were of low quality in the sense that they exhibited serious flaws.

---

[1] https://backstage2.pms.ifi.lmu.de:8080/about/.

Surprisingly, the second question received a negative answer: Neither the quality of the reviews students delivered, nor the quality of the reviews they received correlates with their examination performances. However, the amount of reviews students received does correlate with their examination performances what might reflect the often observed positive correlation between doing homework and examination performances. Indeed, only submitted homework can be reviewed.

Investigating the third question has shown that merely submitting homeworks had no significant impact on examination performance. To investigate this phenomenon further, the submission quality was assessed using an automated testing tool. This revealed that submission quality correlates positively with examination performance (Pearson's correlation coefficient, in the following Pearson's $r = 0.49$). However, this value is not statistically significant.

To answer the fourth research question, a qualitative survey has been conducted at the end of the course. That survey revealed that the general attitude of those course's students who completed the survey toward the novel course format is positive stressing both positive and negative aspects.

This article is structured as follows: Section 1 is this introduction, Sect. 2 is dedicated to related work. Section 3 introduces the course format. Section 4 describes the scientific method of the case study. Section 5 reports on the results regarding the participation and attendance, the quality of peer reviews, the quality of homework submissions, and the qualitative survey. Section 6 discusses the results and makes a comparison to a previous "traditional" course. Section 7 draws a conclusion for improving course design and for further research.

2   Related Work

Peer Review. Following [18], peer review can be defined as a learning activity in which students evaluate, make judgements on, and deliver written feedback on the work of their peers. Several meta analyses report on the learning efficiency of peer review [7,8,24]. Recently, some authors have compared the positive effects of *delivering* peer reviews and *receiving* peer reviews [6,17,18]. The positive impact on learning of delivering reviews is explained by the longer time learners spend on a subject [24] and by the reflection on one's own learning triggered by reviewing the work of others [18].

Among the benefits of peer reviewing for learners, the comparison of different approaches and standard of work and the exchange information and ideas are cited [12,22,25]. Among the negative impacts of peer reviews on learning, the difficulty of making accurate assessments is reported [12].

Peer Teaching. Peer teaching is simply defined as a form of instruction where learners teach each other [11]. Peer teaching is known to improve teamwork abilities and social skills among learners [22] and to contribute to the learners' comprehension [1–3]. Like peer review, peer teaching is has been shown to be beneficial both for learners acting as "teacher" and learners acting as "student" [10] what is explained by the active engagement required by both roles [11]. A difficulty of peer teaching is the choice of suitable peer learning groups or pairs. This difficulty can be overcome by letting instructors decide on the pairings [11] or by relying on previous achievements to form inhomogeneous groups [5].

Skill Theory and Related Models. The evaluation below shows that, in this study, learners' proficiency correlated positively with the quality of given reviews. This effect was *not* found in the meta analyses in [8,24]. Yet, certain theories would predict such an outcome. Fischer's Skill Theory [9] postulates that learners construct a hierarchical framework of skills where high level skills depend on lower level skills [21]. It can be argued that delivering high quality reviews on a topic is a skill that requires skills in that topic. Other theories supporting the findings are the Conscious-Competence Model of Burch that emphasises the importance of being aware of one's own lack of competence in early phases of learning [4] and the Kruger Dunning effect that is often quoted with the following phrase: "we argue that the skills that engender competence in a particular domain are often the very same skills necessary to evaluate competence in that domain" [16, p.1].

3   Course Format

The course format proposed in this article is based on three types of tasks: weekly homework, review, and self-correction (or re-work). These tasks are choreographed for each course "topic" or chapter in three successive one week long phases:

1. At the beginning of a topic's first week, the topic's course material and corresponding homework assignments are published on the learning platform. The students have to deliver their homework within that first week.
2. At the beginning of a topic's second week, that is, after the students delivered their topic's homework, each student is tasked to review the homework of two other students. The students have to deliver their reviews within that second week.
3. At the beginning of a topic's third week, that is, after the students delivered their reviews, "blue prints" or exemplary solutions for the homework assignments are published on the learning platform. The students have to deliver corrections of both their own homework and of the two peer reviews they delivered.

The third phase is introduced to exploit the beneficial effects of self-correction on learning [19].

While a topic is learned over three weeks, every week a new topic is introduced, that is, the afore-mentioned three successive one week long phases of a topic overlap with that of other topics. In other words, with the third week of a course, a student learns a course's topic, reviews the homework of two other students referring to the previous course's topic, and performs a self-correction of her own homework and own reviews referring to course's second to last topic.

This interleaved scheme has been selected so as to exploit the positive impact of timely spaced instruction [23], and shuffled instruction [20].

The learning platform Backstage is specifically tuned to the novel course format as it supports, among others, the afore-mentioned multi-phased choreography requiring almost no supervision thus freeing the teaching staff from time-consuming "administrative" or "organizational" chores.

4   Evaluation Method

An introductory course on functional programming using the programming language Haskell which employed the novel course format was conducted for evaluation in the winter term of 2018 at the Ludwig-Maximilian University of Munich.
Participants. 45 students enrolled in Bachelor computer science programme of whom 12 were female and 33 male attended the course. The students were studying in their second to eighth semester.
Procedure. The course lasted 13 weeks, covered 11 topics that were worked out by the students according to the scheme described in Sect. 3. A topic's homework encompassed two or three exercises, consisting of either a programming task or a set of questions. In total, 27 exercise solutions could be delivered by every student.

Students who missed three consecutive deadlines for a homework delivery or a peer review delivery were removed from the course on the grounds that contributing to the course, both by delivering one's own homework for other students to review and by reviewing the homework of others, is necessary for peers to learn well. This rule was made clear before registration to the course and was accepted by all students who registered for the course. After a student missed two consecutive deadlines, a warning email was automatically sent to the student by the choreography component of the learning platform.

Dataset. After the course, the quality of all student submissions, homeworks, and peer reviews was assessed both by members of the teaching staff and by software specifically designed for this purpose. This quality assessment was performed only for the evaluation reported in this article, its human-performed component is not part of the course format. Its automated component is part of the course format that provides immediate feedback to students.

In order to assess the peer reviews' quality, each review was categorized by members of the teaching staff after the following scheme:

  +FF: "false correctly reported by the reviewer as false"
  −FC: "false wrongly reported by the reviewer as correct"
  +CC: "correct correctly reported by the reviewer as correct"
  −CF: "correct wrongly reported by the reviewer as false"

The correctness of program submissions was assessed automatically using the standard Haskell compiler [14] and by running pre-defined unit tests, that is, tests that compare expected and computed results for a set of inputs. This way, programming submissions were categorized according to the following scheme:

"wrong format" for submissions that were text of PDF files but no Haskell programs, "not compiling" for submissions the compilations of which failed (usually because of syntax errors), "compiling with failed tests" for submissions that compiled (without errors) but failed unit tests, and "tests passed" for submissions that compiled and passed the unit tests, hence that could be considered correct.[2]

These four categories can be considered as steps that have to be consecutively mastered by learners. Indeed, for beginners, the first obstacle to coding is to select the appropriate format, the second obstacle is to write code that compiles (without errors), and the third obstacle is writing code that passes the unit tests. Thus, the automatic categorization scheme reflects levels of skills as proposed by Fischer's skill theory [9].

The students' learning behaviour during the course was assessed as the number of homework and reviews they delivered and when they delivered it.

After the course, an examination referring to the course's topics took place. After that examination, a qualitative survey was conducted to assess the students' attitude towards the novel course format, the learning platform supporting it, as well as the student perception of the course format's usefulness for learning. 18 students who had attended the course and took the course's examination completed that survey.

Of the 45 students, who attended the course, 32 took the course's examination. These students' data forms the dataset of the evaluation this article reports about.

5   Evaluation Results

Participation and Drop Out. Throughout the course, students dropped out. Most of them were removed in application of the rule mentioned at the beginning of Sect. 4 after they missed three consecutive deadlines. Two students freely chose to leave the course after the third week. Figure 1 illustrates the decline of the participation, notably after the third, sixth, and ninth week.

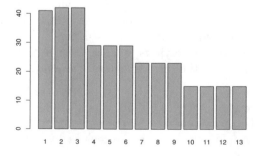

Fig. 1. Numbers of students at each week

---

[2] This assumption is reasonable for short Haskell programs beginners can write.

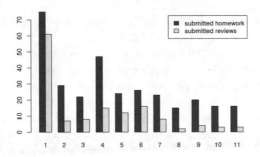

Fig. 2. Numbers of homework and peer reviews submissions at each week

Homework and Peer Reviews Delivered. In total 316 homeworks and 147 reviews were delivered. Similar to course participation, the number of submitted homework and reviews declined throughout the course, as shown by Fig. 2. After the first week, only a fraction of the homeworks were peer reviewed. Peer review participation varied largely, with 18 of 45 students giving 90% of the reviews.

Peer Review Quality. An evaluation of the quality assessment of the peer reviews described in Sect. 4 reveals that most reviews were correct in the sense that they correctly identified either errors or correctness. The relative frequencies of labels is as follows:

+FF: 25% ("false correctly reported by the reviewer as false")
−FC: 22% ("false wrongly reported by the reviewer as correct")
+CC: 47% ("correct correctly reported by the reviewer as correct")
−CF: 6% ("correct wrongly reported by the reviewer as false")

Interestingly, only 6% of the reviews identified errors where they were none and 22% failed to indicate errors.

The Correlations between the frequencies of the labels +CC and +FF where significantly positive (Pearson's $r = 0.44$, $p = 0.05$) and the frequencies of the labels +FF and −FC significantly negative (Pearson's $r = -0.45$, $p = 0.03$). Other correlations between the frequencies of the labels were not significant. This suggests that students good at spotting errors of their peers are also good at identifying correct submissions of their peers and therefore are little prone to give false feedback.

To estimate a student's average review quality, for each student a review score defined as the relative frequency of the number of correct reviews (+CC and +FF) minus the relative frequency of the number incorrect reviews (−CF and −FC) has been computed. The review scores correlate positively with the relative frequency of the number of peer reviews delivered ($r = 0.4$, $p = 0.05$), indicating that good reviewers (in the sense of delivering quality reviews) are more likely to deliver their peer reviews.

Although the participation in peer reviews was low, those students receiving reviews profited from them: Indeed, the relative frequency of the number of received reviews per homework submission correlates positively with the examination performance ($r = 0.44$, $p = 0.03$).

Homework Quality. Of the 316 homework submissions, 232 contained executable code files. The remaining 84 homework submissions either referred to non-coding assignments (40 submissions) or were erroneously submitted in a wrong format (like Word or PDF, 44 submissions).

Of the 232 code submissions, only 129 compiled (without errors). Most of the non-compiling submissions contained syntax errors. Interestingly, of the 129 compiling submissions, only 12 failed to pass the unit tests suggesting that the automatic testing approach makes sense for such a course.

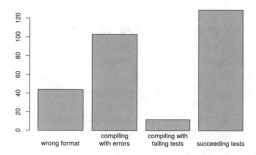

Fig. 3. Numbers of code submissions in the respective categories

Considering the "problem solution steps" mentioned in Sect. 4 shows that most students failed during the first two steps while the last step did not seem much of a hurdle for those students who mastered the previous steps. This is remarkable because it is in the last step (writing code that passes the unit test) that the actual problem is solved. The total frequencies are shown on Fig. 3.

The number of submissions compiling (without errors) of a student correlates positively with the relative frequency of the number of peer reviews that student delivered (Pearson's $r = 0.35$, $p = 0.01$). This indicates that students able to solve the programming assignments are more likely to deliver peer reviews.

The number of submissions compiling (without errors) also correlates with the examination results ($r = 0.44$) but this value is not significant.

Students' Attitudes Towards Peer Review. The perceived usefulness of both delivering and receiving peer reviews was assessed. Most students (44%) indicated that delivering peer reviews was "mostly helpful" for their learning, while on the other hand, most students indicated that *receiving* peer reviews was only sometimes useful. Figure 4 illustrates the perceived usefulness of receiving and delivering peer reviews.

While the received peer reviews are rarely experienced as helpful, the students are relatively confident that their reviews were useful to others (median of 4, on a 6 point Likert scale ranging from "not useful at all" to "absolutely useful").

Students mentioned advantages of the course's peer reviews: The opportunity to see different solutions and of learning from one's peers, and comparing homework

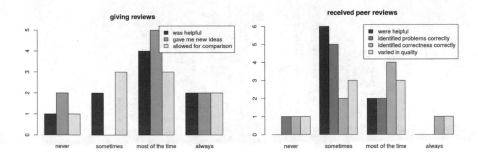

Fig. 4. Perceived properties of given and received peer reviews

standards. Worthwhile noting is the comment: "Peer review gave me evidence that I'm not the only one too stupid to understand the topic." Weaknesses of the peer reviews were also mentioned: low number of reviews received, and low quality of some reviews.

Students' Attitudes Towards Towards Provided Material and Functions. The course's learning material and homework exercises were perceived as very useful for learning (median of 5 on 6-point Likert scale ranging from "not useful at all" to "completely useful"). The online compiler and the unit tests were also perceived as useful (median of 3.5 and 4.5 on the same scale).

Reasons for Drop Out. Students were also asked if they dropped out of the course, and, if so, why. The reasons given were personal reasons like time constraints, and loss of motivation due to a too small number of received reviews.

6   Discussion

Peer Review Quality. The fact that the average quality of the received peer reviews did not correlate with the examination performances is surprising since the importance of the feedback quality for learning has often been stressed in the literature [13]. This surprising fact can be explained as follows. Firstly, it could be due to the small number (32) of students completing the course's examination. Secondly, with the novel course format based on self-learning, reading low quality reviews might motivate to learn more. Furthermore, the students were tasked to self-correct their homework, that is, to re-work.

Homework and Examination Performance. The number of submitted homeworks does not correlate significantly with the examination performances. As a comparison, data from a previous course was examined. That preceding course was held with a teaching staff consisting of 10 tutors who reviewed all homeworks and a professor who hold lectures once a week. That preceding course format had neither peer reviews nor self-correction. 593 students, of which 419 attended the final examination, attended the course. The lecture material and exercises were, except for minor changes, the same in both courses.

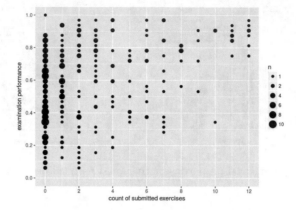

Fig. 5. Relation of examination performance to number of submitted homeworks (aggregated by week) in the preceding course

Figure 5 shows the relation between examination performance and submitted homeworks in the previous course. Two observations can be drawn from the figure: Firstly, students who submitted no homeworks do not necessarily fail in the examination. In fact, these students achieved an average mark of 64%. Secondly, submitting enough homeworks was a sufficient, but non-necessary, condition for examination success, as the almost empty bottom-right triangle of Fig. 5 shows. In the novel course in contrast, submitting enough homeworks was not a sufficient condition for examination success, indicating that the novel format helped students struggling with the course's content less in overcoming their learning problems than the previous course did.

7   Conclusion

Improving The Course Format. The evaluation identified low participation in the peer review process as a shortcoming of the course format. While this may be caused by the novelty of the format, requiring tasks the students are not accustomed to, three measures can be envisaged to provide all students with constant and possibly better peer reviews:

1. Rather than pairing students randomly, proficient students (who are more likely to provide good reviews) could review "struggling students" (who would benefit most from having their homework reviewed) and vice versa. This would increase the reviewing *efficiency* without increasing the teachers' involvement. To identify proficient students, the four submission categories of Sect. 4 could be used. This approach would provide a very natural pairing: Those who are able to write syntactically correct code should be able to help those struggling with that task.
2. Peer review quality could be improved by providing a review scheme, as sort of conceptual scaffolding [15]. Again, the submission categories of Sect. 4 could

be used in asking questions like "Does this submission contain valid Haskell code?" or "Does it compile?"

3. Finally, the social dimension of the course design could be improved. In the case of a missing or unclear peer review, the platform could provide reviewees with means to contact their reviewers directly. This could naturally change the course format from a fixed three-step script (submission, review, rework) to a personalized design where the process of working out a problem, discussing solutions, and reworking solutions takes as many steps as needed.

The proposed improvements rely in part on the discovery of submission categories which in turn relied on automated compiling and testing. This seems to make the use of these techniques in other (non programming) courses impractical. However, it can be argued, that STEM subjects are often based of formal languages (such as algebraic expressions in mathematics or structural formulas in chemistry) that could be interpreted and tested automatically. It is a conviction of the authors that novel course formats requiring less teacher involvement could benefit from such techniques, especially in STEM education, since these techniques not only identify *proficient*, but also *motivated* students.

This article has introduced a novel course format which requires a minimal involvement of teachers. The course format has been evaluated in a case study during a university course in computer science. The proposed format relies on peer reviews and self-correction. The evaluation has shown the effectiveness of the approach and that an insufficient participation in peer reviewing, and hence a lack of reviews, was a problem. Perspectives for overcoming this problem without requiring more teacher work and for applying the format to other subjects have been discussed.

Acknowledgments. The authors are thankful to Elisabeth Lempa for her contribution to assessing the quality of reviews and to coding.

References

1. Bathini, P.P., Sen, S.: Impact of integration through peer instructed lectures. Int. J. Basic Clin. Pharmacol. **6**(6), 1293–1296 (2017)
2. Benè, K.L., Bergus, G.: When learners become teachers: a review of peer teaching in medical student education. Fam. Med. **46**(10), 783–787 (2014)
3. Bester, L., Muller, G., Munge, B., Morse, M., Meyers, N.: Those who teach learn: Near-peer teaching as outdoor environmental education curriculum and pedagogy. J. Outdoor Environ. Educ. **20**(1), 35 (2017)
4. Burch, N.: The Four Stages for Learning any New Skill. Gordon Training International, CA (1970)
5. Carrell, S.E., Sacerdote, B.I., West, J.E.: From natural variation to optimal policy? the importance of endogenous peer group formation. Econometrica **81**(3), 855–882 (2013)
6. Cho, K., MacArthur, C.: Learning by reviewing. J. Educ. Psychol. **103**(1), 73 (2011)

7. Dochy, F., Segers, M., Sluijsmans, D.: The use of self-, peer and co-assessment in higher education: a review. Stud. High. Educ. **24**(3), 331–350 (1999)
8. Falchikov, N., Goldfinch, J.: Student peer assessment in higher education: a meta-analysis comparing peer and teacher marks. Rev. Educ. Res. **70**(3), 287–322 (2000)
9. Fischer, K.W.: A theory of cognitive development: the control and construction of hierarchies of skills. Psychol. Rev. **87**(6), 477 (1980)
10. Gartner, A., et al.: Children Teach Children: Learning by Teaching. ERIC (1971)
11. Goldschmid, B., Goldschmid, M.L.: Peer teaching in higher education: a review. High. Educ. **5**(1), 9–33 (1976)
12. Hanrahan, S.J., Isaacs, G.: Assessing self-and peer-assessment: the students' views. High. Educ. Res. Dev. **20**(1), 53–70 (2001)
13. Hattie, J., Timperley, H.: The power of feedback. Rev. Educ. Res. **77**(1), 81–112 (2007)
14. Jones, S.P., Hall, C., Hammond, K., Partain, W., Wadler, P.: The glasgow haskell compiler: a technical overview. In: Proceedings UK Joint Framework for Information Technology (JFIT) Technical Conference. vol. 93 (1993)
15. Jumaat, N.F., Tasir, Z.: Instructional scaffolding in online learning environment: a meta-analysis. In: 2014 International Conference on Teaching and Learning in Computing and Engineering (LaTiCE), pp. 74–77. IEEE (2014)
16. Kruger, J., Dunning, D.: Unskilled and unaware of it: how difficulties in recognizing one's own incompetence lead to inflated self-assessments. J. Personal. Soc. Psychol. **77**(6), 1121 (1999)
17. Lundstrom, K., Baker, W.: To give is better than to receive: the benefits of peer review to the reviewer's own writing. J. Sec. Lang. Writ. **18**(1), 30–43 (2009)
18. Nicol, D., Thomson, A., Breslin, C.: Rethinking feedback practices in higher education: a peer review perspective. Assess. Eval. High. Educ. **39**(1), 102–122 (2014)
19. Ramdass, D., Zimmerman, B.J.: Effects of self-correction strategy training on middle school students' self-efficacy, self-evaluation, and mathematics division learning. J. Adv. Acad. **20**(1), 18–41 (2008)
20. Rohrer, D., Taylor, K.: The shuffling of mathematics problems improves learning. Instr. Sci. **35**(6), 481–498 (2007)
21. Schwartz, M.S., Fischer, K.W.: Building vs. borrowing: The challenge of actively constructing ideas in post-secondary education. Lib. Educ. **89**(3), 22–29 (2003)
22. Seenan, C., Shanmugam, S., Stewart, J.: Group peer teaching: A strategy for building confidence in communication and teamwork skills in physical therapy students. J. Phys. Ther. Educ. **30**(3), 40–49 (2016)
23. Shaughnessy, J.J.: Long-term retention and the spacing effect in free-recall and frequency judgments. Am. J. Psychol., 587–598 (1977)
24. Topping, K.: Peer assessment between students in colleges and universities. Rev. Educ. Res. **68**(3), 249–276 (1998)
25. Williams, E.: Student attitudes towards approaches to learning and assessment. Assess. Eval. High. Educ. **17**(1), 45–58 (1992)

Algorithm and Software Tool for Multiple LMS Users Registering

Vladlen Shapo[✉]

National University "Odessa Maritime Academy", Odessa, Ukraine
vladlen.shapo@gmail.com

Abstract. Learning Management Systems (LMS) administrator periodically deals with big users number registering. Manual registering is monotonous but may not be assigned to low skilled staff, because such employees must have administrator's rights. It entails possible LMS non-working condition, causes risk of strangers access and increases probability of e-learning courses author's rights violation. Administrator's dull work may be minimized, if to use data, entered at applicants enrollment in any files, prepared manually or exported from software applications. So, it's necessary to develop the template, where to define formation order of significant information components about users, which must be registered; to develop software tool, which will read data from source files and will form text files, ready for import by LMS accordingly to template; to convert data using software tool and to import obtained files into LMS.

In this paper proposed algorithm and software tool, which allowed to automate registering more than 6400 LMS users at once. Further specified software tool often used to minimize LMS users registering time (650–700 people at once). It's especially actual at the studying year start, before exam sessions, training groups formation, etc. It allows to import text files and creates output text files in format, necessary for importing into LMS. Developed software tool may be easily modified for different input data formats and languages and close tasks solving in different educational institutions.

Keywords: Multiple LMS users registering automation algorithm and software tool · LMS user profiles importing

1 Context

Last years a lot of different organizations in the entire world actively use program applications and environments, which allow to create and to exploit learning management systems (LMS) for students knowledge level increasing and employees qualification enhancement. Possibilities, advantages and lacks of most popular LMSs are analyzed in [1, 2]. LMS Moodle became most popular because it's well described in multiplicity of papers, has simple and convenient interface, allows to solve most tasks of e-learning, possesses of numerous possibilities and has some advantages in comparison with competing LMSs. Moodle still developing regularly from 2002 in base variant and by quantity and possibilities of additional modules, created by third party developers, which allow to solve different auxiliary tasks. In Ukrainian conditions

© Springer Nature Switzerland AG 2020
M. E. Auer and T. Tsiatsos (Eds.): ICL 2018, AISC 916, pp. 99–109, 2020.
https://doi.org/10.1007/978-3-030-11932-4_10

LMS Moodle wins also in connection with insufficient financing level of education from state, fear to make investments in corresponding software using own financial resources, high inertia and inadequate qualification level of some employees in educational institutions and following reasons: meaningless to buy application software (program environment) if it will be almost or half empty (without any content or with just a little part of it) during many years. Also Moodle has a wide spectrum of possibilities for working of students, trainees, teachers, course creators and administrators, which organize directly users' work, their registration, user names and passwords issuing and changing, users' rights granting [1, 2].

For most organizations, which use LMS, problem of necessary fast creation of numerous user accounts does not occur, because sufficient changing of human resources happens not often in general, and qualification enhancement of staff members using e-learning is realized also not often and by small groups. But at active LMS using in educational institutions arising a necessity of users variety registering (from some hundreds to some thousands) at very short term (for example, newly entered 1st year full and part-time (studying by correspondence) students enrollment).

2 Purpose

Most Significant difference of National University "Odessa Maritime Academy" (NUOMA) from a lot of another universities is presence of big remote structured subdivisions in different parts of Ukraine: Danube Institute (city of Izmail) and Azov Maritime Institute (city of Mariupol). Also thousands of full or part time students, postgraduate students, advanced training courses trainees, aimed for trainings by shipping or crewing companies, every year pass many months naval training on commercial vessels and need to have access to teaching materials (TM) and teachers' consultations, being far from NUOMA. In the past (15–20 years ago) these possibilities could be realized only by passing paper or electronic TMs to vessels during crew members change (unexpectedly and unreliable), sending electronic TMs via e-mail or downloading TMs from NUOMA's site (main hindrance was high cost or impossibility of data exchange using satellite data transfer systems). These ways were realized during 13 years without any statistics, users registration, efficiency analyzing, etc. Using Moodle (started in test mode in 2009, in full mode started in 2010) has significantly allowed to enhance quality of TMs preparing, and also theirs quantity and actuality, and to prepare teachers and students step by step for regular using of Moodle possibilities wide spectrum. Hence implementation and LMS using in NUOMA is highly actual, and therefore problems of variety users profiles input and removing in short time, briefly described above, are very actual as well.

LMS administrator periodically deals with vast number of new users registering. Absence of automation of this process leads to huge time wasting (manual registering of one user takes 4–7 min of boring mechanical work) or to necessity to entrust this process to additional employees. Manual work leads to appearing of significant number of additional mistakes; in turn it will be a reason for additional administrator's time wasting. Also there is a necessity of automation of additional tasks solving at administrator work. Some of them described in papers [3–5].

Manual users registering is very monotonous work, which must not be assigned to highly qualified specialists like LMS administrators, but to possessing much lower qualification support staff of educational institution. But in this case low skilled employees will get administrator's rights, which entails possible LMS destroying, its long time non-working conditions and complex restoring. As a result it requires intrusion of highly qualified and highly paid specialists. This situation also causes risk of strangers access to LMS from the outside because of low motivation of support staff to keep commercial secret and defend interests of own organization-employer. Also in this case increases probability of violation of author's rights of TMs developers. Possible way to solve this problem is the plain text file formation by support staff, which will be imported to LMS accordingly to template formed by administrator. In this case at users list importing into LMS may occur mistakes, arising at text typing by support staff during imported file formation. LMS administrator has to correct mistakes in imported plain text file also manually and initiate repeatedly the data importing procedure of LMS registered users.

3   Approach

Volume of dull work may be significantly decreased, if to use data, entered in corresponding subdivisions at applicants enrollment. These data may be obtained as files, prepared in any text processor or in any electronic spreadsheets, which contain tables with information about applicants, prepared manually or by exporting from application software, which automates gathering and processing of information about applicants in corresponding subdivisions of educational institution. At existence of such files proposed course of action for LMS administrator is presented below.

1. To save specified files in text formats (manual work; has to be performed many times depending on number of source files).
2. To develop corresponding template, where to define clearly the order of most significant information components formation about users, which must be registered.
3. To develop corresponding software tool, which will allow: to read data from text files, created at paragraph 1; to form corresponding plain text files, which are ready for import by LMS accordingly to beforehand prepared template.
4. To perform data converting procedure using software tool described in paragraph 3.
5. To import obtained files into LMS.

Figures 1a–d present algorithm scheme, which is realized by author in corresponding software tool, which allows to solve the task on automation of large users number entering into LMS at short term. This is especially actual at the studying year beginning, after publication of orders on enrollment, before exam sessions starting or module control performing, training courses attendees groups formation, etc.

Problems analyzed above are extremely actual for wide spectrum of Ukrainian educational institutions, implementing e-learning for additional studying possibilities creating and employees qualification enhancement. In particular, in NUOMA and its subdivisions in Mariupol and Izmail there are great number of full and part time students, which have to be registered in LMS. Author, working as NUOMA LMS administrator from 2009, created software tool, which allows to solve described problems and automated registering more than 6400 LMS users for first time. Time spent for this software tool development is significantly less, than for manual registering. Further application of specified software tool has allowed to minimize time, necessary for LMS users registering (650–700 people at once), during 8 years.

For automation of user profiles importing procedure it's necessary to prepare plain text file accordingly to special template. Its first line contains some standard fields, separated with commas (these fields may be listed in random order, and some of them may not be included in the list); second and all following lines are random number of users profiles, described accordingly to template.

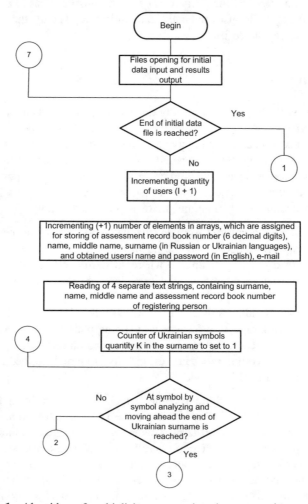

Fig. 1. Algorithm of multiplicity users registration automation program

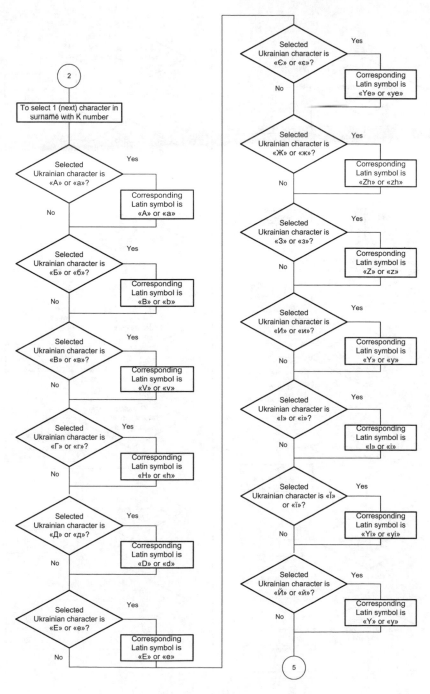

Fig. 1. (*continued*)

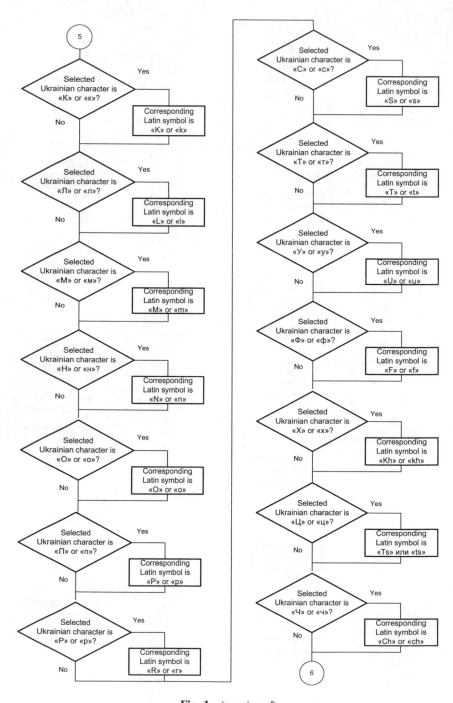

Fig. 1. (*continued*)

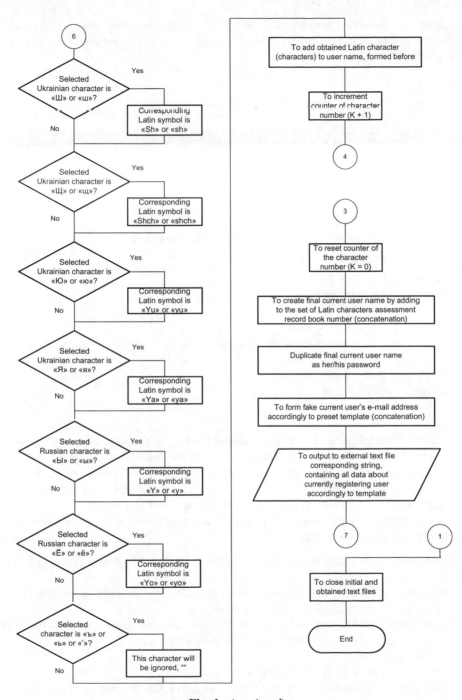

Fig. 1. (*continued*)

Brief description of main template parameters (fields) is shown below. Obligatory fields are described in the Table 1. Fields, included by default, but not obligatory are shown in Table 2.

Table 1. Obligatory fields of template

Field name	Description
Username	User's login
Firstname	User's name (and middle name, if necessary)
Lastname	User's surname and user additional informal ID, if necessary
Email	User's e-mail address

Table 2. Fields, included to the template by default

Field name	Description
Institution	Institution or organization name
Department	Name of department or subdivision
City	Name of the city
Country	Name of the country
Lang	Preferable default interface language (coded by two letters of Latin alphabet: Russian – ru, English – en, Ukrainian – ua, etc.)
Auth	User authentication method
Timezone	Country time zone (for example, +2)

Names of organization or institution, department or subdivision, city, country may be entered in any language. Every line of imported file must contain one record (one user description). Every record consists of data, separated by commas or semicolon. In the case of mismatching of fields number in template, number of data or separators the message about mistake will be displayed and data will not be entered.

As a result of some years real practical work optimal for concrete conditions following template format is formulated:

Username – user name (login), which will be displayed at keyboard typing at LMS admission,
Password – password (not displayed at typing, asterisks are displayed instead),
Firstname – name (name and middle name in any language),
Lastname – surname and additional user ID (specialty name, calendar and studying years of enrollment, department name, form of education, etc.),
Email – user's e-mail,
Lang – preferable by default interface language,
Maildisplay – either to show user e-mail address at LMS working or not (1 – make it visible, 0 – make it invisible for another users),
City – city, where user lives or works (any name in any language),

Country – country (two-symbol standard country code must be chosen from standard list of country codes, for example, DE - Germany, UA – Ukraine, etc.),
Timezone – time zone (time advance or lag with central European time, for example, +2 for Ukraine).

Example of plain text file for importing users' profiles into LMS, containing form template and three users' profiles and screen shot of users' profiles loading procedure into LMS are presented in Figs. 2 and 3.

username, password, firstname, lastname, email, lang, maildisplay, city, country, timezone
teacher1, teacher1, Ivan Ivanovich, Ivanov (t/taccm/prof), net010513@ukr.net, ru, 1, Odessa, UA, +2
teacher2, teacher2, Petr Sidorovitch, Petrov (t/taccm), net020513@ukr.net, ru, 1, Odessa, UA, +2
teacher3, teacher3, Sidor Petrovitch, Sidorov(t), net030513@ukr.net, ru, 1, Odessa, UA, +2
Brunov216123, Brunov216123, Геннадій Іванович, Бурунов (em17/3), brunovg@gmail.com, ru, 1, Odessa, UA, +2
Danylov315023, Danylov315023, Максим И., Данилов (em17/4z), net11012018-6@ukr.net, ru, 1, Odessa, UA, +2

username	password	firstname	lastname	email	lang	maildisplay	city	country	tim
prep1	prep1	Иван Иванович	Иванов (t)	net010513@ukr.net	ru	1		Одесса UA	+2
prep2	prep2	Петр Петрович	Петров (t)	net020513@ukr.net	ru	1		Одесса UA	+2
prep3	prep3	Сидор Сидорович	Сидоров (t)	net030513@ukr.net	ru	1		Одесса UA	+2

Fig. 2. Users' profiles loading procedure into LMS

Upload users preview⊙

username	password	firstname	lastname	email	lang	maildisplay	city	country	timezone
teacher1	teacher1	Ivan Ivanovich	Ivanov (t)	net010513@ukr.net	ru	1	Odessa	UA	+2
teacher2	teacher2	Petr Petrovitch	Petrov (t/taccm)	net020513@ukr.net	ru	1	Odessa	UA	+2
teacher3	teacher3	Sidor Sidorovitch	Sidorov (t/taccm/prof)	net030513@ukr.net	ru	1	Odessa	UA	+2
Burunov216123	Burunov216123	Геннадій Іванович	Бурунов (em17/3)	brunovg@gmail.com	ru	1	Odessa	UA	+2
Danylov315023	Danylov315023	Максим И.	Данилов (em17/4z)	net11012018-6@ukr.net	ru	1	Odessa	UA	+2

Number of preprocessed records: 5

Upload type	Add new only, skip existing users
New user password	Field required in file
Existing user details	Нет изменений
Existing user password	Нет изменений
Разрешить переименования	Нет
Разрешено удаление	Нет
Prevent email address duplicates	Нет
Select for bulk operations	Нет

Fig. 3. Preview of the user list at its loading into LMS

4   Conclusion

Software tool developed using Visual Basic programming language. It allows to import plain text files and creates output plain text files in format, necessary for importing to LMS, which significantly simplifies initial data preparing and further results processing and in hundred times reduces time, spent by LMS administrator for large number of LMS users registering. If necessary, mentioned developed software tool may be easily modified for close tasks solving, among others, for general educational schools, lyceums, gymnasiums and another educational institutions and organizations. Also newly proposed algorithm and software tool may be easily adjusted for using of Greek, Turkish, Polish, French and another non Latin alphabets.

References

1. Shapo, V.F.: Distance learning systems application software analyzing. In: Information Technologies In Learning Process: Proceedings of 4th Scientific-Methodic Seminar, pp. 77–78. South State Pedagogic University, Odessa (2003). Шапо В.Ф. Анализ программных средств систем дистанционного обучения // Информационные технологии в учебном процессе: труды 4-го научно-методического семинара. – Одесса: ЮГПУ. – 2003. – С. 77 – 78. In Russian
2. Shapo, V.F.: Distance learning systems application for educational institutes and specialists competitiveness level enhancing. Network development of management training courses for transport and logistics sector: monograph/University of Paderborn (Germany) and etc., 456 p., pp. 423–437. National Transport University, Kyiv (2008). Шапо В.Ф. Применение систем дистанционного обучения для повышения уровня конкурентоспособности учебных заведений и специалистов // Мережева розробка курсів тренінгу з менеджменту у сфері транспорту і логістики: монографія / Ун-т м. Падерборн (Німеччина) та ін. – К.: НТУ, 2008. – 456 с. – С. 423 – 437. In Russian
3. Shapo, V.F.: Software tool for automation of distance learning system users registering process. In: Problems of Information Society Development: Proceedings of IInd International forum, Kyiv, 12–15 October, 2010. Part I/«Informatio-Consortium» association. – K.: UkrISTEI, 2010, 252 p., pp. 243–249. Шапо В.Ф. Средство автоматизации процесса регистрации пользователей системы дистанционного обучения. // Проблеми розвитку інформаційного суспільства: матеріали II Міжнародного форуму, Київ, 12–15 жовтня 2010 р. Частина I / Асоціація « Інформатіо-Консорціум » . – К.: УкрІНТЕІ, 2010. – 252 с. – С. 243–249. In Russian
4. Shapo, F.S., Shapo, V.F., Volovschikov, V.Y.: Automation of users profiles entering in distance learning system. Electrotech. Comput. Syst. № 1 (77), pp. 109–114 (2010). Шапо Ф.С., Шапо В.Ф., Воловщиков В.Ю. Автоматизация ввода пользовательских профилей в системе дистанционного обучения // Електротехнічні та комп'ютерні системи. – 2010. – № 1 (77). – С. 109 – 114. In Russian
5. Shapo, V.F.: Algorithm and method of distance learning system influence analysis on loading of educational institution information system. In: Problems of Information Society Development: Proceedings of IIIrd International Forum, Kyiv, 20–23 Nov 2012. Part I/

«Informatio-Consortium» association. – K.: UkrISTEI, 2012, 204 p., pp. 184–189. Шапо В.Ф. Алгоритм и метод анализа влияния системы дистанционного обучения на загрузку информационной системы учебного заведения // Проблеми розвитку інформаційного суспільства: матеріали III Міжнародного форуму, Київ, 20–23 листопада 2012 р. Частина I / Асоціація « Інформатіо-Консорціум » , УкрІНТЕІ. – K.: УкрІНТЕІ, 2012. – 204 с. – C. 184–189. In Russian

For Which Type of Students Does the Inverted Classroom Model Work Out?

An Empirical Analysis of Learning Success of Different Types of Students

Nadine Hagemus-Becker[⊠], Ellen Roemer, and Hartmut Ulrich

Hochschule Ruhr West, Duisburger Str. 100, Gebäude 3, Raum 02.117, 45479 Mülheim an der Ruhr, Germany
Nadine.Hagemus-Becker@hs-ruhrwest.de

Abstract. Although the concept of inverted classroom has gained increasing attention in academic teaching, it is still unclear for which type of students inverted classroom models (ICM) really work out. The purpose of our study is to analyse the impact of ICM on the improvement of learning success for different student types. Therefore, our research provides answers to the following research questions: For which type of students does ICM really work out regarding their learning success? Whether and how far do ICM need to be adapted in order to meet the individual needs of heterogeneous student groups? An empirical study with 178 cases was conducted in four courses that have adopted ICM at an University of Applied Science in Germany. Based on five different learning behaviours, three different learning types were identified. Students showing particular learning behaviours need additional, but discriminative support for their learning success. The insights of this paper help to modify the concept of ICM so that it satisfies the needs of different types of learners.

Keywords: Inverted classroom model · Learning success · Student types

1 Introduction

During the last decade, blended learning formats have become more and more popular [1]. According to Bishop and Verleger [2] the model of inverted or flipped classroom is of special interest as a new blended format. The inverted classroom model includes "two parts: interactive group learning activities inside the classroom, and direct computer-based individual instruction outside the classroom" [2, p. 4].

Numerous studies have evaluated the effectiveness of the inverted classroom model. Most of them (a) discuss the effect on students' engagement level [3–7], (b) compare the performance measured by grades or weekly diaries [8–10] or (c) provides insight in individual inverted classroom models [11–13].

Only few studies [14–17] focus on the individual learning styles of students and the learning behaviour in inverted classrooms. They investigate for example, which attitudes support the use of ICM [14]. Moreover, all of them consider students as one

© Springer Nature Switzerland AG 2020
M. E. Auer and T. Tsiatsos (Eds.): ICL 2018, AISC 916, pp. 110–119, 2020.
https://doi.org/10.1007/978-3-030-11932-4_11

homogeneous group. Likewise, Hwang and Arbaugh [14] admit that different attitudes and patterns of learning behaviour lead to various test results and therefore to different learning success.

Therefore, it is still unclear for which type of students inverted classroom models really work out. Moreover, there is a lack of consensus, if, in how far and how inverted classroom models have to be adapted for the individual needs of heterogeneous student groups.

The aim of this paper is to analyse the impact of inverted classroom models on the improvement of learning success for different student types. The contribution of our research is twofold:

- The paper identifies different types of students based on their learning behaviour;
- The paper demonstrates whether and how far inverted classroom models need to be adapted in order to improve learning success of heterogeneous student groups;

To accomplish our aim, the paper is structured as follows. Section 2 provides a literature research focusing on studies that compare the learning success of different student groups in inverted classroom models to identify the research gap. Section 3 outlines the methodology, while Sect. 4 focuses on the data analysis. Section 5 discusses the results and implications. Finally, Sect. 6 outlines avenues for future research.

2   Literature Review

While several studies have investigated the impact of ICM on learning success and outcomes measured by grades [18], only few researchers have investigated different types of students and their learning behaviour.

One exception are Lage et al. [16]. They discuss the capability of inverted classrooms to better meet the needs of different learning types. Basically, Lage et al. [16] state that "a wide range of psychological, sociological, and pedagogical literature has documented that student populations are composed of individuals with distinctly different learning styles. Unfortunately, students do not explicitly select classes based on instructor teaching style; nor can the instructors be expected to change their personality types to accommodate all students" [16, p. 31]. The authors admit that it is difficult to appeal to the learning styles of each and every student. But both lecturers and students benefit from the combination of self-study parts and in-class interactions, as the lecturers can monitor performance and comprehension whereas the students are able to clear up any confusion immediately [16, p. 37]. The authors abstain from supporting their findings by empirical research. Our study fills this gap and suggests different didactic methods for each learning type.

Keller et al. [15] focus on learning strategies in their study on students in distance learning courses. They show that practical and structural skills (e.g., to apply scientific methods to practical problems) are as important as expertise and grades when talking about the broadening of knowledge. Furthermore, Keller et al. [15] underline the importance of cooperation among the students regarding the learning success, especially for students without experiences with self-regulated learning. In contrast to the other approaches, Keller et al. [15] define learning success independent from grades as an indicator for the broadening of the students' knowledge. As Keller et al. [15], our study does not concentrate on measurable grades, but on the individual learning success of different types of students.

McNally et al. [17] also deal with students' learning. In their study, they compare different components of the inverted classroom model in relation to the final grades. Although McNally et al. [17] found differences between those who endorse and those who resist inverted classroom environments (particularly in their engagement), this differentiation based on preferences does not correspond to differences in the final grades in an inverted course. Therefore, McNally et al. [17] assume that preferences alone may not be the most important aspect against which to evaluate an inverted classroom model. Our study deepens the understanding of the students' attitudes by researching the competencies, which might explain the endorsement and the resisting attitudes.

Finally, Hwang et al. [14] monitor students' learning behaviour. They deal with the topic of feedback-seeking activities in inverted classrooms, e.g., ask if the feedback-seeking behaviour and learning performance depend on the students' orientations towards competitive and cooperative attitudes. According to Hwang et al. [14], there are at least three common ways of feedback-seeking such as learning interactions as professors asking questions in-class and outside of class, and checking with fellow students. Their study shows a significant impact on the outcome of multiple-choice test depending on the participation in online discussion forums to seek others' feedback.

To sum up, these findings offer first steps to acknowledge students' diversity instead of viewing them as a one group. However, these studies fail to gain insights in different learning styles based on students' learning behavior. That is what our study provides: Splitting the heterogeneous student group into smaller, but homogeneous subgroups depending on the individual learning style and underlying competencies regarding students' learning behaviour. To know these is necessary in order to adapt inverted classroom models to maximize students' learning success by providing fitting didactic methods.

3   Methodology

The study aimed at identifying different learning types based on learning behavior in order to adapt the inverted classroom model to improve students' learning success. Therefore, we conducted a survey among undergraduate and postgraduate students at a University of Applied Science in Germany. All surveyed students took part in a course that used an ICM format. A total of six courses with 30 students on average were related either to the Business Department or to the Engineering Department. The courses varied in how the inverted environment was incorporated in the teaching practice, e.g. in relation to the length of the particular video sequences.

Participants completed a questionnaire in German after joining the ICM course. The questionnaire contained four different parts, i.e., learning success and learning behavior, the students' attitude towards the course, the overall satisfaction with the course and the learning environment as well as socio-demographic information.

The items for learning success and learning behavior construct were based on Keller et al.'s [15] study and included modified items like 'I improved my ability to find relevant information' or 'I feel more confident with my professional decision' [15, p. 55]. Learning behavior was assessed with seventeen items referring to the learning behavior in the inverted classroom environment. Participants were asked to think about their learning style and indicate their opinion of statements, for example about whether (a) they are able to distinguish more important and less important inputs, (b) they link new information to existing knowledge or (c) they can find a fitting method for the learning matters. All attitudinal scales were anchored on a seven-point scale ranging from "strongly agree" (1) to "strongly disagree" (7), so that lower mean scores represent attitudes that are more positive.

The student data were collected starting in 2015 until 2018. Subsequently, the data were imputed in a spreadsheet and cleaned in the first step, excluding those questionnaires with invalid answers, answer tendencies and a multitude of missing values. In sum, a sample of 185 undergraduate and postgraduate students participated in the survey. Ages ranged from 15 to 39 years (M = 24,47; SD = 3,25). For the data analysis, we used the IBM SPSS 24 software package.

4   Data Analysis

In a first step, we conducted principal component analysis on the items concerning learning behavior. We could reduce 17 items to five factors by using the Eigenvalue criterion and the elbow criterion. The remaining model accounted for 62.3% of item variance. Principal component analysis showed a clear and discriminate factor loadings ranging from 0.912 to 0.512. Based on the items relating to learning behavior, principal component analysis indicated five factors behind the items (see Table 1). Therefore, we summarized the factors by identifying a common ground for each set of items:

In a second step, we used the five factors regarding different learning behaviors to identity clusters, i.e., different types of students. To identify outliers, we used single-linkage cluster analysis. Seven elements could be identified as outliers and were removed from the data set. The final data set consists of 178 cases. Afterwards, we applied the Ward-Method to select the number of clusters. We identified three clusters using the respective dendrogram. Finally, we computed k-means clustering to assign each student to one of the three clusters, i.e., one of the three student types. Table 2 shows the cluster means of the standardized factor values. Lower values (or negative values) should be interpreted as higher outcome of the respective factor (Fig. 1).

In a third step, we used analysis of variance (ANOVA) to profile the different clusters, i.e., to differentiate between student types regarding their learning behaviour. ANOVA revealed that participants in the clusters had significantly different mean scores in relation to their planning competence, their commitment and their distraction from learning. We also conducted ANOVA to test the relationship between the level of learning success between the different student types. Finally, we computed cross tables to explore the three student types in more detail regarding their socio-demographic characteristics.

Table 1. Five factors of learning behaviour

Factor	Items relating to learning behavior
Planning regarding students' learning process	• The most important aspects I write in a list and learn them by heart
	• I write short summaries
	• I learn the most important things by heart to remember them during the exam
	• To check if I understand everything right I repeat the content without my notes or books
	• I plan my learning process before I start
	• I use a timetable for my learning process
Crosslinking of knowledge	• I try to combine the new learning matters with existing knowledge
	• I link new learning matters to existing knowledge
	• I try to repeat the new learning matter in my own words
Commitment	• I push myself even although it is not interesting at all
	• I do not give up even if it is hard to understand
	• The most important documents are always readily at my hand
Distraction from learning	• I am distracted easily
	• I lack concentration when learning
	• I loose myself in details and cannot concentrate on the important things
Selection of learning materials	• I find a fitting method for the learning matters
	• During the course, I can distinguish between more and less important input

5   Results and Implications

The three clusters represent the different types of students. The first cluster contains students with the highest mean score for their ability to commit themselves to the learning process (-0.70), and an even higher mean score for the disposition to lose the focus during the learning process (-0.89). As their lowest mean score describes their ability to plan the learning process (0.15), this cluster summarises students with a **need for support**. These students are very self-motivated but easily distracted. Moreover, they are less able to plan their learning process. In Table 2, the dotted line describes this type of students.

The second cluster shows the highest mean scores for commitment (-0.57), the lowest for distraction (0.95) and a moderate mean score for the planning competence (-0.04). This cluster includes **ambitious** students with a strong commitment and a clear focus on the learning process. Even if their planning competence is only middle-ranged, these students seem to adjust this aspect with their motivation to stay on track. Table 2 shows this cluster by the line with alternating dots and stipes.

Table 2. ANOVA for learning success of the different learning types

Item	Learning type	Mean	SD	SE	Sig.
My expertise raised explicitly	Need for support	1.83	0.694	0.100	0.057
	Pragmatic learners	2.13	0.866	0.111	
	Ambitious learners	1.83	0.706	0.098	
	Overall	1.94	0.777	0.061	
My general educational horizon was broadened	Need for support	2.27	0.818	0.118	0.488
	Pragmatic learners	2.47	0.929	0.120	
	Ambitious learners	2.33	0.879	0.122	
	Overall	2.36	0.879	0.070	
I learned how to apply scientific methods to practical problems	Need for support	2.27	0.939	0.136	0.331
	Pragmatic learners	2.44	0.922	0.118	
	Ambitious learners	2.17	1.061	0.147	
	Overall	2.30	0.975	0.077	
The course motivated me to set a scientific focus	Need for support	2.90	1.189	0.172	0.534
	Pragmatic learners	3.16	1.241	0.159	
	Ambitious learners	3.04	1.283	0.178	
	Overall	3.04	1.237	0.097	
The course motivated me to deal with new subjects	Need for support	2.85	1.215	0.177	0.134
	Pragmatic learners	3.03	1.129	0.147	
	Ambitious learners	2.60	1.089	0.151	
	Overall	2.84	1.150	0.092	
Because of this course I look at professional problems in a different perspective	Need for support	2.60	1.210	0.176	0.697
	Pragmatic learners	2.66	1.124	0.144	
	Ambitious learners	2.47	1.155	0.162	
	Overall	2.58	1.155	0.092	

(*continued*)

Table 2. (*continued*)

Item	Learning type	Mean	SD	SE	Sig.
Because of this course I consider professional problems more critically	Need for support	2.94	1.192	0.172	0.191
	Pragmatic learners	2.80	1.062	0.136	
	Ambitious learners	2.56	0.916	0.127	
	Overall	2.76	1.064	0.084	

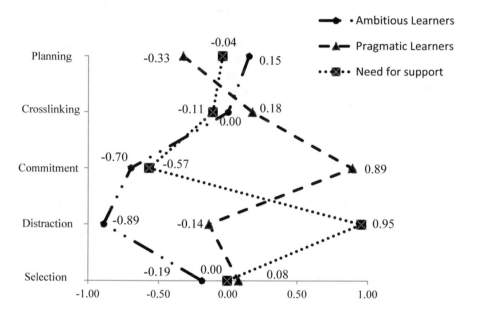

Fig. 1. Three learning types regarding the different competencies

The third cluster describes students with the highest mean scores of planning competence (−0.33), the lowest mean scores for commitment (0.89) and a mean score of distraction close to zero (−0.14). Although these students possess a high capability to plan or manage their learning process, they are not self-motivated at all. Nevertheless, they are able to concentrate on their learning matters. The students summarized in the third cluster are **pragmatic** learners who stuck to the plan. In Table 2, this cluster is represented by the striped line.

To better understand this outcome, the cluster size has to be taken into account: Most of the participants belong to the group of pragmatic learners (61 students), the group of ambitious learners counts 52 students, whereas only 48 students need additional support (17 cases are missing).

A comparison of the socio-demographic factors like gender, age, or duration of study might describe the different student types more in detail. The three clusters do not differ in ratio of gender. That means that all groups contained almost similar percentage of women. Nevertheless, the largest part of all groups study in the fifth or seventh semester, independent of the age. To sum up, the socio-demographic factors do not help to differentiate the types of learners. Apart from these socio-demographic factors, the learning success reveals further insight in the learning behaviour and attitudes of the students.

As Table 2 shows, the self-estimation of the learning progress and therefore the learning success shows no significant differences between the groups. As said before, the learning success does not only concentrate on grades, but follows Keller et al.'s [15] definition as an indicator for the broadening of the students' knowledge. That becomes clear when the students agree to the item that their expertise raised explicitly. The high mean score of this item (1.94) reflects the overall satisfaction with this course and the success of ICM, as all students seem to profit from this learning environment. Moreover, the different student types accord to each other regarding the capability to look at professional problems in a different perspective or the motivation to set a scientific focus. All in all, the students seem to be very satisfied with ICM.

Based on the outcomes, additional support for the students' learning process can be deduced, as the findings from this study provide initial insights on how the five different competencies influence the learning behaviour of the different learning types outside the classroom. This is especially necessary for inverted classroom models, since students' preparation of the in-class sessions is often uncontrolled and therefore unsupported. Hence, the following implications refer especially to the preparation phase.

As ambitious learners seem to benefit most of the inverted classroom model, the given structure supports them in planning their learning process. To meet the needs of ambitious learners and assist them well enough for their individual preparation, a clear and consistent structure of the course is compulsory: The introduction of any inverted classroom course shall contain exploratory elements how to use the video tutorials. This finding is consistent with McNally et al.'s demand for "the importance of tailoring the introduction [...] to better engage those students" [17, p. 292]. A short multiple-choice-test before every in-class session may also help ambitious learners to prepare well enough.

On the one hand, distracted learners are characterised by their tendency to lose their focus easily. On the other hand, they are very committed. Distracted learners need to know how to learn best by providing guidance in how to avoid distraction. This knowledge cannot be assumed, it has to be taught actively. Moreover, short video sequences with distinct topics and additional materials like texts, presentations or other documents shall help these students with a lack of concentration. Small steps lead to quick outcomes, and quick, positive outcomes motivate even more. This finding implies the strength of the inverted classroom model to raise the learning success even of distracted students.

Pragmatic learners also benefit from this method. As well as students with a lack of concentration, unmotivated students like the pragmatic learners benefit from short input sequences, quick learning progress and small tests to structure their learning routines.

Moreover, these students might appreciate virtual feedback as proposed by Hwang et al. [14, p. 282] during the preparation phase. The possibility to pose their questions withdraws them from quick distraction and keeps the motivation high.

The description of the different student types can be linked to McNally et al.'s [17] separation of students who endorse and students who resist inverted classrooms. As our analysis showed, ambitious learners tend to take most out of ICM, whereas distracted and pragmatic learners have to change their natural learning behaviour and habits to follow the structure of inverted classrooms. This might be a hint that at least distracted learners might rather resist ICM.

In accordance with Lage et al., more diversified teaching methods intensify the learning success of every student [16, p. 32], especially outside the classroom. Thus, the combination of short video sequences, a small multiple-choice test before each in-class session and the offer of virtual feedback methods during the preparation phase works as a guideline through the learning process. Furthermore, it supports Lage et al.'s conclusion that "the inverted classroom implements a strategy of teaching that engages a wide spectrum of learners" [16, p. 41].

6 Future Research

Several studies have shown that inverted classroom models have a very positive impact on students' learning success in general. However, we posit that different types of students need support regarding their preparation phase of in-class sessions. Using the insights from this paper, we have identified three student types based on five competencies regarding students' learning behaviour. In this way, tailored support for the different student groups may help further improving students' learning success in inverted classroom modules.

Further research might shed light on the learning process during the preparation phase [19]. When working with video tutorials during the preparation phase in ICM, it is still unclear if the students understand the videos easily and in which parts students may have questions. This is due to the fact that direct feedback, which is given in class by facial expressions or other non-vebal comments of students, is largely missing in the preparation phase of ICM. Therefore, students should be able to provide instant feedback and ask questions regarding their preparation material during the preparation phase.

References

1. Strayer, J.F.: How learning in an inverted classroom influences cooperation, innovation and task orientation. Learn. Environ. Res. **15**(2), 171–193 (2012)
2. Bishop, J., Verleger, M.: The flipped classroom: a survey of the research. Am. Soc. Eng. Educ. (2013)
3. Tsai, C.-W., Shen, P.-D., Chiang, Y.-C., Lin, C.-H.: How to solve students' problems in a flipped classroom: a quasi-experimental approach. Univ. Access Inf. Soc. **16**(1), 225–233 (2017)

4. Subramaniam, S.R., Muniandy, B.: The effect of flipped classroom on students' engagement. Technol. Knowl. Learn., 1–18 (2017)
5. Nouri, J.: The flipped classroom: for active, effective and increased learning – especially for low achievers. Int. J. Educ. Technol. High. Educ. **13**(1), 1–10 (2016)
6. Olakanmi, E.E.: The effects of a flipped classroom model of instruction on students' performance and attitudes towards chemistry. J. Sci. Educ. Technol. **26**(1), 127–137 (2017)
7. Johnson, G.B.: Student perceptions of the Flipped Classroom (2013)
8. Boevé, A.J., et al.: Implementing the flipped classroom: an exploration of study behaviour and student performance. High. Educ. **74**(6), 1015–1032 (2017)
9. Findlay-Thompson, S., Mombourquette, P.: Evaluation of a flipped classroom in an undergraduate business course. Bus. Educ. Accredit. **6**(1), 63–71 (2014). https://papers.ssrn.com/sol3/papers.cfm?abstract_id=2331035
10. Flores, Ò., del-Arco, I., Silva, P.: The flipped classroom model at the university: analysis based on professors' and students' assessment in the educational field. Int. J. Educ. Technol. High. Educ. **13**(1), 1–12 (2016)
11. Betihavas, V., Bridgman, H., Kornhaber, R., Cross, M.: The evidence for 'flipping out': a systematic review of the flipped classroom in nursing education. Nurse Educ. Today **38**, 15–21 (2016)
12. Esperanza, P., Fabian, K., Toto, C.: Flipped Classroom Model: Effects on Performance, Attitudes and Perceptions in High School Algebra (2016)
13. Keck Frei, A., Thormann, G.: Begleitstudie Flipped Classroom ZHAW Informatik: Kurz-Zusammenfassung. https://phzh.ch/Templates/Pages/GenericContentLoader.aspx?controlurl=/Extensions/phzh_forschungs_db/fdb_suche.ascx&idpr=571. Accessed 26 Feb 2018
14. Hwang, A., Arbaugh, J.B.: Seeking feedback in blended learning: competitive versus cooperative student attitudes and their links to learning outcome. J. Comput. Assist. Learn. **25**(3), 280–293 (2009)
15. Keller, H., Beinborn, P., Boerner, S., Seeber, G.: Selbstgesteuertes Lernen im Fernstudium: Ergebnisse einer Studie an den AKAD Privathochschulen. Schriften der Wissenschaftlichen Hochschule Lahr, no. 5 (2004)
16. Lage, M.J., Platt, G.J., Treglia, M.: Inverting the classroom: a gateway to creating an inclusive learning environment. J. Econ. Educ. **31**(1), 30–43 (2000)
17. McNally, B., et al.: Flipped classroom experiences: student preferences and flip strategy in a higher education context. High. Educ. **73**(2), 281–298 (2017)
18. Blair, E., Maharaj, C., Primus, S.: Performance and perception in the flipped classroom. Educ. Inf. Technol. **21**(6), 1465–1482 (2016)
19. Giannakos, M.N., Chorianopoulos, K., Chrisochoides, N.: Making sense of video analytics: lessons learned from clickstream interactions, attitudes, and learning outcome in a video-assisted course. IRRODL **16**(1) (2015)

Circle's Ontology Extended: Circumference and Surface Area of a Circle

Dimitra Tzoumpa[✉], Theodoros Karvounidis[✉],
and Christos Douligeris[✉]

Department of Informatics, University of Piraeus, Piraeus, Greece
dtzoumpa@gmail.com, {tkarv,cdoulig}@unipi.gr

Abstract. In this paper, we present how the use of ontologies and modern information technology tools helped students to connect geometrical meanings with mathematical concepts, based on the properties of circles and regular polygons ontologies, aiming to find the circumference and the surface area of a circle. In order to evaluate this concept, an experiment was set up in a junior high school classroom. The ontology, via abstract and combined thinking, helped the students to have a better understanding of the geometrical meanings and their dynamic interconnections.

Keywords: Semantic web · Ontologies · Geometry · Regular polygons · Circle surface area · Circle segment surface area · Circumference of a circle

1 Introduction

The teaching of Geometry is based on the use of the special features of the shapes it handles and it manages to study space in 2 or 3 dimensions. In a typical classroom setting, the students are asked to deal with a geometric meaning or idea independently and without the visualization of the object under investigation. This means that the student should act simultaneously in two states: (a) empirically and intuitively, exploiting the information coming from the representation of the geometric object, and (b) purely, applying theory and dissociating himself from empirical and intuitive perceptions. This dual demand is the main source of difficulties for learning and teaching Geometry. Nevertheless, Geometry is a field of Mathematics that follows a process of applying experiences, conjectures and productivity. In Mathematics, the students learn how to create and develop personal meanings through hypotheses, refutations, amendments and controls. The choice to apply IT to the teaching process brings up various aspects connected to the roles and the activities of the participants. More specifically, it brings up the need to study the mathematical concepts favored by the IT environment [2], the types of assigned work to the students [3] and, finally, the general teaching process framework [4].

In this paper, we demonstrate the incorporation of Ontologies in the teaching of Geometry, using as an example the case of the circles' circumference and the surface area along with the length of the arc and the surface of a circle's sector. We will present

© Springer Nature Switzerland AG 2020
M. E. Auer and T. Tsiatsos (Eds.): ICL 2018, AISC 916, pp. 120–132, 2020.
https://doi.org/10.1007/978-3-030-11932-4_12

in detail a classroom application where students take advantage of a formerly built circle ontology and connect it with the new fields under investigation.

2 Ontologies and Education

As the educational process changes and evolves, better ways of teaching are proposed. For that purpose, various systems of knowledge organization are set forth. These systems are advanced tools for searching digital information, including all types of organization methodologies, taxonomies, categorizations, ordering, vocabularies, thesauri, semantic networks and ontologies. The last four are more widely used. An ontology sets definitions which are used to describe and represent a field of knowledge. It not only organizes that field but it relates it to another one as well. In this way, an ontology creates reusable knowledge [29]. Ontologies may be applied in numerous ways in education. The usage of ontologies in education has been well- recorded in [3], where an ontology of the ontologies in education was implemented. According to [3], from a technological perspective, ontologies can be applied in the educational framework for knowledge organization, inference, information visualization, navigation, querying knowledge sharing and reusability. From an application perspective, ontologies can be applied as knowledge tools to construct, externalize, communicate and evaluate knowledge and they can describe knowledge based on educational/pedagogical knowledge or according to the structure of the educational system. Modern implementations of e-learning based on ontologies can be grouped into five categories: (i) educational domain knowledge ontologies (ii) knowledge process ontologies (iii) knowledge objects ontologies (iv) Learning Management Systems (LMS) ontologies, and (v) students competence and educational goals ontologies.

3 Why Ontologies in Education?

The significance of involving ontologies in education has been well-documented in the literature. The use of ontologies allows for interoperability between distributed teams on a range of working projects, for convenient formatting for mashups (Semantic Web agents) and for the facilitation of the educational assessment of students through the discovering of hidden links between concepts from the domain of interest, as well as about their relations and their axioms [12].

The proper ways of using digital tools in a Mathematics classroom, with regards to the nature and characteristics of the activities which the students and the educators are asked to get involved, has been a new banner in the context of the broader question of whether and how to integrate digital technologies in a school setting for some years now [1]. The prospect of using technology in the classroom brings out several issues on every aspect related to the roles and the activities of all those involved in teaching, such as: the need for studying the mathematical concepts which are favored by a computational environment [2], the kind of projects assigned to the students [3] and, more generally, the context in which the teaching takes place [4]. The use of technology at all the levels in the design and evolution of a course in a classroom has a catalytic influence on the educational process. The course design and evolution include elements such as the cooperative learning in groups, the change in the traditional role of the teacher and students, and the reinforcement of developing student-oriented didactic models in which the teacher has the ability to actively interfere in the learning process as a counselor and collaborator to the students [5].

4   Application of the Method

4.1   The Goal of This Experiment

Students often find Geometry to be a very hard to understand field of Mathematics, especially when traditional teaching methods are used. These traditional methods include the drawing of shapes, the theory to be applied and the formulas to be used in exercises to demonstrate the students' comprehension of a particular concept. Our goal is to inject computer technology in all the steps of the experiments aiming to help students understand geometrical concepts and their implementation in a better, easier and more efficient way of learning.

The student through these implementations runs a self-knowledge control on his cognition and attempts to sort it in terms of what has learned but has been forgotten, what is already known and can be recalled and, finally, what has to be learned in the near future.

4.2   Some Facts About the Experiment

The experiment took place in a classroom at a junior high school. There were two classes participating, totaling 42 students, equally distributed between boys and girls. The interactive environment of a dynamic Geometry software (GEOGEBRA) [6] along with a programming language to create interactive animations (SCRATCH) [7] were used. The students were working in pairs. They were provided printed sheets with directions that they had to follow.

4.3   Initializing the Process

The first step of the process was aiming to let students present their perspective of what is the meaning of the circle's circumference. All of their opinions referred to what the circumference is indeed and the key idea was about the earth. The earth's rounding edge would be also a good example of a circumference to measure. Their vivid imagination drawn on their printed sheets provided the tool to measure the distance, which was a bicycle! The easiness of programming a bicycle rounding the earth in SCRATCH (Fig. 1) gave a visual understanding of the steps to follow.

An early question that arose was how many circles would the bike's wheel do if it circled the earth. A pair of students displayed to the rest of the class how their first approach would turn into a more mathematical model. A circle moving around another circle (Fig. 2), would provide the measuring of the circumference.

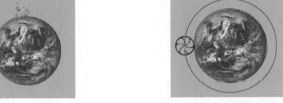

Fig. 1. A bicycle rounding the earth

Fig. 2. A circle rounding the earth

Another pair of students presented the simplified version of the measuring process and rephrased the question to: how many times would the wheel need to move around the other circle.

4.4 Modeling the Process

Although the students did not find the answer to the previous question, they were told to use a compass and draw two circles of the same radius tangent to each other. They also had to draw three more circles of the same radius, forming something like a cross (Fig. 3).

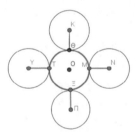

Fig. 3. Four circles tangent to another, all with the same radius

That last exercise had as its target to stimulate more questions or ideas about the outside circle moving around the central one. Although questions were raised, none was answered due to the fact that the scope of this exercise would become apparent at the next step.

4.5 Making It More Interesting and Intriguing

The students drew, using GEOGEBRA, a circle and a tangent line of length equal to the circumference of the circle (Fig. 4). The length of the circumference was automatically calculated by the software.

Fig. 4. The circumference stretched in a line

The line was curved so as to form a circle. As the outside circle was rolling around the circumference circle, its radius was coloring the area spinning around (Fig. 5). At the first stop at right (see Fig. 3), it covered half the surface (Fig. 6). At the next stops of Fig. 3 the rolling radius covered full (Fig. 7), half and full surface again.

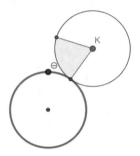
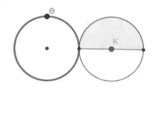
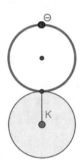

Fig. 5. Rolling the outside circle, as its radius colors the surface rolled around

Fig. 6. First stop at the right

Fig. 7. Second stop at the bottom

What was demonstrated was that the circle rolling around the other circle, made two full spins. But it was not revealed that the center of the outside circle was moving on a circle of double radius than the inside circle (Fig. 8).

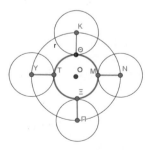

Fig. 8. Rolling the outside circle's center, on a double size radius circle

4.6   Figuring Out the Algebraic Equation

The students used again GEOGEBRA in the case where a circle was rolling and the distance was measured by the software.

They calculated a ratio of distance to radius as the circle was rolling. For a 4.94 radius in half way rolling the ratio reading was 3.14 (Fig. 9). A longer radius of 7.49 in half way rolling, the ratio remained at 3.14 (Fig. 10)

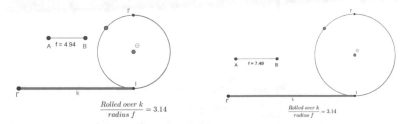

Fig. 9. Ratio in half way rolling   **Fig. 10.** Ratio in half way rolling, but for a longer radius

When the circle was completed rolled, the ratio reading was 6.28 (Fig. 11), no matter what the radius was.

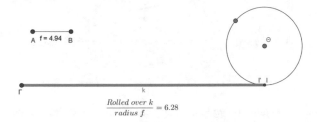

Fig. 11. Ratio of a totally rolled circle

Thus, the students ended up with the appropriate formula to calculate the circumference of a circle of radius r is (Eq. 1)

$$\text{Circumference} = 2 \times \pi \times r \tag{1}$$

5   Circumference to Calculate the Circle's Surface?

The students could not imagine how the circumference would provide the solution to the calculation of the circle's surface. Their new task on the GEOGEBRA that time was to draw a new circle and pivot the top half. Each half had obviously half of the circumference, a length of π × r (Fig. 12), a well-expected result for the students to come up.

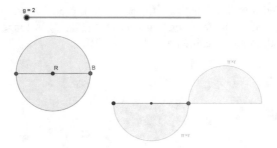

Fig. 12. A circle spread out in half

The next step was to cover the circle with a familiar and well-investigated object so that the surface of this new shape would lead to the calculation of the circle's surface. The already connected shape to the circle from the ontology under investigation were regular polygons (Fig. 13).

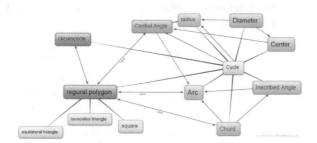

Fig. 13. Regular polygon ontology connected to the circle ontology

The students programmed GEOGEBRA to inscribe regular polygons in the circle, where the number of angles were selected by a runner line. The simplest regular polygon, a square, was firstly selected and the quarters from its diagonals were spread as in Fig. 14.

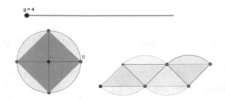

Fig. 14. Inscribed Square spread out in forths

The surface area of the circle was divided into four equal parts. The surface of the four triangles equaled the surface of the square and the rest surface if added would sum up to the surface of the circle. The lengths of the four curved lines would be ¼ of the circumference. Top half was of length $\pi \times$ r. By increasing the number of angles, the inscribed regular polygons in their spread out form, began to demonstrate what the result would be. At the 28th polygon, it was quite obvious (Fig. 15), and still the top side was of length $\pi \times$ r. At the 360th polygon, there was no obvious curved line but it look like a straight line. The object looked as an orthogonal quadrilateral (Fig. 16) although it was still a regular polygon.

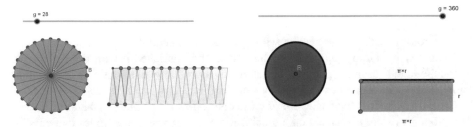

Fig. 15. Inscribed 28polygon spread out

Fig. 16. Inscribed 360polygon spread out slices

Now, the surface area was calculated to be (Eq. 2)

$$\text{Surface} = \pi \times r^2 \tag{2}$$

6   A Part of the Circle

From that point on, the students turned to start learning for themselves based on what had been discovered already. They were asked to draw three circles in the printed sheets, sharing the same point for their center but with different radius. A central angle was given of about 45°. They had to find the length of the arc on each different circle. Setting up the circles in GEOGEBRA, they calculated the three different lengths of each arc. But instead of squeezing for ideas, the students were changing the angle to 90° and 180°. What was obvious was that the arc length was proportional to the angle. When the angle was a right one the arc length was ¼ of the circumference. When the angle was a straight line the arc length was half of it. So the conclusion was reached easily. If the angle is of μ degrees, then the length l of the arc would be (Eq. 3)

$$l = 2 \times \pi \times r \times \frac{\mu}{360}. \tag{3}$$

Applying values to Eq. 3 for the three different circles proved them to be correct for what they had predicted. Calculating of course the surface area of the sector from those different circles was also easy to predict. The area E would be calculated from (Eq. 4)

$$E = \pi \times r^2 \times \frac{\mu}{360} \tag{4}$$

Again, the values were verified against the GEOGEBRA calculations.

7   The Ontology Participation

Students were familiar with the use of Ontologies. They had built from scratch the basic ontology of the circle dealing with inscribed and central angles. They enhanced the ontology by adding regular polygons. Then, they enriched it with right triangles. Throughout the teaching process, the students used these ontologies to recall their knowledge about properties and relations of various geometrical objects. The students had at their disposal a local network of these ontologies for displaying the information. The properties were associated on a graphical design, exposing the triples that exist in the ontologies. At reaching the end of the teaching process, there was time given to the students for connecting the previously learned ontologies [11] to the freshly learned meanings of length and surface area of entire circle or a sector of it. This connection could be described in a Visual Notation of an OWL Ontology as in (Fig. 17).

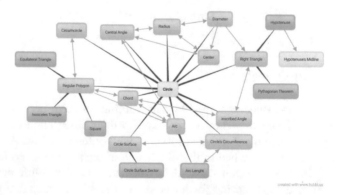

Fig. 17. Circle's ontology expanded

8   Educational Learning Objectives

In order to classify the different objectives and skills set for the students, the teachers used the Bloom's taxonomy [9], as it is shown in Fig. 18), and as expressed verbally by Anderson and Krathwohl [10]:

- Factual type of Knowledge: We mentioned to the students the problem to be addressed (Remember), defining the length of the circumference. We described the model to be created (Understand), the circling of a circle. We wrote a list of the problem parts (Apply). We distinguished the data (Analyze), radius and center were given while surface was unknown. We prioritized the steps to address the problem (Evaluate), from circles of different radius to equals. We designed and represented the hierarchical steps virtually (Create), using SCRATCH to display to movement of bicycle around earth.
- Conceptual type of Knowledge: We identified already known meanings (Remember), like circles, regular polygons and arcs. We sorted and explained (Understand) the meaning of the object creation. We coded individual meanings (Apply) that derive from the problem's solution, finding π to be a constant in our observations. We recorded what has emerged from the solution process (Analyze) helping the students to define the circumference. We validated (Evaluate) our results twice. The students built the model (Create), the cycle rounding earth.
- Procedural type of Knowledge: We defined the relations (Remember) that lead to the solution of the problem. We summarized and explained the process (Understand) using GEOGEBRA for the circle and its radius. We extended the original idea and prepared its implementation (Apply) by blending π in our calculations. We highlighted the new dimensions (Evaluate) of the original problem. We built (Create) the formulas.
- Metacognition type of Knowledge: Students shared (Remember) the results between the two classes. We foresaw (Understand) the future use of this approach. We generalized (Apply) the problem to address other areas. We reviewed (Analyze) our work. We evaluated the criticism (Evaluate) that has been brought to us when viewing the work. We foresaw (Create) the next steps for the model's evolution, what would happen if the bicyclist escapes from earth!

Fig. 18. Bloom's taxonomy

9 Evaluation

The teachers provided the students with a questionnaire in order to measure the effect of this experiment over their preferences. It should be pointed out that the students were quite familiar with the visual programming with SCRATCH as it was taught during their IT lessons, as well as with the GEOGEBRA software which was part of their

Mathematics lessons. The answered questionnaires revealed that the given worksheet was helpful to them to understand in a better way the objectives that were discussed during the teaching process. They preferred working in groups instead of being left alone. The software was really assisting in understanding the various meanings on a visual representation. It was also very enjoying for the students to adjust and change values of variables during software usage and definitely not staying passively attending the teacher's demo. The students also mentioned that this new process improved their collaboration with their classmates; they expressed their belief that that they could achieve better results in Mathematics because they were working in pairs. They also stated that the teaching process became more pleasant. But what really excited them was the help received from the ontology to recall the knowledge of the properties that they needed. Many of the students stated that they wanted the rest of the lessons to be taught implementing the same process. They also stated that other lessons, like Physics and Chemistry, should also implement an ontology! Their preference for the method used is displayed in the following (Table 1) of a Likert scale, where every question out of the twelve in total was given a rate from 1 to 7, where 7 expresses the maximum satisfaction for the method. The average value reached was 84.98%.

Table 1. Students satisfaction

84,98%		Rating						
		1	2	3	4	5	6	7
Question	1	1	1	3	4	7	17	9
	2	0	2	4	3	4	16	13
	3	0	0	2	7	5	15	13
	4	0	3	1	4	3	16	15
	5	0	1	1	2	4	15	19
	6	1	0	1	2	2	18	18
	7	0	0	1	3	4	19	15
	8	0	0	2	1	4	18	17
	9	0	0	1	1	4	17	19
	10	0	2	1	3	5	16	14
	11	0	0	2	1	3	16	20
	12	0	1	0	1	2	17	21

The teachers, on the other hand, observed that all the students participated very actively during the teaching process, even the ones that did not use to be interested in Mathematics. The teachers also observed that the groups of students were working together, exchanging views and thoughts without competing with each other. During the teaching process, the students were giving positive answers showing clearly an improved behavior in class.

In general, the final outcome based on the teachers' views was quite positive as the experiment proved to be very promising. The general impression from this experiment was that if this new teaching process is implemented on a regular basis, the students' learning process quality may significantly improve. Their knowledge would be transformed from fragmented pieces to a clear holistic picture that would be crystal clear in their minds. All the above led the teachers to declare that they would try to repeat this approach in future lessons and in other disciplines.

10   Conclusions and Future Steps

This paper presented the expansion of the circle ontology to include the concept of the circumference and of the surface area aiming to cover more additional fields as a continuation to our previous work [8]. This ontology could be helpful in various ways. It could be used as an intellectual instrument helping teachers organize their teaching lessons in a more flexible way. It could also be used as a communication framework between various application software providers which are dealing with education. Even more, this ontology could serve as a visual browsing connection to the material that is been taught.

The presented experiment was based on the widely accepted assumption that every learning process which refers to a cognitive field is based on specific learning objectives and uses a variety of educational resources. The use of regular polygons' ontology helped students organize the geometrical meanings in their minds. The students were able to understand that there exist geometrical meanings bound together in a strong relationship. The students were set to abstract and combined thinking at the same time.

With this application of the GEOGEBRA software and of SCRATCH, the students entered the role of the scientist who makes assumptions, conducts experiments, formulates his or her own theorems and possibly revises them through refuting them. Ontologies can take this experiment a step further, by allowing examples as this one of regular polygons to interconnect and interact with other examples of geometrical objects like the inscribed angles of a circle. The teaching process that takes advantage of ontologies allows the cooperation not only between the students and the teachers but also among the students. This is achieved through paradigms and exercises embedded inside the ontology.

The implementation of the ontology in the teaching process showed several advantages compared to the traditional teaching modes. Having already used the ontology of the circle and relating it to the right triangles as well as to regular polygons, in this work, we explored the concept of the circumference and of the surface area. Based on the circle's ontology, Trigonometry could become the next field to explore the use of ontologies.

Acknowledgment. The work presented in this paper has been partially funded by National Matching Funds 2016–2017 of the Greek Government, and more specifically by the General Secretariat for Research and Technology (GSRT), related to EU project Medusa.

References

1. diSessa, A., Hoyles, C., Noss, R. (eds.): Computers and Exploratory Learning. Springer, Berlin (1995)
2. Sutherland, R., Balacheff, N.: Didactical complexity of computational environments for the learning of mathematics. Int. J. Comput. Math. Learn. **4**, 1–26 (1999)
3. Hoyles, C.: From describing to designing mathematical activity: the next step in developing the social approach to research in Mathematics education. Educ. Stud. Math. **46**, 273–286 (2001)
4. Nardi, B. (ed.): Context and Consciousness: Activity Theory and Human Computer Interaction. MIT Press (1996)
5. Hoyles, C., Noss, R.: A pedagogy for mathematical microworlds. Educ. Stud. Math. **23**, 31–57 (1992)
6. GEOGEBRA. http://www.geogebra.org/
7. SCRATCH. https://scratch.mit.edu/
8. Tzoumpa, D., Karvounidis Th., Douligeris, C.: Towards an ontology approach in teaching geometry. In: Proceedings of ICL 2016 19th International Conference on Interactive Collaborative Learning 45th IGIP International Conference on Engineering Pedagogy. 21–23 September 2016, Belfast, UK (2016)
9. Bloom, B.S., Englehart, M.D., Furst, E.J., Hill, W.H., Krathwohl, D.R.: The Taxonomy of educational objectives, handbook I: The Cognitive domain. David McKay Co., Inc, New York (1956)
10. Anderson, L.W., Krathwohl, D.R., et al. (eds.): A Taxonomy for Learning, Teaching, and Assessing: A Revision of Bloom's Taxonomy of Educational Objectives. Allyn & Bacon, Boston, MA (2001). (Pearson Education Group)
11. Tzoumpa, D., Karvounidis, Th., Douligeris, C.: Extending the application of ontologies in the teaching of geometry: the right triangle in the circley. In: Proceedings of EDUCON 2017, 25 – 28 April 2017, Athens, Greece (2017)
12. OntoMathPro Ontology: A Linked Data Hub for Mathematics (2014) Knowledge Engineering and the Semantic Web - 5th International Conference. https://arxiv.org/pdf/1407.4833

Hint-Giving Phraseology for Computer Assisted Learning

Marietta Sionti[1(✉)], Ellen Schack[2], and Thomas Schack[1,3,4]

[1] Neurocognition and Action Research Group, Cluster of Excellence-Cognitive Interaction Technology, Bielefeld University, Bielefeld, Germany
{marietta.sionti, thomas.schack}@uni-bielefeld.de
[2] von Bodelschwinghschen Stiftungen Bethel, ProWerk, Bielefeld, Germany
ellen.schack@bethel.de
[3] Faculty of Psychology and Sport Sciences, Bielefeld University, Bielefeld, Germany
[4] CoR-Lab, Research Institute for Cognition and Robotics, Bielefeld University, Bielefeld, Germany

Abstract. In this paper we will discuss work to date on an exploratory study of tutoring patterns and their role for tutoring systems' assistance in initiating creative responses and sustaining learning. Based on classroom discourse analysis, investigation of several studies of computer assisted learning -both for single and peer learning- and our understanding of the *assistance dilemma* of Intelligent Tutoring Systems (ITS), we argue that the students' need for (artificial) facilitator's help emphasizes on the manner of hint-providing, so we should take particular note of the challenges that arise from formalizing these definitions to apply to the style of problem solving and the students' different age groups. This is the first step towards a mobile ITS that utilizes eye-tracking and virtual reality glasses, in order to identify difficulties and offer on time feedback in textual or visual mode, which will be superimposed on a transparent lens in users' field of view. We will conclude with a discussion of implications and future directions.

Keywords: Computer assisted learning (single and peer learning) · Assistance dilemma · Discourse analysis

1 Introduction

The ultimate goal of each tutoring system is to increase productive learning behaviors through supporting both meta-cognitive and cognitive processes. The meta-cognitive process serves as the fundamental learning principle of self-explanation. Its hints will introduce different strategies for improving student's explanations, such as procedural replay, reduction strategy (Socratic dialog), similar reduction strategy (Koedinger and Aleven 2007; Aleven 1997; Rau et al. 2009). Cognitive process is subject to the correct analogy between information or assistance giving and withholding. Koedinger and Aleven (2007; McLaren et al. 2008) call this problem the *assistance dilemma*, which emerges from the difficulty in choosing between lower- and higher-level instructions.

© Springer Nature Switzerland AG 2020
M. E. Auer and T. Tsiatsos (Eds.): ICL 2018, AISC 916, pp. 133–144, 2020.
https://doi.org/10.1007/978-3-030-11932-4_14

Lower-level instructions help students self-act and construct knowledge while higher-level instructions urge them to reach sophisticated knowledge burdened with the risk of non-assimilation. Assigned detailed parameters are used to decide between the two levels.

Peer tutoring systems exhibit higher complexity compared to one-student ITS because its outmost benefit depends on students' interaction and transactivity. The tutor's assistance should not distort interaction or interrupt the intertextual dialogical coherence but owes to elicit transactivity, a highly cited notion for its constructive knowledge acquisition. Ensuring transactivity in peer tutoring systems becomes synonymous with using carefully phrased hints that follow the *accountable talk moves* and *problem based learning* phraseology − two dominant trends in collaborative classroom discourse literature (Resnick et al. 1991; Michaels et al. 2007; O'Connor et al. 2007; Michaels et al. 2008; Sohmer et al. 2009; Resnick, Michaels, and O'Connor, in preparation, Weinberger and Fischer 2006; Gweon et al. 2007; Berkowitz and Gibbs 1983; Ai et al. 2010; Sionti et al. 2012). Based on this phraseology, we will propose linguistic patterns for peer ITS hints, which would respect age groups younger than in previous studies.

Concerning age group, the majority of literature in the computer assisted learning and intelligent tutoring systems − closely affiliated fields − focuses on high school or college students. Piaget (1970) professed that each person can absorb knowledge, given specific ways according to cognitive age. Elementary and middle school children cannot perceive abstract information since they still belong in the stage of concrete operations (7–12 years). In this stage, there is evidence for organized and logical thought but thought is still not very abstract. It incorporates the principles of formal logic at the next period of formal operations (12 years and older). Middle-school children's thinking is still tied to concrete reality, and these students are facing difficulties in understanding and transforming knowledge above their cognitive age. Based on this point, we could explain the above-mentioned differences between Clark and Mayer (2003) and McLaren et al (2006, 2007) concerning the personalized principle; the formers claimed that the informal speech might better facilitate learning than formal speech in a class discourse or an e-Learning environment, while the latter's experiment showed that formal style may be more supportive. In other words, according to Clark and Mayer, instructions and hints should be written in first - or second - person language in a more informal way, while McLaren proves −contrary to his intuition − that a businesslike manner harmonizes with cognitive theories of learning. Both opinions may be right, depending on the age of students, since McLaren's study used college students who had already acquired Piaget's highest intellectual level. The linguistic style of the intervention, regardless of the environment (classroom, e-learning), is of huge importance and responsible for the success or failure of learning procedure, depending on factors such as age.

In the rest of the paper, we briefly present the intended ITS design, which will utilize the proposed phraseology and eventually the linguistic patterns to elicit transactivity through hint-giving and engage students in the most captivating way based on their age.

2 Mobile ITS for Single and Peer Learning

Following on our previous successful design and development of intelligent glasses (Fig. 1) for the help of mentally handicapped and elderly person, we are extending the applications of this pioneering equipment to ITS for school students. This unique mobile assistance system combines techniques from eye tracking, augmented reality and memory research, in order to identify difficulties in actual action processes and offer on time feedback in textual or visual mode, which is superimposed on a transparent lens in users' field of view.

The ongoing single and peer learning ITS is designed to serve as a mobile assistance system for elementary school students' learning and for special groups such as dyslectic or ADHD pupils. For the latter two conditions the v. Bodelschwinghschen Foundation Bethel will contribute their expertise in diagnosis and action assistance, while both Bethel and Uni Bielefeld will attend deriving ethical issues to avoid unforeseen consequences for the society, an important issue when target groups comprise people with above-average vulnerability such as children and people with disabilities (Strenge et al. 2017).

Fig. 1. Mobile assistance system, with problem detection and hint providing on user's field of view

In order to allow for age appropriateness, individual assistance in learning situations and cognitive sequences will be diagnosed in a first step using the structural dimensional analysis motoric method (SDA-M; Schack 2012). The method applied is already well established in cognitive psychology and captures the mental representation structures using a hierarchical splitting procedure. The SDA-M contains four steps: (1) A special split procedure involving a multiple sorting task is used to create a distance scaling among the concepts of a suitably predetermined set. (2) A hierarchical cluster analysis is used to transform the set of concepts into a hierarchical structure. (3) A factor analysis is used to reveal the dimensions in this structured set of concepts.

(4) The cluster solutions are tested for invariance within and between individuals and groups (Fig. 2). The system is then able to identify user specific problems in the execution of actual learning sequences. Based on this information, the actual visual focus and the detected objects in the environment, the system is able to provide adequate feedback, i.e. by displaying situation and user specific help comments and hints.

Age appropriateness and individual cognitive status are key issues in our system. Before building on the above mentioned features, we wanted to establish the most efficient phraseology for the hint giving, especially because there is a lot of ink poured for elementary school students in traditional classroom discourse analysis but the majority of electronic systems' automatic assistance is not based on specific linguistics patterns, which may effortlessly elicit meaningful and productive responses, both for single and peer systems.

 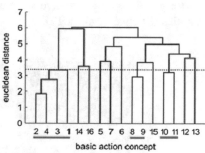

Fig. 2. Structural Dimensional Analysis of Mental Representations: virtual display where participant should select whether the presented concept is relevant or not to the anchor concept (left) and cluster analysis of coherent concepts (right).

3 Linguistic Patterns for Assistance in Intelligent Tutoring Systems

3.1 IRE Model, Accountable Talk Moves and Problem Based Learning

In classroom discourse literature a lot of ink has been poured in defining the linguistic style of instructions that teachers should use in their classes, in order to achieve their goals, such as knowledge acquiring and reasoning cultivation. Several approaches have been developed and this valuable knowledge could be incorporated in e-learning and computer assisted learning, such as intelligent tutoring systems. We will briefly present the major tendencies; the IRE model, the accountable talk moves and incorporate problem based learning questions in the design of our model.

The IRE or IRF model is known as the Discourse analysis or the Birmingham model and it was developed by John Sinclair and Malcolm Coulthard. Its structure reveals that teacher is the person who initiates all actions, because he decides who speaks in every turn and evaluates the knowledge on the basis of the needed or already expressed information. According to Megan (2003), when an evaluation follows a question–answer pattern, it is believed that a known information is asked, while if it was asked by an acknowledgement it may seem like an information seeking question. Students are not eligible to evaluate the other's contribution but they have to subordinate to teacher's managing moves, regardless his good intentions.

According to Michaels, O'Connor and Resnick (O'Connor and Resnick 2009; Resnick, Michaels and O'Connor, in preparation, Weinberger and Fischer 2006; Gweon et al. 2007; Berkowitz and Gibbs 1983) accountable talk (AT) focuses on teacher-led instruction regardless whether the recipient is a small group or the whole class. Teachers engage open questions and topics from different perspectives to elicit students' contribution. Due to specific types, which are dialogically adapted, accountable talk urges students to investigate and acquire knowledge not as subordinates but the primary goal is to hear other's opinion and creatively contribute to alter contribution by using transactive features. Therefore this talk is accountable to the community. Related is the accountability to standards of reasoning, because students can explain their reasoning and provide necessary correction based on logic and evidence. Last but not least, the accountability to knowledge is the ultimate goal of the learning procedure. Every student has pre-existing ideas of the world, although at school he has to formulate them into objective knowledge expressed in scientific terminology and argumentative support. Teacher is responsible to assist student to the conflict of the pre-existing with the academic knowledge, in order to form the latter's noesis.

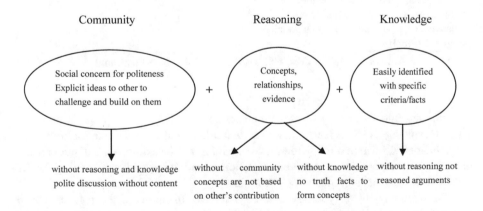

Problem-based learning (PBL) is a collaborative classroom theory, which emphasizes the approach of learning procedure through multifaceted, and realistic problems. Collaboration skills and self-directed learning are its main goals. The tutor helps students to identify their prior knowledge and reach the problem's solution, through providing the appropriate learning scaffolding. Encouragement and understanding are considered as cornerstones to the procedure (Hmello-Silver 2004).

3.2   Phraseology for Peer ITS Hints

The hints in our data combined these three approaches accordingly. The goal of this assistance is to manage intelligibility, motivation and focus (Table 1).

Table 1. Phraseology for peer ITS hints

A. Keeping utterance vivid	B. Managing interaction to facilitate thinking
A1. Revoicing (at) Need clarification (pbl)	B1. Directives to look at materials (at)
A2. Interpretation of a student's statement (at)	B2. Directives to partner talk (at) request directive (pbl)
A3. A student to paraphrase (at)	B3. Monitoring and asking for self-explanation (pbl)
C. Reasoning about content of utterance	
C1 reasoning focused	C2 IRE focused
C1a. Prompting further Participation (at)	C2a. Questions require an answer only (at) Verification (pbl)
C1b. Reasoning application request (at) Example, interpretation (pbl)	C2b. Evaluation an answer as correct and incorrect (at)
C1c. Request for elaboration (at) Disjunctive, concept completion, feature specification, quantification, definition, comparison, causal antecedent, causal consequence, enablement,, expectational, judgmental (pbl)	

(at: based on Accountable talk moves, pbl: based on Problem based learning)

Ordinary and e-learning classes emphasize the vitality of keeping the utterance vivid. Revoicing (A1) - as teacher's move - is fundamental and serves in multiple ways. In the canonical form of a revoicing move, the inference based on student's utterance is preceding, its repetition or even reformulation is following and the opening for student's validation of inference is occurring last.

The revoicing move typology may be divertive. In one case the teacher may be confused at first, but then he gets a clearer sense of what the student knows and doesn't know. The student realizes that the teacher wants to understand her contribution. Over time, this can have a profound effect. The student can accept or reject the teacher's interpretation, which positions the student (at least for that moment) as a legitimate participant in the intellectual enterprise. In another case the teacher is using these students' work to create a more coherent lesson. The student gets the chance to hear a clear version of the reasoning of himself or his peers. The originating students get the chance to expand on their ideas, in public. Finally the teacher may try to bringing out the greater significance.

In Problem Based Learning the need for clarification is part of the task oriented categories of questions. The speaker does not understand something and needs further explanation or confirmation of previous statement Interpretation of a student's statement

(1) *Are you, are you, Jeff are you talking about micro vascular damage that then, which then causes the neuropathy?* (Hmello-Silver 2004).

Repetition (A2) exists in response to a restate request (note that the restate request itself is not included, because the request itself is purely management).

Especially in peer tutoring systems the teacher asks a third student to make request to the first who said it for paraphrasing (A3). The equivalent in problem based learning is group dynamics, which lead to discussions of consensus or negotiation of how group should proceed.

(2) *So Mary, do you know what they are talking about?* (Hmello-Silver 2004).

Although, we did not have many chances to observe assistance according to "managing interaction to facilitate thinking" in our corpus, this group is still important in specific e-learning tasks. A hint could give directives to look at materials (B1). The example that follows is taken from a typical class but it could be used in computer assisted learning as well.

(3) *...So take a moment and work by yourself in your math section of your binder and write down your thoughts on what integers are.* (Hmello-Silver 2004).

In peer tutoring systems, one of the great challenges is to warranty the equal turns of all the participants by giving directives to partner talk (B2) and requesting for collaboration. In classroom discourse a typical question to serve equal turns is:

(4) *What does your partner think about that?*

PBL has similar processes, which is called Request/Directive, e.g.

(5) *why don't you give Jeff a chance to get the board up?* (Hmello-Silver 2004).

The following two managing moves are important for ITS. Monitoring and asking for self-explanation are meta- cognitive functions, first introduced by PBL but also well appreciated in the rest ITS community. Rau et al (2009) support meta-cognitive process of self-explanation and claim that is beneficial for students to explain their solutions "by reference". So their cognitive tutor aims to help students to improve their explanations through dialog. In our data, there is a hint that cultivates these meta-cognitive procedures and achieve optimal learning.

(6) *also the answer is less important than explaining the process of figuring it out*

Reasoning about content of utterance (C) is divided in two larger categories: reasoning focused (C1) and IRE focused (C2). Reasoning focused holds the bulk of interventions that promote learning and knowledge acquisition. Sometimes, when a student is unwilling to continue or has reached a dead end, a hint could prompt further participation (C1a), while in reasoning application request -according to accountable talk moves or example/interpretation following PBL (C1b)- the tutor's question gives a theory instance, by providing an example or interpret a theory,

(7)   *give us an example* or *when have we seen this kind of patient before?*

The request for elaboration (C1c) can be distinguished in

(8)   *is it all the toes? or just the great toe?* (disjunctive)
(9)   *what supplies the bottom of the feet? Where does that come from?* (concept completion)
(10) *could we get a general appearance and vital signs?* (feature specification)
(11) *how many lymphocytes does she have?* (quantification)
(12) *what do you guys know about pernicious anemia as a disease?* (definition)
(13) *are there any more proximal lesions that could cause this?* (comparison)
(14) *what do you know about compression leading to numbness and tingling?* (causal antecedent)
(15) *what happens when it's, when the, when the neuron's demyelinated?* (causal consequence)
(16) *how does uhm involvement of veins produce numbness in the foot?* (enablement)
(17) *how much, how much better is her, are her neural signs expected to get?* (expectational)
(18) *should we put her to that trouble, do you feel, on the basis of your thinking?* (judgmental).

Most of the help which was offered for this study and belonged to C1b and C1c was I question form but few of them had instructional form. According to King (1995) the generic question, in the following table, is the best way to guide critical thinking and induce specific processes and it can easily be used for computer assisted learning hints (Table 2).

Table 2. Fragment from King's questions that best serve the request for application or elaboration (King 1995)

Generic questions	Specific thinking processes induced
Explain why____.(Explain how____)	Analysis
What would happen if____?	Prediction/hypothesizing
What is the nature of____?	Analysis
What are the strengths and weaknesses of___?	Analysis/inferencing
What is the difference between____and____?	Comparison-contrast

Last but not least, in our data we saw IRE focused hints (C2), such as questions that require an assertion or a verification (C2a)

(19) *This number is three intervals or spaces from zero but to the left. What do you think is the opposite of five?* or *are headaches associated with high blood pressure?*

and evaluation of an answer as correct and incorrect (C2b).

(20) *you are right about how many c triangles but you should think about how many of those are in each cup of ice cream*

Even though the IRE model is considered generally insufficient, compared to the collaborative learning techniques, it can still serve a few student-centered approaches, as those that are used in ITS. Especially, the use of the evaluation offers meta-cognitional benefits and justification in the cognitive space (Teasley and Roschelle 1993). Therefore, as soon as the evaluation releases from the grading system it can be used productively by both teachers, students and of course tutoring systems, because evaluation and justification support an idea's validity. Moreover, through incorporation of encouragement and motivation to evaluation, motivation is promoted.

4 Discussion

The need for hint-giving during the learning procedure is rooted centuries ago. In traditional cultures such as the US Indians or in Zen_Buddishm (Japan), the natives give higher level instructions in an irrational way to develop meta-cognitive knowledge and to actively understand the cosmos. Modern society and its instrumental role in help seeking behavior in learning was thoroughly analyzed by Nelson-Le Gall (1985) and specialized the same year in how elementary school students employ help seeking as problem solving strategy (Nelson-Le Gall and Glor-Scheib 1985; Newman 1990; Calarco 2011). In the following years help-seeking took was recognized as a self-regulated learning mechanism which leads to meta-cognitive awareness (Karabenick and Berger 2013) with an emphasis on the motivation for achievement (Newman 1990; Ryan and Shin 2011). This short and by no means exhaustive reference of pedagogic literature indicates that the assistance problem has received major attention in traditional classroom learning. But the rise of ITS and computer assisted learning -both for single student or in a collaborative set up- creates a series of questions associated with the assistance dilemma: Is it preferable to provide hints on demand only or for the tutor to decide the timing of providing hints as needed? (Koedriger et al. 2007). Cognitive tutors give hints mainly after students' requests. According to Baker et al. (2004), students might *game the system* to avoid thinking by either quick guessing or asking for more frequent hints, without full depletion of their resources. In other cases, students might avoid asking for help when needed. Then we come to the second part of our question: How and when can an intelligent tutor detect the need for providing hints? Timing of a tutor's intervention is a rather difficult task, which lays in the dialogue patterns and turns between students and teacher, effective both in traditional and collaborative classes. In the present paper, we suggest the phraseology for hint-giving, while in the next step we need to decide *when* to automatically provide this help. We tend to believe that that simpler measurements – such as the number of turns, number of movements, meaningful and non-meaningful verbal responses – might become indicators for hint needs, which might help us to calculate these variables and propose an algorithm for effective timing allocation of hint giving in single and peer intelligent tutoring systems.

References

Ai, H., Sionti, M., Wang, Y., Rosé, C.P.: Finding Transactive contributions in whole group discussions. In: Proceedings of the International Conference of the Learning Sciences (2010)

Aleven, V., Ashley, K.D.: Teaching case-based argumentation through a model and examples: empirical evalution of an intelligent learning environment. Paper presented at the Artificial Intelligence in Education, Kobe, Japan (1997)

Baker, R.S., Corbett, A.T., Koedinger, K.R., Wagner, A.Z.: Off-Task Behavior in the cognitive tutor classroom: when students "Game the System". In: Proceedings of ACM CHI: Computer-Human (2004)

Berkowitz, M., Gibbs, J.: Measuring the developmental features of moral discussion. Merrill-Palmer Q. **29**, 399–410 (1983)

Calarco, J.M.: "I need help!" Social class and children's help-seeking in elementary school. Am. Sociol. Rev. **76**(6), 862–882 (2011)

Chiu, M.M.: Effects of status on solutions, leadership, and evaluations during group problem solving. Sociol. Educ., 175–195 (2000)

Chiu, M.M.: Group problem solving processes: Social interactions and individual turns. J. Theor. Soc. Behav. **30**(1), 27–50 (2000)

Chiu, M.M.: Analyzing group work processes: Towards a conceptual framework and systematic statistical analysis. In: Columbus, F. (ed.) Advances in Psychology Research, vol. 4, pp. 193–222. Nova Science, Huntington, NY (2001)

Chiu, M.M.: Adapting teacher interventions to student needs during cooperative learning. Am. Educ. Res. J. **41**, 365–399 (2004)

Clark, R.C., Mayer, R.E.: E-Learning and the Science of Instruction. Jossey-Bass/Pfeiffer (2003)

Gweon, G., Rosé, C.P., Albright, E., Cui, Y.: Evaluating the effect of feedback from a CSCL problem solving environment on learning, interaction, and perceived interdependence. In: Proceedings of Computer Supported Collaborative Learning 2007 (2007)

Hmelo-Silver, C.E.: Problem-based learning: what and how do students learn? Educ. Psychol. Rev. **16**(3) (2004)

Howley, I., Chaudhuri, S., Kumar, R., Rosé, C.: Motivation and Collaboration Online. Poster Presented at Artificial Intelligence in Education (AIED 2009). Brighton: UK (2009a)

Howley, I., Chaudhuri, S., Kumar, R., Rosé, C.: Motivation and Collaborative Behavior: An Exploratory Analysis. Poster Presented at Computer Supported Collaborative Learning. Greece, Rhodes (2009)

Karabenick, S.A., Berger, J.L.: Help seeking as a self-regulated learning strategy (2013)

King, A.: Designing the instructional process to enhance critical thinking across the curriculum. Teach. Psychol. **22**(1), 13–17 (1995)

Koedinger, K.R., Aleven, V.: Exploring the assistance dilemma in experiments with cognitive tutors. Educ. Psychol. Rev. **19**(3), 239–264 (2007)

McLaren, B.M., Lim, S., Koedinger, K.R.: When is assistance helpful to learning? Results in combining worked examples and intelligent tutoring. In: Woolf, B., Aimeur, E., Nkambou, R., Lajoie, S. (eds.) Proceedings of the 9th International Conference on Intelligent Tutoring Systems, pp. 677–680. Springer, Berlin (2008b)

McLaren, B.M., Lim, S., Yaron, D., Koedinger, K.R.: Can a polite intelligent tutoring system lead to improved learning outside of the lab? In: Proceedings of the 13th International conterence on AI in Ed, pp. 433–440 (2007)

McLaren, B.M., Lim, S., Gagnon, F., Yaron, D., Koedinger, K.R.: Studying the effects of personalized language and worked examples in the context of a web-based intelligent tutor. In: Proceedings of the 8th International Conference on International TUT System, pp. 318–328 (2006)

Megan, H.: The structure of classroom discourse. In: Handbook of Discourse Analysis, pp. 119–131 (2003)

Michaels, S., O'Connor, C., Resnick, L.B.: Deliberative discourse idealized and realized: Accountable Talk in the classroom and in civic life. Stud. Philos. Educ. **27**, 283–297 (2008)

Nelson-Le Gall, S.: Chapter 2: Help-seeking behavior in learning. Rev. Res. Educ. **12**(1), 55–90 (1985)

Nelson-Le Gall, S., Glor-Scheib, S.: Help seeking in elementary classrooms: an observational study. Contemp. Educ. Psychol. **10**(1), 58–71 (1985)

Newman, R.S.: Children's help-seeking in the classroom: The role of motivational factors and attitudes. J. Educ. Psychol. **82**(1), 71 (1990)

O'Connor, M.C., Michaels, S., Chapin, S.: Small-scale experimental studies of classroom talk: seeking local effects in discourse-intensive instruction. In: Paper Presented at Annual Meeting of the American Educational Research Association (2007)

Piaget, J.: Piaget's theory. In: Mussen, P.H. (ed.) Carmichael's Manual of Child Psychology, vol. 1, 3rd edn, pp. 703–732. Wiley, New York (1970)

Rau, M., Aleven, V., Rummel, N.: Intelligent tutoring systems with multiple representations and self-explanation prompts support learning of fractions. In: Dimitrova, V. Mizoguchi, R., du Boulay, B., Graesser, A. (Eds.) Proceedings of the 14th International conference on Artificial Intelligence in Education, AIED 2009, pp. 441–448. IOS Press, Amsterdam (2009)

Resnick, L.B., Salmon, M., Zeitz, C.M.: The structure of reasoning in conversation. In: Proceedings of the Thirteenth Annual Conference of the Cognitive Science Society

Resnick, L.B., Salmon, M., Zeitz, C.M., Wathen, S.H., Holowchak, M.: Reasoning in conversation. Cognit. Instr. **11**(3, 4), 347–364 (1993)

Resnick, L.B., William, D., Apodaca, R., Rangel, E.: The relationship between assessment and the organization and pratice of teaching. In: McGaw, B., Peterson, P., Baker, E. (eds.) International Encyclopedia of Education, in press, 3rd Edition Teaching of Psychology, vol. 22, pp. 13–17 (2010)

Ryan, A.M., Shin, H.: Help-seeking tendencies during early adolescence: An examination of motivational correlates and consequences for achievement. Learn. Instr. **21**(2), 247–256 (2011)

Sionti, M., Ai, H., Rosé, C.P., Resnick, L.: A framework for analyzing development of argumentation through classroom discussions. In Educational Technologies for Teaching Argumentation Skills, pp. 28–55 (2011)

Teasley, S.D., Roschelle, J.: Constructing a joint problem space: the computer as a tool for sharing knowledge. In: Computers as Cognitive Tools, pp. 229–258 (1993)

Weinberger, A., Fischer, F.: A framework to analyze argumentative knowledge construction in computer supported collaborative learning. Comput. Educ. **46**, 71–95 (2006)

Wittenbaum, G.M., Hollingshead, A.B., Paulus, P.B., Hirokawa, R.Y., Ancona, D.G., Peterson, R.S., et al.: The functional perspective as a lens for understanding groups. Small Group Res. **35**(1), 17–43 (2004)

Fostering 21st Century Skills in Engineering and Business Management Students

2CG®: A Contemporary Teaching Approach

Christina Merl[✉]

TalkShop.cc, Vienna, Austria
cmerl@talkshop.cc

Abstract. This paper demonstrates how a contemporary teaching approach known as 2CG® method, which stands for content- and context-specific generic competency coaching, can help educators foster 21st century skills in engineering and business management students in the framework of English language programs. 2CG® is built around the concept of peer learning and exchange in student-empowered classrooms and has been specifically designed for learning workers who continually need to improve and practice their skills to succeed in the VUCA world where speed, flexibility, innovation and communication are the key to success. 21st century skills as developed with the 2CG® teaching method comprise critical thinking, problem-solving, creativity, communication, collaboration, curiosity, initiative, persistence, adaptability, leadership, and social and cultural awareness. The teaching approach itself is based on the cornerstones of communities of practice (CoP) and teachers take the role of facilitators who nurture their communities with relevant input and impulses. A blended mix of interactive and introspective teaching activities, collaborative learning tasks, artistic impulses and focused feedback is used to empower students to continually develop their competencies and enhance their qualities as well as take responsibility for their learning.

Keywords: Community of practice · Peer learning · 21st century skills · Teaching approach · Learning workers · Collaborative learning · Student-empowered classroom · Artistic impulses · Content and context

1 The Age of STEM and Learning Workers

It is a great time to be in the STEM industry! The number of tech employment opportunities is expected to further increase by 12% by 2024 [1]. We will see a particular increase in demand for the following tech sectors: web development, biomedical engineering, analysis, IT security, oil and gas, and mining [1]. As the tech and engineering industries continue to grow, so will the need of organisations to hire well educated talent, so called "learning workers" [2] – people of all ages who develop and upgrade their skills on the job and thus work themselves into better positions. What sets learning workers apart is their knowledge of how to learn: Rather than acquiring one

© Springer Nature Switzerland AG 2020
M. E. Auer and T. Tsiatsos (Eds.): ICL 2018, AISC 916, pp. 145–156, 2020.
https://doi.org/10.1007/978-3-030-11932-4_15

set of specific skills, learning workers continually develop their skills on the go – they adapt and apply their learning to new situations and issues [3].

Educational programs and educators are challenged to serve these learning workers and provide them with educational experiences that reflect today's digital world and equip them with the skills and competencies that empower them to engage with it. Successful learning workers, among them 21st century engineers, will particularly have to be able to adapt to the different workflows of diverse teams within and across organisational boundaries [4]. They will need to learn continuously and stay abreast of developments in their field of work through effective educational programs, self-study and by participating in professional networks and communities of practice. This development will ask learning workers to increasingly take responsibility for their learning, develop a habit of continuous learning, and be able to filter relevant content and integrate it into their real-world work experience.

1.1   Who Is the 21st Century Engineer?

On their day of graduation, engineering and business management students should not only have domain skills and technical knowledge but also demonstrate social and emotional proficiency. For example, they need to be able to communicate and collaborate in interdisciplinary and cross-cultural teams, listen to multiple viewpoints, and talk to people with diverse professional and cultural backgrounds, from investors and marketing people to product managers, fellow engineers to people who work in manufacturing. "Cross-team functionality and communication skills are skills that are in high demand in the tech industry in 2018", as pointed out in [4]. What is more, 21st century engineers need to be able to understand complexity and act agile in the VUCA world [5] where volatility, uncertainty, complexity and ambiguity rule. They need to make decisions, analyze issues, situations and related facts, data and evidence without being influenced by biases and personal feelings. To do so, 21st century engineers have to master 21st century skills as defined in the *New Vision for Education* of the World Economic Forum [6]: foundational literacies, communication competencies, and character qualities that will equip them with the skills they need to succeed in the VUCA world and evolving digital economy (Fig. 1).

2   Teaching 21st Century Skills

Fostering 21st century skills in engineering and business management students requires a contemporary approach to teaching and supportive learning environments. The present paper presents a modern blended approach to developing and practising 21st century skills within the framework of English language programs. The approach, which is referred to as 2CG® method (content- and context-specific generic competency coaching), assumes that educational programs for 21st century engineers should be rooted in the liberal arts and at the same time need to engage the practical skills students require on the job. It further assumes that effective learning environments for acquiring and mastering 21st century skills have to be experiential, collaborative, technology-enabled and social and empower learning workers to cope adequately with their real-world complexities.

The present paper aims to demonstrate how 2CG® coaching can support engineering and business management students in acquiring and practising communication competencies and character qualities as defined in the *New Vision for Education* of the World Economic Forum [6]. These so called 21st century skills comprise critical thinking and problem-solving, creativity, communication, collaboration, curiosity, initiative, persistence/grit, adaptability, leadership and social and cultural awareness (Fig. 1). The 2CG® method is built around the concept of peer learning and exchange in communities of practice (CoP) and has been applied within the framework of English language classes in bachelor programs for industrial engineering and business management students, among others.

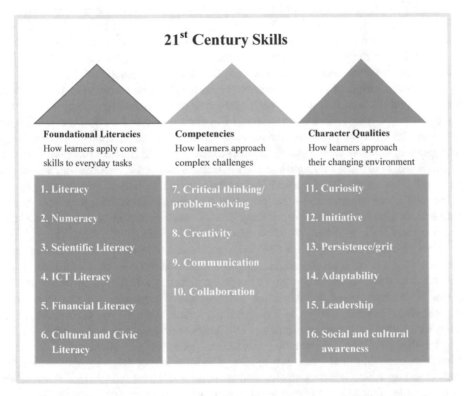

Fig. 1. 21st Century Skills as defined in the New Vision for Education (2015) [6]

2.1 Setting Up Student-Empowered Classrooms

In traditional classroom settings with well designed lessons, students will definitely interact and collaborate, but only when instructed by their teacher. What is often missing in traditional classroom settings are learning opportunities where students can engage authentically and share their real-world stories and experiences. Also, there is hardly enough time in traditional programs to really put the responsibility for learning into the hands of students. Clearly, effective higher learning in the 21st century must be experiential, collaborative, technology-enabled, social and closely aligned to the professional practice of learning workers. The 2CG® approach is based on the idea that communities of practice (CoP) can provide a supportive environment for this new way of teaching and learning. CoPs provide a collaborative, social and tech-enabled learning environment that is built around trust, mutual respect and peer exchange and empowers learners to interact regularly, share their real-world learning experiences and practice and improve their skills and competencies.

CoPs are defined as "groups of people who share a concern or a passion for something they do and learn how to do it better as they interact regularly" [7]. Three characteristics are crucial for CoPs:

The domain: A CoP has an identity as defined by a shared domain of interest. Membership implies commitment to the domain, and therefore a shared competence. While the domain is not necessarily something recognized as "expertise" outside the community, CoP members value their collective competence and learn from each other. In the context of the present teaching case, the domain of interest is "21st century skills with a focus on communication competencies and character qualities".

The community: In pursuing their interest in the domain, members engage in joint activities and discussions, support each other, and share knowledge. They build relationships that further enable them to learn from each other. In the context of 2CG® as presented in this paper, CoP members are industrial engineering and business management students who regularly interact to improve and practise their skills. It has to be noted here that an English class in itself is not a CoP. Attending the same language class only makes for a CoP if classmates regularly interact and learn together, share real-world experiences, solve problems together, build relationships and cross-connections and build trust, even if they do not work in the same organisation and practice their skills in different real-world contexts.

The practice: Members of a CoP are practitioners – in the present teaching case they are students of industrial engineering and business management who have a job and need to develop and improve 21st century skills to effectively engage with the digital economy in their real-world job environment. When interacting in their CoP, these students develop a shared repertoire of resources: experiences, stories, tools, and ways of addressing recurring problems. This takes time and sustained interaction.

Benefits of CoPs for learning workers: CoPs can provide an ideal learning environment as they enable students to take individual and collective responsibility for their learning, which includes sharing and managing relevant knowledge. The teacher takes the role of facilitator and nurtures the community by providing input and impulses that allow students to mature and grow 'as they go'. This is how students are encouraged to build relationships, create cross-connections and develop a shared practice. It should be pointed out here that the social ties between CoP members ensure that they make ethical decisions and support and benefit from, rather than exploit, each other. What is more, weaker students usually learn from their more advanced peers. Students of different cultural and language background who often struggle to connect with their class mates usually get a fair chance to integrate, connect and bond within their CoP.

CoPs definitely have a life cycle (Fig. 2) and only exist as long as members find it useful to interact and further improve their practice. Since bachelor programs for industrial engineering and management students take 6 semesters and offer English lessons in the first 5 semesters, these CoPs may fall apart after that. They may continue to exist in a self-organized form, though.

Life Cycle of A CoP-based 2CG Classroom Community				
(based on Etienne Wenger's concept of 5 CoP stages)				
1 Committing	2 Starting Up	3 Peak Time	4 Winding Down	5 Closing Down
Students enrol for their study program; the teacher designs the course outline and frames the purpose of the CoP.	The purpose of the CoP is defined, students meet in class and negotiate their roles. The teacher takes the role of facilitator; the process i supported by the university.	Operational peak time:Students get input and impulses from their teacher or CoP facilitator, share experiences and knowledge, solve problems together, set up task forces, develop their skills and improve their practice; the teacher nourishes the CoP.	Students are about to complete their studies, the value of their CoP membership may start to diminish – problems have been solved and the practice has improved; the CoP may continue to exist in the form of a loose network.	The purpose of the CoP is fulfilled; the members and/or the organization decide to shut it down.

Fig. 2. Life cycle of a CoP-based 2CG® classroom community

2.2 The Role of Teachers in Student-Empowered Classrooms

In 2CG® classroom settings that are built around CoPs, teachers need to take the role of facilitators and put the responsibility for learning into the hands of students. They have to provide relevant input and impulses and at the same time empower students to contribute content and share experiences. What is more, teachers need to support students in analyzing complex situations, identifying problems that have to be solved, and encourage them to explore new ways of thinking to come up with creative solutions. Teachers can do a number of things to facilitate learning in 2CG® classrooms. For example, they can

- help students to understand the importance of learning continuously in order to stay abreast in their field;
- help students to value different ways of learning, including tech-enabled learning, self-organized learning, collaborative learning and find out what works for them;
- make the classroom a space of trust, mutual respect and support where students are encouraged to share and reflect on their real-world as well as in-class learning experiences;
- ensure that the classroom is a space where sharing is appreciated and rewarded;
- formalize self-organized learning if this makes students feel more comfortable [8];

– empower students to see the bigger picture and figure out the solutions, rather than deliver them every single insight.

2.3   Bringing About a Shift in Students' Mindsets

Usually, students adapt fast to the new teaching style. However, students who are used to an authoritarian teaching style may feel puzzled when they first experience a 2CG® teaching approach and classroom setting. Some may not know what to do with the space they are given to self-organize their learning. While some students will immediately feel empowered by the new way of learning, it may take a little while for others to feel comfortable.

The following anecdote demonstrates what may happen in a 2CG® classroom setting: "At the beginning of the session, the teacher set a learning goal and asked students to collaborate in small teams, identify the problems that needed to be solved and find possible solutions. This instruction required students to self-organize their learning. Suddenly, students became silent and stared at their instructions. Some of them started writing, others looked a bit helpless. It was an awkward situation. The teacher once again pointed out that this was a collaborative team task and encouraged students to exchange their thoughts, think out loud, communicate, ask questions, discuss solutions, etc. Now a couple of students started whispering. To make students feel more comfortable, the teacher invited them to leave the classroom, find themselves a space where they felt safe, work on their task and come back with their solutions after 15 min. The majority of students took this offer seriously and started exchanging and working on their solutions at the table close to the coffee machine, on a park bench outside the building, and in the lounge corner in the corridor. A small bunch of students enjoyed the "break" and definitely did not talk about their team task. When discussing the solutions and reflecting on their task afterwards, one student said, "We are not used to this way of learning. You are the only one here who encourages us to self-organize our learning process. I find that a really pleasant and motivating way to learn." The "lazy" students suddenly understood that they should have come up with solutions and that both, their teacher and their peers had expected them to contribute their ideas, rather than take a break."

As there are always students who may misunderstand the fact that they are given space to self-organize their learning, the teacher is well advised to set clear rules when setting up a student-empowered classroom. Students need to realize fast that they have to commit themselves to their shared practice, engage and make an effort to come up with good ideas and solutions. This usually works and after some initial hurdles students will know that in an experiential learning environment, they need to take responsibility for their learning. Working learners usually prefer this freedom over a command and control system where they fill out worksheets in silence. This is how a classroom becomes a communication and competency lab, a space for developing and practising 21st century skills.

3   Activities for Student-Empowered Classrooms

To facilitate student-empowered classrooms and foster 21st century skills in learners, teachers can make use of a specifically designed, technology-supported mix of collaborative, social and introspective learning activities, including pitch presentations, discussion leads, role play and simulations, poetry reading, reflective writing tasks, analogical thinking activities, impro exercises and much more. Regular interaction, collaborative learning, peer learning and exchange as well as a focused feedback culture can support students in their skills development. So far, the 2CG® method has been applied in the framework of English language programs and places emphasis on aspects such as tone of voice, language register, choice of words, context and style, grammar and mechanics, content delivery, body language, cross-cultural aspects and all the other communication-related competencies and character qualities that fall under 21st century skills. In focused and extensive feedback sessions, students are encouraged to openly share their thoughts and insights with their peers and the teacher. They usually can further internalize their learning by means of tailored reflective tasks that are presented in the form of team tasks or written and oral assignments.

3.1   Artistic Impulses

The 2CG® method places particular emphasis on providing learners with relevant artistic impulses, such as poetry reading, analogical problem-solving activities, role play and impro exercises. These impulses usually increase flexible, creative and critical thinking capacity in students. They also promote skills like curiosity, empathy, cross-cultural understanding, critical thinking and emotional intelligence. When confronted with creative and emotional input, engineering and business management students have to break out of their tech and management bubble and deal with the emotions that are evoked in them. For example, poetry can help teachers to augment students' reality by triggering new thoughts in them, as well as asking them to share and explore different personality traits. For example, The Robo-Boss (Fig. 3), a poem that uses accessible language and mirrors the emotional side of many students' real-world work environments, was used to encourage industrial engineering and business management students to identify communication problems in today's job world and formulate possible solutions. The majority of students stated that the pressure was getting more, both in the job and at university. Some told that socializing was strictly monitored at work and that communication with colleagues had changed as a result thereof. Once they had identified current challenges, students started thinking about what they could do to improve communication at work. Reflecting and talking about all these things in English helps students to improve relevant 21st century skills while discussing real world-related issues.

This is how poetry can be a powerful tool to raise awareness and inspire critical thinking in learning workers – it is the task of the teacher or facilitator to make students share, formulate and communicate their thoughts, make themselves understood and articulate complex issues in their second language. As the American philosopher Richard Rorty [9] put it: "Our reality – our truth – is no more and no less than what can be framed with language."

English 4, WING SS 2018 I cmerl@talkshop.cc

The Robo-Boss (by Lovelyn Andrade, 2018)

Nobody expects you
To be human anymore,
A feeler does not have a place
On the platform of success.
Tuck your emotions
Deep inside your pockets,
The new boss is here,
His business approach,
Does not require emotional intelligence.

This is the future.
This is now.

Leaders are designed to be objective,
Accurate, precise and clear
Sensitivity is going down the drain.
All they care about,
Is that the job is done.
You have a problem?
It's none of their business.
You are only as good as your last success.

The world, as we know it,
Is evolving.
We, human beings,
Are trapped in a fast-paced world,
Our days are spent multi-tasking,
The more you do,
The more is required to do.

We wrestle against time,
To get the job done,
Leaving us too done,
To take time for ourselves.

The new boss is here,
Highly objective and less sensitive,
While I have been begging for caffeine,
He is already solving cases with each tick of the clock.

This is the future.
This is now.

I reach for the cup of coffee,
Left too long and gone cold,
Along with those meals,
I skipped because I was rushing
To get things done
In order for me to enjoy a meal for once.

A message from the new boss
Is clear and brief,
My argument is entertained,
Only to be rejected.
The Robo-Boss is always sure,
There is no negotiation.
I wish, just for once,
Someone pats my back,
For a job well done.

Fig. 3. 'The Robo-Boss' (by Lovelyn Andrade, 2018) was used in the framework of a so called '2CG® reflective pavilion on AI' with industrial engineering students. The poem was also interpreted by a puppet player and filmed: https://bit.ly/2DsO2dk

3.2   Role Play and Impro Activities to Enhance Character Qualities

The 2CG® approach makes use of specifically designed role plays and simulations of real-world settings to help students acquire the ability to cope with complexity, identify problems that need to be solved, make fast decisions and adapt their communication behavior to different contexts while staying connected with a core self. In these activities, students are challenged to become familiar with their role and learn to anticipate what the other players may need and what they expect from them. Role play activities and impro exercises can encourage students to demonstrate authenticity by acting out "the bits of their personality that are relevant for the specific situation" [10]. This two-step process forces them to leave their comfort zone in a double way: First, they need to identify with their role whether or not it matches their personality; second, they need to communicate and make decisions in their second language spontaneously, choose the appropriate language register, make the right choice of words, and apply a proper tone of voice to win acceptance in the given situation – all in their second language. By means of detailed video analysis and reflective activities as well as honest and focused peer feedback, students can create cross-connections to their real-world job experiences and integrate the learning into their professional performance. What is essential here is that all this takes place in a collaborative and social learning environment which is based on trust, mutual respect and sharing (CoP).

3.3 Reflective Activities and Metacognitive Skills

Basically, in order to act and communicate effectively and successfully in their real-world context, working learners – engineering students – need to be able to filter relevant input and integrate it into their actual doing. Reflective and critical thinking tasks can be very useful to practice and improve 21st century skills as they stir students' curiosity, train their habit of looking at things from different angles, and motivate them to create cross-references to their real-world context. Questions like 'How could we solve this better?', 'What makes you think this statement is not to the point?', or 'How can you write this passage more authentically?' will help students train and rely on their inner voice as well as empower them to make and communicate their decisions with confidence. Reflecting on their learning process is one way of training metacognition and ensures that metacognitive skills become part of students' repertoire of critical thinking skills, as suggested in [11].

Here is another example of how students can learn to make useful cross-connections between theoretical input and real-world practice: In the third semester, the bachelor program for industrial engineering and business management students of the institution where the 2CG® method has been applied, focuses on "intercultural communication". Rather than just reading about theoretical cross-cultural concepts and filling out worksheets, which is often done in traditional classroom settings, students need to reflect on their own cross-cultural communication experiences and find out how the theoretical input they get in class could help them "do it better" in their real-world context. What could they have done differently, which behavior was successful and what could they try out next time. The teacher needs to guide students and support them in identifying problems, asking good questions, and finally filtering and integrating relevant content into their real world context. This is how teachers can spark off the learning process and help students increase their self-awareness as learners: "What works for me and what does not?", "What are my strengths and weaknesses and how can I best improve?", "How can I effectively accomplish this task?", "How can I avoid making the same mistakes again?" This is how students learn to focus on their strengths and outbalance the weaknesses, use external resources like the Internet, their classmates or co-workers most effectively, pursue their learning goals and self-organize their learning process [12]. While doing so in the context of an English language class, students need to focus on all four strands of the curriculum (reading, writing, listening, speaking). The process allows them to acquire new vocabulary and apply the foreign language in context. What is more, students are given the opportunity to take risks and make fast decisions in their second language.

3.4 Reflective Written and Oral Assignments

To reinforce the learning, students are asked to reflect on their learning progress in between their face to face interactions in the form of written or oral assignments. These assignments are technology-enabled, meaning that students have to deliver their assignments in the form of videos or text documents within an online learning environment specifically developed for 2CG® classroom settings, which can be integrated into any LMS. They receive focused feedback on grammar and mechanics (linguistic competency), style and tone (sociolinguistic competency), context and meaning (pragmatic competency), content and structure, form and delivery, body language as well as feedback in the form of comments and questions that make them practice specifically selected 21st century skills. These technology-enabled assignments can serve as interactive online learning journal where students can exchange with their teacher and their peers, can raise questions and make notes that support their learning progress.

4   Evaluation in Student-Empowered Classroom Settings

While it may be challenging to measure the exact impact of the teaching activities used in 2CG® classroom settings and attribute them to particular results, educators definitely can trace how students change their practice and improve their skills-related performance along the way. The approach to teaching in collaborative learning environments (CoP) as presented in this paper therefore recommends that traditional forms of assessment, such as vocabulary tests, written and oral assignments that are based on the European Framework of Reference for Languages, be combined with more progressive evaluation techniques, such as focused correction, peer feedback, and interactive technology-enabled feedback. 2CG® teaching is based on the assumption that students can benefit from a synergy of awareness raising and actual error correction. Teachers should provide students with different forms of assessment and train them to give constructive peer feedback. Apart from traditional assessment methods, educators will be able to assess the learning progress of the group as well as of individual students through questions they ask, through the personal stories and anecdotes they share about their professional practice, by means of the stage their CoP is in, and last but not least through their performances related to various teaching activities.

5   Conclusions and Outlook

Successful 21st century engineers and business managers – learning workers – need to be able to adapt to the different workflows of diverse teams within and across organizational boundaries. Educators and educational programs are challenged to offer them learning opportunities that reflect the requirements of the evolving digital economy and empower learners to cope with situations in their VUCA world job environment where speed, flexibility, innovation and communication are the key to success. The contemporary 2CG® teaching approach as presented in this paper is built around the concept of peer learning and exchange in communities of practice (CoP). It puts emphasis on creating student-empowered classrooms and fostering 21st century skills in the framework of English language classes. 21st century skills comprise critical thinking and problem-solving, creativity, communication, collaboration, curiosity, initiative, persistence/grit, adaptability, leadership and social and cultural awareness. The 2CG® method enables teachers and students to shift from traditional content-driven curricula to focusing on context-specific real-world needs of learning workers by empowering students to fully take resonsibility of their learning, to take the initiative to collaborate, and to practise peer-assessment in class.

By giving learners the opportunity to experience their communication behavior through artistic impulses, role play, team-based learning, reflective and cognitive activities, focused correction and regular peer feedback in a student-empowered learning environment the approach empowers them to take responsibility for their individual and collective learning. As a matter of fact, students can be expected to show high engagement in 2CG® based learning environments since the method moves away from teaching abstract concepts and skill sets towards creating settings where students can experience and develop transferable 21st century skills. What is more, teachers and students stay abreast of new developments in their field as they go. It should be pointed out here that due to group dynamics and needs of individual learners, each class or CoP will be different. Teachers who act as facilitators need to adapt the materials they use, take students by the hand and give guidance. It also should be noted that the 2CG® teaching approach will not be fully effective if applied in traditional classroom settings. Students who expect traditional language learning with traditional instructions and worksheets will most probably be dissatisfied in the beginning. It may take them a little while to adapt to the new setting, explore and define their new role as learners and make use of the space they are given in a student-empowered classroom.

2CG® teaching is tailored to the needs of 21st century practitioners – learning workers who have to improve their competencies and character qualities to effectively engage with the digital economy and the VUCA world. It is important that educators do not over-intellectualize their learning program: while reflective activities are vital, the focus must still be on practising and mastering 21st century skills and competencies as required by students in their real-world job environment. To further extend and explore the opportunities and benefits of 2CG® teaching, a number of pilot projects across partner universities, study programs, disciplines and countries will need to be established.

Acknowledgment. The author of this paper would like to express her special thanks to Professor Andreas Pester, Professor Erich Hartlieb, and Professor Michael Auer at CUAS for their kind support and confidence in the effectiveness of the content and context approach to language teaching as presented in this paper. A further thanks goes to the poet Lovelyn Andrade, who gave permission to use her poems for this teaching case. Thanks also to puppet theater experts Angela Sixt and Christian Pfuetze for their artistic support. Last but not least, the author wants to praise her industrial engineering and business management students at CUAS, whose commitment and conributions have delivered invaluable educational insights.

References

1. Modis: 2017 STEM Workplace Trends Survey Results. http://www.modis.com/it-insights/infographics/the-state-of-the-tech-job-market/ (2018)
2. Morgan, J.: The Future of Work: Attract New Talent, Build Better Leaders, and Create a Competitive Organization, 1st edn. Wiley (2014)
3. Forbes: Goodbye to knowledge workers. https://www.forbes.com/sites/jacobmorgan/2016/06/07/say-goodbye-to-knowledge-workers-and-welcome-to-learning-workers/#30cadad82f93 (2016)
4. Forbes Technology Council: Top tech skills in high demand for 2018. https://www.forbes.com/sites/forbestechcouncil/2017/12/21/13-top-tech-skills-in-high-demand-for-2018/#bd9c9a21e5c1 (2018)
5. VUCA 2.0: A Strategy for Steady Leadership in an Unsteady World. https://www.forbes.com/sites/hbsworkingknowledge/2017/02/17/vuca-2-0-a-strategy-for-steady-leadership-in-an-unsteady-world/#2608dc6613d8
6. World Economic Forum: The New Vision for Education: Unlocking the Potential for Technology. http://www3.weforum.org/docs/WEFUSA_NewVisionforEducation_Report2015.pdf (2015)
7. http://wenger-trayner.com/introduction-to-communities-of-practice/. Accessed 31 May 2018
8. Hart, J.: What it is to be a "learning worker" (an interview). http://www.c4lpt.co.uk/blog/2015/06/24/what-it-means-to-be-a-learning-worker/ (2015)
9. Rorty, R.: Contingency, Irony and Solidarity, pp. 4–5, 6, 27, 51–52. Cambridge University Press, Cambridge (1989)
10. Act Like a Leader, Think Like a Leader. Harvard Business Review Press (2015)
11. Smith, S., Conti, G.: The Language Teacher Toolkit, 1st edn. CreateSpace Independent Publishing Platform (2016)
12. https://gianfrancoconti.wordpress.com/2015/06/11/modelling-metacognitive-questioning-in-the-foreign-language-classroom/

Leveraging Collaborative Mobile Learning for Sustained Software Development Skills

Sigrid Schefer-Wenzl[(✉)] and Igor Miladinovic

University of Applied Science Campus Vienna, Favoritenstr. 226,
1100 Vienna, Austria
{sigrid.schefer-wenzl,igor.miladinovic}@fh-campuswien.ac.at

Abstract. The demand for software developers is growing fast, and programming skills are one of the most in-demand skills in the world. The primary goal of our Bachelor program is to provide students sustainable foundations for careers in software engineering. However, teaching and learning software development are challenging tasks.

We designed a new course concept to ensure that students apply software development knowledge in an industry-like project and are continuously engaged in further improving these skills. Our hypothesis was that by creating an interactive, collaborative and open learning environment we motivate students to extend their attitude from only learning for the exam, with short-term learning effects, to learning for a sustained competence growth, with self-motivated and long-lasting effects. With different students' evaluations we were able to validate this hypothesis. In this paper we show the results of these evaluations with particular focus on the sustainability of the gained knowledge.

Keywords: Mobile learning · Collaborative learning ·
Software development education · Sustainable learning

1 Introduction

Sustained learning is one of the main goals in education. Creating learning environments that enable and support sustained learning is a precondition for achieving this goal. However, it is not a trivial task as the sustained learning effects cannot be evaluated easily. It requires a long term observation of students' skills development and effects can only be determined after a longer time period.

Key skills gained with our Bachelor program in Computer Science and Digital Communications include a solid understanding of software development as well as comprehension of modern software engineering principles. Solid software development skills require a continuous and targeted education over several years. In previous years, our students' capability of practically applying programming languages and software engineering principles leaves room for improvement.

© Springer Nature Switzerland AG 2020
M. E. Auer and T. Tsiatsos (Eds.): ICL 2018, AISC 916, pp. 157–166, 2020.
https://doi.org/10.1007/978-3-030-11932-4_16

We therefore currently undergo a substantial update and refurbishment of our study program in order to strengthen these skills.

We started with redesigning a software development course module by applying a mobile learning approach. The main goal was to create a motivating, collaborative and student-oriented learning environment, where the students would be able to individually identify their development potential and to continuously improve it by targeted learning. We designed and optimized all the learning material for the presentation on smartphones and tablets. These devices are also used for software development, enabling interactive testing and making successful progress immediately visible. By using appropriate tools students are also able to communicate and collaborate virtually by their devices. This triple function of smartphones and tablets as mobile learning devices, together with a selected mix of multiple modern didactic methods resulted in an important improvement of learning outcomes, as presented in [24].

In this paper we investigate the sustainability of the gained knowledge. We selected different indicators to measure the sustainability of our course concept. These indicators will be described and discussed in detail. The results show crucial points that need to be considered to achieve sustained learning.

The remainder of this paper is structured as follows. Section 2 discusses related work on mobile learning in higher education with the particular focus on software engineering education. Section 3 presents our mobile learning concept, including challenges and design of the course module. Subsequently, Sect. 4 describes selected indicators for sustained learning and discusses results. Section 5 concludes the paper.

2   Related Work

Rapid techological advancements enable learners to use mobile technology anytime and anywhere to access up-to-date educational resources and to individualize learning [1, 15, 19, 20, 23]. In a recent meta-analysis on mobile learning studies, collaboration opportunities between students as well as ubiquitous learning were identified as the two main perceived advantages of mobile learning [2]. Thus, it facilitates the interaction with other students to share information or work collaboratively on a task or project.

Several authors analyze the challenges in teaching software development topics [3, 9, 10, 18, 21]. Amongst others, these challenges include the vast amount of material that needs to be taught in short time in order to enable students to start their own software projects, the limited quality of most programming textbooks, as well as the dissatisfying learning outcomes of such courses. It is also important that students receive proper feedback on their practical programming tasks in a timely manner, which is often not realized in software development courses [26].

With the advent of mobile devices and especially smartphones, several authors started to argue that the teaching of programming should be done directly on the mobile devices itself [7, 12, 17, 28]. Some authors also advocate for combining traditional, e-learning, and mobile learning strategies in order to successfully teach programming contents [31]. One positive effect of integrating mobile learning in software development education is that it allows implementing and including new technologies to make students more prepared for their future work [5].

Just-in-Time Teaching (JiTT) was introduced in [16]. The main aim of this teaching method is to combine web-based assignments with active in-class hours. Jonsson presented a course concept for integrating JiTT into a programming course [13]. He then compared the outcomes of the course with a programming course teaching the same contents in a traditional lecture form. The share of students passing the course as well as the percentage of good grades increased substantially in the JiTT-based course. Another example for integrating elements of JiTT into software development education has been introduced in [27]. In a pre-lecture phase background information and warm-up exercises are prepared by the students, which are reviewed by the lecturer and then discussed in class, followed by post-lecture activities. In course evaluations, students reported better learning performance due to the introduced JiTT elements.

3   Module Concept

In this section, we present the design for our newly designed Mobile App Development course module tailored to cover the identified shortcomings regarding insufficient programming and software engineering skills. Students take this module in the 4th and 5th semester of their Bachelor study and it is awarded with 18 ECTS (European Credit Transfer and Accumulation System) credits, which corresponds to approximately 450 working hours. This module consists of a lecture (4 ECTS), a tutorial (4 ECTS) and a project (10 ECTS).

3.1   Challenges

When designing our course module concept, we had to address several challenges. One challenge is that students enter our course with very diverse levels of programming skills. This is mainly because we also offer an evening study allowing students to pursue a study while working full time. Thus, our student groups include professional software developers next to students with very limited software development experience. Our combination of collaborative mobile learning and JiTT concepts addresses this issue by enabling individualized learning speed and intensity.

Another challenge is to support students in finding appropriate project ideas for their own software projects. These projects need to be settled in a realistic environment close to industry standards. However, students need to be able to finish these projects with reasonable effort within one semester. Thus, we also want to ensure a continuous progress tracking of the students' projects. We addressed these topics with a hackathon event, regular coaching units throughout the course as well as tool support for project management and collaboration.

The third challenge is to develop self-motivated learning skills of students, which are at very different levels among students. To address this challenge we gradually increased the amount of distance learning elements throughout the module from basic in the lecture to high in the project.

Our fourth challenge is to achieve a sustained knowledge in software development among the students. The sustainability effect can only be measured over a longer time period. For this challenge we introduced continuous assessments during the module to support students' learning through feedback and to increase students' motivation [11]. In addition we offered students coaching on demand after the module by supervising bachelor thesis projects and we repeated the final exam six months after the course.

3.2   Design

Our module concept is based on knowledge levels according to Bloom's Taxonomy [4]. For each part of the course module we defined the targeted knowledge levels (see Fig. 1). Thus, the main aim of the lecture is to provide a solid theoretical background that enables them to apply the taught concepts in tutorial exercises and later on to create their own projects. Moreover, students have to analyze and evaluate other projects. We achieve these levels by gradually increasing the amount of self-motivated learning activities.

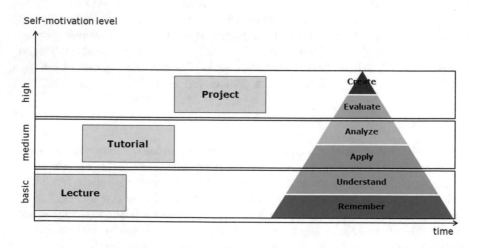

Fig. 1. Addressed knowledge levels

We derived the following learning objectives for our Software Development course module:

– Understand and apply an effective software engineering process, based on knowledge of widely used software process models.

- Employ team working skills including structured planning, time management and inter-group negotiation.
- Capture, document and analyze requirements.
- Translate a requirements specification into an implementable design, following a structured process.
- Implement the software according to the specified software design.
- Design a testing strategy for a software system, employing techniques such as unit testing, test driven development and functional testing.
- Evaluate the final projects by checking compliance with the requirements, and analyze the design and implementation.

In order to achieve the learning outcomes stated above, we designed a collaborative mobile learning module concept by integrating a set of different teaching methods.

In the lecture, we apply an inverted classroom setting [8, 14] combined with elements of JiTT and mobile learning. The learning content is provided in short learning snippets that can be consumed fast and from anywhere via a smartphone or tablet. We also supply short duration videos, usually lasting between two and four minutes, explaining and illustrating complex lecture contents. Students prepare for each lecture by reading and watching learning material as well as doing warm-up exercises. Subsequently, the students reflect each self-study phase in a learning diary. The instructors provide feedback for each learning diary entry before the next in-class lesson and adjust the contents of the lesson accordingly. Based on the learning diary entries, the in-class hours start with clarification of open questions and then focus on further deepening the contents, especially those which were not well understood in the self-study phase. The combination of inverted classroom and just-in-time teaching enables a strong learner-orientation as well as individualization of course contents during class hours. The collaborative mobile learning aspect further supports location- and time-independent learning as well as immediate interaction between students about learning contents. Thus, the lecture part covers the levels *remember* and *understand* of Bloom's taxonomy (see Fig. 1).

The tutorial part of this module is based on and partly intertwined with the lecture. It applies the concepts from the lecture within a student project, i.e. the design and development of a mobile application. We officially start the tutorial part with a hackathon event. A hackathon (hacking marathon) can be described as a problem-focused computer programming event, where people come together to develop prototype applications by collaborating intensively over a short period of time [30]. After the hackathon event, the students work in groups of two to three people on these projects on their own. In addition, they are supported by regular meetings with their instructors as well as a range of software tools for project management, code version management and tools supporting the virtual collaboration of students during the software development process. By applying the knowledge from the lecture on a concrete project, the tutorial part addresses the levels *apply* and *analyze* of Bloom's taxonomy (see Fig. 1).

In the project part we simulate an industry-like software project, where the acceptance of a software is done by persons not being involved in the development. Students have to come up with own ideas and implement them within the defined schedule. To stimulate distance learning, coaching by lecturers is provided on demand in contrast to regular coachings during the tutorial. Depending on the self-motivated learning skills of the group the coaching meetings occurred more or less frequently. The main goal of this course part is on the one hand to introduce students to the nature and complexity of real software projects, and on the other to stimulate the development of self-motivated learning skills of the students through distance learning elements. Each group also evaluates the outcome of other groups via peer-assessment. As shown in Fig. 1, the project part meets the levels *evaluate* and *create* of Bloom's taxonomy.

A detailed description of the course module design and learning outcomes is presented in [24,25].

4   Evaluation and Discussion of Sustained Learning

As presented in [24], the students' software development knowledge increased during our course module by more than 50%. In this paper, we want to investigate the sustainability of this knowledge. First, we analyze how many students chose elective projects and bachelor theses in the area of mobile app development. Second, we repeated the final exam without further notice six month after the end of the tutorial. Third, we present how many students plan to continue a Master study in the area of software development at our university.

4.1   Evaluation

Students can choose elective projects in the 4th and 5th semester of their Bachelor study. These projects are in the areas of mobile app development, embedded systems, IT-security, network systems, and system configuration. Prior to this course module, mobile app development was not offered and only a few students selected software development projects. After introducing this module, from 24 students who attended the Mobile App Development module, 22 students chose their elective project in this area. Moreover, 20 out of these 22 students also selected these topics for their Bachelor thesis.

Six month after the end of the course tutorial, we repeated the final lecture exam without further notice. Thus, students could not prepare themselves for this second exam. We wanted to evaluate whether students could natively answer the questions addressing the levels *remember* and *understand* of Bloom's taxonomy after several months of working in the upper levels of Bloom's taxonomy— by engaging in their elective projects. The exam consisted of three parts. The first part included multiple choice questions on Android and Java fundamentals. In the second part, we asked for the completion of a fragment of code as well as finding errors in a given mobile app code. The third part consisted of software engineering tasks, where students had to write requirements and draw UML diagrams for a certain app idea. The results of the second examination show that no student received less than 80% of points and the majority of students received more than 90%. These results are even better than the results of the original examination.

Those students are currently in the final semester of their Bachelor study. Therefore, they partly started applying for a Master study. At our university we offer several Master studies which are possible after our Bachelor study and one of them is in the software development area—Software Design and Engineering. Although the process of applications at the time of writing this paper has just started, 14 of our mobile app development students already applied for this Master study.

4.2 Discussion

With the introduction and design of a new course module we have noticed a high learning progress in the area of software development by the students during the semester. Going one step further, we wanted to evaluate how sustainable this knowledge acquisition is. However, the sustainability of knowledge is difficult to measure. As explained in Sect. 4.1, we selected three different methods with the following indicators:

- the number of students that elected a subsequent project and bachelor thesis in the area of mobile app development,
- the results of an unannounced, repeated examination six months after the end of the tutorial, and
- the number of students who applied for a subsequent Master study in the area of software development.

The first indicator shows that more than 90% of students stayed in the software development area after the course module. Compared to previous years, with a few students in this area, this is a big change. Given that the introduction of our course module was the only modification of the study program, we explaine this change with the new course module. As described in [24], the students had the possibility of bringing own ideas for project topics in a hackathon event [22,29]. This contributed to the sustained motivation and engagement of students.

The results of the repeated examination confirmed that the students were able to keep the fundamental knowledge achieved during the course module. This is in contrast to bulimic learning [6,32], where the main purpose of the learning is only for the examination. This was possible by continuous assessments of the students supported by learning diaries [24].

The last indicator delivered results which show that the majority of our students are interested in a career in software development. The election of our Master study Software Design and Engineering demonstrates that they want to further deepen their knowledge and specialize in this area.

Summarized, all three indicators show a sustainable motivation of students to engage in the area of mobile app development. One reason for that may be addressing all the levels of the Bloom's taxonomy through our course module. Focusing on the lower levels only could result in a learning for examination only, while focusing on the upper levels could lead to frustration of students, as the fundamental knowledge is missing. Our course module combines several selected teaching methods to ensure continuous learning, including student involvement, goal-oriented teaching, collaboration and continuous assessment.

5   Conclusion

Longer lasting learning effects of particular course concepts are usually not evaluated, although sustainability of knowledge is vital to prepare students for their professional future. In this paper we evaluated sustained learning effects for students of an existing course module. In this course module we applied a mobile learning based teaching method mix in a collaborative and self-motivated learning environment. Students of this course module reflected high learning outcomes immediately after the course. To evaluate sustained learning, we selected three indicators and analyzed them six months after the end of the course. The results demonstrate a sustained competence growth in the area of software development among the students.

In our future work we will continue the observation of this group of students as well as evaluate sustained learning effects of other course concepts at our University.

References

1. Ally, M., Prieto-Blazquez, J.: What is the future of mobile learning in higher education? Revista de Unicersidad y Sociedad del Conocimiento (RUSC) **11**(1), 142–151 (2014)
2. Alrasheedi, M., Capretz, L.F.: Determination of critical success factors affecting mobile learning: a meta-analysis approach. CoRR, abs/1801.04288 (2018)
3. Beecham, S., Clear, T., Noll, J.: Do we teach the right thing? A comparison of global software engineering education and practice. In: Proceedings of the 12th International Conference on Global Software Engineering, Buenos Aires, Argentina (2017)
4. Bloom, B.S.: Taxonomy of Educational Objectives. Cognitive Domain. Addison-Wesley Longman Ltd, Handbook I (1956)
5. Briz-Ponce, L., Juanes-Méndez, J.A.: Mobile devices and apps, characteristics and current potential on learning. J. Inf. Technol. Res. **8**(4), 26–37 (2015)
6. Coffield, F.: Beyond Bulimic Learning: Improving Teaching in Further Education. Inst of Education Press (2014)
7. Esakia, A., McCrickard, D.S.: An adaptable model for teaching mobile app development. In: 2016 IEEE Frontiers in Education Conference (FIE), pp. 1–9 (2016)
8. Gannod, G., Burge, J., Helmick, M.: Using the inverted classroom to teach software engineering. In: 2008 ACM/IEEE 30th International Conference on Software Engineering, pp. 777–786 (2008)

9. Gary, K., Lindquist, T., Bansal, S., Ghazarian, A.: A project spine for software engineering curricular design. In: 2013 26th International Conference on Software Engineering Education and Training (CSEE T), pp. 299–303 (2013)
10. Ghezzi, C., Mandrioli, D.: The Challenges of Software Engineering Education, pp. 115–127. Springer, Heidelberg (2006)
11. Hernández, R.: Does continuous assessment in higher education support student learning? High. Educ. **64**(4), 489–502 (2012)
12. Hsu, Y -C., Ching, Y.-H.. Mobile app design for teaching and learning: educators' experiences in an online graduate course. Int. Rev. Res. Open Distance Learn. **14**(4), 117–139 (2013)
13. Jonsson, H.: Using flipped classroom, peer discussion, and just-in-time teaching to increase learning in a programming course. In: IEEE Frontiers in Education Conference (FIE), pp. 1–9 (2015)
14. Lage, M.J., Platt, G.J., Treglia, M.: Inverting the classroom: a gateway to creating an inclusive learning environment. J. Econ. Educ. **31**(1), 30–43 (2000)
15. Motiwalla, L.F.: Mobile learning: a framework and evaluation. Comput. Educ. **49**(3), 581–596 (2007)
16. Novak, G.M., Patterson, E.T., Gacrin, A.D., Christian, W.: Just-In-Time Teaching: Blending Active Learning with Web Technology. Prentice Hall (1999)
17. Pasamontes, M., Guzman, J.L., Rodriguez, F., Berenguel, M., Dormido, S.: Easy mobile device programming for educational purposes. In: Proceedings of the 44th IEEE Conference on Decision and Control, pp. 3420–3425 (2005)
18. Pears, A. et al.: A survey of literature on the teaching of introductory programming. In: Working Group Reports on ITiCSE on Innovation and Technology in Computer Science Education, ITiCSE-WGR 2007, pp. 204–223. ACM, New York (2007)
19. Ramos, P.R.H., Penalvo, F.J.G., Gonzalez, M.A.C.: Towards mobile personal learning environments (mple) in higher education. In: Proceedings of the 2nd International Conference on Technological Ecosystems for Enhancing Multiculturality (TEEM), Salamanca, Spain, October 2014
20. Roberts, J.B.: Handbook of Mobile Learning, Chapter Accessibility in M-Learning: Ensuring Equal Access, pp. 427–435. Routledge, New York (2013)
21. Robins, A., Rountree, J., Rountree, N.: Learning and teaching programming: a review and discussion. Comput. Sci. Educ. **13**(2), 137–172 (2003)
22. Rosell, B., Kumar, S., Shepherd, J.: Unleashing innovation through internal hackathons. In: IEEE Innovations in Technology Conference, pp. 1–8 (2014)
23. Sarrab, M., Elbasir, M., Alnaelic, S.: Towards a quality model of technical aspects for mobile learning services: an empirical investigation. Comput. Hum. Behav. **55**(Part A):100–112 (2016)
24. Schefer-Wenzl, S., Miladinovic, I.: Game changing mobile learning based method mix for teaching software development. In: 16th World Conference on Mobile and Contextual Learning mLearn 2017, (2017)
25. Schefer-Wenzl, S., Miladinovic, I.: Mobile distance learning driven software development education. In: 8th International Conference on Distance Learning and Education ICDLE 2017, (2017)
26. Staubitz, T., Klement, H., Renz, J., Teusner, R., Meinel, C.: Towards practical programming exercises and automated assessment in massive open online courses. In: 2015 IEEE International Conference on Teaching, Assessment, and Learning for Engineering (TALE), pp. 23–30 (2015)
27. Tao, Y., Liu, G., Mottok, J., Hackenberg, R., Hagel, G.: Just-in-time-teaching experience in a software design pattern course. In: 2015 IEEE Global Engineering Education Conference (EDUCON), pp. 915–919 (2015)

28. Tillmann, N., et al.: The future of teaching programming is on mobile devices. In: Proceedings of the 17th ACM Annual Conference on Innovation and Technology in Computer Science Education, ITiCSE 2012, pp. 156–161. ACM, New York (2012)
29. Topi, H., Tucker, A.: Computing Handbook. Information Systems and Information Technology, 3rd edn. CRCPress (2014)
30. Trainer, E.H., Kalyanasundaram, A., Chaihirunkarn, C., Herbsleb, J.D.: How to hackathon: socio-technical tradeoffs in brief, intensive collocation. In: Proceedings of the 19th ACM Conference on Computer-Supported Cooperative Work & Social Computing, CSCW 2016, pp. 1118–1130. ACM, New York (2016)
31. Venugopal-Wairagade G.: Study of a pedagogy adopted to generate interest in students taking a programming course. In: 2016 International Conference on Learning and Teaching in Computing and Engineering (LaTICE), pp. 141–146 (2016)
32. Zorek, J.A., Sprague, J.E., Popovich, N.G.: Bulimic learning. Am. J. Pharm. Educ. **74**(8) (2010)

Factors and Barriers in Training Financial Management Professionals

Petr Osipov, Julia Ziyatdinova[✉], and Elena Girfanova

Kazan National Research Technological University, Kazan, Russian Federation
posipov@rambler.ru, uliziatd@gmail.com, elena-girfanova@mail.ru

Abstract. The problems of training financial managers in Russian universities appear due to several reasons. Firstly, continuous changes in the business environment and complications of financial management in organizations. Secondly, the demand for creating new requirements for training financial managers under new educational standards. Thirdly, the uncertainty of employer approaches to assessment of key characteristics of a financial manager. Fourthly, differences in ideas about the competencies of a financial manager among employer and universities.

In the last years, the demand for the profession of a financial manager around the world has fallen in comparison with the early 2000s. In regard to this, we decided to analyze the training of bachelor and master degree students in financial management, and demand for them in the labor market. Despite the existing trend of a falling popularity of degree programs in economics and management, the market demand for the graduates is high.

The research goal was to find socially significant factors to influence the demand for financial managers. We used one of the Russian universities as a research environment. In total, 497 students participated in the survey, including graduate bachelor degree students and second year master degree students in management degree programs.

The survey gave the following factors to influence the demand for financial managers in the labor market: university prestige, individual perception of the educational process, student interest for the professional activities, and a degree in management for employment.

At the same time, the survey distinguished a number of barriers for personal fulfillment of the students, including: insufficient attention for non-major courses of the curriculum, absence of on-site practice to get a real life professional experience, and too much theory in major courses.

Keywords: Financial management · Manager · Administrator

1 Introduction

At present, there is a lot of research related to training efficient managers, including abilities and skills required to be employed and to reach the goals set.

The issues of socially significant factors to influence the demand for management graduates are studied in the works of R.A. Abramova and N.B. Zabaznov. They list the

© Springer Nature Switzerland AG 2020
M. E. Auer and T. Tsiatsos (Eds.): ICL 2018, AISC 916, pp. 167–175, 2020.
https://doi.org/10.1007/978-3-030-11932-4_17

following interrelated factors as the most important for a successful employment and practical application of skills: awareness of professional choices, individual perception of the educational process, teaching quality, and real-life practice [2, C. 79].

Another Russian researcher, T.I. Melnikova, in her turn, considers leadership qualities, adaptability to changing market environment and innovations, grounded decisions regarding finances as the socially important factors to influence the demand for financial managers [2, P. 123].

E.V. Yashkova and L.V. Lavrentieva add professional independence, competence, communicative and organizational skills to these factors [10, P. 37].

L.M. Spenser claims that the demand for financial managers is influenced by the following professionally important qualities: organizational skills (including the ability to resolve conflicts, plan and control the actions, self-learning and self-criticism, ability to inspire people) and entrepreneurial skills (including the potential to forecast and evaluate the circumstances, calculate the results and risks) [3, P. 159].

It is also important to develop self-learning skills for personal development of the students, and this can also be added as one of the factors [4].

The majority of the authors agree with the following approach to professional training of managers: a well-structured presentation of fundamental information, improvement of applied skills to help a young employee get adjusted to the working conditions and responsibilities, foresee and plan his own career progress in accordance with the goals set [5].

The analysis of the research papers on the issue of training managers showed the following gaps:

- between the growing and changing demands for the professional training of the future managers from the dynamically developing contemporary market and the real life priority in theoretical information in university curriculum without developing the innovative and entrepreneurial competencies of the future managers as the result of their professional education at a university;
- between the demand for a theoretical justification of the process of developing the innovative and entrepreneurial competencies of the future managers through implementing the pedagogical conditions of efficient training for the future professional managers and the serious differences in the methodological and theoretical approaches to characterization of these conditions;
- between the demand for activating the process of developing the innovative and entrepreneurial competencies of the future managers under condition of constant monitoring the success in their professional training and the lack of methods, technologies, teaching and learning materials for the educational process, aimed at developing the innovative and entrepreneurial competencies of the future managers.

2   What is a Contemporary Manager?

The collapse of the state-run economy resulted in new requirements to training managers and administrators with new skills and competencies for the market economy. Today, the Russian universities implement the US model of training managers, that is, the MBA degrees (Master of Business Administration). The MBA degree is recognized worldwide; the degree holders can work in middle or top management positions of any companies.

This model has been accepted in the Russian Federation due to several reasons. First of all, the United States is the country with the strongest economy and, therefore, highly qualified managers and administrators with hands-on experience.

Secondly, the US system of university education has the best experience in training managers and administrators with a well-developed infrastructure, teaching and learning materials and accreditation standards [6].

The importance of educating managerswas recognized in early 1930s when it became obvious that management is a profession, a field of knowledge, and a science. At the same time, managers and administrators were recognized as an influential social force.

The founders of management as a theory claim that a manager is a professional leader who is responsible for the organizational infrastructure and for the professional activities of the company.

Based on the results obtained through analyzing the model of training managers in the US system of education, including its conceptual, institutional, managerial, organizational and methodological characteristics, we distinguished the following features in the US system of educating managers:

- the importance of professional education of the future manager for his success and success of the organization as a whole;
- in accordance with the concept of a "professional manager", which serves as the theoretical basis for the system of education in management and business, the activities of a manager represent an independent profession with very high requirements for his level of education;
- the initial focus of business schools on demands from the individual and corporate consumers;
- the professional manager education includes two equally important constituents: academic skills (fundamental principles of management, new experience in management, instrument of scientific analysis) and practical skills (applied skills for making efficient decisions, monitoring and regulating the stresses and conflicts in marketing and time management).

3 Approach

The difficulties that the Russian Federation faces in social and economic spheres are related to its integration into the global context, strengthening the market economy, improving and reforming the system of college and university education, innovating the approaches to training educators.

Moreover, the situation in the country contributes to viewing the education in management and administration as a constituent of global learning and character education, the main link in this chain is a manager who matches the requirements of the new economy.

A number of experts predict a so-called 'administrative revolution', which will occur in the nearest future and will be one of the main reasons for Russia to get out of the economic stagnation phase. It is supposed that the key vectors of the Russian social and economic development will include a 'human factor' and professional development of managers [7].

Nowadays it is very important to provide companies and organizations with the new type managers and administrators who have a different model of thinking, who are mobile and capable of reaching the highest results in their profession due to well-developed innovation and entrepreneurship competencies.

This is a complex task for colleges and universities. The Russian government pays a special attention to the issues of training competitive professionals in management and administration.

The Federal law "On Education in Russian Federation" and National Doctrine for Russian Education Development till 2025 list the new requirements to these professions.

The state and society create an elevated demand for qualified professionals in management and administration who are capable of demonstrating creativity in performing their duties, successful and fast solutions of conflicts by using innovative methods aimed at developing entrepreneurship values.

A contemporary manager should be characterized by the following personal and professional qualities:

- entrepreneurial attitude;
- mobility;
- innovativeness;
- self-directed development;
- generating innovative business solutions.

The listed qualities make a manager competitive under the market economy conditions.

At present, the activities of the manager define the results of the organization. These activities include a number of managerial functions which characterize the innovative and entrepreneurial activities in making a rational choice based on economic competencies and similar competencies contributing to efficient solutions of conflicts [8].

The Russian education demonstrates a very rapid development in training managers and administrators. Unfortunately, very few degree programs meet the demands of the society.

As a rule, there is a superficial approach to western ideas and methods without a deep analysis of the educational content. The Russian degree programs simply borrow the pedagogical schemes and readymade theories which have very little in common with the real life managerial tasks.

We can speak about the demand for training managers according the new and improved contemporary standards contributing to a more efficient development of the managers' professional competencies. In this case, their productivity will grow and provide the necessary conditions for the success of the company and country as a whole [9].

There is no official unified ranking of the 'most demanded' professions; the demand for each of the professions depends directly on the situation in business and law.

The analysis of the labor market, however, can show the list of the most popular and demanded professions. According to this analysis, every profession is evaluated according to three criteria:

- the share in the job banks of the HR agencies;
- the demand characteristics;
- the increase in salaries in the last six months.

Managers are in the final list. For all the professions on this list, the employers are ready to pay today more than they paid yesterday, and tomorrow they will pay more than today.

The demand for managers can be explained by the shortage of highly qualified administrators in the country economy.

Many company leaders and owners start to understand that new managerial decisions are in demand for the development of the industrial production today. It is a difficult task to train a highly qualified manager.

The high demand for managers can be explained by the complex character of this profession, because managers can work in any field, including economics, business, industry, state administration, public organizations. Managers have always been in demand.

This qualification includes complex competencies in management, economics, marketing, advertising and psychology. In some cases, managerial positions are occupied by economists and lawyers, who, unfortunately, lack knowledge in administration [6].

4   Characteristics of Russian Degree Programs in Management

The specific characteristics of the Russian education in management include:

- prevailed classical approach to university education based on the German university model with the priority of the state order in developing the educational curriculum; the demands of the employers are often ignored;
- the curricula of BA and MBA degree programs often include courses which do not correlate with the practical activities of the future manager, and, as a result, the student does not gain the knowledge and skills which will be necessary in his future career and will help him to use his potential;
- a demand for creating new models of training a modern professional manager taking into account the experience of the US universities and the Russian national traditions of developing state educational standards;
- a demand for improving the master degree programs in the Russian education and developing programs similar to MBA as the most dynamic, demanded, aimed at a specified consumer;

- movement to the new quality level of training managers based on the lifelong learning model;
- transformation of education in management as the key factor contributing to professional career development;
- application of new technologies based on learning through reflective experience and activities to unite theoretical and practical learning focusing on professional realities which the students will face after graduating from the university;
- enhanced role of independent work, including individual work and team work, in the educational process;
- development of federal state educational standards of the new generation where different degree programs can be merged;
- development of federal state educational standards in accordance with the demands of the employers, including private employers and public organizations which ccan create rules and legislations for the educational system aimed at training competitive professionals who can be greatly demanded be the modern market economy [3].

The analysis of the Russian federal state educational standards shows that the standards of the 1st and 2nd generations were aimed at changing the contents of education; at the same time the education has a unified space and guarantees the minimum volume of the necessary information.

The standards of the 3rd generation, however, give the leading role to the results of the education which correlate to the competence based model of the university graduate.

Although these standards identify the entrepreneurial competencies of the professional managers, they are not enough for training an efficient manager, who can be educated only in case when the professional education is aimed at developing his innovative and entrepreneurial competencies as a system [4].

5 Experiment

We aimed at investigating the socially important factors which influence the demand for the graduates in Financial Management degree programs in a Russian technological university.

In recent years, we observed a decreased demand in degree program in Financial Management.

Therefore, it was necessary to investigate the trends in employment of the program graduates and analyze the curriculum and contents of the courses taught.

Despite the decreased popularity of the degrees in economics and management in general, the professionals in this sphere are still in demand.

The market imposes very high requirements for them in their constant professional development.

In order to identify the factors and barriers of training managers we considered it important to do the following:

- to specify the approaches to definition of characteristics for a financial manager;
- to define the quality characteristics of the labor market in this field (management);
- to develop a matrix of financial manager professional competencies;
- to develop a model to evaluate the demand for professionals in financial management.

These issues can be solved only if we accept a certain approach to education quality where education is considered as a system of hierarchy of socially important characteristics corresponding to the demands and interests of a person, the society and the country as a whole.

When analyzing the socially important factors to influence the demand for the graduates of the degree program in Financial Management at a technological university it is important to consider the following.

The paper uses a practically oriented approach, where the authors analyze the 2015–2017 data from one of the engineering universities which runs financial management degree programs.

At the same time, the authors use a regional approach, investigating the practical experience of a region to distinguish the social and industrial development factors which influence the demand for the financial managers in the industry and other businesses.

The authors compared the numbers of the students in the Financial Management Bachelor degree program during the three academic years between 2015 and 2017.

In the 2016–17 academic year, there was a 10% drop in the total number of students, while the 2017–18 showed a sharp increase by 16%.

These numbers prove that the demand for the financial management program is very unstable and depends on many external and internal factors.

These factors were distinguished based on the analysis of the theoretical literature on the topic, including peer reviewed papers and results of practical investigations of the other researchers.

After analyzing the theoretical prerequisites, the authors developed and conducted a survey for the students who study in the Bachelor and Master degree programs of engineering universities for the purpose of finding their attitudes towards the demand for their profession and their career expectations.

The survey included a number of questions related to the obvious advantages and disadvantages of studying financial management in a degree program at an engineering university, expected career prospects and competitiveness criteria in the labor market. The questionnaire was tested for validity and reliability and proved its efficiency.

The survey was conducted anonymously; around 200 Bachelor and Master degree students participated in it.

The results of the survey allowed the authors to give a better definition to the demanded competencies of a financial manager, characterize the labor market in financial management, to model and predict the demand for financial managers in the region.

6 Results

The results of the research gave the following outcomes:

- Detailed approaches to defining the characteristics of a financial manager.
- Determined quality characteristics of the labor force market in management.
- Developed matrix of financial manager professional competencies.
- Developed model of evaluating the demand for financial managers.

The study showed factors and barriers for the competitiveness of the engineering university graduates in Financial Management degree programs in the regional labor market.

The authors distinguished the following factors contributing to the competitiveness:

- a positive university image, which gives competitive advantages during employment;
- individual perception of the educational process through the use of interactive study tools;
- interest of the students to work in this professional sphere; over half of the university graduates admit that this education helped them to build their careers, at the same time, many young people claim that real-life experience is very important;
- a degree in management serves as a catalyst for a career success.

The authors also define a number of barriers for the competitiveness of the financial managers which included:

- poor training in some of the major disciplines; the first years or education focus on courses to develop general cultural and professional competencies, the courses related directly to management appear only in the third year of the curriculum;
- lack of industrial experience in situ where the future managers could get real-life practice in their profession;
- overwhelming theory and deficient practice (especially in Master degree programs);
- very few opportunities to show presentational skills and to experiment in applying the competencies.

Thus, the outcomes of the survey proved the results of the theoretical investigations and showed that it is necessary to implement practice oriented, problem based and student centered approaches in the university education, which is of primary importance for degree programs in financial management.

At the same time, the survey gave a new picture of a financial manager with new characteristics and competencies demanded, including flexible thinking under conditions of uncertainties in the environment.

7 Conclusions

The study conducted allows the authors to draw the following conclusions: the school graduates choose a university degree program based on the career opportunities, including the status of the profession and labor market demand for it.

Therefore, degree programs in financial management should be very flexible for the changing environment, and finely tuned for the leading industries which is possible in the engineering universities.

If an engineering university aims at improving the competitiveness of its graduates in financial management degrees, the following steps should be taken:

- reviewing the existing degree programs and making them more practically oriented;
- introducing new disciplines into the degree programs, focusing on problem based learning and practical skill development;
- developing leadership and decision making qualities of the financial management graduates so as to help them adapt to the new market demands.

Therefore, by improving the factors and eliminating the barriers in training financial management professionals, an engineering university can increase the competitiveness of its degree programs and its graduates thus contributing to the prosperity of the region.

References

1. Abramov, R.A., Zabaznova, N.M.: Evaluation of the vocational relevance of graduates in higher education. Sociol. Educ. **11**, 79–87 (2016)
2. Melnikova, T.I.: Competence approach to the training of financial managers. Education, 123–128 (2015)
3. Spencer, L.M: Competences at work [Text]. In: Hippo, M. (ed.) Models for maximum performance. Per. with English, p. 384 (2005)
4. Osipov, P.N., Ziyatdinova, J.N.: Pedagogical enhancement of students' personal development in the course of engineering pedagogy. In: Proceedings of the 2014 International Conference on Interactive Collaborative Learning (ICL2014), Dubai, UAE, pp. 307–310 (2014)
5. Yakimova, Z.V., Nikolaeva, V.I.: Assessment of competencies: professional environment and university. Mod. Pract. 13–18 (2015)
6. Abramov, R.A.: Problems of administrative specialties VPO in economic higher education institutions of the Russian Federation. Policy Soc. **9**, 1198–1209 (2015)
7. Kolchina, V.V.: The state of the problem and the conditions for the formation of innovative entrepreneurial skills in pedagogical theory and practice. In: Herzen, A.I. (ed.) News of the Russian State Pedagogical University, 12(91), pp. 196–201 (2009)
8. Ho John, J.: Financial Director of the New Era. Per. with English, p. 304. Vershina (2007)
9. Kolchina, V.V.: Intensification of the process of formation of innovation-entrepreneurial competence among university students - future managers. Sci. Sch. **3**, 78–82 (2013)
10. Kolchina, V.V., Sergeeva, M.G.: Modeling the process of formation of innovation-entrepreneurial competence of future managers. Sci. Rev. Humanit. Res. **9**, 6–12 (2016)
11. Yashkova, E.V., Lavrentieva, L.V.: Professional skills and abilities of a modern manager. Prospect. Sci. Educ. **6**, 37–40 (2017)

Designing Online Environment for Collaborative Learning in a Scientific Community of Practice

Joanna Jesionkowska[(⊠)]

University of Liege, Liège, Belgium
j.jesionkowska@doct.uliege.be

Abstract. The motivation for this research was to find a way to sustain the collaboration between scientists engaged in a training program. The collaborationstarted face to face, but it is envisaged that the training program and the trainers' community will transform to an online format. According to the needs criteria can be proposed to design an online space, that would answer the needs of the members of the community to sustain the process of collaborative learning. Data was gathered from content analysis, observations, interviews, focus groups and online surveys with community coordinators and community members. The model of the community was developed which led to a set of recommendations for an online community of practice. Developing an online community of practice between the members of the community could be an answer to their learning needs and could improve the culture of knowledge sharing and collaboration.

Keywords: Collaborative learning · Situated learning · Community of practice · Metrology in chemistry · Online community

1 Introduction

Science nowadays is highly collaborative. Inter-organizational collaboration is increasing in frequency and importance [1]. Scientific collaborations span fields, institutions, sectors, and countries [2] and have the potential to solve complex scientific problems. Collaboration in science has led to scientific breakthroughs that would otherwise not have been possible [3]. Increasingly, public and private research funding agencies require interdisciplinary, international and inter-institutional collaboration.

Because of their increasing number and importance, it is highly relevant to understand the process of scientific collaborations. The potential of scientific collaboration can be even more evident in European projects, where researchers from different countries come together across language and cultural barriers, with co-operation encouraged through shared budgets and resources.

© Springer Nature Switzerland AG 2020
M. E. Auer and T. Tsiatsos (Eds.): ICL 2018, AISC 916, pp. 176–185, 2020.
https://doi.org/10.1007/978-3-030-11932-4_18

2 Objective

There are many studies analyzing collaboration [4, 5], often they analyze the authorship of papers and the benefits of collaboration for the impact of these research papers (like in: [6]). In this study the focus is on one aspect of scientific collaboration: collaborative learning. It is a study of a group of scientists, a community involved in a training program in the field of chemistry, namely metrology in chemistry. The focal point of the study is on the processes of collaborative learning and collaborative knowledge production in this group.

The training program has flourished for more than a decade but currently it is winding down. The value to the organization and the value of membership isdiminishing as more and more laboratory practitioners around Europe were trained and there is no new training material created. The idea to keep the community alive and sustain the collaborative learning is to transform from a traditional mode of working to online environment. The group members are facing this change whereby they are challenged to evolve from face-to-face to an online community. The study aims at defining the conditions to make such a change successful. Reader will find a set of recommendations that were created for this community to support and sustain the process of collaborative learning in the community.

3 Literature Review

3.1 Scientific Collaboration

The challenges in today's science are often too broad for a single person to cope with. The global scale of many large-scale projects or 'big science' forces scientists to start broader and more inclusive international cooperation [7]. Collaboration in science has led to scientific breakthroughs that would otherwise not have been possible [3]. Increasingly, public and private research funding agencies require interdisciplinary, international and inter-institutional collaboration. The increased number of team science projects has been driven by a variety of factors [8]:

- growing interest in scientific problems that span disciplines,
- advances in communication and transportation technologies that make remote collaborations easier to sustain,
- government policies that encourage collaboration, especially between universities and companies.

The increasing networking on different scales from global to local is having a profound effect on learning and teaching. It makes new forms of collaborative and personalized learning experiences a reality [9]. Learners shift between formal, non-formal, and informal learning. They come together in different social settings and communities.

3.2 Community of Practice

A community of practice (CoP) is a group of people who share a common concern, a set of problems, or interest in a topic and who come together to fulfill both individual and group goals [10]. CoPs often focus on sharing best practices and creating new knowledge to advance a domain of professional practice. According to Wenger [10], people who want to participate in CoPs get ready to share their knowledge, sharpen their expertise, build up interpersonal networks and pursue their interest. Different roles appear among the CoP members and, in certain contexts, the presence of an animator is necessary to facilitate their activities. A CoP emerges spontaneously or/and through a participatory design process [11]. Communities have lifecycles—they emerge, they grow, and they have life spans [12]. Wegner [13] described that CoPs continually evolve through five stages: potential, coalescing, maturing, stewardship and transformation. Communities of practice can generate new ideas, solve problems, promote the spread of best practices and develop people's skills. Although they are self-organizing and thus resistant to supervision and interference, they do require specific managerial efforts to develop them and integrate them into an organization. Only then can they be fully leveraged.

3.3 Situated Learning

The concept of a community of practice is based on social interaction where participants create an environment whereby feelings of isolation are reduced as a strong sense of community develops [14]. It is through this shared sense of community that sharing and deep approaches to learning take place [15]. Viewing learning as situated within cultural activities is the central focus of the situated learning theory. It is a theory on how individuals acquire professional skills, extending research on apprenticeship into how legitimate peripheral participation leads to membership in a community of practice. Situated learning takes as its focus the relationship between learning and the social situation in which it occurs [16]. The emphasis is away from the individual and toward the social setting and the collections of people within such settings.

Lave [16] argues that learning is situated; that is, as it normally occurs, learning is embedded within activity, context and culture. It is also usually unintentional rather than deliberate. Knowledge needs to be presented in authentic contexts — settings and situations that would normally involve that knowledge. Social interaction and collaboration are essential components of situated learning — learners become involved in a 'community of practice' which embodies certain beliefs and behaviors to be acquired. Communities of practice are natural learning groups found in many contexts. Each individual is generally involved in a number of them through their occupation, school, home, community, and leisure activities.

3.4 Collaborative Knowledge Building

Knowledge can be created and shared at the same time. Scardamalia and Bereiter [17] introduced the concept of collaborative knowledge building. The group needs to engage in thinking together about a problem or task, and produce a knowledge artefact (such as a verbal problem clarification, a textual solution proposal or a more developed theoretical inscription) that integrates their different perspectives on the topic and represents a shared group result that they have negotiated. The process of collaborative knowledge creation takes place in several interconnected contexts: local social environment, institutional learning environment, created by the researched higher education institutions, and online, in virtual learning and sharing environments [18].

3.5 Scientific Community

The basic activities of scientists are fundamentally social undertakings [19]. Scientific 'truth' is defined by a consensus of a scientific community. For instance, they jointly make judgments about what papers will be published and what grants will be funded. A scientific community consists of various researchers who have different specialities, each group member has various backgrounds and usually participates in various scientific communities [20].

Scientific users were early adopters and promoters of many of the technologies that lie at the basis of long-distance collaboration, like email and the Internet [21]. That would suggest that scientific collaboration in the digital age should flourish. But research shows that many computer-supported scientific collaboration environments are not living up to their expectations [22]. Only some of them manage to overcome the scaling-up problems, sustain long-distance participation and initiate breakthrough science.

3.6 Collaborative Learning in Scientific Community

Learning is an integral component of scientific collaboration, especially interdisciplinary [23, 24]. Collaboration is viewed as "one of the most effective forms of knowledge transfer" [25, p. 38]. Scientists need to learn from each other to develop a common working understanding regarding their research project and how they can integrate their specialized knowledge to create new knowledge.

Learning is not traditionally discussed or included in research proposals as a research activity [26]. Scientists may recognize that knowledge (both explicit and tacit) is exchanged among collaborators. Learning can be a motivation to initiate collaboration, but it is in the sustainment stage that it may be most challenging [1]. It has become widely accepted, in recent years, that students learn more effectively when they collaborate with others and collaborative learning has positive effects on students [27]. Online collaborative learning can promote the development of self-efficacy, enhance learning motivation and active learning attitudes and lead to improved learning outcomes [28].

3.7 Online Communities

Similarly, online communities can provide a valuable form of professional learning and support. They allow members to gain equitable access to human and information resources that may not be available locally or that budget constraints might limit [29]. They are seen as mitigating the barriers of distance, time and professional isolation, increasing opportunities for knowledge sharing, making interactions more visible, sustaining these interactions and extending their reach [30]. They provide informal professional learning opportunities as well as ongoing support for changes in professional practice. Various types of knowledge capital—learning capital, human capital, social capital, reputational capital and tangible capital—are produced through members' participation. But even when online communities are created with great care and planning, success is often elusive [29]. Overcoming these challenges could be the key for CoPs to survive and to serve organizations in a better way.

3.8   Designing Sustainable Community of Practice

Because communities of practice are voluntary, what makes them successful over time is their ability to generate enough excitement, relevance, and value to attract and engage members. The goal of community design is to bring out the community's own internal direction, character, and energy [13]. Designing for aliveness requires a different set of design principles. The goal of community design is to bring out the community's own internal direction, character, and energy. Hearn and White [31] described criteria to support CoPs:

- Foster instead of control
- Focus on facilitation not technology
- Understand members' needs and capacities
- Recognise the two faces of communities
- Create conditions for two-way learning
- Balance participation and reification
- Be sensitive to the different stages of CoP development.

Using this guide allows to create a community that will be self-sustaining and beneficial for the members and the organisation.

4   Methodology

4.1   Research Design

An exploratory study has been chosen as methodology, whereby the exploration occurs within the context of a case. A multi-method approach was used combining both qualitative and quantitative methods. The analytical approach chosen was sequential data analysis. In this research, the results from one method were used to develop and inform the other method. Data was gathered from content analysis, observations, interviews, focus groups and online surveys with community coordinators and community members. The integration of the qualitative and quantitative findings took place during the analysis and interpretative stages of the research.

4.2   Participants

Participants in this study were trainers and candidate trainers. Participation in the study was completely voluntary. Interviews were proceeded with a discussion with people involved in the program at JRC-Geel. There are 6 people engaged in one way or another in the organization of the events. First interview was an unstructured interview, next five interviews were semi-structured. Each interview was about 30 min. The interviews were digitally recorded and transcribed and sent to the interviewees to be checked and corrected.

After deeper analysis of the data collected from these interviews a questionnaire for trainers (authorized and non-authorized) was created. It was an on-line survey, invitation was sent to all the trainers from the database and to the candidates who attended a training to become trainer. 96 out of 156 trainers answered.

The last group of participants were trainers who participated to the 2016 convention. There were about 30 persons, a mix of new and experienced trainers.

4.3   Data Collection Instruments

For the purposes of this research several instruments were developed. For the qualitative part of the study an interview guide was produced. For the quantitative part, a set of online questionnaires were developed.

The interview guide was built as the basis of an unstructured interview and had a short set of open questions about the main characteristics of the community of practice. These questions focused on the desired purpose and outcome for the community: Who are the participants and how they can be described? The type of participant interactions: What activities must the group undertake to achieve its goals?

The online questionnaire was built based on the findings from previous steps. Questions were organised into four sections: General information, Experience with professional collaborations, Experience with TrainMiC and TrainMiC's future.

4.4   Defining Conditions for the Transformation

A proposal is made here for the conditions that would enable the change from face to face to online community and would sustain collaborative learning of the community members. The investigation was basing at the guide described by Wenger, McDermott and Snyder in their book *Cultivating communities of practice: a guide to managing knowledge* [13]. Taking into account the needs of the community members a set of recommended conditions was examined:

- community's members roles, their expertise (expert, novice…) in the topic,
- a technological solution to communicate, share and collaborate that could generate interactions, energy and engagement,
- a mission and vision, alignment of interests, the major topic areas for community content and exploration, potential categories of activities that community members are likely to want to carry out in the community, communication paths.

5 Results

The results indicated that studied group is a community of practice. The community, domain and practices were identified, as well as the learning cycle of the community. The model of the community was developed which led to a set of recommendations for an online community of practice. The community allows learning and getting new knowledge in the topic of metrology in chemistry. Participants share best practice ideas, which are more genuine and credible than if coordinators would simply impose these solutions. The emphasis in this community is on better understanding and constant improving the knowledge about metrology in chemistry. The community's practice should be seen as learning opportunity. Community should also help and support new members to develop their skills to train others in professional and practical way, so they can use their knowledge and share with others and spreading the idea of metrology in chemistry and this way improving the global knowledge and raising the quality of chemical measurements.

5.1 Recommendations

Basing on the learning cycle of the community a set of conditions to transform the community into an online format was proposed. Those conditions describe the new community that would best sustain collaborative learning and collaborative knowledge creation and sharing for community:

1. **Activities**:

   - sharing and exchanging information on existing and new technical/scientific developments in the field of metrology in chemistry,
   - building a consensus opinion on metrology issues starting from scientific-technical interaction,
   - organising scientific events and seminars,
   - providing training and set up sustainable learning activities,
   - influencing educators and organisations,
   - promoting the harmonisation of measurements and co-ordinate quality assessment activities,
   - performing method development and validation,
   - harmonising practice,
   - participating to standardisation activities,
   - developing common research projects and pilot studies.

2. **Learning, knowledge sharing and communication**:

   - external resources (like publications, reports, standards),
   - work related examples,
   - real-life stories,
   - laboratory exercises,
   - applications of methods and standards,
   - case studies, interpretation of the standard,

- quality assessment and assurance,
- interlaboratory comparison exercises between the members of the community.

3. **Communication and interactions** between the members of the community (and with the content of the community):

- email, telephone, web conferencing to stay in touch in a daily basis,
- newsletter, to keep its members informed on the current activities of the community,
- discussion board,
- training material repository.

4. **Roles and social structure:**

- community leader,
- management board,
- editorial board,
- quality manager,
- community facilitator,
- community administrator,
- community members.

6 Conclusions

This is an example of a community that was built face to face which is changing format to an online space. Developing an online community of practice between the members of the community could be an answer to their learning needs and could improve the culture of knowledge sharing and collaboration. The model of the community, its learning cycle and the set of recommendations could be used by other scientific communities to sustain the collaborative learning of their members.

References

1. Sonnenwald, D.H.: Scientific collaboration. Ann. Rev. Inf. Sci. Technol. **41**, 643–680 (2007)
2. Walsh, J.P., Maloney, N.G.: Collaboration structure, communication media, and problems in scientific work teams. J. Comput. Mediat. Commun. (2007)
3. Cooke N.J., Hilton M.L.: Enhancing the Effectiveness of Team Science. National Research Council, National Academies Press, Washington DC, US (2015)
4. Canals, A., Ortoll, E., Nordberg, M.: Collaboration networks in big science: the ATLAS experiment at CERN. El Profesional de la Información **26**(5), 961–971 (2017)
5. González-Alcaide, G., Gómez-Ferri, J.: The tenets of scientific collaboration: rules and recommendations. Curr. Sci. (00113891) **113**(8), 1505–1506 (2017)

6. Keck, A.-S., Sloane, S., Liechty, J.M., Fiese, B.H., Donovan, S.M.: Productivity, impact, and collaboration differences between transdisciplinary and traditionally trained doctoral students: a comparison of publication patterns. PLoS ONE 12(12), e0189391 (2017). https://doi.org/10.1371/journal.pone.0189391

7. Jirotka, M., Lee, C.P., Olson, G.M.: Comput. Support. Coop. Work 22, 667 (2013)

8. Mindeli, L., Markusova, V.: Bibliometric studies of scientific collaboration: international trends. Autom. Doc. Math. Linguist. 49(2), 59–64 (2015)

9. Conole, G., Klobučar, T., Rensing, C., Konert, J., Lavoué, É.: 10th European Conference on Technology Enhanced Learning. In: Proceedings EC-TEL 2015, Toledo, Spain (2015)

10. Wenger, E.: Communities of Practice. Learning, Meaning, and Identity. Cambridge University Press (1998)

11. Ashwin, P.: Analysing teaching-learning interactions in higher education: accounting for structure and agency. Br. Educ. Res. J. 36(1), 171–172. ISSN 0141-1926 (2009)

12. Kaplan S., Suter. V.: Community of Practice Design Guide, A Step-by-Step Guide for Designing & Cultivating Communities of Practice in Higher Education. Darren Cambridge (2005)

13. Wenger, E., McDermott, R., Snyder, W.: Cultivating Communities of Practice: A Guide to Managing Knowledge. Harvard Business School Press (2002)

14. Farnsworth, V., Kleanthous, I., Wenger-Trayner, E.: Communities of practice as a social theory of learning: a conversation with Etienne Wenger. Br. J. Educ. Stud. 1–22 (2016)

15. Wenger-Trayner, E., Wenger-Trayner, B.: Learning in Landscapes of Practice: Boundaries, Identity, and Knowledgeability in Practice-Based Learning, pp. 13–29 (2015)

16. Lave, J., Wenger, E.: Situated Learning: Legitimate Periperal Participation. Cambridge University Press, Cambridge, UK (1990)

17. Scardamalia, M., Bereiter, C.: Knowledge building. In: Guthrie, J.W. (ed.) Encyclopedia of Education, 2nd edn. Macmillan Reference, New York, USA (2003)

18. Mikhaylov, N., Fierro, I., Beaumont E.: The influences of capital development strategies choice on international management students' collaborative knowledge creation. In: INTED2016 Proceedings, Turkey and Ecuador, pp. 7682–7691 (2016)

19. Collins, H.M.: The meaning of data: open and closed evidential cultures in the search for gravitational waves. Am. J. Sociol. 104(2), 293–338 (1998)

20. Kienle, A., Wessner, M.: Principles for cultivating scientific communities of practice. In: Van Den Besselaar, P., De Michelis, G., Preece, J., Simone, C. (eds.) Communities and Technologies. Springer, Dordrecht (2005)

21. Bos, N., Zimmerman, A., Cooney, D., Olson, J.S., Dahl, E., Yerkie, J.: From shared databases to communities of practice: a taxonomy of collaboratories. In: Olson, G.M., Zimmerman, A., Bos, N. (eds.) Scientific Collaboration on the Internet. MIT Press, Cambridge, MA (2008)

22. Muff, K.: The Collaboratory: A Co-Creative Stakeholder Engagement Process for Solving Complex Problems. Greenleaf Publishing, Sheffield (2014)

23. Maglaughlin, K.L., Sonnenwald, D.H.: Factors that impact interdisciplinary natural science research collaboration in academia. In: Ingwersen, P., Larsen, B. (eds.) Proceedings of ISSI 2005, pp. 499–508. Karolinska University Press, Stockholm (2005)

24. Solomon, N., Boud, D., Leontios, M., Staron, M.: Researchers are learners too: Collaboration in research on workplace learning. J. Work. Learn. 13(7/8), 274–281 (2001)

25. Lambert, R.: Lambert Review of Business-University Collaboration, Final Report (2003)

26. Davenport, T.H.: Process management for knowledge work. In: Handbook on Business Process Management 1. Springer (2005)

27. Du, J., Ge, X., Xu, J.: Online collaborative learning activities: the perspectives of African American female students. Comput. Educ. 82, 152–161 (2015)

28. Liao, Y., Huang, Y., Chen, H., Huang, S.: Exploring the antecedents of collaborative learning performance over social networking sites in a ubiquitous learning context. Comput. Hum. Behav. **43**(11), 313 (2015)

29. Booth, S.E., Kellogg, S.B.: Value creation in online communities for educators (Report). Br. J. Educ. Technol. **46**(4), 684 (2015)

30. Cranefield, J., Yoong, P., Huff, S.: Rethinking lurking: invisible leading and following in a knowledge transfer ecosystem. J. Assoc. Inf. Syst **16**(4), 213 247 (2015)

31. Hearn, S., White, N.: Communities of practice: Linking knowledge, policy and practice. ODI Background Note (2009)

Scalability and Performance Testing of an E-Learning Platform Integrating the WebRTC Technology: Scenario "Authentication"

Cheikhane Seyed[1]([✉]), Jeanne Roux Ngo Bilong[1], Samuel Ouya[1], Mohamedade Farouk Nanne[2], and Ibrahima Niang[3]

[1] Laboratory LIRT, Ecole Supérieure Polytechnique, University Cheikh Anta DIOP, Dakar, Senegal
{ch.hamod, samuel.ouya}@gmail.com, jeanneroux@yahoo.fr
[2] Department Computer Engineering, University Nouakchott AL-AASRIYA, Nouakchott, Mauritanie
mohamedade@gmail.com
[3] Direction Informatique et des Systèmes d'Information, University Cheikh Anta DIOP, Dakar, Senegal
ibrahimal.niang@ucad.edu.sn

Abstract. African countries, which are in the process of economic emergence, are placing increasing emphasis on hybrid education systems that promote socioconstructivism between members of distance education. To support this dynamic, we have set up an E-learning platform based on WebRTC technology. This platform operates in an intranet, and allows users to communicate in real time. The purpose of this paper is to test the scalability and performance of said platform for the "authentication" scenario, in order to ensure its functional capacity, stability, availability and latency.

To achieve our goal, we installed JMeter Apache, which is an open source Java application designed as a graphical interface. It allows to measure and analyze the functional capabilities and performance of a web application. The results of the various tests show that our application is efficient because it reacts to a scalability with a number of users much higher than that which should contain minimally a virtual class. In addition, multiple users authenticate simultaneously to reach the virtual classroom without server connection failure.

Keywords: Load testing · Performance testing · E-learning platform · JMeter apache

1 Introduction

The complexity of distance learning environments, integrating new technologies such as WebRTC for real-time audio/video communications management is becoming more and more a fact. In this context, researchers [1] have designed an e-learning platform dedicated to the context of African universities allowing synchronous learning between teachers and students via virtual classes.

© Springer Nature Switzerland AG 2020
M. E. Auer and T. Tsiatsos (Eds.): ICL 2018, AISC 916, pp. 186–193, 2020.
https://doi.org/10.1007/978-3-030-11932-4_19

With regard to the weak network infrastructure deployed in African universities, the regulation often limits the number of users connected in a virtual classroom to 30 to avoid problems related to server overload and bandwidth saturation.

Today, the number of students wishing to take distance courses via the platform is gradually increasing, hence the need to evaluate the performance of the platform in order to deduce the ability to assume a higher number of users. The question we ask ourselves in this place, is how to minimize or avoid platform failures, due to problems related to system performance and scalability.

According to the authors of the articles [2, 3], To evaluate a system and deduce the performance reflecting the exact reality, it takes weeks or even years of observing the system in order to accommodate a large amount of information.

According to the author of the article [4], most teams struggle because of the lack of professional performance testing methods, while ensuring problems with availability, reliability and scalability, when deploying their application on the real world.

Because of these two discussions, we chose the unique simulation method using the Apache JMeter tools. Our simulation environment is based on five iterations whose corresponding samples are 10, 50, 100, 200 and for a duration of 30 s.

We analyze the resource data obtained by this experiment, then calculate the average query response time, deviation, max and min response times for the total number of pages visited for each iteration.

The throughput and response speed calculated in the CSV files show that the system is reliable despite interaction with the WebRTC server. In addition, the CSV files show that the platform supports up to 200 uses without any error and therefore an increase of about 600% of the initial load.

The remainder of this paper is organized as follows: Sect. 2 describes the JMeter technology and e-learning platform integrating WebRTC, Sect. 3 proposes the architecture of the solution, and Sect. 4 shows the test results and analysis. In Sect. 5, a conclusion and perspectives for future work.

2 Related Work

2.1 Apache JMeter Tool

JMeter is a java-based application, designed to test client/server applications. The choice of this tool is justified by the fact that it is an open source application that is easy to learn. It is versatile of the fact that it allows testing not only web applications, but also FTP applications. In addition, it is scalable with its modular design that allows components to be merged for large-scale test management [5]. It is user-friendly, highly scalable through an API provided and is by far, one of the most used open-source, freely distributed testing applications that the net can offer [6, 7]. Its use consists first in establishing a test scenario describing a visitor's navigation session, then in having this scenario repeated as many times and by as many simultaneous visitors (or threads).

2.2   Presentation of the Platform Integrating WebRTC

To improve the quality of learning, several works were carried out for the implementation of e-learning platforms integrating the WebRTC. This is the case of the work of the authors of the article [8], which have set up a platform on which users will be able to share files Pear-to-Pear, reduce server load, transfer files by fragment to improve the transfer rate decrease the waiting time for users. The authors of article [9] have set up a platform with collaborative content sharing and streaming video technology, using the WebRTC technology. The implementation of this platform is to demonstrate the reliability of the proposed system. In the same logic, the authors of article [10] set up a Model platform to create virtual classes incorporating video, audio and instant messaging. The proposed test platform offers teachers and students, through a browser, the ability to connect with the aim of communicating by chat, audio, video and also transfer files or share their computer screen. For this, it was necessary to implement a WebRTC signaling server to manage real time and use the three WebRTC API: MediaStream API for acquisition and synchronization of audio and video streams, the PeerConnection API for communication between users' browsers, the DataChannel API for file transfer, the Chat and screen sharing [1].

3   Architecture of the Test Environment

The proposed architecture in Fig. 1 represents the issue and the environment of our test, consisting of several components, which mutually interact. The proposed solution explains the process of test plan. This process shows the interactions between the various components of JMeter and tested application.

The following are detailed descriptions of the development of a test plan.

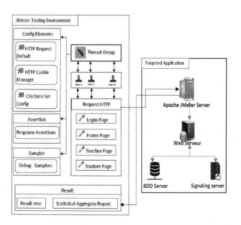

Fig. 1. Architecture of the test environment

- JMeter mainly runs with two components which are: the test plan section is a working space for elements constituting the test plan and execute the Workbench section or work plan, which is an area of unregistered temporary work.
- The Thread Group is configured to set the number of virtual users, iteration and duration of the Ramp up (charge-up time). It is the basic element of our "authentication" scenario. The Thread group evokes the configuration item "parameter http" by default that is responsible for the JMeter connect to the application server through the apache server (one entry in this component: the IP address, port and the path to the application).
- The Thread Group uses the HTTP Request SAMPLERS to run the performance test scenario. This launches the first page of "login" for user authentication. These users have sequentially connection information stored in a CSV file.
- The HTTP REQUEST SAMPLERS launches the redirect page with settings of login and password, using configuration items namely, the CSV data source, the cookie manager, the assertion response to the execution of treatments subsequent, Debug and samplers to check intermediate data values.
- The HTTP REQUEST SAMPLES load the user data file stored locally in CSV form. The HTTP REQUEST SAMPLES opens the corresponding home page depending on the type of users connected (Teacher and Student).
- By running the system under test, the application receives the requests, verifies the authenticity of the user returns the status (successful or unsuccessful) to JMeter via the assertion response component. Finally, JMeter generates results using the listeners.

4   The Test Results

4.1   Presentation of Results

This section presents the results of several test iterations of authentication scenario on our platform integrating WebRTC. These performance tests were conducted on the unique configuration of the JMeter environment.

- Iteration 1: is to increase the test sample to 10 virtual users, while maintaining the parameters set in the first iteration. The execution of the test was successful. See Fig. 2.

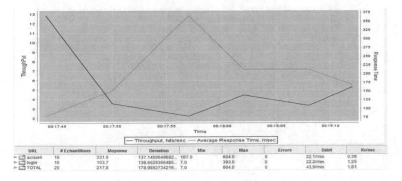

Fig. 2. Test plan for 10 virtual users

- Iteration 2: was successfully performed for 50 users while maintaining the same parameters set in the first iteration. See Fig. 3.
- Iteration 3: has been successfully completed for 100 users while maintaining the same parameters set in the first iteration. See Fig. 4.
- To The platform allows a number of actors to take part in video conferencing for a course on our platform (maximum 30 students). To be realistic we decided to load up to 200 virtual users for the iteration 4. See Fig. 5.

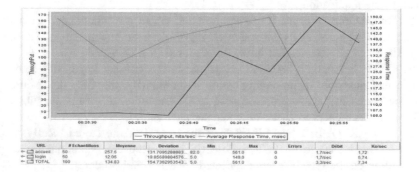

URL	# Echantillons	Moyenne	Deviation	Min	Max	Errors	Débit	Ko/sec
accueil	50	257.6	131.7095288883...	82.0	561.0	0	1,7/sec	1,72
login	50	12.06	19.85689804576...	5.0	149.0	0	1,7/sec	5,74
TOTAL	100	134.83	154.7362953543...	5.0	561.0	0	3,3/sec	7,34

Fig. 3. Test plan for 50 virtual users

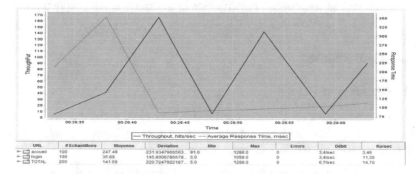

URL	# Echantillons	Moyenne	Deviation	Min	Max	Errors	Débit	Ko/sec
accueil	100	247.49	231.9347966563...	91.0	1288.0	0	3,4/sec	3,46
login	100	35.89	145.8006786678...	5.0	1059.0	0	3,4/sec	11,35
TOTAL	200	141.69	220.7247922187...	5.0	1288.0	0	6,7/sec	14,70

Fig. 4. Test plan for 100 virtual users

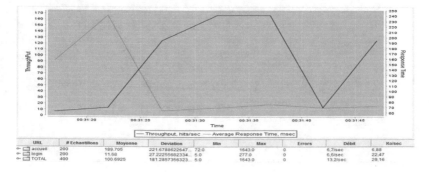

URL	# Echantillons	Moyenne	Deviation	Min	Max	Errors	Débit	Ko/sec
accueil	200	189.705	221.6788622647...	72.0	1643.0	0	6,7/sec	6,88
login	200	11.68	27.22255682334...	5.0	277.0	0	6,6/sec	22,47
TOTAL	400	100.6925	181.2857356323...	5.0	1643.0	0	13,2/sec	29,16

Fig. 5. Test plan for 200 virtual users

4.2   Analysis of Results

The Fig. 6 provides the total readings of all tests performed in the simple setup of JMeter, with a varying number of virtual users.

	Total Echantillons	moyennes	Déviation	Min	Max	Errors	Débit	KO/sec
Iteration2	20	217.8	178.99	7.0	604.0	0	0.83/s	1.61
Iteration3	100	134.83	154.74	5.0	561.0	0	3.3/s	7.34
Iteration4	200	141.69	220.72	5.0	1288.0	0	6.7/s	14.70
Iteration5	400	100.69	181.28	5.0	1643.0	0	13.2/s	29.16

Fig. 6. Report excerpt from csv file

- The total sample represents the total number of requests processed.
- The average is the average response time to a query. It is obtained through the following formula:
  $\mu = 1/n * \sum_{i=1...n} x_i$ (**n** represents the total number of application of a test plan, x_i represents the server's response time to a request). Consider iteration 1 with the respective response times of the following queries:
  x1 = 581.6;   x2 = 206.0;   n = 2.   Hence μ = 1/2 (581.6 + 206.0) = 393.8 ms, observed on the third column of iteration 1.
- The deviation allows measuring the variability or dispersion of a set of data relative to the average value. It is calculated with the following formula:
  $\sigma = 1/n * \sqrt{\sum i} = 1...n (x - \mu)^2$, with **n** le the total number of application x the response time to a request and μ average response time.
- The Min is the smallest response time to a request and the Max meanwhile, represents the highest response time to a query.
- Errors, the state "0" means that the tests are completed without errors.
- The last column of Fig. 6 show that the rate (KB/s) transfer of information does not decrease much, despite the significant increase in the number of requests.

The Fig. 7 shows that the amount of bytes transmitted is invariable, as the user accesses the home page (1967 bytes) or the authentication page (3707 bytes). This may mean that the system is stable and efficient.

The Fig. 8 shows that the more the number of users or the number of applications increases, less latency increases. Hence the observation of significant latency 22: 17: 32.

The detailed report of a csv file test plan shows that latency does not increase in proportion to the increased number of users, and therefore requests. This allows us to infer that the system is performing. See Fig. 9.

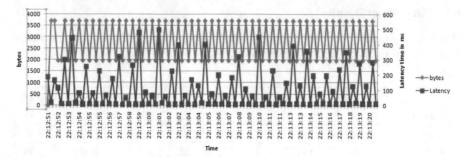

Fig. 7. Throughput and latency over time to 50 Users

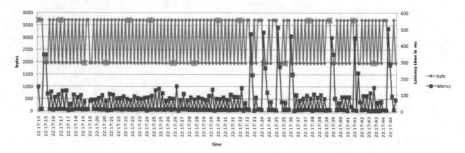

Fig. 8. Throughput and latency over time for 100 Users

Utilisateurs virtuels total	Temps total écoulé	Etiquette	Code Réponse	Message Réponse	Type de données	succès	Bits	Temps de Latence total
10 UV	3130	Login/Accueil	200	OK	Texte	True	56716	2359
50 UV	13527	Login/Accueil	200	OK	Texte	True	283584	10607
100 UV	20157	Login/Accueil	200	OK	Texte	True	567168	14383
200 UV	27332	Login/Accueil	200	OK	Texte	True	1134344	16325

Fig. 9. Summary report testing

5   Conclusion

In this paper, we performed in load testing and performance of an e-learning platform integrating the WebRTC technology. We used several components of JMeter in a single environment to run the tests. We varied the number of virtual users while avoiding stressing the system. The results show the stability, reliability and performance of the tested system. Looking ahead. We will study the impact of learning environments integrating the WebRTC technology.

References

1. Ouya, S., Seyed, C., Mbacke, A.B., Mendy, G., Niang, I.: WebRTC platform proposition as a support to the educational system of universities in a limited internet connection context (2015)
2. Sharmila, S., Ramadevi, E.: Analysis of performance testing on web applications. Int. J. Adv. Res. Comput. Commun. Eng. (2014)
3. Elbaum, S., Karre, S., Rothermel, G.: Improving web application testing with user session data. In: Proceedings of 25th International Conference on Software Engineering, pp. 49–59 (2003)
4. Sarojadevi, H.: Performance testing: methodologies and tools. J. Inf. Eng. Appl. (online) **1** (5) (2011)
5. Nevedrov, D.: Using JMeter to Performance Test Web Services, Published on dev2dev (2006)
6. Kiran, S., Mohapatra, A., Swamy, R.: Experiences in performance testing of web applications with unified authentication platform using Jmeter. In: 2015 International Symposium on Technology Management and Emerging Technologies (2015)
7. Halili, E.H.: Apache JMeter: A practical beginner's guide to automated testing and performance measurement for your websites (2008)
8. Duan, Q., Liang, Z., Wang, L.: Efficient file sharing scheme based on WebRTC. In: 2015 International Conference on Material, Mechanical and Manufacturing Engineering (2015)
9. Oh, H., Ahn, S., Choi, J., Yang, J.: WebRTC based remote collaborative online learning platform. In: 2015 Proceedings of the 1st Workshop on All-Web Real-Time Systems (2015)
10. Ouya, S., Sylla, K., Faye, P.M.D., Sow, M.Y., Lishou, C.: Impact of integrating WebRTC in universities' e-learning platforms. In: 2015 5th World Congress on Information and Communication Technologies (2015)

Task Based Learning to Enhance the Oral Production: Study Case Chimborazo

Adriana Lara Velarde[(✉)], Ruth Infante Paredes, Cristina Páez Quinde,
and Ana Vera-De la Torre

Facultad de Ciencias Humanas y de la Educación, Universidad Técnica de Ambato,
Ambato- Ecuador, Ambato, Ecuador
alara@unach.edu.ec, {rutheinfantep,mc.paez,aj.vera}@uta.edu.ec

Abstract. This research aimed to analyze the relationship between Task Based Learning and the English-speaking skill. This study primarily utilizes an experimental research study. The research was conducted for two months, 6 hours a week. It employed a speaking pre-test and post-test as instruments in order to assess the effectiveness of the implementation of TBL to enhance the English-speaking skill. The population consisted of one hundred forty A2 students who belong to the Language Center of Administrative and Political Science Faculty at Universidad Nacional de Chimborazo. To evaluate the efficacy of this study a rubric was used with the purpose of providing a more detailed evaluation of students' performance. To verify the hypothesis of this research, the Kolmogorov-Smirnov Test was used to measure the sample data and the Wilcoxon Test was used to get the sample results. They show that Task- Based Learning (TBL) fosters the English-speaking skill. It is concluded that students had a significant improvement in their oral production after the implementation of tasks based on TBL....

Keywords: Learning · Task · Based learning · Speaking skill · Fluency · Assessment

1 Introduction

The study of the English language has become one of the principal subjects of students in most of educational institutions around the world. It has also become an essential vehicle of communication for humans in different knowledge areas and in the development of professional and economical life. Communication is a key element in social interaction among people. One cannot do anything without communication. There have been problems with students while communicating in English. There is a low level of oral production in students because tasks performed inside the language classroom are inadequate and the opportunities for

© Springer Nature Switzerland AG 2020
M. E. Auer and T. Tsiatsos (Eds.): ICL 2018, AISC 916, pp. 194–201, 2020.
https://doi.org/10.1007/978-3-030-11932-4_20

using real language inside the classroom are limited. Moreover, activities performed in class do not have a communicative purpose because the main focus is on the form of the language rather than on the use itself. As a result of this situation, there is an ineffective communication and consequently it is not possible for students to keep a conversation using English as a foreign language. The main purpose of this study was to analyze the relation between Task Based Learning and the English speaking skill. This research provides information about how TBL enhances the oral production. It was conducted due to the necessity to help students improve their speaking skill. To develop this research project a pre-test was carried out. It was based on the speaking part of the Preliminary English Test (PET). During the test, students were assessed on what they can do with the language in certain language functions such exchanging personal information, describing pictures and negotiating meaning, and other real life situations. The results obtained from the pre-test showed that there is a limited oral production in students. Therefore, lesson plans were developed based on the principles of TBL. The content of the book that students were using was adapted to this approach. Moreover, the lesson plans included the tasks that TBL proposes and they focused on enhancing students' oral production. The implementation of this proposal lasted two months. Once it had finished, a post-test was conducted with the objective of determining if TBL implementation boosted oral production or not. A rubric was used in order to provide a more objective assessment of students' performance. This rubric evaluated the speaking skill under the following criteria: Grammar and Vocabulary, Discourse Management, Pronunciation and Interactive Communication. The results obtained after the implementation of TBL demonstrate that it surely enhances oral production. Students notably improved their ability to communicate and interact with others.

2   State of the Art

Learning a foreign language has become essential in today's world due to the fact that there is an increasing necessity of using it in several aspects such technology, education, business, medicine, etc. Therefore, English has been taught for many years in classrooms with the purpose of helping students to be able to communicate and interact using this language. Thus, in Ecuador, English is taught since elementary school until higher levels such as careers at university. In this regard, some approaches have been proposed to teach English as a foreign language. Starting by traditional teaching methods, such Grammar Translation, Direct Method, Total Physical Response, and Suggestopedia whose main focus was not the development of communication but the memorization of grammar rules or languages patterns. Reference [1] argues that these methods do not develop language skills naturally instead they promote passive language learning by mastering just certain aspects. By the time, there have been more language learning approaches that have appeared which aim to develop the communicative competence. Thus, some of these approaches are Communicative Approach, Task Based Learning, Content Language Learning among others.

Reference [2] supports that Communicative Approach enhances communication and interaction. Moreover, he affirms that students feel motivated to learning English as a foreign language. Similarly, [3] affirms that Communicative Language Teaching emphasizes that when referring to language it is natural to think about communication since people always use language to negotiate meaning and to accomplish language functions. Therefore, language learning requires an approach which principal purpose is communication. However, nowadays the focus must not only be on communication but on using language in real-life situations.

Taking these insights into consideration it is primarily relevant to apply other methods that can contribute to language skills development. Aside from the Communicative Approach, [4] proposes the use of TBL which tends to develop language by working on meaningful tasks that involve learners in using the language for a communicative purpose.

Task Based Learning (TBL) involves a process in which learners will work on completing a task to achieve an outcome. Learners have to participate, communicate, negotiate meaning, and work cooperatively to carry out a task [5]. Similarly, [6] affirms that TBL is a way of teaching that considers language as the main tool for communication than as an object for study or manipulation. Similarly, [4] highlights that by implementing TBL makes students use the target language through communicative tasks that aim to carry out a specific outcome. In this regard, the main purpose of TBL is to motivate learners to communicate effectively by giving them a purpose to use the language that in this case is developing a task to achieve a final outcome. As it was already mentioned, TBL uses communicative tasks to develop language interaction. Tasks allow learners to use the language as they were doing in a real context [6] hence, there is interaction in every class since learners are communicating, interacting, negotiating meaning in order to achieve a final outcome. In this regard, the role of tasks is highly relevant when motivating learners to develop their oral production by letting them work on completing the task [7].

Reference [4] proposes two kinds of tasks: open and closed. These tasks are classified according to their objective. Open tasks do not have a specific objective due to the fact that learners have the possibility to negotiate meaning among them and decide which final outcome they will present. These open tasks are classified into: listing, comparing, troubleshooting, and sharing personal experiences. Closed tasks are very specific. Learners are required to achieve a specific final outcome. They are classified in: sort and classify, matching, and projects.

The implementation of these kind of tasks fosters students' oral production because in each stage of this process students are required to interact with their classmates and consequently they develop their abilities to communicate.

Buitrago [8] mentions that by implementing open and close tasks contribute to the development of the communicative competence because of the different type of tasks developed in classrooms.

The implementation of a task consists of three phases: pre-task, task cycle and language focus. At the first stage the teacher presents the input. It could be presented a picture, audio, or video to demonstrate the task. Its objective is to prepare students to what they will do in the others stages [9]. The task cycle phase is divided into: task, planning and report. Students plan the activity; then they work on the task and finally they prepare the report [4]. The main purpose in this stage is to give learners the chance to do the activity by using the target language spontaneously to achieve the goal and final outcome by working collaboratively [10]. Finally, the language focus is the third stage of this process. At this level, the teacher makes a brief review of the language used along the activity. According to [7] an analysis of the whole activity is carried out in order to identify mistakes that were committed during final outcomes' presentations. Therefore, the role of the teacher is to provide feedback on students performance about language use.

English language skills are divided into receptive and productive. Thus, speaking is considered as part of the productive skills. It is the capacity that a person has to communicate effectively with others. When referring to a competent person in a language it is often implicit that this person can produce meaningful language patterns to communicate. In this respect, [11] argues that this skill is one of the most used in everyday life in order to share ideas, thoughts, and believes with those who surround them.

It also implies the use of sub-skills such as Grammar. Fluency, Vocabulary and Pronunciation. These sub-skills contribute to the enhancement of the oral production due to the fact that each of them focuses on a specific issue of the communicative competence. Thus, one of the main sub-skills is Fluency because it allows a person to keep a conversation in a natural way. It plays an important role in speaking, as [12] posits that fluency allows the teacher to evaluate students ability to communicate and use the language accurately. Likewise, [13] states that fluency is the ability that a person has in order to make use of the language at a normal flow. Therefore, for a person to be considered fluent in a language, he or she must produce any kind of speech by being intelligible [14].

As fluency is part of the speaking sub-skills, it has to be assessed with the purpose of providing a precise information of how the learner is developing the oral production. In this regard, speaking can normally be assessed by using a rubric. Wolf [15] argues that a rubric is an instrument that teacher employs under a specific criteria and detailed descriptors in order to assess the accuracy of speaking. Similarly, [16] agrees that a rubric is a very useful tool that allows teachers to evaluate the learner's speaking abilities. Moreover, the assessment of the oral production lets the learner know how is he/she using the language and how he/she can improve it.

3   Methodology

This study consisted of an experimental research owing that TBL was applied to a specific group of students aiming to improve their oral production. This research was conducted to 142 A2 students who belong to the Language Center of Administrative and Political Science Faculty at Universidad Nacional de Chimborazo. Students at this institution have different cultural backgrounds in terms of level of education, age, culture, needs, and socioeconomic status. The experiment was conducted for 2 months, 6 hours a week. At the beginning of the process, it was made a diagnosis through a pre-test that pretended to evaluate the level of the oral production of this group before applying TBL.

Table 1. Speaking sub-skills pre- post-test

Subskill	Pre-test	Post-test
Grammar and vocabulary	1,93	2,67
Discourse management	2,12	2,24
Pronunciation	1,94	2,21
Interactive communication	2,02	2,10

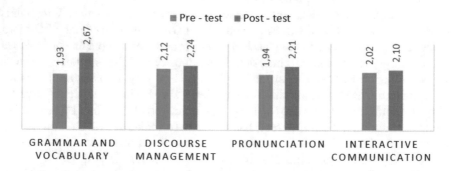

Fig. 1. Speaking sub-skills pre- post-test

Students' responses were recorded with the purpose of analyzing them in detail. The instrument that was used to assess their performance was a rubric. After the implementation of TBL, students showed a significant improvement in their oral production.

Because it was an experimental research the Kolmogorov-Smirnov Test was used due to the fact that it was applied in TBL. Therefore, it was necessary to conduct a pre-test before the implementation and a post-test after its implementation with the purpose of determining if the experiment was successful or not. This test was used because the sample is greater than 50 and the Normality Probabilistic is used to determine what the distribution and symmetry is.

Table 2. One- sample Kolmogorov-Smirnov Test

Details	Difference	
N		140
Normal parameters a, b	Mean	−7,3571
	Standard deviation	3,18511
Most extreme differences	Absolute	,170
	Positive	,159
	Negative	−,170
Test statistic		,170
Asymptotic sig (2-tailed)		,000c

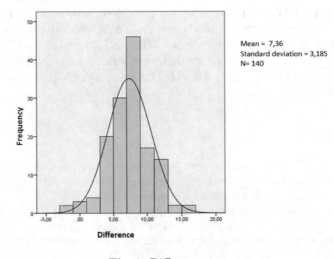

Mean = 7,36
Standard deviation = 3,185
N= 140

Fig. 2. Difference test

Once the normality test was carried out, it was used the Wilcoxon Signed Rank Test in order to know if TBL enhances the English speaking skill and to validate the hypothesis. Therefore, as the Gauss bell shows, it is symmetric which means that TBL it contributes to enhance the English speaking skill. Furthermore, the level of significance is 0,05 therefore the Null hypothesis is rejected whereas the alternative hypothesis is accepted. In other words, TBL enhances the English speaking skill in students at Universidad Nacional de Chimborazo.

Table 3. Wilcoxon Signed Rank Test - ranks

Details	N	Ranks average	Ranks sume
Pos-tes negative ranks	2a	3,50	7,00
Pre-test positive ranks	135b	69,97	9446,00
Ties	3c		
Total	140		

4 Conclusions

The results of this research show that TBL enhances the oral production. It improves the ability of students to interact with others around them. By developing a task, students had the opportunity to practice English. TBL components such as goals, teacher's and student's roles are essential in this learning process due to the fact that each of them accomplishes a specific function. Therefore, they should be considered as an important element when planning, developing, and assessing a class. Moreover, the tasks that TBL propose (closed and open) are useful to develop the English speaking skill since each of them focused on developing the oral production. Furthermore, the speaking sub-skills: Grammar and Vocabulary, Discourse Management, Pronunciation, and Interactive Communication that were assessed in the pre and post test obtained a significant improvement after the implementation of TBL. As future research TBL is needed to be applied to enhance the oral production of students at all levels of the Common European Framework (A1, A2, B1, B2, and C1) of higher education in all the universities in Ecuador.

References

1. Kumar, J.: The Best Method to Teach English Language Retrieved on May 1st, 2018 from https://www.researchgate.net/publication/282974160-The_Best_Method_to_Teach_English_Language/citations (2015)
2. Ahmad, S.: Applying Communicative Approach in Teaching English as a Foreign Language: A Case Study of Pakistan. Portalinguarum. Retrieved on April 25th, 2018 from http://www.ugr.es/portalin/articulos/PL_numero20/12%20%20Saeed.pdf (2013)
3. Torres, M.: Basic Language Methodology: Cooperative Learning Guidebook for Training English Teachers. Riobamba, Unach (2017)
4. Willis, J.: Framework for Task-Based Learning. Longman, Harlow (1996)
5. Munirah, M., Arief, M.: Using task-based approach in improving the students' speaking accuracy and fluency. J. Educ. Hum. Dev. American Research Institute for Policy Development (2015)
6. Ellis, R.: Task-based language teaching: sorting out the misunderstandings (2009)
7. Crdoba, E.: Implementing task-based language teaching to integrate language skills in an EFL program at a Colombian university. Profile Issues Teach. Prof. Dev. (2016)

8. Buitrago, A.: Improving 10th graders' English communicative competence through the implementation of the task-based learning approach. Profile Issues Teach. Prof. Dev. **18**(2) (2016)
9. Sholihagh, U.: Task Based Language Teaching (TBLT) can improve student's writing ability. Magistra (86) (2013)
10. Kalavathi, T.: Task-based language teaching: its implementation to improve speaking skills of rural school students- a case study. IOSR J. Humanit. Soc. Sci. (2017)
11. Ismaili, E.: Information gap activities to enhance speaking skills of elementary level students. Procedia Soc. Behav. Sci. **232**, 612–616 (2016)
12. Aly, E.: Using a Multimedia-Based Program for Developing Student Teachers' EFL Speaking Fluency Skill. Retrieved from ERIC data base on April 28th, 2018 from https://files.eric.ed.gov/fulltext/ED539987.pdf (2013)
13. Marashi, H.: ADHD and adolescent EFL learners' speaking complexity, accuracy, and fluency in English. Iran. J. Lang. Teach. Res. **4**(2), 105–126 (2016)
14. Molina, M.: The use of the 3/2/1 technique to foster students' speaking fluency. i.e.: Inq. Educ. **9**(2) (2017)
15. Wolf, K.: The role of rubrics in advancing and assessing student learning. J. Eff. Teach. **7**(1), 3–14 (2007)
16. Ounis, A.: The assessment of speaking skills at the tertiary level. Int. J. Engl. Linguist. **7**(4) (2017)

Collaborative Content Authoring: Developing WebQuests Using SlideWiki

Maria Moundridou[✉], Eleni Zalavra, Kyparisia Papanikolaou, and Angeliki Tripiniotis

School of Pedagogical and Technological Education, Athens, Greece
{mariam, kpapanikolaou}@aspete.gr, zalavra@sch.gr,
angelastem@master.aspete.gr

Abstract. This paper reports on a study that was conducted to explore the potential of Slidewiki (a newly developed platform supporting crowd-sourced content creation) for collaborative learning in a teacher education context. In this study 143 students (pre-service teachers) used SlideWiki to collaboratively develop a WebQuest according to a set of technological and pedagogical specifications. According to students' responses to questionnaires, reflecting their perceptions, the vast majority of them liked sharing and co-authoring WebQuests with their group in SlideWiki considering that this process is time saving and urges collaborators to equally contribute to the assigned task. They suggested Slidewiki to be extended with a forum to further facilitate collaboration. They also developed a positive attitude towards Open Educational Resources (OERs) and copyright issues considering that sharing their WebQuests with all SlideWiki users motivated them to try harder and improve their work. Finally, they acknowledged the improvement of their knowledge and skills in learning design.

Keywords: Collaborative content authoring · Wikis · Learning design · Teacher education · WebQuest · SlideWiki

1 Introduction

In this paper we explore the potential of wikis for collaborative learning in a teacher education context. In particular, we report on a study in which 134 students (pre-service teachers) working on the design of technology-enhanced learning activities, collaboratively developed Webquests [4], using a specific wiki platform, SlideWiki (http://slidewiki.org) [1]. SlideWiki is a platform supporting the crowd-sourced creation of richly structured learning content, supporting collaborative authoring and allowing the elicitation and sharing of knowledge using presentations [8].

Considering that little technical ability is required to use wikis [9], this technology permits students to focus on collaboration and the exchange of information [11]. Although wikis include features that facilitate collaboration, it does not necessarily follow that their use will ensure or even encourage collaborative learning behaviour [6], as collaboration is not automatically a direct outcome of wiki technology [5]. Moreover, crowd-sourcing as large-scale collaboration around educational content is

© Springer Nature Switzerland AG 2020

M. E. Auer and T. Tsiatsos (Eds.): ICL 2018, AISC 916, pp. 202–214, 2020.
https://doi.org/10.1007/978-3-030-11932-4_21

currently supported only in a very limited way [1]. Applications like Google Docs or Prezi support collaborative authoring of particular type of content, whilst applications like SlideShare, TeacherTube, OpenStudy are commonly used as educational social networks allowing the sharing of various types of content like slides, presentations, diagrams, assessment tests etc., which are mainly created individually.

However, collaborative content authoring is a challenge for the educational community as well as for the teacher education area. Especially WebQuests [4], are considered to provide an excellent organisational model for teacher education [10, 16]. Furthermore, the collaborative development of WebQuests is a promising way of enhancing the comprehension of inquiry-based learning and digital technologies, focusing on the synthesis of pedagogy, content and technology [10].

This study aims at contributing to the evaluation of the newly developed SlideWiki platform. Specifically, we provide evidence about the potential of the innovative characteristics of SlideWiki as a crowd-sourcing platform in promoting community collaboration, authoring, reusing and re-purposing of educational content in a constructivist teacher training context.

2   SlideWiki

SlideWiki (http://SlideWiki.org) is a platform aiming to facilitate the collaboration around educational content [1]. Focusing on educational content creation, the platform is grounded on reusable content authoring and large-scale community collaboration (or crowd-sourcing) [14]. Inspired by the wiki paradigm, which since its inception in the early 2000s has become a ubiquitous pillar for enabling large-scale collaboration, Tarasowa et al. [14] envisioned extending its unstructured, textual content. Aiming to support (semi-)structured content creation (e.g. presentations, questionnaires, diagrams, etc.) and the collaboration of large user communities around such content, they developed the CrowdLearn concept and applied it to SlideWiki. The CrowdLearn concept combines the wiki style for collaborative content authoring -in the form of presentation slides- with the requirements of the Sharable Content Object Reference Model (SCORM) for reusability [15].

For someone to use SlideWiki, a user account needs to be created either by signing up or by using a Google or GitHub login. In SlideWiki users can create slides, arrange them in presentations, and collaborate on them [1]. Actually, the term "deck" is adopted for a collection of slides which users can either create from scratch or import from PowerPoint or OpenDocument Presentation files. In decks, tags can be assigned and a commenting feature is available. Authoring the slides within a deck follows the familiar approach of editing a presentation. A horizontal toolbar including all the basic features for formatting a slide (e.g. defining fonts, arranging paragraphs in textboxes, etc.) is provided, along with a vertical toolbar with tools for embedding images, videos, tables, equations, and other content. Moreover, SlideWiki employs an inline HTML5 based WYSIWYG text editor so that users can see the slideshow output at the same time as they are authoring their slides [8].

SlideWiki allows users to collaborate through managing editing rights in a deck. The deck creator can authorize other users to edit the deck. SlideWiki supports versioning, forking/branching and merging for slides and decks ensuring that every user's personal revisions of slides and decks are always preserved [1]. When a deck or a slide is forked (copied), information about its origin and its creator is displayed in the deck description. Consequentially, SlideWiki attributes three roles to users: (a) 'origin' for the first user who created a deck, (b) 'creator' for the user who forked a deck and (c) 'contributor' for the user who has edit rights to the deck.

SlideWiki is built on the Open Educational Resources (OER) ethos and all content available in it, is licensed under the Creative Commons Attribution Share Alike license (CC-BY-SA) [1]. In this way, users can share, re-purpose and reuse content for their own purposes, meaning that they can revise, adapt and re-mix any slides and decks on SlideWiki.

Figure 1 shows a screenshot of a deck developed during our study. The sidebar on the left (1) includes the deck name and the slide list, which follows the WebQuest model. In the centre (2), there is information regarding the deck, including the creator and the origin, since this deck originated from another deck that had been forked. On the right sidebar (3), one can see the creator as well as the contributors of the deck, along with an activity feed showing the interaction of users with this deck (editing, commenting, forking). At the bottom (4), there are facilities such as tagging and commenting. At the bottom right (5), there is the trademark of the Creative Common License under which the deck is published.

Fig. 1. A SlideWiki deck developed during our study (usernames altered)

3   Empirical Study

Aiming to contribute to the collaborative content authoring area by combining research in wikis and learning design, we used SlideWiki in an undergraduate course in order to evaluate specific functionalities supporting community collaboration, authoring, reusing and re-purposing of educational content, in real educational settings.

The research questions of the study are the following:

- RQ1: How do the students perceive SlideWiki's usability and usefulness?
- RQ2: What are the students' perceptions of SlideWiki about facilitating collaborative content authoring?
- RQ3: What are the students' perceptions of SlideWiki about supporting them in developing WebQuests?

3.1 Methodology

SlideWiki was used in the context of an Educational Technology undergraduate course offered in the School of Pedagogical and Technological Education (ASPETE), Athens, Greece, during the spring semester of the academic year 2017–18. The participants were 134 students of the departments of Civil Engineering Educators and Mechanical Engineering Educators. The research team included the two instructors of the course and two facilitators.

In this course, it is compulsory for students to carry out a learning design activity as their final project. Specifically, they are assigned to collaboratively develop a learning design according to a set of technological and pedagogical specifications. In particular, students are asked to use the WebQuest model of inquiry [4, 10] and integrate technologies, such as web 2.0 tools, into their learning designs.

In this study, the students' project was organised over a seven-week period as a blended learning activity as follows:

- 1st week: Students were informed about the framework of the activity. They were introduced to the SlideWiki platform, created user accounts and got familiar with it.
- 2nd week: Students were asked to explore exemplary learning designs (WebQuests) which were already prepared and published as SlideWiki decks by the research team. They were also asked to comment on one exemplar WebQuest following specific evaluation criteria given by the research team.
- 3rd week: Students were introduced to the final assignment of the course asking them to collaboratively develop a WebQuest as a deck in SlideWiki. They formed groups, mainly of 2 and in a few cases of 3 members, and organised their collaboration in SlideWiki by sharing a deck and defining its contributors.
- 4th - 5th week: Students developed their Webquests in SlideWiki.
- 6th - 7th week: Peer review was organized. Each student commented on the SlideWiki Deck of another group about the appropriateness of technology and pedagogy integration in the design, following specific evaluation criteria given by the research team. The same criteria were used by the course instructors who also commented on the Decks where the WebQuests were published. Finally, taking into account the reviews received by their peers and the instructor, each student group completed its WebQuest.

At this point, it is worth mentioning that the activity was structured by means of *a framework for organising a collaborative learning design activity based on wiki* proposed by Zalavra and Papanikolaou [17]. This framework utilises Salmon's five-stage model [12, 13], a valuable curriculum-building tool for designing and implementing e-tivities based on e-learning tools. It restructures the context of Salmon's model from e-learning to blended learning and proposes how the stages should be implemented specifically for organizing a collaborative learning design activity in a wiki platform. This study also serves as an empirical study of this framework to be analysed in another paper.

At the end of the course, students completed: (a) the System Usability Scale (SUS) questionnaire [3], and particularly the Greek version of it [7], that investigates the likelihood of them reusing the system, and (b) a survey questionnaire assessing the students' learning design experience (evaluation questionnaire).

3.2   Data Collection and Analysis

The data collected consist of: (a) 67 learning designs (SlideWiki decks) developed by 134 students, mostly working in groups of two, (b) 125 SUS questionnaires, and (c) 125 survey questionnaires assessing the students' learning design experience.

The survey questionnaire comprised of 38 questions: 33 Likert-scaled ones (ranging from 1: strongly disagree to 5: strongly agree), 3 open-ended ones, and 1 multiple choice one, organised in four sections focusing on the following aspects:

- *Section A*: SlideWiki's usefulness with regard to working on a specific assignment (Q1–Q11, Tables: 1, 2 and 3)
- *Section B*: SlideWiki's usefulness as an educational tool (Q12–Q17, Tables: 4, 5)
- *Section C*: SlideWiki facilitating collaborative content authoring (Q18–Q29, Table 6)
- *Section D*: SlideWiki supporting WebQuest development (Q30–Q38, Table 7)

In this research we analysed the SUS questionnaires and Sections A and B of the survey questionnaire in order to provide evidence about the usability and usefulness of SlideWiki (RQ1) based on the students' perceptions. Moreover, we analysed the students' responses to the questions of sections C and D of the survey questionnaire in order to provide evidence about the potential of SlideWiki for facilitating collaborative content authoring (RQ2) and supporting students in developing WebQuests (RQ3) correspondingly.

3.3   Results

RQ1: How do the students perceive SlideWiki's usability and usefulness?

SlideWiki was assessed by students along three dimensions: (a) the system's usability, (b) the system's usefulness with regard to working on their specific assignment, and (c) the system's usefulness as an educational tool.

(a) *SlideWiki's usability.*

SUS score interpretation is usually based on data from over 5000 users across 500 different evaluations [2]. The average SUS score from all 500 studies is 68. Thus, a SUS score above 68 is considered above average and anything below 68 is below average. The SUS mean usability score from the 125 students participating in this study, was 62.4 indicating a marginal level of usability for SlideWiki. In the study of Tarasowa et al. [14] where computer science researchers answered the SUS questionnaire, the mean usability score for SlideWiki was 69.62. This difference in the SUS scores between the two studies could be attributed to the users' different background, since the students participating in our study have basic digital skills.

(b) *SlideWiki's usefulness with regard to working on a specific assignment.*

Table 1 summarises the students' responses in statements that address how they perceive SlideWiki's usefulness when working on their assignment. Generally, students perceived Slidewiki to be highly usable: 80% of them agreed or strongly agreed that the system was easy to access and use (Q1) and 69% used it without any assistance (Q2). 79% of students appreciated the ability to access their work anytime from anywhere (Q3) and 61% stated that this ability motivated them to work on their assignment (Q4).

Table 1. Distribution of students' responses in statements Q1 to Q9 of Section A

Statement	1	2	3	4	5
Q1. I could easily access and use SlideWiki	3 (2%)	7 (6%)	15 (12%)	76 (61%)	24 (19%)
Q2. I used SlideWiki for my assignment without needing any assistance	0 (0%)	11 (9%)	28 (22%)	72 (58%)	12 (11%)
Q3. I appreciate that SlideWiki allowed me to access my work anytime, from anywhere	3 (2%)	8 (6%)	16 (13%)	66 (53%)	32 (26%)
Q4. The ability to access my work from anywhere, at any time, has motivated me to work my task	3 (2%)	9 (7%)	36 (29%)	53 (42%)	24 (19%)
Q5. Structuring the content as Slide Deck was practical and intuitive	4 (3%)	17 (14%)	18 (14%)	61 (49%)	25 (20%)
Q6. I enjoyed SlideWiki's feature for easily integrating images, videos and Web 2.0 objects into slides	2 (2%)	3 (2%)	10 (8%)	62 (50%)	48 (38%)
Q7. SlideWiki's feature for authorizing contributors in Slide Decks and providing a history of their actions prompted me to participate actively	2 (2%)	3 (2%)	34 (27%)	67 (54%)	19 (15%)
Q8. I would like SlideWiki's history feature to provide a review of the actions carried out within each slide and the users who performed them	1 (1%)	6 (5%)	32 (26%)	58 (46%)	28 (22%)
Q9. I would like the SlideWiki's history feature to support reverting to previous versions of slides	0 (0%)	3 (5%)	20 (16%)	63 (50%)	39 (31%)

The most popular (88% of students) feature of SlideWiki was the easy integration of images, videos, and Web 2.0 objects into slides (Q6). 69% of students felt that they were prompted to participate actively in their assignment due to SlideWiki's feature for authorizing contributors in Slide Decks and providing a history of their actions (Q7).

However, 68% of students would also like the history feature to provide a review of the actions carried out within the content of each slide and the users who made them (Q8), and 81% would like to be able to revert to previous versions of slides (Q9). Concerning the structuring of content as Slide Decks, while 69% of students find it to be practical and intuitive, a not negligible 17% disagree (Q5).

Students were also asked to name the most useful features of SlideWiki for their assignment (Table 2), as well as the features that interfered with their work (Table 3).

Table 2. Students' responses to the open-ended question (Q10) of Section A

Q10. Which features of SlideWiki do you consider useful for your assignment?

Category of responses	Frequency
Slide Authoring: Integrating images, videos and tables	57 (46%)
Slide Authoring: Adding Hyperlinks	16 (13%)
Online access provided easy access	37 (30%)
Familiar content structure (PowerPoint alike)	31 (25%)
Sharing a deck with my peers and working synchronously	26 (21%)
Being able to see exemplar as well as other groups' WebQuests	10 (8%)
When my deck/slide is copied, my copyrights are preserved	4 (3%)

Table 3. Students' responses to the open ended question (Q11) of Section A

Q11. Which features of SlideWiki do you consider that interfered with your assignment?

Category of responses	Frequency
Slide Authoring: handling content boxes	54 (43%)
Slide Authoring: not enough features for text and/or image formatting	13 (10%)
Slide Authoring: formatting a table	11 (9%)
Slide Authoring: toolbar for formatting text occasionally not visible	4 (3%)
Slide Authoring: not being able to define my own background style	2 (2%)
Platform not responding e.g. when saving a slide	27 (22%)
Platform working slowly	16 (13%)
Overall impression that platform is not easy to use	10 (8%)
Not being able to collaborate synchronously on the same slide	3 (2%)
No support for communication among contributors	2 (2%)
Not being able to attach material (e.g. upload a pdf file)	2 (2%)
None	15 (12%)

Integrating images, videos, and tables into slides was considered useful by nearly half of the students (46%). 30% of the students considered it useful that their work was performed and stored online, enabling them to easily access it from anywhere, at any time. One out of four students favoured that content in SlideWiki is structured in the form of slides, which is familiar to them from presentation software like MS-PowerPoint. Sharing a Slide Deck with peers and working synchronously on it, was considered useful by 21% of students. Fewer students considered other features of SlideWiki useful, namely: adding hyperlinks to slides (13%), being able to see others' Decks (8%), and preserving the copyrights when a Deck is copied (3%).

On the other hand, there were features of SlideWiki that interfered with the students' work. In fact, as shown in Table 3, the students' responses reveal that what interfered with their work was either a platform's malfunction or a missing feature. Malfunctions include: handling content boxes (43%), formatting Tables (9%), platform not responding (22%) or responding slowly (13%). Missing features include: not enough features for text and/or image formatting (10%), not being able to collaborate synchronously on the same slide (2%), no support for communication among contributors (2%), not being able to attach material -e.g. upload a pdf file (2%). It should also be noted that 8% of students said that the platform was overall not easy to use. In contrast, 12% did not view any feature of SlideWiki as interfering with their work.

(c) *SlideWiki's usefulness as an educational tool*

Table 4 summarises the students' responses in statements that address their perceptions of SlideWiki's usefulness when they consider using it in the future. 46% of students would like to use SlideWiki for their assignments in other courses during their studies (Q12). Assuming their future role as teachers, 50% stated that they intend to utilize the WebQuests their fellow students have developed at SlideWiki (Q13), 59% would use SlideWiki to develop educational resources (Q14), and 58% would choose to assign the students to use it (Q15).

Table 4. Distribution of students' responses in statements Q12 to Q15 of Section B

Statement	1	2	3	4	5
Q12. I would like to use SlideWiki for assignments in other courses during my studies	15 (12%)	13 (10%)	39 (31%)	48 (38%)	10 (8%)
Q13. I intend to utilize the WebQuests developed by my peers at Slidewiki in the future	10 (8%)	21 (17%)	33 (26%)	46 (37%)	16 (13%)
Q14. As a future teacher, I would choose to use SlideWiki for developing OERs	9 (7%)	10 (8%)	32 (26%)	55 (44%)	19 (15%)
Q15. As a future teacher, I would choose to assign my students to use SlideWiki	11 (9%)	11 (9%)	30 (24%)	54 (43%)	19 (15%)

The OERs that the students -as future teachers- would develop in SlideWiki are presented in Table 5, Q16. It is of no surprise that 79% of the students would use SlideWiki to develop WebQuests, since they did so when they used the platform for the first time. The fact that 67% of the students would develop presentations in SlideWiki was also an expected finding due to the structuring of content as Slide Decks. It is quite promising that half of the students would use SlideWiki to develop Learning Designs, since these students declare that they intend to use SlideWiki in another context than the already known ones i.e. WebQuests and presentations. Creating learning designs in SlideWiki could be connected to sharing them with the teacher community and/or developing constructivist learning contexts for their students.

Table 5. Students' answers to the multiple choice question Q16 and the open-ended question Q17 of Section B

Q16. As a future teacher, what type of OERs would you develop in SlideWiki?	
Category selection	Frequency
WebQuests	99 (79%)
Presentations	84 (67%)
Learning designs	64 (51%)
Q17. As a future teacher, what would you assign your students to do in SlideWiki?	
Presentations	82 (66%)
WebQuests	45 (36%)
Content authoring group projects	19 (15%)
None	3 (2%)

As shown in Table 5, Q17, presentations is the kind of assignment that most students (66%) would give as future teachers to their students to carry out in SlideWiki. A somewhat confusing answer is the one given by 36% of students who said that they would ask their students in the future, to develop WebQuests in SlideWiki. Obviously, that 36% includes students who misunderstood the question and others who have not got a clear view of what a WebQuest really is. Finally, 15% of the students would use as future teachers the SlideWiki platform as an online space for their students to organize their projects and work collaboratively in groups.

RQ2: What are the students' perceptions of SlideWiki about facilitating collaborative content authoring?

The students' responses in statements addressing their perceptions of SlideWiki about facilitating collaborative content authoring (Table 6) indicate that the platform performed well at that indeed. In particular, 89% of the students appreciated the fact that SlideWiki allowed them to share their work with their group (Q18) and 81% stated that being able to author the assignment at their convenience, facilitated the collaboration (Q19). It is quite promising that although the particular students have only basic digital skills, 44% of them considered the collaboration at SlideWiki to be more effective than face-to-face collaboration (Q20). About two out of three students agreed or strongly agreed that the platform helped them save time (Q21) and facilitated the collaboration in their group (Q22). Moreover, 86% of students enjoyed collaborating within their group on a shared SlideWiki Deck (Q26) and 85% thought that in this way all group members might equally contribute to the assigned task (Q27). SlideWiki also did fairly

Table 6. Distribution of the students' responses in statements Q18–Q29 of Section C

Statement	1	2	3	4	5
Q18. I appreciated that SlideWiki allowed me to share my work with the other group members	2 (2%)	0 (0%)	12 (10%)	71 (57%)	40 (32%)
Q19. Being able to collaboratively author my assignment, at my convenience, facilitated our collaboration	1 (2%)	2 (2%)	21 (17%)	70 (56%)	31 (25%)
Q20. I consider collaboration at SlideWiki more effective than face-to-face collaboration	7 (6%)	18 (14%)	44 (35%)	38 (30%)	18 (14%)
Q21. SlideWiki helped me save time while collaborating within my group	3 (2%)	12 (10%)	25 (20%)	60 (48%)	25 (20%)
Q22. SlideWiki facilitated the collaboration of my group	5 (4%)	5 (4%)	27 (22%)	61 (49%)	27 (22%)
Q23. Collaborating in SlideWiki has enhanced the sense of community in my class	3 (2%)	13 (10%)	28 (22%)	60 (48%)	21 (17%)
Q24. The collaboration could be further facilitated if SlideWiki provided a forum for communication among group members	2 (2%)	7 (6%)	28 (22%)	60 (48%)	28 (22%)
Q25. I was comfortable commenting in SlideWiki Decks developed by others	2 (2%)	5 (4%)	22 (18%)	76 (61%)	36 (29%)
Q26. I enjoyed collaborating within my group in a shared SlideWiki Deck	0 (0%)	4 (3%)	14 (11%)	77 (61%)	31 (25%)
Q27. Collaborating in a shared SlideWiki Deck allowed my group's members to contribute equally to our task	1 (1%)	3 (2%)	15 (12%)	63 (62%)	29 (23%)
Q28. Developing a Webquest in SlideWiki as an OER available to all SlideWiki users motivated me to improve it	0 (0%)	3 (2%)	31 (25%)	61 (50%)	28 (22%)
Q29. Being able to review fellow groups' WebQuests while developing mine stimulated my own creativity	0 (0%)	6 (5%)	27 (22%)	4 (49%)	31 (25%)

well at enhancing the sense of community in the class through collaboration (65%, Q23), at motivating students to improve their work due to its availability as an OER to all SlideWiki users (72%, Q28), and at stimulating the students' creativity by enabling them to review fellow groups' WebQuests while developing their own (74%, Q29). A quite unexpected but welcome finding was that 9 out of 10 students felt comfortable commenting in SlideWiki Decks developed by others (Q25). Finally, 70% of students suggested that collaboration could be further facilitated if SlideWiki provided a forum for communication among group members (Q24).

RQ3: What are the students' perceptions of SlideWiki about supporting them in developing WebQuests?

The students' perceptions concerning SlideWiki's support in developing their Web-Quests are presented in Table 7. The majority of the students stated that they could easily search for educational resources available in SlideWiki (77%, Q30) and easily locate the ones they need (71%, Q31). The fact that SlideWiki publishes content under the Open Creative Commons licence was considered important by 8 out of 10 students (Q32–Q34).

Table 7. Distribution of students' responses in statements Q30 to Q38 of Section D

Statement	1	2	3	4	5
Q30. I can easily search for educational resources available in SlideWiki	2 (2%)	7 (6%)	20 (16%)	70 (56%)	26 (21%)
Q31. I can easily locate the educational resources I need in SlideWiki	1 (1%)	5 (4%)	31 (25%)	66 (53%)	22 (18%)
Q32. I consider it important that SlideWiki provides access to OERs published under Open Creative Commons license	0 (0%)	3 (2%)	20 (16%)	77 (62%)	25 (20%)
Q33. I consider it important that SlideWiki supports creating and publishing educational resources under Open Creative Commons license thus retaining the creator's copyright	0 (0%)	5 (4%)	18 (14%)	78 (62%)	24 (19%)
Q34. I consider it important that SlideWiki supports copying and reusing educational resources under the CC-BY-SA 4.0 license, thus allowing efficient collaboration	0 (0%)	7 (6%)	18 (14%)	72 (58%)	28 (22%)
Q35. I consider it important that SlideWiki supports developing a learning design in the form of a WebQuest as slides in a deck	0 (0%)	6 (5%)	26 (21%)	68 (54%)	25 (20%)
Q36. The exemplar WebQuests available in the SlideWiki platform helped me develop my group's WebQuest	2 (2%)	4 (3%)	25 (20%)	70 (56%)	24 (19%)
Q37. Collaborating with my fellow students/groupmates and co-authoring our WebQuest has enhanced my own knowledge in learning design/developing a WebQuest	1 (1%)	8 (6%)	24 (19%)	69 (55%)	23 (18%)
Q38. I believe that SlideWiki has met the requirement to serve as a platform for the collaborative development of a WebQuest	7 (6%)	8 (6%)	19 (15%)	67 (54%)	24 (19%)

Concerning the development of WebQuests, 73% of students agreed or strongly agreed that SlideWiki has met the requirement to serve as a platform for the collaborative development of a WebQuest (Q38). In particular, 74% considered it important that SlideWiki supports developing a learning design in the form of a WebQuest as slides in a deck (Q35); 75% stated that exemplar WebQuests available in the SlideWiki platform helped them develop their own WebQuest (Q36); and 73% of students said that their knowledge of learning design has been enhanced through the collaboration with peers and the co-authoring of WebQuests (Q37).

4 Conclusions

Harnessing the potential of wikis in education has turned out to be very challenging. Empirical studies using wikis as collaborative learning tools document equivocal findings. This paper reports on a study concerning the SlideWiki platform; focusing on its potential to support collaboration in a teacher education context. Thus, the contribution of the paper is important both in the area of teacher education and as for the development of wikis as crowd-sourcing platforms promoting collaborative authoring.

Indeed, the results of the study revealed that SlideWiki performed quite well at supporting students (pre-service teachers) collaboration to develop learning designs. The fact that almost half of the students considered the collaboration at SlideWiki to be more effective than face-to-face collaboration is quite promising, since these particular students are not experienced in working from a distance. The vast majority of students enjoyed sharing and co-authoring their WebQuest assignment within their group in SlideWiki platform and stated that they saved time, collaborated more easily, and contributed more equally to the assigned task. The students also considered that these features of SlideWiki enhanced the sense of community in the class. However, most students suggested that collaboration could be further facilitated if SlideWiki provided a forum for communication among group members.

The fact that their WebQuests will be available as OERs to all SlideWiki users and not just to their peers motivated students to try harder and improve their work. Moreover, being able to review fellow groups' WebQuests while developing their own, stimulated the students' creativity. Finally, students believed that the collaboration with peers and the co-authoring of WebQuests in SlideWiki significantly enhanced their knowledge of learning design.

Acknowledgment. The research "SlideWiki for collaborative learning design" is implemented and funded through the Horizon 2020 EU research and innovation program, SlideWiki EU Project - EU ICT-20-2015 Project SlideWiki, Grant Agreement No 688095.

References

1. Auer, S., Khalili, A., Tarasowa, D.: Crowd-sourced open courseware authoring with SlideWiki.org. Int. J. Emerg. Technol. Learn. (iJET) **8**(1), 62–63 (2013)
2. Bangor, A., Kortum, P.T., Miller, J.T.: Determining what individual SUS scores mean: adding an adjective rating scale. J. Usability Stud. **4**(3), 114–123 (2009)
3. Brooke, J.: SUS: a "quick and dirty" usability scale. In: Jordan, P.W., Thomas, B., Weerdmeester, B.A., McClelland, A.L. (eds.) Usability Evaluation in Industry, pp. 189–194. Taylor and Francis, London (1996)
4. Dodge, B.: WebQuests: a technique for internet-based learning. Distance Educ. **1**(2), 10–13 (1995)

5. Hadjerrouit, S.: Wiki as a collaborative writing tool in teacher education: evaluation and suggestions for effective use. Comput. Hum. Behav. **32**, 301–312 (2013)
6. Judd, T., Kennedy, G., Cropper, S.: Using wikis for collaborative learning: assessing collaboration through contribution. Australas. J. Educ. Technol. **26**(3), 341–354 (2010)
7. Katsanos, C., Tselios, N., Xenos, M.: Perceived usability evaluation of learning management systems: a first step towards standardization of the system usability scale in Greek. In: Proceedings of the 16th Pan-Hellenic Conference on Informatics, PCI 2012, pp. 302–307 (2012)
8. Khalili A., Auer S., Tarasowa D., Ermilov I.: SlideWiki: elicitation and sharing of corporate knowledge using presentations. In: ten Teije, A. et al. (eds.) Knowledge Engineering and Knowledge Management, EKAW 2012. Lecture Notes in Computer Science, vol. 7603, pp. 302–316. Springer, Berlin (2012)
9. Li, K.M.: Learning styles and perceptions of student teachers of computer-supported collaborative learning strategy using wikis. Australas. J. Educ. Technol. **31**(1), 32–50 (2015)
10. Moundridou, M., Papanikolaou, K.: Educating engineer educators on technology enhanced learning based on TPACK. In: Proceedings of 2017 IEEE Global Engineering Education Conference (EDUCON), pp. 1247–1254. IEEE (2017)
11. O'Bannon, B.W., Lubke, J.K., Britt, V.G.: 'You still need that face-to-face communication': drawing implications from preservice teachers' perceptions of wikis as a collaborative tool. Technol. Pedagog. Educ. **22**(2), 135–152 (2013)
12. Salmon, G.: E-tivities: The Key to Active Online Learning. Routledge, London & New York (2002)
13. Salmon, G.: E-tivities: The Key to Active Online Learning, 2nd edn. Routledge, London & New York (2013)
14. Tarasowa, D., Khalili, A., Auer, S.: CrowdLearn: collaborative engineering of (semi-) structured learning objects. In: Proceedings of the 3rd Russian Conference on Knowledge Engineering and Semantic Web, Saint-Petersburg, Russia (2012)
15. Tarasowa, D., Khalili, A., Auer, S.: Crowdlearn: crowd-sourcing the creation of highly-structured e-learning content. Int. J. Eng. Pedagog. (iJEP) **5**(4), 47–54 (2015)
16. Taylor H.G.: The WebQuest model for inquiry-based learning using the resources of the world wide web. In: Watson, D., Andersen, J. (eds.) Networking the Learner. IFIP—The International Federation for Information Processing, vol. 89. Springer, Boston, MA (2002)
17. Zalavra, E., Papanikolaou, K.: A Framework for Organising a Collaborative Learning Design Activity Based on Wiki. Manuscript submitted for publication (2018)

Study Effort and Student Success: A MOOC Case Study

Jeanette Samuelsen[(✉)] and Mohammad Khalil

University of Bergen, Bergen, Norway
{jeanette.samuelsen,mohammad.khalil}@uib.no

Abstract. Learning was once defined as the function of efforts spent in relation to efforts needed [3]. Provided that effort is closely linked to time, previous research has found a positive relationship between student effort over time and student success, both in university education and Massive Open Online Courses (MOOCs). With the complex environment of tracing and identifying relevant data of student learning processes in MOOCs, this study employs learning analytics to examine this relationship for MITx 6.00x, an introductory programming and computer science MOOC hosted on the edX MOOC platform. A population sample from the MOOC ($N = 32,621$) was examined using logistic regression, controlling for variables that may also influence the outcome. Conversely, the outcome of this research study suggests that there is a curvilinear relationship between effort over time and student success, meaning those who exert effort for the longest amount of time in the MOOC actually have a lower probability of obtaining a certificate than others who exert effort over somewhat less time. Finally, research implications are discussed.

Keywords: Massive open online course (MOOC) · Efforts · Learning analytics · Logistic regression · Total time · Study success

1 Introduction

Academic success through course completion and/or degree attainment is a requirement for many types of jobs. In addition, obtaining a degree is linked to different long-term benefits [15]. Effort is about the "exertion put forth during a task" [20, p. 13]. It may seem intuitive that effort toward academic activities over time will influence student success. However, it is also common for people to view success more as a result of innate abilities, whereby high effort is viewed as a sign of low ability [6].

The relationship between student effort over time and student success in higher education has been examined in the literature. Bowman et al. [2] conducted multiple regression analyses to examine the relationship between perseverance of effort and grade point average (GPA) among undergraduates at Bowling Green State University and the University of Wisconsin at La Crosse. They found a significant positive relationship between the two variables. Strayhorn [23] examined the role of grit on college grades among a subpopulation at a research university, using hierarchical regression techniques. The results indicated that grit was positively related to college

© Springer Nature Switzerland AG 2020
M. E. Auer and T. Tsiatsos (Eds.): ICL 2018, AISC 916, pp. 215–226, 2020.
https://doi.org/10.1007/978-3-030-11932-4_22

grades among the subpopulation. Grit is actually a composite measure encompassing both the behavioral part perseverance and the cognitive aspect passion. However, evidence suggests that behavioral measures are more important than cognitive regarding academic outcomes [23]. Cross [5] examined the role of grit on current student GPA for a group of non-traditional doctoral students in a private university. The results showed a small, but significant relationship between grit and GPA.

A recent trend has been for institutions to host their own Massive Open Online Courses (MOOCs). "A MOOC is an online course with the option of free and open registration, a publicly-shared curriculum, and open-ended outcomes" [17]. Thus, MOOCs allow people with varying levels of time commitment and education levels to participate. Learning analytics studies in MOOCs have found that MOOC participation rate varies to a larger degree than for traditional higher education [4]. The dropout rate is also much higher in MOOCs than in traditional university courses [12, 13].

There has been some research on the relationship between effort over time and success in MOOCs. Researchers at Google examined the "Mapping with Google" MOOC [25], where participants could earn a certificate only through completing a final project. This study found that completing other course activities was positively correlated with earning a certificate. Likewise, researchers analyzing data from the HarvardX course "CB22x: The Ancient Greek Hero", found that taking many actions in the MOOC was positively related to earning a certificate [21]. While these studies offer valuable insights, individual MOOCs often have large differences in instructional conditions, student characteristics and collected data [9]. Thereby, statistical models for MOOCs, and their implications, may not be generalized to MOOCs which occur in different contexts.

This study expands upon previous research by examining the relationship between student effort over time and student success in an introductory level MOOC for programming and computer science, "MITx 6.00x". The insights derived from studies such as this one could benefit not only the research community, but also individuals in pursuit of academic success and its potential positive outcomes, and course instructors/providers who want to help and motivate students to realize their potential.

1.1   Research Question

Based on previous findings in other contexts from traditional university courses and MOOCs [2, 25], this research study will examine the relationship between student effort over time and student success in a MOOC case study. The assumption being that more effort over time will continually lead to higher likelihoods of success (i.e. increasing effort will result in better outcomes) [8]. Following, the research of the current study will tackle the following research question:

What is the type of relationship between students invested efforts over time and their success in MOOCs?

2 Method

2.1 MITx 6.00x

MITx 6.00x was an introductory course to computer science and programming offered by MIT from 2012 to 2013, hosted on the edX MOOC platform (http://edx.org). The course included content such as video lessons, homework questions, assignments, three exams, and a forum. Among resources used were 148 videos, 209 problems, and 31 web pages. So-called chapters gave an overarching structure for a majority of the MOOC content (forums were organized outside of this structure). 14 chapters were released over 15 weeks. Earning a certificate, at no cost, was based on getting a final grade of at least 55%. The final grade depended on the performance of exercises, homework, and exams [22].

2.2 Dataset and Participants

This study uses the HarvardX-MITx Person-Course dataset [19], which is a freely available dataset. The dataset contains de-identified, aggregated information for each individual that participated in MOOCs from Harvard and MIT on the edX MOOC platform; excluding individuals that could not be reliably de-identified. The dataset was loaded into R from a comma-separated values (CSV) file. Data from 16 MOOC offerings were included in the dataset, but through filtering on the course ID of our studied MOOC, the data used in this study contained only observations from the MITx 6.00x MOOC. Information for the variables in the dataset was either derived from the usage of the MOOC (through log data), or self-reported by the participants in an online questionnaire provided upon registration.

In total, 84,511 students originally registered for the course, but due to removal of observations, records from 32,621 participants were analyzed using logistic regression.

2.3 Data Pre-processing

Filtering and corrections were carried out before the dataset was analyzed. Outliers of extreme amount of interactions were removed. Some other observations were filtered out due to inconsistencies. Unrealistically high values and observations with blank values on some of the control variables (indicating participant unwillingness to answer a specific question) were also removed. Some variables were transformed before being used: The dataset variable named year of birth was transformed to age. The dataset variable start time was transformed from date format to day format to ease the analysis process. Avoiding discrimination was also considered, for instance individual countries and regions were recategorized into continents.

2.4 Measures

This study employed logistic regression, a method that has been widely used in the learning analytics and educational data mining fields [1]. The dependent variable for the regression, representing student success, was certified (0: no; 1: yes). As stated before, obtaining a certificate implied getting a final grade of 55% or more. The independent variable, representing effort over time, was number of days active (at least one click in a given day). Both variables were derived from log data. Some control variables were included in the regression: age, gender, level of education, continent and start day. Continuous variables were the following: number of days active, age and start day. The other variables were categorical: gender, level of education and continent.

Measure for Effort Over Time

Deciding to use the number of days active as the most efficient measure metric for effort over time from the log files is based on two reasonings: (a) supporting literature such as the research studies by Khalil and Ebner [13] and Kloft et al. [14]. And (b) exploratory examination of correlations, knowledge of the problem domain, and examining descriptive statistics for the variables as the following.

There were also four other nominees that could represent effort over time, in addition to the number of days active: number of events (interactions with the MOOC), number of video play events, number of chapters accessed, and number of forum posts. Executing Pearson correlation for the five variables ($n = 35,115$ due to removal of observations missing values) made it clear that the number of forum posts had very little correlation with the rest of the variables. Its largest correlation was with number of days active ($r = .26$), and its smallest correlation was with the number of video play events ($r = .16$) (see Table 1). Based on the exploratory examination of the correlations, it did not seem that the number of forum posts was a good indicator of effort over time. In retrospect, it seemed that the number of forum posts might be a measure of social behavior [16].

Table 1. Correlation matrix for candidate variables effort over time variables

	No. events	No. days active	No. video play events	No. chap. accessed	No. forum p.
No. events	1				
No. days active	.87***	1			
No. video play events	.74***	.66***	1		
No. chapters accessed	.81***	.88***	.61***	1	
No. forum posts	.25***	.26***	.16***	.22***	1

$* p < .05; ** p < .01; *** p < .001$

With regards to the number of chapters accessed, it is reasonable to assume that this variable would have a high value for individuals who exerted much effort over time in the MOOC. However, because chapters served as an overarching structure for materials such as exercises, homework and exams, it would also be high for students who earned a certificate despite exerting low effort in the MOOC (for instance students with pre-requisite knowledge who obtained a certificate only through graded exercises, without learning anything new).

The next variable to consider as a measure for effort over time was the number of video play events. Descriptive statistics for this variable suggested video play habits of individuals varied greatly. The maximum number of play events was 8,632, and 583 records contained more than 1,000 video play events ($n = 36,289$ after deletion of observations missing values). A thousand plays of the 148 available videos would imply that each video had been seen almost seven times. Here, a more plausible explanation is that these numbers are a result of pressing pause and play, rewinding the videos, etc. It seemed that this measurement encompassed different types of interactions; hence, it was excluded.

The number of events variable did not seem like a good choice either, as it included video play events and forum posts. In the end, we decided that the most reasonable measurement for effort over time was the number of days active. In fact, Kloft et al. [14] has also identified the number of active days as the most important metric to predict dropout. Although this surrogate measure admittedly is more of a quantitative measure than a qualitative (it is impossible to assess the exact level of effort exerted over the number of days active), it does indicate students' commitment to the MOOC.

Control Variables

As previously mentioned, this study controls for differences in start day, age, continent, and level of education ("Less than Secondary", "Secondary", "Bachelor's", "Master's", "Doctorate"). Start day is an integer signifying which day a student registered for the MOOC, relative to days since registration was made possible (start day 1 would mean that the student registered on the first possible day). This variable could influence certification, as MITx 6.00x was a highly structured course [22]. Considerable research has also suggested that demographic factors may influence student success [23]. In addition, it seems reasonable to assume that level of education could have an impact on individual students' likelihood of obtaining a certificate. Start day, age and continent variables were either transformed or recategorized from original dataset variable to measure used in this study (see Sect. 2.3). Of the control variables, start day and parts of the continent data were based on information derived from the log files. Data for the other measures were self-reported.

2.5   Statistical Analysis

Correlations between candidate measures for student effort over time were assessed initially, to help find one or more appropriate measures to include in the study. Descriptive statistics were utilized to better understand the characteristics of the participants, with regards to the independent, dependent and control variables. Logistic regression was run to examine if there was an association between the independent variable student effort over time (represented by number of days the student was active) and the dependent variable student success (obtaining a certificate for successfully completing the MOOC). We also controlled for additional variables by adding them as additional independent variables to the logistic regression.

Categorical variables were dummy coded for use in the logistic regression. The reference category for gender was female, the reference category for level of education was less than secondary, and the reference for continent was North America.

Measures were taken to assess how well the data met the assumptions of logistic regression, and to make necessary corrections where possible. To check if the continuous predictors were related to the log of the outcome variable, interaction terms for the continuous predictors were tentatively added to the logistic regression and assessed for their significance score after running the regression. This was based on the recommendation by Field et al. [7, p. 344–345]. Actually, it was found that the interaction term for the number of days active was significant, indicating that the assumption had been violated for this variable. To address this violation, the squared term for the number of days active was added to the logistic regression. Running the logistic regression with the square of the number of days active yielded a significantly better model fit than the model that did not include the squared term ($X^2 = 226.43$, $p < .001$, df = 1), suggesting a curvilinear relationship between the predictor and outcome variable. Curvilinear relationships are a quite common occurrence within the social sciences [18].

To check the assumption of absence of multicollinearity for logistic regression, variance inflation factor (VIF) and tolerance values were assessed for the independent variables entered into the logistic regression (number of days active, number of days active squared, and the control variables). Because the number of days active squared had been entered into the regression, it was natural to assume that this term would be highly correlated with the number of days active variable (i.e., itself not squared). However, in the instance of curvilinear relationships between predictor and outcome, multicollinearity can still be accepted [18]. Both number of days active and number of days active squared had a VIF of 9. The VIF value is quite high, nearing the value of 10 which is often especially problematic. The mean VIF was 3.6, which may indicate that multicollinearity can lead to some bias in the model [7].

Due to focusing on a sole MOOC, it seemed reasonable to assume that the data were not related (i.e. errors are independently distributed). Observations were assessed for their DFBETA value to identify influential cases for the logistic regression. No observations were identified as having a substantial influence (DFBETA value above 1). Observations missing values for variables used in the logistic regression were removed before analysis.

2.6   Limitations

Some limitations apply to this study. The full population participating in the MOOC could not be analyzed, for a variety of reasons. When downloading the dataset some observations had been removed, for anonymity reasons. After filtering out data and removal of observations with missing values, we analyzed a sample of only 32,621 students with logistic regression. This may have biased estimates and inflated standard errors since data were not missing completely at random. The use of a squared term in the logistic regression, for number of days active, resulted in a better model fit but also introduced a degree of multicollinearity, which may have somewhat biased the model [18]. Number of days active was admittedly a quite coarse-grained measure for student effort over time, and if available we might have found that one or more other measures (e.g. combined through factor analysis) were better options. Using certification as a measure of student success is limited by the fact that some individuals may not see test scores and certification as a necessity. One more concern is that some information for the control variables were self-reported, which may have introduced bias [2].

3 Results

3.1 Descriptive Statistics

Table 2 presents the means and standard deviations for the continuous variables included in the study, and the percentages for each level of the categorical variables.

From the table, we see that only five percent of the participants earned a certificate (mean 0.05, SD 0.22). This percentage amounted to 1601 of the 32,621 participants. The dropout ratio is as high as reported in many studies like [12, 13]. For certificate learners, the mean number of days active were 66.31.

Table 2. Descriptive statistics for independent (including control) variables and the dependent variable in the study

n = 32,621	M/SD	Percent(%)
Number of days active	9.15/16.43	
Start day	65.29/34.69	
Certification	0.05/0.22	
Age (years)	26.14/7.51	
Gender?		
Male		86
Female		14
Level of education?		
Less than secondary education		3
Secondary education		34
Bachelor's degree		43
Master's degree		19
Doctoral degree		1
Continent?		
Africa		10
Asia		28
Europe		24
North America		29
Oceania		1
South America		7

3.2 Logistic Regression Analysis

A logistic regression was used to predict the relationship between student effort over time and student success, controlling for some demographic variables, level of education, and start time. A likelihood ratio test of the full model against a null model was statistically significant, indicating that predictors can reliably separate between students who obtain a certification and students who do not ($X^2 = 9607.44$, $p < .001$, df = 14) (see Table 3). The model correctly classified 97,3% of the observations.

To interpret how the number of days active was related to obtaining a certificate, the total logit for the different possible values for number of days active (1-138) was calculated, holding the other continuous variables at mean (start day = 65.29, age = 26.14), and the categorical variables at reference group (gender, female; level of education, less than secondary; continent, North America). The individual total logits were then transformed into probabilities, and the calculated probabilities for obtaining a certificate were plotted for the individual number of days (all these operations were coded manually in R, due to limitations in the margins library for R). As seen in Fig. 1, the results suggest there is initially almost a linear positive relationship between the number of days active and the probability of obtaining a certificate, but around day hundred the previously almost linear relationship seems to hit a plateau, and somewhat later the positive relationship actually weakens (the relationship is curvilinear). Here, the model implies that the participants with the most number of days active were actually less likely to obtain a certificate than participants who were active for some-what less number of days (since the probability of earning a certificate is based on total logits it is dependent on the values of the control variables).

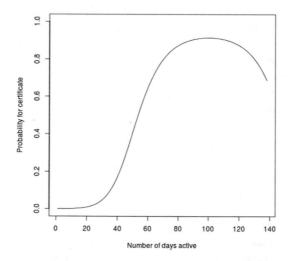

Fig. 1. Calculated probability for obtaining a certificate, related to the number of days active (continuous variables set at means, categorical variables set at reference group)

For the control variables, we see in Table 3 that age is significant, at 0.95 odds ratio, implying that as age increases by one unit, the odds of obtaining a certificate decrease by five percent, when holding the other variables constant. Start day is significant with the odds ratio of 1.01, implying that the odds of obtaining a certificate increases by one percent for each successive day of a participant registering for the

MOOC. For gender, being male has a less positive relationship with earning a certificate than being female (odds ratio 0.62). For instance, if we examine the gender differences in the total probability for certification with 100 days active (thus generally a high probability for earning a certificate, as seen in Fig. 1), holding the other continuous variables at their mean, and the other categorical variables at their reference, the model suggests there is a 4.5% higher probability for earning a certificate for females than for males (91.4% versus 86.9%). For level of education, we see that having less than secondary education (the reference group) is associated with much lower odds for obtaining a certificate than the other education levels. For continents, Africa is significant (odds ratio 0.42), implying lower odds of obtaining a certificate for people from this continent than for those from North America (the reference group). On the other hand, being from Asia is associated with higher odds for obtaining a certificate than being from North America (odds ratio 1.58). Results from the other continents suggest that being from those respective continents are associated with higher odds for obtaining a certificate than being from North America; however, these results are not significant.

Table 3. Logistic regression analysis of the relationship between student effort over time and student success in the studied MOOC

	B (SE)	95% CI for odds ratio		
		Lower	Odds ratio	Upper
No. days active	0.22*** (0.01)	1.23	1.25	1.26
(No. days active)^2	0.00*** (0.00)	1.00	1.00	1.00
Start day	0.01*** (0.00)	1.00	1.01	1.01
Age	−0.05*** (0.01)	0.93	0.95	0.97
Male	−0.47*** (0.14)	0.48	0.62	0.82
Secondary education	1.18*** (0.31)	1.80	3.27	5.99
Bachelor's degree	1.12*** (0.31)	1.67	3.07	5.68
Master's degree	1.27*** (0.33)	1.88	3.56	6.78
Doctoral degree	1.58** (0.49)	1.85	4.84	12.64
Africa	−0.86** (0.28)	0.24	0.42	0.72
Asia	0.45** (0.15)	1.18	1.58	2.11
Europe	0.12 (0.13)	0.88	1.13	1.44
Oceania	0.47 (0.47)	0.62	1.60	3.96
South America	0.10 (0.23)	0.70	1.10	1.72
Constant	−7.81*** (0.40)	0.00	0.00	0.00

Note. R^2 = 0.75 (Hosmer-Lemeshow), 0.26 (Cox-Snell), .79 (Nagelkerke)
Model X^2 (14) = 9607.44, $p < .001$.
* $p < .05$; ** $p < .01$; *** $p < .001$

4 Discussion and Conclusion

To answer our research question *"What is the type of relationship between students invested efforts over time and their success in MOOCs?"*, this research study findings suggest that there is initially almost a linear positive relationship between student effort over time and student success in a MOOC, as was expected. However, interestingly enough, the study indicates that the previously almost linear relationship plateaus over time, and eventually the positive relationship actually weakens. This suggests that those who exerted effort over the longest amount of time actually had a lower probability of obtaining a certificate than others who exerted effort over somewhat less time (as exemplified in Fig. 1). Thus, the study suggests the relationship between effort over time and success is actually curvilinear. The curvilinear relationship was quite surprising given the initial assumption presented in Sect. 1.1 that more effort over time would continually lead to higher likelihoods of success.

Although the previous explanation clarifies the correlation between certification ratio and student efforts over time, there are other variables that affect this correlation. Among the included control variables, we saw that increasing age somewhat negatively influences the odds for student success, increasing start day slightly positively influences the odds for student success, that females have a higher probability of earning a certificate than males, and that those with less than secondary education had much lower odds for obtaining a certificate than their counterparts with more education. Being from Asia is associated with higher odds for obtaining a certificate than being from North America, while the opposite is true for Africa.

Returning to the finding that the relationship between student effort over time and student success is suggested to be curvilinear, one possible reason for this may be related to the concept of achievement goals. Achievement goals are about why someone shows achievement [20, p. 255]. The two types of achievement goals are presentation goals and mastery goals. People who set presentation goals are generally more concerned with proving to others that they are competent and have high ability. We can envision that earning a certificate is one such way of proving competence. On the other hand, people who set mastery goals are more concerned with self-improvement, developing competence, and overcoming challenges through effort. Thus, it may be that some of the people who exert the most effort over the longest periods of time are mastery oriented, i.e., they work hard and master challenging tasks, but may not even be interested in earning a certificate (proving their ability). It has been found that people who set mastery goals are often more internally than externally motivated [20]. Another possible explanation for the finding of the curvilinear relationship between student effort over time and student success could be that some students may just need more time to learn and develop competence in introductory programming and computer science than others, for instance, based on their prerequisite knowledge.

While researchers have found a positive relationship between effort over time and student success, both for university education and MOOCs [2, 25], a curvilinear relationship between student effort over time and student success is to best of our knowledge a unique conclusion in this research study. Given that MOOCs have a higher dropout rate than more traditional university education, and that there are more pressures (e.g. economic pressures) related to completing a university education than a MOOC, it does not necessarily follow that we would have the same findings when researching more traditional university courses. Since it is difficult to generalize statistical models and the implications of this study to other MOOCs, it is unclear if this finding would apply to MOOCs occurring in other contexts as well. However, since this study accounts for student characteristics, we may expect that the results could, at least to a larger extent, be generalized to similar types of MOOCs (introductory programming and computer science MOOCs), provided that instructional conditions and data collection procedures are closely matched.

Control variables also had a significant impact on the outcome variable. The finding that increasing age had a negative relationship with odds for obtaining a certificate could be due to younger people generally being more used to information technology than the older adults. The finding that increasing start day has a slightly positive effect on the odds of obtaining a certificate could actually be because the start day is set to the first day when it was possible to register, instead of for instance the day of the MOOC launch. The finding that gender influences the probability of success is consistent with other findings from more traditional education [24]. In the studied MOOC, we saw that there was a much higher amount of men than women enrolled. This could be due to the subject matters of computer science and programming, which tend to have a higher amount of males than females, both in education and in the workforce [11]. It is perhaps unsurprising that having less than secondary education could influence the odds for success in a MOOC, compared to having more education. The finding that being from the continent of Africa suggests lower odds for student success in comparison to being from North America may for instance be influenced by difficulties with the English language. However, the continent measure is an aggregate, meaning that there may be large differences among individual countries. For Asian learners, the finding that they have higher odds for student success than learners from North America could for instance be influenced by the fact that some of the Asian countries are among the best on all three facets of the PISA performance rankings [10], suggesting quite excellent education systems for some of the countries. Still, it should again be stressed that continent is an aggregate measure.

References

1. Baker, R.S., Inventado, P.S.: Educational data mining and learning analytics. In: Learning Analytics: From Research to Practice, pp. 61–75. Springer, New York (2014)
2. Bowman, N.A., Hill, P.L., Denson, N., Bronkema, R.: Keep on truckin'or stay the course? Exploring grit dimensions as differential predictors of educational achievement, satisfaction, and intentions. Soc. Psychol. Pers. Sci. **6**(6), 639–645 (2015)
3. Carroll, J.: A model of school learning. Teach. Coll. Rec. **64**(8), 723–733 (1963)
4. Clow, D.: MOOCs and the funnel of participation. In: Proceedings of the Third International Conference on Learning Analytics and Knowledge, pp. 185–189 (2013)
5. Cross, T.M.: The gritty: grit and non-traditional doctoral student success. J. Educ. Online **11**(3), 1–30 (2014)

6. Dweck, C.S.: The secret to raising smart kids. Sci. Am. Mind **18**(6), 36–43 (2007)
7. Field, A.P., Miles, J., Field, Z.: Discovering Statistics Using R. Sage, London (2012)
8. Firmin, R., Schiorring, E., Whitmer, J., Willett, T., Collins, E.D., Sujitparapitaya, S.: Case study: using MOOCs for conventional college coursework. Distance Educ. **35**(2), 178–201 (2014)
9. Gašević, D., Dawson, S., Rogers, T., Gasevic, D.: Learning analytics should not promote one size fits all: the effects of instructional conditions in predicting academic success. Internet High. Educ. **28**, 68–84 (2016)
10. Gurria, A.: PISA 2015 results in focus. http://www.oecd.org/pisa/pisa-2015-results-in-focus.pdf (2018)
11. Hill, C., Corbett, C., St. Rose, A.: Why So Few? Women in Science, Technology, Engineering, and Mathematics. American Association of University Women (AAUW) (2010)
12. Khalil, H., Ebner, M.: MOOCs completion rates and possible methods to improve retention-a literature review. In: World Conference on Educational Multimedia, Hypermedia and Telecommunications, pp. 1305–1313 (2014)
13. Khalil, M., Ebner, M.: What massive open online course (MOOC) stakeholders can learn from learning analytics? In: Spector, M., Lockee, B., Childress, M. (eds.) Learning, Design, and Technology: An International Compendium of Theory, Research, Practice, and Policy, pp. 1–30. Springer, Heildelberg (2016). http://dx.doi.org/10.1007/978-3-319-17727-4_3-1
14. Kloft, M., Stiehler, F., Zheng, Z., Pinkwart, N.: Predicting MOOC dropout over weeks using machine learning methods. In: Proceedings of the EMNLP 2014 Workshop on Modeling Large Scale Social Interaction in MOOCs, pp. 60–65 (2014)
15. Kuh, G.D., Cruce, T.M., Shoup, R., Kinzie, J., Gonyea, R.M.: Unmasking the effects of student engagement on first-year college grades and persistence. J. High. Educ. **79**(5), 540–563 (2008)
16. Lackner, E., Khalil, M., Ebner, M.: How to foster forum discussions within MOOCs: a case study. Int. J. Acad. Res. Educ. **2**(2) (2016)
17. McAuley, A., Stewart, B., Siemens, G., Cormier, D.: The MOOC Model for Digital Practice (2010)
18. Mehmetoglu, M., Jakobsen, T.G.: Applied Statistics Using Stata: A Guide for the Social Sciences. Sage (2016)
19. MITx, HarvardX: HarvardX-MITx Person-Course Academic Year 2013 De-Identified dataset, version 2.0. Harvard Dataverse (2014). https://doi.org/10.7910/DVN/26147
20. Reeve, J.: Understanding Motivation and Emotion, 6th edn. Wiley (2014)
21. Reich, J., Emanuel, J., Nesterko, S., Seaton, D.T., Mullaney, T., Waldo, J., …, Ho, A.D.: HeroesX: The Ancient Greek Hero: Spring 2013 Course Report (2014)
22. Seaton, D.T., Reich, J., Nesterko, S.O., Mullaney, T., Waldo, J., Ho, A.D., Chuang, I.: 6.00 x Introduction to Computer Science and Programming MITx on edX Course Report - 2012 Fall (2014)
23. Strayhorn, T.L.: What role does grit play in the academic success of black male collegians at predominantly white institutions? J. Afr. Am. Stud. **18**(1), 1–10 (2014)
24. Zhang, G., Anderson, T.J., Ohland, M.W., Thorndyke, B.R.: Identifying factors influencing engineering student graduation: a longitudinal and cross-institutional study. J. Eng. Educ. **93**(4), 313–320 (2004)
25. Wilkowski, J., Deutsch, A., Russell, D.M.: Student skill and goal achievement in the mapping with google MOOC. In: Proceedings of the First ACM Conference on Learning@ Scale Conference, pp. 3–10 (2014)

Computer Aided Language Learning (CALL)

Defying Learning Traditions: From Teacher-Centred to Student-Centred Foreign Language Education Through Digital Transformation at Sri Lankan Universities

Neelakshi Chandrasena Premawardhena[✉]

Department of Modern Languages, University of Kelaniya, Kelaniya, Sri Lanka
Neelakshi3@yahoo.com, neelakshi@kln.ac.lk

Abstract. The learning traditions of Sri Lanka demand the authoritative role of the teachers. As a result, the learners become comfortable in their passive role and depend heavily on their teachers as the sole provider of knowledge and resources. It has been observed that this dependency impedes the learning process and restricts the potential for self-improvement especially in language education. Thus, shifting from traditional teacher dependent learning to autonomous learning is a major challenge for any student gaining entrance to the university who has experienced frontal teaching for over twelve years in school in Sri Lanka. Nevertheless, transformation from the traditional teacher dependent learning to digital language education has resulted in successfully addressing many issues faced by the teachers and students alike in foreign language education at university level. This paper examines the contribution of authentic material available online to the vast improvement of student performance at the University of Kelaniya and how this digital transformation contributed to shifting from teacher dependent learning to learner autonomy through online resources.

Keywords: Foreign language teaching · Traditional learning styles · Teacher-centred learning · Learner autonomy

1 Introduction

Teacher centred education is prevalent in primary and secondary education in Sri Lanka despite the initiatives taken several decades ago by the state and teacher training institutions to shift to learner centred approach. The learning traditions of Sri Lanka is marked by the authoritative role of the teachers [2, 3]. As a result, the learners become comfortable in their passive role and take no extra effort to improve their knowledge on their own. The inclination to depend on teachers as the sole provider of knowledge leaves little or no room for the learner for independent study or self- improvement. This is a major shortcoming especially in teaching and learning of languages where the learning process is not restricted to the classroom alone. Thus, shifting from traditional teacher dependent learning to autonomous learning is a major challenge for any student gaining entrance to the university in Sri Lanka who has experienced frontal teaching for over twelve years in school. Pilot studies and research conducted since 2004 at the University of Kelaniya in

© Springer Nature Switzerland AG 2020
M. E. Auer and T. Tsiatsos (Eds.): ICL 2018, AISC 916, pp. 229–238, 2020.
https://doi.org/10.1007/978-3-030-11932-4_23

Sri Lanka on foreign language teaching as stated in [4, 5] bear testimony to the positive results achieved in enhancing student performance through Computer Assisted Language Learning (CALL). Transformation from the traditional teacher dependent learning to digital language education has resulted in successfully addressing many issues faced by the teachers and students alike in foreign language education at the university [5].

2   Significance of the Study

Figure 1 encouraging and facilitating learner autonomy is pivotal in language education. Numerous scholars have emphasised the need for learner autonomy for more effective language learning. Cotterall mentions three types of autonomy i.e. philosophical, pedagogical, and practical reasons with regard to learning of languages [6]. She argues that the language learners need to have a choice of what they learn at their own pace and that the learners should be able to engage in their activities independent of the supervision of a teacher [7, 8]. Littlewood as stated in [10] points out that the students will not always have "their teachers to accompany them throughout life" and thus autonomy is an integral part of learning. Benson focuses on three areas including the nature of autonomy, efforts to foster learner autonomy and the relationship between learner autonomy and effective language learning [1]. Little as stated in Ref. [9] describes autonomy "as a capacity for detachment, critical reflection, decision-making and independent action." Scholars recognize the significance of learning outside the classroom. Benson as stated in [1] describes out of class learning as "any kind of learning that takes place outside the classroom and involves self-instruction, naturalistic learning or self-directed naturalistic learning". Thus, it is beyond doubt that transforming from teacher dependent learning to student centred learning benefits both the teacher and the learner.

2.1   Background

University of Kelaniya has the only dedicated department of study in Sri Lanka for the teaching of foreign languages. The Department of Modern Languages offers two types of degree programmes for the Bachelor i.e. a three year degree programme and a four year Honours Degree Programme specializing in one selected subject. Master of Philosophy and doctoral degrees by research in Chinese, French, German, Japanese, Korean and Russian are offered as postgraduate programmes. Hindi Studies has an independent department of study. Prior knowledge of languages for the Bachelor Degree is required only for French and Japanese since many secondary schools offer the two languages at the national examination General Certificate of Education (Advanced Level) which is also the qualifying examination to gain entrance to a state university in Sri Lanka. All the other languages can be offered at beginner level from the first year of study at the university. With two types of Bachelor Degrees offered for students of foreign languages they are expected to reach B1 and B2 levels of the Common European Framework of Reference (CEFR) respectively at the completion of their study programmes. The undergraduates are expected to accumulate a minimum of 30 credits during the academic year with each subject offering 10 credits. According to the Sri Lanka Quality Assurance Framework one credit is equivalent to 2 credits of European Credit Transfer and

Accumulative System (ECTS). Large numbers of students opt for a foreign language in their first year due to its novelty, wider career prospects and scholarship opportunities as exchange students. The numbers exceed 100 for Chinese, German and Korean. However, from the second year onwards many students select a subject of their choice for the Honours Degree Programme which requires a Grade Point Average (GPA) of 3.0 in the selected subject and 2.3 in the other subjects offered in the first year. Thus, the numbers selecting a foreign language for the Honours Degree Programme is limited. They are expected to follow a variety of course modules during the next three academic years and submit a comprehensive Bachelor thesis in the final year in the respective foreign language. The degree is awarded after accumulating a minimum of 120 credits.

3 Purpose

The subject of study in this paper, German Studies (Germanistik), has 5–10 students selecting the Honours Degree Programme every year. Some of these students already have prior exposure to the language at secondary school level whereas many others start learning the language during their first year at the university. Therefore, at the commencement of the study programme, the students may have different levels of language knowledge varying from A1 to A2 levels of CEFR. Furthermore, a certain number of students continue to follow language courses at the Goethe Institute, Colombo and may reach B1 level at the beginning of their Honour Degree Programme from the second year onwards. Thus, the challenges faced by the teachers and students alike to acquire competency in the foreign language are not to be undermined. Shifting from traditional passive learning at school to active learning at the university is also no easy task for the learners [3]. However, integrating CALL into the language classroom since 2004 vastly contributed to enhancing student motivation and performance [5]. This paper examines how digital transformation in foreign language education at the University of Kelaniya has affected the student performance with supporting data available from the Bachelor of Arts Degree Programme in German Studies at the University of Kelaniya. Prior knowledge of the language is not a pre-requisite for the students of German in their first year of study which is offered at two different degree programmes i.e. the three year Bachelor Degree Programme with German as a Foreign Language (*DaF*) as one of the subject options offering 30 credits, and the four year Bachelor majoring in German Studies (*Germanistik*) offering 120 credits. During the first year a student is required to obtain a minimum of 30 credits which is generally achieved by opting for three subjects. As mentioned above, any student obtaining a Grade Point Average (GPA) of 3 in the intended subject for specialisation and 2.3 in the other remaining 20 credits is eligible to enter the Honours Degree Programme specialising in one selected subject from the second year onwards for the next three academic years. It is noteworthy that the GPA of 3 is in the range of 55–59% marks. When analysing the results of previous ten years, it has been observed that this goal is achieved during the first year by over 70% of the students as the learning outcome is to reach language level A1 of Common European Framework of Reference (CEFR). Thus, any student who fulfils this criterion enters the Honours Degree Programme in German Studies that comprise course modules covering a wide spectrum including language, literature, culture, history, teaching methodology, linguistics, political and social history, amounting to a minimum of 90 credits that are to be accumulated within 3 academic years. The challenges occur from the second year onwards if the students select the specialised programme in *Germanistik* as the content and workload is very high with continuous assessments, presentations, take-home assignments and lecture sessions held throughout the week. Among the eligible students who select the Honours Degree Programme are also some who had offered German for their Advanced Level Examination at secondary school, thus reaching A2 of CEFR at the start of their first year of study. Therefore, a knowledge gap exists between absolute beginners at the university who reach A1 in the first year and those with prior knowledge of the language. This is not clearly visible at the first year examinations as the examinations are conducted at A1 level. However, during the high demanding course modules of the Honours Degree Programme the students who start German as an absolute beginner at the university have to work with extra effort to keep abreast with their peers who have already reached A2 level before entering the university. This paper examines how use of online teaching material and authentic material contributed to the vast improvement of the student performance bridging the gap between the absolute beginners and advanced learners and how this digital transformation contributed to shift from teacher dependent learning to autonomous learning through online resources.

4   Approach

A common learning platform for BA Honours Degree students majoring in German Studies (Germanistik) that was created in March 2015 at the beginning of the academic year with access to authentic material available online proved successful in enhancing overall student performance at the university examinations [5]. Over 200 students accessed the learning platform during their studies and worked on additional material provided. However, as mentioned above, the knowledge gap between the students beginning the language at the university and ones with prior knowledge continued to exist. Since 2016 the delivery of lectures for course modules relevant to language and culture entirely depend on online resources with audio and video material, online exercises, interactive exercises and online testing tools available on the online platform created at www.openlearning.com. The platform managed by the author is open to all students and teachers of German at the Department of Modern Language. A separate course is available for each batch of students created to achieve the learning outcomes of the course units conducted by the author. New material is uploaded prior to every lesson with additional resources for independent study hours. When considering the time allocation, only four hours of lecture sessions per week are available for the two course modules in the third year for *GERM 33516 Oral and Written Communication Skills II* and *GERM 33536 Analysis and Interpretation of Non-literary Texts*. For the fourth year students three hours are allocated for the module on *GERM 43515 Discourse Analysis and German Linguistics*. Nevertheless, the material available on the online learning platform gave students ample opportunity to work independently at their own space outside the regular lectures, thus increasing the number of contact hours with the language. The students could follow instructions given on the learning platform and complete the assigned work for the day even during the absence of the lecturer. Performance of two batches of students from the Honours Degree Programme was assessed for this study on their competency at examinations, overall performance and how the online learning platform contributed to shift from teacher dependent learning to autonomous learning.

Figures 1, 2 and 3 demonstrate the sample lessons from the online courses for the modules in third and fourth years. The material is authentic and up to date, mostly comprising texts, audio, video material and interactive exercises to address the learning outcomes of the relevant lesson, main resource being the website of the Deutsche Welle for Teaching German as a Foreign Language at www.dwcom/de. The material provided is well designed and structured to suit the target group ranging from A1-C1 of CEFR. The students are thus exposed to the linguistic as well as cultural resources which equip them with required competencies of the course modules.

Fig. 1. Authentic texts - German university life

Fig. 2. Authentic texts - politeness strategies

Fig. 3. Authentic texts and online exercises

The examinations taken for the sample include written papers of three hours duration for each course module and class tests, continuous assessments, assignments and oral presentations. Transcripts of students were obtained to analyse their performance at all the examinations they hitherto faced. The third year group had one academic year of work with the online learning platform whereas the fourth year batch used the platform for two academic years. Teacher and student feedback as well as performance at examinations during the past three years, continuous assessments, presentations and assignments were taken into consideration during the analysis. The student feedback included responses on the time spent outside designated lecture hours in improving their language skills and what resources were used, their views on the common learning platform, material available, opportunity created for independent learning, IT facilities available, issues faced and suggestions on areas for further improvement. The teacher feedback focused more on their view of student performance and their inclination to work independently in contrast to previous years.

5 Actual Outcomes

The analysis of examination results of the students for a period of three years showed remarkable improvement in the language competency of all the students in the sample. It was evident that the gap between the students who started German at the university and those who had prior knowledge had dramatically reduced. The most significant improvement was observed among the students in the third year who merely had one year of work on the online learning platform. An increase of 15–20% is evident in the marks obtained at the last examinations held in January/February 2018 in their third year in comparison to their second year. Moreover, some students who showed average performance in their second year by obtaining 45–50% of marks previously before being introduced to the online platform achieved above 70% after working for one academic year with digital resources on the platform. The following table gives an insight into the student performance before and after the student experience on independent study through the online platform.

Table 1. Grades obtained at second and third year examinations

Student	Second year (2017)				Third year (2018)			
	GERM 23516		GERM 23556		GERM 33516		GERM 33536	
	Grade	Marks%	Grade	Marks%	Grade	Marks%	Grade	Marks%
1	A+	85–100	A+	85–100	A+	85–100	A+	85–100
2	A+	85–100	A+	85–100	A+	85–100	A+	85–100
3	B+	60–64	A	70–84	A	70–84	A	70–84
4	C	40–45	B+	60–64	A	70–84	A	70–84
5	B	50–54	B	50–54	A	70–84	A	70–84
6	C+	45–49	B	50–54	A	70–84	A	70–84
7	A	70–84	A+	85–100	A+	85–100	A+	85–100
8	A	70–84	A	70–84	A	70–84	A+	85–100

As seen in Table 1 all the students in the sample obtained Grade 'A' by scoring between 70–89% or 'A+' by scoring between 85–100% at the end of module examinations in their third year. Student numbered 1, 2, and 3 in Table 1 had prior knowledge of German when entering the university having offered the subject at the G. C E (Advanced Level Examination – A2 of CEFR). The results in the third year show that the gap between the new learners and the others ceased to exist as seen in the two columns relevant to the performance in their third year.

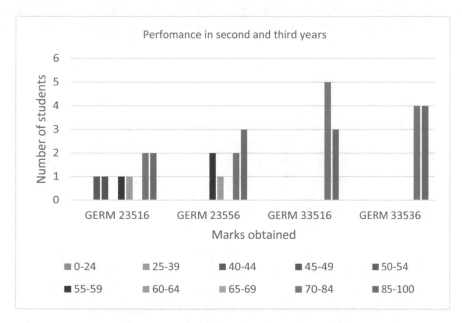

Fig. 4. Performance in second and third years

Figure 4 further illustrates the performance of students in the selected modules in their second and third years. The second year modules were completed in 2017 prior to this batch of students being introduced to the online learning platform. In their third year all the students obtained Grade 'A' or 'A+' by scoring 70% or above in both modules considered for the study at the examinations held in January-February 2018. Furthermore, it was evident that the oral and written competencies and cultural awareness had vastly improved and the enhanced performance was reflected on all course modules including literature and social history in the third year.

The student feedback gave an insight into increased motivation of the students to work independently by addressing their own strengths and weaknesses and the advantage of having selected authentic material for the online learning platform with relevant instructions. Accordingly students spend 3–4 h of independent study per day on average during the week outside the classroom using material on the online platform. This time is spent at the ICT Centres at the university or at university hostels where access to the wireless network of the university is available free of charge. According to the student responses, the lack of internet access or limited access at home due to financial constraints restrict many students from working on the online platform during weekends. All the students were of the view that the new learning style vastly contributed to enhancing their performance and confidence in using the language at any given situation. Students were also motivated to search for more online resources on their own. As areas for improvement the students requested for more access to IT facilities at the university and integrating other course modules of the German Studies Programme also into the learning platform. Feedback from teachers confirmed the vast improvement in the oral and written competency of the selected sample, learner motivation and commitment. It was also revealed during the teacher survey that the teachers spend less time on the online platform than the students due to lack of time resulting from heavy teaching load and commitments to lesson preparation.

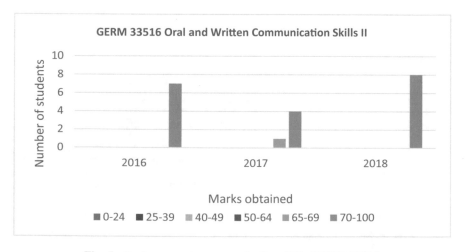

Fig. 5. Performance in communication skills GERM 33516

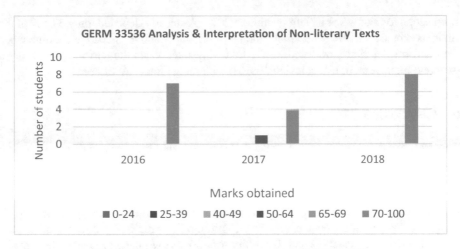

Fig. 6. Performance in text interpretation GERM 33536

Figures 5 and 6 illustrate the performance of students at the end of course examinations of the selected two modules in the third year of study from 2016 to 2018 over a period of three academic years since introducing the online learning platform in 2015. Each batch of students had one year of exposure on the learning platform. In both 2016 and 2018 all the students achieved Grade 'A' or above by scoring above 70% during their performance at the examination.

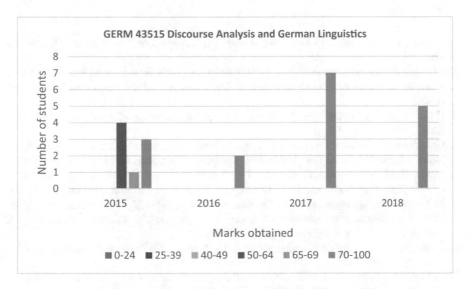

Fig. 7. Performance in Discourse Analysis GERM 43516

Figure 7 demonstrates the performance of students in the final year course module on *Discourse Analysis and German Linguistics* over a period of four years. The students in the years 2016 to 2018 have had the experience of working on the online learning platform for two academic years since its introduction in 2015. As clearly depicted, only 37.5% of the students scored above 70% securing a Grade 'A' Grade or above in 2015 at the time when the platform was first introduced. In the following three years the performance shows a dramatic improvement by achieving 100% in obtaining Grade 'A' of above.

6   Conclusion

The results of the present study conducted over a period of three academic years confirm that the digital transformation in language education has contributed immensely to shift from the traditional textbook book based frontal teaching to learner centred teaching that leads to remarkable improvement in language competency. Albeit the study sample was not large enough to arrive at general conclusions it was a significant step towards deviating from the traditional teacher dependent learning. Working with online resources with less dependence on teachers and the ability to work at own pace motivates the learner. Moreover, the students play an active role in the learning process and believe in their ability to improve their language competency through the material available on the online platform. Thus, the digital transformation in language teaching has immensely benefited the teachers and the students alike and it is envisaged to enhance the facility to all the students offering German as a subject at the university. Language teaching and learning is no longer a tedious and laboured process but a rewarding experience.

References

1. Benson, P.: Teaching and Researching Autonomy, 2nd edn. Longman/Pearson, Harlow, UK (2011)
2. Chandrasena Premawardhena, N.: Lerntraditionen im Vergleich: Sri Lanka und Deutschland, 14th International Congress on Education and Information Technology. Karlsruhe, Germany (2006)
3. Chandrasena Premawardhena, N.: Foundations in Language Learning. Author Publication (2009)
4. Chandrasena Premawardhena, N.: ICT in the foreign language classroom in Sri Lanka: a journey through a decade. In: 10th World Conference on Computers in Education (WCCE 2013), Nicolaus Copernicus University, July 2–5 2013, Torun, Poland, pp. 223–224 (2013)
5. Chandrasena Premawardhena, N.: Developing a common learning platform for foreign language teaching. In: Auer, M.E., Guralnick, D., Uhomoibhi, J. (eds.) Proceedings of the 19th ICL Interactive Collaborative Learning. Advances in Intelligent Systems and Computing, vol. 2, pp. 369–382. Springer, Cham (2017)
6. Cotterall, S.: Developing a course strategy for learner autonomy. ELT J. **49**(3). Oxford University Press (1995)
7. Cotterall, S.: Promoting learner autonomy through the curriculum: principles for designing language courses. ELT J. **54**(2). Oxford University Press (2000)
8. Cotterall, S.: The pedagogy of learner autonomy: lessons from the classroom. Stud. Self-Access Learn. J. **8**(2), 102–115 (2017)
9. Little, D.: Learner Autonomy. 1: Definitions, Issues and Problems. Dublin: Authentik (1991)
10. Littlewood, W.: "Autonomy": an anatomy and a framework. System **2**(4), 427–435 (1996)

Towards the Journey to Accomplish the "Joy of Learning"

Charilaos Tsihouridis[1]([⊠]), Marianthi Batsila[2],
and Anastasios Tsichouridis[3]

[1] University of Thessaly, Volos, Greece
hatsihour@uth.gr
[2] Directorate of Secondary Education of Thessaly, Ministry of Education,
Katerini, Greece
marbatsila@gmail.com
[3] Democritus University of Thrace, Komotini, Greece
tsabtsih@gmail.com

Abstract. In this study we aimed to explore teachers' decision to employ a number of ICT applications to enhance their digital literacy, thus, eliminating the "fear of technology". At the same time we wished to investigate whether factors such as gender or type of school influenced their decision to use the tools. In the study which was conducted in two phases, both quantitative and qualitative research methods were employed. To provide answers for the research questions anonymous questionnaires were delivered to 98 teachers followed by short interviews to help us expand further on the research issues. The results have revealed that the notion of "computer literacy" need is gradually being cultivated in the teachers' minds that have used the tools for both teaching practices and European projects. More interesting results are also discussed with implications for educational authorities and curricula designers.

Keywords: ICT applications · Digital literacy · Teacher training

1 Introduction

Research shows that information and communication technologies have long before penetrated all areas of human actions such as production, work, communication and culture bringing about many radical changes [1]. Education is one of those sectors that have been influenced by their presence to a great extent due to the many features they display that foster teaching and learning processes. For instance, due to their interactive nature, ICT are able to cover gaps in learners' cognitive level, enhancing their knowledge and skills acquisition [1]. Research findings have revealed that ICT encourage e-learning, provide alternative modes of teaching and make lessons more interesting, enjoyable and appealing to learners [2]. Games, videos, educational software and all kinds of educational applications provide a pleasant but at the same time more efficient teaching and learning framework [3]. Over the past years numerous ICT tools, else Web 2.0 tools, have been emphasized by many educators. According to research these tools have been found to foster on-line communication, cooperation and

© Springer Nature Switzerland AG 2020
M. E. Auer and T. Tsiatsos (Eds.): ICL 2018, AISC 916, pp. 239–250, 2020.
https://doi.org/10.1007/978-3-030-11932-4_24

assist teaching and learning procedures with the use of numerous applications [4]. Applications such as "movie maker", "taxgedo", "prezi", google forms, "hot potatoes", "educanon", "socrative", "toondoo", "paddlet", "weebly", "answergarden", "joomag", "inspiration" and others are some of the digital tools which are simple and easy to use in class, supplementary to the lesson, to assist and support teaching and learning.

Based on the extensive offering of ICT tools and their capabilities for educational exploitation, it seems that teachers need to be trained accordingly in order to enhance their digital literacy to be exploited for the fulfilment of their teaching goals. However, according to research, teachers' attitudes towards digital tools and their educational exploitation vary [6]. It has been found that teachers did not use the tools they were trained for, neither for personal nor educational purposes [7] whereas, elsewhere, teachers were rather ineffective in the usage of digital tools despite their preparation to use them in class [8]. In other cases teachers' interest for the use of digital tools was rather low as they did not feel comfortable enough and needed to overcome many personal or practical obstacles to feel secure to use them [9]. Other teachers did not use the digital tools they were suggested as they claimed their knowledge needed expansion [10]. Nevertheless, elsewhere, teachers were very positive in using the digital tools for class purposes [11]. Based on these facts the question seems to remain to be investigated: Are teachers of today adequately digitally literate and if not are they really prepared and ready to "confront" technology and leave aside the "fear of the unknown" for new and more exciting and efficient instructive digital ideas and knowledge paths?

2   Rationale of the Research

New technologies display a lot of advantages to help teachers fulfill these aims. But this is not the only reason to use technology in class. Teachers need to approach the students' digital contemporary world, understand it and exploit it for many reasons. For, when school is interesting for learners it may lead to better educational results, support inclusion policy and foster school drop outs avoidance and exclusion. For these goals to be achieved teachers need to continuously update their ICT skills. Further to the above, during the past years there is a tendency to implement European programs such as e-Twinning or Erasmus+ . However, due to many teachers' low competency in ICT skills or "fear of technology" their participation is not feasible even though they are highly interested in taking part. Based on the aforementioned concerns,, the authors of this paper decided to conduct seminars in order to train primary and secondary education teachers on a series of ICT applications (mentioned above), often used in teaching or in European projects within Erasmus+ or e-Twinning programs. However, the main concern was to investigate the extent to which, and after their training, they finally used these tools and for what purpose/s. Our long-term goal was to detect the reasons for doing or not doing so and whether factors such as type of school (primary or secondary education) and gender had any influence on their decision to use these tools.

3   Methodology

The purpose of the research was to detect whether already trained teachers on the use of digital tools actually employed them in class for educational purposes, why and how. Conducting our research with a mixed method approach (quantitative and qualitative), the main research questions were the following: 1. To what extent have the teachers used the target tools they were trained for and why or why not? 2. To what extent were they satisfied with the use of these tools for educational purposes? 3. How have factors such as gender and type of school (primary/secondary) affected the use of the target tools?

A number of primary and secondary education school teachers voluntarily participated in the research, namely 98 teachers. Forty five of them (45) were male and fifty three (53) were female. Meanwhile, fifty (50) teachers derived from primary education whereas forty eight (48) teachers worked in secondary education in the research area, which is Thessaly, Greece. They all worked in different schools and were randomly selected so as to be representative. A graphical representation of their categorization as regards their gender and type of school is given below (Fig. 1):

Fig. 1. Categorization of the participant groups regarding gender (on the left) and type of school (on the right)

The study was conducted in two phases and a mixed method approach was used. The first phase was implemented in spring 2017 in two successive workshops of three hours each (in groups of 25) and the second part in winter 2017–2018. During the first phase teachers were trained on the use of the tools. In the second phase an anonymous questionnaire was delivered to the teachers, who wished to participate, which comprised two parts (demographic – questions about the use/why/how) and the items were closed type [12] but some answers also had to be given on a 5 point Likert scale [13]. The questionnaire was also stabilized for its validity and reliability and its content and construct validity were examined with a group of teachers. Before proceeding with the actual study all tools were piloted for validity and reliability purposes. Upon completion of the questionnaires, short follow up voluntary interviews were conducted with some teachers who were explained beforehand the purpose of the short interviews-conversations and that their content would remain anonymous and would not be revealed to anyone as it only served the research purposes. The questions followed the same philosophy of the questionnaire but their actual purpose was to expand on the given answers and shed more light into them. The content method and the SPSS statistical package were used for the analysis

4 Data Analysis

The analysis of the questionnaire items was implemented with the use of descriptive techniques and indicators and their graphical representation for a clearer and more explicit picture of the data depicted. Techniques and criteria of induction statistics were also employed for the investigation of the impact of the independent variables (gender and type of school – primary/secondary) on the dependent variables of our research which are: (a) use of digital tools in the educational practice and (b) the degree of satisfaction-effectiveness from the educational use of the target tools. Thus, two sections follow: 1. Research data analysis using descriptive statistics, 2. Research data analysis using inductive statistics.

4.1 Research Data Analysis Using Descriptive Statistics

Question 1: The first question asked the teachers whether they used at least one tool after their training. According to their answers from the 98 participants 81 teachers used at least one tool whereas 17 of them did not use them at all.

Question 2: The second question was about the extent to which the teachers used the tools they had already been trained for. Figure 2 presents the percentage of the target digital tools usage by the participant teachers for their educational exploitation:

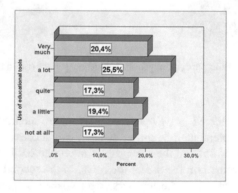

Fig. 2. Percentage of digital tools usage by the participants

According to Fig. 2 data above, it can be observed that the majority of the teachers have used the target tools very much to 25,5% and a lot (20,4%), thus, expecting perhaps to gain better school and learning outcomes. It is also worth noticing that the percentage of the teachers who have not used the tools reaches 17,3% and those who have used it very little (19,4%) despite the fact that they had already been trained on the use of these tools.

Question 3: Possible reasons for some teachers' negative attitude towards the use of the tools, as seen above (Fig. 2) were investigated through question number 3 and are presented in the figure below (Fig. 3):

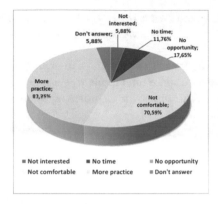

Fig. 3. Reasons for trained teachers not using the target digital tools in the educational process

According to the data in Fig. 3, the majority of the teachers who have not used the target tools did not feel comfortable using them (70,59%) or they would like more practice (82,35%). A number of participants (17,65%) answered that they did not have the opportunity to use them, which according to interview statements meant that they had a tight syllabus content to follow and therefore trying to design scenarios with ICT tools would deviate their purpose and focus [P4, P11]. Some of them (11,6%) also replied they had no time, which is interesting as the possible reason behind this reply (according to the interviews) it was probably the feeling of their not being ready to use them and therefore they needed more time to practice before they could actually "expose" themselves to their learners [P. 9, P16].

Question 4: This question investigated the reasons for using the digital tools, (see Fig. 1 above). The answers are given on the figure below (Fig. 4).

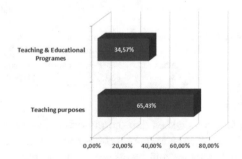

Fig. 4. Reasons for using the target digital tools in the educational process

According to the data in Fig. 4 it can be seen that the majority of the teachers (65,43%) have used the tools for teaching purposes. However, a considerable percentage has used them for both teaching purposes and educational programs. We consider the latter reason quite interesting as it depicts the decision of teachers to expand on their knowledge and apply it to other fields as well. In our case, and according to the interviews statements [P2, P7, P15, P21] the aforementioned programs are implemented within the framework of Erasmus+ or e-Twinning European programs.

Question 5: This was designed to detect ways to exploit the digital tools. There was also an option "other" (due to space limitations it is not presented on the table as there are no answers) for different, unknown or not suggested ways. Their answers are presented in the table below (Table 1).

Table 1. Ways teachers have employed to exploit the digital tools in the educational practice

Tools	Reasons for using the tools				
Movie maker	Video 81,48%	Clips 39,51%	Titles 75,31%	Captions 17,28%	Motivation 86,42%
Taxgedo	Word clouds **95,06%**	Meaning 17,28%	Message **83,95%**	Vocabulary **54,32%**	Motivation **64,20%**
Prezi	Presentations **34,57%**	Digital literacy 7,41%	Variety 22,22%	Group work 11,11%	Motivation **41,98%**
Google forms	Writing **92,59%**	Decision making 8,64%	Peer correction **30,86%**	Joint Story making **27,16%**	Motivation **30,86%**
Hot potatoes	Evaluation 11,11%	Test skills **86,42%**	Vocabulary **54,32%**	Grammar **65,43%**	Motivation 1,23%
Educanon	Questions 7,41%	Reflective thinking 2,47%	Extract parts 16,05%	Digital literacy 3,70%	Motivation 12,35%
Socrative	Quiz 8,64%	Assessment 8,64%	Game based learning 8,64%	Digital literacy 2,47%	Motivation 2,47%
Toondoo	Comic 17,28%	Creativity 17,28%	Writing skills 16,05%	Digital literacy 9,88%	Motivation 17,28%
Paddlet	Riddles 6,17%	Fill-in blanks 3,70%	Crosswords 6,17%	Quizzes 7,41%	Motivation 4,94%
Weebly	Web page for news 6,17%	Web page for my work 3,70%	Webpage for students 6,17%	Web page-forum 7,41%	Motivation 4,94%
Answergarden	Stud.opinion 54,32%	School Survey 19,75%	Poll 9,88%	Digital literacy 8,64%	Motivation 12,35%
Joomag	Online journal 7,41%	Digital literacy 7,41%	Group work activities 3,70%	Cooperation 4,94%	Motivation 6,17%
Inspiration	Digital literacy 2,47%	Mind maps **19,75%**	Communication 0,00%	Group work 0,00%	Motivation 0,00%

Based on the above it can be seen that the majority of the tools were not used to a big percentage with the most popular ones and their uses being movie maker, taxgedo, prezi, google forms, hot potatoes and inspiration. The implications of these findings will be discussed in the discussion section.

Question 6: The last question item in the questionnaire dealt with the extent to which the participant teachers had enjoyed the use of the target tools. Their answers are provided in the figure below (Fig. 5).

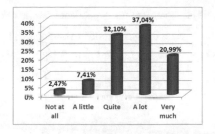

Fig. 5. The extent to which the participants enjoyed using the target digital tools

According to the data of Fig. 5 above, the majority of the participant teachers, who actually employed the target tools for educational purposes, enjoyed a lot the use of the target tools (37,04%), or quite (32,10%), followed by very much (20,99%) and a smaller percentage by a little (7,41%) or not at all (2,47%).

4.2 Research Data Analysis Using Inductive Statistics

To investigate the impact of the factors "gender" and "type of school" (primary/secondary) on the use of the target digital tools, techniques and criteria of inductive analysis will be used. The statistic criteria that will be used are the statistical criterion of factorial analysis of variance. This is a parametric criterion which aims at studying the impact of more than one independent variables on the dependent one and their interaction. Specifically, we shall use (a) the criterion of Univariate Analysis of Variance (One way Analysis of Variance –ANOVA) to study the impact of the independent variables on the dependent variable and (b) the criterion of two-way independent measures –ANOVA to study the interaction of the two independent variables on the dependent variable.

The research data include two factors (independent variables): factor A: Gender with levels 1: Male and 2: Female; factor B: Type of school with levels: 1. Primary Education and 2. Secondary Education.

We consider as dependent variable of the research design the questionnaire item which is given to the participant teachers regarding the use of the target digital tools by the teachers. It should be noted that for all cases the statistical test of dispersions with the criterion of Levene's Test of Homogeneity of Variances gives a value of $p > 0.05$, which indicates that the null hypothesis Ho is not rejected. Therefore, there is no statistically significant difference in the dispersions of the samples and thus ANOVA can be applied in all cases.

Impact of factor A "Gender" on the dependent variable "Use of tools".
We consider the independent variable factor A "Gender" and we will examine its impact on the dependent variable "Use of tools". Null hypothesis: There is no difference between the mean scores of factor A on the variable "Use of the tools" (Ho: m1 = m2). Alternative hypothesis: There is a difference between the mean scores of factor A on the variable Use of Tools (H1: m1 $\neq$ m2). The corresponding calculations of univariate analysis of the independent samples variance are given below (Table 2):

Table 2. Univariateanalysis of variance of the dependent variable use of tools on Web 2.0 tools.

	Sum of squares	df	MeanSquare	F	Sig.
BetweenGroups	0,010	1	0,010	0,005	0,944
WithinGroups	190,521	96	1,985		
Total	190,531	97			

The statistical presentation of the above result is F (1,96) = 0.005, p = 0,944. Since the calculated value of F:0,005 is less than the critical value 3,94, we accept the nullhypothesis and conclude that the variable "Gender" does not affect the use of the digital 2.0 tools. This is also shown on the above error graph wherein the overlap of the error rods is bigger than half the average margin of error (Fig. 6).

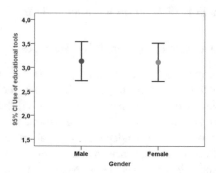

Fig. 6. Error graph for the two research groups of the variable "Gender" (Male - Female) on the variable "Use of Tools"

Impact of factor B "Type of School" on the dependent variable "Use of tools".
We consider the independent variable factor B: Type of School and we will look into its effect on the dependent variable "Use of Tools". Null hypothesis: There is no difference between the mean scores of factor B on the variable "Use of Tools" (Ho: m1 = m2). Alternative hypothesis: There is a difference between the mean scores of factor B on the variable "Type of School" (H1: m1 ≠ m2). The corresponding calculations of univariate analysis of the independent samples variance are given below (Fig. 7, Table 3).

The statistical presentation of the above result is F (1,96) = 0.016, p = 0,900. Since the calculated value of F:0,016 is less than the critical value 3,94, we accept the nullhypothesis and conclude that the variable "Types of School" does not affect the use of the digital 2.0 tools. This is also shown on the above error graph wherein the overlap of the error rods is bigger than half the average margin of error.

Interaction of factor A: "Gender" and Factor B: "Type of School" on the dependent variable "Use of tools".
With the independent variables factor A and factor B we will examine their interaction on the dependent variable "Use of tools". Null hypothesis: There is no impact on the

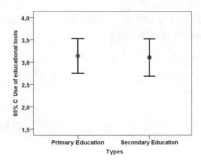

Fig. 7. Error graph for the two research groups of the variable "Type ofSchool" (Primary - Secondary) on the variable "Use of Tools"

Table 3. Univariateanalysis of variance of the variable "type of school" on the "Use of Tools".

	Sum of squares	df	MeanSquare	F	Sig.
BetweenGroups	0,031	1	0,031	0,016	0,900
WithinGroups	190,499	96	1,984		
Total	190,531	97			

mean ranks from the interaction between the two factors on the variable "Use of the tools" (Ho: impact interaction = 0). Alternative hypothesis: There is impact from the interaction between the two factors on the variable "Use of the tools" (H1 impact interaction ≠ 0). The corresponding calculations of univariate analysis of the independent samples variance are given below: (Table 4)

Table 4. Summary of the interaction of the factors

Tests of between-subjects effects				
DependentVariable:				
Source	df	F	Sig.	PartialEtaSquared
CorrectedModel	3	3,797	0,013	0,108
Intercept	1	499,123	0,000	0,842
Gender	1	0,009	0,926	0,000
Types	1	0,016	0,901	0,000
Gender * Types	1	11,365	0,001	0,508

From the above table it can be observed that a statistically significant interaction of the independent variables [Gender * Types] [$F(1,94) = 11,365$, $p = 0,001$, $\eta p2 = 0,508$] has been found, depicting the fact that there are no differences between the two sexes but there are differences as regards the interaction of sexes with the type of school. This is shown on the frequency polygon where the mean ranks of both independent variables are simultaneously depicted and the lines intersect (Fig. 8a). This is also shown on the error graph (Fig. 8b):

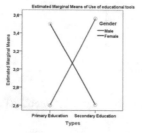
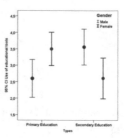

Fig. 8. a) Interaction graph of the two independent variables b) Error graph for the four research groups

The above points discussed are also shown on the above error graphs wherein the overlap of the error rods is bigger than half the average margin of error. Based on this, it can be said that female teachers in primary education have used the digital tools to a greater extent than their male colleagues, whereas the roles are reverse in secondary education with male teachers outbalancing female colleagues in the use and educational exploitation of the target digital tools.

5 Discussion and Conclusions

The present study focused on the investigation of the extent to which already trained primary and secondary education teachers on a number of ICT tools actually used them for educational purposes, why or why not and in what ways. Questionnaires were delivered to 98 teachers followed by short discussions. Interesting results were revealed and a number of implications can be drawn from them. Firstly, out of 98 teachers that had been trained and were questioned 81 used the tools which we consider a very encouraging percentage, and which reveals the teachers' will to be digitally literate to a greater extent than in the past. However 17 teachers did not use the tools with the majority replying that they did not feel comfortable enough (70,59%) or that they needed more practice (82,35%). Based on the answers of the follow up discussions it was pointed out that just one or two seminars for those who have a low level of ICT skills is not enough and that even though they wish to learn more, they forget the operation of the tools very soon if they do not have continuous support. Meanwhile, as some of them admitted, they are embarrassed to ask other colleagues in the fear of being ridiculed and therefore they would like continuous ICT support and training on frequent basis. What is more, they suggested small group based training in order to have more time to practice themselves instead of having to be among five people in front of one computer in a lab as it usually happens. Furthermore, and based on the interview answers, the teachers who replied that had no opportunity to use the tools (17,65%) explained that their daily school schedule was too tight, or that their syllabus content was quite demanding for them to experiment on new ideas [tools]. Expanding on their answers, they admitted that they would like more curricula flexibility and that the society itself (parents mostly who intervene in their work) should be positive towards a more digitally based lesson and not against it, considering it a waste of time or simply a game in the classroom (especially for lower classes). In addition, those who replied they had no time (11,76%) admitted that they felt they would have to devote a lot of time to design an ICT based scenario/lesson and that this would not only be time-consuming but also perhaps not well presented. As they argued, their fear was for their learners to "correct" [P27] their mistakes in case they would show they knew less (especially in secondary education). In the question about reasons for their using the tools the majority (65,43%) replied they did it for teaching purposes. However, it is equally interesting to see a percentage of 34,57%) of the teachers using the tools within Erasmus+ (secondary) or eTwinning (primary) programs. This implies a tendency to improve themselves and their practices, though it is worth investigating whether this is for better educational results or other reasons [i.e. win relevant competitions (sometimes to please the schools or parents) as a few of the participants admitted], [P2, P23]. In the question about the ways the teachers used the target tools, according to the findings, the most popular tools seem to be "movie maker" with a percentage of 81, 48% in creating videos, "Taxgedo" with 95,05% for word clouds and 83, 95% for focusing on a message, Google forms with 92,59% used to practice writing and "hot potatoes" with 86,42% for designing tests. The rest of the tools revealed small percentages in the ways they were used. Crosschecking the questionnaire data and teachers' interviews answers, this fact implies three things: one, teachers used all the tools and therefore, they are gradually beginning "to set aside the fear of the 'digital monster'" [P17]; two, the teachers seem to have adopted tools which are popular on their own and "were already known" [P 4, P18, P25, P21] to some of them (google forms, prezi); three, teachers needed "readymade scenarios" [P1, P19] with the tools integrated in them to help them get started. Whatever the case, we consider the percentage of ways of usage quite interesting as it shows a tendency to "adopt to the digital version of contemporary schools" [P29]. Finally, based on the questionnaire and interviews answers teachers enjoyed the use of the specific tools from a lot (37,04%) to quite (32,10%) and very much (20,99%). The research has some implications for educational authorities, curricula designers and teacher trainers. Based on the findings, educational authorities need to emphasize a mode of continuous support to teachers who many a time feel they are on their own. This should be well organized and based on teachers' frequent needs analysis so as to emphasize gaps in their digital knowledge and not only. What is more, curricula designers should ensure the practical suggestions of ICT based scenarios in the syllabi content, linked to a parallel series of training that will enable the connection of theory to practice. Last, teacher trainers should ensure the implementation of smaller and more frequent training sessions so as to focus on all teachers' clearer and better ICT preparation and understanding so as to help them experience the "joy of learning"!

References

1. Tyagi, S.: Adoption of web 2.0 technology in higher education: a case study of universities in National Capital Region, India. Int. J. Educ. Dev. Using Inf. Commun. Technol. (IJEDICT) 8(2), 28–43 (2012)
2. Grand-Clement, S.: Digital Learning–Education and Skills in the Digital Age. An Overview of the Consultation on Digital Learning held as Part of the Corsham Institute Thought Leadership Programma. U.K.:RAND Corporation, Santa Monica, California, and Cambridge, UK. (2017)
3. Haelermans, C.: Digital Tools in Education-On Usage, Effects and the Role of the Teacher. SNS Förlag, Stockholm (2017)
4. Mikre, F.: The roles of information communication technologies in education, review article with emphasis to the computer and internet. Ethiop. J. Educ. Sci. 6(2), 10 (2013)
5. Khany, R., Boghayeri, M.: The use of Web 2.0 tools in language pedagogy: an Iranian EFL teachers' attitude. In: Proceedings of the Global Summit on Education (GSE2013), Kuala Lumpur, 11–12 March 2013, pp. 151–159 (2013)
6. Zahedi, S.R., Zahedi, S.M.: Role of information and communication techonlogies in modern agriculture. Int. J. Agric. Crop Sci. (IJACS) 4(23), 1725–1728 (2012)
7. Munyengabe, S., Zhao, Y., Sabin, H.: Primary teachers' perceptions on ICT integration for enhancing teaching and learning through the implementation of one laptop per child program in primary schools of Rwanda. Eurasia J. Math. Sci. Technol. Educ. 13(11), 7193–7204 (2017)
8. Lufungulo, E.S.: Primary School Teachers' Attitudes towards ICT integration in Social Studies: A Study of Lusaka and Katete Districts. Master of Education in Social Studies, The University of Zambia (2015)
9. Majoni, A., Majoni, C.: Views of primary school teachers on the use of information communication technology (ICT) in teaching and learning. Glob. J. Adv. Res. 2(11), 1799–1806 (2015)
10. Popa, O.P., Bucur, N.F.: Romanian primary school teachers and ICT. In: The 10th International Conference on Virtual Learning ICVL, Timisoara Romania, 31 October (2015)
11. De Aldama, C., Ignacio Pozo, J.: How are ICT used in the classroom? A study of teachers' beliefs and uses. Electron. J. Res. Educ. Psychol. 14(2), 253–286 (2016)
12. Roussos, P.L., Tsaousis, G.: Statistics in Behavioural Sciences Using SPSS. TOPOS, Athens (2011)
13. Vagias Wade, M.: Likert-type Scale Response Anchors. Clemson International Institute for Tourism & Research Development, Department of Parks, Recreation and Tourism Management. Clemson University (2006)

Does Innovation Need a Reason? The CRS Within the Secondary Education Framework

Charilaos Tsihouridis[1]([⊠]) and Marianthi Batsila[2]

[1] University of Thessaly, Volos, Greece
hatsihour@uth.gr
[2] Directorate of Secondary Education of Thessaly, Ministry of Education, Volos, Greece
marbatsila@gmail.com

Abstract. The present paper aims at investigating the extent to which 127 secondary education teachers integrated the Classroom Response Systems (CRS) online platform of Kahoot in their teaching after they had already been trained to use it. Additionally, the study aimed at investigating the reasons for using it or not, the ways it was used and the extent to which factors such as gender may have affected the teachers' decision to use Kahoot. The study employed a mixed method approach. To this end, a relevant anonymous questionnaire was delivered to the participants followed by two focus group discussions to shed more light into the research issues. The analysis of the questionnaires was implemented with the help of SPSS statistical package whereas the focus group discussions were analysed with the method of content analysis. The study which revealed that the majority of the teachers (30,7%) did not use Kahoot due to fear of incompetency and the need for continuous digital training. The study has implications for policy makers and curricula designers.

Keywords: Game-based learning · Innovation · Classroom response systems · Teacher training

1 Introduction

New technologies display many advantages for education due to the vast applications they offer that provide today's educational practice with a multimodal way of teaching. Through image, sound and interaction they manage to attract learners' attention and interest, thus, being valuable tools in the hands of the teachers. Among the many applications used for teaching purposes "Classroom Response Systems" (CRS) are those that are said to enhance communication [1] among teachers and learners and promote students' active participation in the lesson. It is claimed that the CRS provide opportunities for student engagement and increase their interest in attending their school lessons. Furthermore, it is argued that the CRS offer game based features that enhance student interaction and motivation which lead to the successful implementation of their tasks. Additionally, CRS are believed to promote the development of effective relationships between students and teachers, thus, turning the educational procedure into an enjoyable activity which consequently results in an effective student

© Springer Nature Switzerland AG 2020
M. E. Auer and T. Tsiatsos (Eds.): ICL 2018, AISC 916, pp. 251–262, 2020.
https://doi.org/10.1007/978-3-030-11932-4_25

participation in the classroom activities [2]. What is more, CRS have been found to motivate learners, thus facilitating their learning, and consequently offering positive educational results [3]. It has also been shown that C.R.S help instructors to generate discussion in the classroom, reduce stress, enhance understanding of the lesson and use them for formative assessment purposes [4]. What is more, it is suggested that when CRS are included in curriculum design, they can provide a new dimension for inter-activity in the classroom and allow student and teacher interaction to a great extent. According to research CRS have been used for attendance, gauging comprehension and testing purposes but have also been used to overcome limitations of traditional lectures or improve students' attitudes. Kahoot is one of those classroom response systems (C. R.S) that are offered free and online, and are considered an effective way to introduce new concepts or assess the extent to which those have already been mastered by learners [5]. At the same time, Kahoot is a game-based learning platform that allows students to approach the lesson as a fun game rather than a boring or obligatory process within school duties. It is a free online tool that allows the implementation of quick quizzes to assess students' knowledge in real time [6]. Kahoot has been found to be both an enjoyable and educational platform which impresses learners through the use of quizzes and questions relevant to their learning.

Based on this extensive offering of digital contemporary tools competency it seems crucial that teachers are familiarized with digital technologies and are properly trained to use them in the most appropriate way within their teaching practices so as to help their students learn more effectively. However, to what extent are teachers familiar with digital tools, what difficulties do they encounter and how do they use these tools in their classrooms? Research has shown that teachers have found their use as being positive, partly because they had received long training [7]. Elsewhere, teachers in a research revealed that they did not use the digital tools a lot partly due to difficulties in accessing digital equipment and partly because they felt inadequately prepared or trained [8]. Others report the minimum or absent use of digital tools due to lack of interest, even though, as they argued, they would like to learn how to use them. Research also revealed that there was a large gap between what teachers believed of digital tools and the ways they actually used them [9]. As it was stated this was because, although the teachers believed that the digital tools make lessons more enjoyable, learner-centered and motivate them to get more involved in class, they themselves did not use them much for educational purposes or were not comfortable to change their traditional teaching methods [10]. More research claims that digital tools were used very little by the teachers but the real reason was not actually investigated. Similarly, in another research it was found that the teachers had revealed that they found digital tools important for learning but they needed more support and training to acquire the digital skills necessary to help their students [11].

In previous work we investigated the extent to which 149 teachers, who had been trained to use Kahoot for teaching purposes, had liked the tool and intended to use it in their future classes [12]. The results had shown that they considered it a motivating and interesting tool and they expressed their intention to use Kahoot with their students. This research is a continuation of our previous research with Kahoot. Therefore, it the sections that follow we shall describe our attempt to investigate the extent to which the aforementioned teachers actually used Kahoot or not with their classes after all.

2 Rationale of the Research

Teaching learners today is becoming all the more a demanding process and educators need to be well prepared to respond to the students' needs. A major challenge teachers need to confront is their learners' incredible capability to handle digital applications of the latest technology to such an extent that their knowledge too often surpasses that of their teachers' ICT skills and knowledge. This amazing digital dexterity of the young generation has some impact on teaching and learning. On the one hand, it helps them enrich their background knowledge on many subjects and topics. On the other hand, it raises their expectations for more exciting and innovative teaching methods as opposed to traditional or unattractive teaching tools and/or materials, still very often used in classes, due to digital fear, digital illiteracy or teachers' concern of allocating more time to "innovation" in the expense of syllabi content delivery. Nevertheless, and whatever the case, the majority of teachers strive in order to find ways to attract the learners' attention and participation so as to fulfill their goals for efficient educational results. Based on this need, and our interest to help teachers supplement their knowledge with new applications and tools, a number of 149 secondary education instructors of a variety of subjects had been trained in the use of the on-line CRS platform of Kahoot aiming to explore their views on its usage and features and detect their intention to apply this tool in their future teaching practices. The results of this study have already been presented and briefly discussed in the discussion section [13]. The present research constitutes a continuation of the above study, taking it now a step further. Thus, in the present study the particular teachers were approached again for two main reasons: to find out how many of them actually implemented tasks with Kahoot with their classes, how they did this and for what purposes; to explore the problems or the difficulties they might have faced; whether or not factors such as gender might have affected their decision to use Kahoot in their classes.

3 Methodology

3.1 Purpose and Research Questions

The purpose of the research was to detect whether already trained teachers on the use of Kahoot actually employed it in class for educational purposes, why and how. Conducting our research with a mixed method approach (quantitative and qualitative), the main research questions were the following: 1. To what extent have the teachers used Kahoot and why or why not? 2. To what extent were they satisfied with the use of Kahoot for educational purposes? 3. How have factors such as gender affected the use of the target tool?

3.2 The Sample

A sizeable number of secondary education school teachers voluntarily participated in the research, namely 127 teachers. Fifty three of them (53) were male and seventy four (74) were female, all working at different schools in Thessaly, Greece. This number derived from the original 149 teachers who had already been trained on the use of Kahoot some months before and now they were re-approached for the purposes of the present research. Having taken the permission of the school principals an anonymous questionnaire, which included an explanation of the research, was delivered to schools. The questionnaires were not handed to the teachers themselves but they were left at the schools for those who wished to participate, for ethical reasons and for avoiding to put pressure to anyone by answering it just because they had already been trained. To this end, 127 teachers finally responded. However, and in order to expand on the teachers' answers and have a better insight on the research issues, upon completion of the questionnaire taking, two focus group discussions (14 and 16 per group respectively) took place with randomly selected teachers who voluntarily consented to their participation. Care was taken that the discussions with the teachers were implemented on a day, place and time of common preference (this was decided based on options they were given upon their consent to participate) so as to make them feel comfortable and at ease. A graphical representation of their categorization as regards their gender is given in Fig. 1 below.

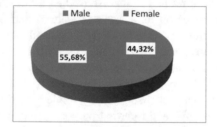

Fig. 1. Categorization of the participant groups regarding gender

3.3   The Research Tools

For the purposes of the research a mixed method approach was employed. Thus, an anonymous questionnaire with closed type questions were used to record the participants' answers, mainly for practical reasons, as well as because of the effectiveness of the processing of codified answers in such types of questions [14]. There were also some questions to be answered on a five point Likert scale. This scale was selected for the grading of the answers as being the most appropriate to measure participants' attitudes and views, ranging from one (not at all) to five (very much) [14]. Upon completion of the questionnaires, the focus group discussions were conducted. The teachers were explained the purpose of the discussions and that their content would remain anonymous and would not be revealed to anyone as it only served the research purposes. They were also explained that they could have a copy if they wished. The topics/questions of the discussions followed the philosophy of the questionnaire items but their actual purpose was to expand on the questionnaire answers and shed more light into the research issues. Finally, the questionnaire research data were digitized and were analysed using the SPSS statistical package. In addition, the content analysis method was used to analyse the content of the focus group discussions, identifying the key words, and classifying them accordingly so as to provide us with possible explanations/clarifications for the research questions [14].

3.4   Structure of the Questionnaire

The questionnaire comprised two parts. The first part included demographic information (gender, specialty). The second part comprised four items. The first question, with answers on a Likert scale, was about the extent to which teachers used Kahoot. The second question asked those that did not use Kahoot why they did so, having to tick more than one reasons if they wished on a given list, plus an option "other" to provide with other reasons, if they would like so. The third question asked those who used Kahoot to select more than one option, if they wanted, from a given list which was about ways they used to exploit Kahoot in class. However, there was also an option "other" for them to provide with more ways if they had thought of or had used in their classrooms. The fourth question asked the extent to which they had enjoyed using Kahoot having to answer on a five-point Likert scale.

3.5   Stabilizing the Questionnaire

When using a research tool, it needs to be stabilized regardless of whether it was already stabilized or created for the needs of a particular research project. In order to check the validity of the questionnaire, the content and construct validity were examined. For the questionnaire content validity, the questionnaire was initially given to a group of instructors who worked at secondary schools, who checked the questionnaire items in relation to its objectives and tried to detect the extent to which its items questioned what they were design to question. The instructors checked the questionnaire and found that there was correspondence between goals and questions and therefore the questionnaire presented content validity. The construct validity of the questionnaire was checked with a group of teachers who answered it, and then by interviewing them in this pilot phase. More specifically, the questionnaire was initially given to a small group of teachers (six teachers), and a small group discussion (four teachers) followed with the participants. According to the results of the pilot survey and based on the interviewees' comments, the questionnaire items were fully understood by the majority of the participants. Thus, and based on these results, the questionnaire exhibited also construct validity and it was developed to be handed in for the main study.

3.6   Research Phases

The study was conducted in two phases. The first phase was implemented in spring 2017 where 149 teachers were trained on the use of Kahoot and had then been asked their opinion on a relevant questionnaire. The second phase, which lasted almost four months, was implemented in winter 2018 and teachers were re-approached to answer an anonymous questionnaire about its use or not. Before proceeding with the actual study all tools were first piloted for validity and reliability purposes and for ethical reasons permission was taken by all parties (principals and participant teachers).

4 Data Analysis

The analysis of the questionnaire items was implemented with the use of descriptive techniques and indicators and their graphical representation for a clearer and more explicit picture of the data depicted. Techniques and criteria of induction statistics were also employed for the investigation of the impact of the independent variable (gender) on the dependent variable of our research which is: "use of Kahoot" in the educational practice. Thus, two sections follow: 1. Research data analysis using descriptive statistics, 2. Research data analysis using inductive statistics:

4.1 Research Data Analysis Using Descriptive Statistics

Question 1: To which extent have you used Kahoot with your classes? Figure 2 presents the percentage of Kahoot usage by the participant teachers:

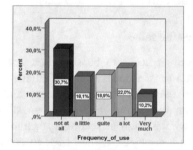

Fig. 2. Percentage of Kahoot usage by the participants

According to Fig. 2 data above, it can be observed that the majority of the teachers did not use Kahoot at all (30,7%) whereas 22,0% used it a lot, 18,1% quite, 18,1% a little and only 10,2% used it very much.

Question 2: The reasons for those who did not use Kahoot are presented in Fig. 3 below:

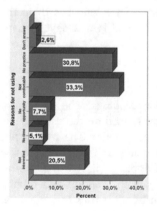

Fig. 3. Reasons for trained teachers not using Kahoot in the educational process

As seen in Fig. 3 above the majority of the teachers (33,3%) did not use Kahoot because they did not feel comfortable with its use while a percentage of 30,8% felt they needed more practice. It is worth mentioning the percentage of those who declared not interested (20,5%). Finally, the percentage of those who did not have the opportunity or time to use Kahoot was 7,7% and 5,1% respectively, which we do not consider high. The implications of these results will be discussed in the discussion section.

Question 3: The reasons for using Kahoot are presented on Fig. 4 below.

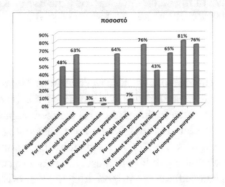

Fig. 4. Reasons for using Kahoot in the educational process

According to the data in Fig. 4 it can be seen that the majority of the teachers (81%) have used Kahoot for student enjoyment purposes which we consider really interesting. Similarly, high percentages of use are also displayed for competition and motivation purposes (both at 76%), followed by classroom tools variety purposes (65%), game-based learning purposes and formative assessment with slight differences (64% and 63% respectively), diagnostic assessment with 48% and student autonomy learning with 43%. The rest of the options (mid-term assessment, final school year assessment and student digital literacy) received very low percentages. The implications of these results will be discussed further down.

Question 4: To which extent are you satisfied with the educational exploitation of Kahoot? The answers are presented in Fig. 5 below.

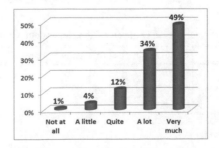

Fig. 5. Teachers' satisfaction from the educational exploitation of Kahoot

Based on the data of Fig. 5 the majority of the teachers who used Kahoot were very much satisfied (49%) followed by those who liked the process a lot (34%) and a smaller percentage (12%) who quite liked the use of the tool. However, there was a very small percentage of those who replied that they liked it a little (4%) or not at all (1%).

4.2   Research Data Analysis Using Inductive Statistics

To investigate the impact of the factor "gender" on "the use of Kahoot", techniques and criteria of inductive analysis will be used. The statistic criterion that will be used is the statistical criterion of factorial analysis of variance. Specifically, we shall use the criterion of Univariate Analysis of Variance (One way Analysis of Variance – ANOVA) to study the impact of the independent variable on the dependent variable. The research data include one factor (independent variable): factor A: Gender with levels 1: Male and 2: Female. We consider as dependent variable of the research design the questionnaire item (1) which is given to the participant teachers regarding the use of Kahoot by the teachers. It should be noted that for all cases the statistical test of dispersions with the criterion of Levene's Test of Homogeneity of Variances gives a value of $p > 0.05$, which indicates that the null hypothesis Ho is not rejected. Therefore, there is no statistically significant difference in the dispersions of the samples and thus ANOVA can be applied in all cases.

Impact of factor A "Gender" on the dependent variable "Use of Kahoot".
We consider the independent variable factor A "Gender" with levels - conditions: 1. Male, 2. Female and we will examine its impact on the dependent variable "Use of Kahoot". First it is considered necessary to formulate the research hypothesis:

Null hypothesis: There is no difference between the mean scores of factor A on the variable "Use of the Kahoot" (Ho: m1 = m2). Alternative hypothesis: There is a difference between the mean scores of factor A on the variable Use of Kahoot (H1: m1 $\neq$ m2).

The statistical presentation of the result on Table 1 and Fig. 6 below is F $(1,86) = 9,738$, $p = 0,002$. Since the calculated value of F: 9,738 is bigger than the critical value 3,94, and the null hypothesis is rejected, we conclude that the gender affects the use of Kahoot. Therefore, men outbalance women in their preference in using Kahoot. This is also shown on Fig. 6 with the error graph below where the overlap of the error rods is bigger than half the average margin of error.

Table 1. Univariate analysis of variance of the dependent variable use of Kahoot

	Sum of squares	df	MeanSquare	F	Sig.
BetweenGroups	9,366	1	9,366	9,738	0,002
WithinGroups	82,714	86	0,962		
Total	92,080	87			

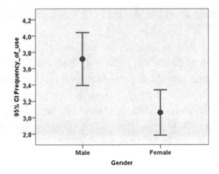

Fig. 6. Error graph for the two research groups of the variable "Gender" (Male - Female) on the variable "Use of Kahoot".

5   Focus Group Discussions Results

Two focus groups were conducted for the purposes of the research instead of one to have analytical views from more participants. The discussions focused on the same issues dealt with in the questionnaire but an effort was made to expand on them, in the hope of receiving valuable data which would give a better insight and understanding to the research questions. Due to the fact that there were two groups we will refer to them as P1a, P2a, P3a for the participants of the first group, and P1b, P2b, P2b and so on for those of the second group. Thus, based on the discussions data, the teachers revealed that they felt very eager to use the tool they were trained for as they considered it easy and innovative [P2a, P7b]. They justified their answer explaining that they needed to do this in order to offer variety to their classes as they were gradually beginning to "lose them" [P2a, P7b, P15b]. They admitted however that they were not sure whether their learners would "take the tool seriously" [meaning seeing as learning tool too and not only as a game] [P7b] as students were not familiar with such methods in class but they were used to a linear approach of content delivery [P15a]. What is more, as they claimed, they were also concerned about their colleagues' "acceptance" [P2b, P7a, P9b] and especially at schools with teachers of older age.

In the question why one would not use Kahoot the teachers admitted that they felt insecure [P1b, P4b, P8a, P12b], or that they had forgotten how to use it [P4b, P5b, P6a] because summer had mediated (since their training) and they were shy to ask others about its use [P4b, P6a]. What is more, they expressed their doubt whether it would finally serve any purpose if they integrated it in their lessons. This concern mainly originated from female participants who admitted they were afraid this "innovation" [P6b, P11b, P8a] might "cause chaos" [P11b] in the school and this would have a "negative impact" [P8a, P1b] on the way the principal or the colleagues thought about them as teachers. In the question why one would not have time to use Kahoot they answered that this was simply a matter of good preparation at home and this would mean less time to allocate to personal or social life [5b, 12b, 6a]. Based on this, it was suggested that the Ministry of Education should provide material with game-based learning techniques which would remove their fear of "deviating from syllabus" on one hand and on the other they would allow them to use these materials without having to think "how exactly to do it" [9a, 12a, 14b].

When asked about why someone might not have the opportunity to use Kahoot they seemed not to be able to discuss it (probably because none of them had given such an answer) even though one person argued that he meant to use it but had not done so yet. In the question how the teachers used Kahoot very interesting points were revealed. Firstly, their wish to motivate their learners was almost unanimous [P1a, P4a, P3a, P12a, P2b, P8b, P1a, P9a]. According to their responses they also felt it was high time they "added more arrows in their quiver" [P12a, P2b] so as to "keep learners in their class" [P9a]. Formative assessment was also something they found very useful [P8b, P3a, P14b] because their learners were very negative towards traditional testing and this was an obstacle to their work. It is worth mentioning however that even though they would like to use such tests in the end of the school year – even supplementary to the traditional ones if necessary [P3a, P15b] - this was forbidden by the state and the national curriculum. As they explained they would welcome some flexibility of choice in the way their learners perceived teaching and testing, arguing that this is needed if one wants to support students' different modes of learning (auditory, visual, kinesthetic and so on) [P1a, P1a, P2b]. Meanwhile, in being asked whether they enjoyed the use of Kahoot, they admitted they did but would have enjoyed it more had they had more time to allocate to it as their classes last not more than 45 min at the most. As they explained, time is an important factor in order to organize a digitally based scenario or activity due to the fact that often enough technology "betrays" them and "destroys their plans" [P4a, P2b, P15b]. Last, it is worth mentioning that none of them actually suggested more ways to use Kahoot. When asked why they expressed their lack of confidence on "digital ideas" [P10a, P13b) and suggested that before "inventing new ideas" [P10a, P13b) they should make sure they could "implement the 'offered' ones" [P10a, P13b]. When asked to expand on this, they explained that they need to feel digitally competent enough before designing something new and suggested that digital literacy training should be on frequent basis and based on the needs of each discipline [P10a, P13b, P14b]. It is worth noticing as well that there men used Kahoot more than women. In asking them to expand on this women revealed that they were afraid they might not be able to "control the fuss" [P6b, P11a] especially with quite naughty students and large classes.

6 Discussion and Conclusions

The present study constituted the continuation of a previous study based on which a number of 149 teachers were trained on the use of the CRS tool, known as Kahoot. Thus, the next phase of that research is the present research which entailed the re-approaching of the particular teachers to detect the extent to which they had actually implemented tasks using Kahoot with their classes. The purpose was also to investigate why or why not, in what ways they used Kahoot for its educational exploitation and whether they enjoyed this process. A number of 127 teachers finally responded to an anonymous questionnaire followed by two randomly and voluntarily selected focus group discussions. According to the results the majority of the teachers (30,7%) did not use Kahoot at all, where a percentage of 22,0% used it a lot. When asked why they didn't use it the majority replied they felt insecure or that they needed more practice (33,3% and 30,8% respectively). Those who used Kahoot replied that they used it for many reasons with the most popular choices the need for student enjoyment (81%), for motivation purposes (76%), for competition purposes (76%), for classroom variety purposes (65%), for formative assessment (63%), for diagnostic assessment (48%) and for student autonomy learning (43%). The focus group discussions revealed teachers' need for more training, fear of incompetency, difficulty to use the tool due to excessive syllabi content but also the need to update their digital literacy. The results comply with those of the literature review according to which teachers do not use digital tools in class due to insufficient training or lack of confidence [14, 15, 16, 17, 18, 19]. Furthermore, the gender was a factor which affected their choice of using the tool. Especially men seemed to like the tool more because they felt they could control their classes easier than female colleagues. The study has implications for policy makers and curricula designers who ought to re-consider curricula orientation and allow for more teacher flexibility so as to help teachers adjust to the demands of an on-going digital literacy and will allow them to adapt to the digital task-based teaching needs and learning approaches of the contemporary education. What is more, it is revealed that curricula designers need to consider the teachers' needs and base their educational policies on them. It seems that perhaps governments may "gradually forget" in a way that learning is an ongoing process and that teachers, together with learners, also need knowledge expansion, continuous support and autonomy opportunities. Teachers may be seen as continuous students after all, who need to be well trained if we want them to be the multipliers of knowledge in schools.

Teaching seems to be a great challenge and a difficult task for almost the majority of the teachers who desperately and restlessly seek ways to improve their methods and approaches. ICT and their latest applications have a lot to offer to assist this difficult task offering a vast source of ideas and tools. Game-based learning is an uprising technique and if it is well planned, it can prove to be a valuable tool in the hands of the teachers for the benefit of learners and education in general. What is more, if we wish to really make a difference in education we should not be afraid to experiment and try new ideas and methods. Teaching, although a magical process that may last a few exciting moments of enjoying effective outcomes, is also an everlasting and stressful fight both inside and outside the classroom aiming at new paths to knowledge that never seize to amaze us with the outstanding results they may offer. After all, it is these paths of knowledge that give us more moments of joy than the final destinations, for, as Cavafy once said [15] "Keep Ithaka always in your mind – arriving there is what you are destined for – but do not rush throughout the journey at all – Better if it lasts for years, so you are old by the time you reach the island, wealthy with all you have gained on the way" "Ithaka" by Cavafy [15].

References

1. Muncy, J.A., Eastman, J.K.: Using Classroom Response technology to create an active learning environment in marketing classes. Am. J. Bus. Educ. **2**(2), pp. 213– 218 (2012)
2. Beatty, I.D., Gerace, W.J., Leonard, W.J., Dufresne, R.J.: Designing effective questions for classroom response system teaching. Am. J. Phys. **74**(1), 31–39 (2006)
3. Duncan, D.: Clickers in the Classroom: How to Enhance Science Teaching Using Classroom Response Systems. Pearson/Addison Wesley, San Francisco, CA (2005)
4. Eastman, J.K.: Enhancing classroom communication with interactive technology: how faculty can get started. Coll. Teach. Methods Styles J. First Quart. **3**(1), pp. 31–38 (2007)
5. Owusu, A., Weatherby, N., Otto, S., Kang, M.: Validation of a classroom response system for use with a health risk assessment survey. Poster session at the 2007 AAHPERD National Convention and Exposition, Baltimore, Maryland (2007)
6. Diaz, C., Trejo, C.: Kahoot: The Student-Teacher Interactive Classroom Tool. Intera, NY (2015)
7. Heiss, B.M.: The effectiveness of implementing classroom response systems in the corporate environment. A Thesis Submitted to the Graded College of Bowling Green (2009)
8. Tsihouridis, C.,Vavougios, D., Ioannidis, G.: Assessing the Learning Process Playing with Kahoot–A Study with Upper Secondary School Pupils Learning Electrical Circuits. Interactive Collaborative Learning, ICL, Budapest, September, 2017
9. Lufungulo, E.S.: Primary School Teachers' Attitudes towards ICT integration in Social Studies: A Study of Lusaka and Katete Districts. Master of Education in Social Studies, The University of Zambia (2015)
10. Majoni, A., Majoni, C.: Views of primary school teachers on the use of information communication technology (ICT) in teaching and learning. Glob. J. Adv. Res. **2**(11), 1799–1806 (2015)
11. De Aldama, C., Pozo, J.I.: How are ICT used in the classroom? a study of teachers' beliefs and uses. Electron. J. Res. Educ. Psychol. **14**(2), 253–286 (2016)
12. Varol, F.: Elementary school teachers and teaching with technology. Turk. Online J. Educ. Technol. (TOJET) **12**(3), 85–90 (2013)
13. Batsila, M., Tsihouridis, C.: "Let's go… Kahooting"–teachers' views on C.R.S. for teaching purposes. In: International Conference on Interactive Computer Aided Learning, Budapest. 27–29 September 2017
14. Roussos, P.L., Tsaousis, G.: Statistics in Behavioural Sciences Using SPSS. TOPOS, Athens (2011)
15. Keely E., Sherrard, P., Cavafy, C.P.: Collected poems. In: Savidis, G. (ed.) Revised Edition. Princeton University Press, New Jersey (1992)

New Tasks for a Dyslexia Screening Web Application

Nikolaos C. Zygouris[1(✉)], Filippos Vlachos[2],
Antonios N. Dadaliaris[1], Evangelos Karagos[1],
Panagiotis Oikonomou[1], Aikaterini Striftou[1], Denis Vavouguios[2],
and Georgios I. Stamoulis[3]

[1] Computer Science Department, University of Thessaly, Lamia, Greece
{nzygouris,dadaliaris,ekaragkos,
paoikonom,astriftou}@uth.gr
[2] Department of Special Education, University of Thessaly, Volos, Greece
{fvlachos,dvavou}@uth.gr
[3] Electrical and Computer Engineering, University of Thessaly, Volos, Greece
georges@uth.gr

Abstract. Dyslexia is the most common learning disability and it is characterized by a persistent failure to acquire reading skills despite normal intelligence, adequate cognitive abilities, conventional instruction and sociocultural opportunities. The main target of the present study was to extend our previous dyslexia screener web application by constructing and embedding new tasks. An extended battery of tests was implemented, which included sentence comprehension, auditory working memory, syllable composition and word explanation skills. Moreover, another aim of the study was to incorporate these tests into our existing web application. Statistical analysis revealed that dyslexic children presented lower correct responses ($p < 0.01$) and larger time latency ($p < 0.05$) in comparison to their average peers in all four tasks of the web application. Additionally, the main purpose of delivering a screening procedure via internet is that the users are doing something useful while perceiving that they are "playing a game". The present study does not replace a seven task battery of tests with a four task one. One the contrary, it enriches an already used seven task battery with four more tests in order to design a more complete dyslexia web application screener.

Keywords: Dyslexia · Web application · Screening reading disabilities

1 Introduction

Reading is a complex process as its achievement requires repetitive and consistent practice [1]. This process draws on phonological awareness [2] and on orthographic knowledge [3]. However, some children present a marked impairment in the development of reading skills defined by difficulties in accurate and/or fluent word reading and spelling [4]. It is assumed to be of neurobiological origin [5]. Furthermore, it can be described as a serious and specific failure in reading in spite of typical general cognitive development (such as intelligence) and adequate access to conventional

© Springer Nature Switzerland AG 2020
M. E. Auer and T. Tsiatsos (Eds.): ICL 2018, AISC 916, pp. 263–271, 2020.
https://doi.org/10.1007/978-3-030-11932-4_26

instruction [6]. Dyslexia is identified in about 10% of the scholar children [6]. The prevalence of dyslexia in Greece has been estimated at 5.52%, which is consistent with data from other countries with "pure" orthographies [7].

The most well established and studied deficit that dyslexic children face is the phonological deficit. Phonological processing is widely accepted as the core cognitive process underlying most dyslexic students' reading difficulties. Neuropsychological studies have demostrated that word reading depends mainly on the left lateralized cortical network including the frontal, temporal, parietal and occipitotemporal brain areas [8]. Grapheme – Phoneme conversion is established in the temporoparietal cortex, while the visual word area is established in the occipitotemporal lobes [8]. The occipitotemporal region has a strong relationship with Broca's area which is associated with articulation [9]. Moreover, it has been suggested that dyslexia is related with anatomical and functional abnormalities in brain areas such as the parietal operculum, which is less asymmetrical in dyslexics in comparison to typical achieving readers [10]. Furthermore, dyslexics present over activation in the right posterior areas in contrast to their average peers, and more dyslexic pupils display a preference for a right hemisphere thinking style compared to their peers who adopted a left hemisphere thinking style [11].

Although considerable research in dyslexia has been devoted to a phonological deficit, less attention has been paid to other aspects of language processing. Recent studies suggest that dyslexia is the outcome of multiple risk factors and children with language difficulties at school entry are at high risk [12]. Likewise, cognitive skills testing generally confirms that many dyslexic children present impairments in working memory, which causes difficulties in recalling and controlling information [12] or deficits in processing auditory temporal information [13]. Furthermore, dyslexia causes difficulties in vocabulary breadth. Several recent studies suggest that dyslexics present lower scores in vocabulary tests in comparison to typical achieving readers (e.g. [14]). In addition, a number of studies suggest that children with dyslexia present deficits in syllable awareness [15] and a lot of studies indicate variance in reading comprehension tasks, particularly with students with reading difficulties [16].

The use of computers in screening dyslexic children is now well-established in many European countries such as the U.K. There is a number of different applications available that can be implemented in early identification and screening in children, adolescents and adults [17]. Digital media has been identified as a promising tool to address dyslexia [18]. In our previous study [19] a web application battery was implemented including seven tasks in order to screen children with dyslexia. In more details, a task of word reading, visual discrimination, working memory, reading and auditory discrimination, and two tasks on orthographic abilities were designed and implemented.

Given that the aforementioned recent studies [11–16] have reported that students with developmental dyslexia may manifest difficulties in some skills that we did not include in our first research protocol, this study aimed to extend our previous study [19]. The novel feature was that we constructed a new battery of tests that can be delivered by computer in order to screen children's literacy, auditory working memory, word composition and vocabulary skills. Moreover, another purpose of the study was to incorporate these tests in our previous web application. The hypotheses of the present study was that children that are already diagnosed by paper-pencil tests as dyslexic will also achieve (a) significantly lower scores and (b) larger latencies in the new tasks of the web application.

2 Method

2.1 Participants

A total of fifty two right handed children (26 male and 26 female, age range 8–11 years old M = 11.15 SD = 1.03) participated in this study. The dyslexic students (N = 26, 13 male and 13 female age range 8–11 years old M = 11.23 SD = 0.91) had a statement of dyslexia after assessment at the Centre of Diagnosis, Assessment and Support of Lamia Greece, as it is required by Greek Law. The control group (N = 26. 13 male and 13 female age range 8–11 years old M = 11.08 SD = 1.16) was formed by pupils who attended the same classes with dyslexics. They presented typical academic performance according to teacher ratings. Additionally all children that participated in the present study did not have a history of major medical illness, psychiatric disorder, developmental disorder or significant visual or auditory impairments according to their medical reports of their schools. The participants of the comparison group were matched for age and gender with dyslexics (1 dyslexic: 1 control).

2.2 Materials and Procedure

Literacy, auditory working memory, word composition and word vocabulary skills were examined in dyslexics and typically achieving children, using a battery that consisted of four tasks that evaluate some critical skills relevant for dyslexia screening through the use of the aforementioned web application. During the test procedure correct responses and latency was measured. In more detail we used: (a) a sentence comprehension task; children had to decide if the meaning of the presented sentence was correct or incorrect (Fig. 1), (b) An auditory working memory task consisting of twenty two sequence of numbers, where the first sequence included three numbers; the second four numbers; the third five numbers; the fourth six numbers; the fifth seven numbers; the sixth eight numbers. Participants were asked to report the numbers with the use of a 0–9 numerical pad that was displayed. They had the opportunity to correct their answer if they report a number at fault. It is worth to mention that if children could not remember two series in a row or three series in general the test was stopped (Fig. 2), (c) A syllable composition task in which children had to discriminate and compose a correct word, by putting its letters in the correct order (Fig. 3), (d) A vocabulary test during which children had to select the correct meaning of a word by identifying the correct picture that gives the meaning of the word (Fig. 4).

Fig. 1. Sentence comprehension task. Children had to identify if the sentence presented a logical meaning (push the green button) or not (push the red button).

Fig. 2. After hearing, a sequence of numbers children had to use the numerical pad in order to choose the numbers that they had heard

Fig. 3. Children had to discriminate and compose the word by putting its letters to the correct order.

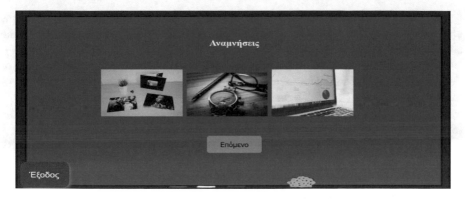

Fig. 4. Children had to identify the meaning of the word by choosing the correct picture.

2.3   Implementation

The implementation of the aforementioned web-based application followed the classic pattern of dividing the front-end and back-end functionality into separate design flows. One major decision was the usage of a minimal set of basic web development technologies, taking into account maintenance and re-usability issues, concerning our code, and the potential smooth expansion of its functionality when and if needed.

Front-end web development focused on the implementation of the interaction between the end user and the application data, while back-end web development practically added utility to everything created in the front-end, since our application requires constant communication between a client and the main server that hosts it.

The front-end's toolbox consisted of HTML5, CSS3 and JavaScript. HTML5 is the latest version of the core technology markup language HTML of the World Wide Web. It was used to define the structure of each page. One of its biggest advantages is the support of the latest multimedia codecs which enhance the streaming or playback of high definition audio and video. Additionally, it has error-handling capabilities, meaning that older browsers can safely ignore new HTML5 contracts and features without damaging the overall content. CSS3 was used for the beautification of the graphical environment. The graphical representation of each test is of paramount importance, since the user must remain undistracted throughout his/hers effort to complete it. The application contains a few subtle color changes in order to avoid making the test a tedious ordeal. Finally, the application's functionality and mechanics were implemented using JavaScript and jQuery. JavaScript is a key component of every modern web browser and is used to develop client-side scripts that can interact directly or indirectly with the user choices. This interaction is the core that generates the dynamic behavior of this application.

Concerning the back-end, the two main components used were PHP and MySQL. PHP handled the communication between client and server, feeding data generated by the user to the hosting server and defining the stability and duration of each session. The data was stored to a database which was created using MySQL. The role of

MySQL was twofold since apart from the implementation of the main database, it was also used to generate the appropriate queries that returned the values needed for the statistical analysis.

Hosting service reliability and availability are essential to the proper functioning of the screening application. The periodical usage of the application where a plethora of online screening tests are performed simultaneously points out that the Cloud is the appropriate hosting plan. Cloud hosting is flexible, reliable and cost-efficient while its infrastructure is secured.

Alongside the security issues faced by the cloud providers, the web application has been enhanced using secure encryption and decryption functions into the MySQL queries. Furthermore, data exchanged between the browser and the server has been encrypted using the HTTPS protocol. Finally, several network vulnerability scanners like Nessus where used in order to identify and fix any potential exposures to unauthorized users like browsable web directories and misconfigurations.

3   Results

Descriptive statistics were performed in order to obtain mean scores and standard deviations of participants in all four tasks (see Table 1). Analysis of Variance (ANOVA) was conducted to compare the scores of children with dyslexia and control group. ANOVA revealed that children with dyslexia had statistically significant ($p < 0.01$) lower mean scores of correct answers in all four tasks compared to their average peers that participated at the control group (Table 1).

Table 1. Mean scores and standard deviations of children with dyslexia vs the control group correct responses and the associated statistical significance in all tasks.

Tasks	Groups				
	Dyslexic		Control		
	Mean	SD	Mean	SD	F
Sentence comprehension	14.08	2.61	17.00	1.60	23.75**
Auditory working memory	8.08	2.28	13.38	3.66	39.46**
Syllable composition	1.23	1.07	4.62	4.69	12.86**
Vocabulary	11.23	4.75	14.92	3.08	11.05**

note**$p < 0.01$

Moreover, time latency was examined in three out of four tasks, as in auditory working memory test time was not measures, so it was excluded from the statistical analysis. As it was revealed by the statistical analysis children with dyslexia needed statistically significant ($p < 0.01$) more time in order to complete the tests in comparison to typically achieving children. The results are presented in Table 2.

Table 2. Mean scores and standard deviations of children with dyslexia vs. the control group time latency and the associated statistical significance in all tasks.

Tasks	Groups				
	Dyslexic		Control		
	Mean	SD	Mean	SD	F
Sentence Comprehension	8.26	4.53	4.51	1.80	35.63**
Syllable Composition	2.46	0.57	1.46	0.60	35.09**
Vocabulary	4.11	3.41	2.42	1.38	5.50*

note* $p < 0.05$ **$P < 0.01$

In sum, the aforementioned statistical analyses revealed that children with dyslexia had significantly lower scores of correct answers in all tasks of the web application and needed more time in order to respond in comparison to children that formed the control group. It is worth to keep in main that in auditory working memory, task time latency was not measured.

4 Conclusion

The web application screener that was used in the present protocol revealed significant differences in the four tasks that were constructed. The dyslexic and non-dyslexic groups were found to differ substantially in sentence comprehension, auditory working memory, syllable composition and word explanation skills. It was found that the mean scores of dyslexic were significantly lower in all tests answers, result that verifies our first hypothesis. Furthermore, the time latency were significantly higher compared to the control group, which verifies our second hypothesis. Our results confirm that the four tasks examined in the present study could be used to differentiate dyslexics from typically developing students. Another important component of the present study was that all children used a web application screener, which presents that the internet can be a useful tool in order to screen easily children with dyslexia and typical readers.

The web application paradigms if they are properly constructed and presented aims to engage users into a screening activity which produces common good to the child itself, to its parents and teachers. The main idea of implementing a web application screener is that the user does something useful while "playing a game" via the internet [20].

Additionally, web application screeners can be very useful for children as they have the ability to lure "players" into performing more accurate an assessment of dyslexia as it was shown by the present research protocol. If the user enjoys the application he/she will probably continue to play putting more effort in order to complete all the needed tasks [18]. Moreover, the context of learning to use the computer can provide the child with an experience of mastery and a sense of control, which can help to minimize user frustration or loss of confidence when working on tasks once accomplished with ease.

However, it has to be clarified that the present study does not replaces a seven tasks battery of tests [19], with a four tasks one. One the contrary, it enriches an already used seven tasks battery with four more tests in order to design a more complete web application in order to provide first-pass screening services and referral. Furthermore, standardization on a large-scale sample representative of the general population is necessary before widespread adoption.

In addition, it must be highlighted that strict psychometric and educational assessment that is placed by expert professionals that forms the multidisciplinary team cannot be replaced by a web test. They are responsible not only on making the diagnosis but also, construct the intervention program that a child with dyslexia and school have to follow.

References

1. Waldie, K.E., Wilson, A.J., Roberts, R.P., Moreau, D.: Reading network in dyslexia: similar, yet different. Brain Lang. **174**, 29–41 (2017)
2. Joubert, S., Beauregard, M., Walter, N., Bourgouin, P., Beaudoin, G., Leroux, J.M., et al.: Neural correlates of lexical and sublexical processes in reading. Brain Lang. **89**(1), 9–20 (2004)
3. Loveall, S.J., Channell, M.M., Phillips, B.A., Conners, F.A.: Phonological recoding, rapid automatized naming, and orthographic knowledge. J. Exper. Child Psychol. **116**(3), 738–746 (2013)
4. Vlachos, F.: Dyslexia: a synthetic approach to causal theories (in Greek). Hell. J. Psychol. **7**, 205–240 (2010)
5. Zygouris, N.C., Avramidis, E., Karapetsas, A.V., Stamoulis, G.I.: Differences in dyslexic students before and after a remediation program: a clinical neuropsychological and event related potential study. Appl. Neuropsychol. Child, 1–10 (2017)
6. Gabrieli, J.D.: Dyslexia: a new synergy between education and cognitive neuroscience. Science **325**(5938), 280–283 (2009)
7. Vlachos, F., Avramidis, E., Dedousis, G., Chalmpe, M., Ntalla, I., Giannakopoulou, M.: Prevalence and gender ratio of dyslexia in Greek adolescents and its association with parental history and brain injury. Am. J. Educ. Res. **1**(1), 22–25 (2013)
8. Pugh, K.: A neurocognitive overview of reading acquisition and dyslexia across languages. Dev. Sci. **9**(5), 448–450 (2006)
9. Cohen, L., Dehaene, S.: Specialization within the ventral stream: the case for the visual word form area. Neuroimage **22**(1), 466–476 (2004)
10. Robichon, F., Levrier, O., Farnarier, P., Habib, M.: Developmental dyslexia: atypical cortical asymmetries and functional significance. Eur. J. Neurol. **7**(1), 35–46 (2000)
11. Vlachos, F., Andreou, E., Delliou, A.: Brain hemisphericity and developmental dyslexia. Res. Dev. Disabil. **34**, 1536–1540 (2013)
12. Thompson, P.A., Hulme, C., Nash, H.M., Gooch, D., Hayiou-Thomas, E., Snowling, M.J.: Developmental dyslexia: predicting individual risk. J. Child Psychol. Psychiatry Allied Discip. **56**(9), 976–987 (2015)
13. Banai, K., Ahissar, M.: Poor frequency discrimination probes dyslexics with particularly impaired working memory. Audiol. Neurotol. **9**(6), 328–340 (2004)
14. Ben-Artzi, E., Fostick, L., Babkoff, H.: Deficits in temporal-order judgments in dyslexia: Evidence from diotic stimuli differing spectrally and from dichotic stimuli differing only by perceived location. Neuropsychologia **43**(5), 714–723 (2005)

15. Corkett, J.K., Parrila, R.: Use of context in the word recognition process by adults with a significant history of reading difficulties. Ann. Dyslexia **58**(2), 139–161 (2008)
16. Gori, S., Facoetti, A.: How the visual aspects can be crucial in reading acquisition: the intriguing case of crowding and developmental dyslexia. J. Vis. **15**(1), 8–8 (2015)
17. Collins, A., Lindström, E., Compton, D.: Comparing students with and without reading difficulties on reading comprehension assessments: a meta-analysis. J. Learn. Disabil. **51**(2), 108–123 (2018)
18. Singleton, C., Horne, J., Simmons, F.: Computerised screening for dyslexia in adults. J. Res. Read. **32**(1), 137–152 (2009)
19. Gaggi, O., Palazzi, C.E., Ciman, M., Galiazzo, G., Franceschini, S., Ruffino, M., et al.: Serious games for early identification of developmental dyslexia. Comput. Entertain. (CIE) **15**(2), 4 (2017)
20. Zygouris, N., Vlachos, F., Dadaliaris, A., Oikonomou, P., Stamoulis, G.I., Vavougios, D., et al.: The implementation of a web application for screening children with dyslexia. In: Auer, M.E., et al. (eds.) Interactive Collaborative Learning, Advances in Intelligent Systems and Computing, pp. 415–423. Springer International Publishing, Cham (2017)
21. Roccetti, M., Marfia, G., Semeraro, A.: Playing into the wild: a gesture-based interface for gaming in public spaces. J. Vis. Commun. Image Represent. **23**(3), 426–440 (2012)

IGIP Multilingual Glossary

Tatiana Polyakova[✉]

Moscow Automobile and Road Construction State Technical University,
Moscow, Russia
kafedral01@mail.ru

Abstract. International Society for Engineering Pedagogy (IGIP) uses Global English as the working language of its conferences and publications. The analysis of ICL/IGIP 2017 Conference Presentations showed that English is a foreign language for more than 80 per cent of its participants. This fact causes certain problems for using and understanding Engineering Pedagogy terminology. To solve these problems IGIP launched a project aimed at compiling a multilingual Engineering Pedagogy glossary. "IGIP Multilingual Glossary" consists of the English core – basic Engineering Pedagogy terms and their definitions followed by translations into the target languages of the countries participating in IGIP activities as its National Sections. The paper describes the structure of the Glossary, the word entry, the Chapters of the main Subject Part, the problems that occurred during the work, and the ways of their solutions. The IGIP Working Group "Languages and Humanities in Engineering Education" is responsible for this project. The electronic version of the Glossary will be uploaded to IGIP site and it will give the opportunity to add new terms as soon as they appear in the professional communication.

Keywords: Engineering pedagogy · Terminology · Multilingual glossary

1 Context

The process of globalization required a global language for international communication. For political, economic, technological and other reasons including the spread of Internet this function is fulfilled by English that nowadays plays the role of lingua franca and is known as Global English, World English, Common English or General English.

The English language is used in many countries all over the world. Firstly, English is spoken nationally in Great Britain, Ireland, the USA, Canada, Australia, New Zealand, South Africa, and several Caribbean countries. For most people of these countries English is their mother-tongue. Secondly, in a great number of the countries English traditionally has a special status and it is used officially as a medium of communication by the government, in law courts, state institutions, and educational institutions. They are India, Pakistan, Singapore, etc. The population of these countries considers English to be their second language. Thirdly, there are countries that historically were not connected with English but the language is learnt by a great number of people in educational institutions for international communication, for example, in

© Springer Nature Switzerland AG 2020
M. E. Auer and T. Tsiatsos (Eds.): ICL 2018, AISC 916, pp. 272–282, 2020.
https://doi.org/10.1007/978-3-030-11932-4_27

China, Poland, Russia, Germany, Spain, and other countries. English is a foreign language for them.

As a result of the unprecedented spread of English in the world in the second part of the 20th century, at the beginning of the 21st century one fourth of the world population spoke English. It is necessary to underline that that three people out of the four using English for communication are not native speakers [1].

English as a means of international communication is the language of international relations, business, science, education, tourism, the media, advertising, international safety, and pop music. English gives access to the latest information. As early as in 1981, 85 per cent of scientific publications in the sphere of biology and physics were written in English [1]. By the way, this paper is also in English. English is used for air traffic control ("Airspeak") and as a means of international communication of the sea ("Seaspeak"). These variants of English have a restricted number of words and standard sentence patterns. Boeing found out that 11 per cent of airplane fatal crashes from 1982 to 1991 happened due to misunderstanding between the pilots and dispatchers. For example, in 1977, the collision between two Boeings at Tenerife was caused by unclear English pronunciation and terminology [1].

But what is Global English? Certainly it is not one of the variants of English. It is not American English, British, Australian or Indian English. There is an idea [1] that Global English rises from the merger of General American English and Standard British English with a mixture of other varieties. One of the consequences of the global spread of English is its moving away from one standard, which made the linguists coin the terms "Englishes" and "the English languages". The only thing we are sure of is that Global English has American English spelling thanks to the software produced in the USA. We cannot predict the ways of its further development as we there are no precedents in the history of human civilization. But it is evident that Global English no longer belongs to one country and its changes are influenced by its usage of not native speakers.

At present IGIP also speaks English and annual conferences are organized without the assistance of interpreters. However, in the history of IGIP there were different periods as far as the working language is concerned. As stated in [2], IGIP was founded during the First International Symposium on Engineering Pedagogy, which took place in Klagenfurt (Austria) in 1972. At that time the working language of the Society was German as the majority of IGIP members were from German-speaking countries. For that reason they exchanged their views without any language barriers. The Symposium Proceedings, IGIP official journal "Report" were published in German. No wonder that the main, most cited and referenced work on Engineering Pedagogy "Ingenieurpädagogik" was also written by A. Melezinek in German [3].

In the 1990s, as stated in [4], there was the transition period when the symposium speakers made presentations either in German or English. The keynote speeches were simultaneously interpreted. The papers in Symposium Proceedings were also published either in German or in English and only the abstracts were sometimes given in the other language, for example [5]. For the majority of the participants who did not have the communicative competence in the two languages it did not give the opportunity to evaluate the contents of all the publications. The official journal of IGIP and even the IGIP book "Who is Who" [6] were published both in English and German. As more and more educators were getting interested in the ideas of Engineering Pedagogy, at that time the book "Engineering Pedagogy" by A. Melezinek was translated into Hungarian, Czech, Slovenian, Russian, Ukrainian, Bulgarian, and Polish.

English received the status of the working language of IGIP and its conferences in 2006 during the Symposium in Tallinn (Estonia) [2]. Since that time the Conference Proceedings and all Conference presentations have been in English.

2 Goal

The usage of English as IGIP working language helped to overcome most evident difficulties connected with the language barrier and to achieve better understanding between IGIP members. The analysis of the presentations of the 20th International Conference on Interactive Collaborating Learning and 46th International Conference on Engineering Pedagogy "Teaching and Learning in a Digital World" (27–29 September 2017, Budapest, Hungary) showed that the speakers represent 50 countries from five continents (Table 1).

Only 8 per cent of the presentations were made by representatives of the countries where English is the mother-tongue of the most part of the population. They are from Australia, Great Britain, Canada, New Zealand, Ireland, and the USA. The presentations made by the authors who are from the countries where English traditionally has a special status and is used as a second language is equal to 3.1 per cent. They are from India, Pakistan, and Sri-Lanka. The majority of the presentations (78.6 per cent) were written by the authors for whom English is a foreign language. But this portion is larger as many of the 10.3 per cent of the presentations of international teams of authors are also from the countries where English is learnt as a foreign language (Table 2).

As a result, many IGIP activities' participants may have certain problems with the Engineering Pedagogy terminology usage in the process of professional communication. The problems are connected with improper usage of the terms, mispronunciation of the terms, not accurate interpretation of the terms' meaning, etc. The problems and their examples are described in more detail in [4].

Of course there are a lot of glossaries, encyclopedias, and dictionaries on the problems of education that are available now. But the main thing is that they do not cover Engineering Pedagogy terminology. Besides, a lot of glossaries that are in use are monolingual, for example [7], and do not help much if it is necessary to translate a term. There are multilingual glossaries as well. The well-known English, German and Russian Glossary on the Bologna Process [8] contains very important detailed explanations of the meaning of the terms introduced due to the Bologna reforms and used in a great number of documents regulating them. The explanations are in the three languages but very often the glossary does not give equivalents of the terms in the other languages. For example, it is not possible to find either an English or a German equivalent for the Russian word "аспирантура" (the post graduate course aimed at doing research, writing and defensing a dissertation which is now the third cycle of higher education). A very good bilingual manual [9] does not provide equivalents in many cases either.

Table 1. Presentations of the 20th international conference on interactive collaborative learning and 46th international conference on engineering pedagogy

№	Country	The number of presentations	Per cent
1	*2*	*3*	*4*
1	Algeria	3	1.3
2	Australia	3	1.3
3	Austria	19	8.5
4	Belgium	1	0.4
5	Brazil	3	1.3
6	Bulgaria	2	0.9
7	Canada	3	1.3
8	Czech Republic	5	2.2
9	Ecuador	3	1.3
10	Estonia	4	1.8
11	France	3	1.3
12	Germany	8	3.6
13	Great Britain	4	1.8
14	Greece	6	2.7
15	Hungary	16	7.1
16	India	4	1.8
17	Indonesia	6	2.7
18	Ireland	1	0.4
19	Israel	2	0.9
20	Italy	2	0.9
21	Japan	5	2.2
22	Macedonia	1	0.4
23	Malaysia	1	0.4
24	Mexico	1	0.4
25	New Zealand	1	0.4
26	Norway	1	0.4
27	Oman	3	1.3
28	Pakistan	1	0.4
29	Peru	1	0.4
30	Poland	6	2.7
31	Portugal	12	5.4
32	Qatar	1	0.4
1	*2*	*3*	*4*
33	Romania	4	1.8
34	Russia	24	10.7
35	Saudi Arabia	1	0.4
36	Serbia	2	0.9
37	Slovakia	14	6.3

(*continued*)

Table 1. (*continued*)

№	Country	The number of presentations	Per cent
38	South Korea	1	0.4
39	Spain	3	1.3
40	Sri Lanka	1	0.4
41	Sweden	2	0.9
42	Switzerland	1	0.4
43	Thailand	1	0.4
44	Turkey	3	1.3
45	Ukraine	6	2.7
46	USA	6	2.7
47	Several countries (including Egypt, Nigeria, Uganda)	23	10.3
Total		224	100.0

Table 2. Correlation of the presentations and the countries of their authors

Presentations	The number of presentations	Per cent
Presentations by the authors from the countries where English is the mother-tongue	18	8.0
Presentation by the authors from the counties where English is the second language	7	3.1
Presentations by the authors from the countries where English is a foreign language	176	78.6
Presentations by the authors from different countries	23	10.3
Total	224	100.0

In order to solve the above mentioned terminology problems the author of the paper suggested the idea of compiling a concise dictionary of Engineering pedagogy terminology. This idea was supported by the IGIP Executive Committee and the IGIP Working Group "Language and Humanities in Engineering Education" was chosen to be responsible for this project.

The goal of the project is to compile a concise dictionary of Engineering Pedagogy terminology under the name of "IGIP Multilingual Glossary". The Glossary is a list of terms in Engineering Pedagogy with corresponding definitions aimed to facilitate communication among IGIP members, to understand professional information and to be understood. The Glossary should consist of the English core comprising basic Engineering Pedagogy terms and their definitions to be followed by translations into the target languages according to IGIP National Sections, for example, Russian, German, Hungarian, Portuguese, Estonian, and other languages of the countries participating in the IGIP activities.

The structure of "IGIP Multilingual Glossary" provides Preface, the section "Guide to use the dictionary", the main English Subject Part, and references. The Preface describes the aim of the dictionary, the target audience and the basic statements of the lexicographical concept applied. The Guide should give clear instructions on the dictionary usage. The main Subject Part should consist of Chapters that contain selected Engineering Pedagogy terms united by the subject of the Chapters. Each entry-word comprises a term in English, its transcription, and a term definition. If necessary some lexical comments may be added.

In order to achieve the goal the following objectives were set:

- Defining the Chapters
- The selection of terms for each Chapter
- Definitions of terms
- Reviewing, editing and adding terms
- Translation of terms into target languages (National Sections)
- Electronic version
- Uploading to IGIP platform.

The solution of the first objective implies defining the main topics of Engineering Pedagogy to be reflected in the glossary. Then for each Chapter it is necessary to select most relevant terms used in oral and written professional communication. The definitions are compiled on the basis of deep analysis of glossaries, encyclopedias, dictionaries, textbooks on pedagogy, scientific papers on the problems of education including IGIP Proceedings. As soon as a Chapter is completed, it should be reviewed by IGIP members who are specialists in this particular field. For example, the Chapter connected with the classical terminology of the Klagenfurt Engineering Pedagogy School has been reviewed by Tiia Ruutman (Estonia) and Dana Dobrovska (Czech Republic) who are A. Melezinek followers. According to their recommendations the Chapter was edited. The reviewers came to the conclusion that no other terms should be added. At the next stage, when all the Chapters are ready, the process of their translation into target languages begins. This process can go simultaneously or asynchronously. For example, the terms have been already translated into Russian, but the translation into other languages can take a long time as it depends on the enthusiasm of IGIP National Sections. Moreover, as soon as a new country organizes a National Section it has the opportunity to start this work. If there is a difference in the interpretation of a notion, definitions of the terms in the target languages may be added.

3 Approach

"IGIP Multilingual Glossary" presents Engineering Pedagogy terminology that absorbed the terms of general pedagogy, contains the terms of engineering education systems, engineering education, technical teachers training, research, and the terms of international integration. Frequently used in discussions and papers word combinations have also been included.

The Glossary is compiled according to the lexicographical concept developed for a series of learners' concise terminology dictionaries published by Moscow Automobile and Road Construction State Technical University. The concept is described in detail in [4].

According to the concept a term can consist of more than one word. In Engineering Pedagogy we can find two-word terms like "teaching process" or "European dimension". There are three-word terms like "teaching/learning process" or four-word terms like "European Higher Education Area". Even it is possible to find terms consisting of five words: "European Credit Transfer and Accumulation System". Following this interpretation of a term the selection of lexicographical units for the Glossary embraced not only one-word terms but multicomponent terms as well.

It is also necessary to underline that a term can be a component of the basic stock of a language but it enters it through a certain terminology system. Typically a term is characterized by belonging to one terminology system, it is monosemic, it has a definition, and it shows absence of expressiveness and stylistic neutrality. It is common knowledge that monosemy is necessary for effective professional communication. A term is considered to be the most informative unit of a language. It reflects scientific and professional knowledge of the objects and phenomena, and its definition determines its most relevant characteristics.

The work on the Glossary has revealed that the situation with Engineering Pedagogy terminology is quite different. Obviously it is connected with the international character of the Society's activities. Firstly, we mainly deal with Global English that reflects the diversity of engineering education systems and Engineering Pedagogy theories. Secondly, Engineering Pedagogy terms very often refer not to one but to more than one terminology systems. In sciences, mathematics or civil engineering there exists the universal system of notions that are monosemic and quite clear to all the specialists no matter what country they are from. This universal terminology system is just expressed by means of different languages. That is why when engineers fail to understand professional communication or to explain some phenomena they write formulas, turn to sketches, plans, diagrams, etc. and succeed in understanding.

As a result, due to belonging to various terminology systems, sometimes an Engineering Pedagogy term may lose its monosemy; there appear a lot of synonyms of some terms; the terms may have more than one definition or definitions may determine different relevant characteristics as Engineering Pedagogy terminology reflects scientific and professional knowledge of different nations. This situation required solving problems of two types. The problems of the first type are connected with the selection and presentation of English terms in the glossary and the problems of the second type arise with the translation of English terminology into target languages.

While selecting the terms it became evident that due to different national traditions in educational theories in scientific literature published in English the key notion of all pedagogical sciences is expressed by a number of terms. They are "pedagogy", "pedagogics", "education", "education science", "educational science", "pedagogical science'. In the Glossary all of them are placed in the corresponding Chapter with definitions in English and explanations in Russian (of course, explanations if necessary can be given in any other target languages). These explanations can help to prevent common mistakes in translation from English into Russian. For example, the term "comparative education" is used to denote an established academic field of study that examines education in one country (or group of countries) by using data and insights drawn from the practices and situation in another country, or countries. Comparative education also includes the comparative study of educational systems. But interpreting the name of this science into Russian the translators choose the broad meaning of the word "education" that is defined as the process of learning and acquiring information. It is the reason of the mistakes in translation of this term from English into Russian.

At the same time one term can denote more than one scientific notion. Ignoring this fact may lead to misinterpretation of such terms in professional communication. This case can be illustrated with the help of comparatively new concept of "international education". On the basis of the analysis of various glossaries, dictionaries, and scientific papers it turned out that at the moment the definition of this term is debated. There are two general meanings used by specialists in the sphere of Engineering Pedagogy. The first meaning refers to education that transcends national borders through the exchange of people, for example, students studying at oversees campuses as part of a study abroad program or as part of a student exchange program. The second meaning refers to a comprehensive approach to education which purposefully prepares students to actively participate in the global world by developing international attitudes, international awareness, international-mindedness, international understanding and by providing appropriate forms of assessment and international benchmarking.

Then there are cases when two Engineering Pedagogy terms are commonly used as synonyms which sometimes makes paper authors in a difficult position of unawareness which of them to choose in this or that situation. It concerns the terms "assessment" and "evaluation". In many glossaries they are interpreted as complete synonyms and the terms "self-assessment" and "self-evaluation" are used to prove this point of view. But careful analysis of pedagogical works helped to find out that there also exists the opinion that these two terms differ from each other to some extent. "Evaluation" is defined as the processes of description, analysis, and judgment of educational programs, practices, institutions, and policies for a range of purposes. "Assessment" is the term used most often in relation to student learning or performance and relates to the activity or task designed to show what a person knows or can do. So "assessment" constitutes only a part of evaluation in connection with a concern for the outcomes of a course, curriculum, or program. This particular point of view that explains the usage of these terms has been accepted for "IGIP Multilingual Glossary" because according to linguistics parallel usage of complete synonyms does not last a long time and results in either dropping out from communication of one of them or acquiring a slightly different shade of meaning.

So in order to solve the problems of the first type it was necessary to include all the synonyms in independent word entries and provide them with distinct definitions. In order to show the difference of their interpretation necessary explanations were added.

The problems of the second type are connected with translation of the terms into Russian as one of the target languages. It turned out that there are English terms that do not have equivalents in Russian. There are no nationally accepted terms to denote education provided by educational or training institution, structured in terms of learning objectives, time or support and leading to certification ("formal education") opposed to structured education that is provided by an educational or training institution but not leading to certification ("non-formal education") or to learning resulting from daily life activities related to work, family or leisure that is not structured and not leading to certification ("informal education"). The term "post-secondary education" also illustrates the same problem.

Then, there are many cases when terms definitions do not completely coincide in different languages as they reflect various theories or different educational systems. The term "higher engineering education" illustrates this case as for Russian educators engineering education is always associated with higher education. The term "further education" as post-secondary education that is distinct from education offered in universities and that may be at any level above compulsory secondary education has also a different shade of meaning.

We meet a similar problem trying to express the terms typical for Russian Engineering Pedagogy in English. For example, the term "principle" is one of basic notions of Russian pedagogy that determines the main requirements to the teaching/learning system as a whole and to its components. The English terminology has a different connotation.

And last but not least, looking for a definition in a target language we can find different variants of it given by specialists. These parallel definitions are used in literature by representative of various scientific schools existing in one country. The key terms of "competence"/"competency" and "skill" or "habit" serve as most typical examples.

The translation of Engineering Pedagogy terms into a target language required finding proper equivalents, suggesting possible equivalents in case of missing analogies, working out definitions of many terms in the target language on the analysis of scientific literature stressing the significant characteristics of these terms. Besides, in case of serious differences on the terms interpretation explanations were added to describe the origin of these differences and different points of view.

4 Actual and Anticipated Outcomes

Following the aim of "IGIP Multilingual Glossary", the English Subject Part is the heart of the glossary. Taking into consideration the main body of the Engineering Pedagogy terminology the following Chapters have been defined:

1. Education
2. Main Educational Sciences and notions
3. Engineering Education
4. The Terminology of the Klagenfurt Engineering Pedagogical School
5. Structure of Higher Engineering Education
6. Curriculum Development
7. Teaching Process Formats in Engineering Education
8. Methods and Techniques in Engineering Education
9. Technical Aids of Teaching in Engineering Education
10. Learning Outcomes and their Assessment
11. Professional Development of Technical Educators
12. The Bologna Process in Engineering Education
13. Engineering Pedagogy Research.

The Chapter "Education" contains the terms denoting the concept of education, the main types of educational systems and education. The Chapter "Main Educational Sciences and Notions" consists of two parts and includes the basic terms of educational sciences and the basic pedagogical concepts. In the Chapter of "Engineering Education" the terms are grouped as key terms, the terms of initial and secondary technical education, and the terms of higher engineering education. After introduction of the terms connected with educational systems, the Chapter "The Terminology of the Klagenfurt Engineering Pedagogical School" covers, what we may call, "classical Engineering Pedagogy terms". Chapter "Structure of Higher Engineering Education" contains the terms connected with organizational structure of higher engineering education and Engineering Degrees. The Chapter "Curriculum Development" is extremely important as it has played a significant role in IGIP work having the Working Group with the same name and devoting a lot of presentations devoted to the issues of curriculum development at IGIP Conferences. These Chapters are followed by "Teaching Process Formats in Engineering Education", "Methods and Techniques in Engineering Education", "Technical Aids of Teaching in Engineering Education", and "Learning Outcomes and their Assessment". The Chapter "Professional Development of technical educators" is divided into two parts: "Academic and Administrative Positions" and "Types of Professional Development". The Chapter "The Bologna Process in Engineering Education" is necessary as the Bologna Process introduced great changes in engineering education. As Engineering Pedagogy is not only a discipline but a science as well it was necessary to include the Chapter "Engineering Pedagogy Research". The first part of it shows various scientific approaches and theories in Engineering Pedagogy. The second part deals with the terms denoting research methods in Engineering Pedagogy and reflects the variety of them used in different countries.

5 Conclusions

IGIP uses English as the working language of the Society. Due to the international character of the Society activities it is Global English. For the majority of IGIP Conferences participants (more than 80 per cent) English is a foreign language. It causes problems of Engineering Pedagogy terminology usage and may result in ambiguity in the process of oral and written professional communication.

In order to solve this problem IGIP started a project aimed at compiling "IGIP Multilingual Glossary". The work on the Glossary showed that Engineering Pedagogy terms belong to more than one terminology systems that reflect diversity of national engineering education systems and Engineering Pedagogy approaches and theories. As a result, there are a lot synonyms, one term may have different meanings or different shades of meaning, etc.

Deep analysis of the works on Engineering Pedagogy and those related to Engineering pedagogy allowed to select the terms, formulate their definitions and give necessary explanations on their usage.

"IGIP Multilingual Glossary" do not only help IGIP members with linguistic problems but the Glossary can be considered to be the first step on the way to harmonization and standardization of Engineering Pedagogy terminology which is significant for research and may contribute to further development of the Society.

References

1. Crystal, D.: English As a Global Language, p. 212. Cambridge University Press (2003)
2. Prikhodko, V., Polyakova, T.: IGIP. International Society for Engineering Pedagogy. The Present, past and future. Moscow: Technopoligraphcenter, p. 143. Приходько, В.М., Полякова, Т.Ю. (2015) IGIP. Международное общество по инженерной педагогике. Прошлое, настоящее и будущее. М.: Техполиграфцентр, с. 143
3. Melezinek, A.: Ingenieurpädagogik: Grundlagen einer Didaktik des Technik-Unterrichte. Springer-Verlag, s. 152 (1977)
4. Polyakova, T.Y., Prikhodko, V.M., Rementsov, A.N., Bogacheva, E.: Engineering pedagogy terminology. Teaching and learning in a digital world. In: Proceedings of the 20th International Conference on Interactive Collaboration Learning, vol. 2, pp. 195–204. Springer, New York (2017)
5. Proceedings of 34th IGIP International Engineering Education Symposium Design in Education in the 3rd Millennium. Froniers in Engineering Education, 12–15 Sept 2005, Mor Ajans, Istambul, Turkey, p. 904 (2005)
6. 25 Jahre IGIP: Who is Who. Leuchtturm-Verlag, p. 284 (1997)
7. Higher education terms glossary. https://www.hotcoursesabroad.com/study-abroad-info/once-you-arrive/higher-education-glossary
8. Glossary on the Bologna Process. English-German-Russian. HRK – German Rectors' Conference. Bonn, p. 196 (2006)
9. Kolesnikova, I., Dolgina, O.: A Handbook of English-Russian Terminology for Language Teaching, p. 223. Cambridge University Press, Russia-Baltic Information Center, S.-Petersburg (2001)

Educational Virtual Environments

Addressing the Cultivation of Teachers' Reflection Skills via Virtual Reality Based Methodology

Kalliopi-Evangelia Stavroulia[(✉)] and Andreas Lanitis

Visual Computing Media Lab Cyprus University of Technology,
Limassol, Cyprus
{kalliopi.stavroulia, andreas.lanitis}@cut.ac.cy

Abstract. Virtual Reality (VR) based approaches have emerged lately as a new learning and training paradigm. This paper addresses the possibility to cultivate teacher's reflection skills using a VR based approach targeting to maximize their professional development. Reflection has been considered as a skill of paramount importance for teachers and an integral part of their professional development. The research aimed to investigate possible difference in cultivating reflection skills between participants who used a VR system and the control group who participated in a real classroom setting. The results indicate that the process of reflection is essential for the participants of all groups. However, the participants using the virtual classroom environment scored higher to the reflective items that the control group that tended to be undecided. Moreover, the VR system gave the participants the opportunity to enter the students' virtual body and understand the different perspectives affecting at a higher level the reflection process than the control group.

Keywords: Virtual reality · Teacher education · Reflection skills

1 Introduction

Teaching is a dynamic and complex profession that requires highly skilled teachers able to act to the various situations that arise within the school context. Thus, teacher's professional development and high-quality expertise is of paramount importance and the key that will eventually maximize the quality of the education provided by an educational system.

Reflection is defined as an everyday process, an important human activity that allows as to step back from our experiences, recapture them, explore them in depth and evaluate them to improve our actions in future similar situations. Through reflection we find ourselves in place to develop new strategies and to take the appropriate course of action. The process of reflection in teaching practice can be a powerful tool contributing significantly in maximizing learning of teachers and their professional expertise. The development of critical reflection skills is of paramount importance in teacher education [1] because education is a moral and ethical act and thus must be in question by educators [2]. Thus, the ability to reflect is considered an integral part of

© Springer Nature Switzerland AG 2020
M. E. Auer and T. Tsiatsos (Eds.): ICL 2018, AISC 916, pp. 285–296, 2020.
https://doi.org/10.1007/978-3-030-11932-4_28

teacher professionalization and one of the basic standards that teacher candidates must develop to achieve self-development.

This paper presents the possibility to improve teacher training and reflection skills via using VR environments as part of teacher training methodology that will allow in-service and pre-service teachers to experience an entirely new side of training. VR can provide teachers an absorbing, realistic and interactive virtual classroom world, allowing them to virtually experience real life class scenarios. Within the virtual world teachers can experience real-life situations, allowing them to test their knowledge and decide the appropriate course of action without real risk of harming students. The aim of the experiment is to investigate the possibility of using a VR-based technique for the cultivation of reflection skills. Additionally, the experiment aims to identify any differences between the groups that used the VR system and those who were trained in a real classroom setting without the use of VR technology. The VR tool used in the experiment addresses specific teachers' competences derived from an extensive documentation of existing Teachers' Competence Models and significant guidance by experts who pointed specific competencies of primary importance to teachers.

2   Literature Review

2.1   Reflection in Teacher Education: Significance and Challenges

Reflection is not a new concept but can be traced back to ancient times coming from the Latin word "reflectere" that means to turn back [3]. Socrates and Plato were the first to use the process of reflection. Nevertheless, although the concept has been around for a long time, the educational system became capable of reflection in the 18th century when education was treated as part of the total social system [4]. Over the past several decades, reflection and reflective practice has been considered as an important part in the acquisition of knowledge and skills and has been widely adopted in the field of both initial and lifelong education [5].

There is no single definition of the concept of reflection but a variety of classifications representing the different viewpoints of the theorists that used the term. Dewey [6] defined reflection as an "active, persistent, and careful consideration of any belief or supposed form of knowledge in the light of grounds that support it, and the further conclusions to which it tends". Hinett and Varnava [7] state that reflection is an everyday process, "an important human activity in which people recapture their experience, think about it, mull it over and evaluate it". Seibert [3] and Bengtsson [8] state that Reflection is a learning mechanism that includes the process of stepping back from an experience and through extensive consideration get a better and deeper understanding of a phenomenon.

Reflection has been considered as a skill of paramount importance for teachers and an integral part of their professional development [8]. Dewey [9] argued that teacher education programs should emphasize on the promotion of reflection skills that will help teachers understand the connection between theory and practice maximizing their effectiveness. Reflection can promote the development of new knowledge regarding good teaching practices that goes beyond the role of the university and needs the active participation of the teacher [10].

Reflection on teaching practice concerns all teachers, educators or tutors of all educational levels, who are led by their need to understand and evaluate their lesson, so that they can identify possible problems and mistakes [11]. Student-teachers might reflect on their practice aiming to improve their teaching skills, newly appointed teachers could use reflective practice to increase their knowledge and understanding maximizing their professional development, while experienced teachers through reflection are able to understand in depth various educational issues [12].

Through reflection teachers can explore more critically their teaching practice allowing them to enhance their professional knowledge. Nevertheless, a factor that must be taken into consideration is that teachers' life as students and all those years of experience gained during this period of their life constitute the roots of their professional development in the teaching path [13, 14]. Those experiences affect their beliefs regarding the teaching profession that lead their future practice affecting negatively in many cases the acquisition of reflective skills by pre-service teachers. Thus, reflective practice is essential to facilitate future teachers understanding and change of beliefs [14].

The significance of reflection lies in its relationship with "professional knowledge and practice" [11]. Reflective practice is a promising method to be used in education allowing people to recapture their experience and use it as an opportunity to learn [7, 15]. The strong relationship of reflection with practice promotes experiential learning allowing people to learn from their experience leading to the development and maintenance of professional expertise [5, 11, 12]. Thus, reflection can enhance the professional development of an individual affecting implicitly the society as a whole [15].

Even though the process of reflection is quite promising in the field of education, several challenges have been encountered that need to be addressed. One significant issue is that reflection is not characterized by conceptual clarity and as a result it has been used differently by different scientists and practitioners [5, 15]. Furthermore, there is high possibility that in many cases reflective activities promote predetermined outcomes turning the reflective process to a memorization process preventing learners self-reflect and brainstorming [5]. Moreover, in many cases the reflective activities do not target learning. It is important to design the reflective activities within the context of specific aims and consequences, otherwise the expected outcomes might not be accomplished [12]. In many cases reflection is not being encouraged by the learning context the trainer has set and as a result the trainees do not focus on their personal experience and do reveal their misunderstanding and uncertainties, undermining their learning [5]. Equally important is the fact that the aims and boundaries of reflection following an activity must be set, otherwise it is possible to observe ethical dilemmas affecting negatively the relationship between the teacher and a student [5].

Reflection must become a top priority of teacher education programs, as through the reflective process teachers will be able to evaluate their experiences and throughs, especially those coming from their classroom practice and analyze different viewpoints increasing their awareness and understanding [13–15]. As a result, reflection has been set as a goal of many teacher education programs as there are indications that the integration of reflection within the curriculum resulted in more empowered beginning teachers [14]. Moreover, reflection is included in most teacher competence models and frameworks these days [16, 17].

Nevertheless, despite the fact the reflection is a key component in most competence models, teachers do not learn how to critically evaluate and improve their own practice [18]. There are indications that teacher education programs promote superficial cultivation of teacher's reflective skills [19, 20]. Moreover, teacher education focuses on the promotion of reflection regarding the teaching practice and the ways to improve the teaching strategies to respond to the standardized curriculum ignoring the importance of reflection to the promotion of teacher's professional development [10]. Furthermore, results indicate that candidate teachers do not possess the ability to reflect even after receiving relevant education and as a result they face difficulties in critically reflect on their teaching practices [2]. This lack of teacher's reflective skills indicates a failure of teacher education programs [21]. Thus, there is a need to enhance teachers' reflection skills that will promote experiential learning, allowing teachers to learn from their experience leading to the development and maintenance of professional expertise [12].

Taking all the above into consideration reflection is a key point in the professional development of teachers. It is an undeniable fact that notable changes take place in the society and for this reason teachers must allocate the ability to respond creatively in any given situation. This can be accomplished through the reflective practice. Thus, teachers must become reflective practitioners, reflect upon their practice, seeking to detect their strengths and weaknesses and learn through the experiences of their everyday classroom lives. Using reflecting practices will enable teachers to invent new strategies of action and make decisions based on prior experiences. Moreover, by recognize their own errors and weaknesses teachers can have an explicit picture of their progress and see what needs to be improved. Reflection is the means that will enable them to increase their self- awareness, develop critical thinking and evaluation skills. In addition, reflection encourages metacognition and self-empowerment, enhances creativity and contributes in the improvement of communication, which will help them respond to the new social challenges of the 21st century.

2.2   VR-Based Approach in Teacher Education to Enhance Reflection Skills

The last decades, there is a growing interest in using VR based learning within teacher education and the results seem promising regarding issues such as bullying and recognition of student's vision disorders [22, 23]. Moreover, the use of VR environments is very useful in domains complex and difficult to master, and as such it is well suited for teacher professional development. VR has the potential to become an innovative and essential tool in teacher training offering teachers the room for experimentation but without the risk of harming real students. Equally important is the fact that the knowledge gained using VR can be applied to real life and thus teachers will be able to apply the knowledge gained in their real classroom. Moreover, such a VR system can offer users the opportunity to record their performance, reflect on it and experiment again maximizing their teaching performance. Taking all the above into consideration, we propose the use of a VR tool for the development of effective future teachers that will be successful in the classroom, while constant training within the virtual environment will ensure their survival in todays' digital and multicultural classrooms.

This research aims to give an innovative Virtual Reality-based approach to teacher education and the related training methodology that will allow in-service and pre-service teachers to experience an entirely new side of training. Moreover, the use of VR in teacher training responds to a real need from the European Union for improving the development of teaching practices and attracting more high-quality candidates to the teaching profession. The aim of this research is to offer a new paradigm in teacher training, an alternative safe and low-cost environment that allow users to learn from their mistakes without consequence for the students. The proposed virtual reality tool aims to provide strong support for teachers' professional development through high-quality scenarios that address real needs of teachers. Moreover, the VR tool addresses specific teachers' competences as outcome, after an extensive documentation of existing Teachers' Competence Models and significant guidance by experts who pointed specific competencies of primary importance to teachers.

The implementation of the VR based approach followed a full designed cycle, a five-phase process like this of the ADDIE model used by instructional designers and training developers including Analysis, Design, Development, Implementation, Evaluation. The five phases -Analysis, Design, Development, Implementation, and Evaluation- represent a dynamic, flexible guideline for building effective training and performance support tools. The development of the application is linked to both strong theoretical foundations in education derived from the literature, and real teachers' problems based on extensive literature analysis and survey.

Reflection via the use of VR will allow teachers to become "researchers in the practice context" [24] constructing knowledge of the problematic situations that they will experience. The development of the virtual worlds in response to real teacher training needs is a crucial component that makes VR a reliable mean for experimentation. Additionally, VR based approach allows the user to construct knowledge from facts and improve their teaching practice via recalling their own experiences and reflecting on them [14, 25]. Furthermore, VR offers the possibility to experience a range of professional situations while it also includes the element of repetition. A professional practitioner, experiences various situations repeatedly and through reflection he/she can criticize and think back the situations he/she experienced maximining his/her knowledge and understanding and be prepared for future situations [24]. Thus, a teacher within a VR environment can encounter a certain situation again and again and through experimentation and constant practice to be able to respond in a similar situation in the real classroom context.

3   Method

The aim of the experiment was to identify the potential of using a VR based approach to enhance the cultivation of reflection skills. Moreover, the research aimed to identify possible differences regarding reflection skills between the groups that used the VR system and the control group that was trained within a real classroom setting without the use of VR. To achieve the above aims two research questions were posed:

- Does the use of VR promote the cultivation of reflection skills?
- Is there a difference in the cultivation of reflection skills between participants who used a VR system and those who were trained in a real classroom setting?

3.1   The Framework

As part of the experiment two groups were exposed to two different classroom scenes:

- The control group, the Physical Space group (PS.group) experienced the scenario in a physical classroom with real students.
- The second group, the Virtual Reality group (VR.group) experienced the scenario in a virtual classroom environment (see Fig. 1).

Fig. 1. The control group on the left, the virtual classroom environment on the right

The scenario dealt with multiculturalism and verbal bullying and its selection was the result of an extensive literature review research, survey and interview conducted with experienced in-service teachers and academic experts regarding the most important problems that teachers face within today's classrooms. The scenario begins with the teacher introducing a new foreign student called Lynn to the classroom. Following her introduction to the class, Lynn receives verbal bulling from her classmates. The user-teacher was given the opportunity to experience the problem from two different perspectives: through the eyes of the student Lynn (Perspective I) and through the eyes of the teacher (Perspective II). The scenario and the dialogues were the same at both perspectives, changing only the camera position allowing the user to experience the two different viewpoints.

The VR application was developed with the Unity© game engine. The 3D avatars (teachers and students) were created using the online software Autodesk® Character Generator and the 3D models (school and objects including desks, chairs etc.) were develop using Maya Autodesk student version. Moreover, the Head Mounted Display (HMD) HTC VIVE was used as a means of viewing the application.

3.2   The Sample

A total number of 33 participants, 22 male and 11 female, took part in the experiment all from higher education sector. Most of the respondents (63.6%) were aged from 25 to 29 years old, 27,3% were aged from 30 to 39 years old, two of the participants were aged from 50 to 59 years old and one was aged from 18 to 24 years old. Regarding teaching experience there are differences between the two groups. In the first group half of the participants (54.6%) had teaching experience from 1 to 2 years, while 36.4% claimed to have no teaching experience at all. On the contrary in the second group 72.8% of the participants claimed to have teaching experience from 1 to 5 years. As far as the third group is concerned 45.5% of the participants had teaching experience from 1 to 5 years, while 36.4% has experience from 6 to 10 years. The results indicate that most of the participants were unfamiliar with the use of virtual reality.

3.3   The Research Tool and the Process

For the measurement of reflection, a questionnaire was developed based on several instruments concerning reflection as none of the already existing instruments responded to the needs of the current research. The questionnaire was pre-tested along with the virtual world by academic experts, and their comments and support lead to several modifications to achieve a clear outcome for the participants. The experiment took place in December 2017 in Geneva. Before taking part in the experiment all partici- pants were informed about the experiment and its purpose, assurances of confidentiality and non-traceability regarding their participation were given along with explanations regarding the right to withdraw at any time and for no reason. After voluntary informed consent was obtained, the scenario was explained to the participants with more details and the experiment began. After the end of the experiment the participants were asked to complete the questionnaire.

4   The Results

After the data has been collected and analyzed with the use of SPSS software (Sta- tistical Package for Social Sciences). Before the presentation of the results there is a need to clarify the names of the two groups participated in the experiment and explained in the framework section above: VR.group (Virtual Reality group) and PS.group (Physical Space group).

Cronbach's Alpha coefficient of internal consistency was quite good for the items of the reflection scale (Cronbach s Alpha = 0.88). The results indicate that reflection is considered an important process for all of the participants as the vast majority tend to score from agree to strongly agree in both groups (VR.group M = 5.18 SD = .54, PS.group M = 5.27 SD = .47). From the data it can be concluded that participants' experience within the virtual classroom can contribute to the change of their beliefs and ideas regarding multiculturalism and bullying as the majority tended to score from agree to strongly agree (VR.group M = 5.00 SD = .41). On the contrary, participants of the control group tend to be more neutral and undecided (PS.group M = 4.36 SD = .51). The output states that participants of both groups disagree with teachers disinterest in analyzing student's needs as they tended to score from disagree to strongly disagree (VR.group M = 4.68 SD = .48, PS.group M = 4.36 SD = .51). The results suggest that the group that used the VR system argued that such an experience can change the way teachers attend the needs of their students, as the participants tended to score from agree to strongly agree (VR.group M = 5.23 SD = .43). The majority of the participants in the control group tended to be more undecided and neutral indicating their training experience in the real classroom setting did not have the same impact as the VR experience to the other two groups (PS.group M = 4.36 SD = .51). From the data it can be also concluded that both the VR and the real-time classroom training experience can help teachers change the way they will react on disruptive behavior among my students, nevertheless many of the participants tended to be neutral especially those of the control group (VR.group M = 4.82 SD = .73, PS.-group M = 4.36 SD = .51). Moreover, the results indicate that training both via VR and real-time can help teachers discover faults in their interaction with students that they previously believed to be right as the participants tended to score from agree to strongly agree, however, many of the participants tended to be neutral and undecided in both groups (VR.group M = 4.82 SD = .79, PS.group M = 4.55 SD = .52). Based on the questionnaire responses it can be concluded that the participants who used the VR system tended from agree to strongly agree that VR experience can offer them the ability to enter their student's position, while the participants of the control group tended to be more neutral regarding entering their student's position (VR.group M = 5.55 SD = .51, PS.group M = 4.36 SD = .51). Furthermore, the results suggest that VR can promote teacher's ability to provide improved support to students of other racial and ethnic groups. The majority of the participants that used the VR system tended to score from agree to strongly agree (VR.group M = 5.00 SD = .62), while the participants of the control group scored lower tended from disagree to agree, indicating that their training in the real classroom did not seem to promote at high level their ability to support more students belonging to racial and ethnic minority groups (PS.-group M = 4.09 SD = .70).

The results from the tests of normality (namely the Kolmogorov-Smirnov Test and the Shapiro-Wilk Test), revealed that most of the items are below 0.05, therefore, the data significantly deviate from a normal distribution and non-parametric tests were used for the analysis.

Mann-Whitney U test indicated statistically significant differences between the two groups in some of the reflection scale items (see Fig. 2). There was a statistically significant difference between the VR and the control group regarding the impact of the VR experience and the real-time classroom training on challenging teachers' firmly held ideas about the issues of multiculturalism and bullying (U = 54.5, p = .003). Moreover, Mann-Whitney U test revealed that there was a statistically significant difference between the two groups regarding changing the way that teachers attend the needs of the students (U = 34, p = .000). Another statistically significant difference between the two groups was found regarding teacher's ability to enter the student's position (U = 20, p = .000). Furthermore, Mann-Whitney U test revealed a statistically significant difference between the VR and the control group regarding teacher's ability to support more efficiently students belonging to racial and ethnic minority groups (U = 45, p = .001).

A Spearman's rank-order correlation was run to determine the relationship between some items of the reflection scale. There was statistically significant evidence of a strong positive correlation between teachers' change of reaction on disruptive

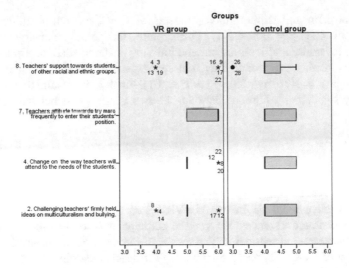

Fig. 2. Mann-Whitney U test showing the differences between the two groups

behaviors among the students and teachers' ability to discover faults in their interaction with students that they previously believed to be right (r_s = .830, n = 33, p = .000 < .01) (see scatterplot at Fig. 3). Moreover, there was a moderately positive correlation between the way teachers will attend to the needs of the students and support more students of racial and ethnic minority groups (r_s = .524, n = 33, p = .002 < .01). The way teachers will attend the needs of the students was also moderately correlated to teachers' effort to enter more frequently their students' position (r_s = .476, n = 33, p = .005 < .01). From the data it can also be concluded a moderately positive correlation between teachers' effort to enter more frequently their

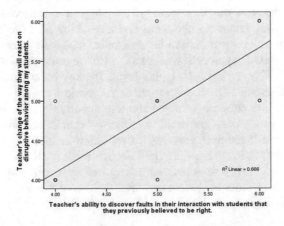

Fig. 3. The relationship between teachers' change of reaction on disruptive student behaviors and their ability to discover faults in their interaction with students for both groups

students' position and support more students of racial and ethnic minority groups (r_s = .445, n = 33, p = .009 < .01). Furthermore, the participants argued that their experience regarding multiculturalism and bullying challenged some firmly held ideas and because of this experience they will support more students of other racial and ethnic groups (r_s = .496, n = 33, p = .003 < .01). Finally, the results do not reveal any statistically significant difference between the two genders and between the different age groups.

5   Discussion

The current research aimed to explore the possibility of promoting and cultivating teacher's reflective skills via the use of a VR system. Moreover, another objective was to identify possible differences between the participants who used the VR system and those who were trained without it. The results of the research are promising regarding the potential of using VR based methodology in teacher training regarding the cultivation of reflection skills.

   The results indicate that the process of reflection is significant for the participants of both groups. Nevertheless, experimental results indicate that the control group that did not use the VR system scored lowered in all items of the reflection scale and tended to be more undecided regarding the effectiveness of their training within the real classroom setting. On the contrary, experimental results indicate that the VR experience can contribute to a higher level to the change of beliefs and ideas regarding multiculturalism and bullying, change the way teachers attend the needs of their students and react on disruptive behavior among the students. Moreover, the VR experience offered the participants the opportunity to enter the student's position and experience bullying through the student's viewpoint, challenging some firmly held ideas and promoting more support to students of racial and ethnic minority groups. Within the VR environment the camera was placed in a way allowing the participants to see themselves as being in the body of the student, offering a more immersive experience and deeper understanding of the problem. This could not be achieved in the real classroom setting where the participants felt more like themselves, resulting in more neutral scores. It seems that although there are indications that reflection skills were cultivated at all groups, the VR based approach was more effective encouraging reflection and understanding of different views at a higher level. The key point consists in the ability of VR to offer the user the opportunity to view a scenario from the eyes of a different person, making their experience a unique learning opportunity. The great variation of different point of views does not seem possible to be achieved in a real classroom setting, and as a result multiple perspectives reflection cannot be achieved. Overall, there are several indications regarding the effective use of VR for the cultivation of reflection skills. However, further research is required to investigate the difference on participant's reflective thoughts before and after the use of the VR system in order to identify its effectiveness.

Acknowledgment. Authors acknowledge funding from the European Union's Horizon 2020 Framework Programme through NOTRE Project (H2020-TWINN-2015, Grant Agreement Number: 692058).

References

1. Hammerness, K., et al.: How teachers learn and develop. In: Darling-Hammond, L., Bransford, J. (Eds.), Preparing teachers for a changing world: What teachers should learn and be able to do, pp. 358–389. San Francisco: Jossey-Bass (2005)
2. Avgitidou, S., Hatzoglou, V.: Educating the reflective practitioner in the context of teaching practice: The contribution of journal entries. In: Androusou, A., Avgitidou, S. (eds.) Teaching Practice During Initial Teacher Education: Research Perspectives, pp. 97–124. National Kapodistrian University of Athens, Athens (2013). (in Greek)
3. Seibert, W.K., Daudelin, W. M.n.: The Role of Reflection in Managerial Learning: Theory, Research and Practice. Quorum Books, Westport, USA (1999)
4. Luhmann, N., Schorr, K.-E.: Problems of Reflection in the System of Education. Waxmann, New York (2000)
5. Boud, D., Walker, D.: Promoting reflection in professional courses: the challenge of context. Stud. Higher Educ. **23**(2), 191–206 (1998)
6. Dewey, J.: How We Think: A Restatement of the Relation of Reflective Thinking to the Educative Process. DC Heath and Company, Boston (1933)
7. Hinett, K., Varnava, T.: Developing Reflective Practice in Legal Education, p. 51. UK Centre for Legal Education, Coventry (2002)
8. Bengtsson, J.: What is reflection? On reflection in the teaching profession and teacher education. Teach. Teach. **1**(1), 23–32 (1995)
9. Dewey, J.: The relationship of theory to practice. In: Borrowman, M. (ed.) Teacher Education in America: A documentary History. Teachers College Press, New York (1966)
10. Zeichner, K., Liu, K.Y.: A critical analysis of reflection as a goal for teacher education. In: Handbook of Reflection and Reflective Inquiry, pp. 67–84. Springer, New York (2010)
11. Harrison, J.: Professional development and the reflective practitioner. In: Dymoke, S., Harrison, J. (eds.) Reflecting Teaching and Learning. A Guide to Professional Issues for Beginning Secondary Teachers, pp. 7–44. SAGE, London (2008)
12. Pollard, A.: Reflective teaching, 2nd edn. CONTINUUM, London (2005)
13. Körkkö, M., Kyrö-Ämmälä, O., Turunen, T.: Professional development through reflection in teacher education. Teach. Teach. Educ. **55**, 198–206 (2016)
14. Yost, D.S., Sentner, S.M., Forlenza-Bailey, A.: An examination of the construct of critical reflection: Implications for teacher education programming in the 21st century. J. Teach. Educ. **51**(1), 39–49 (2000)
15. Procee, H.: Reflection in education: a Kantian epistemology. Educ. Theor. **56**(3), 237–253 (2006)
16. Caena, F.: Literature review Teachers' core competences: requirements and development. Education and training, 2020 (2011). http://ec.europa.eu/dgs/education_culture/repository/education/policy/strategic-framework/doc/teacher-competences_en.pdf
17. European Commission: Supporting the Teaching Professions for Better Learning Outcomes (2013). European Commission website http://eur-lex.europa.eu/LexUriServ/LexUriServ.do?uri=SWD:2012:0374:FIN:EN:PDF
18. Hagger, H., McIntyre, D.: Learning Teaching From Teachers. Realizing the Potential of School-Based Teacher Education. Open University Press, Maidenhead (2006)

19. Calderhead, J.: Reflective teaching and teacher education. Teach. Teach. Educ. **5**(1), 43–51 (1989)
20. Conway, P.F.: Anticipatory reflection while learning to teach: from a temporally truncated to a temporally distributed model of reflection in teacher education. Teach. Teach. Educ. **17**(1), 89–106 (2001)
21. Poulou, M., Haniotakis, N.: Teacher candidates' reflection relative to practice: a first attempt (in Greek). Educ. Sci. **1**, 87–97 (2006)
22. Manouchou, E., Stavroulia, K.E., Ruiz-Harisiou, A., Georgiou, K., Sella, F., Lanitis, A.: A feasibility study on using virtual reality for understanding deficiencies of high school students. In: Proceedings of the 18th IEEE Mediterranean Electrotecnical Conference (MELECON 2016), 18–20 Apr 2016, Limassol, Cyprus (2016)
23. Stavroulia, K.E., Ruiz-Harisiou, A., Manouchou, E., Georgiou, K., Sella, F., Lani-tis, A.: A 3D virtual environment for training teachers to identify bullying. In: Proceedings of the 18th IEEE Mediterranean Electrotecnical Conference (MELECON 2016), 18–20 Apr 2016, Limassol, Cyprus (2016)
24. Schön, D.A.: The Reflective Practitioner: How Professionals Think in Action. Ashgate, England (1991)
25. Romano, D.M., Brna, P.: Presence and reflection in training: support for learning to improve quality decision-making skills under time limitations. Cyber Psychol. Behav. **4**(2), 265–277 (2001)

Contribution to the Setting of an Online Platform on Practical Application for the Science, Technology, Engineering and Mathematics (STEM): The Case of Medical Field

Kéba Gueye[(⊠)], Ulrich Hermann Sèmèvo Boko,
Bessan Melckior Degboe, and Samuel Ouya

LIRT Laboratory, Higher Polytechnic School, University of Dakar,
Dakar, Senegal
mamekeb@gmail.com, bulrich91@gmail.com,
samuel.ouya@gmail.com, bessan@degboe.org

Abstract. The objective of this paper is to contribute to the improvement of distance education in medicine by offering a platform for practical work. To do this, we combine the intelligence of WoT with the power of WebRTC. This platform, based on the WebRTC Kurento multimedia server and the Web of Things (WoT), allows teachers and students to do remote labs. Kurento Media Server (KMS) allows you to create media processing applications based on the pipeline concept. The Web of Things (WoT), considered a subset of the Internet of Things (IoT), focuses on standards and software frameworks such as REST, HTTP and URI to create applications and services that combine and interact with a variety of network devices. To prove the relevance of our approach, we described a scenario where the teacher initiates a medical consultation TP with any patient on which sensors are placed. Patient data is visible in real time for all students who follow the teacher's comments/explanations and also interact. However, our experimental results may be relevant for other STEM disciplines.

Keywords: Practical work · E-learning · Medicine · WoT · KMS · WebRTC

1 Introduction

Nowadays, the ability to do practical work online has become vital in science, technology, engineering and mathematics (STEM). That's why in recent years, research on remote or online labs has become very popular.

Previous research has shown that laboratory experiments in engineering disciplines have a real impact on a student's practical knowledge [1, 2]. Therefore, considerable attention has recently been given to remote laboratories in STEM. Several authors have

© Springer Nature Switzerland AG 2020
M. E. Auer and T. Tsiatsos (Eds.): ICL 2018, AISC 916, pp. 297–307, 2020.
https://doi.org/10.1007/978-3-030-11932-4_29

concluded that real laboratory experiments play an important role in STEM [3, 4]. The main problem is how to provide effective STEM training in e-learning.

In the literature, several approaches have been proposed to implement laboratories [5, 6]. Each approach solves a specific problem [7]. There are some STEM disciplines such as medicine and biology where it is difficult to do e-learning due to lack of remote laboratory equipment. The literature on remote laboratories in medicine and biology is poor. Our contribution applies first to distance education in medicine. However, our experimental results may be relevant for other STEM disciplines under certain conditions.

The paper aims to contribute to the establishment of a platform of practical work in the distance education of medicine. This platform, based on the WebRTC Kurento multimedia server and the Web of Things (WoT), allows teachers and students to do remote labs. Kurento Media Server (KMS) allows you to create media processing applications based on the pipeline concept. The Web of Things (WoT), considered a subset of the Internet of Things (IoT), focuses on standards and software frameworks such as REST, HTTP and URI to create applications and services that combine and interact with a variety of network devices. To prove the relevance of our approach, we described a scenario of practical experiences in medicine. Students and professors could interact interactively on the platform.

2   Related Works

In the literature, studies have shown that web-based infrastructures can be used to conduct clinical consultations. The authors [8–11] propose solutions based on old communication technologies, not including real time and requiring the installation of third-party software. For this study, we focus on recent work, some of which uses WebRTC.

The authors [12] focus their work on audio/video communication and data exchange between the patient and the caregiver. In addition, patients have medical sensors for taking vital parameters. To provide specific services to type I diabetic patients, [13] offer an e-health platform. The latter involves medical sensors and a robot. The robot interacts with the patient to collect the data and sends it to the health care staff. Article [14] provides an API for integrating sensor data into a web application.

3   Technology

The first part presents the technology used to establish communication between the different users of the system. Then, the Web of Things allows controlling and interacting with different intelligent objects of architecture.

3.1   Kurento Media Server (KMS)

Traditional WebRTC applications are standardized so that browsers can communicate directly without the mediation of third-party infrastructures. This is sufficient to provide basic multimedia services, but features such as group communications, stream recording, streaming, or transcoding are difficult to implement. For this reason, the most interesting applications require the use of a multimedia server.

There are many other services we can offer with media servers: augmented reality, computer vision and alpha blending. These services can add value to applications in many scenarios such as e-health, e-learning, security, entertainment, games or advertising. Kurento Media Server (KMS) is an evolution of traditional media servers that provides a modular architecture where other features can be added as modules.

Kurento is an open source WebRTC multimedia server that allows you to create media processing applications based on the pipeline concept. Media pipelines are created by interconnect modules called Media Elements. Each Media Element provides a specific feature. KMS contains Media Elements capable of recording and mixing streams, computer vision, etc.

From the point of view of the application developer, Media Elements are like Lego pieces: just take the necessary elements for an application and connect them according to the desired topology. This type of modularity is new in the field of RTC multimedia servers.

Kurento Media Server offers the capabilities of creating media pipelines through a simple JSON-RPC-based network protocol. However, to further simplify developer work, a client API that implements this protocol and directly leverages Media Elements and pipelines is provided. Currently, the Java and JavaScript client API is ready for developers [15].

Taking into account the integrated modules, the Kurento Toolbox is detailed in Fig. 1.

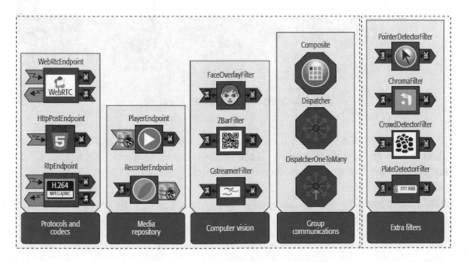

Fig. 1. Kurento Elements toolbox

3.2   Web of Thing

The Web of Things (WoT) is a specialization of the Internet of Things (IoT). On the one hand, it provides an abstraction of the connectivity of smart objects. On the other hand, WoT adds a standard web standards-based application layer to simplify the creation of IoT applications. In IoT, the communication protocols are multiple (MQTT, CoAP, ARMQP …), which creates groups of users.

The main interests of using WoT instead of IoT are the simplicity of development using APIs, standardization and simple coupling. The idea is that all intelligent objects can communicate using a Web language through an API. This API can be present in the intelligent object itself or in an intermediary that can act on behalf of the intelligent object [16]. This has become possible with the improvement of embedded systems.

4   Presentation of the Proposed Architecture

To exploit the advanced features of the webrtc, the proposed architecture integrates Kurento Media Server (KMS). During a WebRTC multimedia session, the solution provides access to the information collected by the sensors and sends it to the other end of the communication in real time. Figure 2 describes the architecture of the proposed system.

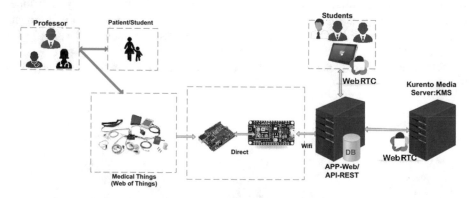

Fig. 2. Proposed system architecture

For a good understanding of the architecture, a scenario applicable in the field of telehealth (health services, online, information, training, social networks, serious games) is proposed. This scenario consists of making Practical Work (PT) possible as the clinical examination.

Distance education in medicine, biology, etc. is faced with many technical and logistical difficulties. Indeed, the students are geographically dispersed and do not have virtual laboratories for Practical Work.

The proposed solution allows the teacher to initiate a live medical consultation TP via the KMS-IoT platform. The goal of TP is to teach students how to perform a clinical exam. To do this, the professor uses the connected medical devices (stethoscope, thermometer, electrocardiogram, sphygmomanometer, etc.) to collect the constants of a student taken as a patient. This data is directly transmitted to the application and then shared in real time with the other students via the platform. With the help of a connected computer, tablet or smartphone, any student can view the constants and follow in real time the comments/explanations of the teacher. Each student can also interact by asking questions.

5 Implementation

A platform using Node.js and KMS is implemented in this paper. On the one hand, it allows establishing a multimedia communication between two users by simply using their browser. On the other hand, it allows users to access data from predefined connected objects. The proposed architecture consists of three distinct entities: Internet of Things, API and Web Application.

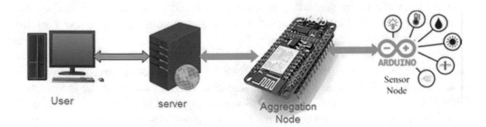

Fig. 3. Architecture WoT

5.1 Internet of Things

The first part is the WoT part. Each endpoint is considered a gateway to its set of smart objects. In addition, each user has control over these objects. The NODEMCU ESP8266 aggregation node is not responsible for reading the sensors. It simply provides a gateway between the user and the sensor network, and then performs data analysis. The sensor node is the lowest level of a sensor network. It is responsible for gathering information from sensors, performing user actions, and using communication mechanisms to send data to the aggregation node (see Fig. 3).

The ESP8266 gateway can then communicate with the sensors using one of the well-known communication protocols (Lora, Zigbee, Bluetooth, WIFI…). In the current platform, a DHT11 humidity and temperature sensor is used. The latter is connected to the NodeMCU gateway (ESP8266) which sends the sensor data using WIFI.

Fig. 4. Wiring the ESP8266 with the DHT11

In the case of e-health scenarios, all we need is portable medical sensors. They can communicate via any protocol, since the WoT summarizes the complexity of the connectivity of objects.

Using the current architecture, an implementation of the remote clinical examination is possible, where a doctor communicates with students using Kurento Media Server. The teacher has access to a set of sensors. Then he can send the information collected by these sensors to students in real time using the KMS-IoT platform. Finally, these data can be analyzed and commented by the actors. Figure 4 shows the wiring of the Node MCU Gateway with the DHT11 temperature and humidity sensor.

5.2   Api

We have developed a REST API capable of retrieving information collected by a connected medical device and storing it in a MongoDB database. MongoDB belongs to the NoSQL family Document-store, developed in C ++. It is based on the concept of a key-value pair. The document is read or written using the key. MongoDB supports dynamic queries on documents. Since this is a document-oriented database, the data is stored as JSON, BSON style [17].

According to recent work [18–20], NoSQL database systems are non-relational databases designed to provide great accessibility, reliability and scalability to huge

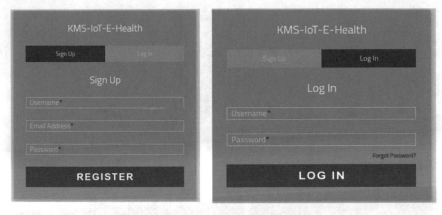

Fig. 5. Authentication and login on the KMS-IoT-E-health

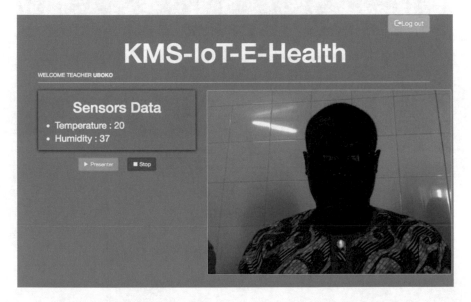

Fig. 6. Communication between teacher and students: teacher side

data. NoSQL databases can store unstructured data such as e-mails and multimedia documents. MongoDB has many security risks that can be overcome by a good, secure cryptographic system [21].

5.3 Web Application

To set up the web application, we use the NodeJs and Kurento Media Server technologies. This platform allows teachers and students to register and authenticate themselves to access Kurento Media Server features. Once connected, students can

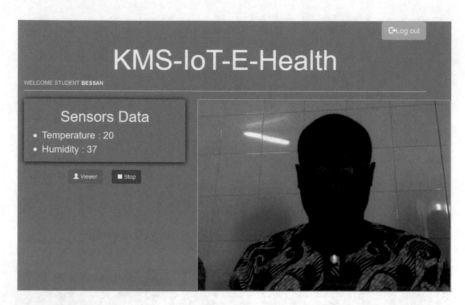

Fig. 7. Communication between teacher and students: student side

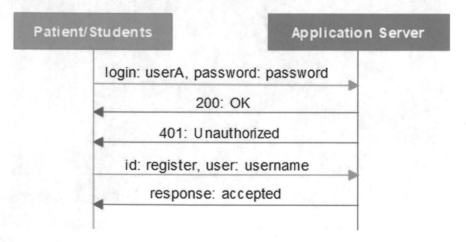

Fig. 8. Authentication flow

view the sensor data and the media streams of the teacher in charge of the TP. Figure 5 shows the authentication and login principle on the KMS-IoT platform.

The web application can also collect information from the database and display it. Connected users can then view sensor data. Figures 6 and 7 show that actors can access the temperature and humidity sensor information. The same mechanism is applicable to any other sensor.

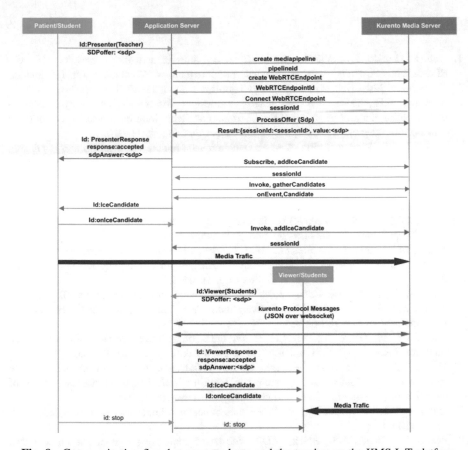

Fig. 9. Communication flow between students and the teacher on the KMS-IoT platform.

Figures 8 and 9 show the authentication and communication steps between the students and the teacher on the KMS-IoT-E-Health platform.

6 Conclusion

In this article, we propose the platform KMS-IoT which allows to do TP in the distance education of medicine. This platform is built around a WebRTC Kurento multimedia server, an API, a Web application and connected objects. This allows STEM students, in the context of e-learning, to access remote labs and conduct hands-on experiments. Adopting the proposed approach could change distance learning activities for students in STEM disciplines such as medicine and biology.

Acknowledgment. The authors kindly thank colleagues who helped them to achieve this paper, especially the members of RTN laboratory.

References

1. Coti, C., Loddo, J.V., Viennet, E.: Practical activities in network courses for MOOCs, SPOCs and eLearning with Marionnet. In: International Conference on Information Technology Based Higher Education and Training, Lisbon, 11–13 June, pp. 1–6 (2015)
2. Elawady, Y.H., Talba, A.S.: A general framework for remote laboratory access: a standarization point of view. In: IEEE International Symposium on Signal Processing and Information Technology, Luxor, 15–18 December, pp. 485–490 (2010)
3. Hashemian, R., Riddley, J.: FPGA e-Lab, a technique to remote access a laboratory to design and test. In: Proceedings of IEEE International Conference on Microelectronic Systems Education: Educating Systems Designers for the Global Economy and a Secure World, San Diego, CA, 3–4 June, pp. 139–140 (2007)
4. Lee, T.H., Lee, H.C., Kim, J.H., Lee, M.J.: Extending VNC for effective collaboration. In: Proceedings of IFOST-2008-3rd International Forum on Strategic Technologies, Novosibirsk-Tomsk, 23–29 June, pp. 343–346 (2008)
5. Tawfik, M., Salzmann, C., Gillet, D., Lowe, D., Saliah-Hassane, H., Sancristobal, E., Castro, M.: Laboratory as a service (LaaS): a novel paradigm for developing and implementing modular remote laboratories. Int. J. Online Eng. **10**, 13–21 (2014)
6. Willems, C., Meinel, C.: Users clients players problem terminal assignment B. Queuing queue terminal assignment problem C. Other methods A. Time slotting time reservation. Int. J. Online Eng. **4**, 179–185 (2008)
7. Bochicchio, M., Longo, A.: Hands-on remote labs: collaborative web laboratories as a case study for it engineering classes. IEEE Trans. Learn. Technol. **2**, 320–330 (2009)
8. Magrabi, F., et al.: Home telecare: system architecture to support chronic disease management. In: Conference Proceedings of the 23rd Annual International Conference of the IEEE Engineering in Medicine and Biology Society, vol. 4 (2001)
9. Lau, C., et al.: Asynchronous web-based patient-centered home telemedicine system, vol. 49, 12 (2002)
10. Zheng, H., Davies, R.J., Black, N.D.: Web-based monitoring system for home based rehabilitation with stroke patients. In: 18th IEEE Symposium on Computer Based Medical Systems (CBMS'05), June 2005, pp. 419–424
11. Chiang, C.Y., et al.: An efficient component-based framework for intelligent home-care system design with video and physiological monitoring machineries. In: Fifth International Conference on Genetic and Evolutionary Computing, Aug 2011, pp. 33–36
12. Pierleoni, P., et al.: An innovative webrtc solution for e-health services. In: IEEE 18th International Conference on e-Health Networking, Applications and Services (Healthcom), Sept 2016, pp. 1–6
13. Al-Taee, M.A., et al.: Web-of-things inspired e-health platform for integrated diabetes care management. In: IEEE Jordan Conference on Applied Electrical Engineering and Computing Technologies (AEECT), Dec 2013, pp. 1–6
14. Azevedo, J.A., Pereira, R.L., Chainho, P.: An api proposal for integrating sensor data into web apps and webrtc. In: Proceedings of the 1st Workshop on All-Web Real-Time Systems, ser. AWeS'15, pp. 8:1–8:5. ACM, New York, NY, USA (2015). http://doi.acm.org/10.1145/2749215.2749221
15. Garcia, B., Lopez-Fernandez, L., Gallego, M., Gortazar, F.: Kurento: The swiss army knife of WebRTC media servers. IEEE Commun. Stand. Mag. **1**(2), 44–51 (2017)
16. Guinard, D., Trifa, V.: Building the Web of Things. Manning Publications Co (2016)

17. Truică, C.O., Boicea, A., Trifan, I.: CRUD Operations in MongoDB. In: International Conference on Advanced Computer Science and Electronics Information, pp. 347–348 (2013)
18. Chopade, M.R.M., Dhavase, N.S.: Mongodb, couchbase: Performance comparison for image dataset. In: 2017 2nd International Conference for Convergence in Technology (I2CT), Mumbai 2017, pp. 255–258 (2017)
19. Jose, B., Abraham, S.: Exploring the merits of nosql: A study based on mongodb. In: 2017 International Conference on Networks & Advances in Computational Technologies (NetACT), Thiruvanthapuram, 2017, pp. 266–271 (2017)
20. Patil, M.M., Hanni, A., Tejeshwar, C.H., Patil, P.: A qualitative analysis of the performance of MongoDB vs MySQL database based on insertion and retriewal operations using a web/android application to explore load balancing—Sharding in MongoDB and its advantages. In: 2017 International Conference on I-SMAC (IoT in Social, Mobile, Analytics and Cloud) (I-SMAC), Palladam 2017, pp. 325–330 (2017)
21. Kumar, J., Garg, V.: Security analysis of unstructured data in NOSQL MongoDB database. In: 2017 International Conference on Computing and Communication Technologies for Smart Nation (IC3TSN), Gurgaon, India, 2017, pp. 300–305 (2017)

e-Pedagogical Practice Assessment in a Higher Education Comparative Context

Sayed Hadi Sadeghi$^{(\boxtimes)}$, Nigel Bagnall, and Michael J. Jacobson

The University of Sydney, Sydney, Australia
ssad2473@uni.sydney.edu.au,
{nigel.bagnall,michael.jacobson}@sydney.edu.au

Abstract. In this study the pedagogical understanding of e-practice in a higher education comparative context, using an Australian and an American institute was investigated. The theoretical framework focus was on e-learning practice in the area of pedagogy. The studied e-pedagogical sub variables were learner-centre interactivity, socio-communication, assessment, e-resources and e-Environment. Participants were postgraduate students, lecturers and staff engaged with online learning and teaching. Comparing the answers of Australians and Americans in the three levels of institute position showed that there were significant differences in evaluation of the e-pedagogical practice factor between Australian and American administrative staff, where Australians significantly evaluated the e-pedagogical practice factor higher than Americans. The results of analysis of variance (ANOVA) also revealed that there was significant difference in evaluation of the e-pedagogical practice factor between Australian and American lecturers indication that Americans evaluated this factor more significant than Australians. Moreover, an ANOVA test revealed that there was significant difference in evaluation of the e-pedagogical practice factor between Australian and American students, showing that American students evaluated this factor significantly higher than Australian students.

Keywords: E-learning · Comparative context · Pedagogical practice

1 Introduction

The process of education, which addresses the fertility of thoughts, is one of the main concern in social cominucation and nowadays it's getting more important at the same time as improvement of digital humanities [1, 2]. The goal of most researchers in the process of education and e-learning field is to provide a comprehensive pattern of practice that can be applied to all students and all learning environments based on the success and quality of distance education and e-learning [3]. While, cognitive and social tools are considered as the main factors of effective education, the quality of both the pedagogical and technological content is also viewed as an important issue influencing learning practice [4–7]. Based on some definitions have focus on the technology base rather than pedagogy it has been stated that describing -e-learning in terms of the enabling technologies is not useful as it has no distinguish between the types of design features for various e-learning approaches and more important between

© Springer Nature Switzerland AG 2020
M. E. Auer and T. Tsiatsos (Eds.): ICL 2018, AISC 916, pp. 308–320, 2020.
https://doi.org/10.1007/978-3-030-11932-4_30

different paradigms for teaching and learning [8]. According to the evidence practice based on pedagogy can support e-courses by developing a model for the learning and teaching process [9–11].

A large number of researchers have directed their attention to the field of e-pedagogical practice. These studies provide a variety of models, guidelines, critical success factors and benchmarks put forward as the best e-pedagogical practice in order to enhance and assure quality in higher education institutes. The ultimate aim of this current research is to improve the quality of online learning courses in Australian and American institutes. The pedagogical factor, which addresses the process of learning and teaching in terms of how learning and teaching is done, is at the core of e-learning environments [12]. Based on literatures the pedagogical factor is considered to be the most critical in practice, which has five sub-factors [11, 13–15].

2 The Pedagogical Factors

2.1 Learner-Centre Interactivity

Learners' interactivity is a main critical success strategy in e-learning [16, 17]. Student success can be significantly affected by active engagement in such practices as learning interactivity and integrating past experiences [18]. Interaction and discussion are at the core of the learning and teaching process that can create opportunities for networking and encourage dialogue between and among all the actors in an online learning classroom. According to the studies by Chickering et al. [17], Phipps et al. [19], Chou [20] and Zhao [15], the main focus of learner-centre interactivity is on learner-centre practices and activities, interactive networks and discussion in the classroom between and among all the actors.

2.2 Socio-Communication

Socio-communication effectiveness has been defined in variety of ways [21]. Socio-communicative orientation concerns one's approach towards others and how one perceives him/herself, and is much less descriptive of how a person actually behaves [22]. Socializing and building a concept of community attracts and retains students in online learning settings [23]. A competitive environment, effective communication, facilities and opportunities for good communication and social interactive tools are counted as influential factors in the success of e-learning [9, 15, 22, 24].

2.3 e-Environment

Environmental learning facilities allude to locations, contexts, settings and cultures in which students learn. Students who keenly support the collaborative wiki tool are successful at using it to complete unit tasks in a flexible online environment [25]. Creating and improving a sense of space and feeling at home could be important elements in reducing the dropout rate between online students [12]. A flexible environment system and environmental learning facilities are the main factors of an effective learning environment [25, 26].

2.4   e-Assessment

Assessment can focus on the students' progress, learning community, teacher practices, e-learning systems and organization. The assessment of learning in online programs requires policies, practice and tools that are clear, valid, reliable, and can be automatically administered and scored [27]. Assessment in e-learning can be carried out in different modes by teachers, peers, in terms of self-assessment as well as the students' portfolios [12]. The results of assessment in online programs could be affected due to problems of classroom feedback, academic honesty, plagiarism and feedback [6, 28, 29].

2.5   e-Resources

Having adequate contents is a necessary priority in academic setting. It was achieved by Hostager [30] that adequate learning resources and services have a positive effects on the student grades in e-learning programs. E-learning providers are expected to provide a variety of e-resources to support learning practices and activities [6, 14, 19, 31]. Organizing resources incorporates different practices in a classification system and will store organized content of e-learning programs [32].

3   Method

3.1   Participants and Design

A total of 231 participants from an Australian Institute and an American Institute were recruited to take part in this study through an online invitation email asking for volunteers. To check the normality of the distribution and homogeneity of variance in this sample the researcher applied several tests including Boxplot and Kolmogorov–Smirnov. The results indicated 16 cases as outliers (7 Australians and 9 Americans), so they were excluded from the main analysis. There were 129 female and 86 male participants out of 215 candidates with the age range of 20–30 years (number of participant (n) = 99), 30–40 years (n = 68) and 40–50 years (n = 48). Among the participants 149 students, 45 lecturers and 21 administrative staff had contributed. The type of previous experience of participants in the e-learning educational system was categorized as both blended and online (n = 155) or fully online (n = 60). In this study the primary dependent and independent variables were the pedagogical practice and positions of participants, respectively.

3.2   e-Pedagogical Practice Questionnaire

The questionnaire self-constructed was used by the researcher as an instrument. Exploratory factor analysis was applied to test the validity of the constructed questionnaire. Participants answered each question by using the Likert scale (1 = Extremely

Poor, 2- Poor, 3 = Average, 4 = Above Average (good), 5 = Excellent). It is worth mentioning that three versions of the e-pedagogical practice questionnaire were presented to participants based on their positions. The factor of e-pedagogical practice has 5 sub factors elicited by 13 questions: student-centred interactivity, socio-communication, learning environment, assessment and learning resources [10–12, 30].

Table 1. Sub-factors, Items and questions of e-pedagogical practice.

Sub-Factors	Items
Learner-centre interactivity	Learner-centred practices
	Interactive network classroom
	Using the blackboard discussion board
Socio-communication	Effective communication
	Facilities and opportunities
	The social interactive tools
	Competitive environment
e-Environment	Flexible environment system
	Environmental learning facilities
e-Assessment	Classroom constructive feedback
	Academic honesty plagiarism policy
	Feedback on assessment results
e-Resources	Access to e-resources

3.3   Procedure

After obtaining ethical approval, the study was conducted by creating an online questionnaire of e-pedagogical practice using Lime Survey software. The link to the questionnaire then was sent to the e-learning centre of health sciences in Sydney Institute. The e-learning coordinators of the institute sent the link of the survey to lecturers, administrative staff and students who were engaged with online courses. The participants responded to the questionnaire voluntarily.

4   Results of Current Status of e-Pedagogical Practice

The e-pedagogical practice factor was measured by five sub factors namely: learner-centred interactivity, socio-communication, assessment, learning resources and learning environment. In this section, the results of each sub factor based on academic position of participants in Australia and America in case of e-learning courses were reported. At the end, the total results of all sub factors of the main factor of e-pedagogical practice were reported.

4.1   Learner-Centre Interactivity

The means and standard deviation of the learner-centre interactivity sub-factor based on academic positions of Australian participants is reported in Table 2. As can be seen in this table, the highest mean regarding this sub factor belongs to administrative staff (Mean (M) = 11.12, Standard Deviation (SD) = 1.12). After administrative staff, lecturers reported this factor next to the highest score (M = 9.70, SD = 2.27) and the lowest one was reported by the students (M = 9.07, SD = 1.45). The analysis of variance (ANOVA) was applied to investigate the possibility of differences in the evaluation of learner-centred interactivity between students, lecturers and administrative staff. The results showed that academic position had significant effect on evaluation of learner-centre interactivity in Australian participants [F (2, 98) = 6.22, p = .003]. The least significant difference (LSD) multiple comparison test between the three academic positions revealed that administrative staff reported this factor more significantly higher than students and lecturers. However the evaluation by lecturers and students of this sub-factor were the same. The results showed that all Australian participants believed that learner-centre interactivity e-practice is above average.

Table 2. Mean, SD, and F value of evaluation of pedagogical sub factors.

Sub factors	Country	Students		Lecturers		Staff		F	P
		M	SD	M	SD	M	SD		
Learner-centre interactivity	AUS*	9.07	1.45	9.70	2.27	11.12	1.12	6.22	.003**
	USA	10.39	1.13	11.10	1.01	11.23	1.01	6.21	.003**
Socio communication	AUS	12.71	1.55	12.70	2.07	14.50	1.69	4.16	.01*
	USA	13.84	1.40	13.06	1.06	12.84	1.77	6.21	.003**
e-Assessment	AUS	10.18	1.21	9.85	1.59	11.87	1.35	7.21	.001**
	USA	10.59	1.11	11.39	1.31	10.00	1.41	6.78	.002**
e-Resources	AUS	3.54	0.71	4.37	0.67	4.12	0.64	11.41	.00***
	USA	4.37	0.58	4.18	0.55	4.00	1.00	2.40	.09
e-Environment	AUS	7.52	0.89	7.75	1.06	8.25	1.03	2.37	.09
	USA	8.04	0.86	7.70	0.73	7.53	0.51	3.28	.04*
e-pedagogical Practice	AUS	43.04	4.01	44.35	5.29	49.87	3.31	9.43	.00***
	USA	47.26	3.11	47.45	2.60	45.61	2.81	1.91	.15

Note: *p < .05, **p < .01, ***p < .001. AUS* = Australian

In America, as can be seen in Table 2, the highest mean of answers to this sub factor belongs to administrative staff (M = 11.23, SD = 1.01). After administrative staff, the lecturers reported this factor (M = 11.10, SD = 0.91) with the next highest mean and the lowest score was reported by the students (M = 10.39, SD = 1.13). To investigate the possibility of differences on evaluation of learner-centred interactivity between American students, lecturers and administrative staff, ANOVA was applied. The results revealed that there was a significant main effect of academic position on evaluation of learner-centre interactivity in American participants [F (2, 115) = 6.21, p = .003]. An LSD multiple comparison test between the three levels of academic positions determined that students reported significantly of this factor lower than administrative staff and lecturers. However the evaluation of this sub factor between lecturers and administrative staff was the same. The results showed that all American participants believed that learner-centred interactivity is above average.

Comparing the answers of Australians and Americans in the three levels of academic position showed that there were no significant differences on evaluation of this sub factor between Australian and American administrative staff; where in both country the administrative staff evaluated this sub factor higher than students and lecturers [F (1, 20) = 0.05, p = .82]. However the results of ANOVA revealed that there was significant difference in evaluation of this sub factor between Australian and American lecturers [F (1, 44) = 7.94, p = .007]. Comparing the means of both countries samples indicated that American lecturers evaluated this sub factor significantly higher than Australian lecturers. Further, an ANOVA test showed that there was a significant difference in evaluation of this sub factor between Australian and American students [F (1, 148) = 39.03, p = .00]. It was conclude by comparing the means of both countries samples that American students evaluated this sub factor significantly higher than Australian students. The results showed that in both countries, students, lecturers and administrative staff believed that learner-centre interactivity was above average.

4.2   Socio-Communication

Table 2 reports the means and standard deviations of the socio- communication sub-factor based on the academic position of Australian participants. As can be seen in this Table, the highest mean regarding the socio-communication belongs to administrative staff (M = 14.50, SD = 1.69) and the students reported this sub factor (M = 12.71, SD = 1.55) next to the highest score, where the lowest score was reported by the lecturers (M = 12.70, SD = 2.07). To investigate the possibility of existing differences in evaluation of socio- communication between students, lecturers and administrative staff, ANOVA was applied. According to the results there was a significant effect of academic position on evaluation of learner-centre interactivity by Australian participants [F (2, 98) = 4.16, p = .01]. It was obtained by LSD multiple comparison test between the three academic positions revealed that administrative staff reported this factor significantly higher than students and lecturers. However the evaluation by lecturers and students of this sub-factor was the same. The results showed that all Australian participants believed socio- communication is above average.

According to the Table 2, the highest mean of the answers to this sub factor belongs to students (M = 13.84, SD = 1.40). After students, the lecturers reported this factor (M = 13.06, SD = 1.06) next highest and the lowest score was reported by the administrative staff (M = 12.84, SD = 1.77). The results of ANOVA showed that there was a significant main effect of academic position on evaluation of this sub factor by Australian participants [F (2, 115) = 6.21, p = .003]. An LSD multiple comparison test between the three levels of academic positions revealed that students reported this factor significantly higher than administrative staff and lecturers. However the evaluation of this sub factor by lecturers and administrative staff was the same. The results showed that all American participants believed socio-communication to be above average.

Comparing the answers of Australians and Americans in the three levels of academic position showed that there were significant differences in evaluation of this sub factor between Australian and American administrative staff; Americans evaluated this sub factor higher than Australians [F (1, 20) = 4.46, p = .04]. Also the results of

ANOVA revealed that there was no significant difference in evaluation of this sub factor between Australian and American lecturers [F (1, 44) = 0.58, p = .45]. An ANOVA test showed that there was significant difference in evaluation of socio-communication sub factor between Australian and American students [F (1, 148) = 39.00, p = .00]. The results showed that in both countries, students, lecturers and administrative staff believed socio-communication to be above average.

4.3   e-Assessment

The mean and standard deviation of the assessment sub factor based on the academic position of Australian participants is reported in Table 2. As can be seen in this table, the highest mean regarding this sub factor belonged to administrative staff (M = 11.87, SD = 1.35). After administrative staff, the students reported this sub factor (M = 10.18, SD = 1.21) as next highest and the lowest score was reported by the lecturers (M = 9.85, SD = 1.59). To investigate if there are any differences in evaluation of this sub factor between students, lecturers and administrative staff, ANOVA was applied. The results showed that there was a significant effect of academic position on evaluation of assessment on Australian participants [F (2, 98) = 7.21, p = .001]. It was observed by LSD multiple comparison test between the three academic positions that administrative staff reported this factor significantly higher than students and lecturers. However the evaluations by lecturers and students of this sub-factor were the same. The results showed that all Australian participants believed assessment to be above average.

In America, as can be seen in Table 2, the highest mean of answers to this sub factor belonged to lecturers (M = 11.39, SD = 1.31). After them, students reported this factor (M = 10.59, SD = 1.11) next highest and the lowest score was reported by the administrative staff (M = 10.00, SD = 1.41). To investigate if there are any differences in evaluation of socio- communication between American students, lecturers and administrative staff, ANOVA was applied. The results showed that there was a significant main effect of academic position on evaluation of this sub factor by Australian participants [F (2, 115) = 6.78, p = .002]. An LSD multiple comparison test between the three levels of academic position revealed that lecturers reported this factor significantly higher than administrative staff and students. However the evaluation of this sub factor by students and administrative staff was the same. The results showed that all American participants believed assessment is above average.

Comparing the answers of Australians and Americans in Student, lectures and administrative staff showed that there were significant differences in evaluation of this sub factor between Australian and American administrative staff; Australians evaluated this sub factor higher than Americans [F (1, 20) = 8.97, p = .007]. The results of ANOVA revealed that there were significant differences in evaluation of this sub factor between Australian and American lecturers [F (1, 44) = 12.69, p = .001] which indicated that Australian administrative staff evaluated the assessment sub factor significantly more highly than American administrative staff. An ANOVA test showed that there was a significant difference in evaluation of this sub factor between Australian and American students [F (1, 148) = 4.67, p = .03] illustrating that American students evaluated this sub factor significantly higher than Australian students. The results showed that in both countries, students, lecturers and administrative staff believed assessment is above average.

4.4   e-Resources

Also, Table 2 reports the means and standard deviations of the learning resources sub factor based on the academic position of Australian participants. As can be seen in this table, the highest mean regarding this sub factor belongs to lecturers (M = 4.37, SD = 0.67). The administrative staff and (M = 4.12, SD = 0.64) students (M = 3.54, SD = 0.71) reports next highest and lowest score, respectively. To investigate the possibility of differences in evaluation of this sub factor between students, lecturers and administrative staff, ANOVA was applied. The results showed that there was a significant main effect of academic position on evaluation of this sub factor by Australian participants [F (2, 98) = 11.41, p = .00]. An LSD multiple comparison test between the three academic positions revealed that students reported this factor significantly lower than administrative staff and lecturers. However the evaluation by lecturers and administrative staff of this sub-factor was the same. The results showed that in Australia, students believed that learning resources are above average. However lecturers and administrative staff assessed this sub factor as excellent.

The highest mean of answers to e-resources was achieved by students in American institute (M = 4.37, SD = 0.58). Lecturers reported this factor next highest (M = 4.18, SD = 0.55) and the lowest score was reported by the administrative staff (M = 4.00, SD = 1.00). The ANOVA results showed that there was no significant main effect of academic position on evaluation of this sub factor by American participants [F (2, 115) = 2.40, p = .09], which leads to the conclusion that in America, students, lecturers and administrative staff have believe that learning resources are available at an excellent level.

Comparing the answers of Australians and Americans in the three levels of academic position showed that there were no significant differences on evaluation of this sub factor between Australian and American administrative staff [F (1, 20) = 0.09, p = .75]. Also the results of ANOVA revealed that there was no significant difference in evaluation of this sub factor between Australian and American lecturers [F (1, 44) = 0.77, p = .38]. An ANOVA test showed that there was significant difference in evaluation of this sub factor between Australian and American students [F (1, 148) = 60.62, p = .00] illustrating that American students evaluated this sub factor significantly higher than Australian students. Comparing the results of Australian and American participants showed that only Australian students assessed this sub factor as above average while American students and lecturers and Australian lecturers and administrative staff believed learning resources are excellent.

4.5   e-Environment

TAs can be seen in Table 2, the highest mean of e-Environment sub factor respectively belongs to administrative staff (M = 8.25, SD = 1.03) lecturers (M = 7.75, SD = 1.06) and students (M = 7.52, SD = 0.89). The results of ANOVA showed that there was no significant main effect of academic position on evaluation of this sub factor by

Australian participants [F (2, 98) = 2.37, p = .09]. The results showed that in Australia, students and lecturers believed the learning environment was above average, whereas administrative staff assessed this sub factor as at an excellent level.

In America, as can be seen in Table 2, the highest mean regarding this sub factor was reported by students (M = 8.04, SD = 0.86), where lecturers (M = 7.70, SD = 0.73) and administrative staff (M = 7.53, SD = 0.51) noticed the nest highest and lowest score, respectively. It was obtained from ANOVA analysis that there was significant main effect of academic position on evaluation of e-Environment sub factor by American participants [F (2, 115) = 3.28, p = .04]. An LSD test showed that students evaluated this sub factor significantly higher than lecturers and administrative staff. There were no differences in evaluation by lecturers and administrative staff of this sub factor. The results showed that in America, students believed the learning environment is at an excellent level. However lecturers and administrative staff assessed this sub factor as only above average.

By comparing of the Australians and Americans answers from the three levels of academic position in institutes it was concluded that there were significant differences in evaluation of this sub factor between Australian and American administrative staff [F (1, 20) = 4.43, p = .04]; Australians significantly evaluated this sub factor higher than Americans. However, the results of ANOVA revealed that there was no significant difference in evaluation of this sub factor between Australian and American lecturers [F (1, 44) = 0.02, p = .87]. An ANOVA test showed that there was significant difference in evaluation of this sub factor between Australian and American students [F (1, 148) = 13.36, p = .00]; American students evaluated this sub factor significantly higher than Australian students. Comparing the results of Australians and Americans showed that Australian staff and American students assessed this sub factor at an excellent level while American lecturers and staff and Australian lecturers and students believed the learning environment is above average.

According to the reported data of the means and standard deviations of the e-pedagogical practice factor based on the academic positions of Australian participants in Table 2, administrative staff (M = 49.87, SD = 3.31), lecturers (M = 44.35, SD = 5.29) and students. As can be seen in this table, the highest mean of the e-pedagogical practice factor belonged to. After them, the reported the e-pedagogical practice factor as high and the lowest score was reported by students (M = 43.04, SD = 4.01). The ANOVA results for investigation of the possibility of differences in evaluation of the e-pedagogical practice factor between students, lecturers and administrative staff, showed that there was significant main effect of academic position on evaluation of the e-pedagogical practice factor by Australian participants [F (2, 98) = 9.43, p = .00]. The LSD test revealed that administrative staff evaluated this factor significantly higher than students and lecturers. However there was no difference between the evaluation of students and lecturers of this factor. The results showed that all Australian participants believed e-pedagogical practice is above average.

According to the results in Table 2, the highest, the middle and the lowest mean of answers to the e-pedagogical practice factor belongs to lecturers (M = 47.45, SD = 2.60), students (M = 47.26, SD = 3.11) and administrative staff (M = 45.61, SD = 2.81), respectively. To investigate if there are any differences in evaluation of the e-pedagogical practice factor between American students, lecturers and administrative staff, ANOVA was applied. The results showed that there was no significant main effect of academic position on evaluation of the e-pedagogical practice factor by American participants [F (2, 115) = 1.91, p = .15]. The results showed that all American participants believed e-pedagogical practice is above average.

Comparing the answers of Australians and Americans in the three levels of academic position showed that there were significant differences in evaluation of the e-pedagogical practice factor between Australian and American administrative staff [F (1, 20) = 9.92, p = .005]; Australians significantly evaluated the e-pedagogical practice factor higher than Americans. However, the results of ANOVA revealed that there was significant difference in evaluation of the e-pedagogical practice factor between Australian and American lecturers [F (1, 44) = 6.63, p = .01]; Americans evaluated this factor significantly higher than Australians. To continue, an ANOVA test showed that there was significant difference in evaluation of the e-pedagogical practice factor between Australian and American students [F (1, 148) = 51.95, p = .00]; American students evaluated this factor significantly higher than Australian students. The results showed that in both countries, students, lecturers and administrative staff believed e-pedagogical practice to be above average (see Fig. 1).

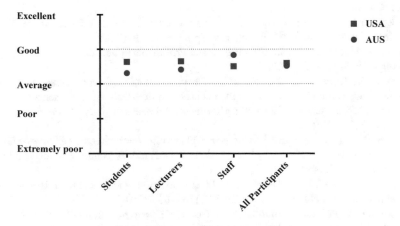

Fig. 1. Mean level of pedagogical e-practice.

5 Conclusion

The five sub factors of e-pedagogical practice including learner-centre interactivity, socio-communication, assessment, learning resources and learning environment were studied and the results illustrated that the, American students evaluated all these sub factors higher than Australian students. Furthermore, American lecturers evaluated the learner-centre interactivity and assessment sub factors higher than Australians. It was concluded from the collected results that there were no differences between the evaluation of American and Australian lecturers of remain e-pedagogical practice sub factors. Based on highly evaluated of the learner-centre interactivity and assessment sub factors by American lecturers compared with Australian lecturers and also the highest evaluation of the main factor by American students it was observed that pedagogical practice of e-learning in America is more fitted to the needs of learners than the Australia. Australian administrative staff evaluated the sub factors of assessment and learning environment higher than Americans, whereas American administrative staff evaluated the sub factor of socio-communication higher than Australians. This results shows that engaging socially with the e-learning system and adoption of an e-learning system with a social perspective is considered more important by American administrative staff. There were no differences in evaluation of the rest of the e-pedagogical practice sub-factors between those participants.

Obviously, the collected results in this study are based on the two institutes from America and Australia, which make it difficult to generalize [33–36]. However, given the size and level of deviation it could be justified to conclude that a significant difference is evident in the pedagogy of e-learning practices between the two countries.

References

1. Sadeghi, S.H.: Pathology of Learning in Cyber Space: Concepts, Structures and Processes, Vol. 156. Springer (2018)
2. Sadeghi, S.H.: Training in cyberspace. In: Pathology of Learning in Cyber Space. Studies in Systems, Decision and Control, vol. 156. Springer, Cham (2019)
3. Boud, D., Molloy, E.: Feedback in Higher and Professional Education: Understanding it and Doing it Well. Routledge (2013)
4. Cottrell, S., Donaldson, J.H.: Exploring the opinions of registered nurses working in a clinical transfusion environment on the contribution of e-learning to personal learning and clinical practice: Results of a small scale educational research study. Nurse Educ. Pract. **13**(3), 221–227 (2013)
5. Gerjets, P.H., Hesse, F.W.: When are powerful learning environments effective? The role of learner activities and of students' conceptions of educational technology. Int. J. Educ. Res. **41**(6), 445–465 (2004)
6. Kala, S., Isaramalai, S.A., Pohthong, A.: Electronic learning and constructivism: a model for nursing education. Nurse Educ. Today **30**(1), 61–66 (2010)
7. Wilkinson, A., Forbes, A., Bloomfield, J., Gee, C.F.: An exploration of four web-based open and flexible learning modules in post-registration nurse education. Int. J. Nurs. Stud. **41**(4), 411–424 (2004)
8. Jacobson, M.J., Jacobson, M.J., Kim, Y., Lee, J., Kim, H., Kwon, S.: Learning sciences principles for advanced e-learning systems: implications for computer-assisted language learning. Multimed. Assist. Lang. Learn. **8**(1), 76–115 (2005)
9. Herrington, J., Oliver, R., Herrington, A. (2007). Authentic Learning on the Web: Guidelines. Flexible learning in an information society. 26–35
10. Thurmond, V.A.: Considering theory in assessing quality of web-based courses. Nurse Educ. **27**(1), 20–24 (2002)
11. Sadeghi, S.H.: E-Learning Practice in Higher Education: A mixed-method Comparative Analysis. Springer (2017)
12. Masoumi, D.: Quality in E-learning in a Cultural Context: The case of Iran. Department of Education, Communication and Learning; Institutionen för pedagogik, kommunikation och lärande (2010)

13. Chickering, A.W., Gamson, Z.F.: Seven principles for good practice in undergraduate education. AAHE Bull. **3**, 7 (1987)
14. Finger, G., Jamieson-Proctor, R., Watson, G.: Measuring learning with ICTs: an external evaluation of Education Queensland's ICT curriculum integration performance measurement instrument. Paper to be presented at AARE. 5 (2005)
15. Zhou, Y.: Towards Capability Maturity Model of e-Learning Process (2012)
16. Paechter, M., Maier, B., Macher, D.: Students' expectations of, and experiences in e-learning: Their relation to learning achievements and course satisfaction. Comput. Educ. **54** (1), 222–229 (2010)
17. Rodríguez-Ardura, I., Meseguer-Artola, A.: E-learning continuance: The impact of interactivity and the mediating role of imagery, presence and flow. Inf. Manag. **53**(4), 504–516 (2016)
18. Chickering, A.W., Ehrmann, S.C.: Implementing the seven principles: Technology as lever. AAHE Bull. **49**, 3–6 (1996)
19. Phipps, R., Merisotis, J.: Quality on the Line: Benchmarks for Success in Internet-Based Distance Education (2000)
20. Chou, C.: Interactivity and interactive functions in web-based learning systems: a technical framework for designers. Br. J. Educ. Technol. **34**(3), 265–279 (2003)
21. Spitzberg, B.H., Hurt, H.T.: The measurement of interpersonal skills in instructional contexts. Commun. Educ. **36**(1), 28–45 (1987)
22. Frymier, A.B.: Students' classroom communication effectiveness. Commun. Q. **53**(2), 197–212 (2005)
23. Marshall, S.: eMM Version Two Process Assessment Workbook. Victoria Institute of Wellington, Wellington (2006)
24. Reeves, T.C., Reeves, P.M.: Effective dimensions of interactive learning on the World Wide Web. Web-based Instr. 59–66 (1997)
25. Raitman, R., Augar, N., Zhou, W.: Employing wikis for online collaboration in the e-learning environment: case study. In: Information Technology and Applications, ICITA, Third International Conference. IEEE (2005)
26. Achtemeier, S.D., Simpson, R.D.: Practical considerations when using benchmarking for accountability in higher education. Innov. High. Educ. **30**(2), 117–128 (2005)
27. Thompson, M.M., Braude, E., Canfield, C.D., Halfond, J., Sengupta, A.: Assessment of KNOWLA: knowledge assembly for learning and assessment. In: Proceedings of the Second. ACM Conference on Learning@ Scale. ACM (2015)
28. Gáti, J., Kártyás, G.: Practice oriented higher education course definitions and processes. In: 5th International Symposium on Computational Intelligence and Intelligent Informatics (ISCIII). IEEE (2011)
29. Wahlstedt, A., Pekkola, S., Niemelä, M.: From e-learning space to e-learning place. Br. J. Educ. Technol. **39**(6), 1020–1030 (2008)
30. Hostager, T.J.: Online learning resources do make a difference: mediating effects of resource utilization on course grades. J. Educ. Bus. **89**(6), 324–332 (2014)
31. FitzPatrick, T.: Key success factors of e-learning in education: a professional development model to evaluate and support e-learning. Online Submission (2012)
32. Sadeghi, S.H.: E-Learning Instructional design practice in American and Australian institutions. In: International Association for Development of the Information Society (2017)

33. Sangrà, A.: A new learning model for the information and knowledge society: the case of the Universitat Oberta de Catalunya (UOC), Spain. The International Review of Research in Open and Distributed Learning 2(2) (2002)

34. Ragin, C.C., Becker, H.S.: What is a Case?: Exploring the Foundations of Social Inquiry. Cambridge Institute Press (1992)

35. Gomm, R., Hammersley, M., Foster, P.: Case Study Method: Key Issues, Key Texts (2000)

36. Flyvbjerg, B.: Five misunderstandings about case-study research. Qual. Inq. 12(2), 219–245 (2006)

Understanding of Time-Based Trends in Virtual Learning Environment Stakeholders' Behaviour

Martin Drlík[⊠]

Constantine the Philosopher University in Nitra, Tr. A. Hlinku 1,
949 74 Nitra, Slovakia
mdrlik@ukf.sk

Abstract. The analysis of data collected from the interaction of users in the virtual learning environment attracts much attention today as a promising approach for advancing the current understanding of the learning content development, learning process in general as well as VLE stakeholders' behaviour. The learning analytics research has not frequently focused on analysing of time-based trends in VLE stakeholders' behaviour or their preferences in the same VLE over different years of deployment, as well as on analysing of temporal trends in the selection of different activity types over a typical period. Therefore, the paper deals with several methods, which can be used for analysing VLE stakeholders' behaviour over several academic years. The paper introduces a case study, which shows that several analytical and data mining methods can give useful insight into the changing behaviour of the stakeholders of the VLE over a longer period. Finally, the paper summarises the obtained results and discusses possible implications and limitations of the applied approach from different perspectives in the context of the management of the virtual learning environment, VLE stakeholders and educational content improvement at the institutional level.

Keywords: Learning analytics · Log analysis · Virtual learning environments · User behaviour · Time-based trends analysis

1 Introduction

It is not very surprising that there are more and more educational data being collected about us. The data comes from different resources; whether there are stored in the virtual learning environments (VLE), MOOCs, generated using mobile devices or participating in social media. Higher educational institutions (HEI) see the educational data, which arose from the students and faculty activity, as a source, which analysis should be useful for the stakeholders (persons with an interest or concern in educational process at the university) at every level of the organisation, from individual students and their tutors to educational researchers, as well as to the HEI's management [1].

Nowadays, there are plenty of approaches and methods covered by several research disciplines, more or less well-founded by statistical and mathematical background, which try to analyse educational data with the aim to support learning as well as

© Springer Nature Switzerland AG 2020
M. E. Auer and T. Tsiatsos (Eds.): ICL 2018, AISC 916, pp. 321–332, 2020.
https://doi.org/10.1007/978-3-030-11932-4_31

decision making processes. Learning analytics (LA) is one of such research disciplines, which gains more and more attention in the field of education. LA is closely related to the educational data mining and academic learning research disciplines. LA applies modern statistical techniques and methods to the system level of the learning process, targets the institutional level, and tries to support the decision-making processes of the educational institutions [2]. It is adoption and implementation at the institutional level is still in its infancy.

Considering the results of several solid review papers, it is clear, that most institutions are still in a preparatory or early stage of LA adoption [2]. In other words, HEIs are aware of LA opportunities, but surprisingly, they use only some basic reports. The majority of HEIs have not yet moved beyond basic reporting [3]. They are predominantly in the stages marked as Aware or Experimentation according to the general LA sophistication model [4].

The LA research has not frequently focused on analyzing of time-based trends in VLE stakeholders' behaviour or their preferences in the same VLE over different years of deployment, as well as on analysing of temporal trends in the selection of different activity types over a typical period. Therefore, the paper contributes to this debate and deals with several analytical and LA methods, which can be used at the HEIs belonging to the Aware and Experimentation phase of the maturity LA deployment.

The paper has the following structure. After a short review of the related work, the subsequent sections of the paper introduce a case study, which shows how several analytical and data mining methods can give useful insight into the changing behaviour of the stakeholders of the VLE over a longer period. Particular steps of the used methodology are described with the aim to explain the complexity of the practical adoption of LA methods on data stored in the VLE. Finally, the paper summarises the obtained results and discusses the possible implications and limitations of the applied approach from different perspectives.

2   Related Work

Regarding the focus of the paper, e.g., assuming suitable approach to the effective analysis of educational data stored in the VLEs at the HEIs, Adejo and Connolly [5] narrowed the widely accepted LA definition [6] to its use in the context of HEI and VLE. They understand the LA as a collection, storage, and analysis of data from learning management systems in order to gain useful and intelligent insights for decision making that will have a lasting impact on learners and the HEI. In other words, the main intention of LA is for gathering stakeholders' data for the development of models, algorithms, and processes that can be further used and generalised to improve stakeholders' performance.

Moreover, the LA outcomes can be applied not only to the evaluation of the student's performance, but they can help to review the study programs, syllabus, even the educational institutions. The statistical and data mining methods frequently used in academic learning, as a part of the LA, are described in [7].

Romero et al. published a review of the clustering methods used in the LA [8]. They identified the most frequent tasks, where the clustering methods are successfully applied. Navarro et al. [9] used educational data from VLE to provide a comparison of clustering algorithms for LA. Similarly, Preidys and Sakalauskas analyzed students' activities in VLE [10]. Special focus on time characteristics of educational data from VLE can be found in [11]. All these studies confirmed the importance and suitability of LA methods like clustering of educational data for support of the decision of the HEI's management.

3   A Case Study of an Analysis of VLE Stakeholders' Logs

The university belongs to the universities, which are in the preliminary stage of maturity of LA deployment according to the maturity LA adoption model. The university management has not yet fully uncovered the possibilities of the data-driven era. Educational data analysis is partially realized in the form of occasional reports, pivotal tables, and simple charts. Considering the individual dimensions of LA defined in [12], the current state can be summarized as follows.

Educational data is systematically collected in the academic IS and VLE for more than ten years. It has the structured (relational databases) and unstructured form (e-learning courses and scholarly publications content). While the structured data is regularly evaluated, transformed and loaded into the form of the data warehouse, the unstructured data stays in a raw form.

Instruments in the meaning of tools, which are applied in the LA context, have been already implemented. VLE Moodle is maintained at the university level provides up-to-date LA tools and interface. However, its effective use depends on the systematic use of other instruments, like guidelines, following the rules of using VLE and the knowledge of the trained stakeholders. Objectives of LA adoption exist only in the conceptual form described in several papers [13]. Their materialization requires the support of all HEI management levels. Moreover, the stakeholders at all levels have a little knowledge, as well as vague expectations related to the possibilities of LA. Consequently, it is obvious, that these weaknesses constrain the discussion about further internal limitations of LA adoptions. Besides this joyless state, the discussion about the LA adoption at the university has already begun, because the effort to effectively utilize the knowledge hidden in the educational data is growing.

3.1   Dataset Characteristics

The dataset used in this study describes individual accesses of stakeholders to the activities (modules, resources, interactive and collaborative activities) of e-learning courses during ten academic years (2007–2017).

The teachers and students of all faculties of the university use the VLE predominantly in the blended learning form. The average number of unique daily logins is more than 1000 students (Fig. 1a). The stakeholders' logs from e-learning courses used at the university during 2007–2018 were selected. Unique 20840 stakeholders enrolled and actively participated in 1200 e-learning course instances during the mentioned period.

Figure 1 depicts the sum of the stakeholders' activity logs as well as the number of unique stakeholders logged in the VLE during the observed period. While the number of the unique stakeholders can be considered stable, it seems, the activity by means of stakeholders' activity logs is growing. However, this simple observation will be evaluated more precisely later due to the fact that the log storage mechanism used in the VLE has changed. On the other hand, the number of unique logins remains quite stable despite the general decrease in the number of university students during the last decade.

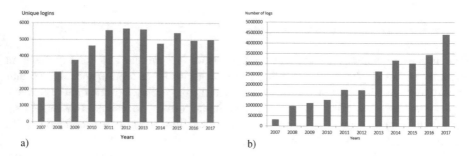

a)   b)

Fig. 1. a) The number of unique logins in the university VLE. b) Total count of logs per year.

3.2 Data Preparation

VLEs usually have the same software architecture as other web-based systems. For that reason, many VLEs store information about their users not only in server log files but also in relational database systems. Not only the data creating the content, defining the structure or the state of the systems but also metadata and a huge amount of data about the VLE stakeholders' activities can be found there.

Data preparation consists of several steps, such as data cleaning, user identification, session identification, path completion [14]. The known structure of the tables partially simplifies the process of data collection by querying a database system. A set of SQL queries can be prepared, which returns all the attributes, which describe the stakeholders' activity. For the purpose of this case study, SQL scripts were written, which selected the attributes *ID, userID, courseid, timecreated, module, action* from the tables mdl_log, mdl_logstore_standard_log. These tables represent the primary storage of logs in the VLE Moodle. The final SQL queries also required joining other relational tables to improve the over semantics of the results.

Since the log system of the VLE Moodle changed, the records about the stakeholders' activity in the e-learning course opened after the academic year 2015 had to be mapped into the structure of the previous academic years. The attributes *userid, courseid, compoment, action, timecreated* were mapped to the appropriate attributes of the mdl_log table.

The records were cleaned from unnecessary items. The entries about users with the role other than the student or teacher were removed. Finally, 22,435,175 entries were accepted to be used in the next steps of data preparation. The prepared log file is not directly suitable for many analytical methods and data mining methods [15, 16]. Regarding the aim of the paper, the data was further preprocessed in the following several steps:

1. SQL queries were launched with the aim to aggregate individual log records according to attributes *role, years, courseid,* userid, *module, action.*
2. Two files with respect to the roles of students and teachers were created.
3. The modules and actions were mapped to the more meaningful logical structures called *activities,* which will be explained later.
4. The crosstables were calculated to transpose the individual records of activities to the columns.
5. Each row of the crosstable belongs to the aggregated logs of an individual stakeholder across all identified activities.
6. The additional columns with the sum of similar activities were added.

The final log files contained 241 rows, which correspond to the individual teachers and 7935 records about the student's activities.

3.3   Comparison of New and Re-opened Courses

The first insights into the behavior of the VLE stakeholders can be gained without the application of data mining methods. This approach corresponds to the Aware and Experimentation phase of the above-mentioned Learning analytics sophistication model defined by Siemens et al. [4].

The e-learning courses attended by the stakeholders were created in the VLE of the university. The university management decided to centralize the VLE of the university during the year 2007. The VLE Moodle was selected as a final university e-learning platform. The available e-learning courses were created considering the outcomes of several educational projects focused on the e-learning courses development, which were realized at the university.

Table 1. The number of new courses created in individual academic years.

Year	04	05	06	07	08	09	10	11	12	13	14	15	16	17
Number of new courses	11	307	18	35	44	21	25	63	81	90	131	74	63	58

Table 1 uncovers the first interesting findings. It shows that the majority of courses, already used by the teachers, were created before the central platform has been launched. These courses were imported from previous VLE instances.

The closer look based on the log analysis extends the finding. These courses are actively involved in the education process each year. The visualization shown in Fig. 2 has the main contribution to the university management and administrator of the VLE because it indicates the current perception of the position of the e-learning courses in the educational process. It shows how many courses, created in the given year, are repeatedly re-opened in the consequent years.

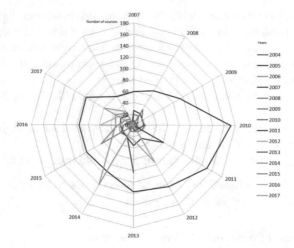

Fig. 2. The number of re-open courses during the observed period.

The university managers could state that the majority of teachers prefer to repeatedly re-open ones created courses. The year 2005 confirms this statement. At the same time, it generates new questions related to the structure of these courses, preferred types of activities, as well as stakeholders' behavior, which will be discussed in the next section.

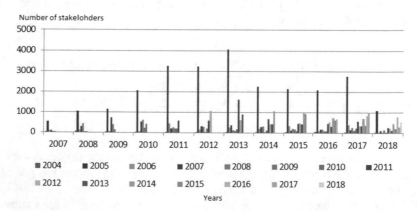

Fig. 3. Comparison of the number of enrollments into courses created in particular year.

Considering the VLE administrator point of view, this finding is important due to the requirement of backward compatibility of the system. During the observed period, the VLE has undergone with many upgrades, which also related to the core functionality, changes in GUI requirements, availability with other devices, etc. If the courses contain obsolete approaches, the overall maintenance of the system requires additional effort. If the stakeholders' enrolment into the individual course is taken into account, the view on the issue is slightly changed (Fig. 3). The chart shows that the maximum of enrolled students to the courses created in 2005 culminated in 2013. It can be assumed, that the improvements to the VLE regarding the GUI and core functionality could cause the decrease of enrolments in the older courses.

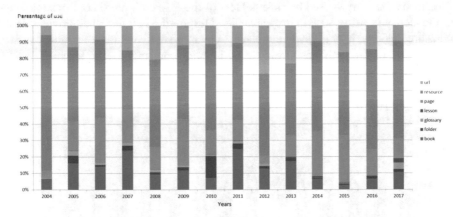

Fig. 4. The comparison of different types of resources used in the courses during the years.

3.4 Changes in Resources and Activity Structure

The structure of the activities and resources during the observed period has also slightly changed. Figure 4 shows that the teachers prefer to upload different kinds of resources to the courses. The development of the online content in the form of activities *book*, *lesson*, *page* is less popular. This finding is partially in accordance with the overall trend in e-learning, where the teachers become more the curators of the subject oriented curated learning content than the authors of the resources themselves.

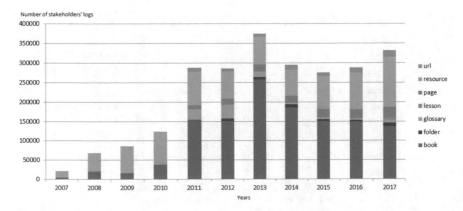

Fig. 5. The real use of different types of resources based on the stakeholders' activity.

However, if the real stakeholders' activity is taken into account, the preferences of the stakeholders in using available resources were changed. Figure 5 confirms the fact that the considerable groups of stakeholders still use the already created online content during their stay in courses.

Finally, the same approach can be applied to the analysis of the structure and types of interactive activities, used by stakeholders of the VLE. Figure 6 confirms the long-term tendency in using VLE at the university. Despite the fact the VLE provides many interesting options, how to involve students to be more active in the e-learning courses; the spectrum of the activities already used is quite unchanged. The teachers usually include *quizzes*, *assignments* and *forums* in their newly opened courses.

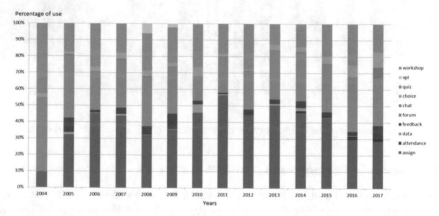

Fig. 6. The comparison of different types of activities used in the courses during the years.

Similarly, as in the case of the resources, the analysis of the stakeholders' logs brings a different view on their real utilization at the VLE (Fig. 7). As a result, the

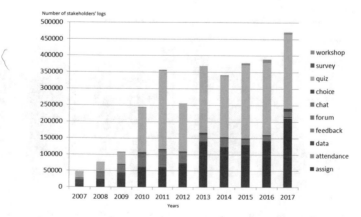

Fig. 7. The real use of different types of activities based on the stakeholders' activity.

dominancy of the activities *quiz* and *assignment* increased. On the other hand, the effort of course developers, who added other types of activities to the courses, like a *forum*, *chat*, *feedback*, has not been reflected yet.

3.5 Identification of Clusters of Stakeholders

The previous section uncovers some differences between the frequency of using different kinds of educational resources and interactive activities and their real use by the VLE stakeholders over the years. This section involves some data mining clustering techniques to identify other potentially useful aspects of the analysis log files about the stakeholders' activity in the VLE. The main question is whether the VLE stakeholders could be clustered according to their activities stored in the log files and whether this behavior has changed over the years.

As was already mentioned, before any application of a statistical or analytical method, the data has to be preprocessed using some of well-known data preparation steps [17]. Special attention was therefore paid to the attribute *activity*. Romero et al. define the abstraction *activity* as groupings of records that are related in some way [8]. Sherd [18] defines an *activity* as an abstraction of discrete behavioural activity of the stakeholder stored in the VLE Moodle log file. Subsequently, the *activity* can represent a more semantically meaningful set of actions belonging to the module used in the e-learning courses. Page views, sessions, tasks, modules and related actions of the VLE in this case, are typical examples of these abstractions.

In this case study, the fact the original variables *module* (39 items) *action* (50 items) create too many categories, was taken into account. Many values would lead to the lower counts of accesses. The higher number of the categories of activities would cause worse interpretation of the results. For that reason, an abstraction, called *activity*, was also introduced.

A group of the VLE and e-learning experts had been asked to divide the combination of available *modules* and *actions* into the three *activities*:

- *resources* – contains all modules and actions, which are connected with the resources available in the courses (view resource, view url, print book, etc.),
- *assignments* – contains the modules and actions, which evaluate in some manner the knowledge of the stakeholders and provide feedback to teachers (assign view, assign grade, chat, post discussion, open attempt, etc.),
- *activities* – includes all other modules and actions related to improving interactivity and engagement of the students in the e-learning courses (workshop, database, chat, vpl, choice, wiki).

These new *activities* were calculated from the already prepared log files with the stakeholders' aggregated records. Only active stakeholders, with the activity above the defined threshold, were selected to ensure, the clusters will better represent the typical groups of VLE stakeholders. The log files where divided based on the interval of years (2007–2013, 2014–2017). The facts that the university management changed the e-learning strategy, the GUI of the VLE has changed, and the system has been upgraded, were taken as the reason, why the academic year 2013/2014 was selected as a boundary. Subsequently, an agglomerative clustering method, which belongs to the unsupervised data mining methods, was applied to the pre-processed data files with the aim to estimate the most suitable count of possible clusters. Final selection of three of clusters has also been confirmed by the elbow function (Fig. 8).

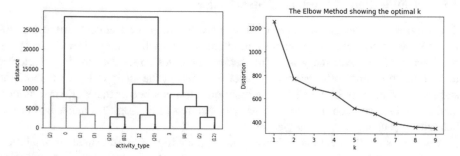

Fig. 8. Identification of suitable count of clusters using agglomerative clustering and elbow function.

Finally, unsupervised data mining method K-means clustering was used on the normalized dataset to identify possible groups of stakeholders based on their activities in VLE (Fig. 9). Points represent the number of logs of each stakeholder in categories *resources*, *assignments* and *activities*. In case of the teachers' logs, the results showed the disproportion between the groups of teachers based on their experience or attitude to involve the VLE possibilities in the education process. Most of the teachers use the VLE only as a repository of the resources. Only the minority of teachers provides a

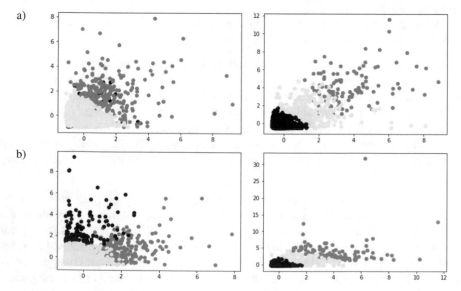

Fig. 9. The comparison of identified clusters in different periods. a) resources-assignments, b) assignments-activities dimensions. Axes represent normalized distances between points.

large spectrum of interactive modules (assignments, quizzes, workshops, forums) in their courses and is able to motivate students to participate in them actively.

Figure 9a depicts the visualization of the clusters for the students' data in the both observed periods of years. The changes in behavior by means of the different count of identified clusters in the case of students uncover the influence of the minority of teachers, who use other interactive modules in courses. While in the first observed period only the groups of students used *resources* and *assignments* were identified, the second observed period visualizes the sole cluster of students, who actively participated in other *activities*. Figure 9b provides an additional point of view to students, who use *assignments* and other *activities*. It shows that the activity of students in these modules highly depends on modules' availability in the courses and the effort of the teacher, who decided to involve students to be more active in the online environment.

4 Discussion and Conclusions

The approach used in the paper showed the potential of application of different analytics method on educational data stored in the form of logs in the university VLE. The proposed analysis of the log files confirmed that the VLE stakeholders' records could provide additional information to the managers about the real use of the e-learning courses. The paper provides some examples, how the expectations of the course developers, often teachers, differ from the real use. It provides an example, how to explore the logs using exploratory data mining methods. Identified clusters of stakeholders should be further researched using other data mining methods to ensure their usefulness. However, they provide a suitable starting point for discussion about the groups of VLE stakeholders and their behavior.

This approach also has the limitations, which results from the used K-means clustering method itself. The abstraction of the *activities* represents the second potential limitation. The division of the attributes *modules* and *actions* to the *activities* was done by experts. However, their final shape should be further discussed with regard to the detailed analysis of their meaning in the VLE's environment. Nevertheless, the results can be useful for different VLE stakeholders. Teachers will be able to identify and consequently modify resources and activities according to their real use. Course developers could obtain better knowledge about the changing trends in the stakeholders' behaviour. They will be able to identify, what kind of activity is preferred by different kinds of stakeholders, which activities are used together, what changes in the course structure in means of used kind of activities lead to the higher visit rate. Management of the university could analyse the preferences of the VLE stakeholders over time, manage the process of development of new or reopened courses. Moreover, they will be able to better focus on teachers, who have not yet uncovered all the potential benefits of the portfolio of activities available in the VLE.

Acknowledgement. This work was supported by the Cultural and Educational Grant Agency of the Ministry of Education of the Slovak Republic under the contract KEGA-029UKF-4/2018 and by the project "IT Academy – Education for 21st Century" under the contract ITMS 312011F057.

References

1. Sclater, N.: Learning analytics. The current state of play in the UK higher and further education. JISC. Effective Learning Analytics. Using data and analytics to support students. Effective Learning Analytics. JISC (2014)
2. Ferguson, R., et al.: Research evidence on the use of learning analytics - implications for education policy (2016)
3. Bichsel, J.: Analytics in Higher Education: Benefits, Barriers, Progress, and Recommendations. EDUCAUSE Center for Applied Research (2012)
4. Siemens, G., Dawson, S., Lynch, G.: Improving the Quality and Productivity of the Higher Education Sector. Office of Learning and Teaching, Australian Government, Canberra, Australia (2014)
5. Adejo, O., Connolly, T.: Learning analytics in higher education development: a roadmap. J. Educ. Pract. **8**, 156–163 (2017)
6. Larusson, J.A., White, B.: Learning Analytics. From Research to Practice. Springer, New York (2014)
7. Lang, C., Siemens, G., Wise, A., Gašević, D.: The handbook of learning analytics. Soc. Learn. Analyt. Res. (2017)
8. Romero, C., Ventura, S., Pechenizkiy, M., Baker, R.S.J.D.: Handbook of Educational Data Mining. Chapman & Hall/CRC (2010)
9. Navarro, A.M., Moreno-Ger, P.: Comparison of clustering algorithms for learning analytics with educational datasets. Int. J. Interact. Multimed. Artif. Intell. **5**, 1–8 (2018)
10. Preidys, S., Sakalauskas, L.: Analysis of students' study activities in virtual learning environments using data mining methods. Ukio Technologinis ir Ekonominis Vystymas **16**, 94–108 (2010)
11. Mlynarska, E., Greene, D., Cunningham, P.: Time Series Clustering of Moodle Activity Data. AICS 2016 (2016)
12. Greller, W., Drachsler, H.: Translating learning into numbers: a generic framework for learning analytics. Educ. Technol. Soc. **15**, 42–57 (2012)
13. Skalka, J., Drlík, M., Švec, P.: Knowledge discovery from university information systems for purposes of quality assurance implementation. In: 2013 IEEE Global Engineering Education Conference (EDUCON), pp. 591–596 (2013)
14. Munk, M., Drlík, M., Benko, L., Reichel, J.: Quantitative and qualitative evaluation of sequence patterns found by application of different educational data preprocessing techniques. IEEE Access **5**, 8989–9004 (2017)
15. Munk, M., Kapusta, J., Švec, P.: Data preprocessing evaluation for web log mining: reconstruction of activities of a web visitor. Procedia Comput. Sci. **1**, 2273–2280 (2010)
16. Munk, M., Benko, L.: Using entropy in web usage data preprocessing. Entropy **20**, 67 (2018)
17. Munk, M., Drlík, M.: Chapter 10 - methodology of predictive modelling of students' behaviour in virtual learning environment A2 - Caballé, Santi. In: Clarisó, R. (ed.) Formative Assessment, Learning Data Analytics and Gamification, pp. 187–216. Academic Press, Boston (2016)
18. Sheard, J.: Basics of statistical analysis of interactions data from web-based learning environments. In: Romero, C., Ventura, S., Pechenizkiy, M., Baker, R.S.J.d. (eds.) Handbook of Educational Data Mining. CRC Press, A Chapman & Hall Book (2011)

The IT Based Internal Combustion Engines Integrated Teaching Complex

M. G. Shatrov, T. Yu. Krichevskaya, A. L. Yakovenko,
and Alexander Solovyev[✉]

Moscow Automobile and Roads Construction State Technical University
(MADI), Moscow, Russia
dvs@madi.ru, iakovenko_home@mail.ru,
soloviev@pre-admission.madi.ru

Abstract. The article considers composition and structure of the Internal Combustion Engines Integrated Teaching Complex (IC Engines ITC) for preparation of students in the field of modern piston engines. The following advantages of its use in the teaching process are demonstrated: raising the quality of preparation, expansion of methods of knowledge transfer to the students, ensuring their realization in the form of practical skills, granting the opportunity to learn independently. Destination, interface and specific features of using individual components of the IC Engines ITC: lecture course, virtual laboratory, computer aided design systems of the engine and the systems of teaching process quality control are presented.

Keywords: Piston engine · Informational technologies · IC Engines ITC · Virtual laboratory

1 Introduction

The use of innovative informational technologies in the teaching process makes it possible to improve the preparation quality, expand the channel of knowledge transfer to the students by ensuring their realization in the form of practical knowledge, granting him the opportunity to study independently. The article considers technology of the development of the Internal Combustion Engines Integrated Teaching Complex (IC Engines ITC) and its application in teaching process at the Department of Thermal Science and IC engines of MADI.

2 The Tasks of the Development of the IC Engines ITC

The aim of the development of the IC Engines ITC is to ensure a modern world level of energy-ecological preparation of qualified personnel in the field of IC engines. The preparation programs should correspond to the state teaching standards for the different levels - from pupils of schools and colleges to bachelors and masters, as well for postgraduate students. A wide range of applications of IC engines is considered: automobile, water and marine transport, small aircrafts, forestry and agriculture.

© Springer Nature Switzerland AG 2020
M. E. Auer and T. Tsiatsos (Eds.): ICL 2018, AISC 916, pp. 333–343, 2020.
https://doi.org/10.1007/978-3-030-11932-4_32

The Complex is an instrument for preparation of high-qualified teachers. It enables to expand the range of their didactic tools and techniques, and for young teachers – to ensure mastering by them of IC engines teaching methods and achieve the required level of expertise.

The IC Engines ITC enables to improve the teaching quality, save time and reduce the scope of routine uncreative actions of students, reduce expenses for creation and servicing of high-technology experimental equipment required for carrying out laboratory works.

3   Composition and Structure of the IC Engines ITC

The conception of the Complex creation is systematization of information on IC engines using the domestic teaching traditions and the world experience. The IC Engines ITC bases on innovative informational technologies and computer technique (see Fig. 1). The Complex united on a single platform all traditional elements of the teaching process: lectures, laboratory practicum, design and quality control. Application of this Complex ensures the further sophistication of the methods of using this type of didactic instruments in the teaching process.

Fig. 1. The title window of the IC Engines ITC

3.1   Lecture Course

The lecture course makes it possible, using multimedia equipment, to get information on composition and structure of the IC engine, its components and processes taking place in it [1].

The aim of the course is to grant the student information in the form of a lecture on the basic aspects of description of piston internal combustion engines (see Figs. 2 and 3). It may be used for different levels and targets of preparation both in case of independent and remote work of the student, in fragment form or to the full extent and when the teacher delivers the lectures. The trajectory of studying the lecture is selected by the student by himself or is set by the teacher.

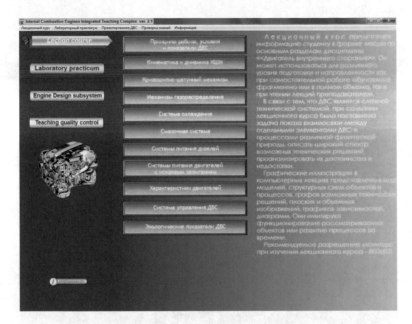

Fig. 2. The content of the lecture course

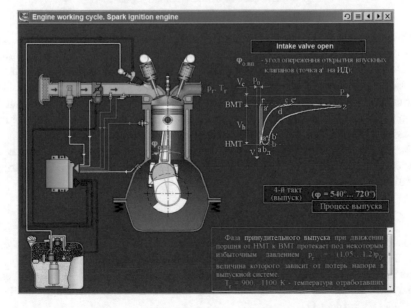

Fig. 3. The window of the lecture on engine working cycle

3.2   Laboratory Practicum

The laboratory practicum, imitating operation of the IC engine, ensures running of the whole complex of experimental works carried out when obtaining its characteristics (see Figs. 4, 5, 6 and 7).

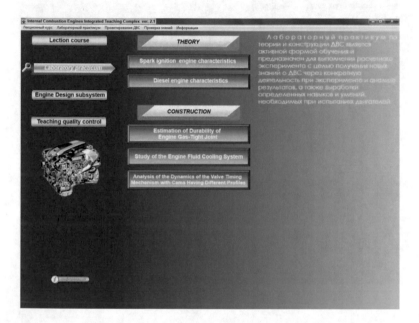

Fig. 4. The content of the laboratory practicum

The Practicum is an active teaching form. It is destined for carrying out a calculation experiment whose aim is getting new knowledge about IC engines though concrete activity during experiment and analysis of the results obtained [1]. For this, a database of meaningful tasks that the student can solve was developed. During experiment, one can investigate extreme situations, for example, IC engine operation in high mountain region, Extreme North, etc. Finally, the student gets new knowledge, as well as elaborates skills and experience required for testing IC engines.

Carrying out of laboratory works in such a way is cost-effective, ensures ecologically clean experiment, and unites individual and collective approaches during its realization.

3.3   Computer Aided Design System for IC Engines

Computer Aided Design (CAD) system for IC engines ensures fulfillment of the external (pilot) development stage of the IC engine forming its conception. The most complicated and important issues of the engine are realized in the system. IC engine CAD system ensures both course design and engine development when fulfilling the final student's project [1].

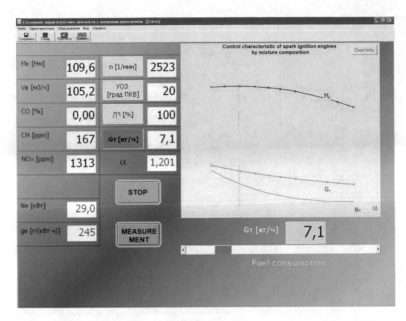

Fig. 5. Window of the laboratory work "Control characteristic of spark ignition engines by mixture composition"

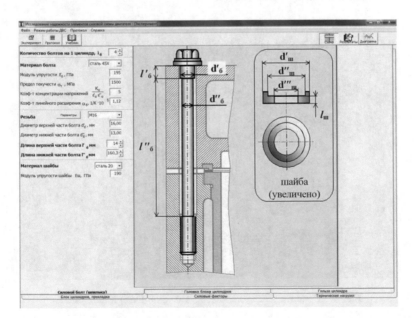

Fig. 6. Window of the laboratory work "Estimation of Durability of Engine Gas-Tight Joint"

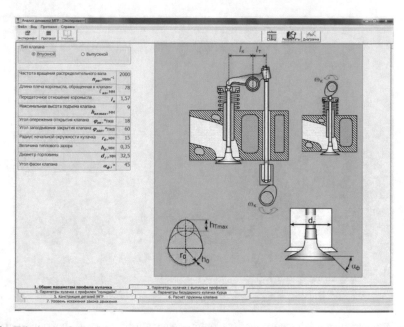

Fig. 7. Window of the laboratory work "Analysis of the Dynamics of the Valve Timing Mechanism with Cams Having Different Profiles"

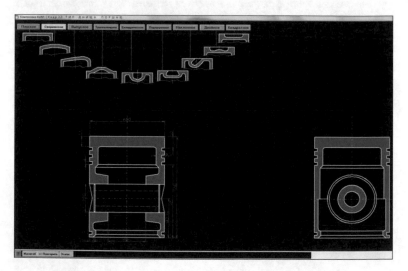

Fig. 8. Window of the Geometrical Engine Design subsystem

The system consists of two object-directed subsystems: geometrical and calculated engine design.

In the Geometrical Engine Design subsystem (see Fig. 8), the most complicated and important issues of arrangement and engineering study of the connecting rod gear and timing gear, as well as their coordinated operation as parts of the engine are realized.

In the Calculated Engine Design subsystem, calculation of working processes, dynamics and strength of the parts, productivity and efficiency of the engine systems and their components are realized.

3.4 Teaching Quality Control

Teaching quality control (see Figs. 9 and 10) is used for estimation of teaching quality of a student during current and final tests. It is important in case of independent estimate by the student of the level of his mastering of the sections of the discipline and selection of the required correction of the teaching trajectory.

The system bases on the banks of test questions systematized by all the aspects of

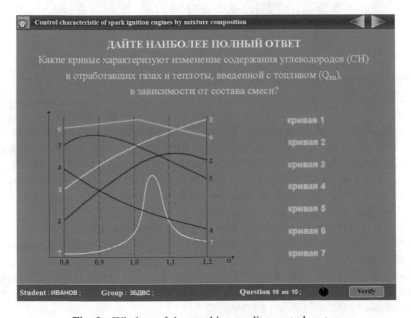

Fig. 9. Window of the teaching quality control system

description of the design, processes and characteristics of the engine. The questions are arranged according to the established scheme and the sequence of questions is determined by their level of complexity and training program.

Fig. 10. Window of the teaching quality control system

4 The History and Modern State of the Art of the IC Engines ITC

The development of the components of the IC Engines Integrated Teaching Complex started in the 90th of the XX century. Today, the Complex is widely used for energy-ecological preparation of students of all mechanical specialties of MADI, as well as the pupils of the Moscow schools. It ensures studying of the IC engine courses by students in more than 150 universities in different regions of Russia, CIS countries, Europe, Latin America, Asia and Africa.

For the development of the IC Engines textbook/complex, the team of the Department was awarded the Prize of the Government of the Russian Federation in the Field of Science and Technique.

The complex is an open developing system, which is rapidly upgraded and permanently supplemented with the results of new scientific research and world achievements in the field of IC engines.

5 Experience of Application of the IC Engines ITC

More than 20-year experience of application the IC Engines ITC in the teaching process at the Department of Thermal Science and IC Engines of MADI demonstrated that the decision taken was correct. As far as computer technologies develop rapidly and opportunities granted by the designers of hardware and software grow, accumulation of the experience of operation of the Complex, its individual components were updated many times. The Complex continues to develop rapidly.

The lectures on engine mechanisms and systems were finalized completely including lectures on ecology (toxic emissions and noise (see Fig. 11)).

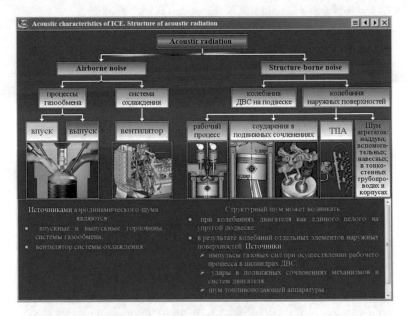

Fig. 11. The window of the lecture on noise

Fig. 12. The window of the IC Engines CAD

At present, the lectures on working processes of IC engines are at the finishing stage. A new version of the IC Engines CAD on the base of 3D modeling is developing actively (see Figs. 12 and 13). One of the most important tasks is the further sophistication of the methods of using the IC Engines ITC for teaching.

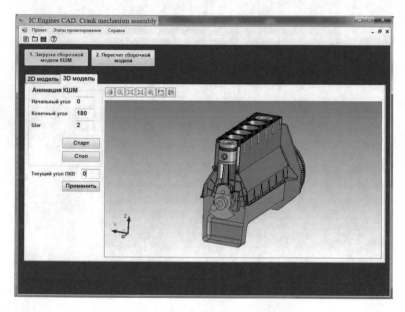

Fig. 13. The window of the IC Engines CAD

6   Conclusions

The technology of the development of the IC Engines ITC on a single platform was created which includes all elements of the teaching process (lectures, laboratory works, design and quality control). It insures getting knowledge and skill training when studying a complicated technical system - a piston internal combustion engine.

The first version of the Thermal Science ITC was prepared based on this technology. It is permanently developing and its ninth version is finalized at present time.

The methods of using the IC Engines ITC both for traditional organization of the teaching process and in case of individual work of the student were elaborated.

IC Engines ITC in the teaching process of different educational institutions by different categories of students demonstrated that it ensures the level of mastering of the discipline IC Engines set by the State standards

The Complex enables to decrease considerably the time of energy-ecological preparation of young teachers. Educational institutions save considerably expenses for didactic provision of the teaching process, as well as stiff requirements for environment protection and saving hydrocarbon fuels are followed efficiently.

Reference

1. Lukanin, V.N., Shatrov, M.G., Krichevskaya, T.Y. et al.: Internal Combustion Engines. In 3 books. Book 3. Computer Practical Course. In: Lukanin, V.N., Shatrov M.G. (eds.) Modeling of Processes in Internal Combustion Engines. College textbook. The 3rd modified and updated edition. Moscow. High School, 414 p 2007. (In Russ)

Scheduling Synchronous Tutoring Sessions in Learning Activities

Amadou Dahirou Gueye[1]([⊠]), Pape Mamadou Djidiack Faye[2],
Bounama Gueye[1], and Claude Lishou[3]

[1] Alioune Diop University, Bambey, Senegal
{dahirou.gueye,bounama.gueye}@uadb.edu.sn
[2] Virtual University, Dakar, Senegal
papedjidiack.faye@uvs.edu.sn
[3] Cheikh Anta Diop University, Dakar, Senegal
clishou@ucad.sn

Abstract. This paper deals with a solution allowing digital universities or distance learning structures to extend the functionalities of their distance learning platform to improve not only the management of course session activities but also the organization and synchronous scheduling of these sessions. Indeed, the major problem that is often encountered at the level of distance learning platforms is the management of course hours. This paper proposes a solution for the FOAD Moodle platform to develop a plugin offering teachers the possibility to dynamically choose their course availability and learners the possibility to visualize in real time the week's schedules. The developed plugin is integrated as a new educational resource into the Moodle platform to extend its functionality. This plugin which constitutes the originality of this paper bears the name of Planning.

Keywords: Distance learning platform · Synchronous programming · Plugin development · Planning

1 Introduction

In a context marked by a new generation of highly connected students who prefer digital tools for learning, informing themselves and keeping in touch with the outside world, higher education institutions in developing countries are finding, through distance learning, the means to strengthen their role and place in the democratization of access to higher education. Today, we realize that Open and Distance Learning (ODL), which initially interested professionals who wanted to complete their training, welcomes young people as soon as they obtain their baccalaureate between the ages of 17 and 18, this is the case of the Université Virtuelle du Sénégal (UVS) [1]. These young people need to be well trained in areas such as communication, STEM (Science Technology Engineering and Mathematics) [2] for practical skills. To this end, the Government of Senegal also pays particular attention, on the one hand, to the promotion of STEMs (Science Technology Engineering and Mathematics) in all components of the education system and, on the other hand, to the development of distance

© Springer Nature Switzerland AG 2020
M. E. Auer and T. Tsiatsos (Eds.): ICL 2018, AISC 916, pp. 344–352, 2020.
https://doi.org/10.1007/978-3-030-11932-4_33

education in higher education. This is reflected in the new higher education reforms that emerged from the Presidential Council held in 2013 [3].

Today, traditional public universities in Senegal also integrate distance education into their mode of operation. This new type of education is based on distance learning platforms. Many distance learning platforms exist and integrate online self-assessment and evaluation tools. This is the case of Moodle which is used by many universities [4, 5]. Now, with distance learning platforms, the learning process becomes independent of time and place because one can stay anywhere and at any time to learn. This method of training requires means of communication based on electronic technology but also on good organization to avoid overlapping of course hours. Research shows solutions to improve the functionality of training platforms for better course management [6, 7]. However, the major problem that is often encountered in distance learning platforms is the management of course hours. Sometimes two teachers try to teach at the same time (for a given time) for the same or different groups. So we know that a group cannot receive several teachers at the same time. Also a teacher cannot teach with different groups at the same time. Students who do not master their class hours can sometimes connect to hours that have not been scheduled for their group. The teacher can connect to unplanned hours. To solve all these gaps, we would like to set up a plugin for course planning.

The rest of the paper will be organized as follows:

Section 2 presents the modularity of Moodle. Section 3 presents a study of table structure in Moodle. In Sect. 4, we propose to develop a plugin that will manage the students' time schedule and its integration into Moodle. Section 5 presents the results of our implementation.

2 Study of the Modularity of Moodle

Many of Moodle's features are modular.

Activities are the most important modules. They are placed in the "mod" folder. There are several modules by default: assignment, survey, test, forum, consultation and resource, etc. Each module is placed in its own subfolder and consists of various mandatory elements (as well as other scripts specific to each module) [8].

The Blocks:
Blocks are rectangular information areas that can be displayed in the left or right column, and stack one on top of the other. It is also possible to display them only on the left, or on the right, and for the dashboard to appear in the central section of the course space. It is also possible with some third party themes to make them appear in the central section of courses [9].

Themes:
Themes (or "skins") define the appearance of a site. Some simple themes are provided in the Moodle distribution. You can also create your own theme with your favorite colors, logo, styles and graphics.

Database Schematics:
Once the database and its tables are defined, the intentionally simple SQL language used by Moodle should work correctly with a large number of database brands. It can happen that the automatic creation of new tables in a database is problematic. To enable this automation in each type of database, database schemas can be created, which include the SQL commands needed to create for a particular database the tables that Moodle uses [10].

Course Formats:
Moodle currently offers three different course formats: weekly, thematic and informal. These formats are a little more dependent on the rest of the code (and therefore less modular), but it is still quite easy to add new ones [11].

Moodle APIs (Application Programming Interface):
Moodle has a number of core APIs that provide tools for Moodle scripts.

They are essential when writing Moode plugins.

These APIs are: Access API (Access), Data Handling API (dml), File API (Files), Form API (Form), Log API (Log), API Navigation (Navigation), API of page (page) output), String API (string), Upgrade API (upgrade), Moodlelib API (core) and other APIs [12].

3   Study of Table Structure Under Moodle

The study on tables in Moodle revealed about 250 tables. The database being very complex we concentrated mainly on the table containing the users, the table containing the roles and finally the table identifying the group of each student.

We have an overview of the different tables that will allow us to create our module (Fig. 1).

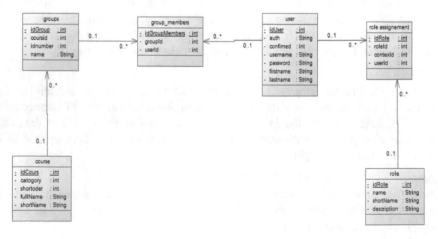

Fig. 1. Extract from tables managing users under Moodle [13]

In the Moodle platform a user must belong to a group to be able to participate in a course. The relationship between these different tables can be translated as follows: A role can be assigned to one or more users. A user can be a member in one or more groups. One or more groups can register for one or more courses.

It is in this context that we propose a class diagram of the different tables that we use for the realization of our plugin.

4 Proposal of Our Solution

Moodle is a modular platform. Its architecture allows developers to easily add new features. These new features will be available as plugins.

4.1 Use Case Diagram

Our system will have two actors, a teacher and a student. First, the teacher has the possibility to authenticate himself. Then, he can plan either a course or tutorial or practical work by checking the time and day that suits him. Students have the ability to authenticate themselves and view their schedule according to the group they belong to, as shown in Fig. 2.

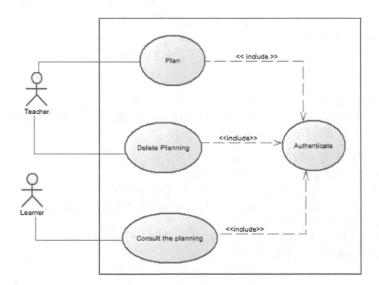

Fig. 2. Use case diagram

4.2 Class Diagram for Course Planning

After having made a detailed study on the various tables which interest us, we proposed the addition of a new "planning" table to dynamically manage our schedule.

The class diagram in Fig. 3 shows the internal structure of our plugin.

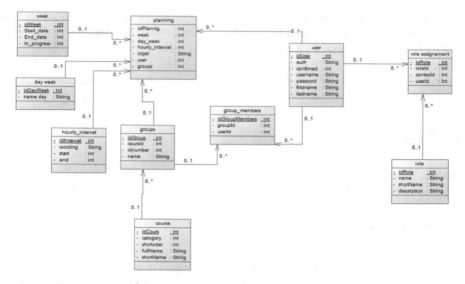

Fig. 3. Adding tables to manage planning

The relationship between these different tables can be explained as follows:

- A user can plan or view the schedule according to his role.
- A user can belong to one or more groups
- The schedule can be planned for one or more groups
- A user can have one or more roles.

The planning table:

In this table we have the schedule of all courses created at the Moodle platform level. The attributes that make it up are:

- id: Identifier specific to the planning
- week: the identifier of the current week
- day_week: the day for which a course is scheduled
- interval_week: the interval of the week for which the timetable remains valid
- course_title: Course name or title
- user: identifies the user if he has the right to view the timetable or not
- group: identifies the user's group.

5   Proposal of Our Plugin for Scheduling Class Hours

The plugin development sequence diagram is as follows: (Fig. 4)

To achieve this, we have created the following files:

Display: In this file, we retrieve the identifier of the connected user to know if it is a teacher or a learner. If it is a teacher, the file that manages the timetable planning is

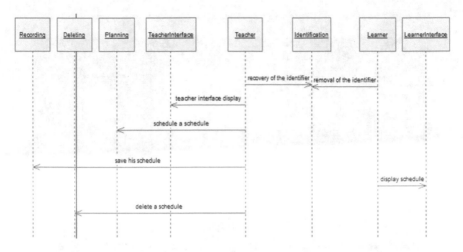

Fig. 4. The plugin development sequence diagram

called because the connected user is a teacher. Otherwise, if the user is a learner, the file that manages the learners is called.

Choice_time_day: This file manages the choice of days and hours of classes. It includes the algorithm that allows the teacher to choose the day and time of his course.

Employment: This file manages part of the layout of the schedule.

Save: This file manages the algorithm that allows to save the time schedule at the database level.

TeacherSheet: It is this file that manages all teachers to know what days and hours are available in the timetable. If there is a free day to plan a schedule or if a group is not already scheduled by another teacher.

StudentFile: This file contains the algorithm which allows, according to the identifier of the student and the group to which he belongs, to display his schedule.

Form: This file manages user identification at the platform level.

6   Implementation

This interface allows us to add our plugin as an activity resource (Fig. 5).

Figure 6 shows our added plugin as an activity module.

The planning interface can be displayed in two ways depending on whether the user is a teacher or a student. If the user is a teacher the first thing he will have to do is to first authenticate himself then give the name of his course and check the hours that suits him.

We have an overview of the interface of a teacher who wishes to plan his lesson hours: (Fig. 7).

Fig. 5. Adding our plugin as an activity module

Fig. 6. Added plugin as an activity module

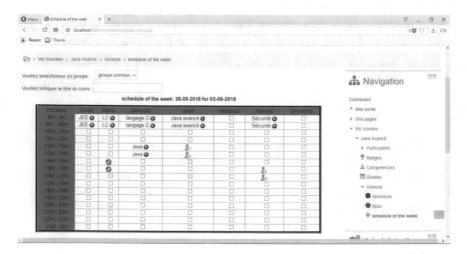

Fig. 7. Teacher's home page for course Planning

The student once connected will see his own interface to know what are the hours of lessons that have been scheduled by the teachers (Fig. 8).

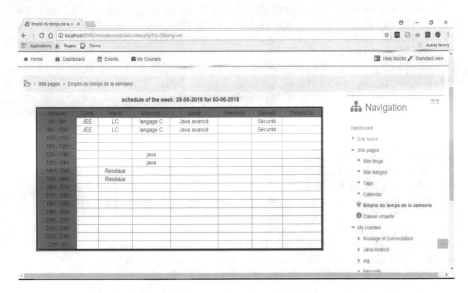

Fig. 8. Student home page for time use visualization

7 Conclusion

This paper shows the possibility of developing a plugin for scheduling class hours among teachers and visualizing the timetable among students. We believe that with this plugin, there will no longer be time overlaps during online course scheduling. This will allow you to have real-time scheduling of lessons and tutorials from the training platform. This paper can be extended in the future for better absence management and better monitoring of synchronous tutoring activities.

Acknowledgments. We thank CEA-MITIC (http://www.ceamitic.sn) for contributing financially to this paper. CEA-MITIC is a consortium that brings together academic institutions from Senegal and the region, research institutions and national, regional and international companies involved in the ICT sector.

References

1. UVS: Université Virtuelle du Sénégal (2015) http://www.uvs.sn/
2. Nite, S.B., Margaret, M., Capraro, R.M., Morgan, J., Peterson, C.A.: Science, technology, engineering and mathematics (STEM) education: a longitudinal examination of secondary school intervention. In: Frontiers in Education Conference (FIE), 2014 IEEE, pp. 1–7, 22–25 Oct 2014

3. Conseil présidentiel sur l'enseignement supérieur et la recherche (2013). http://www.cres-sn. org/sites/default/files/cnaes_decisions19aout-l.pdf
4. Aydin, C.C., Tirkes, G.: Open source learning management systems in e-learning and Moodle. In: Education Engineering (EDUCON), 2010 IEEE, pp. 593–600, 14–16 (2010)
5. Blas, M.: Teresa and Serrano-Fernández, The role of new technologies in the learningprocess: Moodle as a teachingtool in Physics. Ana Computers and Education, vol. 3, pp. 35–44. March 2009 (2009)
6. Zorrilla, M.E., Álvarez, E.: MATEP: monitoring and analysis tool for e-learning platforms. Published In: Advanced Learning Technologies, 2008. ICALT 2008. Eighth IEEE International Conference on, IEEE, Conference Location: Santander, Cantabria, Spain (2008). https://doi.org/10.1109/icalt.2008.33. July 2008
7. Faye, P.M.D., Gueye, A.D., Lishou, C.: Proposal of a virtual classroom solution with WebRTC integrated on a distance learning platform. In: Proceedings of the 19th ICL, Volume: 544, Publisher: Springer International Publishing, eBook ISBN: 978-3-319-50337-0, Series ISSN: 2194-5357 (2017). https://doi.org/10.1007/978-3-319-72965-7. Edition Number: 1, Copyright: 2017
8. [Online] Available. http://camri.uqam.ca/moodle/lang/fr/docs/developer.html
9. [Online] Available. https://docs.moodle.org/2x/fr/Blocs
10. [Online] Available. http://camri.uqam.ca/moodle/lang/fr/docs/developer.html
11. [Online] Available. https://docs.moodle.org/19/fr/Formats_de_cours
12. [Online] Available. https://docs.moodle.org/dev/Data_manipulation_API
13. [Online] Available. http://www.examulator.com/er/

A Process of Design and Production of Virtual Reality Learning Environments

Lahcen Oubahssi, Oussema Mahdi, Claudine Piau-Toffolon[⊠],
and Sébastien Iksal

Le Mans Université, LIUM, EA 4023, Avenue Messiaen,
72085 Le Mans Cedex 9, France
{lahcen.qubahssi, oussema.mahdi, claudine.piau-
toffolon, sebastien.iksal}@univ-lemans.fr

Abstract. Facilitating human learning is one of the uses of virtual reality (VR). Users interact within original and dynamic situations of learning in an integrated learning environment called Virtual Reality Learning Environment (VRLE). Nevertheless, the design of these environments is still considered as a complex task. We intend to study and propose technical and methodological solutions to help teachers to design (adapt or reuse) their pedagogical situation with a scenario-based approach and to operationalize it in a VR learning environment. In this work, we defined a design process allowing teachers to generate their VRLE. Then, we instantiated the proposed design process with a pedagogical situation as an illustrated example.

Keywords: Pedagogical scenario · Learning design ·
Educational virtual environments · Virtual reality learning environments ·
Pedagogical situation

1 Introduction

Virtual reality is more than a new concept: it is emerging as a new medium with its own characteristics. Some of these characteristics include the ability to allow individuals to live an immersive experience, carry out a senso-motor activity in an artificial world, and interact with each other and with events that are unavailable or unrealistic due to distance, time, cost, or safety factors. These characteristics find a big interest in the scope of learning and are integrated in educational environments. Mikropoulos, and Natsis [1] define a Virtual Learning Environment (VLE) or Educational Virtual Environment (EVE) as a *«virtual environment that is based on a certain pedagogical model, incorporates or implies one or more didactic objectives, provides users with experiences they would otherwise not be able to experience in the physical world and redounds specific learning outcomes»*. To illustrate, in this research paper we propose a VRLE where learners may study the phenomenon of relativity of movement and the gravitational attraction. In this example, learners may visualize and interact with the three-dimensional virtual representation of the moon, the earth and the sun. They will, *"experience the virtual environment in real-time, visualize abstract concepts, articulate their understanding of phenomena by constructing or manipulating the virtual*

© Springer Nature Switzerland AG 2020
M. E. Auer and T. Tsiatsos (Eds.): ICL 2018, AISC 916, pp. 353–364, 2020.
https://doi.org/10.1007/978-3-030-11932-4_34

environments and visualize the dynamic relationships between several variables in a virtual environment system" [2]. However, the VRLE design is a complex activity. The difficulties can be technical and cognitive at the same time [3, 4]. According to the model of technology integration, VRLE's design should combine three sources of knowledge: technology, pedagogy and content [5]. To fully describe the learning experience, designers have to describe a pedagogical scenario, its operationalization and the control of activities in the target virtual learning environment e.g the environment where the pedagogical situation is described. According to our literature study, we noticed that numerous VRLE environments depend, most of the time on a specific domain and context of learning. The scenario model has to be planned at the early stages of the environment's design where all possible situations have to be envisaged. This is often difficult for teachers. The existing environments are more intended for specialists in the virtual reality field. It is not easy for teachers to define and adapt the scenarios models according to their learning situations. Teaching is a design activity that may be considered as *"the intelligent center of the whole teaching-learning life-cycle"*, open to transformation by the learners [6] and supported by process and tools [7]. There are few methods and tools helping and giving support to teachers for designing their own solution [8] and it is particularly true for VRLE. Therefore, designing a virtual learning situation is a complex task for the teachers and solutions are required to help them to design, to reuse and to spread their pedagogical scenarios in the VRLE. The research questions of this study are relative to the activity of design and operationalization of the pedagogical scenarios by the teachers - designers in the target VRLE. The main question is the following: how to help the teachers to express and formalize their learning situations, which are not dependent to a virtual reality-based environment? Once the pedagogical needs are formalized, how to operationalize/to spread them in VRLE by respecting the pedagogical intentions of teachers and by limiting the semantic losses? Our research efforts aim at developing methodological and technical tools to answer these questions. In this way, we propose an iterative design teacher-centered approach. Teacher's design practices are iterative, reflection may occur before, during and after a unit's implementation in a participatory approach [8, 9]. The rest of this paper is structured as follows. The second section presents some VRLE tools and environments selected according to our research interest. We analyze these tools according to design quality criteria [10]. In the next part, we propose a VRLE design process framework. To illustrate our approach, we report in the last section an experiment consisting of using an on-purpose pedagogical situation based on a virtual reality environment. Finally, we present some concluding remarks and the next steps of this project.

2   Review of VRLEs Literature

In this literature review, we take a closer look at some VRLE and focus on their architectures, design models and learning scenario models. We examine the key strengths and limits of these proposals according to the research issues we mentioned in the previous section.

2.1   The Different VRLEs Design Models

The usual approach is to start with technical considerations before addressing pedagogical issues. For example, Trinh et al. [11] provide models for the knowledge
explanation of virtual agents populating virtual environments. This knowledge focuses
on the structure and dynamics of the environment as well as procedures that teams can
perform in this environment. This makes it possible to ensure the different semantic
constraints in VR: (1) internal properties of the spatial object, (2) spatial relationships
between a set of spatial objects, and (3) semantic of spatial interactions (for example,
before and after the state of the spatial tasks). Chen et al. [12] propose a theoretical
framework to guide VRLEs' design. This framework is divided into two subsets. The
first is called "macro-strategy". It refers to the overall design of the VRLEs and
involves (1) identification of learning objectives (skills, knowledge, etc.) and the
relationship between these objectives; (2) identification of pedagogical scenarios
allowing the learner to acquire the targeted learning; (3) identification of the help
provided to the learner (resource information, tools, etc.) to facilitate the acquisition of
targeted learning. The second subset is called "micro-strategy". It refers to the pedagogical scenarios adaptation according to the type of VRLE that one wishes to design.
Chen and Teh [13] propose some improvements of the virtual environment pedagogical
design model proposed in [12]. Ritz [14] provides guidelines for best practices in
integrating immersive virtual reality, especially Cave Automatic Virtual Environment
(CAVE), into teaching. These guidelines will address a practical need by informing and
supporting educators in adapting instructional design to emerging technology. We note
that the proposed design models are not easy to achieve for non-computer specialists.

2.2   VRLEs Learning Scenario Models

Many studies in the field of VRLEs have addressed the issue of modeling pedagogical
situations in virtual environments. For example, Sehaba and Hussaan [4] propose a
system that allows personalizing for each patient the running of virtual games for the
evaluation and rehabilitation of cognitive disorders. Marion et al. [15] propose a
learning scenario model POSEIDON able to integrate a VRLE in the learning process.
The approach is based on meta-modeling ensuring the modeling genericity, regardless
of the nature or domain of VRLEs. The authors use a meta-model that provides an
abstract representation of virtual environments both generic and machine readable.
Fahim et al. [16] ensured that the generic side of the POSVET pedagogical scenario
model using the MASCARET meta-model, allows to reuse pedagogical scenarios on
different platforms. We note that, MASCARET is a meta-model describing virtual
environments and agents that evolve in this environment. This meta-model is based on
UML that describes the structure of the environment (entities, positions), the entities
and behaviors of agents [17]. The main POSVET advantage is to allow the adaptation
of pedagogical activities and to offer to learners a control on their learning. This work
aims at adapting the pedagogical scenario to the learners' needs but doesn't offer
solutions for assisting the teachers in their design process. Chen et al. [12] as cited

previously, propose a theoretical framework which identify four principles of peda-gogical scenarios' realizations: (1) the conceptual principle that guides the learner towards the information to be considered; (2) the principle of metacognition that explains to the learner how to think during learning; (3) the procedural principle that indicates how to use the information available in the VRLEs; (4) the "*strategic*" principle that allows the learner to analyze the learning task or problem to be solved. According to Le Corre et al. [18] a pedagogical scenario in the VRLEs allows to organize the training for a pedagogical purpose, however the scenario is designed for any learner without considering the individualities, which can slow learning. These authors [18] identified some weaknesses of the Intelligent Tutorial System (ITS) PEGASE for learning with virtual reality [19] and identified its lack of con-nection with the pedagogical scenario, its lack of modularity and its lack of individ-ualization. To fill these weaknesses, they proposed an ITS called CHRYSAOR based on POSEIDON. This new proposal allows to define a pedagogical scenario and in order to perform it with MASCARET, it will be considered as a knowledge base for the agents. Based on the study of these research works, we noticed that the pedagogical models are planned at the early environment's design stages and all the possible pedagogical situations must have been considered in advance.

2.3   Architecture of VRLEs

In this part we studied different VRLEs' software architectures. Lanquepin et al. [20] propose a platform called HUMANS (Human Models based Artificial eNvironments Software), a generic framework designed to build custom virtual environments. This approach involves the dynamic computation of situations that varies according to pedagogical goals, moreover, it is not easy to handle by non-computer specialists. It is interesting to note that this platform proposes a set of software covering the VRLE life cycle from the design to its exploitation by the learners and trainees. Gerbaud et al. [21] offer a technical infrastructure not for trainers but for engineers seeking to develop VRLEs, by reusing existing components. A first study of these architectures led us to note that they do not address the problem of the design (adaptation or reuse) and operationalization of the scenario models directly by the trainers according to their pedagogical situations.

3   Analysis of Existing VRLE Tools

From the previous literature study, we covered VR tools and environments according to the two main objectives of this research work: the tools that help the teachers to design and generate their own VRLE and design their pedagogical activities. We selected some criteria from the literature [10] to characterize VR tools and design process models. Tables 1 and 2 illustrate the results of the study we conducted.

Table 1. Support for teachers to generate VRLE

	Define a design process model			Teaching oriented
	Repeatable	Reusable	Deployable	
GVT	Yes	Yes	Partial	No
ARVAD	No	Yes	Partial	Yes
HUMANS	Yes	Yes	No	Partial
MASCARET	Yes	Yes	No	No
VTS Editor	Yes	Yes	Yes	No

Table 2. Support for designing pedagogical scenario

	Define a scenario model				Propose an editor
	Generic	Adaptable	Reusable	Operationalizable	
GVT	Yes	Yes	Yes	Yes	Partial
ARVAD	No	Yes	Yes	Yes	Partial
HUMANS	Partial	Partial	Yes	Yes	Yes
MASCARET	Yes	Yes	Yes	No	No
VTS Editor	No	Yes	Yes	Yes	Yes

3.1 Support for Teachers to Generate a VRLE

We study in this part the question concerning the support for teachers to produce their own VRLE. What is the process? Is this process repeatable,[1] reusable or deployable [22]? Is the process teacher-oriented? Table 1 presents the resulting analysis that ensures this objective, for the selected tools: VTS Editor, GVT, ARVAD, HUMANS and MASCARET. VTS Editor [23] allows generating simulation-type serious games. The designer creates scenarios with an intuitive graphical editing mode based on block settings (QCM, Q&A, images, virtual characters, scoring, statistics, etc.). The designer can get a preview of his current creation and change what he has already designed. The VTS design process is both repeatable and reusable. Nevertheless, from the first use of VTS, we find out that it is intended for pedagogical engineer and e-learning project managers. Economically, VTS is very expensive compared to other simulation tools. Generic Virtual Training (GVT) [21] is a platform to create virtual environments for procedural training. It aims at improving technical training overall, in terms of productivity and qualification, and exploiting new potential, design, monitoring and capitalization of educational pathways. The GVT platform is based on visual meta-phors. This concept is important in our research situation because it focuses on interactions with objects using a menu of icons representing possible interactions between the object and the user. GVT design process focuses on three elements: modeling an actor's activity, modeling the collaborative scenario and setting up an

---

[1] The necessary process discipline is in place to repeat earlier successes on projects with similar applications.

action selection mechanism. Its model allows reproducing or realizing new pedagogical situations oriented in the industrial sector, but their deployment is partial.

The main focus of ARVAD's project [24] was on the design activity of pedagogical scenario models by teachers themselves and their operationalization in a VR environment. The ARVAD project aim is rather similar to this current research work but the public concerns exclusively LUSI (Local Units for School Inclusion) students to enable them to be more autonomous in their personal and professional life. As it is a work in progress, the design process proposes a reusable but not repeatable (generic) approach. Also, the effort required for operationalization still remains semi-automatic to deploy a new scenario, the teacher having to set variables in text files.

The major disadvantage of GVT and ARVAD environment is that they offer pedagogical models which depend on a particular learning field and training context.

MASCARET is a MultiAgent System for Collaborative, Adaptive & Realistic Environments for Training [17]. It defines an application design process repeatable and reusable and ensures a high level of abstraction when designing. It provides a language that allows an expert to define both the environment and the activities which are performed in that environment and provides operational semantics to each language concept, which automatically creates the simulation in a VR application and is seen as a knowledge base of the agents who perform the activities in the environment. The HUMANS (HUman Models based Artificial eNvironments Software) platform is a generic framework designed to build virtual environments. It *"can be adapted to different application cases, technological configurations or pedagogical strategies"* [20]. As GVT, VTS Editor, the MASCARET and HUMANS design process model allows realizing and repeating pedagogical situations oriented simulation. However these environments do not allow their deployment (except VTS) and cannot be easily handled by teachers.

3.2   Support for Designing Pedagogical Scenarios

We study in this section our second question related to the possibility of support offer to the teachers in designing VR-oriented pedagogical model. In our research work, the pedagogical model is represented by a pedagogical scenario in the form of a workflow of activities. Therefore, we are asking if models of scenario in a VRLE are generic, if they can be transformed in computational model and if they can be reused and adapted to the pedagogical situations. Table 2 presents the resulting analysis of different tools for designing pedagogical scenarios. VTS editor [23] offers a graph-based scenario editor. The notion of the graph allows teachers to design various and non-linear scenarios, with complex interactions. The graph is a set of icons called "Blocks". They are connected by logical links, represented by lines. Each type of block has its own operation, making it possible to enrich the course of the scenario. Blocks are regular structures that are instances of reusable styles. The Scenario Graph editor contains all the scenes that compose the scenario currently edited. In VTS editor, the pedagogical scenarios are adaptable and reusable for other pedagogical simulation, but they are not generic because they do not allow producing non-oriented simulation situations. The HUMANS platform offers via its SELDON module an approach for adaptive scenario [20]. The SEDLON model is extrinsic, this means that scenario is seen as an additional step in framing an existing virtual environment, and not as an integral part of the design process for that environment. Its scenario model is partially adaptable, reusable, and operationalizable, but it is intended for direct, behavioral, or motivational control activities, semi-autonomous virtual characters, or instantaneous changes in simulated system states would interfere with behavioral consistency. However, this approach does not offer a scenario editor. The ARVAD [24] environment proposes models of scenario that may offer to the teachers the possibility to define their own scenarios according to the learner's profile and the pedagogical situation. These scenarios are reusable and adaptable with new pedagogical situations but they are not generic. An editor is partially developed. It facilitates the design and parameterization of scenarios in virtual environments for learning travel autonomy for LUSI class.

In MASCARET [17], the scenario's model is a virtual agent-oriented. This model is based on four concepts: the organization, the role, the agent and the element of behavior. The organization serves as a structuring factor, providing a framework for interactions of the agents who are part of it. There are two types of organization: social organization and physical organization. Roles are the responsibilities of an agent in the organization. These responsibilities are defined by a set of behavioral elements that must be adopted by the agent playing the role. An agent must have the capabilities to use these behavioral elements (a role therefore imposes prerequisites).

GVT [21] includes mainly a reactive environment composed of behavioral objects, an interaction engine to manage complex interactions (STORM), a scenario engine to manage the course (LORA) and pedagogical engine to guide the learner. The LORA (Language for Object Relation Application) model is designed for pedagogical activities used in an industrial context. In this context, procedures and in particular maintenance procedures are very strict (actions have to be performed in exactly the given order), long and complex. GVT scenario model is reusable thanks to its generic model STORM used to describe reusable behaviors for 3D objects and reusable interactions between those objects. It is also adaptable thanks to its scenario language LORA which allows non-computer scientists to author various and complex sequences of tasks in a virtual scene.

4 Proposition of an Engineering Process Teacher-Oriented: From Design to Generation of VRLE

Our goal is to propose a solution to support and guide teachers and trainers in producing VRLEs adapted to their needs. We define a process of several steps from the definition of the learning situation to its deployment/operationalization in the VR environment (Fig. 1) based on the scenario design process model in [25]. At the beginning of the process, trainers expressed their needs according to their learning situation, with the help of a virtual reality scenario model. Therefore, this step involves in the formalization of the learning situations. A good way to formalize teacher's needs is to use a pattern-based approach. A pattern-based formalization, considering its semi-structured data, allows teachers-designers to express their pedagogical needs without extensive loss of semantic information while representing their pedagogical intention with a pattern-based editing tool [26, 27]. Then, we suggest creating pedagogical

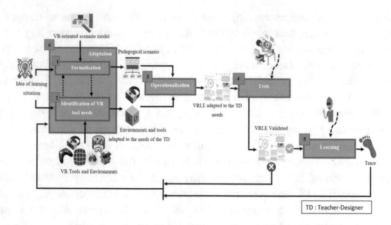

Fig. 1. A process of design and production of VRLE

scenario that defines an orchestrated sequence of learning activities within this for-
malism. The second step consists in identifying the virtual reality needs (the 3D
environment and the virtual reality tools to use). In this step, teacher-designer or
community of teachers chooses and adapts a virtual reality environment where a
pedagogical scenario is instantiated. The questions we have to deal with are: (1) which
architecture we shall use to create this service (2) How we insure the interoperability of
the various 3Ds environments? (3) How we shall face the limits of compatibility of the
technical components? At this stage, the difficulty is to provide virtual-reality tools and
environments teachers-friendly that teachers may use by themselves for specifying their
learning needs. The third step "Operationalization" consists in deploying a pedagogical
scenario in the selected virtual reality environment. The main activity consists in
operating the scenario on the chosen environment. With this step the generation of a
new VRLE or evolution of existing one is possible, based on teacher's needs. We shall
study in this step the feasibility of a service, which will allow the operability of any
scenario on any adapted 3D environment. The fourth step "Tests" focuses on the
simulation and testing activities to adapt the selected VRLE. Finally, after the learning
step, we propose to analyze the tracking data recovered from the test and learning steps
to anticipate as possible the future adaptations and modifications of VRLE. Within this
last step, we enforce an adaptation process from the tests and learning tracks to adapt to
the teachers' needs.

In the next part, we illustrate this process on a case study. First, the teacher comes
with an idea of learning scenario, expresses and formalizes it thanks to an editor. The
system generates a structured and adaptable/reusable scenario as a pattern. Then the
teacher selects an already existing 3D environment. The adaptation service has to apply
the necessary features to return the compatible environment and send it to the integration's service. The latter will instantiate the scenario on the chosen environment and
generate the new adapted VRLE.

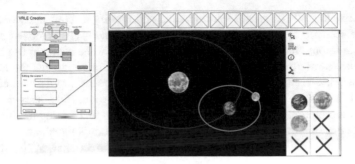

Fig. 2. Overview of the scenario editor

5   A Process Instantiation of a Pedagogical Situation

A learning situation (also known as a pedagogical situation), is a set of conditions and circumstances that can lead a person to build knowledge. An instantiation of pedagogical situation from our design process was made with a physics teacher in French college. He can be considered as non-expert designer since, on the one hand, he has not participated in the design of the first version of models and tools that we propose, and, on the other hand, he is not an advanced user of computer sciences. A teacher's design of a pedagogical situation involves the expression of a need in a disciplinary context and a goal to be achieved for the learners. The expressed need is to show students the phenomenon of relativity of movement and the gravitational attraction. The pedagogical objectives are the following: Observation of the Moon's movement relative to the Earth (circular movement); Observation of the Moon's movement relative to the Sun (curvilinear movement); Observation of the Moon's movement relative to itself (rotation-synchronous revolution); Understand the formula of gravitational attraction force ($F = G * M$ Earth $* M$ Moon $/$ Distance^2).

After this work, the teacher can begin the design process by the formalization stage in which we propose a VR-oriented scenario editor that allows the teacher's ideas to be formalized in a computational language (Fig. 3). The editor allows the definition of the scenario and its scenes. The teacher can either create his own personalized scenario or

```
<EVAH>
    <name> space travel </name>
    <field> physics </field>
    <level> second </level>
    <course> relativity of movement and gravitational attraction </course>
    <teaching_purpose> understand that the nature of movement depends on the chosen frame of reference </teaching_purpose>
    <phase id="1">
        <name> in Space </name>
        <skill> notice </skill>
        <mission id="1"> observe the movement of the earth in its orbit about the sun </mission>
        <mission id="2"> observe the movement of the moon in its orbit about the earth </mission>
        <mission id="3"> observe the movement of the moon in its orbit about the sun </mission>
        <mission id="4"> observe the movement of the earth in its orbit about the earth itself </mission>
        <tool id="1"> casque RV </tool>
        <tool id="2"> manette RV </tool>
        <environment>
            <object id="1" type="artifact">
                <name>Space</name>
                <size height="1024px" width="1024px" />
                <trajectory>Null</trajectory>
            </object>
            <object id="2" type="pedagogical">
                <name>earth</name>
                <size height="50px" width="50px" />
                <position x="60" y="34" z="57"/>
                <speed value="100"/>
                <trajectory>circular</trajectory>
            <object id="3" type="pedagogical">
                <name>moon</name>
                <size height="10px" width="10px" />
                <position x="120" y="85" z="200"/>
                <speed value="100"/>
```

Fig. 3. XML file modeling the formalization and the virtual environment identification stage

choose an existing VR-oriented scenario model and modify its content according to the needs. The scenario is a series of scenes where each scene is characterized by a name, skills and missions. Indeed, the teacher specifies the scene he wants to use in order to reinforce the learner's knowledge or skill and define the missions that the learner will have to do. The activities of the scenario will be the exercises proposed in each scene. There will be four scenes: space, Earth, Sun and Moon. The learner will have hypotheses representing the different possible movements and it is up to him to choose which one corresponds to the true phenomenon of the moon's movement in relation to the earth, the sun and in relation to itself. With each correct answer, the student can move on to the next exercise. Finally, the editor generates an XML file. This file will be used in the next operationalization phase.

After the formalization phase, the teacher identifies his needs for VR tools and objects. A VR tool is a VR-specific hardware interface, which have been booming and the public's craze for interactive virtual worlds [28]. We propose in this step a module for the selection of the VR tools by the teacher. He will also need to identify also all VR objects. VR objects can be pedagogical objects or artifact objects. A pedagogical object is a semantic unit of learning resources. It can be an exercise, a definition, examples, etc. Each pedagogical object can gather elementary components like an image named "component" (or "asset"). It may also be composed of other pedagogical objects. At first, the teacher selects VR tools he needs from the list. In our case the teacher will only need a VR headset. In a second step, he defines VR objects. The virtual environment corresponding to this teaching situation is space. The Earth, the Moon and the Sun represent the VR objects of this environment. It is possible to move the planets by simply "drag and drop" in the scene. It will also have a parameter button to handle the background of the scene. Subsequently, the teacher modifies certain parameters: he sets the size, speed and trajectory of each object. Thus, he animates the scene and sets the pedagogical objectives. Indeed, the idea of the teacher is to allow the learner to move on one of these objects to observe which objects revolve around another object.

Below (Fig. 2) is an example of the XML file resulting from the part of the scenario where the objective of the task asked to the students is to observe the movement of the Sun relative to the Moon (curvilinear movement). Subsequently, we identify a phase of operationalization resulting from the adequacy between the two previous phases (formalization and identification). The goal of the operationalization stage is to generate a VRLE. Once the operationalization is done, the teacher enters the stage of simulations and tests. This phase allows the simulation of the generated VRLE. Next, the teacher tests whether his learning objectives are achieved: Has the student been able to understand the notion of gravitation? Did the student also know how to calculate the gravitational attraction force that depends on mass and distance? Then the teacher compares the test results with the desired results during the learning phase and may apply an adaptation stage from the test traces when the results do not correspond to the pedagogical needs. The adaptation will be either at the level of the formalization of the needs, or at the level of the identification of tools and VR objects.

6 Conclusion

In this article we proposed a design process framework model of VRLE for teachers/designers. Our challenge is to facilitate the design of pedagogical scenarios and their integration/operationalization/deployment in various virtual reality environments by the teachers themselves without being constrainted by the technical difficulties, which are related to the use of technology in a virtual reality environment. This work is in progress. Future works will be dedicated to the design of at least two learning situations and the development of a technical solution and the instantiation of the process on these learning situations. A tool, a graphic-editor-like will be developed to support the design process and provide a "proof-of-concept" of the proposed approach. Last but not least, extra effort will be required to implement, evaluate and improve the approach in other pedagogical situations in the design process editor.

References

1. Mikropoulos, T.A., Chalkidis, A., Katsikis, A., Emvalotis, A.: Students' attitudes towards educational virtual environments. Educ. Inf. Technol. **3**(2), 137–148 (1998)
2. Chen, C.J.: Theoretical bases for using virtual reality in education. Theme. Sci. Technol. Educ. **2**(1–2), 71–90 (2010)
3. Carpentier, K., Lourdeaux, D.: Generation of learning situations according to the learner's profile within a virtual environment. Commun. Comput. Inf. Sci., 245–260 (2014)
4. Sehaba, K., Hussaan, A.M.: Adaptive serious game for the re-education of cognitive disorders. AMSE J. Adv. Model. Ser. Model. C **3**(73), 148–159 (2013)
5. Mishra, P., Koehler, M.J.: Technological pedagogical content knowledge: a framework for teacher knowledge. Teach. Coll. Rec. **108**(6), 2017 (2006)
6. Goodyear, P.: Teaching as design. HERDSA Rev. High. Educ., 27–50 (2015)
7. Hernández-Leo, D., Agostinho, S., Beardsley, M., Bennett, S., Lockyer, L.: Helping teachers to think about their design problem: a pilot study to stimulate design thinking. In: 9th annual International Conference on Education and New Learning Technologies EDULEARN17, Barcelona, Spain (2017)
8. Bennett, S., Agostinho, S., Lockyer, L.: The process of designing for learning: understanding university teachers design work. Educ. Technol. Res. Dev. **65**(1), 125–145 (2017)
9. Tan, E., Könings, K.D.: Teachers as participatory designers: two case studies with technology-enhanced learning environments. Instr. Sci. **43**(2), 203–228 (2015)
10. Zeiss, B., Vega, D., Schieferdecker, I., Neukirchen, H., Grabowski, J.: Applying the ISO 9126 Quality Model to Test Specifications – Exemplified for TTCN-3 Test Specifications", pp. 231–244 (2007)
11. Trinh, T.-H., Querrec, R., De Loor, P., Chevaillier, P.: Ensuring semantic spatial constraints in virtual environments using UML/OCL. In: VRST '10 Proceedings of the 17th ACM Symposium on Virtual Reality Software and Technology, pp. 219–226. Hong-Kong, Nov. 22–24, 2010
12. Chen, C.J., Toh, S.C., Fauzy, W.M.: The theoretical framework for designing desktop virtual reality-based learning environments. J. Interact. Learn. Res. **15**(2), 147–167. Norfolk, VA: Association for the Advancement of Computing in Education (AACE) (2004)
13. Chen, C.J., Teh, C.S.: Enhancing an instructional design model for virtual reality-based learning. Australas. J. Educ. Technol. **29**(5), 699–716 (2013)

14. Ritz, L.T.: Teaching with CAVE virtual reality systems: Instructional design strategies that promote adequate cognitive load for learners. SMTC Plan B Papers **5** (2015)
15. Marion, N., Querrec, R., Chevaillier, P.: Integrating knowledge from virtual reality environments to learning scenario models-a meta-modeling approach. In: Proceedings of the 1st International Conference on Computer Supported Education, PP. 254–259 (2009)
16. Fahim, M., Jakimi, A., El Bermi, L.: Pedagogical Scenarization for Virtual Environments for Training: Towards Genericity, Coherence and Adaptivity. Int. J. Adv. Eng. Res. Sci. **3**(12), 96–103 (2016)
17. Buche, C., Querrec, R., De Loor, P., Chevaillier, P.: Mascaret: Pedagogical multi-agents system for virtual environment for training. J. Distance Educ. Technol. **4**(2), 41–61 (2004)
18. Le Corre, F., Hoareau, C., Ganier, F., Buche, C., Querrec, R.: A pedagogical scenario language for virtual environment for learning based on UML meta-model. Application to Blood Analysis Instrument. In: Proceedings of the 6th International Conference on Computer Supported Education (CSEDU), pp. 301–308. Spain (2014)
19. Buche, C., Bossard, C., Querrec, R., Chevaillier, P.: PEGASE: A generic and adaptable intelligent system for virtual reality learning environments. Int. J. Virtual Real. **9**(2), 73–85 (2010)
20. Lanquepin, V., et al.: HUMANS: a human models based artificial environments software platform. In: VRIC 2013, Mar 13, Laval, France, pp. 59–68 (2013)
21. Gerbaud, S., Mollet, N., Ganier, F., Arnaldi B., Tisseau, J.: GVT: a platform to create virtual environments for. In: Virtual Reality Conference, Reno, NE, USA (2008)
22. The process approach in iso 9001:2015, International Organization for Standardization
23. Ruzzu, G.: Blog VTS Editor authoring software for Degital Learning. Serious factory, 8 December 2017. https://www.seriousfactory.com/virtual-training-suite/?lang=en
24. Oubahssi, L., Piau-Toffolon, C.: Virtual learning environment design in the context of orientation skills acquisition for LUSI class. In: CSEDU 2018, the International Conference on Computer Supported Education, Funchal, Madeira, Portugal, pp. 47–58 (2018)
25. Emin-Martinez, V., et al.: Towards teacher-led design inquiry of learning. eLearning Papers, no. 36, pp. 1–12 (2014)
26. Tadjine, Z., Oubahssi, L., Piau-Toffolon, C., Iksal, S.: A process using ontology to automate the operationalization of pattern-based learning scenarios. In: Communications in Computer and Information Science (CCIS), pp. 444–461. Springer (2016)
27. Mor, Y.: Embedding design patterns in a methodology for a design science of e-Learning. In: Kohls, C., Wedekind, J. (ed.) Problems Investigations of E-Learning Patterns: Context Factors Solutions (2010)
28. Fuchs, P., Moreau, G., Guitton, P.: Virtual Reality: Concepts and Technologies. CRC Press (2011)

From Old Fashioned "One Size Fits All" to Tailor Made Online Training

Daša Munková[1(✉)], Michal Munk[2], Ľubomír Benko[3], and Jakub Absolon[1]

[1] Department of Translation Studies, Constantine the Philosopher University in Nitra, Nitra, Slovak Republic
dmunkova@ukf.sk, absolon@asap-translation.com
[2] Department of Informatics, Faculty of Natural Sciences, Constantine the Philosopher University in Nitra, Nitra, Slovak Republic
mmunk@ukf.sk
[3] Institute of System Engineering and Informatics, University of Pardubice, Pardubice, Czech Republic
lubomir.benko@gmail.com

Abstract. Nowadays, post-editing of machine translation output represents a significant element in the translation market and industry. Subsequently, the preparation of future translators must cover not only all routine methods but must be cost-effective, efficient and in accordance with human resources available. That is the reason we use Internet-based technologies more and more. New emerging technologies are very often driven by the marketing power of companies developing and selling applications. Each of us experienced dozens of fantastic features available in teaching software and applications. The core skill of the online educator is to find a balance between our needs and ability to use technology. Since translation demand keeps growing every day, a large number of translators use various technical tools including translation memories, terminology management tools or Machine Translation (MT) technologies and thus increase their productivity and meet this high demand. The post-editing of MT should be only done by a person who is familiar with this method and knows exactly what, how and how much needs to be edited in the text. Otherwise, the sense of post-editing is losing importance, as the work of post-editor would not be more effective as a translator's, who translates the text traditional way "from scratch". The contribution of the paper is to create an online educational system tailored to translators' needs; an online system in which students translate and revise a text, post-edit machine translation output and also assess the quality of the translation.

Keywords: Education · Online system · Translation · Post-editing · Evaluation

1 Introduction

Today's world and modern technologies are being developed very quickly. The stage of their development and the degree of applicability can be shown later in the practice. New technologies have an immense influence on many industries and branches,

© Springer Nature Switzerland AG 2020
M. E. Auer and T. Tsiatsos (Eds.): ICL 2018, AISC 916, pp. 365–376, 2020.
https://doi.org/10.1007/978-3-030-11932-4_35

inclusively translation industry. Computers have already become a part of the 21st-century lifestyle, and most of the people use them because they can make their work easier and faster. There is a great number of companies and professionals using machine translation and considering it an elementary feature of the whole output. Despite distinguishable imperfections of machine translation, its popularity is still increasing. IT corporations invest their money into development and improvement of translation technologies and bring up the products which aim is to generate the adequate translations and to create their own translation memories for further use.

Modern translators must keep the pace with more and more demanding market and use all technologies available. From state of the art text editors with predictive typing which gives you automatic suggestion of what you are going to type, through online terminology databases and automatic quality assurance application to MT engines using AI. Subsequently, the preparation of a professional translator of the future must cover not only all the mentioned fields but must be cost-effective, efficient and in accordance with human resources available. That is the reason internet-based technologies are used more and more.

The latest technologies are very often driven by the marketing power of companies developing and selling applications (e.g. SDL Trados, Memsource, MemoQ, etc.). Each of us experienced dozens of fantastic features available in teaching software and applications. Unfortunately, we are not able to use all of them, like most of us regretfully do not make use of different features of our cameras. However, the core skill of the online educator is to find a balance between students' needs and technologies available. To make a theoretical lecture in PC classroom or to use an interactive whiteboard for watching video is wasting the resources. We would like to reveal our best practical experience with the online training of translators. In preparation of contemporary translators ready to compete on the 21st-century translation market, and, based on practical experience from both, developing a training course for university students of translation studies and translation professionals, three different methods are used in the online training of translators: webinar, one-to-one online tuition, and self-paced online exercise. The webinar is perfect for one purpose activity like software training; to educate professionals/students/novice translators theoretically or to show new features available in software e.g. style guides, or terminology resources. However, it is more demanding from the side of the organizer and presenter (it is highly recommended to have two presenters - one for introduction and the other one as a teacher with necessary expertise). One-to-one online training is used for distance training of computer-aided translation tools. It is the fast, cheap, and easy way how to offer personalized tuition to a translator, but it is still training how to use some computer-assisted tool. Self-paced online exercises are used for the specific preparation of translators (e.g. Post-Editing of Machine Translation (PEMT) course).

However, all of these online learning types are only web applications or certain parts of some learning management system (such as LMS Moodle). So far, no one has focused on or tried to design a comprehensive system that, on one hand, would copy a similar interface of professional computer-aided tools (e.g. SDL Trados or MemoQ) or web applications (e.g. Memsource) which are familiar to students, and on the other hand, meets the requirements of learning and teaching translation. We attempted to design such an online learning system that would meet the requirements of online education and as well would be tailored to the translator's needs, who should be trained and prepared to translation market.

The aim of our contribution is to describe a design and creation of an online OSTPERE system in which students do not only translate (as in other translation technologies) but also correct translations, post-edit machine translation, but also assess the quality of translation (machine or human translation), and to show one application of OSTPERE system for translation evaluation, where students translations or post-edited machine translations are evaluated using automatic metrics of accuracy and error rate of MT evaluation.

2 Post-editing of Machine Translation Output

Post-editing (PE) is not a brand new discipline as it might seem for the first sight. It covers the tasks of editing, modification, or correction of the machine-translated output. Munkova [1] claims that machine translation designated for publication inevitably needs an intervention of a translator.

Gradually, post-editing found its position in translation and arose greater interest in professionals. Most of them distinguish several PE types, according to the extent of corrections done by post-editor in MT output. Allen [2] discusses the factors determining PE and he defines inbound and outbound translation related to the types and levels of PE. He distinguishes two levels for the inbound translation: MT with no PE, and rapid PE. For the outbound translation, he defines three PE levels: no PE, minimal PE and full PE. International Standard specifies two PE types: light and full PE [3]. There are two types of quality distinguished here: 'a good enough quality' is defined as comprehensible and accurate (it conveys the meaning of the source text, the text does not have to be grammatically or stylistically flawless but semantically correct, no information is added or omitted). The second level is of 'publishable quality', which is similar to the quality of the human translation (translation is comprehensible and accurate; grammar and stylistics need to be appropriate).

In translation, a focus is given on adequate variant of translation- in PE, a post-editor evaluates correctness or incorrectness of the variant suggested by the MT system. An important part of post-editor work is that he always needs to make quick decisions – she/he needs to evaluate the importance of machine-translated suggestions in the text and to decide about possible omissions. This ability is further important for PE efficiency in terms of speed and decision making [4].

Witczak [5] conducted an experiment on translation students in their master's program examining their opinion on PE. There were 21 students involved. The students were assigned to post-edit two texts (newspaper and a medicine leaflet). After PE, they filled in the questionnaire-based review which showed that most of the students (11 out of 21 students) were in favor of PE. The study showed that PE can accelerate man's work, and saves time. They added it was an easy and quick work and the results may depend on the type of the MT output being post-edited.

Carl et al. [6] compared productivity in human translation and PE of MT output and gaze activity. Translators (all with Danish L1 and English L2) were asked to post-edit three English-to-Danish MT texts produced by Google Translate. The results showed that the PE was somewhat faster and more efficient than the human translation.

Koehn and Germann [7] focused on the impact of MT quality on human PE. They studied the correlation between four different MT systems and post-editors' productivity and behavior. There were four English-German bilingual post-editors involved, with no professional experience in translation and different language skills. They found out that the best system showed 20% better productivity than the worst system. In the case of a better system, they spent less time on editing, and on figuring out translation problems. They also found out that the differences between post-editors are much larger than the differences between MT systems.

De Almeida and O'Brien [4] studied speed and keyboard/mouse activity and the correlation between post-editing and translator's previous experience. The experiment traced 6 translators post-editing IT texts from English (all participants) into French (3 participants) and Spanish (3 participants). It showed that the more-experienced translators were faster and more precise in post-editing but on the other hand, they did the highest number of stylistic corrections which were not necessary to do. It also found out that keyboard and mouse activity is a highly individual factor in PE.

3   Online System for Translation, Post-editing, Revision, and Evaluation

An interactive online system (OSTPERE – Online System for Translation, Post-editing, Revision and Evaluation) was firstly created only for manual evaluation of MT output, namely for MT post-editing done by translators and for the needs of errors analysis of machine translation output.

3.1   Design and Description of the OSTPERE

The first version of the OSTPERE system was introduced by authors Munkova et al. [8, 9]. Later, it was improved and adjusted based on the requirements of post-editors and linguistic experts in the field of Slovak language. The designed system requires alignment of the source/input texts, therefore the HunAlign tool [10] was used and the output (target text) had a table format (.xls).

The system offers an online interface available anytime through a web browser. It was created using the PHP framework CodeIgniter and as a data storage MySQL database was used. The system workflow is shown in Fig. 1. Functional requirements for the designed system were defined for a better description of the relations between the system and user:

- Uploading and processing of files with aligned texts in xls format;
- Selection of document for post-editing of MT output;
- Selection of document for translation (human translation);
- Classification of basic MT errors by post-editor (language, accuracy, terminology a style);
- Revision of the post-edited documents by linguists;

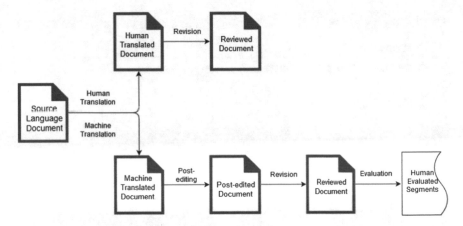

Fig. 1. Illustration of the workflow of MT evaluation using OSTPERE

yntax and

- Automatic evaluation of MT output using metrics of automatic evaluation.

After uploading and processing documents, the administrator can assign the documents to corresponding translators or post-editors. The administrator can also assign all documents of one specific domain to the corresponding post-editor. The system contains documents of various styles and genres. The working interface (Fig. 2) for translator and post-editor was inspired by the professional translation tools and programs, such as SDL Trados Studio. The created online system displays 25 segments of the current working document. The segments are translate or post-edit by the user (translator/post-editor) on the right side and on the left side, the corresponding source segment is located. The system also monitors another important parameter during the human translation or PE of MT output. It records thinking and editing time of the user. The thinking time means the time from the moment the user initiates the contact (through a mouse click) with the working segment to the moment of starting editing the segment. The editing time means the time from the moment the user starts typing (editing) the working segment to the moment he/she confirms the edit with a keyboard shortcut. These two-time intervals can help us to better analyze the cognitive load through the time needed for translation or PE.

When PE of MT output (split into segments) is done and the user made changes, the post-editor has to classify and determine the extent of occurred errors of MT output (Fig. 3). After saving the post-edited segments in the database, a post-edited segment will be used as a reference for automatic error identification in MT. The interface for human translation is similar to post-editor one but does not have the error classification.

A similar interface is designed for linguists (of Slovak language). The experts are responsible for correcting the post-edited segments from stylistics and grammatical aspects. The checked segments are ready for evaluation of MT errors.

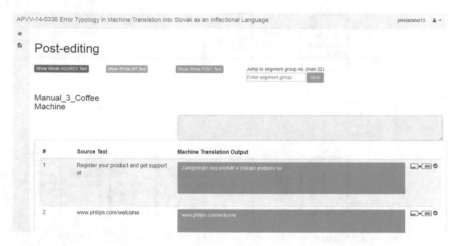

Fig. 2. Preview of the post-editing work environment

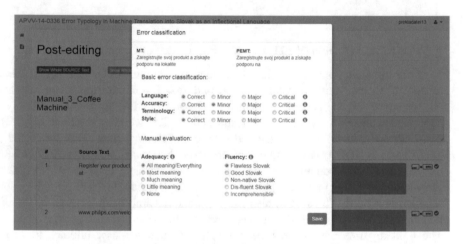

Fig. 3. Preview of MT error classification for post-editors

In perspective of the Slovak language, the interface (Fig. 4) is tailored to linguistic experts for identification and classification of MT errors (error analysis). Both identification and classification are recorded for further deeper analysis.

Up to the present, the system contains more than 550 documents of journalistic, administrative or scientific style. More than 370 of the documents are written in English (source language) and around 150 in German as a source language. This number of documents matches around 62,000 of sentence segments (around 570,000 MT words and 720,000 post-edited MT words).

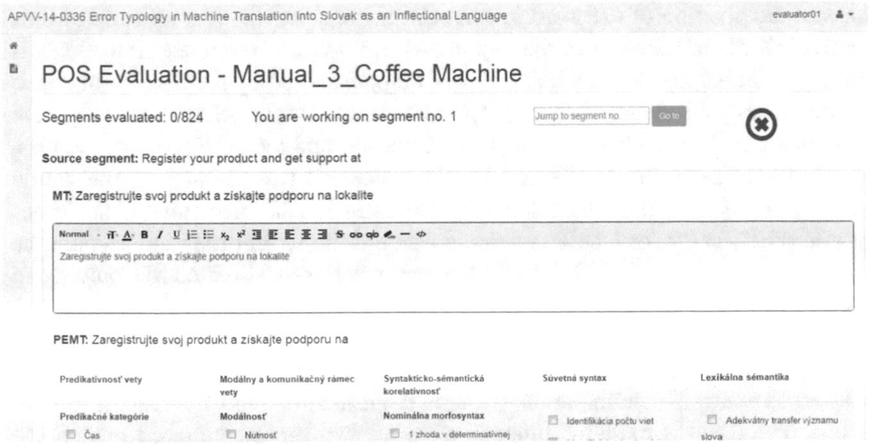

Fig. 4. System environment preview for machine error rate evaluation

3.2 Learning Through OSTPERE

Later, the system was tailored to students (future translators) needs and specialized in training translation skills, post-editing, revision, and also the evaluation of machine translation output from/to foreign languages (English and German) into/from mother tongue (Slovak). The OSTPERE system can be used both, for education (training translation competence 24/7; students do not translate "per paper", their translations are stored in a database, on one place and available anytime not only to students but also to their teachers) and for evaluation of students' translation or revision or MT post-editing (based on the evaluation results teacher can focused on errors or issues found in translation or in MT post-editing). Students can access to the system anytime, they just need an access to computer and internet connection. The flexibility of working time is important in translation market as well as in translation studies because translation itself is an art, it is a creative work, not a work in a factory with fixed working time; some translators prefer work at night, others in the morning or afternoon.

4 The Student Assessment Using the Automatic Evaluation Metrics of Post-edited Machine Translation Output

In this section, we introduce one of the possibility how to use OSTPERE system for the evaluation of MT post-editing performance of students. We assess post-edited machine translation based on the metrics of automatic MT evaluation. In this study, we focused only on automatic metrics (metrics of accuracy and edit rate) based on lexicon methods and edit distance. We dealt with metrics of accuracy (Precision, Recall, f-measure, and BLEU) which are based on the closeness of the hypothesis (MT output/ PEMT output/ student's translation) with the reference (teacher's translation or post-edited MT output as a "gold standard" measure of translation quality), similar to bag-of-words, i.e.

regardless of the position of the word in a sentence. Precision (P) is a measure of how many correct words are present in the hypothesis in regard to reference. Recall (R) is the number of correct words in students' PEMT output divided by the number of words of reference (the gold standard), i.e. proportion of all words in reference that are correct. F-measure (F1) is a harmonic mean of precision and recall. F-measure (F1) is a harmonic mean of precision and recall. BLEU represents two features of translation quality, adequacy and fluency, by calculating words or lexical precision [11]. Metrics of error rate (PER, WER, or CDER) take into account the word order and account the Levenshtein distance between MT output/human translation/post-edited MT output and a reference [12].

To assess the metrics of automatic evaluation of students' post-edited MT outputs, we used exploration techniques. Exploration analysis methods serve for the discovery of structures, hypothesis determination, extreme identification, and for visualization of phenomena. The issue is to show different ways of data representation, to recognize regularities and irregularities, structures, patterns and individualities. In the exploration process, we look for interesting configurations and relationships within data.

For each PEMT sentence, post-edited by students, 10 metrics of automatic evaluation of translation quality have been calculated to express the quality in terms of accuracy and error rate of translation.

We used icon charts to visualize multidimensional data, where each icon corresponded to one sentence, was characterized by scores of the metrics of automatic evaluation. The icons representing sentences of the MT are shown in the graph from left to right corresponding to the sequence of sentences in the text.

Using these charts, it is possible to identify sentences with the same quality of post-editing, to divide them into groups, but also to explore sentences different from others. Icon graph can visually detect extremes- sentences with significantly different scores compare to other. This approach can help us to identify problems or issues, which make the student the most in post-editing and focus on texts (sentences) in which students can train and gain better post-editing skills.

As we mentioned earlier, using OSTPERE system in education brings benefit to students in form of online learning and improving ICT skills (which belongs to translator competence- technical competence) and to teacher in form of a novel approach how to evaluate the quality of students- translators' performance based on the automatic metrics of MT evaluation regardless translation direction.

Based on the metrics of accuracy (automatic metrics of MT evaluation) the 3rd, 18th, 24th, and 29th post-edited sentence (post-edited by students) were assessed the best, achieved the highest scores (Fig. 5). On the contrary, the lowest scores of metrics of accuracy were reached for the 8th, 10th, 33rd, and 35th sentence. The individual components of the measure of translation accuracy (Precision, Recall, and F-measure) are almost symmetrical in all sentences.

Based on the metrics of error rate (automatic metrics of MT evaluation) the 3rd, 18th, 24th, and 29th post-edited MT sentence were assessed the best (Fig. 6) and vice versa, the highest score of metrics of error rate was reached for the 8th, 10th, 31st, 33th, and 35th post-edited sentences. Individual components of metrics of translation error rate (PER, WER, and CDER) show a slight asymmetry in some sentences (cases).

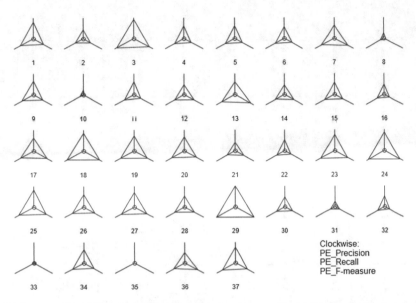

Fig. 5. Scores of Precision, Recall, and F-measure of post-edited MT (Clockwise – values are displayed in the clockwise direction)

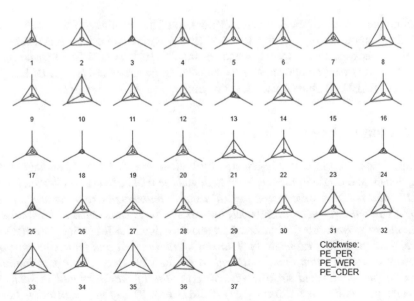

Fig. 6. Scores of PER, WER, CDER of post-edited MT (Clockwise – values are displayed in the clockwise direction)

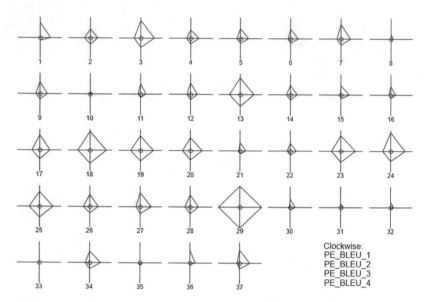

Fig. 7. Scores of BLEU of post-edited MT (Clockwise – values are displayed in the clockwise direction)

Based on the metrics of the automatic evaluation of the accuracy of the translation post-edited by students (Fig. 7), the 18th and 29th sentence were assessed the best. Conversely, the lowest scores were reached for the 8th, 10th, 31st, 32nd, 33th, and 35th sentence. The individual components of the translation accuracy (BLEU_1, BLEU_2, BLEU_3, and BLEU_4) show a significant asymmetry in some sentences.

5 Discussion and Conclusion

The extent of post-editing is highly-affected by both source and target language. For example, Sentence 8 origin in Slovak: *Z tohto dôvodu je preklad, resp. prekladateľstvo neodmysliteľnou a dôležitou súčasťou medziľudskej komunikácie ako v bežnom styku, tak aj pri získavaní informácií potrebných na osobnostne a spoločensky obohacujúci život jednotlivca v spoločnosti.* Translated by machine into the following form (EN): *Therefore, the translation, respectively. Translating an integral and important part of interpersonal communication than in normal traffic, as well as to obtain information necessary for personally and socially enriching the life of the individual in society.* Subsequently post-edit into the form (EN), PEMT(ref.): *Therefore, translation (and hence translation studies) is an integral and important part of interpersonal communication in everyday situations, as well as in gathering the information necessary for the enrichment of an individual's personal and social life in our society.* PEMT (Student1): *Therefore, translation or translating is an integral and important part of international communication in everyday contact, as well as in obtaining information necessary for enriching personal and social life of the individual in society.* PEMT

(Student2): *Therefore, the translation, respectively translating, is an integral and important part of interpersonal communication in everyday life as well as the tool to obtain necessary information for personal and social enrichment of the individual's life in society.* The results indicate, that the main problem of students' performance measured by automatic metrics lays in an ambiguity, i.e. one string of words corresponds to a few different representations. It means that teachers must pay more attention to lexical and syntactical ambiguity, in the case of post-editing of machine translation of journalistic text from Slovak into English. The similarity in translation languages and differences between them has an impact on quality of machine-translated output and consequent post-editing.

The training of translators is very specific, but students (of any subject) still require support and guidance (e.g. using an effective online learning system). We believe our approach will be inspirational for all kinds of areas.

Acknowledgment. This work was supported by the Slovak Research and Development Agency under the contract No. APVV-14-0336 and Scientific Grant Agency of the Ministry of Education of the Slovak Republic (ME SR) and of Slovak Academy of Sciences (SAS) under the contracts No. VEGA- 1/0809/18 and VEGA- 1/0776/18.

References

1. Munková, D.: Prístupy k strojovému prekladu (modely, metódy a problémy strojového prekladu). Univerzita Konštantína Filozofa, Nitra (2013)
2. Allen, J.: Post-editing. In: Somers, H. (ed.) Computers and Translation: A Translator's Guide, Amsterdam & Philadelphia: Benjamins, pp. 297–317 (2003)
3. ISO/CD 18587:2014. Translation Services – Post-editing of machine translation output – Requirements. International Organization for Standardization (2014)
4. De Almeida, G., O'Brien, S.: Analysing post-editing performance: Correlations with years of translation experience. In: Proceedings of the 14th Annual conference of the European Association for Machine Translation, (EAMT 2010) (2010). http://www.mtarchive.info/EAMT-2010-Almeida.pdf. Accessed 5 Jan 2018
5. Witczak, O.: Incorporating post-editing into a computer-assisted translation course. A study of student attitudes. J. Transl. Educ. Transl. Stud. **1**, 35–55 (2016)
6. Carl, M., Dragsted, B., Elming, J., Hardt, D., Jakobsen, A.L.: The process of post-editing: a pilot study. In: Proceedings of the 8th International NLPSC workshop. Special Theme: Human-Machine Interaction in Translation, pp. 131–142 (2011)
7. Koehn, P., Germann, U.: The impact of machine translation quality on human post-editing. In: Workshop on Humans and Computer-Assisted Translation, Gothenburg, Sweden, pp. 38–46 (2014)
8. Benko, Ľ., Munková, D.: Application of POS tagging in machine translation evaluation. In: DIVAI 2016 : 11th International Scientific Conference on Distance Learning in Applied Informatics, Sturovo, May 2–4, 2016. Wolters Kluwer, ISSN 2464-7489, Sturovo, pp 471–489 (2016)
9. Munková, D., Kapusta, J., Drlík, M.: System for post-editing and automatic error classification of machine translation. In: DIVAI 2016 : 11th International Scientific Conference on Distance Learning in Applied Informatics, Sturovo, May 2–4, 2016. Wolters Kluwer, ISSN 2464-7489, Sturovo. pp 571–579 (2016)

10. Varga, D., Németh, L., Halácsy, P., Kornai, A., Trón, V., Nagy, V.: Parallel corpora for medium density languages. Proc. RANLP **2005**, 590–596 (2005)
11. Munkova, D., Munk, M.: An automatic evaluation of machine translation and Slavic languages. In: IEEE 2014: IEEE 8th International Conference on Application of Information and Communication Technologies, IEEE, 2014, pp. 1–5 (2014). https://doi.org/10.1109/icaict.2014.7035992
12. Munkova, D., Munk, M.: Automatic metrics for machine translation evaluation and minority languages, MedCT 2015: Mediterr. In: Conference on Information and Communication Technology, Springer, Cham, Saidia, 2016, pp. 631–636 (2015). https://doi.org/10.1007/978-3-319-30298-0_69

Implementation of an Adaptive Mechanism in Moodle Based on a Hybrid Dynamic User Model

Ioannis Karagiannis[✉] and Maya Satratzemi

Department of Applied Informatics, University of Macedonia,
54006 Thessaloniki, Greece
{karagiannis,maya}@uom.edu.gr

Abstract. Learning styles summarizes the concept that individuals have different learning preferences and they learn better when they receive information in their preferred way. Even though learning styles have been subjected to some criticism, they can play an important role in adaptive e-learning systems. In order to overcome the drawbacks of the traditional detection method, educational data mining techniques have been implemented in these systems for the automatic detection of students' learning styles. The purpose of this paper is to present the implementation of an adaptive mechanism in Moodle. The proposed mechanism is based on a hybrid dynamic user model that is built with techniques that are based both on learner knowledge and behavior. An evaluation study was conducted in order to examine the effectiveness of the proposed mechanism. The results were encouraging since they indicated that our extension affected students' motivation and performance. In addition, the precision attained by the proposed automatic detection approach was rather positive.

Keywords: Learning management systems · Adaptive systems ·
Learning styles · Automatic detection · User modeling

1 Introduction

Learning Management Systems (LMS) have been widely used either for e-learning courses or for blended learning. A causal factor is that LMS offer various sets of tools to support teachers in creating and managing online courses. LMS have a major disadvantage though. Namely, educational resources are the same for all students, ignoring their individual characteristics, such as learning styles. Learning styles are the different methods that students use to acquire and process information [9]. In the literature there are many works on the significance of learning styles and their impact on the learning process [1, 12, 20]. Learning styles allow instructional designers to adapt their teaching styles and educational material to their students' preferences to enhance learning [12]. Learning style detection is important because it helps to improve learning performance, enhance motivation, and reduce the learning time [18].

The purpose of this paper is to present an adaptive mechanism that was implemented in Moodle. This mechanism adapts the presentation and the proposed navigation within a course, to students' different learning styles and knowledge level. The

© Springer Nature Switzerland AG 2020
M. E. Auer and T. Tsiatsos (Eds.): ICL 2018, AISC 916, pp. 377–388, 2020.
https://doi.org/10.1007/978-3-030-11932-4_36

aim of our study was to investigate whether our adaptive mechanism was able to increase students' motivation and help them to improve their grades on the mid-term exam. Besides that, we investigated whether the precision attained by the proposed automatic detection approach is high.

The remainder of the paper is organized as follows. Related work is presented in Sect. 2. This is followed by a section where the proposed system is presented. After Sect. 4, where the evaluation study is presented, in Sect. 5 are the conclusions.

2 Related Work

2.1 Learning Styles

Students differ from one another in many ways, and the way they receive and process information determines their learning style [9]. Although most researchers accept that students have different learning preferences, the use of learning styles has caused a controversy. Proponents of using learning styles believe that each student has a specific learning style or preference, and he/she learns best when information is presented in this style [1, 3]. On the other hand, adherents of the learning style theory believe that there is no real scientific basis for the proposition that a learner actually has a certain optimal learning style, there is a reliable and valid way to identify it and learning is improved if teaching is matched to learning styles [16]. However, researchers still believe that the theory of learning styles continues to offer something useful and the criticism that has been raised is invalid [20].

There is a wide variety of learning style models that have been put forward. The Felder-Silverman Learning Style Model (FSLSM) [9] is used far more than any other in AEHS. Dorca et al. [7] argued that FSLSM stood out because it combined different main learning styles models and, therefore, it describes learning styles in much more detail. FSLSM has four dimensions each with two scales: active/reflective, sensing/intuitive, verbal/visual and sequential/global, according to the way students process, perceive, receive and understand information. The resulting preferences are considered as tendencies, since even those learners with a strong preference for a particular learning style can at times act differently [14]. The Index of Learning Styles (ILS) was developed in order to identify learning style preferences in FSLSM [10]. Every learner has a personal preference for each dimension, which is expressed with a value of between -11 and 11 (including only odd values).

2.2 Automatic Detection of Learning Styles

Traditionally, learning styles are mainly identified using questionnaires that students are asked to fill out. Even though these questionnaires present good reliability and validity for identifying learning styles [11], they have been subjected to some criticism [12, 16]. To begin with, the particular type of detection can be biased as it depends on students' judgment [20]. Moreover, it is done only once while the learning styles can change over time [20]. To overcome the aforementioned drawbacks, research has focused on approaches for the automatic detection of learning styles that are based on the analysis of students' behavior data such as time spent on different activities. Automatic approaches do not require learners to waste their time on completing a questionnaire and therefore, they are free from the problem of inaccurate self-conceptions of students at a specific time [8]. Moreover, they can easily be applied to dynamically update students' user model regarding their learning styles.

There are two main approaches to automatically detect a student's learning style: data driven and literature based approach [14]. The first approach applies data mining algorithms to students' behavior data in order to build a model that imitates the questionnaire that is used by the respective learning style model. Data driven approach has the advantage of being more accurate than literature based approach, mainly due to the use of real data for the detection of students' learning styles [20]. Several data mining algorithms have been used to detect students' learning styles such as Bayesian networks [2, 13, 14], neural networks [3, 5], decision trees [6] and genetic algorithms [3]. The literature based approach depends on the idea that learners with a particular learning style preference behave in a predefined way. This approach uses students' behavior to obtain hints about their learning style preference and, after considering the theory of the adopted learning style model, a simple rule-based method is applied to calculate the learning style from the number of matching hints. Several literature based approaches have been proposed [7, 8, 14, 17, 18]. Most of the aforementioned works employed FSLSM to represent a student's learning style and attained promising results regarding their precision. However, some FSLSM dimensions were not considered in some of the works and, therefore, their validity can be challenged.

2.3 Adaptive Learning Systems

Provision of same instructional conditions to all students can be pedagogically ineffective [1]. Several systems have been developed to provide content that fits students' individual learning preferences. These systems reflect users' characteristics in a user model and apply that model to adapt instructional aspects of the system accordingly [4]. Different frameworks have been used for the development of adaptive systems such as previous knowledge and student background [19]. Developers reported that student's learning styles were the most useful framework though [19] because they can be used to adapt the content presentation to the learner [3].

The user model in such systems can be built with techniques based either on learner knowledge or on user behavior and can be either static or dynamically updated to describe learner's current state. The adaptation techniques that can be used are classified into two categories according to the type of adaptation provided: adaptive presentation and adaptive navigation support [4]. Taking these techniques under consideration, researchers design their proposed system's architecture.

Many systems have been developed to adapt the presentation of a course to students' learning styles. WELSA [18] is an adaptive educational hypermedia system that implements dynamic student modeling and adapts educational resources to users' learning styles ignoring learners' knowledge. Yang, Hwang and Yang [21] developed the AMDPC system which is an adaptive educational hypermedia system that considers users' learning styles to adapt the presentation of the educational material and cognitive styles to adapt the system interface. The validity of the pilot study can be questioned though, since the two groups of students used actually a different system. Other researchers [14, 17] developed add-ons for Moodle in order to extend it with adaptivity because commonly used learning management systems are somewhat limited in the amount of personalization they can offer [1]. Both [14, 17] implemented dynamic student modeling but they totally ignored learners' knowledge. Bachari, Abelwahed & Adnani [2] developed the LearnFit system which is an extension to Moodle that recommends resources to learners based on their learning styles. The user model is initialized according to the results obtained by the ILS and then updated by using the Bayesian model. Although the proposed approach was beneficial to students, it is quite complicated and time consuming for the instructor.

3   Proposed System

3.1   Architecture

After taking under consideration related works, it was decided to extend Moodle in order to provide courses that suit students' learning preferences. The architecture of the proposed system is illustrated in Fig. 1. Our system is innovative as it adopts a hybrid dynamic user model. The proposed user model is called hybrid as it is built with techniques that are based both on behavior and knowledge of the learner regarding both static and dynamic student modeling modules. Taking into account the advantages of FSLSM and its wide use in adaptive systems, we decided to adopt the particular model to represent students' learning styles in the proposed system.

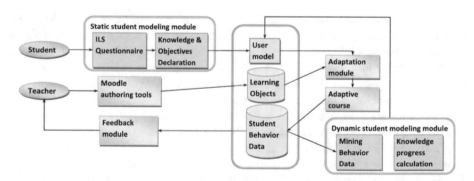

Fig. 1. Architecture of the proposed system

Automatic approaches for student modeling has the advantage of achieving better results than questionnaires without requiring additional effort by the students. However, they can only be applied when a significant amount of data is available. This means that should a researcher wish to apply an automatic approach for student modeling, it would have to be applied after the students have become involved in studying and doing the Moodle course so as to create a significant amount of interaction data. To overcome this drawback, the static student modeling module (Fig. 1) can be used so that the system can adapt the course from the very beginning. To achieve this, the particular module requests that students respond to ILS and also declare their educational objectives right after their first login. Although the user model is built from the very beginning, it has to be dynamically updated in order to reflect students' current state, surmounting the disadvantages of static approaches. As can be seen in Fig. 1, dynamic student modeling module applies a data mining algorithm to users' behavior data comprising the number of visits to each type of learning object, as well as their duration. Besides mining behavior data, the dynamic student modeling module calculates knowledge progress as described in detail in our previous work [15].

3.2 Automatic Detection of Learning Styles

Traditionally, data-driven approaches for the automatic detection of learning styles implement a specific data mining algorithm. This means that should a researcher wish to apply a different algorithm in the system, it would have to be implemented from scratch. To achieve flexibility, the proposed approach runs Weka online. Weka is a collection of tools and algorithms for data mining tasks. It contains tools for data pre-processing, classification, clustering, association rules, and visualization. The algorithms can either be applied directly to a dataset or called from Java code. The direct application of the algorithms to a dataset can be implemented either via the graphical interface or the command line. The proposed approach makes use of the latter option.

In order to enable communication between Moodle and Weka, we implemented an extension in Moodle and installed Weka on the server. The extension stores the command that has to be executed in Weka in a string variable. Next, the respective command is executed by passing this string as input to "system" function. "System" is a php function that executes an external program and displays the output. In our case, the output of executing the command in Weka is redirected to a file created on the server. The main advantage of our approach is that should a researcher wish to apply a totally different algorithm in the system, he would only have to change the respective command line which stores the Weka command in the string variable. Thus, automatic detection of students' learning styles can be achieved in a flexible way.

Relevant Patterns for FSLSM. The first step of the proposed approach for automatic detection of learning styles is to determine the behavioral patterns that are relevant for FSLSM's dimensions. The proposed adaptation mechanism [15] used seven different types of learning objects, namely outlines, content objects, videos, solved exercises, quizzes, open-ended questions and conclusions. Therefore, the behavioral patterns should be related to these types. It was decided to use the total time that a student studies a specific type of learning object and the number of visits to it. Next we present patterns of behavior that were decided to be used and their description:

- outline_dur, outline_vis: relative time spent and number of visits on outlines
- content_dur, content_vis: relative time spent and number of visits on content objects
- video_dur, video_vis: relative time spent and number of visits on videos
- quiz_dur, quiz_vis: relative time spent and number of visits on quizzes
- conclusion_dur, conclusion_vis: relative time spent and number of visits on conclusions

- solved_dur, solved_vis: relative time spent and number of visits on solved exercises
- quiz_review_dur, quiz_review_vis: relative time spent and number of visits on reviewing quizzes' results
- open_dur, open_vis: relative time spent and number of visits on open-ended questions

After taking under consideration the literature regarding the FSLSM [9] and other related works [8, 14], relevant patterns for identifying learning style for each dimension of FSLSM were found. The aforementioned findings are summarized in Table 1. The "+" and "−" indicate a high or low occurrence of the particular behavior from a student with the respective learning style. Taking the example of active learners, the findings in Table 1 indicate that such learners don't spend a lot of time to read content objects while, the opposite behavior is expected from the reflective learners.

Table 1. Relevant patterns for each dimension of FSLSM

Pattern	Active	Reflective	Sensing	Intuitive	Visual	Verbal	Sequential	Global
content_dur/vis	−	+	−	+	−	+		
outline_dur/vis	−	+					−	+
solved_dur/vis	−	+	+	−			+	−
video_dur/vis					+	−		
quiz_dur/vis	+	−	+	−				
quiz_review_dur/vis	−	+						
open_dur/vis	+	−	+	−	−	+		
conclusion_dur/vis	−	+	+				−	+

As can be seen in Table 1, each learning style dimension consists of a relatively low number of patterns, compared to the number of patterns of related works [14, 17]. A greater number of patterns is expected to signify a higher precision in the automatic detection of learning styles, given that a user model with many variables implies describing the students' behavior in more detail [18]. However, in the works mentioned above the patterns proposed demanded specialized tracking mechanisms that cannot be applied to every LMS. Our proposed method aims to adopt a significantly simpler user model that is generic and can be applied to more LMS.

Data Preparation. The second step of the proposed approach for automatic detection of learning styles is to prepare the data that are going to be used as input in the decision tree algorithm. Data relevant to each one of the patterns presented in Table 1, are extracted from Moodle's database. In order for the data to be ready as input in a decision tree algorithm, firstly a new table has to be created consisting of a different row for each student. Each one of these rows stores cumulated information extracted from the aforementioned data as well as the student's learning preference for each dimension as detected from answering ILS. After the new table's construction is completed, data regarding time and number of visits have to be transformed from absolute to relative values. Next, a different csv file is created from the specific table for each dimension of FSLSM. Each one of these files is created by extracting from the table only the patterns that are relevant for the specific dimension, as well as the equivalent learning style preference. Subsequently, each csv file should be converted into an arff file which is more appropriate for Weka. In order to perform this conversion, Weka is used in the background as described above. Finally, each file is randomly divided into a training and test dataset by using the "RemovePercentage" class which can be found in the "weka.filters.unsupervised.instance" package of Weka.

Application of the Decision Tree Algorithm. The third step of the proposed approach for automatic detection of learning styles is the application of the decision tree algorithm which, in our system, is J48. J48 uses the concept of information entropy to build a decision tree from a set of training data. Four decision trees were created, one for each dimension of FSLSM. Taking the example of the active/reflective dimension, the following lines of code were used in order to apply the J48 algorithm.

```
$train_string = 'java -cp /path/to/weka.jar weka
 .classifiers.meta.FilteredClassifier -F weka.filters
.unsupervised.attribute.RemoveType -W weka.classifiers
.trees.J48 -t '.$pathname.'/train_active.arff -i -T
$pathname.'/test_active.arff -p 1 > '.$pathname.'/ re-
sult_test_active.csv';
$train = system($train_string);
```

"-cp" defines the class path of "weka.jar". The "FilteredClassifier" class is used in order to perform an "on the fly" classification which means that data are passed through a filter before they are used as classifier input. The "RemoveType" filter is used via "-F" in order to remove but not delete users' id from the dataset. The "-W" parameter is used to declare the specific classification algorithm that is used which, in our case is J48 and the relevant class can be found in "weka.classifiers.trees" package. Subsequently, the "-t" and "-T" parameters are used to define the training and the test dataset, respectively. The "–p 1" parameter is used to output the predictions about the attribute which is in the first column of the respective test dataset, which in our case is the learning style preference variable. Prediction is redirected via the ">" symbol and is stored in a new csv file, namely result_test_active.

Update of User model. The last step of the proposed approach is to update the user model. After running the J48 algorithm, the respective predictions for all the students are stored in four new csv files, one for each dimension of the FSLSM. Each csv file consists of one row for each student. In turn, each row consists of the user's id, the predicted and the actual learning style preference in terms of the respective dimension. Thus, each student's learning style is described in a different row in each of the four csv files. In order to update the user model with these results, it was decided to create a new table in Moodle's database. Data are extracted from the resulting csv files and these values are stored in the respective attributes of the new table.

3.3   Adaptive Course

The mechanism that is proposed to adapt a Moodle course to a student's learning style and knowledge is based on a framework that was thoroughly presented in our previous work [15]. According to this framework, the sequence of a course's learning objects can be adapted to students' learning styles. To achieve this, an adaptation matrix is used in order to calculate a ranking score for each learning object. These scores determine their positioning within the respective section of the course. In addition, the presentation of the course can be adapted to students' knowledge in terms of the proposed method for progress calculation.

In the present paper, we should only exemplify how a course can differ among learners with different learning styles regarding the position of the learning objects in the particular course. Taking the example of visual/verbal learners, Fig. 2 illustrates the proposed sequence of the resources. Visual learners, who learn best from visual representations of the underlying concepts, would appreciate videos, which should be presented both before and after the content. The video before the content provides the stimulus for the student to become actively involved in the section. Verbal learners learn best from text. Solved exercises, which are more suitable for this type, should be presented in both the area before and after content.

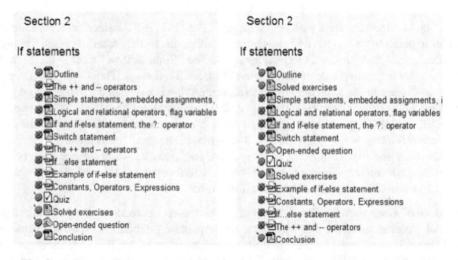

Fig. 2. Proposed sequence of educational resources for a visual and a verbal learner

4   Evaluation Study

4.1   Methodology

In order to assess the precision of our approach regarding the automatic detection of learning styles in Moodle, as well as the effectiveness of the proposed adaptive system, an evaluation study was conducted during the winter semester of the 2016/17 academic year in the context of the Procedural Programming introductory course taught in our department. Course's total length is 13 weeks and it consists of a 2-hour weekly lecture and a 2-hour weekly laboratory where students practice and solve a problem. The students are required to take a mid-term exam worth 30% of their final grade. The study was conducted over the first six weeks of the course, up to the mid-term exam. During this time, students were presented with five sections on fundamental concepts of procedural programming.

Two groups based on students' preferences were formed, the experimental and the control. Having been informed about the differences between each group, students were assigned to a group based on their student number. The random formation of the two groups is a threat to the validity of the study. In order to mitigate this threat, we investigated the two groups' average grades for another similar course which is also taught in the first semester. None statistically significant difference was found and, thus, it can be assumed that the two groups were equally engaged in their studies and any differences in the study results were probably due to the proposed mechanism.

Both groups attended the same lectures and laboratories. In order to evaluate the proposed system, two different courses using the same learning material were created in Moodle. The experimental group used the first course where the adaptive version described here was implemented. The control group used the second course where the standard version of Moodle was applied. At this point, it should be mentioned that the results and the analysis regarding learning style detection are relevant only for the experimental group.

Overall, 96 students participated in the study and took the mid-term exam. They were equally assigned to each group. On completion of the 5 sections of the respective Moodle course but prior to the mid-term exam, both groups of students had to answer a questionnaire evaluating the system usability and motivational appeal. The questionnaire consisted of five-point Likert type questions, ranging from 1 'strongly disagree' to 5 'strongly agree'. The efficiency of the proposed mechanism was examined using student performance on the mid-term exam.

For measuring the precision of the proposed approach for automatic detection of learning styles, the following formula (1) proposed in [13] is used.

$$Precision = \frac{\sum_{i=1}^{n} Sim\left(LS_{predicted}, LS_{ILS}\right)}{n} \cdot 100 \tag{1}$$

In (1): n is the number of students. The function Sim compares the predicted learning styles ($LS_{predicted}$) with those that are detected by the ILS questionnaire (LS_{ILS}), and returns 1 if both parameters are equal, 0.5 if one represents a balanced learning style and the other represents a preference for one of the two poles of the dimension, and 0 if they are opposites.

4.2   Results

As can be seen from the results in Table 2, students rated the system usability highly overall. Slight differences exist between the mean values of the two groups for some questions. None of these differences was statistically significant though. Consequently, the system usability is not affected when the adaptive mechanism is embedded in Moodle, which was one of our main goals when designing the system.

Table 2. Student feedback on system usability

Pattern	Experimental group		Control group	
	M	SD	M	SD
Pages loaded fast	4.3	0.739	4.1	0.872
Navigation was easy	4.2	0.869	4.3	0.742
System was user-friendly	3.9	0.957	3.9	0.958
Links were definite	4.5	0.757	4.2	0.915
System was considered adequate for novices	4.1	0.900	4.1	0.913
Easily familiarized myself with the system	4.4	0.957	4.3	0.857

The findings regarding system motivational appeal are presented in Table 3. As can be seen from these results, all students were generally satisfied with the use of Moodle and the quality of the educational resources. The findings regarding the questions on whether students were motivated to study more and found Moodle helpful in making learning easier showed that the experimental group gave higher means than the control. Both differences were found to be statistically significant ($p < 0.05$).

Table 3. Student feedback on the system's motivational appeal

Pattern	Experimental group		Control group	
	M	SD	M	SD
System motivated me to study more	4.4	0.777	4.1	0.958
System helped me to learn easier	4.4	0.618	4.1	0.813
I'm generally satisfied from system usage	4.1	0.852	4.0	0.760
Quality of educational resources	4.1	0.543	4.1	0.621

The implementation of the proposed technique also improved students' grades on their mid-term exam. The experimental group performed better on the mid-term exam in comparison to the control group, with average grades of 18.13 and 15.13 out of 30, respectively. A two-tailed Mann-Whitney U test revealed that this difference was statistically significant ($p = 0.011 < 0.05$). Therefore, the conclusion can be reached that the proposed mechanism did contribute to enhancing student performance.

Table 4 presents the findings on the precision that was attained by different automatic approaches including the proposed. Regarding the effectiveness of the proposed approach for automatic detection of learning styles, the precision attained was 86%, 80%, 75% and 80% for the four dimensions of FSLSM, respectively. The aforementioned results are quite promising and indicate that the proposed approach is suitable for detecting students' learning styles in regards to all the dimensions of FSLSM. As can be seen in Table 4, our approach attained higher precision in all dimensions with the exception of the visual/verbal dimension. Although Dung & Florea [8] attained higher precision regarding the visual/verbal dimension compared to our approach, the overall precision of their approach is lower.

Table 4. Precision attained by different approaches

Approach	Active/Reflective	Sensing/Intuitive	Visual/Verbal	Sequential/Global
Data-driven approaches				
Proposed	86%	80%	75%	80%
[13]	58%	77%	–	63%
[14]	62.5%	65%	68.75%	66.25%
Literature based approaches				
[8]	72.73%	70.15%	79.54%	65.91%
[14]	79.33%	77.33%	76.67%	73.33%
[18]	65%	75%	76.25%	77.5%

5 Conclusions

Although learning styles have been subjected to some criticism over the last years, the adaptation of educational material to suit students' learning styles is proven to be beneficial for them. Many systems have been proposed to adapt the educational material and several approaches have been applied for the automatic detection of learning styles. However, the integration of learning styles and adaptive learning system still requires further researches and experiments [20].

Trying to contribute to research, we propose an adaptive mechanism that can be implemented in Moodle based on a data driven approach for the automatic detection of learning styles. The proposed approach differs from other works in that it calculates students' learning styles by considering not only their responses in the ILS but also their interaction with Moodle in order to surmount the disadvantages of each approach. The added value of the present work is that it uses Weka in the background in order to enhance system flexibility. Although the results of the study were quite positive, further research is needed to validate the attained precision of the automatic detection approach, mainly because data-driven approaches strictly depend on the available data in order to build an accurate classifier [8]. Our future work will focus on the inclusion of new types of learning objects that promote collaborative learning.

References

1. Akbulut, Y., Cardak, C.S.: Adaptive educational hypermedia accommodating learning styles: A content analysis of publications from 2000 to 2011. Comput. Educ. **58**, 835–842 (2012)
2. Bachari, E., Abelwahed, H., Adnani, M.: E-learning personalization based on dynamic learners' preference. Int. J. Comput. Sci. Inf. Technol. **3**(3), 200–216 (2011)

3. Bernard, J., Chang, T.W., Popescu, E., Graf, S.: Learning style identifier: Improving the precision of learning style identification through computational intelligence algorithms. Expert Syst. Appl. **75**, 94–108 (2017)
4. Brusilovsky, P.: Adaptive hypermedia. User Model. User Adap. Interact. **11**, 87–110 (2001)
5. Cabada, R.Z., Estrada, M.L.B., García, C.A.R.: EDUCA: a web 2.0 authoring tool for developing adaptive and intelligent tutoring systems using a Kohonen network. Expert Syst. Appl. **38**(8), 9522–9529 (2011)
6. Crockett, K., Latham, A., Whitton, N.: On predicting learning styles in conversational intelligent tutoring systems using fuzzy decision trees. Int. J. Hum Comput Stud. **97**, 98–115 (2017)
7. Dorça, F.A., Lima, L.V., Fernandes, M.A., Lopes, C.R.: Comparing strategies for modeling students learning styles through reinforcement learning in adaptive and intelligent educational systems: An experimental analysis. Expert Syst. Appl. **40**(6), 2092–2101 (2013)
8. Dung, P.Q., Florea, A.M.: An approach for detecting learning styles in learning management systems based on learners' behaviours. In: International Conference on Education and Management Innovation, pp. 171–177. IACSIT Press, Singapore (2012)
9. Felder, R.M., Silverman, L.K.: Learning and teaching styles in engineering education. Eng. Educ. **78**(7), 674–681 (1988)
10. Felder, R.M., Soloman, B.A.: Index of Learning Styles Questionnaire. http://www.engr. ncsu.edu/learningstyles/ilsweb.html
11. Felder, R.M., Spurlin, J.: Applications, reliability and validity of the index of learning styles. Int. J. Eng. Educ. **21**(1), 103–112 (2005)
12. Feldman, J., Monteserin, A., Amandi, A.: Automatic detection of learning styles: state of the art. Artif. Intell. Rev. **44**(2), 157–186 (2015)
13. García, P., Amandi, A., Schiaffino, S.N., Campo, M.R.: Evaluating Bayesian networks' precision for detecting students' learning styles. Comput. Educ. **49**(3), 794–808 (2007)
14. Graf, S.: Adaptivity in learning management systems focusing on learning styles. Ph.D. dissertation, Faculty of Informatics, Vienna University of Technology, Austria (2007)
15. Karagiannis, I., Satratzemi, M.: Enhancing adaptivity in moodle: framework and evaluation study. In: Auer, M., Guralnick, D., Uhomoibhi, J. (eds.) Interactive Collaborative Learning ICL 2016. Advances in Intelligent Systems and Computing, vol. 545, pp. 575–589. Springer, Cham (2017)
16. Kirschner, P.A.: Stop propagating the learning styles myth. Comput. Educ. **106**, 166–171 (2017)
17. Liyanage, M.P.P., Gunawardena, K.S.L., Hirakawa, M.: Using learning styles to enhance learning management systems. ICTer **7**(2), 1–10 (2014)
18. Popescu, E., Badica, C.: Creating a personalized artificial intelligence course: WELSA case study. Int. J. Inf. Syst. Soc. Change **2**(1), 31–47 (2011)
19. Thalmann, S.: Adaptation criteria for the personalised delivery of learning materials: a multi-stage empirical investigation. Australas. J. Educ. Technol. **30**(1), 45–60 (2014)
20. Truong, H.M.: Integrating learning styles and adaptive e-learning system: current developments, problems and opportunities. Comput. Hum. Behav. **55**, 1185–1193 (2016)
21. Yang, T.C., Hwang, G.J., Yang, S.J.H.: Development of an adaptive learning system with multiple perspectives based on students' learning styles and cognitive styles. J. Educ. Technol. Soc. **16**(4), 185–200 (2013)

Ontology-Based Database for Chemical Experiments: Design and Implementation

Baboucar Diatta[1]([⊠]), Adrien Basse[1], and Samuel Ouya[2]

[1] Department TIC University Alioune Diop Bambey, Bambey, Senegal
{baboucar.diatta,Adrien.basse}@uadb.edu.sn
[2] Department Computer Engineering,
University Cheikh Anta Diop Dakar Sénégal, Dakar, Senegal
Samuel.ouya@gmail.com

Abstract. Virtual laboratory teaching helps develop and acquire skills. Chemistry ontologies address chemical elements and their interaction. We seek to implement ChemOnto, ontology that makes it easy to develop chemical virtual laboratories that will help learners get acquainted with hands-on tools and procedures. We will study existing ontologies on chemical elements by searching, prioritizing, and linking key terms related to experimentation and safety rules. We will implement an ontology describing chemical experiments with the inception of an interface for database usage, safety rules, and skills involved. We will feed our practicum and safety rule database. Anticipated results are: Safety rules are known and skills involved are acquired. For each practicum, our ontology proposes a set of skills to be acquired over time along with various actions to complete them.

Keywords: Ontology · Virtual laboratory · Lab work · Chemistry

1 Introduction

In science education, experimental activities are of paramount importance because they help learners to build their own knowledge. Works like [10, 15] also indicate that students feel more motivated to learn chemistry when they are doing practical classes. However, it is not guaranteed that laboratories are always available to carry out practical activities. The integration of virtual practical experiences into a chemistry course helps students perform well in a practical laboratory and can contribute to improving learning skills [14]. Virtual laboratories allow for several prior manipulations before entering a physical laboratory. such virtual manipulations help learners to become familiar with the material and master the procedures to follow in carrying out practical work [4]. They reinforce the know-how and make it possible to reduce learners' time of presence in laboratories; reduce product losses and pollution that it causes.

The representation of knowledge in a virtual laboratory has always been a concern that has become more important in recent years [2]. Learners need information about the identifiable elements of a virtual laboratory in order to adapt their behavior and make coherent choice of actions [16] for the realization of a virtual manipulation. To

© Springer Nature Switzerland AG 2020
M. E. Auer and T. Tsiatsos (Eds.): ICL 2018, AISC 916, pp. 389–397, 2020.
https://doi.org/10.1007/978-3-030-11932-4_37

help learners clearly identify the elements of a virtual laboratory with which they will interact in order to achieve a set of consistent tasks, we have opted for ontology. Our aim in this article is to use an ontology to describe the manipulations to be performed, the products and materials used as well as the safety instructions to be observed.

Section 2 contains a revision of the literature on ontologies. Section 3 describes the ontology we have proposed. Section 4 shows an example of how to use this ontology and Sect. 5 is the conclusion.

2 Related Work

The ontologies represent a huge opportunity and bring great benefits to e-learning systems. Their implementation is seen as a better solution for organizing and visualizing didactic knowledge, and for this knowledge to be shared and reused by different educational applications [5]. In this review of literature, we will start with the ontologies used for virtual labs, then the ontologies used in the domain of chemistry. For the ontologies of virtual Labs [11] provide a platform architecture for e-learning that is an online learning management system with metadata. This system is an ontology for the e-learning process, such as teaching methods, learning styles, learning activities and the curriculum. This system helps students, administrative staff and teachers set up and manage the course data and to review the learning content. This system of architecture will be able to gain in adaptability, scalability of performances and in reutilization of concepts. [12] presents an ontology of a formative evaluation, called Onto Note F, which allows teachers to add formal semantics to the formative feedback annotations that they put on students' laboratory reports in the lab Book system. [1] presents a design of an accessible learning object model by using an ontological approach that will define the structures of semantic content and relationships between the learning objects.

In the field of chemistry, most ontologies deal with the structure, characteristics of chemical elements or the annotation of chemical products applied to biology. [9] analyses the different categories of structural classes in chemistry, presenting a list of models for the found characteristics in the class definitions. Then, compare these models of class definition to tools that allow the automation of hierarchical construction within the ontology. In [6], the chemical ontology applied to the biology, chebi (chemical entities of biological Interest) provides a classification of chemical entities such as atoms, molecules and ions. The Chebi ontology contains about 15000 chemical entities and regroups them based on common structural characteristics. Chebi is widely used as a database of chemical entities that can be queried by functional annotations in the ontology of roles. Other work, such as [19], is a first step towards the design of an ontology for classifying chemical compounds into functional groups. The work presented in [19] represents a preliminary step towards the description, the reasoning and the interrogation of the structure and the function of the molecules. [7] proposes an ontology that will serve as a powerful tool for the search for chemical databases and the identification of the key functional groups responsible for biological activities.

This article does not focus on the structure of chemical elements such as atoms, molecules and ions. It presents an ontology to describe the manipulations to be achieved, the products and materials used and the safety instructions to be observed.

3 Virtual Lab Ontology

In this paper, we have developed ontology to describe the major steps of an experiment, products and materials used, as well as safety instructions required for chemical laboratories. After defining the ontology objectives, we present the most representative methodologies used in ontology. The several methodologies that have been developed in different construction steps include [3, 8, 13, 17, 18]. We choose Noy and McGuinness [13] method to build our ontology because it is better suited to the context of this work. From Noy and McGuinness methodology we choose the three following step:

- Enumerate important terms of the ontology: The definition of important terms has been possible thanks to the checking of documents such as a book of the first year chemistry lab work and scientific articles such as [7, 19]. All those terms have been validated by an expert of the domain that is responsible for the first year chemistry course of our university. This phase highlights key terms used in the proposed system.
- Define classes and class hierarchy as well as properties of classes and relations: Organization of terms by using classes, relations and properties according to hierarchical organization principles. Free, open source ontology editor and framework Protégé[1] is used from this step.
- Create instances: Populated with individual instance class of the ontology define in step 2.

Our goal is to build our ontology by following the abovementioned steps.

3.1 Enumerate Important Terms

In this part of paper, we will proceed to define the key concepts used. That will make it possible to describe the different components of a chemistry laboratory, the safety instructions and the step of experience. Our first source for collecting knowledge is a book of the first year chemistry lab work of our university. We also take into account scientific articles such as [7, 19]. A part of terms used in this step is shown in the Fig. 1.

---

[1] https://protege.stanford.edu/.

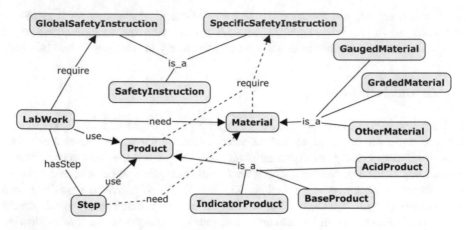

Fig. 1. Part of important terms

3.2   Classes, Properties and Relations

This section illustrates a brief description of some classes, hierarchy of class (see Fig. 2), data properties defined for each class, objet properties and relations among elements. Figure 3 represents a part of properties and provides a brief description of data and objet properties.

– *Labwork*: Define lab work
– *Material*: This class defines material needed for a lab work. This class contains three subclasses: *GradedMaterial, GaugedMaterial* and *OtherMaterial*. In the *GradedMaterial* subclass we have a list of graded material, *GaugedMaterial* contains chemical gauged material and in *OtherMaterial* we have all chemical that is not graded or gauged.
– *Product*: List product we will use for our lab work. It contains four subclasses: *Acid, Base, Indicator,* and *OtherProduct*.
– *Step*: This class shows a list of tasks required for every lab work. Each task is formed by pratical activities.
– *SafetyInstruction*: Describe the safety instructions. It contains two subclasses : *GlobalSafetyInstruction* and *SpecificSafetyInstruction. GlobalSafetyInstruction* contains safety instruction for all activities in the laboratory and *SpecificSafetyInstruction* regroups the instructions associated whith specific material or product

Part of Data and Object properties.

– *title*: this objet property represents the title of lab work;
– *goal*: It represents the goal;
– *theory*: This objet property represent the theory;

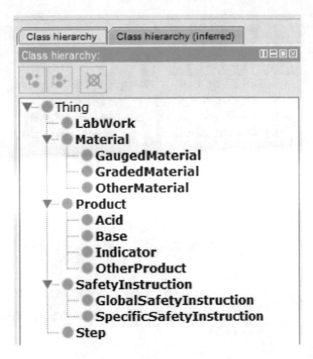

Fig. 2. Class hierarchy of the ontology

- *name*: It represents a name of each element of laboratory;

- *description*: It gives a small description of each element of laboratory;

- *equation*: This object property represents the equation of mixture;

- *formula*: This objet property represents the chemical formula of product

- *phValue*: It represents the ph value of the mixture

- *need*: Represents a relation between *LabWork* and *Material* or between *Step* and *Material*

- *use* : Is a link between *LabWork* and *Product* or between *Step* and *Material*

- *isUsedBy*: Describes inverse of objet property *use*

Instances.

We have created many instances of lab work. Figure 4 shows a part of created instances. The instances that we have created include:

- Acidimetrie and Alcalimetrie: Are instances of Labwork
- Burette: Is an element of class Material
- NAOH: Represents an instance of Product

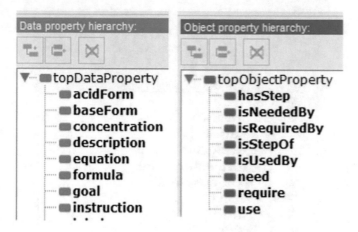

Fig. 3. Part of properties

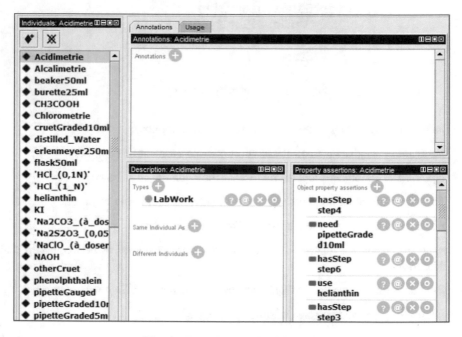

Fig. 4. Part of Individual lab work

4   Use Case

Virtual chemistry lab is a working environment where learners can work in a virtual environment similar to real practice lab. In this section of our paper, we describe the application of ontology that we use. Our goal is to illustrate how to exploit this ontology in order to achieve a better organization of chemical virtual lab work. We use Fuseki[2] server and zend framework[3] to implement our interface. Figure 5 shows a list of virtual lab for chemical experience. For each lab work we show the title, the goal and the theory. Other properties are available through a button.

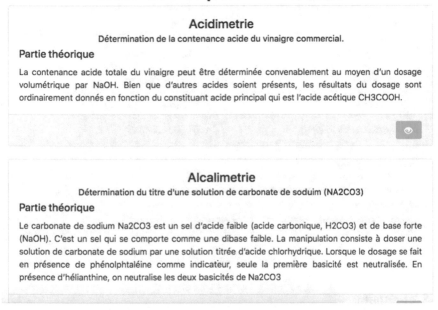

Fig. 5. List of lab works

The system provides information such as title of lab work, goal, description and a list of tasks required for lab work. Figure 6 provides a lab work named "Acidimetrie" with list of materials, products and tasks required.

---

2 https://jena.apache.org/documentation/fuseki2/.

3 https://framework.zend.com/.

TP Acidimetrie

Détermination de la contenance acide du vinaigre commercial.

Partie théorique

La contenance acide totale du vinaigre peut être déterminée convenablement au moyen d'un dosage volumétrique par NaOH. Bien que d'autres acides soient présents, les résultats du dosage sont ordinairement donnés en fonction du constituant acide principal qui est l'acide acétique CH3COOH.

Partie expérimentale

Produits et matériels requis

Produits	Matériels
Phénolphtaléine	1 Burette de 25 ml
Eau distillée	3 Becher de 50 ml
Héliantine	1 Erlenmeyer de 250 ml
Solution de NAOH (0.1N)	1 Pipette de 10 ml

Mode opératoire

Faire bouillir dès le début 100 ml d'eau distillée et laisser refroidir
Pipeter 5 ml de vinaigre commercial que l'on verse dans une fiole de 50 ml

Fig. 6. "Acidimetrie" lab work

5   Conclusion

Technology can be used to enhance the field of teaching. Much research has been done in the field of virtual laboratories to enable practical activities that complement theoretical activities. The proposed ontology will play a significant role in the representation of different concepts of a laboratory. This ontology aims to facilitate development of chemical virtual lab, which will help learners get acquainted with hands-on tools and procedures. Our ontology proposes for each lab work, the skills that will be acquired at term, the safety instructions to be respected, as well as the various actions to carry out a manipulation. The ontologies give an explicit definition of conceptualization on a specified domain of knowledge. This approach blazes the trail for further exploration. Based on the current approach, our future work seeks to study the sequencing of lab work in the form of training paths with support for the learner's progress.

References

1. Abdellaoui, H., Mohamed, M.A.B., Bacha, K., Zrigui, M.: Ontology based description of an accessible learning object. In: 2013 Fourth International Conference on Information and Communication Technology and Accessibility (ICTA), pp. 1–5. IEEE (2013)
2. Aylett, R., Cavazza, M.: Intelligent virtual environments-a state-of-the-art report. In: Eurographics Conference, Manchester, UK (2001)
3. Bachimont, B., Isaac, A., Troncy, R.: Semantic commitment for designing ontologies: a proposal. In: International Conference on Knowledge Engineering and Knowledge Management, pp. 114–121. Springer, Berlin, Heidelberg (2002)
4. Dalgamo, B.: The potential of virtual laboratories for distance science education teaching: reflections from the initial development and evaluation of a virtual chemistry laboratory In: Dalgarno, B., Bishop, A., Bedgood, D. (eds.) Proceedings of the Improving Learning Outcomes Through Flexible Science Teaching, Symposium.-The University of Sydney, October 3, p. 90 (2003)
5. Dalipi, F., Idrizi, F., Rufati, E., Asani, F.: On integration of ontologies into e-learning systems. In: 2014 Sixth International Conference on Computational Intelligence, Communication Systems and Networks (CICSyN), pp. 149–152. IEEE (2014)
6. De Matos, P., et al.: Chemical entities of biological interest: an update. Nucleic Acid. Res. **38** (suppl_1), D249–D254 (2009)
7. Feldman, H.J., Dumontier, M., Ling, S., Haider, N., Hogue, C.W.: CO: A chemical ontology for identification of functional groups and semantic comparison of small molecules. FEBS Lett. **579**(21), 4685–4691 (2005)
8. Fernández-López, M., Gómez-Pérez, A., Juristo, N.: Methontology: from ontological art towards ontological engineering (1997)
9. Hastings, J., Magka, D., Batchelor, C., Duan, L., Stevens, R., Ennis, M., Steinbeck, C.: Structure-based classification and ontology in chemistry. J. Cheminformatics **4**(1), 8 (2012)
10. Josephsen, J., Kristensen, A.K.: Simulation of laboratory assignments to support students' learning of introductory inorganic chemistry. Chem. Educ. Res. Pract. **7**(4), 266–279 (2006)
11. Kaur, P., Sharma, P., Vohra, N.: An ontology based E-learning system. Int. J. Grid Distrib. Comput. **8**(5), 273–278 (2015)
12. Mokeddem, H., Desmoulins, C., Rachid, C.: A formative assessment ontology for students' lab reports in lab book. In: 2015 IEEE 15th International Conference on Advanced Learning Technologies (ICALT), pp. 253–255. IEEE (2015)
13. Noy, N.F., McGuinness, D.L.: Ontology development 101: a guide to creating your first ontology (2001)
14. Ramos, S., Pimentel, E.P., Maria das Graças, B.M., Botelho, W.T.: Hands-on and virtual laboratories to undergraduate chemistry education: toward a pedagogical integration. In: Frontiers in Education Conference (FIE), 2016 IEEE, pp. 1–8. IEEE (2016)
15. Temel, H., Oral, B., Avanoglu, Y.: Kimya ogrencilerinin deneye yonelik tutumlari ile titrimetri deneylerini planlama ve uygulamaya iliskin bilgi ve becerileri arasındaki İliskinin degerlendirilmesi. Cagdas Egitim Dergisi **264**, 32–38 (2000)
16. Tutenel, T., Bidarra, R., Smelik, R.M., Kraker, K.J.D.: The role of semantics in games and simulations. Comput. Entertain. (CIE) **6**(4), 57 (2008)
17. Uschold, M., King, M.: Towards a methodology for building ontologies (1995)
18. Uschold, M., Gruninger, M.: Ontologies: Principles, methods and applications. Knowl. Eng. Rev. **11**(2), 93–136 (1996)
19. Villanueva-Rosales, N., Dumontier, M.: Describing chemical functional groups in OWL-DL for the classification of chemical compounds. In: OWLED, Vol. 258 (2007)

Proposal of a Dynamic Access Control Model Based on Roles and Delegation for Intelligent Systems Using Realm

Jeanne Roux Ngo Bilong$^{(\boxtimes)}$, Cheikhane Seyed, Gervais Mendy,
Samuel Ouya, and Ibrahima Gaye

Cheikh Anta Diop University, Dakar, Senegal
{Jeanneroux.ngobilong, gervais.mendy}@ucad.edu.sn,
{ch.hamod, samuel.ouya}@gmail.com,
{Jeanneroux.ngobilong

Abstract. Delegation is an element of administration that remains important in access control systems. Although widely used, delegation is very little taken into account in security policies because of its complexity. The models proposed so far are extensions of the RBAC model. Role-based access controls documentation does not reveals sufficient studies of delegation requirements for role and task. To address this problem, we propose a hybrid model called Role and delegation Based Dynamic Access Control (RDBDAC), which dynamically manage user role updates and task delegation, taking into account parameters such as the level of trust and temporal context. We show that our approach is flexible and sufficient to handle all delegation requirements. For a better expressivity of our model, we use non-monotonic logic T-JClassicδε which make it possible to specify non-monotonic authorizations and a better representation of the temporal aspects specific to a given delegation. For the model application, we used Realm, a role-based access controls management tool. However, it has some shortcomings for information system administrators in terms of dynamically updating roles assigned to different actors. To solve this problem, we interfaced a middleware between the Realm tool and the users, to facilitate the management of the update of the roles on a virtual university platform.

Keywords: Access control · Role · Delegation · Trust level · Realm

1 Introduction

Access controls are of paramount importance in the management of smart structures involving the health, financial and educational sectors [1–3]. In addition, environments further integrate various miniaturized devices as well as mobile communication technology. This allows services to be deployed anywhere, anytime and for anyone. This evolution imposes new security requirements and challenges on these dynamic, intelligent-context sensitive environments [1]. The access control models such as MAC (Mandatory Access Control), DAC (Discretionary Access Control), RBAC (Role Based Access Control) and TMAC proposed so far do not take into account the

© Springer Nature Switzerland AG 2020
M. E. Auer and T. Tsiatsos (Eds.): ICL 2018, AISC 916, pp. 398–409, 2020.
https://doi.org/10.1007/978-3-030-11932-4_38

dynamic aspect of access controls [4], nor the management of obligations, recommendations, permissions, prohibitions, nor the specific rules of the organization. These are static-access control models. In an effort to improve access control policies in dynamic and intelligent environments, researchers have worked on dynamic access control models such as OrBAC, GeoRBAC (Geographic Role Based Access Control), CRBAC (Context Role Based Access Control), AdOrBAC, PolyOrBAC [1]. Each of these models represents an extension of RBAC, but is not entirely satisfactory because they do not all at once support dynamic role administration, task and trust delegation, and temporal context. In our work, we will implement a hybrid model that will manage dynamic delegations of roles and tasks taking into account the trust level and context. The rest of our work is organized as follows. Section 2 presents the state-of-the-art access control models. Section 3 deals with the description of the proposed model. In Sect. 4, we implement the model. Section 5 concludes our paper with an opening for future work.

2 Access Control Model

2.1   Static Access Control

2.1.1   Discretionary Access Control
Discretionary Access Control (DAC) policies are based on the concepts of subjects, objects and access rights. Access rights to each piece of information are manipulated by the information owner. This access control model is flexible because a subject with access rights can grant access rights to any other user. The granting or revocation of privileges is regularized by a decentralized administrative policy [5].

Limits: difficulty of administration and limitation of the access to the objects according to the identity of the user.

2.1.2   Mandatory Access Control (MAC)
The MAC model has a security policy which is set and managed by an authority, and cannot be modified by users. This excludes problems related to information leaks (using Trojans) observed in the DAC model. This is mainly due to not allowing users to interfere with the access control policy [5]. Unlike discretionary access control policies, subjects of a mandatory access control policy do not own the information which they have access to. Moreover, the operation allowing the delegation of rights is controlled by the rules of the policy. Subjects no longer have control over the information they handle. The subject has access to information only if authorized by the system [1].

Limits: Vulnerable to hidden channels, does not taking into account the administration component in role management, does not take into account delegation issues and level of trust.

2.1.3   Role-Base Access Control (RBAC)
The role-based access control model, or RBAC, is seen as an alternative approach to mandatory access control (MAC) and discretionary access control (DAC). Its security

policy does not apply directly to users [2]. The RBAC model is centered on the role. The latter represents in an abstract way a function or a profession within an organization, which associates the authority and responsibility entrusted to a person who plays this role (for example, Professor, Director, Engineer, Technician…). Each role is assigned permissions (or privileges), which are a set of rights corresponding to the tasks that can be performed by that role. A role can have multiple permissions, and a permission can be associated with multiple roles. Just as a subject can have several roles, a role can be performed by several subjects [5].

Limits: No role delegation and the trust level is not taken into account.

2.2 Dynamic Access Control

Dynamic access control models are generally extensions of the RBAC model. In this section, we will focus on the context-sensitive access control models. Organization based access control (OrBAC).

2.2.1 OrBAC Model

In any organization, the administrator is responsible for managing each user's access to a resource, applying security rules. But managing access rights becomes complex as the number of users, resources and activities increases. In this context, the OrBAC model solves this problem by creating abstract entities (Role, View, activity) separated into concrete entities (Subject, Object, Action). The objective of this separation is to apply the security rules to abstract entities, and to each such entity, a concrete entity is associated. OrBAC defines four types of safety rules: Permission, obligation, prohibition and recommendation [6, 10].

Limitations: No delegation or level of trust aspect.

2.2.2 TRUSTBAC Model

The Trust-RBAC model is based on users' trustworthiness evaluation. This value is deduced dynamically based on the identified change in user behavior [7]. The main entities of the model are:

- Users: they are assigned to roles after evaluation of their trust level.
- Roles: they are associated with the trust level required for their assignment.
- Permissions: are associated with the trust level required to activate permission to a user [1].

In the virtual classroom context, the user's assignment to the trust level is determined by three factors - the educational skill, the user's past behavior, the user's knowledge (the identifying information presented by the user) and the recommendation provided by others about the user. The system validates all these factors to decide on the trust level. A trust level is a set of real numbers between −1 and +1. A user, at a given time with a particular session, has a confidence level [9].

2.2.3   TRBAC Model

The TRBAC model is obtained by introducing the time factor into the role-based access control model. The activation and deactivation of roles is temporal. For this purpose, a role can be active for a certain period and not active on others, thanks to triggering role. This model is very suitable for applications with strong temporal constraints, such as systems integrating workflows where the notion of time is important [14, 15].

Table 1. Models Synthesis

Models	Advantages	Limits
DAC	• Flexible	• Updating the security policy is complex and costly • Vulnerable to Trojan horses and hidden channels. • Doesn't distinguish users from subjects.
MAC	• rigid • distinguishes users from the subject	• Costly update of the security policy • Doesn't express prohibitions, recommendations or obligations • No information flow between different hierarchical levels • Doesn't consider the temporal context in the definition of security policy
RBAC	• Includes the advantages of DAC and MAC.	• No role delegation • Trust level not taken into account • Doesn't express prohibitions, recommendations or obligations • Doesn't consider the temporal context in the definition of security policy
ORBAC	• Easy update of the security policy • expresses permissions, prohibitions, recommendations and obligations in the form of rules • No conflicts between the different rules.	• Vulnerable to hidden channels • Lack of administration aspect • No role and task delegation • Trust level not taken into account
TRUSTBAC	• Includes the advantages of RBAC • the confidence level validates the assignment of a subject to a role	• Doesn't express prohibitions, recommendations or obligations • No role and task delegation • Does not consider the temporal context in the definition of security policy
TBAC	• Includes the advantages of RBAC • The activation and deactivation of roles are temporal	• Doesn't express prohibitions, recommendations or obligations • No role and task delegation

2.3   Model Comparison Synthesis

We made a comparative study between the most famous models. The benefits and limitations of each of these models are summarized in Table 1. The static access control models, the most advanced of which is RBAC, have a large limitation due to its non-dynamicity. Several models, such as TrustBAC, TRBAC, have been proposed with the aim of partially improving it, but none of them to our knowledge integrates the parameters concerning the level of trust, the delegation and the context. We introduced Dynamic access control models, the most advanced of which is OrBAC. Unfortunately, the latter, despite its dynamic side and its ability to manage permissions, prohibitions, obligations, recommendations, doesn't take into account some important parameters already existing in the RBAC model extensions. The limitations observed in the comparison table justify our choice to propose a dynamic hybrid model that will take into account the strengths of the previous models to make it a reliable, flexible and easy to implement model.

3   Proposed Model RDBDAC

3.1   Description of Proposed Model

The Role and delegation based dynamic access control model (RDBDAC) is a hybrid model composed of several parameters inspired by each model described in Sect. 2 of our work. The central elements of our model are delegation, the level of trust and the temporal context. We describe our model in a virtual university environment where teachers can temporarily delegate tasks to tutors assigned to them by the organization's administrator. We can define the concept of Teacher as a member of staff who formally performs their function by default, with access privileges related to their role. He teaches during working hours. We define the Teacher as follows:

$$\text{Teacher} \equiv \text{Staff\_Member} \sqcap \text{Licence\_assigment} \sqcap \text{Role\_Assigment} \sqcap \delta\text{Permission}$$

$$(1)$$

The guardian concept refers to a staff member who is officially assigned to his default role. He may nevertheless benefit from a delegation of tasks over a given period by the teacher. The definition of guardian is as follows:

$$\text{Tutor} \equiv \text{Staff\_Member} \sqcap \text{Role\_Assigment} \sqcap \delta\text{Permission} \qquad (2)$$

It is not possible to assign the license to the guardian because the guardian does not have the privilege to delegate his tasks, not even to his counterpart.

3.1.1   Assignment of License and Role

- Licensing and role of the teacher :

When assigning a Teacher to a role, the administrator also assigns a delegation license to delegate tasks over a period of time. The assignment is defined as follows:

$$\delta Licence\_Assigment \sqsubseteq Assignee L.assigment \sqcap Licence L.assigment \sqcap$$
$$\delta Privileges L.Action \sqcap Cible L.Objet \tag{3}$$

$$\delta Role\_Assigment \sqsubseteq Assignee L.assigment \sqcap Role L.assigment \sqcap$$
$$\delta Privileges L.Action \sqcap Cible L.Objet \tag{4}$$

- Role assignment for the tutor:

When the administrator assigns tutors to teachers, he sets the value referring to the tutor's trust level. The teacher will only have a view of the tutors whose confidence level is valid.

$$\delta Role\_Assigment \sqsubseteq Tutor R.assigment \sqcap Role R.assigment \sqcap$$
$$Trust\_level R.assigment \sqcap \delta Privileges R.Action \sqcap Cible R.Objet \tag{5}$$

3.1.2   Task Delegation

There are two types of delegation views: the license delegation view and the role delegation view. The license delegation view gives access to the Partial delegation. It gives to the assignee the possibility of delegating tasks to the beneficiary for a specified period. The trust level constraint must be valid for delegation. Example: In the case of unavailability, the guardian inherits the tasks delegated to him by the teacher. The tutor can then teach, evaluate learners and modify other authorized objects. These tasks will apply the exception on the tutor. We then represent the partial delegation by the following axiom

$$\varepsilon Permission \sqsubseteq Use L.Licence\_Delegation \sqcap Trust\_level L.Trust \sqcap Grantee L.Grantor \sqcap$$
$$Privilege L.Action \sqcap Target.Object \sqcap Duration L.Time$$

$$\tag{6}$$

The role delegation view allows for full delegation of role. This type of delegation is defined by the following axiom:

$$Empower \sqsubseteq Use RD.Role\_Delegation \sqcap Trust\_level RD.Trust \sqcap Assignee RD.Grantor$$
$$\sqcap Assignment RD.Role$$

$$\tag{7}$$

3.1.3   Revocation

Revocation is the process of recovering the license or delegated role [8]. We represent it axiomatically as follows:

$$\delta Permission \sqsubseteq UseL.License\_Delegation \sqcap AssigneeL\_Assignee \sqcap$$
$$PermissionD.GD\_Revoke \sqcap DurationEndL.Licence\_Delegation \tag{8}$$

3.1.4   Illustration with UML Model

The sequence diagram below illustrates the delegation process, from assigning role and licenses to revoking delegation (Fig. 1).

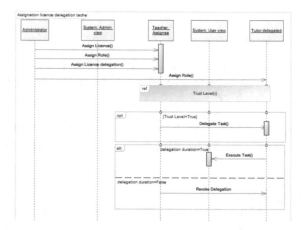

Fig. 1. Sequence diagram

Description of the sequence diagram:

1. License assignment by the administrator to teacher.
2. Role assignment by the administrator to users.
3. Delegation of license to the assignee.
4. If the trust level is true, the assignee delegates one or more tasks to the tutor.
5. If the duration of the delegation = true, then the task delegation is revoked.

3.2   Comparison Between RDBDAC and State-of-the-Art Models

Table 2 shows that our model is more comprehensive, more flexible in terms of defining access control security policies and easier to implement.

Below, the representation of the proposed model (Fig. 2).

Table 2. Comparison between RDBDAC and other models

Criteria of comparison	DAC	MAC	RBAC	TRUSTBAC	TRBAC	ORBAC	RDBDAC
Access control	✓	✓	✓	✓	✓	✓	✓
Contextual Rules	x	x	x	x	✓	✓	✓
Centralized administration	x	✓	x	x	x	x	✓
Trust level	x	x	x	✓	x	x	✓
Dynamic	x	x	x	✓	✓	✓	✓
Delegation/revocation	x	x	x	x	x	x	✓
Permission, recommendation, prohibition, obligation	x	x	x	x	x	✓	✓

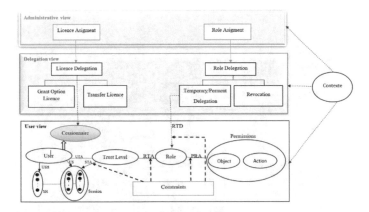

Fig. 2. Proposed model RDBDAC

4 Implementation of the Proposed Model

4.1 Realm Tool

Realm is a Tomcat server tool that integrates a security mechanism based on authentication. It helps to protect resource access to the server, by applying security constraints. Realm's operating principle consists in managing a list of users, with roles. This operating principle is similar to the concept of access control in operating systems with user group management and access privilege. A security constraint will therefore be to define which user or role has access to which protected resource. Realm is a particular element in the Tomcat configuration, as it can be inserted as threads of different containers: Engine, Host or Context. In the internal tree logic of the server, if a Realm is defined for a Context, it will eventually overwrite that of the parent Host. A Realm implements the Java interface org.apache.catalina. Realm. Realms whose goal is to manage users, differ from each other by their management mode (XML file, database) [15].

4.2   Solution Architecture

The architecture of the proposed solution shows the principle of dynamic delegation of tasks from the lecturer called assignee to a tutor. Following the configuration of XML files via Realm, a database is set to define the set of roles, users, permissions via objects and actions to be taken on these objects. A middleware is interfaced to allow different users interact with the database and PLC. When a user connects to the middleware, they are redirected to their view to perform the operations assigned to them. The implementation of this architecture allows to have an intelligent e-learning environment or intelligent classroom. To show the feasibility of our approach, we implemented a platform in the context of the Université Virtuelle du Sénégal. We used the JEE programming language to develop this platform and for better access control management, we used the Realm tool (Fig. 3).

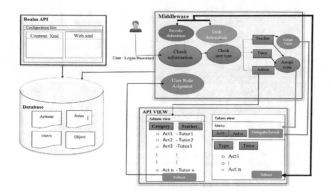

Fig. 3. Solution architecture

Figure 4 shows the home page to administer the entire application, including configuration and access control management. To do this, the administrator dynamically assigns roles or tutors to different users based on their profiles.

Fig. 4. Home page to administer the platform

Figure 5 below is designed to manage the two main access control management functions, such as assigning roles and guardians to teachers. In this figure precisely, the administrator selects the teacher Ahmed Bah and gives him the authorization on the tasks that he will be able to access and he validates.

Fig. 5. Assigning roles to teacher Ahmed Bah

Figures 6 and 7 shows the list of the different tutors registered in the database. Only tutors who have a level of confidence with the mention 'delegateable' are assigned to the teacher. The teacher can then delegate tasks to them for a given period.

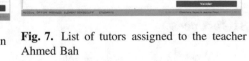

Fig. 6. List of tutors delegateable to be assign to teacher Ahmed Bah

Fig. 7. List of tutors assigned to the teacher Ahmed Bah

Figure 8 shows the teacher's view. He accesses it after being authenticated. Here, teacher Ahmed Bah has the possibility, on the one hand, to perform the seven tasks assigned to him by the administrator and, on the other hand, to delegate or revoke the tasks to the different tutors assigned to him.

Fig. 8. View of Professor Ahmed Bah

5 Conclusion

In this work, we propose a hybrid RDBDAC model for access control in intelligent systems. The model has been designed taking into account the strengths of the RBAC, ORBAC, TRBAC and TRUSTBAC models. The proposed model focuses on the parameters of confidence levels, delegation and temporal context. An administrator view allows you to assign roles to tutors and teachers. Licensing teachers allows exceptions to be managed once the teacher has delegated tasks to the tutor. Revocation of delegation is also dynamic, because the guardian loses the permissions to act on a given object, once the time allocated to the delegation is exhausted. We used the non-monotonic logic T-JClassicδε to correctly express the exceptions and the temporal context. The model has been implemented with Java EE and the Realm tool. The results show that our model is more flexible, more comprehensive and easier to implement than previous models. As our model does not take into account the geo-temporal parameter and the abstract and concrete views present in the OrBAC model, we have in perspective to set up a model that will take into account these elements mentioned above and apply them in the field of e-health. Another perspective will be to compare all existing access control models and propose a standard reference access control model.

References

1. Zerkouk, M.: Modèles de contrôle d'accès dynamiques (Doctoral dissertation, University of Sciences and Technology in Oran) (2015)
2. El Kalam, A.A., et al.: Or-BAC: un modèle de contrôle d'accès basé sur les organisations. Cahiers francophones de la recherche en sécurité de l'information **1**, 30–43 (2003)
3. Bettaz, O., Boustia, N., Mokhtari, A.: Dynamic delegation based on temporal context. Procedia Comput. Sci. **96**, 245–254 (2016)

4. Abakar, M.A.: Etude et mise en oeuvre d'une architecture pour l'authentification et la gestion de documents numériques certifiés: application dans le contexte des services en ligne pour le grand public (Doctoral dissertation, Saint Etienne) (2012)
5. Ennahbaoui, M.: Contributions aux contrôles d'accès dans la sécurité des systèmes d'information (2016)
6. Ghorbel-Talbi, M.B., Cuppens, F., Cuppens-Boulahia, N., Bouhoula, A.: Managing delegation in access control models. In: International Conference on Advanced Computing and Communications, 2007. ADCOM 2007, pp. 744–751. IEEE (2007)
7. Ray, I., Mulamba, D., Ray, I., Han, K.J.: A model for trust-based access control and delegation in mobile clouds. In: IFIP Annual Conference on Data and Applications Security and Privacy, pp. 242–257. Springer, Berlin, Heidelberg (2013)
8. Zhang, L., Ahn, G.J., Chu, B.T.: A rule-based framework for role-based delegation and revocation. ACM Trans. Inf. Syst. Secur. (TISSEC) 6(3), 404–441 (2003)
9. Chakraborty, S., Ray, I.: TrustBAC: integrating trust relationships into the RBAC model for access control in open systems. In: Proceedings of the eleventh ACM symposium on Access control models and technologies, pp. 49–58. ACM (2006)
10. Miege, A.: Definition of a formal framework for specifying security policies. The Or-BAC model and extensions (Doctoral dissertation, Télécom ParisTech) (2005)
11. El Kalam, A.A., Deswarte, Y.: Security model for health care computing and communication systems. In: IFIP International Information Security Conference, pp. 277–288. Springer, Boston, MA (2003)
12. Artale, A., Franconi, E.: A survey of temporal extensions of description logics. Ann. Math. Artif. Intell. 30(1–4), 171–210 (2000)
13. Barka, E., Sandhu, R.: A role-based delegation model and some extensions. In: Proceedings of the 23rd National Information Systems Security Conference, Vol. 4, pp. 49–58 (2000)
14. Bertino, E., Bonatti, P.A., Ferrari, E.: TRBAC: A temporal role-based access control model. ACM Trans. Inf. Syst. Secur. (TISSEC) 4(3), 191–233 (2001)
15. Wiggers, C., et al.: Professional Apache Tomcat. Wiley (2004)

"Software Reconfigurable Hardware" in IoT Student Training

Doru Ursutiu[1(✉)], Cornel Samoila[2], Andrei Neagu[3], Aurelia Florea[4], and Adriana Chiricioiu[5]

[1] Romanian Academy of Scientists AOSR, Transylvania University of Braşov, Braşov, Romania
udoru@unitbv.ro
[2] Technical Sciences Academy ASTR, Transylvania University of Braşov, Braşov, Romania
csam@unitbv.ro
[3] Transilvania University of Brasov, Braşov, Romania
andrei.neagu@student.unitbv.ro
[4] Miele Tehnica, Braşov, Romania
aurelia.florea@miele.com
[5] Transylvania University of Braşov, Braşov, Romania
adriana.chiricioiu@yahoo.com

Abstract. Fast development of IoT technologies require new engineering skills pushing universities towards developing and/or improving their systems and tools used in assisting courses and laboratories goals.

Many companies (Digilent, National Instruments, Cypress, etc.) have started developing new devices related to these requirements, devices necessary to be adopted by universities in their new concepts regarding engineering education.

In this context, one important idea in the development of new systems will be connected with the modern class and ideas of the Software Reconfigurable Hardware. In our development, after a careful analyse of the literature, we selected the use of the new ultra low power PSoC 6 MCU from Cypress Semiconductors as their architecture offers the processing performance needed by IoT devices, eliminating the tradeoffs between power and performance.

Keywords: Reconfigurable hardware · LabVIEW · PSoC Creator

1 Introduction

The actual release of the new Cypress ultra low power PSoC 6 MCU was well connected with the university ideas to develop flexible "software reconfigurable hardware" at low prices as well as easy to be adapted for laboratory needs and creative developments in student training.

Combining this device with the same company Analog-Coprocessor was easy to be done and understood by our students and led us to the necessary knowledge to develop a new reconfigurable platform for IoT development.

The idea of flexible hardware development was permanently in attention of universities [1, 2] but now with the decrease in prices of development kits and generally

© Springer Nature Switzerland AG 2020
M. E. Auer and T. Tsiatsos (Eds.): ICL 2018, AISC 916, pp. 410–416, 2020.
https://doi.org/10.1007/978-3-030-11932-4_39

speaking of MCU's or PROCESORS prices – it must be the main strategy used in adding new technologies, new software in development of students' knowledge and expertise.

The new PSoC 6 MCU offered us a low price and a pertinent solution in the development of new classes of IoT devices with low power, flexibility in programming and built-in security – all actual and necessary requirements in IoT development. In combination with the relative easy to use and understand software configurability and one ultra-low-power 2.7″ E-ink display (with a thermistor, a 6-axis motion sensor and a digital microphone on board) the PSoC 6 BLE PIONEER KIT (CY8CKIT-062-BLE) was selected for preliminary developments [3].

The Cypres Analog-Coprocesor Pioneer Kit (CY8CKIT-048 PSoC) presented in Fig. 1a was selected due to the easy implementation of the sensor interfaces for our students, and one small prototyping board based one same family (CY8C4A45PVI), presented in Fig. 1b was build.

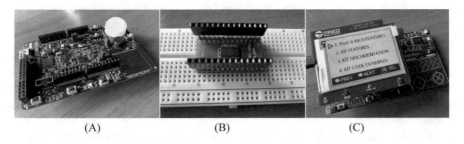

| (A) | (B) | (C) |

Fig. 1. (a) PSoC Analog-Coprocesor, (b) Prototype (c) CY8CKIT-062-BLE PIONEER KIT

PSoC 6 BLE KIT (presented in Fig. 1c) with an Arm® Cortex®-M4 and Arm® Cortex®-M0 +, 1 MB of Flash, 288 KB of SRAM, 78 GPIO, 7 programmable analog blocks, 56 programmable digital blocks, Bluetooth Low Energy (BLE), a serial memory interface, a PDM-PCM digital microphone interface, and industry-leading capacitive-sensing with CapSense™ offered us all the necessary background for our developments.

2 Students' Point of View About the Selected System

The reason why I chose to work with PSoC Analog-Coprocessor Pioneer Kit (CY8CKIT-048 PSoC), as a student, is that it had everything I needed in one place at a convenient price. I started being passionate about this area right before choosing my graduation thesis, and the multiple sensors combined with the easy programming analog coprocessor offered me flexibility and an easy understanding of the system.

In my opinion, working with this kit reunites my major specialization with real working applications, allowing me to go fully practical. I got to build my own prototype, along with the help of my Professor who first introduced me to this area, which will allow users to measure and control different aspects of work environment (temperature, light, humidity) but also many other parts of our lives, such as home safety or health care. This could be a low-cost industrial solution for a proper resource management, maximizing the profit with minimum investment.

It is an important transition for a student from the theory we are studying in college or university to the practicability of real life. Trying to create something ready to be used can be challenging in many aspects.

The final goal is to provide systems for homes and businesses that can detect precise parameters of the environment that will fundamentally change the way we understand and finally consume resources. This will also provide information about the right environment not only for humans but also for the equipment we use to make our lives easier, such as server rooms. An important part of having a server room is providing a good ambient room temperature and humidity level for optimal performance and reliability.

The directions we are following for this project are consisting, firstly, in the understanding of the hardware in its deeper level, then creating an easy and intuitive software which can be used by non-specialized users as well, and then creating our own hardware, starting with temperature, humidity and light sensors that will connect through a universal connector.

The graphical programming is a plus in developing the software, which allows us to simply understand the functionality of the hardware and this way we are able to come with different ideas about how the final user-interface will look like. The main goal of the software is to be as intuitive as possible, allowing the user to choose the needed sensorial module.

As a final part of this project's evolution, everything described above will be transformed in a reliable wireless system in order to offer unlimited flexibility through Bluetooth technology. Moreover, implementing a database for data recording will provide further statistics.

3   Development and Targets

We started to use these ideas (software/hardware) for bachelor and extended to master and Ph.D. students. The continuity offered to the students on increased mobility in creating original systems and obtaining better grades (at Diploma, Dissertations and in future at the Ph.D. Thesis).

Using one reconfigurable system, the investment price decreases, but at the same time the combination graphical programming IDE presented in Fig. 2 (PSoC CREATOR 4.1 for Analog Coprocessor and the new 4.2 for PSoC 6.0) with reconfigurable hardware (like in case of PSoC Analog Coprocessor and PSoC 6 systems) results into increased in quality and variability.

For three years now, we have been working with MIELE Company and Ministry of Education (and a strong team of electronics factories in Romania) in a partnership to implement this system of reconfigurable devices and graphical programming (LabVIEW, Cypress Creator etc.) for children and students training based on a pilot implementation in order to change the curricula in schools and electronics faculties

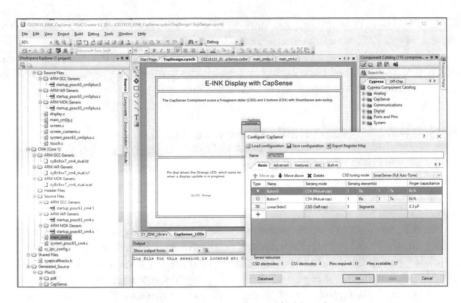

Fig. 2. PSoC Creator 4.2 graphical interface

according with the fast evolution of new technologies in electronics and computer technologies.

The strategic partnership is growing and witnessing industry and academia work together in order to achieve common objectives.

This collaboration is critical as higher performance expectations and demands are being placed on graduates, raising questions about how to adapt the content of academic programs and pedagogical methods to best match the future needs.

University and business in this strategic partnership are aware of the current state of educational problems and share now a common vision and are looking together at the future trends in the electronic industry – contributing to the challenge of changing the training modalities and influencing the curricula.

We believe that the main duty of one academic teacher must be to adapt his teaching methodologies and resources to the actual trends in industrial development and this must be started in school by attracting children to the new technologies.

All the involved parties agreed that the future of electronics can be ensured only by defining a common mission and building a partnership to support its achievement. This strategic partnership is key to building incremental trust and proficiency through a variety of projects and exchanges.

Multiple ideas were generated, working groups were created, an action plan is being developed and the implementation stage will follow. The main directions refer to closing gaps between theory and practice, financing, increasing human resources quality, and consolidating the dialogue and collaboration between academia and industry.

This action is very well regarded at national level and it was included in the national education project coordinated by the President of Romania, "Romania

Educata" – project what will run during 2018–2030; (http://www.presidency.ro/ro/angajamente/detalii-romania-educata/?y=2018&m=3&d=20).

As a result, we aim at a visible and tangible evolution for both the academia and the industry, with a focus on encouraging and sustaining performance, research, innovation and alignment with the trends of the electronic industry.

4 Software Reconfigurable Hardware Tools

Because we thought of easy understandable systems suited for high schools students and first and second year bachelor students we must start by developing simple tools so that the students should understand the huge possibilities of the modern developed embedded systems.

As we presented in first chapter, we focused on imagining, developing and introducing simple PSoC Analog Coprocessor examples, easy to be used for reconfiguring the analog part of the system. After this stage was done the same using the new PSoC 6 architecture retains the strengths of the PSoC programmable fabric, unlocking possibilities for an evolving IoT market. Along with the easy-to-use PSoC Creator IDE and support for a vast ecosystem of partners, it enables next-generation IoT application development in a timely and predictable manner.

For the beginning we developed one simple LabVIEW [4] application which offers students the ability to select the application (module), already developed, and based on their fast ablity to reprogram the embedded system (based on Cypress PSOC integrates). This LabVIEW application Panel and Diagram presented in Fig. 3 can be used in both situations for PSOC Analog Coprocesor and for the new PSoC 6.0.

To be able to do this system reconfiguration using the presented application, learners need first to update the Cypress software to the last versions and if they receive the message from Fig. 4 to make an firmware update.

Cypress KitProg2 (see Fig. 4) uses two types of programming/debugging interfaces: PPCOM and CMSIS-DAP. The PPCOM interface supports Cypress tool chains such as PSoC Creator and PSoC Programmer. CMSIS-DAP is an alternative programming/debugging interface in which the KitProg2 can be used with third-party tool chains to program/debug the target. This mode is selected when you press and release the mode switch for less than two seconds.

In our case after firmware update, using the mode SW3 (press it and afterwards release it) from the Analog Coprocesor board will switch to the Mass Storage mode, when the student sees one the system the "KitProg2 H:" directory (inside can see STATUS.TXT information) and the PSoC can to be programmed with HEX file. Only PSoC 4/6 Families are supported in this version. Dragging and dropping of a HEX file shall trigger the programming of the target device. In case of an incorrect HEX file or a programming error, firmware of the target device shall be erased.

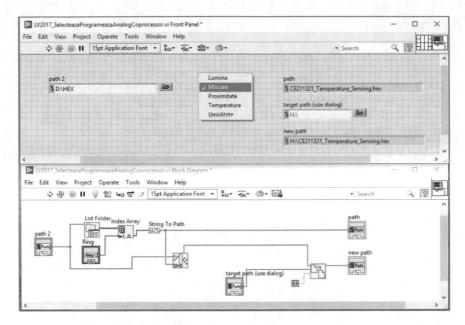

Fig. 3. Front Panel and Digram of the LabVIEW application with selector for modules

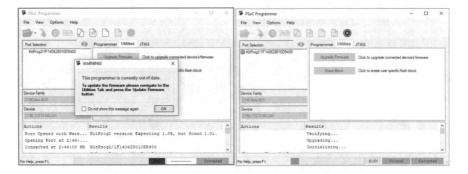

Fig. 4. Cypress Programmer 3.27.1 firmware update

5 Conclusions/Recommendations/Summary

We think that collaboration between schools and universities based on new technologies must be one of our first steps in order to adapt the educational system to the new challenges in Engineering Education.

At the same time "Software Reconfigurable Hardware" and IoT children/student training can be a fast and productive approach already tested in our field of Electronics.

Curricula changes must animate all the teachers starting from high school teachers up to the Ph.D. schools.

References

1. Masselos, K., Voros, N.S.: Introduction to reconfigurable hardware. In: Voros, N.S., Masselos, K. (eds.) System Level Design of Reconfigurable Systems-on-Chip. Springer, Boston, MA (2005). https://doi.org/10.1007/0-387-26104-4_1, Print ISBN 978-0-387-26103-4
2. FITkit informations. http://merlin.fit.vutbr.cz/FITkit/en/uvod.html
3. PSoC6. http://www.cypress.com/documentation/development-kitsboards/psoc-6-ble-pioneer-kit
4. LabVIEW. https://www.viewpointusa.com/labview/what-is-labview-used-for

Multitasking and Today's Physics Student: A Global Perspective

Teresa L. Larkin[1]([⊠]), Diana Urbano[2], and Benjamin R. Hein[3]

[1] Department of Physics, American University, Washington, DC, USA
tlarkin@american.edu
[2] Department of Physics Engineering, University of Porto, Porto, Portugal
urbano@fe.up.pt
[3] Shady Grove Presbyterian Church, Derwood, MD, USA
ben@shadygrovepca.org

Abstract. This paper focuses on a study conducted with undergraduate students at two institutions of higher education – American University in Washington, DC and the University of Porto in Portugal - undertaken in spring 2018. Student participants in the U.S. were enrolled in a first-level introductory physics courses taken primarily by students who are not math or science majors. The Portuguese students were enrolled in a first-level introductory physics course designed for engineering majors. At both institutions, the first level courses focus on basic classical mechanics. Using a survey constructed by the authors, questions related to multitasking and student perceptions of its impact on learning in physics were given to student volunteers taking these introductory courses. Survey questions include those designed using a Likert-type scale as well as a small number of open-ended questions. Upon comparison of a small subset of survey results, this paper will provide a discussion of what factors relating to multitasking our student participants feel are promoting or hindering them both inside and outside of the physics classroom. Through a comparison of these preliminary results across both institutions, we hope to gain a deeper and more global insight into possible reasons students engage in multitasking both inside and outside of the physics classroom. Implications for future studies will also be shared.

Keywords: Generation Y · Generation Z · Learning in introductory physics · Millennials · Multitasking

1 Introduction

The widespread use of the Internet and other recent technological innovations (such as smart phones) have made today's generation of students the first to have the potential for literally around-the-clock exposure to information. In fact, recent research suggests that the constant potential for exposure to information via the Internet highway has caused these students to become a part of the most socially connected generation ever to enter our classrooms. However, if one digs a little deeper into the hidden message in this statement it becomes clearer that while our students are constantly in touch with others through texting, emailing, and through various social media outlets, a lot of the

© Springer Nature Switzerland AG 2020
M. E. Auer and T. Tsiatsos (Eds.): ICL 2018, AISC 916, pp. 417–426, 2020.
https://doi.org/10.1007/978-3-030-11932-4_40

"connections" are done while sitting alone in their dorm rooms, their apartments, and yes – even in our classrooms! In addition, students today are often busy "connecting" with others while simultaneously studying and performing other tasks. Our students are multitaskers whether they realize it or not.

Because of the unique challenges caused by constant exposure to information technology, several questions will be raised in this study. For example, does this around-the-clock potential exposure to information help or hinder students' ability to learn physics? Does this continual opportunity for constant exposure to information via the Internet tacitly encourage students to feel like they must multitask in order to survive their academic experience? What is the reason(s) that students multitask? Furthermore, does multitasking serve to prohibit or enhance student learning in physics? Questions such as these provide the motivation for framework of the current study. Before this study is outlined, an overview of relevant literature will be presented.

2 Relevant Literature

Who are today's students? Today's students are typically those born in the 1990s. Some of these students may fall into what has been coined as the millennial generation and are sometimes referred to as Generation Y (GenY), Generation Me (GenMe) or the iGeneration (iGen) [1]. The millennial generation is loosely defined to be those born in the 1980s and 1990s. In terms of the iGen notation, Twenge suggests that the letter i could mean reference to things such as the individual or even the Internet. Twenge further suggests that as a label, iGen is concise, broad, and relatively neutral [2]. She notes that the iGen label has been portrayed as being bland, which could be a strength as a label needs to be inclusive enough to capture a wide range of people, yet be neutral enough to be accepted by them. The title Generation Z (GenZ) has been loosely used to refer to those individuals born in the late 1990s and early 2000s. Neither the GenY or GenZ labels ever really caught on largely because most individuals in those age categories don't like being labeled with a letter that follows from the preceding generation. To clarify, the older generational references include the Baby Boomer generation which approximately includes those born between 1946 and 1964, followed by Generation X or GenX which roughly includes those born between 1965 and 1980. GenX has also been dubbed with other titles like the 13th Generation and Post Boomers.

The term millennial generation was coined by Neil Howe, a historian, economist, demographer, and foremost expert on generational trends. To come up with a name to use when referring to GenZ, Howe and his colleagues conducted an online contest in 2006 in order to find a title to use for the generation that came after the millennial or GenY generation [3]. The name that was the winner of this contest was Homeland Generation. He suggested that the readers of his books "sensed a worldwide cultural movement towards nationalism, localism, and an increased identification with one's roots. The word also fits since this generation of children is literally kept more at home than any earlier generation of kids, thanks to the protective, hands-on child-raising style of Gen-X parents" [3, p. 2]. Most students in today's classroom were born in the late 1990s. Should these students be called GenY or GenZ, millennials, or the homeland generation? Because the age boundaries are rather loosely defined, and because we are literally at the boundary between the GenY and GenZ generations, we will refer to today's students as millennials in this paper.

Because today's students have such easy and continual access to online Internet platforms, one question that rises to the surface is what impact does this constant access have on today's students? These Internet platforms include everything from Google to YouTube to other social media platforms such as Instagram, Snapchat, and Facebook. Vidyarthyi, for example, has suggested that over the past decade the use of these social media platforms has served to reduce average attention spans of users from 12 min to 5 min [4]. This fact is further supported by a study on media-induced task-switching conducted by Rosen, Carrier, and Cheever [5]. Their study involved the observation of 263 middle school, high school, and university students studying for 15 min in their homes. These researchers also used a questionnaire that focused on such things as study strategies, task-switching preferences, and technology attitudes. They found that the study's participants averaged less than 6 min on task before they moved on to another one. Rosen, Carrie, and Cheever noted that the task-switching had a good deal to do with technological distractions such as social media and texting. To combat the issue, they suggest that educators might provide short technology breaks during class to help alleviate some of the distractions. Regardless of the Internet platform, the social media frenzy and digital culture of today has also been linked to increased user multitasking [6–8]. The jury is still out in terms of what effects multitasking may have on student learners, specifically, student learners in university physics classrooms.

Seemiller and Grace discuss the myth of multitasking as it pertains to Generation Z [6]. These authors suggest that this generation of students is extremely comfortable splitting their time as well as their focus between multiple screens. They suggest that these students are very comfortable conducting multiple Internet-based activities such as Internet searches, posting information to their social media sites, and watching a video – all while simultaneously writing a paper. As Seemiller and Grace suggest, there is a caveat to this behavior. They refer to this caveat as the myth of multitasking which essentially contends that multitasking can be conducted effectively. Arguably, millennial students also exhibit this type of digital behavior.

The myth of multitasking should not be taken lightly. We have been unable to find any research which suggests that multitasking leads to enhanced and increased learning in students. Turkle notes that "when psychologists study multitasking, they do not find a story of new efficiencies" [7, p. 163]. In fact, she suggests that multitasking leads to decreased performance on any of the tasks the multitasker is attempting. It seems pretty obvious that an individual cannot concentrate on two or more things simultaneously and complete any of them as well as they could if they had simply concentrated on a single task. Baron, for example, speaks to this issue and argues that "a cascade of multitasking studies continues to indicate that one of the major issues is interruption. The intrusive stimulus breaks our concentration on the initial task at hand, and performance on that task degrades." [8, p. 218]. Turkle also speaks to the fact that for many, multitasking feels good and that neurochemicals in the brain serve to reward the body by inducing a multitasking "high." Turkle suggests that the end result is that this perceived high tends to trick the multitasker into thinking they are being outstandingly productive. She further suggests that in the years ahead, there will be much to try and sort out and understand. Turkle goes so far as to say that the insurgence of technology-based applications made us love it. Now we need to figure out what effect this insurgence has, and is having, on today's learners.

Multitasking and student learning do not appear to be very good companions. Yet multitasking seems to be the way of today's students who are deeply immersed in the digital culture. Carr suggests that toggling between tasks can actually add significantly to our cognitive load, which may in turn obstruct our thinking and increase the chance that we may skip over or misinterpret important information [9]. Where student learning is concerned, Carr's suggestion cannot be taken lightly. We are concerned that the constant distractions presented by social media and other Internet-based platforms can serve to detract our students from the learning tasks at hand. These constant distractions may slow down the learning process and make any one task take much longer to complete. In addition to the time loss, these distractions may also make it more difficult for a student to refocus their attention and get back to the learning task at hand [10].

It is of interest to explore the impact that technology-driven multitasking has on student learning both in and out of the classroom. It is also of interest to assess students' perceptions regarding their multitasking behavior as well as the reasons they tend to engage in multitasking in the first place. Awwad, Ayesh, and Awwad investigated the use of laptops in the classroom in order to examine whether laptop usage in the classroom was a distracting educational tool while learning was taking place [11]. Through a questionnaire they surveyed a random sample of students at the United Arab Emirates University's Colleges of Engineering, Science, and Information Technology. The results of their investigation suggest that students were using their laptops for both academic and non-academic purposes, and hence, they served as a distraction to learning in the classroom.

Sana, Weston, and Cepeda also looked at laptop multitasking in the classroom [12]. Their study included forty-four undergraduate students enrolled in an introductory psychology course at a large, comprehensive Canadian university. The participants were not only psychology students, but students representing a range of undergraduate disciplines. All participants were asked to use their laptops in the classroom. Half of the students were told to use their laptops to take notes, just as they would in a normal class. The other half were told to also complete 12 online unrelated tasks at some point during the lecture. One unsurprising outcome of this study was that the comprehension of the participants was impaired when they performed multiple tasks while they were learning. The study's participants were also given a short questionnaire which included two 7-point Likert scale questions related multitasking. The first question assessed the impact of the multitasking on student learning of the lecture material. The second question assessed the extent to which the participants felt their multitasking served as a hindrance to the learning experience of their classmates. Interestingly, the results from these questions revealed that the assigned multitaskers felt that their multitasking behavior would only slightly hinder their classmate's learning. In actuality, the observed effect size from peer distraction was almost twice that of the observed self-distraction effect size. Sana, Weston, and Cepeda do suggest that these results need to be taken with a degree of caution and that future studies are warranted.

Through the use of a survey, the present study is preliminary and focuses on whether or not students perceive themselves to be influenced when others are multitasking in the classroom. We also look at the reasons students tend to multitask when studying physics. In the section that follows we outline the framework for the study and provide an overview of the physics courses and student participants at both institutions.

3 Framework of Study

This study was conducted with two different populations of introductory physics students at two different universities – American University (AU) in the U.S. and the University of Porto (U. Porto) in Portugal. One aim of the study was to address the questions outlined in the introduction through a common survey given to students at both institutions. A second aim was to learn what the students perceived as the reason (s) that they engaged in multitasking while studying physics.

The survey used in this study was designed by the authors and the questions formulated were an outgrowth of current research related to multitasking and student learning [5, 11, 13, 14]. Following some basic demographic items, the survey consisted of items related to three main areas. The first area simply asked students how often they were using their computers or smart phones both inside and outside of the classroom. The second area looked at different activities, not related to physics that students engaged in both inside and outside of the classroom. Finally, the third area included some general items that pertained to studying and learning physics in general. Students responded to each item using a 5-point, Likert scale. On this scale, 1 indicated a feeling of strong disagreement and 5 indicated a feeling of strong agreement with the statement presented for each item.

Before presenting a summary of the data collected, the subsections that follow provide a brief description of each course. In addition, a general picture of the student participants in each course will be presented.

3.1 Course Descriptions

The current study was conducted in spring 2018 in two first-level introductory physics courses at two different universities across two continents—AU and U. Porto. As first-level courses, both involved topics that fall under the umbrella of classical mechanics. General topics included fundamental topics such as basic motion concepts, Newton's Laws, energy, and momentum. The primary difference between the two courses was the level of mathematics that was used to present these topics.

At AU, the specific course was entitled Physics 100, Physics for the Modern World. Physics 100 is a first-level introductory algebra-based physics course that students choose to take to satisfy the university's general education requirements for graduation. Most students take this course during their first or second years of study; however, a few students wait to take the course until their third or fourth years. At U. Porto, the specific course was entitled Physics I. Physics I is a first-year; second-semester introductory calculus-based physics course for students in the Environmental Engineering Program. In most of the lectures in the Portuguese course, the Peer Instruction method was used in the classroom [15]. by having students answering to quizzes. These quizzes were presented on Moodle and students could connect using their computers or smart phones.

3.2   The Student Participants

In spring 2018, there were 52 students enrolled in Physics 100 at AU. Of these 52 students, 49 volunteered to take the anonymous survey. Because Physics 100 is a general education course, the background of the participants was very diverse and includes students studying in a wide range of disciplines. At U. Porto, there were 49 environmental engineering students enrolled in Physics I. Of this number, 42 students volunteered to take the survey. A breakdown of the participants at the two institutions is illustrated in Table 1.

Table 1. The student participants

	Number of female participants	Number of male participants	Total number of participants	Age average
American University (AU)	18	31	49	19.7 ± 1.1
University of Porto (U. Porto)	29	13	42	18.6 ± 0.76

The section that follows highlights a subset of the data obtained during the spring 2018 semester. For the purpose of this paper, we give attention to the data related to distractors that might influence students as they study; and, simultaneously encourage multitasking. In addition, we focus on some of the reasons that encourage student multitaskers to multitask.

4   Data and Results

For the purposes of this paper, we will concentrate on a subset that includes 4 items from the third area of the survey which focused on studying and learning in general. These 4 items included:

1. I am easily distracted during class when one or more of my classmates are texting or using their computer for a non-class-related activity.
2. I often multitask (i.e. perform other non-related tasks) while studying physics because I feel I have to.
3. I often multitask (i.e. perform other non-related tasks) while studying physics because I feel I learn more that way.
4. I often multitask (i.e. perform other non-related tasks) while studying physics because I believe I perform better that way.

Students completing the survey recorded their responses to each item using a 5-point Likert scale. Again, the rankings on this scale ranged from 1 to 5 where 1 indicated a strong disagreement with the item's statement and 5 indicated a strong agreement with it. A ranking of 3 meant a student was undecided about the statement.

Tables 2 and 3 provide a summary of the data collected for the above 4 items. Table 2 shows a breakdown of the results, by gender, for students at AU and Table 3 shows the same breakdown of results for students at U. Porto. Average (mean) scores (AS) and standard deviations (SD) for responses to each item are presented in Tables 2 and 3.

Table 2. Average responses from students at AU

Item number	Female students AS ± SD	Male students AS ± SD	All students AS ± SD
1	2.83 ± 1.24	2.32 ± 1.11	2.51 ± 1.17
2	2.50 ± 1.29	2.61 ± 1.15	2.57 ± 1.19
3	2.06 ± 1.00	2.26 ± 1.23	2.18 ± 1.15
4	2.33 ± 1.37	2.29 ± 1.35	2.31 ± 1.34

Table 3. Average responses from students at U. Porto

Item number	Female students AS ± SD	Male students AS ± SD	All students AS ± SD
1	3.07 ± 1.14	3.08 ± 1.19	3.07 ± 1.16
2	2.17 ± 0.71	2.15 ± 0.69	2.17 ± 0.70
3	2.14 ± 0.83	2.23 ± 1.17	2.17 ± 0.93
4	2.00 ± 0.80	2.23 ± 1.23	2.07 ± 0.95

In order to statistically compare the average (mean) scores obtained for the two groups of students we performed an independent samples t-test. The t-tests suggested that there was a statistically significant difference between the average responses to item1 between the AU and U. Porto students. For items 2–4, our results show no statistically significant differences between average responses. In addition, we found no statistically significant differences between average scores of the male and female students within or across either group of students.

In addition to looking at average responses and standard deviations we also present the data for each group of students in percentage as well as media and mode formats [16, 17]. Since a ranking of 3 meant a student was undecided about a particular item's statement, we chose to present the data by breaking it down into the percentage of students that chose to record a response less than 3, equal to 3, and greater than 3. Doing so may assist the interpretation of student responses by allowing us to uncover and possible trends or tendencies across institutions. We also present the data in both median and mode format. The median is the middle number of the data set. The mode is simply the number that appears most often. A summary of student responses in percentage format as well as the corresponding medians and modes are presented in Table 4.

Table 4. Responses presented in percentage and Median–Mode format

Item number	American University				University of Porto			
	Percentages			Median–Mode	Percentages			Median–Mode
	<3	=3	>3		<3	=3	>3	
1	51.0	32.7	16.3	2–3	38.1	19.0	42.9	3–4
2	53.1	22.4	24.3	2–2	71.4	26.2	2.38	2–2
3	65.3	20.4	14.3	2–1	71.9	16.7	11.9	2–2
4	59.2	22.4	18.4	2–1	66.7	26.2	7.1	2–1

In the section that follows we provide a synthesis and discussion of the results elicited from the data presented here. In doing so, we have looked for emergent themes and common items among the study's participants across both institutions. We are particularly interested in reasons behind why students multitask when studying.

5 Discussion of Results

The first survey item asked students to indicate whether or not they were easily distracted by one or more of their classmates who may be texting or using their computers to engage in a non-class related activity. Results of an independent samples t-test reveal a clear difference between AU. and U. Porto students. However, our results did not reveal any difference between gender groups in each country. Based on these results, it appears that the Portuguese student participants were somewhat more easily distracted by the texting and/or computer usage of their classmates during lecture. One potential reason for this difference is that students at AU were discouraged from using their computers during class, while students in classroom at U. Porto were not. As a result, it may be that there were fewer distractions for the students in the AU classroom. In the AU classroom, distractions most likely were due to their classmates'usage of smart phones during lecture. Another possible reason the Portuguese students appeared to be somewhat more easily distracted during class was that one of the lectures at U. Porto occurred very late in the afternoon, at a time when many of the students were probably more tired, and thus perhaps more easily distracted. Table 1 also indicates that the average age of student participants at AU was 19.7 while the average age of those at U. Porto was 18.6. Hence the average student at AU was in their second year of college, while the average student at the U. Porto was in their first year. For college students, a one year difference in age may help explain why the AU students were not as easily distracted as those at U. Porto. It may be that because they were a little older, they had gotten a little better at managing distractions in the classroom.

Items 2–4 involved potential reasons why individual students might multitask while studying physics. As indicated in Tables 2 and 3, average Likert scale rankings for students at both institutions were less than 3 for each of these items. This was also true when looking at the results by gender. These results may suggest that students at both institutions had similar feelings in terms of multitasking while studying. Looking at Table 4, we can see that over 50% of the students at both institutions ranked these three items either with a 1 or a 2. To compare each item average across the two groups of students independent samples t-tests were performed. Results of these t-tests revealed no significant differences between the average scores.

It is interesting that the results tend to suggest that there are more similarities between students taking physics in two different classrooms at different places on the globe than there are differences. Perhaps the student participants provided responses with a feeling that physics is hard and requires complete focus. Alternatively, it may be because literally all of today's students enrolled in our physics classes have grown up with nearly constant access to computers and other Internet-based technologies. As a result, perhaps they are simply oblivious to possible distractions these items may cause in a college classroom. And, perhaps that's a good thing.

6 Summary and Implications for Future Studies

The study reported on here looks at the effects of technology-based classroom distractions and multitasking on physics learning. Student perceptions regarding both classroom distractions and multitasking were elicited through a multiple-item survey developed by the authors. A small subset of survey items provided the framework for the present study. Student participants were those taking introductory physics at AU in the U.S. and at U. Porto in Portugal.

Based on our preliminary results the Portuguese students appeared to be more easily distracted by their classmates who were texting or using their computers in the classroom. Possible reasons for this difference include the fact that the average age of the AU students was a full year higher than the U. Porto students. In addition, the students in the AU physics class were discouraged from bringing computers to class while those in the U. Porto class were not. Hence, there were potentially less distractions in the U.S. classroom.

Of note is the fact that the sample sizes used (49 U.S. students and 42 Portuguese students) were sufficient to allow us to make some initial comparisons, but they were not large enough to allow us to draw definitive conclusions. To confirm our preliminary findings, we plan to expand the present study to include students from other physics courses at both institutions.

Our preliminary results did not reveal that there was a significant difference between average scores for any of the survey items related to multitasking. Based on the preliminary results for these items, there appear to be more similarities in student responses than differences between them across both institutions. More than 50% of students at both institutions indicated that they tended to multitask while studying physics. Perhaps this is because of the perception that physics is a difficult subject. Or, perhaps the items themselves were not as specific as they could have been.

In addition to expanding the sample size involved in the study, we may also modify the items for a future study to include some more specifically related to multitasking during lecture. As presented in Sect. 4, the three items we looked at in the present study that related to multitasking were all focused on multitasking while studying physics.

We feel it is of related interest to also look at multitasking behaviors of students within the formal classroom setting. Items that specifically focus on technology-drive multi-tasking may also provide the basis for an additional study.

References

1. Twenge, J.M.: Generation Me: Why Today's Young Americans Are More Confident, Assertive, Entitled–and More Miserable Than Ever Before. Atria Books, An Imprint of Simon & Schuster, New York (2014)
2. Twenge, J.M.: iGen: Why Today's Super-Connected Kids Are Growing Up Less Rebellious, More Tolerant, Less Happy–And Completely Unprepared for Adulthood* (*And What That Means for the Rest of Us). Atria Books: An Imprint of Simon & Schuster, New York (2017)
3. Howe, N.: Introducing the Homeland Generation (Part 1 of 2). Forbes. (2014). https://www.forbes.com/sites/neilhowe/2014/10/27/introducing-the-homeland-generation-part-1-of-2/#6b94fecd2bd6. Accessed 21 May 2018
4. Vidyarthyi, N.: Attention Spans Have Dropped from 12 to 5 Minutes: How Social Media is Ruining our Minds [Infographic]. (2011). http://www.adweek.com/digital/attention-spans-have-dropped-from-12%20minutes-to-5-seconds-how%20social-media-is-ruining%20our-minds-infographic/. Accessed 21 May 2018
5. Rosen, L.D., Carrier, L.M., Cheever, N.A.: Facebook and texting made me do it: media-induced task-switching while studying. Comput. Hum. Behav. **29**, 948–958 (2013)
6. Seemiller, C., Grace, M.: Generation Z Goes to College. Jossey-Bass, San Francisco, CA (2016)
7. Turkle, S.: Alone Together: Why We Expect More from Technology and Less from Each Other. Basic Books, New York (2011)
8. Baron, N.S.: Always On. Oxford University Press, New York (2008)
9. Carr, N.: The Shallows: What the Internet is Doing to Our Brains. W. W. Norton & Company, New York (2011)
10. Larkin, T.L., Hein, B.R.: Friend or foe? multitasking and the millennial learner. In: Auer, M. E., Guralnick, D., Simonics I. (eds.) Teaching and Learning in a Digital World, Proceedings of the 20th International Conference on Collaborative Learning (ICL 2017), vol. 2, Chapter 79, pp. 702–212. Springer International Publishing, Cham, Switzerland, AG. (2018). https://doi.org/10.1007/978-3-319-73204-6_79
11. Awwad, F., Ayesh, A., Awwad, S.: Are laptops distracting educational tools in classrooms. Procedia Soc. Behav. Sci. **103**, 154–160 (2013)
12. Sana, F., Weston, T., Cepeda, N.J.: Laptop multitasking hinders classroom learning for both users and nearby peers. Comput. Educ. **62**, 24–31 (2013)
13. Burak, L.J.: Multitasking in the university classroom. Int. J. Scholarsh. Teach. Learn. **6**(2), (2012)
14. Fried, C.B.: In-class laptop use and its effects on student learning. Comput. Educ. **50**, 906–914 (2008)
15. Mazur, E.: Peer Instruction: A User's Manual. Prentice Hall, Upper Saddle River, N. J. (1997)
16. Sullivan, G.M., Artino Jr., A.R.: Analyzing and interpreting data from Likert-Type scales. J. Grad. Med. Educ. **5**(4), 541–542 (2013)
17. Norman, G.: Likert scales, levels of measurement and the "Laws" of statistics. Adv. Health Sci. Educ. Theory Pract. **15**(5), 625–632 (2010)

A Design Framework for Building a Virtual Community of Practice

Olga Fragou[✉]

Hellenic Open University, Patras, Greece
fragou0@gmail.com

Abstract. Sustaining competitiveness in global market is dependent on organizations' capacity to innovate; revelation of tacit knowledge and its incorporation in organisations' structures and processes, comprise an important factor in
exploring and implementing innovation. Virtual communities of practice (CoPs)
and virtual learning communities are becoming widespread within Higher
Education Institutions (HEIs) thanks to technological developments which
enable increased communication, interactivity among participants and incorporation of collaborative pedagogical models, specifically through information
communications technologies (ICTs). This paper presents a design framework
for building a Virtual Community of Practice (VCoP) incorporating company
stakeholders, academics, researchers and school practitioners as prospect
members of the community, who interact on the basis of knowledge sharing on
IT technologies and engineering: the design characteristics, important concepts,
structures and preliminary data are presented. Design research sets the
methodological framework used, while, in the context of applying ethnographic
practices, qualitative data have been collected to support the proposed design
mechanism.

Keywords: Knowledge management · Instructional design · Communities of
practice · Design framework

1 Introduction

Knowledge Management is known as *"the discipline of enabling individuals, teams,
and entire organisations to collectively and systematically capture, store, create, share
and apply knowledge to better achieve their objectives"* [1]. Community of Practice
(CoP) as a strategy sets a real breakthrough for managing knowledge on implementation challenges: it involves groups "of people who share a concern, a set of problems,
or a passion about a topic, and who deepen their knowledge and expertise in this area
by interacting on an ongoing basis" [2], sustaining thus innovation [3]. The most
important components that distinguish them from traditional organizations and learning
situations are summarised as follows: (1) different levels of expertise that are simultaneously present in the community of practice; (2) fluid peripheral to centre movement
that symbolizes the progression from being a novice to an expert; and (3) completely
authentic tasks and communication. Within global organizations and emerging forms
of organizations, interacting face to face on a regular basis is costly and time

© Springer Nature Switzerland AG 2020
M. E. Auer and T. Tsiatsos (Eds.): ICL 2018, AISC 916, pp. 427–441, 2020.
https://doi.org/10.1007/978-3-030-11932-4_41

consuming; thus, since Information and Communication Technologies (ICT) can transcend space and time, organizations are increasingly interested in exploiting their capabilities to support CoPs in more technology based schemata such as the Virtual Communities of Practice (VCoPs).

Virtual Communities of Practice (VCoPs) are widely used as a Knowledge Management tool in a number of multinational corporations where they are now the norm rather than the exception [4]. A VCoP may use a large array of traditional media (phone, teleconference, etc.) and more or less sophisticated technological tools such as email, videoconference, newsgroup, online meeting space, common accessed database, Website, Intranet, [5] to establish a common virtual collaborative space. While there is a wealth of literature in exploring the theoretical and conceptual issues related to CoP, very few empirical studies have been undertaken to show how a CoP is designed and works. Aiming towards this direction, this paper is structured as follows: the Literature review section provides insights into design issues of Communities of Practice; the Problem section addresses important issues regarding the design components of the specific Virtual Community of Practice; in the Methodology section, important decisions and methodological tools have been addressed regarding the design of the VCoP community; the Design Framework proposed and Data collected to support it are presented in the respective sections of the paper; the Results section, presents preliminary findings, whereas the Conclusion section finishes with suggestions and future research.

2 Value of CoPs, Tacit Knowledge and Virtual CoPs

A Virtual Learning Environment through the orchestration and use of its tools by the stakeholders can shape up the creation and dissemination of best practices, using raw material our of stakeholders' exchange for value creation, creating understanding as a basis to act and further bring about knowledge, competencies and competitiveness. For creating a knowledge-based organization- a process perspective of knowledge is important to be adopted, providing paradigms, tools and supporting initiatives [3, 5]. Important pillars of how tacit knowledge [4] is incorporated in organisational structures imply sharing of experiences (socialization), articulating tacit knowledge in specific concepts (externalisation), reconfiguring existing information through sorting, adding, combining, and categorising of explicit knowledge which may lead to new information (combination) and learning by doing, embodying explicit knowledge into tacit. *Communities of Practice (CoPs)* are the sum of both stakeholder interest and the development of individuals within the community, while they also emphasize process development over market or product development: they have been recognized as a powerful model for teacher professional development, functioning as a supportive and collaborative learning environment which enables the emergence and discussion on good practices. Based on constructivism [6], they involve ill structured problems, learning in social and physical context of real world problems, shared goals which are negotiated between instructors, learners and between learners, cognitive tools, as well as an instructor's role as facilitator or coach [2, 3]. They shape a process by which collective knowledge is produced through newcomers' and old members' interaction, but they also compose *a physical or virtual entity,* reified in the group that is formed to carry out this process.

Virtual Communities [4] use networked technology, especially the Internet to establish collaboration across geographical barriers and time zones: they are defined as designed communities using current networked technology, whereas Communities of Practice emerge through the ways their participants use the designed community. While traditional communities are *place based* and have memberships according to norms, virtual communities exist according to identification to an idea or task rather than a place; technology implementation implies that virtual team environments require skills, operation in technology and synchronous, asynchronous communication. The combination of content (i.e. text, images, animation etc.), scaffolding (i.e. with respect to web based technology), plus text based communication shape suitable environments for Communities of Practice: because most collaboration is text-based, norms are reduced, enabling introverted participants to share their ideas on an equal footing with extroverts. Technology and its usability impose a critical success factor for Virtual CoPs which need to make good use of Internet standard technologies such as listserv, bulletin boards, and accessible web technology: designing CoPs in organizational structures is based on the role of relevant social groups, the nature of technology and communication used. Difficulties with access and ICT skills in relation to online discussions and e-based learning are acknowledged in the existing international literature [3–5]. Mooted benefits of networking and communication include rendering physical location unimportant and isolation from the peer group is less problematic when academics are scattered geographically or work in small institutions: virtual CoPs can grow up based on interests rather than on physical proximity enabling collaborations, sharing of specialist interests and affording access to mentors and like-minded individuals [4]. What seems to shape Virtual Communities of Practice is the social interactions which often impose their dynamic interactivity in the processes of the community; experts prompt to wrap Virtual CoPs design around social interactions up to the same level as usability [7] and urge an attitude shift from "Designing Web-supported Communities" to "Designing For Web-supported Communities" [7]. An approach as such leads to avoiding the pitfall of imposing an external structure on the group instead of allowing meaningful structures and topics to emerge from group membership. Based on the Wenger's framework on CoPs (2002) important characteristics of CoPs' methodology, are: *a) mutual engagement, b) joint enterprise, and c) shared understanding* [2]. Members engage in action whose meanings they negotiate, while working together pushes practice forward, spurs actions and invites new ideas: in the shared repertoire routines, words, tools, stories, gestures, symbols, actions or concepts are promoted through which the community escalates.

The advent of technology has lead to emergence innovative paradigms of learning such as Ubiquitous learning [9]. The design of an ubiquitous learning environment, which bridges formal (e.g., school), non formal (e.g., museum and library), and informal (e.g., free time) learning, on the basis of developing Virtual Communities of Practice is a challenging endeavor. The Virtual Community of Practice discussed in this paper has been targeted at designing and developing an integrated learning environment [10], web supported, involving academics, researchers, secondary school teachers, practitioners and corporate sector CEOs, students to designing and developing educational scenarios, code material, educational content using state of the art technology for exploring Ubiquitous computing, Mobile computing and Internet of Things in science and STEM education ("UMI-Sci-Ed: Exploring Ubiquitous Computing, Mobile Computing and Internet of Things in Science Education"). The concept "ubiquity of learning" describes how learning takes place everywhere and how a plethora of environments offer different, both pre-planned and spontaneous, learning opportunities [11]. The STEM education approach [6] seems to gain ground over the last (5) years, an approach which is very much linked to the discipline of engineering: the social practice of conceiving, designing, implementing, producing and sustaining, complex technological products, processes or systems seems dominant in professional engineering culture. Effective science dominated engineering curricula do not just aim to achieve technological depth in the engineering disciplines but also enrich and broaden students' background; to help engineering students think about the bigger picture, allow them to see that what they do is affected by global and social trends and thus, could create unintended consequences.

2.1   The Problem

Organizations, as contexts of developing CoPs and VCoPs play a critical role in their nurturing and sustainability; literature often *provides a –one- size- fits- all* for organisations interested in forming, developing, sustaining CoPs and VCoPs [2, 3]. This literature tends to assume that all communities are similar; however, in order to carefully design, develop and nurture them to the full potential there has to be a full understanding of the organisations regarding this schema. CoPs' and VCoPs' members experience different environments because of the media through which they primarily interact and therefore face dissimilar realities that are better studied separately. For instance, building mutual knowledge, trust among members, and the sense of belonging which all increase the likelihood of comprehension, open exchanging and sharing, may be more difficult through computer mediated interaction [12]. Based on this problematic, the research questions of this paper has been shaped as follows: *(a) which are the basic stages of constructing a design framework for a VCoP involving school, corporate stakeholders and education policy makers?, (b) which are the most important components in designing a VCoP framework as such?*

3   Methodology

3.1   Designing a Virtual Community of Practice Methodology

A large part of the research over the past decades has been analytical, trying to create a theoretical underpinning for designing and understanding the current practices of designers. Knowledge of design resides both in practice and in academic research –and the problem is that there is little common ground or communication between the two. Most studies focus on members' behaviors on VCoPs, using these communities [7]. This paper does not focus on users' behaviours of using VCoP but rather on the design considerations for developing a VCoP. In order to produce the Virtual Community of

Practice framework we have deployed a design research methodology [14] supported by the following: a) instructional design tools to reveal important components of the framework [9], b) technological infrastructure building, based on the CoPs literature [10], c) phases of designing and developing Communities of Practice [2]. Iterative design methodologies are based on the underlying assumption that requirements, the problem and the solution contexts can be only understood over time: prototyping is essential, products emerge throughout the process and quality steadily improves. By designing an instructional design process we have defined a set of basic components and steps that describe sequencing of creating a U-learning ecology [13]: this has been the generic framework on which the proposed VCoP design framework has been developed. The design research methodology allowed for iterative loops in design: Action Design Research [8] has been described as a design research method for generating design knowledge through building and evaluating IT-artifacts in an organizational setting. The ADR methodology focuses on two major challenges: (1) addressing a problem situation encountered in a certain organizational setting through intervention; (2) building and evaluating an IT-artifact, which addresses the class of problems typified by an encountered situation.

The design and development of the VCoP has been based on the following steps: (a) selection of the stages for developing the VCoP based on CoPs literature review, (b) initial adaptation of the initial CoPs' schema to specific organization, (c) needs analysis for VCoPs' infrastructure content, (d) review of CoPs' infrastructure requirements, (e) design and development of VCoPs infrastructure resulting in UMI-Sci-Ed CoP's platform [9, 13], (f) initial design and development of material used. Defining the virtual environment's (platform) characteristics has been an important step in the sense of: (a) defining VCoP's platform idiosyncracy regarding other CoP's platforms, (b) creating the learning environment for that purpose, supported by suitable technological tools. To explore how the stakeholders perceive of VCoPs' function and to conduct a preliminary needs analysis, we have collected and analyzed data via interviews. The research sample has been corporate CEO's, secondary school teacher/academics, as educational scenario designers, and secondary school teachers as prospect VCoPs' members and practitioners who shall make use of UMI-Sci-Ed platform services and UMI technologies. Research protocols (semi structured interviews) have been designed and conducted by the author to further support the initially designed framework. Data have been collected through ethnographic practices and specifically by using semi structured interview protocols.

3.2   The VCoP Design Framework

Ubiquitous computing, Mobile computing and the Internet of Things (IoT), collectively mentioned as UMI technologies, seem to emerge as a new learning paradigm imposed by advent of technology such as the 4G networks spread. The Horizon 2020 UMI-Sci-Ed project (http://umi-sci-ed.eu/) focuses on the investigation of the introduction of UMI technologies in education, putting these state-of-the-art technologies in practice, so as to make attractive the prospect of pursuing a career in domains pervaded by UMI [9]. Students through a mentoring mechanism are provided with training material, IoT hardware kits and software tools to explore UMI technologies through hands-on activities: several model educational scenarios are introduced that incorporate UMI technologies, in order to cultivate relevant competences on high school students, and as a means of exchange of experience among practitioners, school teachers and academics. The UMI-Sci-Ed project situation, prompted to explore ways in which the "developers" of knowledge could partner with "end users" of knowledge and enhance awareness towards UMI technologies, attitudes and best practices on using this type of technology in STEM education. The VCoP framework proposed makes use of the conceptual framework of Communities of Practice [2], involving its three important dimensions: *domain, community and practice,* pillars of which are presented in Schema 1. Mutual Engagement, Joined Enterprise and Shared Repertoire are also important pillars of the proposed framework as important CoPs' pillars. The (5) stages of CoPs as set by CoPs' methodology are the following: (a) Potential, (b) Coalescing, (c) Maturing, (d) Stewadership, (e) Transformation [2, 15]. According to the timeline of the design and development of the VCoP, the design framework presented in this case focuses on the stages of Potential and Coalescing, as the data collected out of these stages are going to further shape the Maturing and Stewadership stages of the VCoP as the project proceeds.

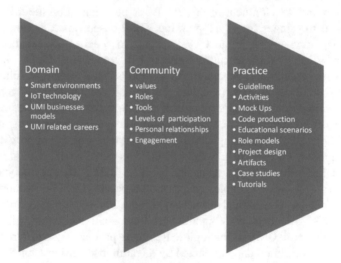

Schema 1. Domain, community & practice in VCoP

As a preparatory phase we have designed and developed a Special Interest Group (SIG) on a local basis so as to: (a) design in micro initial structures, processes and material of the VCoP, (b) start technological tools' selection and launching/pilot testing of UMI-Sci-Ed Communities' of Practice platform. Schema 2 presents design components of the SIG as well as material and documents developed for this purpose.

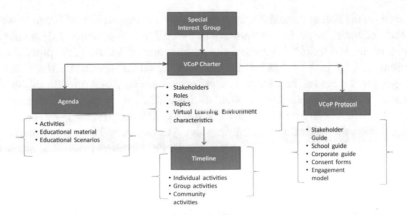

Schema 2. Components of VCoP 's SIG and material/documents supporting it

The use of design artifacts has been a strategic decision for designing and developing the learning environment of the VCoP. For that purpose design artifacts such as the CTI Educational Scenario Template (designed by the author [9, 13]) have been used in a twofold purpose: to use it as a *half baked structure* of UMI Educational Scenarios and use it to structure the collaborative space of the designed and developed UMI-Sci-Ed platform. In an effort to initially conceptualize and structure the ubiquitous learning experience, we have selected and used the schema of *learning ecology* [13], structured but flexible enough due to the seamlessness of relationships among the actors and items involved. Designing and developing an environment for VCoP based on the use of educational scenario has been a core task during the first stage of building the VCoP, proposing in parallel important roles and tools. The most important components of the proposed VCoP design framework are presented in the following Schema 3.

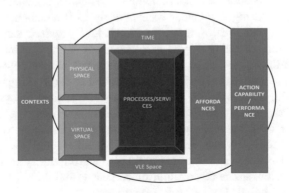

Schema 3. Components of the design framework proposed

Based on the literature methodology for building Communities of Practice we have followed as basic steps for designing the VCoP the following: (a) developed a map/charter for the VCoP, (b) created the architecture of VCoP, (c) explored ways in facilitating activity in VCoP and d) built an evaluation framework for the VCoP. The developed chart includes the background information, vision and mission of the VCoP, important goals and principles as well as identification of potential members. As important parameters to design and develop the VCoP have been the perceived value of the community, the value that VCoP brings to organization and individual members, as well as important outcomes, targeted out of the CoPs' members interaction, such as the engagement in designing educational scenarios and project involving UMI technologies on STEM education topics. Roles have also been assigned in the sense of defining the Administrator, Moderator, and Contributors of the community; a possibility of shifting roles is expected as the VCoP proceeds through development phases (i.e. Coalescing, Maturing etc.). In order to design the **architecture** of the VCoP we have: (a) conducted a survey of existing CoP literature and technological platforms, (b) taken into consideration the charter, the mode of interaction and collaboration required to build the infrastructure of the community at the initial level, (c) developed a first version of the VCoP with technological tools proposed for both preliminary and main phase of VCoPs [8], (d) designed collaborative activities for VCoP members to use during UMI-Sci-Ed platform interaction at a pilot stage. Leadership roles have been pre assigned to academics by the Core Group-though changes in leadership are expected through the interaction process.

UMI-Sci-Ed software platform [9] has been developed using open source technologies such as HTML, PHP, JavaScript and MySQL. It involves: (a) **Content Management Services**, supporting various forms of content management from typical file organization in folders to metadata annotated resource filtering including also a UMI-Sci-Ed repository with artifacts, best practices diagrams etc. (b) **Project coordination services**, services that implement the project management module (*i.e.* creation of a project, allocation of tasks including documents, organizing activities with relevant information using a calendar of events by CoP members, setting visibility rights of the activity etc., (c) **Member feedback and research services** supporting feedback in the form of rating a type of content, providing comments, and finding information according to the ratings and access frequency of their colleagues. Polls and surveys services are provided to facilitate participation to the workings of a community task from a broader group of users. Different types of questions are supported such as select options, likert-scale, and date and text fields. Analysis of the results is also provided (number of submissions per component value, calculations, and averages), (d) **Social media sites services**, Facebook and Linkedin could also be exploited by CoPs for their collaboration and interaction, (e) **Supporting utility services** such as log in, access rights settings, characterization of content visibility, notification reception, web metric reports, submit UMI App to execution.

As Communities of Practice (CoPs) comprise social arrangements in which individuals learn by participating in activities, emerging artifacts such as the products, technology, media and processes that are created by its members are also important in shaping the complexity of a **VCoP digital space**. Open Educational Resources, designed and developed Guides (i.e. CoPs' Stakeholder Guide, Pilot Guide for

Secondary School Teachers in UMI technologies, etc.), diagrams, case studies, webinars, PPT presentations etc.) has been the first basis educational material for VCoPs' members. VCoPs' activities have been proposed and designed for the UMI-Sci-Ed project's pilot phase employing constructivist techniques (i.e. ill structured problems) so as to enable learning to take place in a community of practice.

4 Preliminary Data Collection

4.1 Data Collection

The design of VCoP framework has been the result of applying design research methodology principles, such as the loop like design process, with iteration and evolvement based however on the use of core design mediating schemas such as the learning ecology and the educational scenario schema [13]. In order to support the design framework, at its very first stage with stakeholders' input the author has designed (3) interview protocols and (3) focus groups protocols, conducted and analysed qualitative data regarding: (a) the secondary school teachers' attitudes on proposed UMI technologies and practices, (b) the corporate stakeholders' attitudes, and potential contributions in the VCoP, (c) the designers' of educational scenarios views on using the UMI-Sci-Ed platform. Semi structured interview protocols have been designed by the author to address important issues of the respective research scheme. Content analysis has been applied to trace initial codes, look for patterns, themes, sequences, differences. We have used the inductive approach since there has not been enough knowledge of the research scheme/phenomenon [15, 16]. The categories are derived from the data in inductive content analysis: this included open coding, creating categories and tracing abstraction in terms of formulating a general description of the research topic through generating categories [17].

4.2 Data Analysis

(9) semi structured interviews (3 from academics, 3 from CEOs and 3 from secondary school teachers) have been designed and conducted by the researcher. Field notes have been kept. The data collected have been transcribed, coded and analyzed. Indicative data are presented in the following sections of the paper. Qualitative research aims at generating in-depth knowledge of the phenomena under investigation and is thereby a suitable approach particularly for exploratory work on topics that are new and for which much previous scientific knowledge does not exist [19, 20].

4.3 Results

The involvement of multiple stakeholders governs the development of systems that are intended for organization-wide use. Furthermore, in many organizations the IT function is assigned with the task of managing these projects on behalf of the organization's management and decision makers. The integration of needs, resources, information, and objectives among the providers and users stimulates service co-creation processes in the frame of the service dominant logic [18]. In adopting the service orientation, it is acknowledged that the knowledge and capabilities of the essential stakeholders is key to the effective realization of system development, and hence the IT function is assigned with the task of integrating and coordinating knowledge that is distributed organization-wide [19]. To gain stakeholders' insights regarding pillars of the proposed framework, interviews and focus groups have been designed and conducted in VCoP stakeholders. Feedback emerged regarding (a) the preferred use of resources and educational material, (b) the way platform's learning environment would be expected to support designed scenarios on UMI technologies, (c) topics important to structure the VCoP. Interviewees have been CEO's of IT companies, academics as educational scenarios designers and secondary school teachers as future users of the UMI-Sci-Ed platform. Data from the (3) designed interview protocols are presented as follows.

Secondary school teachers' attitudes and experiences have been explored (research protocol #1) regarding the following axes: (a) teachers' professional background, (b) teachers' confidence in using UMI, (c) potential ways of using UMI in classroom, (d) potential media of teachers to be informed about UMI, (e) potential ways of using UMI technologies for STEM, (f) the way UMI technologies could support the way STEM is taught, (g) their experience in teaching STEM, (h) teachers' participation in professional communities, (i) which teaching methods do Greek teachers use in their classrooms, (j) the STEM activities running in their schools, (k) the value of STEM education, (l) how STEM education can be improved.

Regarding the potential of using state of the art technology such as UMI in their practices and classroom, the teachers mentioned that as an infrastructure it is not difficult to follow in the classroom- however, a more technical part is needed so as to support the teachers in programming and compiling components of UMI technologies as they are not used to interacting with that kind of technology (Inter 2,4,5). Regarding the ways that UMI could help teaching STEM, as the most profound advantage has been stated that teachers would shift their teaching from theory to practice, being able also to first practice themselves of what they are going to teach in their students (Inter 1,3,4); furthermore the multidisciplinary character of STEM has also been stated as important, being able to bridge many domains and disciplines. Regarding the value of STEM education, most of the teachers pinpointed its practical use as well as the holistic picture that it provides to students engaged with STEM activities and the constructionist approach that it allows to the curriculum.

IT companies' CEOs experiences have been explored (research protocol #2) regarding the following axes: position in company and jurisdictions, the company's field of activities and types of products involved, information on the size of the company and employees' work on UMI technology, basic and secondary jurisdictions of those employees, the educational/technical background of corporate employers active in UMI domain, up to what point their education complies with the market needs, certifications that are of interest in the field, estimation on the skills required for UMI domain in the prospect of 5–10 years, ways of being informed about market needs and ways of covering these through human resources, participation in professional networks, envisioned ways of contributing to the CTI VCoP Community of Practice. Table 1 presents important data.

Table 1. CEO's attitudes on predicting important job positions in UMI technologies and skills required

Item1	Job descriptions & UMI domain	Skills & UMI domain
CEO#1	services directors, mobile developers for Android, IOS and bugging developer, web developer and web designer, tester	Team spirit, project management skills, use of project management tools, interaction with clients, problem solving, technical skills in the context of programming skills that they have to know. The programming skills (developing applications in Google for Android (using Java) and in Apple using Objective C). Bugging technologies imply JS platform and JS programming in javascript code
CEO#2	web developers, web designers and mobile developers and graphic designers	Communication skills, interacting with the customer (as "technical people are not clear in explaining their domain in customers"), collaborating in the context of a team, preparing end user interfaces, documented accordingly
CEO#3	Programmer in mobile applications using Android, IoS and Xamarin, Analyst, Architect Solutions, Project Manager, Software Engineer, Machine Learning	Java, working on Xamarin and CsR, objectives c20e, Joomla and PHP, sometimes R and Python, Vapor, SQL, and Logo. The employee's personality, the ability to work in a team

As far as the CEO's#1 insights on the market developments regarding UMI are concerned, it has been stated that there is a definite turn on mobile applications, whereas IoT applications through sensors are expected to become a reality in every house as well as smart cars. The interviewee's response on the ways that the company could contribute to the building of VCoP community has been to provide know how on mobile applications and IoT and products of this work could be inserted in the UMI-Sci-Ed project deliverables. As far as the CEO's#2 insights regarding the company's contribution in the VCoP are concerned, the following have been stated as possible ways of intervention: providing know how on technology used, provide insights on experienced problems regarding projects and mobile applications, and contribute in a reporting aspect in the sense of providing theoretical analysis of relationships among technologies, how these technologies become tools according to the user domain, and providing information on development or amendment of state of the art technologies. Furthermore, it is possible to contribute in events, analysis, informing students on new technologies and providing a selection of tools that they have to know, what they should search for, if they wish to work on this domain. Regarding the CEO'#3 s insights on the evolution of UMI market in the future 5–10 years it has been stated that mobile development is very important whereas the sensors' use creates rapid increase in companies' customers since they cater for different needs.

The ES designers' interview protocol (#3) (1 teacher, 2 academics) comprises of (12) questions in the following axes: their attitudes on the educational scenarios (pedagogical value, expected difficulties, students' knowledge, attitudes, students' motivation, assessment issues, types of preferred educational material for Educational Scenarios and CoPs) ES' functions that platform serves, companies' support on the CoPs regarding the Educational Scenarios etc.

Exchange of knowledge and communication has been the important pillars that the teacher/ES designer stated as important functions of the UMI-Sci-Ed platform. Companies could support the ES by providing know how, skills, examples, on problem solving and how an educational scenario like that and its content could be exploited in the market. It has been stated that this ES cannot be implemented without the platform: communication involving knowledge exchange and problem solving is the post important pillar. IT companies could support this ES with technical managers regarding project management skills, and engineers who actually work in IoT platform development, working on sensors and microprocessors and companies which support statistics data.

Regarding the UMI-Sci-Ed's platform role on supporting the ES, 2 academics answered that there is a preference on problem solving activities and not just in communication: the justification has been that communication and knowledge exchange is easier to be achieved however, this ES implies the use of standards and protocols and an already existing knowledge on students' behalf. As far as the company sector's CEOs that could provide support on the CoPs activities the expertise of a technical manager or marketing expert would be important in order to attract students and experts that will be able to provide demonstrations: technical managers that could provide technical information on state of the art technologies.

4.4   Discussion

The proposed design framework as shaped by the learning ecology schema [13] allowed for emergence of important parameters of the learning environment to arise: the use of Blog, Wiki and Forums shaped the Personal and Collaborative Space of the VCoP. There has been design of on line and off line activities for VCoPs, based on constructivism and inquiry learning. The size of the VCoP has been taken into consideration for the initial design: a local pilot community expected in later stages to merge with other local communities. The VCoP agenda has been finalized according to secondary school teachers' feedback, academics' feedback and CEO's feedback on important issues regarding the VCoP domain (UMI technologies). Affordances are expected to emerge throughout different forms of learning (mobile and on/off line). Themes and structures of CoPs educational activities have been designed, whereas the secondary school teachers' feedback resulted in preferred formats of educational material used in the VCoP videos, tutorials, guides, workshops etc.). Corporate stakeholders' feedback led to the expansion of existing corporate network, as the startup companies included in the initial VCoPs' schema have established collaboration with other organizations such as "Greek International Business Association" - (http://www.seve.gr/) or "Association of Informatics Companies North Greece" (http://www.sepve.org/web/guest/home), Hellenic Association of Mobile Application Companies (http://www.hamac.gr/), Association of Informatics Companies in North Greece (http://www.sepve.org/web/guest/home), "Greek International Business Association" - (http://www.seve.gr/).

As far as the corporate stakeholders' presence is concerned in building the VCoP for UMI-Sci-Ed, we have proposed for the pilot phase (Potential and Coalescing of Wenger's framework) the following processes for attracting UMI corporate professionals in the CTI Community of Practice for Local Pilots:

- Creating an initial pool/directory of startups and medium size IT companies
- Establishing initial communication using Skype and telephone contact
- Presenting the context and goals of UMI-Sci-Ed project and how this interferes with the CEOs' presence in the project
- Running an interview regarding CEO's insights on UMI domain
- Setting an initial agenda of topics on which these IT companies could support the community
- Collaboration/discussion and expert identification on delivering relevant content to students
- Strategic presentation of company's presence in the UMI-Sci-Ed CoPs platforms
- Identification of relevant IT products of these companies that could be disseminated through the UMI-Sci-Ed platform
- Design and develop (by the company stakeholder) a Blog for each company expressing interest
- Setting the CTI CoP Corporate activity content for Local Pilots
- Mediate the collaboration between teachers and CEOs to finalize the content and target of CTI CoP #1 Corporate activity content for Local Pilots
- Engage CEOs in interaction with Upper High School students through CEO's presentations on VCoPs' agenda topics

In order to deliver these processes we have used a) Skype meetings, b) telephone contacts and c) the use of UMI-Sc-Ed platform as a space for providing opportunity for IT companies' stakeholders. Our target has been to start using the UMI-Sci-Ed platform, instead of using broadly social media channels (for dissemination) such as FB, so as the Secondary School teachers from GEL/EPAL (Upper Secondary Schools in Greece) and IT Corporate sector partners to engage through the VCoPs' designed infrastructure (UMI-Sci-Ed CoPs' platform).

4.5   Evaluation

Regarding the VCoP evaluation process we have designed and used based on a proposed framework: (a) formative evaluation, which means evaluation of the implementation plan and (b) summative evaluation, to measure the efficacy and impact of VCoP, at a later stage. As a first basis evaluation the System Usability Scale [19] has been used: this involves a 10point Lickert Scale ranging from Strongly Disagree to Strongly Agree. The items in the SUS cover a variety of aspects of system usability such as the need for support, training and complexity. The broader evaluation framework has been based on critical success factors for developing a CoP such as activities that address details of practice, rhythm and mix of activities, regular communication and interaction among members, flexible and feasible format etc. [20, 21] as well as value creation in CoPs.

5 Conclusion

Implications in observing, understanding, analysing, evaluating knowledge sharing are important in terms of practitioners' tacit and explicit knowledge- thus exploitable by organisations. Designing and developing instructional design tools that support complex learning paradigms which involve the use of UMI technologies involves perceiving the technology-user interaction as a transactional process, which shapes the fully elaborated learning environment. The VCoP framework presented at this paper, encapsulates important pillars, processes, instructional design tools and actors of designing and developing CoP oriented infrastructures: it discerns levels of human-technology interaction, strategic and supportive components of building a VCoP as important pillars to develop web supported Communities of Practice. However, apart from the technological dimension, the proposed framework could be used by educators and trainers as a means to further conceptualize and understand the learning process when involving Knowledge Management schemas such as Communities of Practice, supported by technological tools such as the Virtual Learning Environments, adapting their professional practice on important pillars of the framework. A set of design considerations have been highlighted, important so as to form guidelines for researchers and developers who want to design VCoP, taking into account roles, profile information, infrastructure creation and navigation, content and the diverse character of community. Human ethics approval has been granted to contributing members so far; with the dawn of on line social networks, practitioners steadily and increasingly utilise these networks in their learning habits. Data collected during the Potential and Coalescing phases of Wenger's framework (2002), are expected to further shape organisation, material, practices and interaction among VCoP members in the next phases of the UMI-Sci-Ed project.

Acknowledgment. Dr Olga Fragou is currently an Academic Tutor in Hellenic Open University. The research schema and protocols presented in this paper have been designed and conducted by the author during her research work (2016–2017) at Computer Technology Institute and Press- Diophantus (http://www.cti.gr), for Horizon 2020 Project "UMI-Sci-Ed: Exploring Ubiquitous, Mobile and Internet of Things Technology in Science Education", GA 750183. The information and views set out in this publication are those of the author and do not necessarily reflect the official opinion of the European Union. Neither the European Union institutions and bodies nor any person acting on their behalf may be held responsible for the use which may be made of the information contained therein.

References

1. Young, R.: Knowledge management tools and techniques manual. Asian Prod. Organ., 3–14 (2013)
2. Wenger, E., McDermott, R., Snyder, W.M.: Cultivating Communities of Practice: A Guide to Managing Knowledge. Harvard Business School Press, Boston, MA (2002)
3. Swan, J., Scarbrough, H., Robertson, M.: The construction of "Communities of Practice" in the management of innovation. Manag. Learn. **33**, 477 (2002). https://doi.org/10.1177/1350507602334005.available,   https://www.researchgate.net/publication/247748275_The_

Construction_of_Communities_of_Practice'_in_the_Management_of_Innovation. Accessed 26 May 2018

4. Ardichvili, A., Page, V., Wenthling, T.: Motivation and barriers to participation in virtual knowledge-sharing communities of practice. J. Knowl. Manag. **7**(1), 64–77 (2003)

5. Barrett, M., Cappleman, S., Shoib, G., Walsham, G.: Learning in knowledge communities: managing technology and context. Eur. Manag. J. **22**(1), 1–11 (2004)

6. Squire, K., Johnson, C.: Supporting distributed communities of practice with interactive television. Educ. Tech. Res. Dev. **48**(1), 23–43 (2000)

7. Barab, S.A., Duffy, T.: From practice fields to communities of practice (CRLT Technical Report No. 1–98) (2000). Retrieved from http://crlt.indiana.edu/publications/duffy_publ3.pdf

8. Delaney, K., O'Keeffe, M., Fragou, O.: A design framework for interdisciplinary communities of practice towards STEM learning in 2nd level education. In: Auer, M., Guralnick, D., Simonics, I. (eds.) Teaching and Learning in a Digital World. ICL 2017. Advances in Intelligent Systems and Computing, vol. 715. Springer, Cham (2018). https://doi.org/10.1007/978-3-319-73210-7_86

9. Goumopoulos, C., Fragou, O., Chanos, N., Delistavrou, K., Jaharakis, J., Kameas, A.: The UMI-Sci-Ed platform: integrating UMI technologies to promote science education. In: Proceedings of 10th International Conference on Computer Supported Education (CSEDU) (2018). ISBN 978-989-758-291-2

10. Kumpulainen, K., Lipponen, L.: Productive interaction as agentic participation in dialogic enquiry. In: Littleton, K., Howe, C. (eds.) Educational Dialogues. Understanding and Promoting Productive Interaction, pp. 48–63. RoutledgE, LONDON (2010)

11. Pan, S.L., Leidner, D.E.: Bridging Communities of practice with information technology in pursuit of global knowledge sharing. J. Strat. Inf. Syst. **12**, 71–88 (2003)

12. McKay, J., Marshall, P.: The dual imperatives of action research. Inf. Technol. People **14**(1), 46–59 (2001)

13. Fragou, O., Kameas A., Zacharakis, I.D.: An instructional design process for constructing a U learning ecology. In: Proceedings of 2017 IEEE Global Engineering Education Conference (EDUCON 2017), pp. 1797–1806. IEEE (2017)

14. Laurel, B.: Design Research: Methods and Perspectives. M.I.T. Press, Cambridge (2003)

15. Hsieh, H.-F., Shannon, S.: Three approaches to qualitative content analysis. Qual. Health Res. **15**, 1277–1288 (2005)

16. Gerbic, P., Stacey, E.: A purposive approach to content analysis: designing analytical frameworks. Internet High. Educ. **8**, 45–59 (2005)

17. Badinelli, R., Barile, S., Ng, I., Polese, F., Saviano, M., Di Nauta, P.: Viable service systems and decision making in service management. J. Serv. Manag. **23**(4), 498–526 (2012). https://doi.org/10.1108/09564231211260396

18. Peppard, J., Ward, J.: Beyond strategic information systems: towards an IS capability. J. Strat. Inf. Syst. **13**(2), 167–194 (2004)

19. Brooke, J.: SUS: a "quick and dirty" usability scale. In: Jordan, P.W., Thomas, B., Weerdmesster, B.A., McClelland, I.L. (eds.) Usability Evaluation in Industry, pp. 189–194. Taylor & Francis, London (1996)

20. Ranmuthugala, G., Cunningham, F.C., Plumb, J.J., Long, J., Georgiou, A., Westbrook, J.I., Braithwaite, J.: A realist evaluation of the role of communities of practice in changing healthcare practice. Science **6**, 49 (2011)

21. Wenger, E., Trayner, B., de Laat, M.: Promoting and assessing value creation in communities and networks: a conceptual framework. Rapport 18, Ruud de Moor Centrum, Open University of the Netherlands (2011)

Work-in-Progress: SenseExpress, It Sounds Greek to Me but I Can Imagine How You Feel

Hippokratis Apostolidis, Fotini Siakkagianni, Stergios Tegos,
Nikolaos Politopoulos, and Thrasyvoulos Tsiatsos[✉]

Department of Informatics, Aristotle University of Thessaloniki, GR-54124
Thessaloniki, Greece
{aposti,fsiakkag,stegos,npolitop,tsiatsos}
@csd.auth.gr

Abstract. This study presents and evaluates an application exploring real-time sentence-based emotion recognition. This application is called SenseExpres and is used for the detection of emotions from sentences comprising Greek words. SenseExpress development was based on the Synesketch software library, which is built upon Ekman's basic emotion theory. Greek sentences are classified into six emotional types defined by Ekman (happiness, sadness, anger, fear, disgust and surprise). The evaluation of this application was based on users' self-report and Technology Acceptance Model (TAM). More specifically, 108 users participated in the evaluation activity. Every user was asked to provide 15 simple Greek sentences as input to the application. The self-report evaluation showed that 50.9% of the participants were satisfied with the resulted emotions. Moreover, TAM evaluation resulted in a significant interrelated model.

1 Introduction

The growth of web technology offered many ways in information exchange and communication between humans. The written way of human communication is becoming very frequent, usually expressed through emails, chats, forums and mobile messages (SMS). Furthermore, the interpretation of textual messages via emotion recognition is becoming a crucial aspect of human communications [1]. Affective computing is an emerging area dealing with text-based emotion awareness among other emotion detection methodologies.

Although this research focuses on sentence level emotion recognition, our main aim is the utilization of this technology for educational purposes and the amplification of learning tools supportive mechanisms. Emotions related to academic learning, teaching, taking exams and achievement (e.g., enjoyment of learning, pride of success, or test-related anxiety) are often called as 'academic emotions' [2]. Affective awareness focused on students' academic emotions may prove a considerable support in educational activities. Our Greek sentence-based emotion recognition open source library used by SenseExpress (Fig. 1) application could be applied in Web 2.0 tools such as chats, forum, blog, wiki as well as in real-time educational applications, such as conversational agent systems engaging in natural language interaction with the learners [3].

© Springer Nature Switzerland AG 2020
M. E. Auer and T. Tsiatsos (Eds.): ICL 2018, AISC 916, pp. 442–449, 2020.
https://doi.org/10.1007/978-3-030-11932-4_42

Fig. 1. SenseExpress login screen

SenseExpress open source library is based on Synesketch [4] which detects six basic emotions (happiness, sadness, anger, fear, disgust and surprise) in English sentences using the WordNet affect lexicon. It processes Greek sentences utilizing a rule based logic, Ekman's [5] emotional model and a Greek affect lexicon [6] that consists of 386 roots of Greek words and their emotion weights. First, SenseExpress parses the user sentence (Fig. 2). Then, it attempts to locate common root words with the roots kept in Greek affect lexicon database. This procedure is based on the fundamental application hypothesis that common root words express the same emotion. Finally, the resulted emotions in descending weight order are displayed.

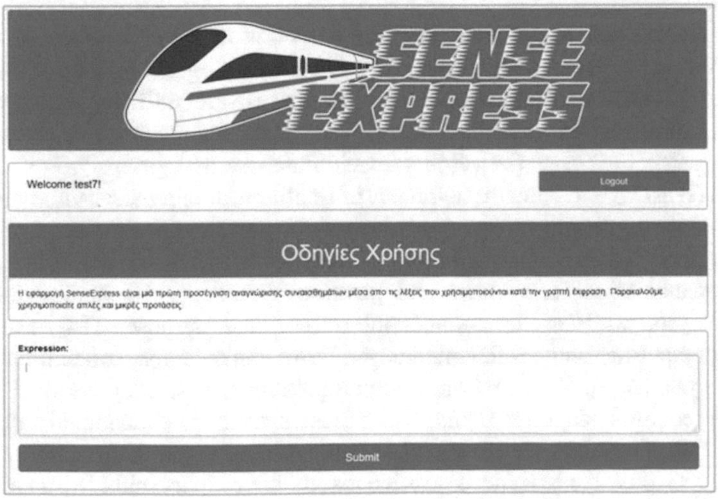

Fig. 2. SenseExpress expression input

In this paper, we outline the theoretical background of our study, our research motivation and a user activity. The evaluation of the activity was based on two methods:

- A self-report assessment. In this case, if the user replied that h/she was agreeing with the resulted emotions, displayed after the processing of each given sentence, the response was added to total agreements otherwise it was added to total disagreements.
- The user acceptance of the SenseExpress application, which was applied by exploring Technology Acceptance Model (TAM) [7].

Finally, the conclusions and the future steps are presented.

2   Related Work

2.1   Affective Computing in Text

In the last decades, affective computing has emerged as a research field in computer science that explores all the possible ways to detect, represent and express users' emotions, moods and personality characteristics [8]. Emotions are considered to be a key element in human interactions; in many situations, they have a significant influence in the relationships among humans and they determine human communication and behavior in personal and social level [9, 10]. All human representations and their modalities, such as facial expressions, gestures, action tendencies, physiological activity and subjective experience have been analyzed. Still, the most convenient access to emotions is through language, especially analyzing and detecting emotions in the text [11].

In computer science, automatic emotion detection in the text has become remarkably interesting and important. Many affective computing methodologies, which are focused on constructing affective lexicon resources, are utilizing emotional words (words that correlate with emotional states). WordNet usually constitute a starting point for building affect lexicons [12]. It is a lexical database that groups words into sets of synonyms (synsets) and contains a number of relations among these synonym sets. It is widely used by applications for text analysis. We used the Greek edition of WordNet [13] as we parse Greek sentences. Moreover, text emotion detection has been applied in social communication for reviews and position statements, in opinion mining and market analysis on Web for commercial or political purposes [14]. Furthermore, affective computing and sentiment analysis can enhance the capabilities of other computer systems, such as real-time communication tools, adaptive and intelligent systems or recommendation systems. Using these systems in educational environments to facilitate sharing ideas and thoughts, argumentation, negotiation and conflict resolution it is very possible to achieve positive results in group learning [8]. The basic idea in a well designed educational tool, such as a robustly developed CSCL environment, is to achieve knowledge distribution and group work by supporting emotion comprehension, collaboration and constructive interactions among peers.

3   Research Goals

The research motivation of this study is the development of an application interface (SenseExpress) based on an open-source affective-awareness library applied on the Greek language. Furthermore, this study is trying to evaluate this application considering:

- its reliability (RG1),
- users' perceived acceptance of SenseExpress (RG2).

In order to explore the second research question, we apply Technology Acceptance Model (TAM) [7]. Thus, the second research issue is shaped by the following research hypotheses according to TAM:

- Users' perceived ease of use has a positive effect on users' perceived usefulness
- Users' perceived usefulness has a positive effect on attitude towards using
- Users' perceived ease of use has a positive effect on attitude towards using
- Users' perceived usefulness has a positive effect on behavioral intention
- Users' attitude towards using has a positive effect on behavioral intention

4   Method

This section refers to the methodology followed for SenseExpress evaluation.

4.1   Participants

The study employed 108 undergraduate students (82 males and 26 females, with a mean age of 18.3 years, SD = 1.21), who attended a course in the domain of computer science. All the students took part in a self-report assessment but only 90 of them responded to the TAM evaluation questionnaire.

4.2   Instruments

In this section SenseExpress evaluation instruments are presented.

Self-report.

This is a self-estimation of every participant indicating h/her perceived satisfaction of the resulted emotions. Every user response indicated by a satisfaction or dissatisfaction flag was kept in a database under the user's code.

Technology Acceptance Model (TAM).

User acceptance of technology concentrates the interest of many researchers of various scientific fields. Among many models that have been proposed to explain the use of a technological system, Technology Acceptance Model (TAM) [7] is the most known and popular. This model is applying two central determinants:

- Perceived usefulness, which refers to "the degree to which a person believes that using a particular system would enhance h/ her performance" and it consists of six items [7].

- Perceived ease of use, which refers to "the degree to which a person believes that using a particular system would be free of effort" and it consists of six items [7].

4.3   Materials

Throughout the evaluation activity, every participant had to submit 15 simple Greek sentences through the application interface of SenseExpress.

4.4   Procedure

SenseExpress evaluation activity was organized in three phases:

- First, participants were informed about the application and they were given detailed instructions about using SenseExpress.
- In the second phase, which lasted nineteen days, students were asked to type at least 15 short Greek phrases or sentences and evaluate each of the results of the application. Every time the user gave a sentence one or more implied emotions were displayed in descending order according to every emotion weight. Besides, the participant was asked to choose if h/she was satisfied or not with the result. When the student was not satisfied with the result, h/she was asked to feel a textbox with the main implied emotion of h/her sentence as well as one to five complementary textboxes with words mainly but not necessarily used in h/her sentence closest to h/her preferred emotion.
- Lastly, in the third phase, for a period of ten days each student was asked to answer a TAM questionnaire, in order to evaluate their experience about using the application.

4.5   Evaluation Results

After applying data analysis on the aforementioned evaluation methods, we came into the following considerations.

Data Analysis

- Self-report user responses were processed in order to form a percentage of total agreement and a percentage of total disagreement.
- We applied confirmatory factor analysis (CFA) in order to examine the factor structure of the Greek version of the questionnaires measuring the latent variables included in TAM model. Finally, we applied path analysis, a structural equation modeling (SEM) technique, in order to test the hypotheses of TAM model. Path analysis and CFA were conducted in SPSS AMOS Version 20.

Results of Data Analysis Based on Every User Self-report.

Many Greek sentences were submitted to the application of SenseExpress (2337). The percentage of overall user agreement with the resulted emotions was 50.9%. More details are showed in the following Table 1.

Table 1. SenseEpress self-report results

Description	Users	Percentage %
Satisfied with the result	1189	50.9
Not satisfied with the result	1148	49.1
Total	**2337**	**100.0**

Results of TAM Evaluation

We employed the confirmatory factor analysis (CFA) in order to examine the factor structure of the Greek version of the questionnaires measuring the measured variables that we included in the path model based on TAM. More specifically, we applied confirmatory factor analyses to compare the apriori factor structures, implied by previous theory and empirical research. The initial models fit the data very well.

The Cronbach's alpha internal consistency estimates, obtained for each of the aforementioned questionnaire variables included in the examined model, were found greater or equal to 0.86. A SEM analysis was applied in order to test the relations between TAM dimensions. The confirmed relations are displayed in Table 2. The confirmed path model is displayed in Fig. 3 with good overall fit, $\chi^2 = 50.099$, $\chi^2/df = 1.252$, CFI = .979 και RMSEA = .053.

Table 2. Relations between TAM dimensions

A/A	Factors	Estimate	S.E.	C.R.	P
1	PU (Perceived Usefulness) ← PE (Perceived Ease of Use)	.182	.121	1.500	.134
2	AT (Attitude Towards Use) ← PU (Perceived Usefulness)	1.059	.183	5.787	***
3	AT (Attitude Towards Use) ← PE (Perceived Ease of Use)	.048	.119	.400	.689
4	BI (Behavior Intention) ← AT (Attitude Towards Use)	1.088	.137	7.965	***

5 Discussion

This study tried to evaluate an application dealing with emotion recognition while parsing simple Greek sentences. According to the evaluation results we can support the following considerations for each research goal:

RG1 - SenseExpress reliability: The percentage of satisfied users (50.9%) is not high but we must take in mind that Greek language has a very complex morphology [15, 16]. Hence, the ultimate goal of recognizing emotion through Greek sentences constitutes a long-term effort. However, the resulted percentage could be considered as quite encouraging under the prism of this preliminary approach.

RG2 – user's perceived acceptance of the presented application: TAM model was verified from our statistical evaluation procedure discussed in the aforementioned

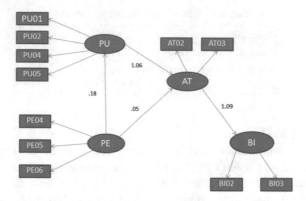

Fig. 3. Emerging TAM model

section. From the derived model we can support that the user who perceives Sen-seExpress as a useful application has also a positive attitude towards it and intends to actually use it. In more detail, participants' perceived usefulness and their attitude towards using the presented application seem to be the most crucial factors driving to behavioral intention.

6 Future Steps

This article presents a broad approach of a Greek affect lexicon being accessed by a real-time emotion recognition application, which receives Greek or English sentences as input and classifies them to the Ekman's model six basic emotions [5]. The future expansion of our research will focus to the enrichment of the lexicon. Furthermore, our Greek sentence-based emotion recognition application is planned to be used in a series of modern web applications and learning tools, such as conversational agents and dialogue support systems, emphasizing the value of collaborative learning.

References

1. Ishii, K.: Implications of mobility: the uses of personal communication media in everyday life. J. Commun. **56**, 346–365 (2006)
2. Pekrun, R., Goetz, T., Titz, W., Perry, R.P.: Academic emotions in students' self-regulated learning and achievement: a program of qualitative and quantitative research. Educ. Psychol. **37**(2), 91–105 (2002)
3. Tegos, S., Demetriadis, S., Tsiatsos, T.: A configurable conversational agent to trigger students' productive dialogue: a pilot study in the CALL domain. Int. J. Artif. Intell. Educ. **24**(1), 62–91 (2014)
4. Krcadinac, U., Pasquier, P., Jovanovic, J., Devedzic, V.: Synesketch: an open source library for sentence-based emotion recognition. IEEE Trans. Affect. Comput. **4**(3), 312–325 (2013)
5. Ekman, P.: Cross-cultural studies of facial expression. Darwin Facial Expr. Century Res. Rev. **169222**, 1 (1973)

6. Panatsias, I., Apostolidis, H., Tegos, S., Tsiatsos, T., Demetriadis, S.: Detecting emotions in Greek sentences: towards an open-source affect-awareness framework. *10° Πανελλήνιο και Διεθνές Συνέδριο «Οι ΤΠΕ στην Εκπαίδευση»* (2017)
7. Davis, F.D.: Perceived usefulness, perceived ease of use, and user acceptance of information technology. MIS Q. **13**, 319–340 (1989)
8. Reis, R.C.D., Rodriguez, C.L., Lyra, K.T., Jaques, P.A., Bittencourt, I.I., Isotani, S.: Affective states in CSCL environments: a systematic mapping of the literature. In: 2015 IEEE 15th International Conference on Advanced Learning Technologies (ICALT), pp. 335–339. IEEE, July 2015
9. Goleman, D.: Emotional Intelligence. Bantam Books, New York (1995)
10. Zohora, S.E., Khan, A.M., Srivastava, A.K., Nguyen, N.G., Dey, N.: A study of the state of the art in synthetic emotional intelligence in affective computing. Int. J. Synth. Emot. (IJSE) **7**(1), 1–12 (2016)
11. Shapparava, C., Mihalcea, R.: Affect detection in texts. In: Calvo, R.A., D'Mello, S., Gratch, J., Kappas, A. (eds.) The Oxford Handbook of Affective Computing, Chapter 13, pp. 184–203. Oxford University Press, New York (2014)
12. Miller, G.A.: WordNet: a lexical database for English. Commun. ACM **38**(11), 39–41 (1995)
13. Greek WordNet: Retrieved 15 of December 2015 from http://wordnet.okfn.gr/ (2013)
14. Cambria, E.: Affective computing and sentiment analysis. IEEE Intell. Syst. **31**(2), 102–107 (2016)
15. Stephany, U.: Verbal grammar in modern Greek child language. In: Dale, P.S., Ingram, D. (eds.) Child Language: An International Perspective. University Press Park, Baltimore (1981)
16. Andreou, G., Karapetsas, A., Galantomos, I.: Modern Greek language: acquisition of morphology and syntax by non-native speakers. Read. Matrix **8**(1) (2008)

Engineering Pedagogy Education

Improving the Quality of Training in Building Information Modeling

Elena S. Mishchenko, Pavel V. Monastyrev[✉],
and Oleg V. Evdokimtsev

Tambov State Technical University, Tambov, Russia
lenochkami@yandex.ru, {monastyrev68,gent_tam}@mail.ru

Abstract. The article considers the issues of improving the quality of specialists' training in the field of Building Information Modeling. The analysis of the development of information modeling and the regulatory base of the Russian Federation in this sphere was carried out. The analysis of data on the job market in the construction sector, the results of the employers' survey, and the experience in the implementation of the educational programs in Russian universities showed the need to improve the quality of education in the field of BIMtechnology. The results of the employers' survey revealed their interest in further co-operation in the realization of educational programs. The main principles of formulating the learning outcomes by combining the requirements of federal state educational standards, professional standards and employers' requirements are considered.

Keywords: Improving the quality of educational programs ·
Building information modeling · BIM-technology

1 Context

At the end of the twentieth century, the information overload and the challenges of processing the increasing flow of information in architectural design led to the emergence of a fundamentally new approach based on the creation of a building information model containing all the information about the construction site. One of the first programs (Building Description System) with the rudiments of information modeling appeared in the early 1970s. The complication of buildings and structures, the high pace of construction and the need to reduce design time, internationalization and international cooperation in design as well as many other factors led to the rapid improvement of computer design tools. At present, this approach to design is called Building Information Modeling (BIM) [1], and the technologies and software tools are called BIM-technologies. The essence of information modeling is that each participant in the construction process receives the amount of information necessary for solving a particular problem [2]. The consequence of the foregoing is the mandatory use of BIM in many countries of the world [3], which is confirmed by the diagram in Fig. 1.

Over the past ten years, BIM-technologies have made a big difference in the design process in the construction industry, and this process continues to gain momentum. With the help of information modeling, it is possible to solve problems that were

© Springer Nature Switzerland AG 2020
M. E. Auer and T. Tsiatsos (Eds.): ICL 2018, AISC 916, pp. 453–459, 2020.
https://doi.org/10.1007/978-3-030-11932-4_43

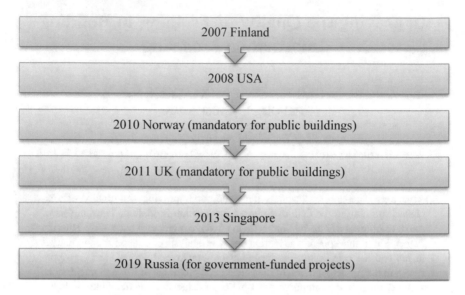

Fig. 1. Timeline of BIM application by countries

previously considered to be unsolvable. With the introduction of BIM-technologies, the approaches to the designer supervision, planning of construction costs, risk manage-ment, etc., have changed dramatically. These technologies play a very important role in the design of unique buildings and structures, as well as hazardous facilities, whose projects are subject to compulsory government expertise. All of the above logically leads to an understanding that in the current conditions, absolutely new educational programs in architecture and construction are required to work with the building information model.

2   Purpose or Goal

The purpose of the study is to improve the quality of training in the field of Building Information Models. The objects of the research are higher education programs (bachelor's and master's cycles) and supplementary education programs (advanced training, professional retraining, etc.). To achieve the goal, both research and applied objectives were set. The former included identification of trends existing in the inter-national and regional labor markets, understanding of the needs of employers in spe-cialists competent in building information modeling, while the latter comprised gathering information necessary for the development and improvement of educational programs. For example, the introduction of the concepts of "smart home", "smart city" and reality modeling, which are currently widely used in building practices, will necessitate the creation of very flexible educational programs in the field of building information modeling to capture changes in the world trends in this area [4]. Another goal of the research was to determine the most effective educational forms (traditional, on-line training, etc.) in the field of BIM-technologies.

3 Approach

To improve the existing educational programs, develop and implement the new ones in the field of building information modeling, with regard to the employers' requirements, the managers of the construction industry companies cooperating with the Tambov State Technical University (TSTU) were surveyed. The survey involved employers in design, construction, and technical maintenance of buildings and was organized throughout 2018. The respondents were the employers either interested in hiring TSTU graduates or currently employing graduates/undergraduates on a temporary or permanent basis. It should be noted that the respondents included those who employ undergraduates, and those who offer jobs only to graduates.

In total, employers had to respond to 30 questions. In accordance with the goals set, the questions were divided into three thematic parts:

- Part A: basic information about respondents and their attitude to information modeling;
- Part B: competences relevant for professional activities of graduates;
- Part C: proposals for combining the requirements of educational programs and the labor market.

The analysis of the survey results identified the employers who are might be interested in further cooperation in the realization of the Master's programs. The survey involved 22 employers engaged in construction, design and operation of buildings (Fig. 2).

To determine the required graduate qualifications, the analysis of the professional standards in the field of architecture and civil engineering was made.

4 Actual or Anticipated Outcomes

The survey results showed the need for the educational programs and the need for training competent professionals in the field of building information modeling.

It should be noted that about one third of respondents (27%) were middle-sized employers (51 to 250 employees), the rest were small-sized companies (11 to 50 employees). Half of the surveyed had from 11 to 25 automated workstations (a PC with the necessary software installed on it). When responding to the question about the scope of BIM-technologies in their company, 77% of employers said that they were not used. A significant part of the respondents (18%) said that BIM technologies were insufficiently used. However, 41% of employers admitted their willingness to use BIM-technologies in their companies.

The main causes of the limited application of BIM-technologies in companies were the lack of qualified personnel, the relatively high cost of licensed software, and the insufficient legal support in the field of building information modeling. The responses to the questions about the types of specialized software used in the company, the skills required to work with the software revealed the employers' preferences in computer tools (Fig. 3).

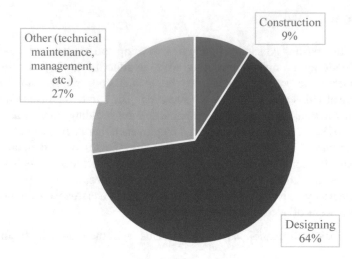

Fig. 2. Employers' profile

More than 68% of employers indicated the need for the third-party assistance in the implementation of BIM-technologies in their organization and noted the need to improve the employees' skills in building information modeling. The data on the optimal duration of the BIM-technologies training course (144 h - 37% of respondents) and frequency (once a year or once every 5 years - 35% of respondents) were obtained.

In Part B, the respondents formulated the learning outcomes required for Building Information Modeling.

The majority (91%) of the respondents confirmed the importance of acquiring competences in BIM-technologies by university graduates; 89% of respondents indicated the importance of mastering the basic and specific packages (CAD, BIM-technologies, BIM-management, etc.) in educational programs. In the employers' opinion, skills in architectural design and construction as well as organizational skills play an important role (Fig. 4). Soft skills include Decision Making Skills, Communication Skills, and Teamwork. Thus, in the development of educational programs, it is necessary to focus on group work and project work, so as to ensure the development of the above-listed skills.

The least relevant competence turned out to be "Good Command of English". Probably, this was due to the fact that the enterprises and organizations participating in the survey did not have the experience and the need for interaction with foreign partners. Nevertheless, from year to year the number of companies and organizations working internationally is increasing. Therefore, more than 68% of respondents stressed that this competence was compulsory (most important, important, less important) for graduates.

Since adaptation in a new working environment is one of the primary challenges new employees have to deal with, it is important to understand how long it takes for a young professional to become involved in the activities of the organization or enterprise. The employers believe that the optimal duration of internship (practice) can be

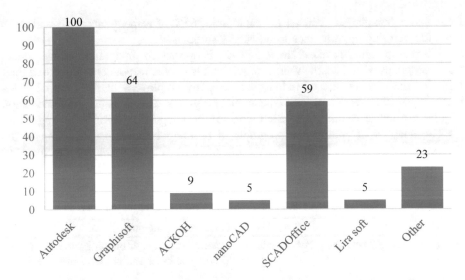

Fig. 3. Employers' responses (%) to the question: What software packages do graduates have to be able to work with to get a job with your company?

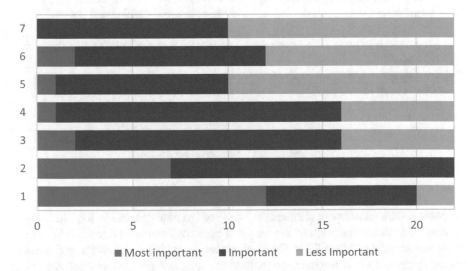

■ Most important   ■ Important   ■ Less Important

Fig. 4. Employers' ranking of graduates' skills and competences (1 - Deep knowledge of architectural and construction design; 2 - Deep knowledge of engineering design; 3 - Deep knowledge of technology, organization and management in construction; 4 - Deep knowledge of the construction economics; 5 - Skills in project management; 6 - Skills in building construction; 7 - Skills in technical maintenance of buildings) from 1 to 3 months per year. At the same time, 63% of respondents were willing to provide internship opportunities to university students.

When responding to questions is Part C - Proposals for integration of the educational programs with the labor market requirements - some respondents expressed their wish for the organization of training of company employees in the field of BIM-technologies, the creation of joint educational programs, as well as the implementation of joint research and industrial design works in building information modeling.

The analysis of educational programs in building information modeling in the Russian Federation has resulted in the classification of programs by the place of implementation; technology training; duration and labor intensity (Table 1).

Table 1. Educational programs in building information modeling in the Russian Federation

Organization	Educational program
Peter the Great St. Petersburg Polytechnic University	Automated design of buildings and structures Theoretical bases and practices of using BIM-technologies
The National Platform for Open Education Project http://npoed.ru/	Building Information Modeling - BIM
Kazan State University of Architecture and Civil Engineering	Information systems and technologies in architecture, construction and geo-engineering
National Research Moscow State University of Civil Engineering	Automated design of building structures using a software package (AutoCad)
Peoples' Friendship University of Russia	Computer Modeling in Civil Engineering
"INFARS" Group of Companies	BIM-manager and BIM-project coordinator responsibilities in innovative BIM-technologies
University of East London, Faculty of Architecture, Computer Science and Engineering	BIM-technologies in architecture and construction design

Federal state educational standards (FSESs) of the new generation require applying a competence-based approach in the design of basic professional educational programs of higher education. The FSESs [5, 6] of higher education (bachelor's and master's degree cycles) in Civil Engineering include the areas of professional activity which graduates need to master as well as the requirements to the learning outcomes. The graduates of this educational program are expected to acquire transversal, general professional and professional (based on professional standards) competences.

The educational organizations have to formulate the requirements to the learning outcomes in terms of professional competences on the basis of the corresponding professional standards, i.e. professional competences have to match the generalized labor functions and labor functions of the selected professional standards. Competence design is a critical element in the development of an educational program, and the formation of professional competence is a high priority for the university.

5   Conclusions/Recommendations/Summary

The educational programs in building information modeling of Russian universities, job vacancies and qualifications in the construction sector and the results of the employers' survey have been analyzed.

The requirements of the federal state educational standards, as well as the employers' responses are the basis for the improvement of the existing educational programs and new ones. The analysis of professional standards in the field of architecture and civil engineering made it possible to interface the labor functions (actions) and professional competences specified in FSES in order to formulate the learning outcomes. The design of future educational programs will be based on the learning outcomes corresponding to the competences of FSES and those developed by the university.

References

1. Talapov, V.V.: Informatsionnoe modelirovanie zdanij-sovremennoe ponima-nie: CADmaster № 4(54) 2010 (oktyabr'-dekabr'): 114–121 (2010)
2. Polterovich, V., Tonis, A.: Innovation and Imitation at Various Stages of Development: A Model with Capital. Working Paper #2005/048. M.: New Economic School (2005)
3. Talapov, V.V.: Tekhnologiya BIM: sut' i osnovy vnedreniya informatsionnogo modelirovaniya zdanij –M.: DMK-press, 2015. 410 s
4. Barison, M., Santos, E.: BIM teaching: current international trends. Gestão & Tecnologia de Projetos, **6**(2), pp. 67–80 (2011)
5. Federal'nyj gosudarstvennyj obrazovatel'nyj standart vysshego obrazovaniya po napravleniyu podgotovki 08.04.01 Stroitel'stvo (uroven' magistratury), utver-zhdennym prikazom Minobrnauki Rossii ot 30.10.2014 g. № 1419
6. Federal'nyj gosudarstvennyj obrazovatel'nyj standart vysshego obrazovaniya – ma-gistratura po napravleniyu podgotovki 08.04.01 Stroitel'stvo, utverzhdennym prikazom Minobrnauki Rossii от 31.05.2017 г. № 482

A Digital Step-By-Step Transformation Towards a Flipped Classroom

Egon Teiniker$^{(\boxtimes)}$ and Gerhard Seuchter

FH JOANNEUM, Kapfenberg, Austria
`egon.teiniker@fh-joanneum.at`

Abstract. Teaching software architecture and design in a part-time bachelor program implies numerous challenges. Part-time students (all already working) start with heterogeneous knowledge, various practical experiences and different ages which leads to diverse learning types. Additionally, the number of students in our bachelor program has been doubled over the past five years. In this paper we present a step-by-step transformation process toward a flipped classroom model based on several digitalization techniques which are also used in industrial practice and open source communities. We recommend such a gradual transformation which is based on evaluation and feedback to reduce risks. Also, efforts needed for this transformation can be distributed throughout many years.

Keywords: Blended learning · Flipped classroom · Digital education · Part-time study · Software design

1 Introduction

In our Software Design part-time bachelor program, we use a blended learning [1] environment which we introduced 14 years ago. This matured learning platform left room for further improvements because of the challenging circumstances. To ensure the education quality for a growing number of students in such a heterogeneous setting, we decided to introduce a flipped classroom model [2]. The existing blended learning practices and resource limitations led us to the following research questions:

- Is it possible to implement a practical step-by-step transformation towards a flipped classroom model?
- How can digital techniques and tools be applied to improve this change process?

For a concrete setting, we selected three related lectures: Software Design in the 3rd semester; Design Patterns and Software Architectures in the 4th semester, which are taught for 10 years in our part-time bachelor program [3]. Two of

© Springer Nature Switzerland AG 2020
M. E. Auer and T. Tsiatsos (Eds.): ICL 2018, AISC 916, pp. 460–471, 2020.
https://doi.org/10.1007/978-3-030-11932-4_44

them include practical parts in which students have to solve design problems and implement design patterns. We therefore have the challenge to convey abstract concepts such as design principles and design patterns together with practical implementation aspects. At the same time, the techniques and tools used should be as close as possible to the industrial practice.

2 Related Work

Before describing the chosen transformation process, we briefly discuss important concepts which influenced our work. We start with a definition of digital education and the major activities which are related to that transformation process in higher education. After that, we give an overview about the flipped classroom model on which our transition is based. Finally, we describe a variant of the inductive learning model which we use to teach software design in a practically oriented way.

2.1 Digital Education

In January 2018, the European Commission updated its Digital Education Action Plan which has three priorities to meet the challenges and opportunities of education in the digital age: Making better use of digital technology for teaching and learning; Developing digital competences and skills; Improving education through better data analysis and foresight [4].

The University of Oxford defines digital education in its Digital Education Strategy (2016–2020) as follows:

Digital Education is understood to mean the employment of technology in the creation and curation of teaching materials in digital form, the design and delivery of teaching, and the engagement and interaction of students with learning through the medium of digital technology. [5]

In addition, the Digital Education Strategy document suggests: Development of remote teaching; Engaging of a wider audience by using Open Educational Resources (OERs) [6] and Massive Open Online Courses (MOOCS) [7]; Development of an infrastructure for the use of students' own devices; Support of an enabling infrastructure for online assessment; And a lecture capture platform as a core service. It is common sense that only converting existing education material into digital formats is not the right approach. Technology should not drive didactics it should be the other way around. Therefore, advanced didactic models are necessary to use digital technologies to create a benefit for the whole learning process.

2.2 Flipped Classroom

The Inverting Education (also known as Flipped Classroom or Inverted Classroom) ideas are not new. One of the first publications of this concept describes the application and results of using the inverted classroom at the University of Miami [8]. The starting point was the frequent mismatch between an instructors teaching style and students learning styles which results in students learning less and being less interested in the topic. The further development of technologies, in particular communication via the Internet and the widespread distribution of mobile devices, is constantly opening up new possibilities for learning outside the lecture theater. Gannod [9] and Helmick [10] describe experiments with podcasts and online courseware. Numerous other applications of that didactic model have been described since then [11]. The Inverted Classroom Mastery Modell (ICMM) [12] consists of the following activities:

- **Knowledge transfer:** The transfer of knowledge takes place online using digital media such as PDF files, e-books and videos. This self-directed learning gives students great freedom in terms of the place and timing of learning.
- **Deepening of knowledge:** In classroom teaching, the in-depth acquisition of skills takes place interactively, assuming the teacher's role as the one of a consultant.
- **Mastery:** The described methodology is extended by autonomous knowledge tests, which show the students the progress in their learning process. These knowledge tests take place prior to classroom teaching, often in the form of multiple-choice tests which are offered online.

Using this didactic model, the classic frontal lecture in the lecture hall is completely eliminated. The transfer of knowledge is entirely transferred to digital techniques and supplemented by online collaboration (forums, chat, etc.). By using different digital media, more student learning styles can be covered as well. The classroom instruction is dedicated to the application and deepening of already learned knowledge. It should be noted that the students acquire this knowledge before the attendance units. Mastery tests help students to determine if they already have acquired the necessary knowledge. It should be mentioned that this form of teaching places higher demands on the students. Self-directed learning requires a greater degree of autonomy and organization. For educators, the digitization of educational materials means a considerable amount of extra work, which is often difficult to overcome, especially in IT subjects with rapidly changing content.

2.3 Teaching Software Design

Software design is an abstract concept that can be described by design principles [13] and design-patterns [14]. The term design pattern goes back to Christopher Alexander who specified this term in the context of building architecture and planning as:

> Each pattern describes a problem which occurs over and over again in our environment, and then describes the core of the solution to that problem, in such a way that you can use this solution a million times over, without ever doing it the same way twice. [15]

The concept of describing a model solution for a common problem turned out to be very useful in the software domain as well. Even on a larger scale, architectural patterns have been successfully introduced [16,17]. Using the idea of inductive learning [18], the abstract concept of design and architectural patterns can be taught in a practical way [19]. Inductive learning is based on examples: It begins with a code review of a concrete design pattern implementation; Then, we attempt to extract the underlying pattern from the implementation. This is a fundamental element of the learning process and the lecturer can help students to find the right abstractions; Finally, students use the just-learned design pattern abstraction to solve similar problems in practice.

This inductive approach is quite similar to the process of mining design patterns for the first time. Design patterns do not just emerge, they are extracted from a proven solution in software design and described in a semiformal way [14].

3   A Step-By-Step Transformation Based on Digitalization

The shift from our classic blended learning platform to the new didactic method of flipped classroom could not be implemented abruptly because neither the resources nor the assured knowledge of a workable solution were available. Therefore, a gradual transformation has been initiated with short feedback loops. After implementing each transformation step we observed the impacts and considered improvements in the next step. It turned out that digitalization was the solution of many problems we observed.

3.1   Primarily Blended Learning Environment

In the early years of the part-time bachelor program, our blended learning method was as follows: The classroom lectures took place in blocks at the end of the week (Thursday, Friday and Saturday). Online lessons were almost always done synchronously (we used a combination of Moodle and Skype for Business), students and teachers sat in front of their computers at the same time and were connected via a conference channel. The split of classroom and online education was around 40%–60%.

Basically, the same teaching methodology was used for online and classroom teaching. Due to the smaller number of students there was enough time to explain the material directly during a classroom unit. Room utilization was not a problem for a maximum of two groups, and in addition to the laboratory units, lectures were also held more often in the lecture hall. Compared to classroom teaching, the online lessons differed only in that we could not use certain media, for example, spontaneous sketches on a blackboard were not possible. Nevertheless, the online lessons were well received by the students, not least because of the avoided trip to the College. The assessment of the practical part was done through home assignment examples, which had to be delivered by students throughout the semester.

3.2   Towards a Flipped Classroom Model

In order to achieve the goal of a flipped classroom model for our part-time bachelor program, a gradual transformation of the teaching methods was undertaken. The following list summarizes the steps taken so far together with the observations we have made:

1. **Shifting knowledge transfer to synchronous online unit:** Following the concept of a flipped classroom, knowledge transfer is shifted to synchronous online units. Slide sets and demo examples (concrete implementations of design patterns) are used. By the means of synchronous instruction, students are always able to ask questions. The presented documents are also stored on Moodle in a digital format. Relevant literature is provided online via e-books, which are linked to the Moodle page.

   Observation:
   - Students passively follow online lessons and interactions are kept to a minimum.

2. **Moving classroom instruction to the laboratory:** The classroom units are used to deepen and apply the learned knowledge. The students solve practical problems in group work or through the method of pair programming [20] which is an agile software engineering practice in which two developers work together on a single computer to solve a problem. Problem-oriented exercises promote the acquisition of skills, while the teacher can assume the role of a consultant and answer students questions immediately.

   Observation:
   - Unfortunately, students often take too little knowledge from the online lessons to be able to approach the practical tasks actively. As a result, parts of the knowledge transfer must again take place in the classroom.

3. **Lab exams at the beginning of each classroom unit:** At the beginning of each classroom unit, a lab exam is held where students have to solve a concrete problem autonomously within a given time. Immediately after the exam, the model solution and possible alternative solutions will be discussed together. The achieved points are included in the overall assessment up to 40%. These practical exams, which are distributed throughout the semester, ensure that students are prepared for the classroom, and give accurate feedback on the current level of the students learning process.

   Observations:
   - Switching from assessed homework exercises to practical lab examinations resulted in a much more differentiated picture of students' competences. In laboratory lab exams, the differences in prior knowledge come to full effect, making the results very heterogeneous.
   - Students complain that they cannot gain enough practical experience in the online units to prepare for lab exams. The demo examples discussed are helpful but not well suited for practice.

4. **Autonomous knowledge checks by test exams:** In addition to the programming examples discussed in the online lessons, the students are also given test exams. These are practical programming tasks which are checked using automated tests. In addition, sample solutions are available to compare the different solutions. These knowledge checks can be performed either asynchronously or as part of synchronous online units. The synchronous approach has the advantage that questions can be asked immediately when problems arise. The asynchronous approach allows students to optimally classify the learning time depending on their previous knowledge.

   Observations:
   – These test exams allow students to tailor their learning activities to their prior knowledge and learning style.
   – In order to reduce the effort for creating examples, the following lifecycle for examples has been introduced: Examples are initially developed as lab exams. Old lab exams are provided together with the model solutions as test exams. Particularly descriptive test exams can be turned into demo examples which are used for knowledge transfer.

5. **Improving knowledge transfer through Inductive Learning:** Part-time students come from real life in companies and sometimes find it difficult to use a deductive approach where they start with abstract concepts before moving on to concrete examples. Inductive approaches are based on examples and extract the underlying concept from them. In doing so, code reviews techniques can be used.

   Observations:
   – The inductive approach works very well in the area of software development and software design, as the abstract concepts can be shown using small examples. Here we can use source code review techniques from industrial practice in the classroom.
   – Activities such as code review work also well within synchronous online units.

The described observations of each transformation step show very impressively how important this gradual approach is. After each step, the impacts could be analyzed and improvements made to the next step through discussions with students.

3.3 Digitization as Facilitator of the Transformation Process

Due to the fact that the flipped classroom model has been developed as an answer to the increasing use of multimedia in education, the solutions to the observed problems are often based on the consistent application of digital concepts. The following digitization techniques have been applied and evaluated over the last five years:

Recording of synchronous online units: Many students take advantage of the opportunity to record online units via Skype for Business. As a result, they can stop the recordings in the middle or repeat units many times to compensate different learning speeds.

Communication via Social Media: Facebook groups are used as an additional communication channel between students and to the lecturer [21]. This medium is used for the distribution of supplementary material (YouTube videos, photos of panel paintings, links to public articles) and for the informal exchange of views on the contents of the course. Facebook groups are used by students mostly via their smart phones.

Bring Your Own Device (BYOD): To solve the space problem in the laboratory, a BYOD approach is used. Students come with their own computers, thus, lab sessions can now be held in seminar rooms or lecture halls as soon as the problem of power supply has been solved.

Examples in Git repositories: All examples used in our software design lectures are published on a Git repository. By using such a distributed source code version control system (VCS), changes and extensions to the examples can be immediately passed on to the students. Following our inductive learning model, we provide two different types of examples suitable for self-study and in-class usage:

- **Demo examples** are indented for code reviews in the lectures. They represent concrete examples including automated tests to study the runtime impacts as well.
- **Exercises and model solutions** are incomplete programs that should be extended by students to make them work. During an exercise, automated tests guide students into the right direction. The model solution is used to compare student solutions with a reference implementation.

Reverse engineering of examples: Inductive learning means learning from examples which starts with a code review and the visualization of structural aspects of a program. Tools like Object Aid can be used to generate UML class diagrams from source code, as shown in Fig. 1.

From such a visualized structure of a concrete example, we can easily infer the underlying design pattern by a step-by-step abstraction process.

Automated Testing Framework: From the agile practice of Test-Driven Development (TDD) [22] we learned the importance of using automated tests in software development.

Every single test case describes a usage scenario of the implemented example. These test cases can be executed many times to study the runtime behavior of a particular functionality. Automated tests are also used as a feedback for students to proof their implementations. For every functionality a student has implemented, another test turns from red to green (Fig. 2). With that feedback students can see that they are on the right track towards the problem solution.

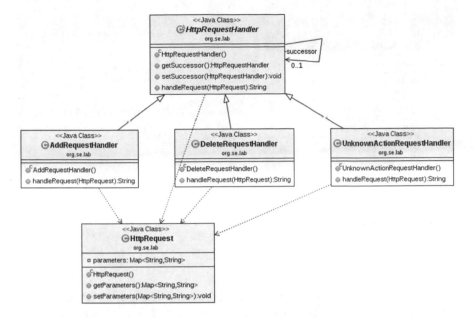

Fig. 1. Generated class diagram from a Chain of Responsibility design pattern example source code.

Virtual Laboratory: We provided a virtual lab for students [19]. This virtual lab is based on a Linux VM image and contains a Java Development Kit (JDK), open-source libraries, Java application- and database servers, Web browsers, and free Integrated Development Environments (IDE) which can be shared without licensing costs. This gives us a standard environment in which all examples can be prepared and used. Another advantage is that students can also use this virtual laboratory for the self-directed learning process at home.

Self-study resources: We extended our Learning Management System (LMS) to be an entry-point for self-directed learning. In addition to the classical PDF documents, we embedded links to Git repositories, FTP server (hosting the virtual machine images), Facebook groups, and public articles. We as well bought related e-books and linked them with the LMS, thus, students can access all material online starting from a single point.

Note that, while we were using a Moodle instance as a LMS, we also decided to focus on methods and tools used in software engineering enterprises. Virtual machines, Git repositories, and automated tests are important techniques for digital supported education in our flipped classroom model but at the same time they are commonly used in industrial practice and open-source communities.

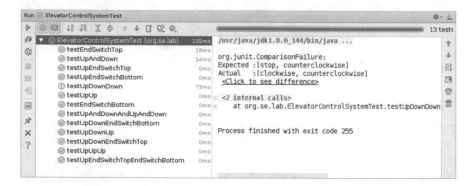

Fig. 2. The automated testing framework is used to check the functionalities of the implementation of an example.

4   Results

Institutional evaluations and direct feedbacks from students indicated that we succeeded in introducing the flipped classroom model in a step-by-step transformation. Whilst the number of students in our part time bachelor program continuously has increased from 36 in 2013 up to 72 in 2018, we were simultaneously able to improve the quality of education.

During the gradually implementation of the flipped classroom model, we have observed the following dependencies between didactic concepts and digitalization techniques:

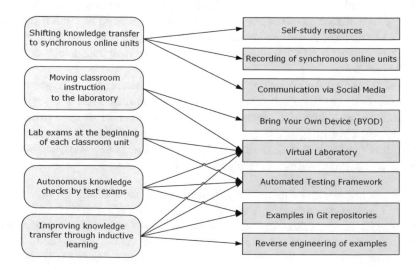

Fig. 3. Transformation steps and their enabling digitalization techniques.

- **Shifting knowledge transfer to synchronous online units** can be improved by giving the students the opportunity to record online sessions, by creating additional communication possibilities via social media, and by providing a central entry point for self-study from where all resources can be accessed by a single click.
- **Moving classroom instructions to the laboratory** to apply the online knowledge can be supported by a DYOD policy, and the usage of a virtual lab which makes students independent of laboratory equipment.
- **Lab exams at the beginning of each classroom unit** motivate students to work seriously on their self-directed learning process. During the lab exam, a set of automated test cases executed by a testing framework guide students the way to the right solutions. Also students can use their familiar virtual lab to solve the given problem.
- **Autonomous knowledge checks by test exams** enable students to determine their learning progress. A library of exercise and model solution examples, equipped with automated test cases, shared via Git repositories can be implemented and executed in the virtual lab.
- **Improving knowledge transfer through inductive learning** is based on demo examples checked out from Git repositories, reverse engineering techniques and automated test cases to analyze the structure and the runtime behavior of these examples, and the virtual lab which provides all the tools needed to analyze and run the examples.

From the dependency analysis between didactic concepts and digitalization techniques, shown in Fig. 3, we can see that the most commonly used digitalization techniques are the virtual lab, the testing framework and the examples in Git repositories.

In order to illustrate the current state of our transformation process, a common usage scenario from a students perspective should be outlined:

1. Students go to the Moodle page of the course (self-study resources) and find a link to the current timetable. They download the PDF handouts and follow the synchronous online lessons via Skype for Business, record the unit and use direct conference communication to ask questions.
2. After the online session, students communicate via social media and listen to the recorded online units. Some of them use e-books to deepen their knowledge. They also check out the examples from Git repositories, use the virtual lab to analyze and run the examples and try to solve test exams. Students are guided by the automated tests and compare their solutions with existing model solutions. The time required and the location of these learning activities can be controlled by the students according to their learning style and prior knowledge.
3. From time to time, students come to the college and start the in-class lectures with a lab exam where their knowledge can be applied and credits for the overall assessment can be earned. Students solve the lab exams with their own devices using the virtual lab. After the lab exam, the sample solution

will be discussed in class. Subsequently, further problems are carried out by the students in small groups. The lecturer assumes the role of the consultant.

Figure 4 shows the relationships between self-directed learning and classroom lectures on the one hand and the necessary digitalization techniques on the other. Digitalization is the prerequisite for self-directed learning, but we also observed that classroom lectures can benefit from these techniques as well.

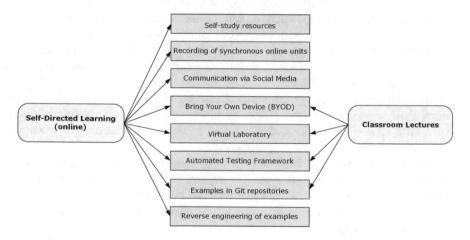

Fig. 4. Self-directed learning needs all of the presented digitalization techniques but classroom lectures can also benefit from them.

Finally, we have implemented an exemplary library of design and architectural patterns iteratively over a period of 10 years as a teaching tool on part-time bachelor programs [19]. Starting in 2016, we published these examples on GitHub [23–25]. The virtual lab can be downloaded for free via a link on these GitHub projects. Thus, the most commonly used digitalization techniques (the virtual lab, the testing framework and the examples on GitHub) are shared as an Open Educational Resource (OER).

5   Conclusion

Driven by experience and feedback we have established a flipped classroom model based on blended learning and digital supported education techniques. Continuous evaluation has shown that this approach is able to support classes in a part-time bachelor program facing a growing number of students with heterogeneous precognitions.

In a next step, we are going to produce different types of videos to support the online knowledge transfer. So far, students recorded their own copies of synchronous online units to adapt the learning rate to their own needs. We assume that specialized videos for theoretical content and example reviews can be more effective for students.

References

1. Bonk, C.J., Graham, C.R., Jay, C., Moore, M.G.: The Handbook of Blended Learning: Global Perspectives, Local Designs. Pfeiffer (2006)
2. Roehling P.V.: Flipping the College Classroom: An Evidence-Based Guide. Palgrave Pivot (2017)
3. FH JOANNEUM: Software Design (2018). https://www.fh-joanneum.at/software-design/bachelor/en/
4. European Commission: Digital Education Action Plan (2018). https://ec.europa.eu/education/initiatives/european-education-area/
5. Digital Education Strategy 2016–2020, University of Oxford (2016). https://www.digitaleducation.ox.ac.uk/strategy
6. Guidelines for Open Educational Resources (OER) in Higher Education, UNESCO and Commonwealth of Learning, United Nations (2015)
7. Baturay, M.H.: An overview of the world of MOOCs. Proc. Soc. Behav. Sci. **174**, 427–433 (2015)
8. Lage, M.J., Platt, G.J., Michael, T.: Inverting the classroom: a gateway to creating an inclusive learning environment. J. Econ. Educ. **31**, 30–43 (2000)
9. Gannod, G.C.: Work in Progress - Using Podcasting in an Inverted Classroom. In: 37th Frontiers in Education Conference (FIE), Milwaukee, USA (2007)
10. Helmick, M.T.: Integrated online courseware for computer science courses. In: 12th Conference on Innovation and Technology in Computer Science Education, New York, USA, pp. 146–150 (2007)
11. Bretzmann, J.: Flipping 2.0, Bretzmann Group LLC (2013)
12. Zeaiter, S., Handke, J.: Inverted Classroom - The Next Stage. Spectrum (2017)
13. Martin, R.C.: Agile Software Development - Principles, Patterns, and Practices. Prentice Hall (2002)
14. Gamma, E., Helm, R., Johnson, R.E., Vlissides, J.: Design Patterns: Elements of Reusable Object-Oriented Software. Addison-Wesley (1994)
15. Christopher, A.: A Pattern Language: Towns, Buildings, Construction. Oxford University Press, New York (1977)
16. Buschmann, F., Meunier, R., Rohnert, H., Sommerlad, P., Stal, M.: Pattern-Oriented Software Architecture: A System of Patterns, vol. 1. Wiley (1996)
17. Fowler, M.: Patterns of Enterprise Application Architecture. Pearson (2002)
18. Prince Michael and Felder Richard: The many faces of inductive teaching and learning. J. College Sci. Teach. **36**(5), 14–20 (2007)
19. Teiniker, E., Seuchter, G., Farrelly, W.: An open educational resource for teaching software design. In: 10th International Conference on Education and New Learning Technologies, Palma de Mallorca, Spain (2018)
20. Beck, K., Andres, C.: Extreme Programming Explained - Embrace Change. Addison-Wesley, 2nd edn (2004)
21. Gaar, W., Teiniker, E.: Improving model-based collaboration by social media integration. In: 27th Conference on Software Engineering Education and Training. Klagenfurt, Austria (2014)
22. Beck, K.: Test Driven Development: By Example. Addison-Wesley (2002)
23. Teiniker, E.: Software Design by Examples (2018). https://github.com/teiniker/teiniker-lectures-softwaredesign
24. Teiniker, E.: Design Patterns by Examples (2018). https://github.com/teiniker/teiniker-lectures-designpatterns
25. Teiniker, E.: Software Architectures by Examples (2018). https://github.com/teiniker/teiniker-lectures-softwarearchitectures

International Network Conference: New Technologies of Interaction for the Development of Engineering Education

Vasiliy Ivanov, Svetlana Barabanova$^{(\boxtimes)}$, Mansur Galikhanov,
Alla A. Kaybiyaynen, and Maria Suntsova

Kazan National Research Technological University, Kazan, Russia
vgiknitu@mail.ru, sveba@inbox.ru,
{mgalikhanov, alhen2}@yandex.ru,
emci2008@gmail.com

1 Engineering Education: Current Trends

Today's global processes of post-industrial society and digital economy development are closely related engineering education. Vigorous growth of technology and communications tools implies mobility and reducing national barriers in professional interaction. Globalization processes lead to integrating and unifying national economies, policies, and cultures. New social, economic and political relations are being formed among countries. Goals, strategies, and conditions of the international capital, technology, goods and services transfer are being changed. New understanding of time-space constraints is being developed [1, p. 41]. In the context of growing negative technology impacts, training the engineers is of key importance for ensuring stable social development. Contemporary engineering education issues are entering now into the fundamental social, political, and humanistic dimension. The ideas have become dominant that education should serve the purposes of a sustainable and dynamic society that is suffering some serious shocks of technogenic and social-political nature [2, p. 60].

Along with this large-scale task, training contemporary engineers must be performed considering the compliance of their education with the requirements and needs of employers. Developing the research activities of universities is also closely related to industrial concerns and orders. In contemporary Russia, in fact, the state does not hold being a universal consumer and work-giver for universities' graduates and their scientific product. In the global practice, multiple examples are well known for the successful participation of business partners in defining the syllabus and supporting scientific research, in developing business activities of students, in establishing endowment funds with the cooperation of graduates, and in building up the university corporate environments. The leading universities of Europe, Americas, and Asia construct the strategies of their development based on lasting cooperation with sponsoring companies and work-givers. In Russia, this cooperation takes specific features due to the sanction policies of Western countries and due to limited competition in most economy sectors [3, p. 48]

These issues often became a subject matter of national and international education conferences. However, interest in achieving the above purposes permanently stimulates

© Springer Nature Switzerland AG 2020
M. E. Auer and T. Tsiatsos (Eds.): ICL 2018, AISC 916, pp. 472–482, 2020.
https://doi.org/10.1007/978-3-030-11932-4_45

developing the innovation forms of partnership between education and business, and these are large corporations and research universities that set the pace here. The practices of international societies also contribute to the above [4, 5].

2 Transformation of Collaboration in Engineering Education

Today, high-technology enterprises in all over the world experience a deficit of the qualified new-generation engineering personnel. Labor market requires from the graduates of engineering universities to master many competences, such as business competences, the ability of independent life-long learning, and the ability to focus on problem solving and continuous development, not on accumulating their knowledge. Close interaction and interpenetration of fundamental and applied studies, as well as the inter- and multidisciplinary nature of the new science-driven technologies to solve comprehensive problems in conventional, allied, and new sciences – all this requires the new paradigms of engineering activities and, correspondingly, engineering education, as well as the new forms of communication between universities and the representatives of business community.

Respectively, a pronounced trend common for many countries is observed today to integrating efforts of governments, scientific and research communities, and business communities, aimed at identifying the areas for further development of engineering education, considering rapidly changing social and economic environments [6]. As another dominant trend in the engineering education development, we can also name the active involvement into this process of the representatives of business and the world's leading industrial corporations. Finally, in Russia, education and science are specific systems closely related to the extremely complicated hierarchy named "Russian social-political and cultural-economic system" [7].

This is why the formation and development of Russian technical and technological universities offering multilevel and multiarea education in Russia allowed seeing in a new light the issue of syllabus and the quality of training professionals for engineering activities within the context of the changing needs of economy and society. Universities of the new type are meant to ensure the fundamentality and amplitude of training, strengthen the professional focus of education, and develop in students the ability to work in interdisciplinary teams and implement comprehensive projects. It, in turn, presupposes the availability of a professional team of professors and administrators capable of organizing training focused on the requirements of the business community and advanced educational technology and of using electronic/digital training. Thus, the problem of studying advanced practices in these areas is updated, as well as new approaches to and new forms of interaction are required to discuss the system-wide problems of training engineers within the conditions of global society, involving relevant work-givers. One of the ways to solve it in Russia is regular holding network conferences supported by title sponsors from among the leading oil-and-gas companies.

This idea may seem not to be new on a scale of the global educational community. Various professional communities, such as IGIP, ASEE, IFEES, SEFI, etc., annually hold conferences united by a certain key concept or the purposes of their activities.

The members of such organizations or just interested participants discuss the issues of engineering education at such conferences. The events, such as annual conferences held by ASEE or WEEF, are traditionally supported by their industrial partners. However, in this paper, we are going to try and present, what the critical differences and key innovations are in the Russian event named SYNERGY, emphasize its importance with both Russian and global engineering education, and summarize some outcomes.

It is no mere chance that SYNERGY practically immediately gained the reputation of an innovative form of developing engineering education and enhancing the methodological competence, as well as the professional and educational qualifications, of its participants. And all this takes place in the unique conditions of collaborative activities performed by professionals from different areas of science and engineering, by being involved, in person or virtually, in the system of a great variety of events held within the framework of SYNERGY. The matter is that one of the opportunities to create the innovative format of the international network conference (supported by large manufacturers) in Russia was the annual methodological workshop that has been held by Moscow Automobile and Road Construction State Technical University (MADI) since 2001.

In 1990s, the search for the new forms of self-organizing the engineering educators and integrating the engineering education community of Russia led, supported by IGIP, to creating a system of national engineering education centers, to some extent a blueprint of networking. Functions of a unifying center of that system were fulfilled by the annual inter-university tutorial workshop held by MADI and supported by the Russian Monitoring Committee of IGIP with its active centers at MADI, Kazan National Research Technological University (KNRTU), and other Russian universities.

Today, the scientific and methodical experience exchange in the area of training engineers has developed to the level of new organizational and technological forms and methods suggested by the time.

Through the organizational and financial support of PJSC Gazprom, Europe's largest energy provider, the International Science and Practice SYNERGY Network Conference successfully has been held for three years, i.e., in 2016–2018, involving the company's 17 flagship higher educational institutions. Its ingenious network format ensured the participation of many representatives of PJSC Gazprom's flagship universities, affiliates and the corporation itself, as well as joint motivated discussions on the best practices and contemporary trends in engineering education, considering the needs of the real sector, primarily those of petrochemical industry and international standards. It became obvious at assessing the very first outcomes that SYNERGY has opened a unique opportunity to comprehensively and objectively discuss the state of, issues of and trends in the national engineering education within the context of global trends.

The idea of the conference came up at the international engineering education forum in Florence, Italy in September 2015. At the suggestion of the Russian Association for Engineering Education and KNRTU, the innovative for Russia format was elaborated for a distributed network conference that would unite holding sessions and video-conferences in several research and educational centers of Russia, with the participation of and assisted by the leading international societies of engineering education. The final session was decided to be held in Irkutsk, at the bank of the Baikal Lake as a unique ecosystem.

The conferences are conventionally organized and attended by the International Federation of Engineering Education Societies (IFEES), International Society for Engineering Pedagogy (IGIP), Global Engineering Deans Council (GEDC), Russian Association for Engineering Education, National Training Foundation, Gazprom's flagship universities, etc.

Recognized engineering-education experts, representatives of the leading global companies, legislative and executive authorities, and other interested parties are invited to participate in the conference. Its agenda includes plenary sessions, round tables, expert workshops, panel discussions, video conferences, and webinars. Proactive participation of researchers and professors from Russian and foreign universities from the USA, Austria, Belgium, Germany, Czech Republic, China, Portugal, Belorussia, Kazakhstan, etc., has become a tradition. The geography of the conferences encompasses the Volga Region, Siberia, Central Russia, Southern and Northwestern Federal Districts, and Yakutia.

PJSC Gazprom, the title sponsor of the conference has stood with its ideas and concept. By the way, the largest Russian energy provider supported such a network-based scientific and practical conference for the first time in its practice of active interacting with universities, primarily with the corporation's flagship universities and KNRTU.

3   Current Content of Networking Cooperation

International Science and Practice SYNERGY Network Conference focused on urgent issues and bottlenecks of engineering education in Russia.

For SYNERGY'2016 Interdisciplinarity in Engineering Education: Global Trends and Management Concepts, interdisciplinarity as the global trend in the development of engineering education was chosen as the central focus. The purpose of the event was studying the global and national experiences in organizing and managing the training of professionals to work in interdisciplinary teams and projects. The organizers were sure that working within such teams and projects would ensure the synergetic effect in their implementation. One of the conference goals was also implementing highly efficient methods into the system of training and re-training engineering staff, interaction of industrial companies and universities in such processes, and the influence of interdisciplinarity upon the competitiveness of engineers [6].

Given where the conference was held, the areas of the conference discussions and speeches included interdisciplinary projects relating to resource-efficient technologies and sustainability, including those focused on the preservation of the Baikal Lake natural resources. Due attention was paid to the students' place in interdisciplinary projects.

Intensive preparatory work allowed attracting to the conference the interested international and domestic partners and join the forces of the conference founding parties and sponsors, including the Ministry of Education and Science of the Russian Federation, Association for Engineering Education of Russia (AEER), National Training Foundation, KNRTU, European Society for Engineering Education (SEFI),

International Society for Engineering Education (IGIP), and International Federation of Engineering Education Societies (IFEES).

Network format allows joining the efforts and experience of some hundreds of our colleagues from around Russia and from around the globe. The SYNERGY'2016 was held on May 24–July 13. It was held in several stages and in different Russian towns, hosted by five leading engineering universities of Russia and by D. Serikbayev East Kazakhstan State Technical University.

"The topic of the conference is serious and very important," Ladislav Musilek, the member of the SEFI Administrative Council, noted. "Young people today are not always interested in related engineering areas, which does not assist much in developing the society or solving global and regional problems."

Within the plenary session at KNRTU, as well as discussions and workshops, the participants discussed global trends in organizing and managing educational projects in the areas of industrial engineering, efficient use of resources, and sustainable development. In the course of the session, organizers supported by local authorities and by the directors of manufacturing enterprises arranged a round table under participation of the representatives of business community (about 40 people in total), devoted to the urgent matters of inventive and rationalization activities.

At the final session held at Irkutsk National Research Technical University on July 11–13, 2016, about seventy conference participants had been fruitfully worked at plenary sessions, at the expert training workshop on managing university environments for performing interdisciplinary projects, and participated in discussions and round tables for three days.

SYNERGY'2017 New Standards and Technologies of Engineering Education: Capabilities of Universities and Needs of Petrochemical Industry had already included a Round Table on Engineering Pedagogy as one of its events (titled Human Resourcing of Petrochemical Complex: Issues of Developing Engineering Pedagogy), which is becoming an integral part of the International Science and Practice SYNERGY Network Conference.

The conference had already been held at Gazprom's flagship universities in 7 cities: St. Petersburg, Moscow, Kazan, Ufa, Tyumen, Ukhta, and Tomsk from September through December. Its formats were: Plenary sessions, round tables, expert workshops, panel discussions, and video conferences. A large-scale round table, Human Resourcing of Petrochemical Complex: Issues of Developing Engineering Pedagogy, was held at KNRTU on September 7–8, 2017. It included the discussions on the pressing needs of enterprises for qualified engineers and on changes within the system of higher engineering education in Russia and in all over the world. Given the importance of the event for the local economy and its basic industries, it was included into the annual agenda of Tatarstan Oil, Gas & Petrochemical Forum.

Over thirty speakers – well-known researchers and scientific leaders in engineering activities – spoke to the participants of the Round Table; in total, over 130 representatives of scientific and educational community and business from Russia, the USA, and other countries, as well as leading domestic and foreign experts in engineering education participated in the meeting. The speeches touched the university innovations in training the engineers, including project-based and online learning, training schoolchildren in engineering vocations, and accrediting educational programs.

The final session of SYNERGY'2017 was also held in Kazan on December 4–6, 2017. Over 150 contribution reports and conveyances were presented by Russian and foreign experts. In the five sections of the conference, the innovation in engineering education [8], engineers' competencies and training quality [9], the system of training professors and instructors [10], and the need for the early engineering vocational orientation of schoolchildren were discussed. They proactively exchanged their advanced experiences and best practices in training engineers and developing the human resources of enterprises. That was also about the interaction of educational and vocational standards, the development of open online education, the Federal Online Examination for bachelor program graduates, the teaching/learning models of the professor-student cooperation, and much more. 13 flagship universities of PJSC Gazprom participated in the conference.

The topics of SYNERGY'2018 (Integrative Training of Line Engineers for Increasing the Labor Efficiency of Petrochemical Enterprises) demonstrate the continuous topicality of the subject matters to be discussed on the conference: considering the work-givers' interests: It will be focused on training the workers for petrochemical industry (due to conducting the WorldSkills Championship in Kazan in 2019). The second topic is determined by the enterprises requiring the proactive training of personnel, and it is related to better performance at work.

4 Promoting the Cooperation of Universities, Authorities, and Business Community: New Opportunities for Engineering Education

According to their participants, the contents of SYNERGY events and talks at final, plenary sessions, largely contributed to enhancing the qualifications of the professors and instructors of the participating universities, developing psychology and educational competences, and growing the teaching skills in accordance with the international standards.

The attractive format combining participation in person and telepresence ensures the long-term informal interaction among the representatives of PJSC Gazprom's flagship universities. The processes have been intensified regarding the comprehensive analysis of the best practices of Russian and foreign technical universities, scientific-pedagogical schools, and individual researchers and educators. Unlike other conferences, the talks and presentations within the SYNERGY events are specifically practice-focused – Gazprom's flagship universities analyze the possibilities of translating the foreign experience in their universities, discuss the applicability of certain conceptual and methodological approaches at universities, jointly focused on supporting through personnel, science and technology, the efficient functioning of the most important sectors of the Russian economy.

Once again: SYNERGY, in complete accordance with its name, opened new opportunities to communicate with colleagues, study the best practices of the leading industry-focused universities of the country, primarily, due to the general sponsor of the conference. Permanent participating in network sessions helps actualize the topics of talks and reports given by researchers, professors, managers, and experts, as well as evolve new actors from educational and engineering communities. Positive feedback of the managements of PJSC Gazprom's flagship universities to our initiatives contributed to the corporations' decision on making the SYNERGY conference annual. KNRTU is appointed to be responsible for this area.

Growing international issues of political and organizational nature are known to affect adversely the academic mobility of researchers, professors, and students at Russian universities. Moreover, approaches based on mechanically taking the best foreign practices in Russian conditions do not always allows achieving the positive practical results. Therefore, Russian universities demonstrate a special interest in national events and in the opportunities of creative discussions with their colleagues from Russian and foreign engineering universities, especially industry-specific ones related to petrochemical, oil-and-gas producing and refining industries. The statistics of the SYNERGY events throws this into sharp relief.

At SYNERGY 2017, Yury Pokholkov, the President of the Association for Engineering Education of Russia, having emphasized in his reports the contradiction between the engineers training quality and the requirements set by employers, proposed to develop the measures aimed at enhancing the prestige of the engineering activities, introducing a certification system for professional engineers, and even adopting a law on engineering professions.

In his turn, Sergey Yushko, the Rector of KNRTU, emphasized in this talk that it was time to start thinking of how to be helpful for the partnering enterprises. "Training managing personnel is highly focused on now, while there are very few good engineers and workers. However, a today's worker is actually an engineer, since modern machines are so complex and diversified. This is why young people want to study majoring in engineering professions." He also emphasized that just providing knowledge to the students was wrong, since the skills necessary for business were updated very fast. Therefore, upon having completed the basic training, young people starting to build their careers return to the university very soon in order to gain new knowledge and competences. It is well exemplified by the practice of implementing additional vocational programs of the young professionals of Gazprom Transgaz Kazan at KNRTU.

Marat Akhmetzyanov, the Deputy Director General of the above company for Corporate Security, told about the company's fruitful collaboration with the university: "Our relationships became permanent and really efficient upon entering into the collaboration agreement in 2012. Since then, 573 people have been trained, we have been developing and implementing joint continuing education programs offered in face-to-face or online forms, and a "Gazprom Class" has been functioning at the KNRTU residential vocational school. Over 50 students annually intern at our company (on-the-job trainings and pre-graduation internships), and a specialized department has been opened. According to the program of cooperative scientific studies, the university is acting as a contractor for four projects simultaneously."

Rafinat Yarullin, the Director General of OAO "Tatneftekhiminvest-holding", participated in the plenary session of the conference. Having presented his review of the industry trends, he turned to the contemporary requirements for engineers. "Universities must educate innovatively trained people. For this purpose, general and higher education must be strong," the holding head said.

Irina Arzhanova, the Executive Director of the National Training Foundation (Russia) continued talking about the targeted training of engineering personnel in the partnership with business. "University today is not just walls and a degree certificate. It is much more the competences that make its graduates successful in their lives, which competences can be gained both distantly or using network forms," she emphasized.

Andrey Verbitsky, the member of the Russian Academy of Education, presented the conceptual idea of training engineers, based on his psychological and pedagogical Context Education Theory, and proposed to develop, under the support of the Ministry of Education and Science of Russia, a system for implementing advanced technology in engineering education.

Nina Aniskina, the President of the Continuing Professional Education Union, spoke about different public accreditation models of educational programs that must be used in engineering personnel development, including rating systems.

Viacheslav Prikhodko, the Chairman of the Russian IGIP Monitoring Committee, touched on the topic of the necessity to upgrade the qualifications of professors at engineering universities in accordance with international standards and issuing the relevant IGIP certificates.

An efficient tool of training engineers ready for solving production tasks has become chemical engineering actively using digital prototyping, Professor Sergey Diakonov, the KNRTU Rector Advisor, said in his presentation.

Professor Vasily Senashenko from the University of Peoples' Friendship (Moscow, Russia) presented a thought-provoking conception about the hybridity of contemporary education, including engineering education, having emphasized the spontaneity and volatility of its forms.

At the section of Engineering Pedagogy and Engineering Education moderated by Professors Hanno Hortsch, Vladimir Kondratyev, and Vasiliy Ivanov, the experience was discussed regarding implementing new teaching technology, training engineers for digital economy, developing the infrastructure of interdisciplinary activity-based teaching and learning, and much more.

At the section of the System of Standards and Accreditation moderated by Professors Gang Chen and Svetlana Barabanova and by Senior Lecturer Liliya Ryazapova, the opinion was expressed that it was necessary to ensure, first of all, the accreditation of professors and instructors, not programs or graduates. Many recommendations were given, such as not to loop educational standards on vocational ones only, but work proactively, etc. Training personnel within the framework of engineering post-graduate schools was also touched on.

At the section of the Engineering Education Ratings and Quality, moderated by Professors José Quadrado and Roza Bogoutdinova, it was a discussion about developing the innovative capacity of a professor who must not be prevented from creating.

Professor Phillip Sanger, the moderator of the section of Developing the Human Resources of Enterprises and Project-Based Training of Engineers, noted that the programs of collaboration between Sibur and Gazprom and universities presented were demonstrating new, non-ordinary approaches and technologies.

Representatives of the colleges and schools of Tatarstan and pre-university departments of universities, KNRTU professors guided by Lyubov Ovsiyenko and Farida Shageyeva shared their best practices in early industry-oriented profilization and engineering orientation of schoolchildren and proactive attracting them to the university and introduced the audience to the interesting projects on working with gifted children.

The participants of the round table and the plenary session of the SYNERGY conference noted that human resourcing the industry-focused enterprises was a complicated systemic problem, solving which would require taking legislative, economic, organizational, and pedagogical measures, improving the mechanisms of public-private partnership, and the availability of clear strategy and tactic, acknowledged by business, scientific and university communities.

One of the restrictions of industry development is disbalance within the vocational education system, which determines the deficit of labor resources at enterprises and on industrial sites. On the labor market, we can observe, on the one hand, the excessive number of academically-trained people and, on the other hand, the deficit of "blue collars" having primary or secondary vocational education. University educational programs do not often take into consideration the specific nature of individual enterprises, which results in educating professionals that do not meet the requirements of employers, while enterprises have problems with personnel adaptation. One of the main conditions for the necessary human resourcing the enterprises is achieving the high quality level of engineering education.

According to the systemacity principle in considering the ways of enhancing the quality of engineering education, the influence of both external and internal factors should be considered. External factors include global and national trends in economy, engineering, and engineering education; national policies; legislative framework; financial resources; business susceptivity to innovations; availability of clear forecasts regarding the labor market's needs for graduates; quality of training high-school graduates; public image of engineer; and some others.

Among external factors determined by consumers, the following should be identified: The level of technologic paradigm implemented at the enterprise, research intensity of products manufactured, requirements for young professionals and working conditions offered to them, the level of interaction with universities, etc.

Internal factors (university factors) include personnel policy; competences of faculty members; quality of educational programs and educational environment; the extent of integrating education, research and innovative activities; graduates' competitiveness on labor market; the level of interaction with research organizations and businesses; the efficiency of the university quality management system and management system, etc.

In these conditions, the interested participation of business in solving the problems discussed seems to be of special importance. A vivid example of responsible partnership is sponsoring by PJSC Gazprom the events that allow the representatives of universities, including Gazprom's flagship universities, to meet those of business community to develop their relationships and efficient discussing the engineering education issues for the benefit of the sustainable development of Russian economy.

Elaborating during SYNERGY specific recommendations for universities, educational authorities, for PJSC Gazprom is its key outcome. In 2017, the conference participants adopted a final instrument that particularly recommends developing on the national level a comprehensive program for modernizing engineering education, reduce the red-tape procedures in organizing research and engineering activities, and develop a system of incentives to attract businesses to funding vocational education and participating in training engineering personnel.

The practice of advanced developing the engineering education in developed countries proves conclusively that the efficiency of the process is determined by "the extent of mutual interest of its subjects and depends on their coordination and responsibility in joint activities, as well as on focusing the research, educational and business structures on achieving the commonly important goals" [4, 33].

5   CDIO Concept – SYNERGY's Basis and Outcome

Summarizing their joint activities, the conference participants elaborate practically important recommendations addressed to flagship universities, to the Russian Ministry of Education, and to PJSC Gazprom. The can considerably change the process of training engineers. The talks given by the leading researchers from participating universities are being published in the leading Russian journals of vocational education.

The conflict between the high requirements set by the state and society for the level of research, engineering, and psychological-pedagogical competence of those teaching at engineering universities and no tool that would be capable of ensuring the desired outcomes is being gradually overcome in the interaction practice of the leading Russian universities, primarily flagship ones of PJSC Gazprom, through the networking system within SYNERGY.

Most sessions of the conference are held in the online mode, the speeches being broadcasted. Some SYNERGY events are held as webinars.

Statistics of the conference events proves that Russian researchers and business representatives trust in its scientific and practical importance and highly estimate its create format. One of the most important outcomes of the conferences is extending the business and scientific contacts of the representatives of PJSC Gazprom's flagship universities, the increased publishing activities of professors, including publications in foreign titles, such as collections of works published by international engineering societies IGIP (International Society for Engineering Education) and ASEE (American Society for Engineering Education), both included into the Scopus international citation database. For two years the Synergy has been existed, about 1,000 people have participated in its events, and about 300 papers have been published, including 30 papers published in the Russian rating journals and over 40 papers in the titles included in Scopus and Web of Science.

The authors of the present research strongly believe that the previous international and national conferences have never aimed at or achieved such goals as the ones set and achieved by the organizers and sponsors of the SYNERGY conference, namely: Development of public-private partnership, modernization of syllabus, advanced training and capacity building of conference participants, promotion of new educational technologies, implementing digital format in the context within a large country, such as Russia, and extending partner relationships with customer companies. Thus, the CDIO conception proves its universal applicability in very different aspects of scientific and educational activities.

Proven experience in holding an international network conference on engineering education, supported by the largest energy provider as its sponsor, has vividly demonstrated the fruitfulness of its innovative format chosen. Ideas and recommendations elaborated by our colleagues allowed us to draw the conclusion that the synergy of efforts made by the interested participants of the innovative development process of training engineers at universities within the context of global trends enables creating a single environment of higher technical and technological education.

References

1. Starikova, L.D., Khokhlova, N.V.: Special features of building individual educational trajectories using electronic contents within the aspect of globalization and open education. Pravo i obrazovaniye [Law and Education] **4**, 41–47 (2018). (in Russian, abstract in Eng.)
2. Ivanov, V.G., Gorodetskaya, I.M., Kaibiyaynen, A.A.: Engineering education for a resilient society. Vysshee obrazovanie v Rossii [Higher Education in Russia] **12**, 60–69 (2015). (in Russian, abstract in Eng.)
3. Zinkovsky, M.A.: Practice-focused formation of the Russian business in the conditions of sanction policies of Western countries. Pravo i obrazovaniye [Law and Education] **4**, 48–52 (2018). (in Russian, abstract in Eng.)
4. Sazonova, Z.S.: MADI-IGIP tutorial workshop: history and outlooks. Vysshee obrazovanie v Rossii [Higher Education in Russia] **2**, 30–38 (2015). (in Russian)
5. Melezinek, A., Prikhodko, V., Zhurakovsky, V., Fedorov, I.: International society for engineering education. High. Educ. Russ. **3**, 53–59 (2004) (in Russian)
6. Ivanov, V. G., Barabanova, S.V., Galikhanov, M.F., Lefterova, O.I.: Advanced training of engineers in research university: traditions and innovations. In: Book Advances in Intelligent Systems and Computing, vol. 544, pp. 353–361 (2017)
7. Sapunov, M.B., Tkhagapsoev, Zh.G.: Culture of critical discourse in higher education and science (paging the magazine). Vysshee obrazovanie v Rossii [Higher Education in Russia] **3**, 20–27 (2018) (in Russian)
8. Ivanov, V.G., Zhurakovski, V.M., Barabanova, S.V., Galikhanov, M.F., Suntsova, M.S.: New trends in training engineers in Russia. In: 2015 ASEE International Forum. Seattle, Paper ID. 14374. Washington, 14 June (2015)
9. Barabanova, S.V., Ivanov, V.G., Zinurova, R.I., Suntsova, M.S.: On legal support for engineering activities: a new managerial project. In: ICL2017–20th International Conference on Interactive Collaborative Learning, pp. 1087–1096. Budapest, Hungary, 27–29 Sept (2017)
10. Sanger, P., Ivanov, V., Barabaniova, S., Ziyatdinova, J.: Training the trainer: an integrated university/industry program of improving Russian industrial trainers 2014. In: ASEE 121 Annual Conference, Paper ID. 9334. Indianapolis, Indiana, 15–18 June (2014)

Changes in Preparation of Future Teachers of Vocational Subjects in a Confrontation with FEP in the Czech Republic

Jan Válek and Petr Sládek[✉]

Department of Physics, Chemistry and Vocational Education, Faculty of
Education, Masaryk University, Brno, Czech Republic
{valek, sladek}@ped.muni.cz

Abstract. Currently the Framework Educational Programme (FEP) is updated
in the Czech Republic. The Framework Educational Programme and its content
of each subject in each branch should correspond to the contemporary perception of the world of contemporary young people. The changes should be based
on the needs of students as well as on the requirements of practice. In most cases
only the range of expected outputs and curricula is increased, but without any
reduction of other unnecessary content. Adjusting of links between the different
disciplines, if any, is minimal. The aim of the paper is to present different views
on the changes of the Framework Educational Programme, especially for
vocational secondary schools. We focus our attention to the study program of
vocational subjects/practical training teacher with respect to the changes of
FEP. With varying output requirements of students of secondary vocational
schools, the demands on the future teachers of vocational subjects/practical
training change too. We must answer on question: How long in advance, how,
and what is the way for the university preparing future teachers of vocational
subjects/practical training when designing and to realizing the innovation of
their study program? Through a meta-research survey, we summarize the data
that present other views on the Framework Educational Programme.

Keywords: FEP · Professional competence · Practical training · Teachers

1 Introduction

Basic Curriculum Documents in the Czech Republic are the Framework Educational
Programme (FEP) and the School Educational Programme (SEP). Thus, the general
educational objectives in the Czech Republic are set by the FEP, which is a binding
document for teachers' creation of SEP. The FEP establishes mandatory education
requirements for each degree and branch of education. The FEP as basic pedagogical
documents (for different levels of education) is approved and issued by the Ministry of
Education, Youth and Sports of the Czech Republic.

FEP should also ensure students permeability between schools across the Czech
Republic and also has teachers provide a framework in which students will learn. The
system of curriculum documents thus indirectly follows that of the years 1990–2007
(Educational Programme/Curriculum); which also determined the core curriculum but

© Springer Nature Switzerland AG 2020
M. E. Auer and T. Tsiatsos (Eds.): ICL 2018, AISC 916, pp. 483–494, 2020.
https://doi.org/10.1007/978-3-030-11932-4_46

the school had little opportunity to adapt it to current circumstances and needs of local character.

Although the FEP authors promised to be different from the old Educational Programs/Curriculum (these documents were valid before the introduction of the FEP, i.e. by 2007) and the declared freedom for schools and teachers, we often find that the opposite is true. Consequently, the current FEPs are often very similar to, if not identical to, old programs/curricula. Given that before 1989 (when the Velvet Revolution took place in Czechoslovakia), there was a different political situation in Czechoslovakia at that time, and centralized management was taking place, then the curriculum at that time was rather conservative. Changes in policy have led to the release of rules and responsibilities, and schools can partially intervene in programs.

Contemporary trends in both primary and secondary education are formulated in the FEP for each type of school. They emphasize key competencies and cross-cutting themes, respecting the new education strategy, not only in the Czech Republic, but the Europe-wide Copenhagen Declaration of 2002.

In preparing future teacher of Vocational Education Training (VET Teacher) and teacher of Vocational Specialized Subjects (VSS Teacher) for Secondary Schools, we should always rely on valid and approved FEPs for Secondary Vocational Education (SVE), corresponding to ISCED 3–4.

As mentioned above, the FEP is a document that is binding on the organization and content of teaching different professions in different schools. Currently we have 281 (!) valid FEP for Secondary Vocational Education (FEP SVE). Is it at all possible, however, that we prepare VET or VSS teachers in one university study program that implement a curriculum according to one of the 281 FEP SVE, at the same time?

2 Approaches to Creating Curriculum Documents

In the past, many documents and traditions from Austrian and German neighbors have traditionally been taken over in the Czech lands, and so has it been in the case of education. Today, however, it appears also other directions in Czech education. Thus, we are in the face of two approaches to content and learning objectives.

A more traditional approach is not only in the Czech Republic focusing on holistic development, content and intentionality of education - *Bildung*. In the process of education, work is carried out with the vision of the state/objectives in which the student should be located. It can be said that he is educated (he can demonstrate this by controlling the area in which he is educated). This approach/tradition includes not only the "educated by someone" page but also the "educate yourself" page, which is very important to realize. It is, in essence, also an unfinished process, i.e. with a view to lifelong learning. The emphasis is on educational results [1].

Against this, at least in the last decade, the *Learning* approach is applied. There is a partial weakening of the role of the state as a guarantor of the curriculum and the influence of the market is strengthened. We can also describe this relationship as the seller-buyer. Or, education is bought as a commodity. The content of education is gradually reduced, and content curricula are weakened in favor of curriculum only in the framework of the defined or very detailed (serving testing) curricula. The *Learning* approach is rather progressive and works with what will come now or in the near future. Learning outcomes are expressed as an assertion that a graduate of learning knows or knows something specific. One put the emphasis on learning outcomes [2].

Currently, curriculum issues are addressed in the Czech Republic by curricula revisions, and it is not clear whether there is a change in education from the type of input control to management derived from outputs, i.e. from *Bildung* to *Learning*.

A similar process is also required in higher education, where more emphasis is placed on "learning outcomes".

3   Timeframe for FEP Adjustments

In view of the expected/ongoing revisions and adaptations of the FEP, further questions need to be raised, mainly.

For how long does the change in the curriculum of a university study program reflecting the changes in the FEP SVE affect the teaching process of students at secondary vocational schools and subsequently in their own practice?

In addition, who should initiate changes to the FEP: Practice? Secondary vocational schools? Universities preparing VET/VSS teachers?

Let us look at the time horizon of applying the changes in the FEP SVE. If we consider the fastest possible way, we get to 7–11 years. This happens when the change of system comes from universities and we are interested in when the transfer to practice in labor market will take place. So, let's just think of the shortest possible case: In the beginning, the university's workplace will reflect current changes in society and in practice, and will initiate changes in the curriculum of the VET/VSS teacher study program. Before it can admit students to this modified program, it will take one year (new students can start in the next academic year). The duration of bachelor study program is three years. Those teachers immediately after graduating go into school practice. It will take another three years for vocational secondary school students (apprentice) to graduate. In total it takes $1 + 3 + 3 = 7$ years.

The whole process is prolonged when VET bachelor's degree program students will immediately start a follow-up Master's program for two years after their bachelor's degree graduation. If we consider matura graduation programs at vocational secondary schools i.e. 4 years we obtain in total $1 + 3 + 2 + 4 = 10$ years. The most extreme is the case when we take in account additional 9 years for a pupil already undergoing changes of FEP for 1st to 9th grade. Here we are getting to an unbelievable 19 years of transfer process new requirements - labor market practice.

As we can see, this system is not very flexible. And we did not consider that the requirement for change at university level can be initiated from a secondary vocational education or labor market. Here it is necessary to add some time to the analysis of the state, which could last at least five years, in order this is not to be a sudden and isolated decision. In total it is expected that the change in FEP will take 12–16 years. More in Table 1.

According to the National Accreditation Bureau for Higher Education in the Czech Republic, future accreditation of university study programs should be more general, which of course is not quite possible in the case of VET study program. There is always a strong emphasis on concrete curriculum and practical examples.

Table 1. Time horizon for editing FEP in years

		Variant A	Variant B	Variant C	Variant D	Variant E	Variant F
I	Changes in the university curriculum	1	1	1	1	1	1
II	University ISCED 5A	3 + 2	3	3 + 2	3 + 2	3 + 2	3 + 2
III	Primary & Lower secondary education ISCED 1 + 2A	9					9
IV	Upper secondary education (Vocational education) ISCED 3		3	3	3 + 2	4	3 + 2
V	$\sum$	15	7	9	11	10	20

Note 1: In row II, Bachelors are preparing for three years, follow-up Masters are preparing for two years.
Note 2: In row IV we consider either an ISCED 3A (4 years) or an ISCED 3C (3 years) or an ISCED 3C + 2 years.

4 Implementation of the FEP in the Czech Republic

Simultaneously with the creation of the Framework Educational Programme, a transformation of a system of education fields occurred in the secondary vocational education. This step was inevitable, because the system needed to better take into account not only labor market demands but also changes in production technologies and work activities. Thus, after the modification, the courses were reduced from almost 720 to 281. The last framework educational programs were issued in 2012, which can be considered as the year of the launch of the education under the FEP.

Prior to this change period, which started in 2007 at FEP for 1st to 9th grade of elementary education and ended in 2012 at FEP SVE (secondary vocational education), a similar solution was outlined, but not so robust. Already after 1989, schools have been able to create and implement their own educational plans for the education of their students. This is also connected with the increased competencies of the directors in terms of determining the individual subjects, which the school will devote to the increased time allocation or the modification of the contents of the subjects. This step had to be backed up by the basic study literature, so it was the school's authority what textbooks and other materials will be used [3].

Before the introduction of the FEP in the Czech Republic, the Educational Programs were valid. For elementary education, they were: Elementary School; Grammar School; Civic School and National School and of them based on curricula [4].

For the secondary vocational education (SVE) there were about 720 educational programs, as we have mentioned above, so it is not possible to list them here.

The individual education programs for elementary education show some differences, but the basic students' knowledge on the output for each level is comparable. Here it is necessary to mention that it has not always been explicitly stated what the objectives of individual subjects or even the overall elementary education; but the content has been given [4– 6].

The basic themes of the FEP were introduced in 2001 in the National Program for Development of Education in the Czech Republic (so-called White Book, Program Document of the Government of the Czech Republic). The approval of the FEP took place in 2004. Pilotage of selected parts of the school education program took place in September 2004 at sixteen elementary schools, where it was taught according to the FEP, or their own school educational program (SEP). A pilot study was completed in June 2006. The teaching according to the FEP began in September 2007 for the students from 1st and 6th grade elementary school, teaching at the secondary vocational schools then began in 2009 [4–6].

Regarding the implementation of the FEP, it was logically structured. Thus, it progressed gradually and copied the situation as pupils of elementary schools gradually passed to secondary schools. The individual stages and years of introduction of FEP into Czech education are shown in Table 2.

Table 2. The course of the reform - Publishing of FEPs

	Approval of a FEP	At school, they learn in the first years under the FEP and SEP
Pre-primary education Kindergartens ISCED 0	March 1, 2005	September 1, 2007
Primary & lower secondary education ISCED 1 + 2	August 31, 2005	September 1, 2007
Upper secondary education (General education) ISCED 3A	July 24, 2007	September 1, 2009
Upper secondary education (Vocational education) ISCED 2C, 3A, 3C, 4A, 4C	August 31, 2007	September 1, 2009

Source: Ministry of Education, Youth and Sports of the Czech Republic
Note 1: For row with Upper secondary education (Vocational education): Only graduates have been progressively introduced for selected vocational training areas.

Individual FEP for SVEs have been published in waves since 2007 [7]. The Ministry of Education, Youth and Sports issued a separate FEP for each branch of education listed in the Government Regulation on the System of Education in Basic, Secondary and Higher Vocational Education (281 SVE FEPs were issued).

5   Demographic Developments and Implications for School Practice

As teachers, for whatever grade, from nursery schools to secondary schools, according to the Ministry of Education, Youth and Sports of the Czech Republic, is a regulated profession, it is necessary to plan the profession properly. As a rule, a specific university degree and branch of education is required. The number of teachers also go hand in hand with the demographic development in the state, which determines the required capacity (personnel and material) that are part of the educational system.

What should be done if there is a change in the numbers of enrolled students in schools? For example, the last major population wave was in 2008–2010, when 118,000 children were born on average in Czech Republic, which represents an increase of about 28,000 (one quarter!) compared to 1999. This difference must also be reflected in educational system. From today's point of view, it is not clear if the educational system has been sufficiently prepared for this as regards the number of qualified and approved teachers. If we consider that children born in 2008 were enrolled in elementary schools at their seven years, i.e. 2015–2017, they would enter the secondary school after nine years of compulsory school attendance in 2024–2026. They then join the university in the years 2028–2030.

We reiterate the question: Is educational system for this population wave ready? And how long in advance does it take for the state (county, municipality) to start solving this situation? Considering the above-mentioned time horizon for elementary, secondary and higher education, it can easily be calculated that the state has addressed this situation already for the generation of teachers born between 1990–1992 and the university graduated between 2015–2017. Those teachers immediately after graduation enter the school and teach in the 1st year of primary school a population wave born in 2008–2010. If we want teachers to have at least five years of experience, then they were born in 1985–1987 and graduated between 2010–2012 This means that soon (no later than a year later) after obtaining statistical data on live-born children state must respond (five years to prepare teacher at university).

For secondary schools, the situation is still seemingly acceptable in the sense that we could get a teacher with the year of birth 1999–2001; in the case of five years of practice, we would need teachers born between 1994–1996. In the case of vocational subjects (not general education), however, it is necessary to count on their professional practice in the appropriate branch (not only the teacher's practice), then the time for the reaction is proportionally shortened. The individual years are presented in the Table 3.

The opposite problem occurs on the part of teachers. Here we will have the problem in about 10 years that there will be a shortage of them. Why? Since there were an average of 175,000 children born between 1970 and 1980 in Czechoslovakia every year, this is the most active group in the working age (Source: The Czech Statistical Office). If we consider that getting into schools as teachers in 1990, it is now (2018) we expect 10 more years of their teaching career (we assume that the teacher will work for 40 years). For teachers from the end of 70's we expect about another 20 years of

Table 3. Distribution of years for teaching children from population boom in Czech Republic at 2008–2010

	Year of birth	Starting attendance at primary & lower secondary education	Starting attendance at upper secondary education	Starting attendance at University
Population boom	2008–2010	2015–2017	2024–2026	2028–2030
Primary & lower secondary education teachers for population boom	1990–1992	1997–1999	2006–2008	2010–2012
Primary & lower secondary education teachers for population with 5 years of practice	1985–1987	1992–1994	2001–2003	2005–2007
Upper secondary education teachers for population boom	1999–2001	2006–2008	2015–2017	2019–2021
Upper secondary education teachers for population boom with 5 years teaching practice	1994–1996	2001–2003	2010–2012	2014–2016
Upper secondary education teachers for population boom with 5 years teaching practice and 5 years' experience in the field	1989–1991	1996–1998	2005–2007	2009–2011

Note: 1: For teachers, the starting year for a university is set at five years earlier because they need to graduate before they enter the school.

Note: 2: The same color indicates years in the table, which are related to the preparation of teachers and the enroll of students in schools.

teaching. In contrast, in academic year 2015/2016 have enroll university population-weak vintages, while in the first classes of primary school (ISCED 1) coming baby boom from 2008 to 2010. These groups will meet at grade 2 of elementary school (ISCED 2) in 5 years, which will be a problem as schools will not be able to provide quality education with qualified teachers because they will not exist in labor market. The data is presented in Fig. 1.

Another serious problem is the fact that teachers are expected to always give 100% performance no matter what. So even despite the low number of teachers. It is clear from psychohygiene that if a teacher is overloaded by either an administration or a large number of students in a class, he/she cannot perform 100% of the performance. This is another aspect that few people are aware of (Table 4).

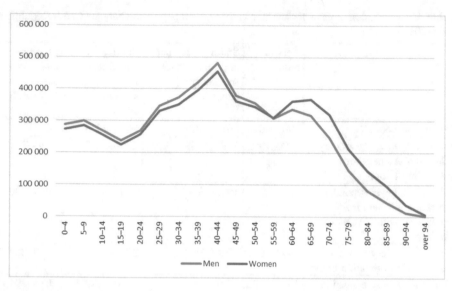

Fig. 1. Population composition of the Czech Republic as at 31st December 2017

Table 4. Number of employed Lower secondary and Upper secondary education teachers in Czech Republic in years 2006–2018

School year	Lower secondary education		Upper secondary education	
	Full time employee	Non-qualified full time employee	Full time employee	Non-qualified full time employee
2006/2007	34,931	5,026	47,452	7,056
2007/2008	33,453	5,149	47,124	6,850
2008/2009	31,963	4,751	46,735	6,428
2009/2010	30,782	4,362	46,489	6,324
2010/2011	30,227	3,974	45,385	5,759
2011/2012	29,700	4,174	43,876	6,580
2012/2013	29,294	3,490	41,789	5,094
2013/2014	29,244	3,120	40,214	4,131
2014/2015	29,240	2,466	39,070	3,124
2015/2016	29,391	1,490	38,385	1,686
2016/2017	29,807	1,527	38,069	1,437
2017/2018	30,552	1,649	38,114	1,369

Source: Ministry of Education, Youth and Sports of the Czech Republic

6 What Should be in the VET and VSS Teacher Study Programs?

Framework Educational Programmes are documents designed to establish a common framework for schools and teachers. This concept emphasizes key competencies important to a person's daily life, their connection to educational content and the use of acquired knowledge and skills in the practical life of each and every one of us. There is a departure from the encyclopedic concept of curriculum documents. The concept itself brings a lifelong learning model. Among other things, the FEP also supports the pedagogical autonomy of schools and the professional responsibility of teachers for the results of their education [4–6, 8].

We must not forget about another aspect that is reflected in today's society and hence in Education, which is Industry 4.0. Its impact on vocational training will not be small. Unfortunately, there are no systematic studies concerning the influence of the Fourth Industrial Revolution on technical education. It is understandable because this phenomenon appeared only three years ago and the first information appeared in journals and on the Internet in 2014/2015. On the Web of Science, after entering the query "Education Industry 4.0" in the "Education - Educational Research," we get to May 31st, 2018 a total of 20 articles containing the keywords above. If we look specifically at the context of articles, we find that our topic suits only five contributions. We will therefore rely on the results of theoretical analyzes.

It is evident that with advancing robotics, automation and the introduction of smart factories the innovative processes in the field of technical education at all types of schools will have to be more dynamic. The strategic importance of innovation will take place mainly in professional technical education at secondary schools and universities. Main areas of innovation can be summarized as follows [9]:

- Ability to solve problems, flexibility and adaptability.
- Ability to propose innovative solutions, creativity, systematic thinking.
- Ability of sectoral and intersectoral communication, team work.
- Ability to withstand the workload and tense situations

Just as the professional public does not agree, who is the "buyer" of the FEP (for any level of education), and the individual university departments preparing VET and/or VSS teachers will never completely overlap in their fields of study curricula and the preparation of VET/VSS teachers. Some say that the customer is the student of others that the teacher.

Some disciplines are taught similarly (in a similar scale or arrangement in each semester, but the rest is different.) We expect that in the course of newly prepared re-accreditation, they will become more unified as the Framework Requirements for Teacher Study Programs entered into force; as the profession of teacher is the regulated profession (Czech Ministry of Education, Youth and Sports of October 5th, 2017). These requirements were fixed unilaterally by ministerial officials without consulting the university guarantors corresponding study programs.

We also believe that it is necessary to focus beyond the basic questions "Can we properly prepare the future VET and VSS teachers in confrontation with the FEP SVE" to focus also on:

- How many VET and/or VSS teachers do we need in practice?
- What specializations are they supposed to teach?
- What are the numbers of graduates?
- What are their opportunities on the labor market?

From the above aspects we should answer the question: *How to construct the study program for future VET and VSS teachers?*

As we mentioned above new Framework Requirements for Teacher Study Programs entered into force in the Czech Republic. Looking at the curriculum of programs offered by individual university workplaces in the Czech Republic, we find that they are similar in some areas. Typically, the department agreed in didactics courses, which are always represented by a minimum of three semesters. Similarly, there is always a subject or more subjects focused on work with Didactic Technology and Information Communication Technologies. Usually, Management is also offered. Often the Economy or, for example, the basis of work safety and health protection and fire protection in school practice. Less often Environmental Studies, Law, or School Policy. Rarely subjects focusing on the natural sciences (Physics, Chemistry or Biology) and only at the Pedagogical Faculty of Masaryk University, Mathematics, and subjects focused on Transport (Source: Catalogs of study programs and of study subjects of Czech universities).

Specializations in Technical branches or Business and Services, sometimes Health Services occur most often from the third, not later than from the fourth semester. This division is based both on tradition and on the other hand it looks logical to the generic designation of large groups of vocational education. Looking at the current professions, or rather the professions that youngsters want to do, it is suggested that a different division in specific cases should be created.

Anyway, everyday work with the rising immersion of the population into the digital world and social networks must be included in the equipment of the teacher (not just VET or VSS teacher).

We see that we are slowly plotting, the composition of the subjects that should be in the curriculum. It is perhaps more important for students to pass them such information and to teach them such skills that enable them and their students to succeed in practice.

7 Discussion and Conclusions

As noted above, the Framework educational Programme is a document that is updated both in terms of current and mainly future societal needs, new teacher experience and supposed pupil needs, but the process of transferring change into practice is lengthy.

The FEP and its content of each subject in each branch should correspond to the contemporary perception of the world of contemporary young people. However, it can fundamentally change the current negative attitude towards science, for example? The document itself probably not, it is necessary to actively involve teachers. Here we can encounter a problem of the natural renewal of natural science teachers [10, 11].

The fact that the final version of these documents could have been interfered with by people who did not have any professional education in the given disciplines, we are all aware of, but we believe together that this happened not in a great extent.

We believe that the requirements for the content of education at primary and secondary schools and universities should be set by experts who know the current state of science, the development of knowledge and the need for practice rather than the educators. The composition and methodical processing should then be created by the university preparing future teachers along with experienced teachers from practice. However, it should not be forgotten what the aim of primary and secondary education is, i.e. to prepare students for everyday life, to be employable in the labor market and to be able of lifelong learning. This should ensure a good structure of FEP.

Another one positive aspect for graduated VET and VSS teachers should be mentioned. At a time when the economy is prosperous state there is a lack of these teachers at secondary vocational schools, but when the economy is (not only the Czech Republic but also in other European Union countries) in recession, many experts will move to secondary education. So, we can say, the school is a rescue plan for many of them.

Another thing to be mentioned is that one of the objectives of the FEP is to motivate the student to a lifelong learning. This goal is easier to define than it does, but we believe that when looking at statistical data, the number of university graduates in the Czech Republic has increased from 30,102 (2001) to 72,057 (2017). Similarly, a number of applicants and graduate of retraining courses increases.

Following the retraining courses and the current developments in technology (Industry 4.0), we realize that it will be more important to equip students/graduates with the core competencies that will help them manage situations that we currently do not even know that could happen. At the same time, the role of the teacher will gradually change from the main source of knowledge for students and school sur-roundings (like in the 18th century) to a guide and manager of education (in the 21st century). The teacher will no longer have to be a specialist in a particular branch but will have to be a good organizer of students' activities during their stay at school.

References

1. Fuhr, T.: Bildung. In: Laros, A., Fuhr, T., Taylor, E. (eds.) Transformative Learning Meets Bildung: An International Exchange, pp. 3–15. SensePublishers, Rotterdam (2017)
2. Biesta, G.: Beyond Learning: Democratic Education For A Human Future. Paradigm Publishers, Boulder (2006)
3. MŠMT: Zpráva o vývoji českého školství od listopadu 1989: (v oblasti regionálního školství). Praha (2009). http://www.msmt.cz/uploads/VKav_200/zprava2009/zprava_vyvoj_skolstvi.doc
4. Šimoník, O.: Úvod do didaktiky základní školy. MSD, Brno (2005)
5. Tupý, J.: Tvorba kurikulárních dokumentů v České republice: historicko-analytický pohled na přípravu kurikulárních dokumentů pro základní vzdělávání v letech 1989–2013. Masarykova univerzita, Brno (2014)
6. Filová, H: Výběr z reformních i současných edukačních koncepcí: (Zdroje inspirace pro učitele). (J. Svobodová, Ed.). Brno: MSD (2007)

7. Rámcové požadavky na studijní programy, jejímž absolvováním se získává odborná kvalifikace k výkonu regulovaného povolání pedagogických pracovníků. (2017). Rámcové požadavky na studijní programy, jejímž absolvováním se získává odborná kvalifikace k výkonu regulovaného povolání pedagogických pracovníků. Praha: MŠMT. http://www. msmt.cz/vzdelavani/dalsi-vzdelavani/ramcove-pozadavky-na-studijni-programy-jejichz-absolvovanim

8. Walterová, E.: Kurikulum: Proměny a trendy v mezinárodní perspektivě. Masarykova univerzita, Brno (1994)

9. Pecina, P., Sládek, P.: Fourth Industrial Revolution and Technical Education. Valenica: IATED-int assoc Technology Education & Development, pp. 2089–2093 (2017)

10. Sládek, P., Válek, J.: (Ne)kvalifikovanost učitelů – létající učitelé. In H. *Cídlová XXIV. Mezinárodní konference o výuce chemie DIDAKTIKA CHEMIE A JEJÍ KONTEXTY. Sborník příspěvků z konference 20.–21. 5. 2015.* (pp. 187–192). Brno: Masarykova univerzita (2015)

11. Sládek, P., Válek, J.: Létající fyzikáři, In M. *Randa Moderní trendy v přípravě učitelů fyziky 7.* (pp. 204–212). Plzeň, Západočeská univerzita v Plzni (2016)

12. Přehled vydávání RVP SOV po vlnách, Národní ústav pro vzdělávání. (2017). Přehled vydávání RVP SOV po vlnách, Národní ústav pro vzdělávání. http://www.nuv.cz/t/prehled-vydavani-rvp-sov-po-vlnach. Accessed 21 May 2017

13. Harmonogram, MŠMT ČR: Harmonogram, MŠMT ČR (2013). http://www.msmt.cz/vzdelavani/skolstvi-v-cr/skolskareforma/harmonogram. Accessed 21 May 2017

The Main Trends in the Development of Engineering Education: The Role of the University Teacher in Systemic Changes

Khatsrinova Olga[✉], Svetlana Barabanova, and Khatsrinova Julia

Kazan National Research Technological University, Kazan, Russia
{khatsrinovao, khatsrinoval2}@mail.ru, sveba@inbox.ru

Abstract. The article is devoted to the development of competences of the teacher of an engineering university in accordance with the trends of modern education. Therefore, it is necessary to improve the qualifications of teachers in the direction of solving emerging new educational problems. He should understand new types of relationships with students, use new teaching technologies. Teaching teachers facilitates understanding of systemic changes and, as a result, the implementation of a new department of the educational process.

Keywords: University teacher · Tendencies in the development of education · Advanced training of teachers · Management of the learning process

1 Introduction

Today there are new trends in socio-economic development, which determine the requirements for the organization of learning processes. Modern education becomes more human-centered than it is for simple transfer and assimilation of knowledge. But the traditional pedagogy of higher education lags behind the requirements of the professional sphere. Therefore, the problem of professional training and development of the competence of the teacher of higher education comes first, because knowledge of the modern level can give a teacher with a range of characteristics.

The engineer becomes the main character of the 21st century. This tendency is also already looked through in an education system, in strengthening of a role of engineering education and applied, practical component of educational programs. It is simple to present a new role of the engineer in future social fabric. The engineer will be called for the solution of matters of principle of the organization of a way of life of society. He will choose safe and steady development strategies of society from a set of possible options, many of which bear threats, risks and accidents. The engineer will bear responsibility for the made decisions and to answer global challenges and threats. In such realities, an opportunity to prepare graduates with the qualification corresponding to the order of business and time becomes tasks of higher education institutions, reacting to dynamics of changes of inquiry in time frames of process of preparation; to guarantee quality of training of graduates according to demand, at the scheduled time [2].

© Springer Nature Switzerland AG 2020
M. E. Auer and T. Tsiatsos (Eds.): ICL 2018, AISC 916, pp. 495–502, 2020.
https://doi.org/10.1007/978-3-030-11932-4_47

Professional activity of the teacher of higher education institution is specific owing to its polyfunctionality. It is among the professions demanding from the worker of a number of qualities. This activity assumes competent and effective realization by the teacher of numerous professional functions. He has to know the taught subject, develop in itself(himself) broad cognitive interests and high informative requirements; to improve flexibility, depth, criticality and independence of thinking; to form orientation to universal values, etc. The teacher of higher education institution has to be organized and benevolent in interaction with students; to have an initiative and responsible approach in work. The professional competence of the teacher of the higher school in general is open synergetic system which components can change or be added in connection with development of pedagogical science and the changing requirements to the content of professional activity of the teacher from the state, student teaching, employers, the public, students.

1.1 The Purpose

The purpose of this study is to identify new types of training for an engineering university teacher. The types of advanced training presented by different researchers for the training of university teachers should be enriched by the specifics of engineering activity, a special emphasis should be placed on the development of the methodological component. The competency model should be, if possible, a full ranked list describing the personal qualities, behavior, knowledge, skills and other characteristics necessary to achieve quality standards and work efficiency. The teacher ceases to be the sole guardian of truth and knowledge. He must act as project manager and colleague.

In our work we rely on the ideas of sociocultural dynamics of society and A.P. Bulkin, P.S. Gurevich, N.D. Kondratyev, A.A. Pinsky, P.A. Sorokin, A. Toynbee's education. Training of the teacher of higher education institution has to be carried out, proceeding from multidimensional idea of it as about the active subject of educational process. The organization of educational activity of the student and designing of the educational environment becomes the main objective of the teacher. Activity of the teacher and activity of the student are interconnected (partnership is their cornerstone), at the same time the teacher in general comes for maintenance of activity of the student. Also we relied on competence-based approach in an education system. Realization of competence-based approach in education sets new tasks for the teacher of the higher school: development of new working programs, new estimated means, designing of occupations in a format of competence-based approach. The generalizing characteristic of the teacher of the higher school is given to Z.F. Esareva. "Teacher, – she notes, is a scientist who seized scientific methods of training and education, skillfully uses technical means of teaching, continuously improves the skills, actively participates in research work, in public life" [3, p. 51].

The analysis of these definitions shows that activity of the teacher of higher education institution is elaborate and includes variety of aspects of its manifestation (educational, educational, organizing, methodical, scientific, innovative, etc.) which differ in a form, contents, ways of implementation and functional orientation. In the set these aspects of professional activity of the teacher of higher education institution are aimed at the solution of the main purpose – reproduction of professional staff, socialization of the younger generation, production of knowledge and their transfer.

The analysis of the works devoted to reforming of the Russian higher school, problems of reproduction of research and educational personnel [4, 8], allowed to define a number of important tendencies which have significant effect on the nature of professional activity of the teacher of engineering higher education institution. It is possible to carry to such tendencies: – activization of Bologna Process and formation of the all-European system of the higher education; adaptation of domestic engineering school to integration into world educational space: the integration in domestic education of the European tendencies with identity of the Russian culture caused by participation of Russia in Bologna Process:

- integration of higher education institutions (creation of the federal universities), change of the status of higher education institutions (creation of the research universities);
- localization of processes of regionalization: increase of influence of system of the higher education on regional growth, strengthening of ties of higher education institutions with regional labor market and economic clusters, creation of regional university complexes;
- change of structure of research and educational personnel: aging of research and educational personnel, feminization of science and education, outflow of shots from science, secondary employment in labor market (combining jobs);
- transition to a new educational paradigm: from transfer of knowledge to formation of need for knowledge, emergence of new competences of professional activity.

It is obvious that the allocated tendencies affected the professional functions which are carried out by the teacher. Under the influence of these tendencies the teaching profession gains new lines, the new requirements connected, in particular, with the necessary level of information culture, social mobility, new professional competences (knowledge of foreign languages, professional mobility, professional communication) are imposed to it. High-quality changes in professional activity of the teacher, in effect, are some kind of answer to calls of new society. Corresponding changes find the reflection in the outlook changing at the teacher and the system of values, strategy and practice of his professional activity.

Y.V. Podpovetna defined a number of features of professional activity of the teacher of higher education institution. Are referred by scientists to such features: the various nature of the activity including besides pedagogical, the research party demanding existence of special abilities; accurate criteria of professional compliance (existence of academic degrees and ranks, scientific awards, number of the published works in the reviewed editions, etc.); accurate criteria of professional efficiency (performance of an academic load, methodical and organizational and pedagogical work, organization of conferences, etc.); prevalence of a communicative component in activity; specificity of means of activity (knowledge, abilities, culture, moral shape and individual style of the teacher); specificity of result of pedagogical activity ("materializovannost in people", "otodvinutost in time", difficult measurability and comparability) [5].

It is also necessary to note that professional and pedagogical activity of the teacher of higher education institution is implemented in several directions: organizing, methodical, scientific, innovative, etc. However often these kinds of activity do not form complete pedagogical system. Though they are structural components of professional competence of the teacher of engineering higher education institution. As a result the community of teachers is divided into "teachers scientists" and "experts teachers" (translators of knowledge) who perform narrowly directed functions.

2 Approach

Today it is possible to allocate three groups of contradictions of formation and development of competences of teachers of modern conditions of their activity, self-education and professional development: the sociocultural, all-pedagogical and personal and creative [6] Predictive purpose of development of competences can be considered professional and pedagogical culture as the high level of development of competences of the teacher and ability to realize professional and pedagogical activity at the level of professional skill. Let's open and will add them. Sociocultural contradictions find the expression in search of ways of overcoming crisis of pedagogical culture. Are the reasons of such crisis: lag of pedagogical system from the changing society, tendencies of development of engineering activity, underestimation of a sociocultural role of engineering education, absence of necessary pedagogical bank of information, etc. Crisis of pedagogical culture demonstrates existence of discrepancy between conditions of surrounding social reality, sociocultural processes and development of competences of teachers of higher education institutions. Permission of these contradictions helps to develop idea of a desirable image of the teacher of higher education institution – the scientist - the teacher, the carrier of high morality, culture, erudition. In the conditions of prolonged economic and moral crisis the efficiency of educational influence of the teacher decreases as the double task is assigned to it: it is designed not only to create the identity of future expert, to develop bases of his universal and professional culture, but also to keep and protect the personality from moral degradation.

However for this purpose the teacher has to realize values of the professional and pedagogical activity, be guided by the corresponding principles, commensurate the professional, social and private life with that moral ideal which carrier it is designed to be. A specific place is held by a contradiction between need of development of scientific and pedagogical capacity of the higher school and lack of sufficient conditions for optimum use. In the conditions of the amplifying commercialization of the highest, in particular engineering education, it is scientific - the pedagogical potential of the teacher in many respects is unclaimed as does not bring fast commercial result, that is arrived.

Other contradiction – between objectively existing common cultural traditions and experience and their insufficient account in professional activity of teachers of the higher school. This contradiction indicates a problem of influence of professional and pedagogical culture of the teacher on development of pedagogical culture of society. It is not casual all outstanding teachers scientists pointed to an important role of public and educational, propaganda activities. The teacher of the higher school has to assume responsibility for the approval of norms and values of pedagogical culture, both in the professional activity, and in the social environment surrounding it. In modern conditions the ability of the teacher to use information technologies, technical means of training, the standardized diagnostics of achievements of students, i.e. external means, than ways of pedagogically caused communication with students developed by him, their motivations on educational activity, creation of the comfortable psychological environment, stimulation of their professional formation as the identity of the professional is more and more appreciated.

The intensity of knowledge and development of new values of pedagogical culture is defined by change of priorities in the theory and practice, the level of readiness of technologies of educational process and also the professional focused installations of the teacher, his motivation, degree of satisfaction with pedagogical activity. In permission of this contradiction the important place is taken by orientation of the identity of the teacher, pedagogical collectives on harmonious mastering and development of values - the purposes, values knowledge, values technologies, on purposeful formation of values relations and values qualities.

Recognition of new pedagogical values stimulates creative search of teachers, promotes the adoption of perspective approaches, technologies, the systems of the relations in pedagogical process. In cases when there is no systematic updating of an arsenal of pedagogical values, the probability of emergence of stereotypic actions of elements of pedagogical conservatism and stagnation increases. Professional deformations are inherent in a considerable part of the teaching case of engineering higher education institutions [6].

Transition to other organizational and pedagogical relationship is connected at teachers with the difficulties having the various nature. Model of educational process, obvious at change, there is a difficulty of his understanding as the system having other properties and communications between components. Considerable problems of the reflexive and methodical plan - lack of the sufficient knowledge and abilities necessary for design and realization of nonlinear model of educational process and realization of new roles with possession of information and pedagogical technologies were allocated. Besides, it is psychologically difficult for teachers to accept other system of values of education, the student as equal participant of educational process, to understand and accept the directions of changes. It can be explained with decrease in intellectual level of students.

For the last decades students significantly changed in social, psychological, physiological aspects. According to experts of Russian joint stock company, they lower creative abilities, vigor of thinking, strong-willed qualities, the emotional discomfort increased, requirement of screen stimulation of informative processes and activity in general is accurately shown. According to the international research of school students of PISA for 2012, GPA of progress of the Russian school students (in comparison with the leading countries of Europe, Japan, China, South Korea) below an average (494) on 12 points.

Allocated internal (personal and creative) and external (sociocultural and all-pedagogical) contradictions are interconnected and are driving forces of development of professional and pedagogical culture of the teacher of the higher school. They allow to define the directions of development of professional competence of the teacher of higher education institution.

The analysis of practice of increase in pedagogical competence of teachers of engineering higher education institutions of TsPPKP at the Kazan national research state technological university and some other the centers allows to allocate a number of problem points in professional competence of the teaching case of engineering higher education institutions. At the same time results of the research [7] conducted by us according to which 64% of the interviewed teachers prefer to improve the skills in the traditional way – through courses are very indicative, 12% - would like to have the mentor (these are generally beginning teachers).

Thus, the overwhelming number of teachers (76%) prefer that they were trained. Only 28% expressed in favor of self-education and only 6% stated intention to study remotely on the Internet and to gain additional knowledge.

The research was conducted in four directions: difficulties of system thinking, reflexive, technique - technological, acceptances new. The low level of proficiency in information technologies (at teachers of technical disciplines it higher, than humanitarian) - 74% and new pedagogical technologies of - 64% came to light. Possession of a limited set of estimated means (tendency to decrease: teachers of technical, humanitarian disciplines) - 56%. On distribution of time for kinds of activity the priority in planning of own activity, but not educational activity of students was designated at teachers. So, planning of educational activity of students comes down to formulation of questions on discipline, definitions of tasks for practical and independent works. Planning of activity of the teacher is connected about planning of content of education, portions of its presentations, but, as a rule, there is no judgment of technological strategy of giving of a training material, work with it.

As shows experience, among the studying teachers various ways of activization which are widely used in foreign practice of training of adults are effective. The greatest activization of students is promoted by the following methods of training. Analysis of practice: students study real situations as teachers carry out different types of pedagogical activity, or roles as they react to various situations. "Groups of criticism": the group is divided into two or more subgroups and participates in assessment of the presentation which is carried out by the third group. At the same time the task of the first group consists in revealing the positive moments, and a task of the second subgroup - in the statement of critical remarks. The role of the teacher consists in synthesis of comments of both groups, and formation of knowledge is carried out on the basis of development of critical thinking. Role-playing game: when students have an opportunity to apply knowledge and skills in the situations demanding fast decision-making. Potential of a game increases as a lot of things teachers are a staff of one higher education institution, and the considered situations are connected with real practice. Method of analysis of cases: students in groups or independently analyze situations. Debate: the presentation of the clashing opinions two people or two groups for the purpose of clarification of their positions. Formation of knowledge becomes more active when people represent the antiput points of view, at the same time it is not obligatory that they share these points of view. - Problems from experience. Participants of group write down problem situations against which they came up in practical pedagogical activities.

Peculiar "bank" of the problem situations or questions relevant for participants of this group is created. Further students are offered to discuss possible solutions of these problems in common. The role of the teacher of system of professional development comes down to generalization and systematization of knowledge, to additional comments whenever possible of use of this knowledge in new situations. Application of these methods of training promotes activization of formation of pedagogical knowledge and abilities of young teachers of technical college. The motivation when studying a course increases when knowledge and abilities can be applied at once in practical pedagogical activities.

3 Conclusion

The results of the analysis of teachers' difficulties showed that the difficulties identified are not just their personal problems, but they are connected with the general value shifts in education, with the contradictions of theory and real practice, a subconscious understanding of this contradiction by teachers and, as a result, a feeling of dissatisfaction. Teacher training for systemic changes in the educational process of the university assumed a gradual removal or at least a reduction of contradictions. After training in the Center for Advanced Training of University Teachers, the greatest growth was noted for the indicator of personal interest in changing the nature of management (27% on average), systemic understanding of the changes (by 24%) and the use of reflexive procedures (16%). Methodical support of teachers, carried out in methodological seminars and individual consultations, allowed to increase personal interest in changing the nature of management (on average 51%), systemic understanding of the changes (by 46%) and the use of reflexive procedures (by 34%).

It is also possible to claim that self-development of the teacher in methodical activity is directed to management of the developing system of continuous professional development of pedagogical shots, initiation of pedagogical creativity and development of modern educational technologies; the didactic and methodological support of introduction of new content of education and training, i.e. is connected with methodical maintenance of professional activity of teachers. Besides, self-development of the teacher of higher education institution in methodical activity holds a specific place in a control system of educational institution, is closely connected with its main objectives and functions and is caused by activization of the personality and creative activity of the teacher. Self-development of the teacher of higher education institution in methodical activity happens in the course of exchange of experience with colleagues: concerning application in educational process of modern information technologies; on introduction in educational process of innovative educational technologies, to holding imitating, business, organizational and activity and other games, the solution of production tasks and concrete situations, etc.; during approbation new and modernizations of the operating laboratory works; at discussion of lectures and other studies attended at colleagues; by preparation for scientific and practical seminars, conferences and so forth; in the organization of work of methodical seminars for beginners, unexperienced teachers, etc. [9]. Thus, we came to understanding that self-development of the teacher of higher education institution in methodical activity is connected, as a rule, with development and adaptation of ready pedagogical innovations (the ideas, development, projects, etc.), their introduction and generalization, however, without analysis of regularities of existence and mechanisms of emergence.

References

1. Vorobyova, I.M.: Strengthening of a role of engineering education and practical component of educational programs in technical college. In: The Young Scientist, vol. 11, pp. 1304–1307 (2015). https://moluch.ru/archive/91/19565/. Accessed 29 May 2018
2. Sadovnichiy, V.A.: The higher school of Russia: traditions and present. The report at the VII congress of the Russian union of rectors, Almamater (The bulletin of the higher school), vol. 12, pp. 7–12 (2002)
3. Esareva, Z.F. Features of activity of the teacher of the higher school, 278 p. (1974)
4. Bard, G.A., Nesterov, A.A.: Trapicin XIU. Quality management of educational process, Saint Petersburg, p. 9 (2001)
5. Podpovetnaya, Yu.V.: The system and synergetic STRATEGY of management of humanely focused scientific and educational process of the university. Bull. Moscow State Reg. Univ. (Online Magazine) **3**, 89–93 (2011)
6. Filippov, V.M.: Education for new Russia, The Higher Education in Russia, vol. 1, p. 11 (2000)
7. Hatsrinova, O.Yu.: Methodical competence of the teacher of engineering higher education institution and its development (education guidance),160 p. Publishing Center "Shkola", Kazan (2015)
8. Fedorova, T.S.: Pedagogical technologies: collection of educational projects. In: Fedorova, T.S., Neudakhina, N.A. (eds.) Viola. state. техн. un-t of I.I. Polzunov, 102 p. Publishing house of ALTGTU, Barnaul (2015)
9. Olga, K., Ivanov, V.G.: Psychology and pedagogical maintenance of formation of career competence of future engineers. Adv. Intell. Syst. Comput. **715**, 287–292 (2018). https://doi.org/10.1007/978-3-319-73210-7_34

Networking Between Engineering University and Enterprises in Future Students Training

Alla A. Kaybiyaynen[1(⊠)], Vladimir V. Nasonkin[2],
Dmitry V. Bondarenko[2], Andrei V. Nazarov[1], and Gennady F. Tkach[3]

[1] Kazan National Research Technological University, Kazan, Russia
alhen2@yandex.ru, digger_45@rambler.ru
[2] Federal Centre for Educational Legislation, Moscow, Russia
nasonkin@mail.ru, bond_22@inbox.ru
[3] Peoples' Friendship University of Russia (RUDN University), Moscow,
Russia
olga-mandusova@yandex.ru

Abstract. This article shows the experience of engineering university in working with future students through the organization of network cooperation with partner enterprises. The article also analysis the efficiency of such forms of networking as a creation and development of children technopark, method of professional tests of engineering talented children, participation in professional skills competition of young professionals WorldSkills Junior, competitions of school children innovative projects etc.

Keywords: Networking between university and partner enterprises ·
Children technopark ·
Professional skills competition of young professionals WorldSkills junior ·
Career guidance activities · Specialized classes ·
Competitions of innovative projects of schoolchildren

1 Introduction

The experience of world best engineering universities shows the efficiency of network cooperation between university and partner enterprises.

The current research highlights the specific role of the university context in networking activities, and in particular, the development of particular types of networks, namely, social and business [1].

However, today it is necessary to develop such cooperation not only in teaching bachelors and masters at the university, but also at the stage of attracting technically talented children, developing their interest in engineering professions. This will allow carrying out more qualitative and targeted training of specialists for industry and society.

But worldwide the youth interest to engineering professions is decreasing and as a result a very little amount of young people choose engineering education [2, 3].

© Springer Nature Switzerland AG 2020
M. E. Auer and T. Tsiatsos (Eds.): ICL 2018, AISC 916, pp. 503–513, 2020.
https://doi.org/10.1007/978-3-030-11932-4_48

An important trend in modernizing and improving today's engineering education is the system of schoolchildren career guidance and pre-university training. The existing attitude of parents and school students to the engineering profession can be changed by combined efforts of school teachers, university professors, and the leading experts of industrial enterprises.

Today one of the most important challenges is a provoking children interest in exact sciences and engineering professions from their school days. According to this fact, engineering universities worldwide try to incline the school students to select technical subjects and engineering professions as a future career. In this respect, we need to develop new approaches and innovative solutions.

Nowadays one of the important trends is strengthening of engineering education practical orientation, introduction of network learning forms in college-university-industry interaction with addition of primary school into this network chain using the education experience [4].

The main aim of this research is analysis, synthesis and extension of experience of networking between university and partner enterprises in attracting prospective students through the system of children technoparks, the organization of innovative projects competitions, participation in competitions of professional skill under the program of young professionals *WorldSlills Junior* and others. The goal is to determine the effectiveness of new forms of networking between University and industry through increasing children's interest to the engineering professions, developing their orientation to technical education, entering the engineering universities.

The goals are achieved through the unity of research tasks and joint actions program of the university and partner enterprises in the selection and targeted training of future engineering students.

We supposed that involving the university partner enterprises in various forms of schoolchildren professional training would allow increasing the attractiveness of engineering professions for the latter ones.

We also think that one of the most efficient forms of working with future engineering university students is creating a technopark for children, proactively supported by partner enterprises.

Specific forms of how the involved enterprises may participate in these activities are as follows: the organization of professional laboratories, the implementation of a professional samples system, the promotion of projects and startups of the most talented children, the organization and holding of competitions of projects focused on the industry area etc.

The implementation of the network cooperation program allows achieving significant results already at the middle of the project. First, these are the victories of children involved in the system of pre-university training of future university students, in professional skills competitions in partner enterprises. Secondly, it is the popularity growing of the children technopark. Thirdly, these are new innovative projects of school children, who were supported and took over the implementation by partner enterprises, etc.

2 School—University—Enterprise: Best Practices

The contemporary theory of engineering pedagogy and the experiences of universities from different countries show a special commitment to the task-oriented and project-organized training (see K. Benjamin, T. Jones, A. Kolmos, S. Meyers, P.G. Larsen, E. Lindsay, J. Raven, J. Stevenson, and F. Flemming) [5, 6].

Today, the model of life-long vocational education is characteristic of many countries, such as UK, USA, France, etc.

We can also distinguish the advanced education model that continues and develops the models of practice-oriented and project-organized training.

For example, in Great Britain much attention is paid to career-oriented work with children, which positively affects their future employment opportunities and forming their personal career paths. Schools in the UK developed simulation modeling, such as business games or industry simulations; monitoring manufacturing processes; and establishing mini-companies. The system of education in the Netherlands system is characterized by a permanent link between secondary and higher education, as well.

A system of dual education is widely known and successfully applied in Western Europe, especially in Germany. According to this training model, at least 60% of the training time is allocated for practical training on training simulators or metallurgic machines in the college, as well as for internships in enterprise shops, and only 40% of the training time is devoted to the theory.

In general, the model of higher engineering education providing its link to businesses has developed in many European countries, such as UK, France, the Netherlands, Denmark, and Scotland. Companies participate in designing curricula in the Netherlands; joint-training programs with a large number of practical training hours are being developed in Denmark; the "professional licenciate" degree has been introduced in France to consider the labor market demands; etc.

As to the specific forms of working with their future students, different universities take different measures to prepare their future students for admitting colleges and polytechnic universities for engineering training programs. As a rule, this is done in form of preparatory courses for a specific bachelor's program. For example, the University of Queensland, Australia has a preparation course for the Bachelor of Science in Engineering. University of Technology Sydney where practical industry-involving training is seriously delivered, holds Open Day TS with over 200 talks, tour and activities on offer. The university's website has a special section for future students with the opportunity to explore course and career options and plan the future "uni you."

Some universities in Russia are implementing programs for the early engineering professional guidance of school students, the former ones involving business, within the school-university-enterprise system. The experience is accumulated in preparing future students for applying to the engineering education programs made in connection with the partner companies.

In the university's partner schools, there is an under professional and professional training offered in special classes. An important trend is also the integration of science and education within the school-university system. Special research and educational activities are arranged for schoolchildren, aimed at developing the motivation for engineering activities.

3 Network Forms of Cooperation Between University and Enterprises

The practice of networking between engineering universities and partner enterprises in the scientific and educational spheres is currently developing rapidly throughout the world.

Multiple forms of network cooperation of the university with the enterprises allow to improve the quality of training engineers and to enhance the attractiveness of engineering professions for schoolchildren – the future students of the engineering universities.

The benefits of such collaboration with a university are obvious for businesses, since continuous training the children for their future engineering activities, starting from their school years, will allow them to get a highly-qualified professional knowing the special aspects of working at this specific enterprise.

The following are some concrete forms of network interaction of universities and enterprises in educational and scientific spheres in the successfully developing preparation of bachelors and masters at the Kazan National Research Technological University (KNRTU).

The forms of network interaction are the following:

- *Creation of joint bachelors and masters educational programs.* These are the new programs or the improved ones which were created by the university teachers with the help of the representatives of the enterprises. For example, joint programs for the preparation of masters for the largest Russian company PJSC Gazprom: "Gas and chemical technologies for the production of raw materials for polymers", "Vacuum and compressor equipment for physical installations" and many others. The University participation in implementation of innovative projects of the enterprises, and the creation and development of new industries. For example, a preparation of a group of students focused on a specific project for their subsequent work at the new product line.
- *Creation of basic university departments at targeted branch enterprises.*
- *Writing textbooks, articles and monographs in conjunction with specialists from partner enterprises.*

4 Network Forms of Cooperation Between University and Enterprises for Future Students

The named forms of network cooperation relate to the basic scientific and educational activity of the university and primarily to the work with the students. However, we consider the work with the future students as one of the important and efficient forms of university-enterprise collaboration. It should be held at the stage of attracting technically talented children to study at the universities and to work for the future employer, developing their interest in engineering professions.

The purpose of such work is not only informing schoolchildren and attracting future students to a certain university. In the first line, it is their career guidance, getting acquainted with the engineering environment where they probably will work, with the very essence of engineering activities, and with their specific tasks. Second, while implemented their programs with schoolchildren, the university and partner enterprises set and solve clear educational tasks, such as knowledge transfer and developing in children a cognitive interest in engineering areas.

The following are among the most effective forms of such cooperation:

1. Creation of children technoparks at the universities.
2. Conducting career-oriented activities (academic competitions, contests) with partner enterprises.
3. Creation of specialized enterprise-classes in vocational schools and colleges at universities.
4. Professional skills competitions, including those by the program for young professionals WorldSlills.
5. Competitions of innovative projects of schoolchildren carried out with the support of partner enterprises.

4.1   Creation of Children Technoparks at the Universities

Today children technoparks in Russia are being created with the aim of reviving the prestige of engineering and scientific professions and training of the personnel reserve for a technological breakthrough of the country on the global scale. This is possible now within the framework of the creation of regional territorial clusters with the active assistance of partner enterprises in Russia.

For example, an application of one of the biggest Russian technological universities for the creation of children technopark "Quantorium" was featured among the winners of the competitive selection of the Ministry of Education and Science of Russia for implementing the measures financed from the state budget to develop the scientific, educational and creative environment in educational organizations and the system of additional education for children. It was created on the basis of Nizhnekamsk Engineering Center (NEC) operating within the regional cluster "Innokam" in the Republic of Tatarstan, a territory of advanced development. "Quantorium" was established in Nizhnekamsk where the largest petrochemical enterprises of Tatarstan are located.

Technopark is designed to serve as a social lift for talented in scientific and technical creativity young people and to create a new form of additional education for children.

The goals of "Quantorium" are to identify and support gifted in engineering sciences children, to develop their orientation to technical education and working professions that are in demand at branch enterprises. Another goal is to give children knowledge beyond the school curriculum (the ability to understand special literature, to speak publicly, to be proficient in Technical English, etc.), to form the ability to technological search.

"Quantorium", (769.5 m^2) is located in the NEC building. About one thousand high school students study here. Well-equipped modern classes represent 6 streams ("quanta") of the technopark: "Nanokvant", "Neuroquant", "Geokwant", "Energikvant", "Robokvant" and "Industrial Design". There are also several laboratories equipped with modern technological equipment in the technopark. Each of the six "quanta" has its own educational program. Teachers of the University assigned to the quanta determine the subject matter of research, advise schoolchildren, give lectures, and help the students to join the problems of scientific research solving at the university and enterprises.

The children technopark works directly with the schools in the city, the technical study groups being its primary stage. On the other hand, the center is directly connected with the educational activities of the university and its branch in Nizhnekamsk.

Specific forms of participation of enterprises in the work of the technopark are:

- creation of specialized laboratories;
- conducting a system of professional samples;
- promotion of projects and start-ups for the most gifted children;
- organization and holding of projects competitions on industrial subjects of the enterprises.

Technopark "Quantorium" is connected with large industrial enterprises of Tatarstan.

Another important task of the technopark is to promote the projects and startups of the schoolchildren. All projects and startups are oriented towards the problems of the industrial enterprises of the regional cluster "Innokam". This will help the students to learn the basics of engineering. The best students will become the residents of the technopark. They will have the opportunity to work on the unique equipment, develop their projects to the level of commercialization with the support of the leading enterprises. The Minister of Education and Science of the Republic of Tatarstan noted at the opening of the technopark that the opening of the "Quantorium" is a great impetus for the further development of high-tech creativity which creates a new environment for the development of gifted children.

4.2   Holding of Career Guidance Activities with Partner Enterprises

A vivid example is holding of a job fair and a "Gazprom Day" at this technological university. The university is one of the 11 main universities for the Russia's largest energy company. This is a special annual event to attract future university engineers to the company. Students from the partner schools of the university are invited to the event. The representatives of the company's subsidiaries communicate with the students; talk about the enterprises, their staffing needs and about the prospects for young professionals and career growth during the job fair. Students and schoolchildren give future employers their completed questionnaires and ask questions of interest. According to the specialists of the personnel services of companies, students of this university and schoolchildren are very active, show great interest in the potential employers.

4.3   Creation of Specialized Classes in Vocational Schools and Colleges at Universities by the Enterprises

A few years ago, a vocational school with in-depth study of chemistry for gifted children was created at Kazan technological university. Here, with the assistance of the three largest technological and energetic world companies and one Russian company special classes were equipped.

Talented vocational school students conduct their scientific research there and also study modern instruments (for example, the automated management of an enterprise). The technological University has become an associated member of the "Gazpromclass" network across Russia. The company's enterprises provide opportunities for internships for the students of both higher and secondary levels of education.

4.4   Professional Skills Competitions

Competitions on the WorldSkills standards are currently being conducted at the regional, national and international levels. Their tasks include popularization of working professions among young people, identification of the best young professionals, upgrading of the status and standards of professional training and qualification around the world.

At the one of the technological universities the focused preparation work for the WorldSkills championships has been conducting for the last 4 years. This work is especially important in connection with the WorldSkills World Cup in Kazan in 2019. Our city already has experience of holding national championships in 2014 and 2015, regional competitions in 2016 and 2017.

KNRTU was one of few Russian universities to start WorldSkills competitions among university students: the first regional competitions was held on November 2017.

Since 2016, the university has introduced a new model of career-guidance work with the future students with the participation of partner enterprises on the basis of so-called "professional testings" for pupils from 7th to 11th school forms, for the vocational school for gifted children, for colleges at the university, and also for the Department of secondary education of one of the technological university.

"Professional testings" are organized on the sites of partner enterprises, where children have the opportunity to join the blue-collar job, to practice and immerse themselves into the profession. This allows students of colleges and vocational schools to gain professional skills during the studying, and this helps them to become good engineers in the future, to learn the industry from within and "make friends" with it. In addition, partner enterprises prepare students' teams from vocational school and college for the participation in professional skills competitions, try to find tutors among their specialists. Also the Regional competitions are held at the sites of the partner enterprises. This work gives good results.

In February 2017, a large regional stage of the WorldSkills competitions in 40 competences took place in Kazan. Students of the university's technical college participated in four competences. The partners of the Poligraphic Technology in Press competence were the Tatarstan's Agency for Press and Mass Communications and three Kazan tipographies, where the students did their internships. Moreover, the university presented a new competence for the competition - Industrial Design, which has been developed by the technology college of the university.

One of the students of the Faculty for Secondary Vocational Education was the second in the Manufacture of Polymers competence. The university's professors had also been among the experts of that competition.

Within the JuniorSkills format (participants aged 10–17), there were the students of the university-supported vocational school for gifted children with the in-depth studies of chemistry. Their team was the second in the Laboratory and Chemical Analysis 14 + competence.

The students of the Technology College also participated in the championship competences, such as Laser Technology, Restaurant Services, and Engineering CADs.

In total, thirty students of the technological university, including those studying at the technology college and chemistry school, participated in regional contests in 2017.

As a comparison, only 8 representatives of the university participated in the WorldSkills competitions in 2016.

On November 27–29, 2017 the regional WorldSkills competition took place in Kazan, and there were already 114 categories in it: 91 categories for participants aged 16–22 and 23 categories for children aged 14–16 (JuniorSkills).

One of them, Vadim Polyakov, the student of our university, became the WorldSkills World Cup Winner in Abu-Dhabi in 2017. He is majoring in Refrigeration Engineering.

In 2017, Kazan National Research Technological University (KNRTU) became a member of associated partners of the Young Professionals (WorldSkills Russia) Union - the Agency for Developing Professional Unions and Working Personnel. This opens new opportunities for the university's collaborating with partner enterprises in training schoolchildren and college students for both the WorldSkills championships and their further conscious choice of engineering professions and improvement of their vocational competences.

As soon as in six months upon holding the first university championship at KNRTU, in June 2018, the second qualifying WorldSkills Russia university championship is being held. It will offer 7 categories, namely: Refridgeration Engineering & Air Conditioning, Fashion Design, Enterpreneurship, Engineering CADs, Engineering CADs JUNIORS, Laboratory Chemical Analysis, and Industrial Design. Competitions among juniors aged 14–16, i.e., schoolchildren and students, are held at the university for the first time, and they will be for the Engineering CADs category only.

The key outcome of the championship is expected to be the increased quality of engineering university education, propagating the practice-focused teaching and learning, and perfecting the university programs taking the Russian and international requirements for professional competences into account.

4.5   Competitions of Innovative Projects of Schoolchildren Carried Out with the Support of Partner Enterprises

In the attracting of gifted children to the project activity in engineering, the university is actively introducing the model "University - Resource Center for Project Training". Partners of the University from business are actively involved in the organization of project competitions. In 2018, a so-called innovative competition "Innovative test range "Tatarstan - the territory of the future" was held for the 6th time. It allows school children, together with their studies, to gain an early experience of project work in teams, to do real research work in the scientific centers of the university. Within six months the school children passed several stages (training, business days at enterprises, organization of the team and the development of their own project), receiving as a result the support for their perspective business ideas from the partners of the competition (public structures, banks, industrial companies).

The main areas of the competition are technologies and materials of the future, as well as communications and ecosystem. Every year the number of participants of the project grows: in 2018, about 150 projects were created by teams of schoolchildren competing to be winners (53 competed in 2016–2017 and 45 in 2015).

Business partners of the contest are several big technoparks and IT companies of the region and also the public administration.

They really commend the high level of the engineering developments made by school students. The experts put a focus on their prominent scientific status, actuality, and novelty, as well as a significant potential for establishing start-ups based on those projects. Many works are invited to participate in other large regional competitions.

In the competitive direction, we can note the holding of the annual multidisciplinary interregional academic competition for school children "The Future of Great Chemistry", which reveals the classical abilities of high school students for polytechnic training.

For example, the contest of research and creative works "Nobel hopes" is held for senior students, successfully supporting the aspiration of children to research: under the guidance of university teachers and tutors from industrial enterprises, school students write serious and original works.

5   Program Results

Realization of this program with the support of partner enterprises shows good results in career guidance of school children in our university with different directions of attracting children to project activity in engineering field. It is first of all the high percentage of submitting school students who participated in these projects in different professional-oriented engineering departments of university. Following the results of our university admission process in 2017 among all submitted students there are more than 60% who came after one of the special school projects. The number of participated school students is growing. For example, in 2017 there were 65 thousands of school children, from primary to high school, participated in our programs which is vastly larger than last year.

The implementation of the network cooperation program allows achieving significant results already at the middle of the project. First, these are the victories of children involved in the system of pre-university training of future university students, in professional skills competitions in partner enterprises. Secondly, it is the growing the popularity of the children technopark. Thirdly, these are new innovative projects of school children, who were supported and took over the implementation by partner enterprises, etc.

For several years, the university has been realizing a special program aimed at developing in the students of partner schools and industry-specific colleges vocational competences for their future employment at individual industry-focused enterprises.

According to the program, the school/college students prone to or talented in engineering areas and natural sciences were clustered in several groups. Students from those groups studied more in-depth the school subjects relating to engineering and natural sciences. Some educational events/contests were arranged within the program development.

Then, a special course was taught to the students of one of the groups, which training was aimed at developing their vocational competences for their future profession. This selected group participated in the regional WorldSkills program.

The program implementation resulted in the significant increase in the percentage of the graduates from those groups who had entered the university upon having participated in the contests and special events.

Below aresome specific figures.

Over 50% (110 people) of the 200 program participants attended two or more events.

50 of them won in the regional WorldSkills competitions, i.e., 25% of the program participants.

75% of participants, i.e., 150 people, decided to study at the university, majoring in engineering areas.

The program also allowed us to attract new partner enterprises and expand the network of partner schools. Generally, the percentage of school graduates that entered KNRTU upon having attended such contests and events, increases significantly. As to partner schools and industry-specific colleges, we should note that over 80% of graduates who have participated in such events choose our university for their further studies.

The main results also are shown in Table 1.

Table 1. Amount of University Students – winners of special programs for school students.

	2014	2015	2016	2017
Program participants	135	150	180	200
Participants of more than one event	40	72	90	110
Became engineering students	47	83	110	150

Schoolchildren who participated in special career guidance events arranged by the university and partner enterprises, showed a great interest in and awareness of their future profession in comparison with regular applicants, according to the data from the questionnaire they fill out when enter the university. This is especially about those who took part in the technopark activities, in the WorldSkills championships, in the "Innovative Site – Tatarstan – Territory of the Future" competition, and in the competitions of schoolchildren's innovative projects, supported by partner enterprises.

Systematic joint actions of the university and several key partner enterprises in attracting technically talented children, their targeted training (within the framework of the children technopark and other special events of the university), organization of professional skills competitions and other forms of cooperation allow achieving high results, in terms of quality, of future students and their motivation to receiving the high-quality professional training.

Participation in such joint actions of the university and partner enterprises leads to informed choosing by potential students the specific engineering programs, studying areas, faculties and departments of the university, and engineering profession.

The research findings can be used by engineering universities in practicing the networking cooperation with partner enterprises, aimed at efficient and task-focused training their future students, and help them in choosing the preferable programs of engineering studies.

References

1. McAdam, M., Marlow, S.: A preliminary investigation into networking activities within the university incubator. Int. J. Enterp. Behav. Res. **14**(4), 219–241 (2008). https://doi.org/10.1108/13552550810887390
2. Ivanov, V.G., Poholkov, Y.P., Kaybiyaynen, A.A., Ziyandinova, Y.N.: Ways of development of engineering education for the global community. Vysshee obrazovanie v Rossii [Higher Education in Russia] **3**, 67–79 (2015). (in Russia)
3. Kaybiyaynen, D.-A., Kaybiyaynen, A.: University as a center of project-based learning of school students. In: Proceedings of International Conference on Interactive Colaborative Learning (ICL 2015), pp. 1018–1021 (2015). https://doi.org/10.1109/icl.2015.7318168
4. Kaybiyaynen, D.A.: Network cooperation in the training of engineering elite for regional economies. In: Proceedings of 2014 International Conference on Interactive Collaborative Learning (ICL 2014), Dubai, United Arab Emirates, 3–6 Dec 2014, pp. 616–618 (2014)
5. Chuchalin, A., Minin, M., Kulyukina, E.: Foreign universities experience in development of professional and transferrable competencies of engineering programme graduates. Vysshee obrazovanie v Rossii [Higher Education in Russia], vol. 10, pp. 105–115 (2010). (in Russia)
6. Lindsay, E., Munt R., Rogers, H., Scott, D., Sullivan, K.: Making students engineers. Eng. Educ. J. High. Educ. Acad. Eng. Sub. Centre. **3**(2) (2008)

Depressive Symptomatology of University Students Via a Web Application

Madalena Soula[1], Nikolaos C. Zygouris[2(✉)],
and Georgios I. Stamoulis[1]

[1] Electrical and Computer Engineering, University of Thessaly, Volos, Greece
{msoula,georges}@uth.gr
[2] Computer Science Department, University of Thessaly, Lamia, Greece
nzygouris@uth.gr

Abstract. Depression is a serious mental health disorder that affects approximately 2–5% of school-age children and adults. In addition to the emotional suffering, depression often disrupts cognitive functions and leads to the deterioration of academic performance. In depressed adults, short and long-term memory as well as attention are the most affected cognitive functions. There are only a limited number of studies on cognitive functions in university students with depressive symptomatology. However, deficits in short-term memory and meta-memory have been reported. Impaired ability to concentrate on task performance is one of the major problems in child and adult patients with depression. A number of studies suggest that university students neglect or even give up on theirs studies after their first year as students. Furthermore, a large number of students finish their studies taking more time than they normally would. A line of researchers recommends assessing their psychological characteristics. Several research protocols have been implemented in order to assess psychological disorders among students, taking into account gender differences, different levels of education, different scientific fields and age.

Keywords: Depression · University students · BDI

1 Introduction

The term "Depression" is used to identify a wide range of emotional states that may be either normal or pathological. It is describing a normal mood or feeling (e.g. significant loss, death), a symptom (e.g. stress, psychiatric problem) and psychopathological disorder (e.g. Dysthymia, Major Depressive Disorder). In general, Depression can be defined as a pathological disorder accompanied by significant reduction in the sense of personal value as well as a painful consciousness of slowing mental, psychomotor and organic processes. Furthermore, depression is used in order to describe a syndrome caused after a painful event. This syndrome includes a combination of symptoms such as sleeplessness, loneliness and nervousness and they normally resolve within a reasonable time. Depression is attributable to the disorder in which the abovementioned symptoms are elevated in intensity, persist over time, and have a specific etiology, course, and prognosis.

© Springer Nature Switzerland AG 2020
M. E. Auer and T. Tsiatsos (Eds.): ICL 2018, AISC 916, pp. 514–519, 2020.
https://doi.org/10.1007/978-3-030-11932-4_49

Throughout the second half of the 20th century several patient-rated assessment scales for detecting Depression were proposed. Among popular scales the Beck Depression Inventory [4] was chosen to be used in the present study. Beck and colleagues developed the 21-item BDI in 1961 in order to aid clinicians in the assessment of psychotherapy for depression. Later, this Inventory was reformulated to the BDI-II [3]. The English version of BDI-II has been translated and validated in 17 languages so far, and it is used in Europe, the Middle East, Asia, and Latin America [1, 10, 14]. BDI has been used more extensively than other measures in published research because of it is easily applicable [2] and highly reliable among adults [9] and students [12].

The main aim of the present study was to evaluate the prevalence of depressive symptomatology among Science and Humanities students. Another target was to assess the reliability of a well-established and studied psychological paper-and-pencil test when it is delivered via a web application. The research protocol included students of Economics, Electrical and Computer Engineering, Special-Education and Early Childhood Education. The aforementioned students were divided into two large groups in order to compare their emotional characteristics. Science students included both students of Economics and Electrical and Computer Engineering and humanities students included both students of Special-Education and Early Childhood Education.

2 Methods

The Beck Depression Inventory (BDI, BDI-1A, BDI-II) consists of 21 multiple choice questions in which the individual carries out a census of himself/herself. Each response is scored on a scale of 0–3. Higher overall scores indicate increasing severity of depressive symptoms associated with suicidal desire [7]. The current version of the Beck Depression Inventory (BDI-II) is designed for individuals aged 13 and over, and is composed of items relating to symptoms such as hopelessness and irritability, cognitions such as guilt or feelings of being punished, as well as physical symptoms such as fatigue, weight loss, and lack of interest in sex.

The standardized cutoff scores are as follows:

0–13: minimal depression
14–19: mild depression
20–28: moderate depression
29–63: severe depression

A sample of 119 students from the Economics, Electrical and Computer Engineering, Special-Education and Early Childhood Education departments was used. The sample of the population consisted of individuals of both sexes aged 18–26 years. More specifically, the average age for the 53 males was 21.98 (SD = 2.03) and the average age for the 66 females was 21.15 (SD = 2.09). The entire sample participating in this research did not present any brain injury, learning disabilities and psychopathological disorders as a result of semi-structured interviews that were taken before participating in the survey. The participants were asked to complete the Greek version of the Beck Depression Inventory [8] online through a web application.

Their answers were stored in a two-dimensional table, which had 24 attributes placed in 24 columns. The first three attributes were about the field of studies, gender and age. The remaining 21 attributes corresponded to the 21 questions of the BDI assessment. A part of the web application is shown in Figs. 1 and 2. Each row of the table contained the choices each student made so that every row corresponded to the choices of one individual.

The answers to the BDI were matched with the appropriate value according to the scale of Greek version of the scale so that all the data was converted to a numerical value.

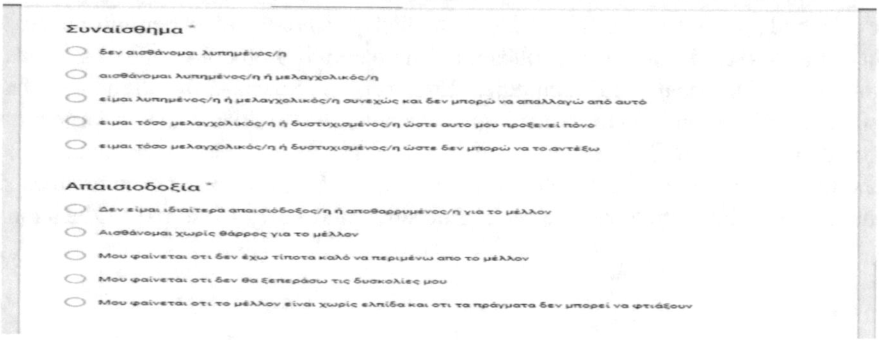

Fig. 1. Part of the web application

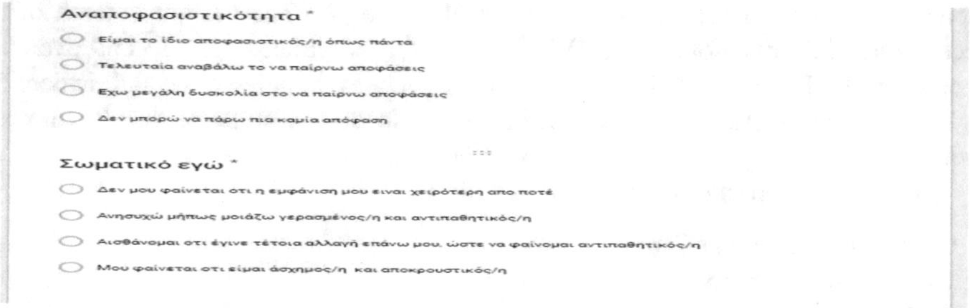

Fig. 2. Part of the web application

3 Results

Descriptive statistics were performed in order to obtain mean scores and standard deviations of the participants over all the questions.

The first statistical analysis that was performed utilized descriptive statistics and resulted in the mean scores and standard deviations for the participants as shown in Table 1.

Table 1. Mean scores and standard deviations of the participants according to the department of studies.

Department	Question 9	Question 18	Question 20
Electrical & Computer Engineering	0.78 (SD = 0.57)	0.87 (SD = 0.77)	0.39 (SD = 0.66)
Economics	0.39 (SD = 0.58)	0.59 (SD = 0.77)	0.65 (SD = 0.78)
Special-education	0.75 (SD = 0.61)	0.46 (SD = 0.66)	0.83 (SD = 0.92)
Early-childhood education	0.61 (SD = 0.70)	1.00 (SD = 0.91)	1.00 (SD = 1.09)
F	2.423*	2.490*	3.356*

Note: *$p < 0.05$

A second statistical analysis was conducted in order to divide the participants according to their field of study. Two groups were formed: the first group named "School of Sciences" consisted of Economics and Electrical and Computer Engineering students and the second group named "School of Humanities" consisted of Special-Education and Early Childhood Education students. Analysis of Variance (ANOVA) was conducted to compare the scores of the students according to the groups that were formed by the aforementioned procedure. ANOVA revealed that participants had statistically significant differences ($p < 0.05$) in four questions. More specifically, on the question about wishes of self-righteousness the group "School of Humanities" (mean = 1.58, SD = 0.82) presented higher results in comparison to the "School of Sciences" (mean = 1.26, SD = 0.77). It seems that the students that participated in the first group had higher scores (F = 4.445, $p < 0.05$) in comparison to the second group of students (Table 2).

Table 2. Mean scores and standard deviations of the participants according to the school of studies.

Department	Question 8	Question 13	Question 20	Question 21
School of sciences	1.26 (SD = 0.77)	0.96 (SD = 0.82)	0.47 (SD = 0.70)	0.63 (SD = 0.80)
School of humanities	1.58 (SD = 0.82)	1.26 (SD = 0.79)	0.88 (SD = 0.98)	0.35 (SD = 0.61)
F	4.445*	3.635*	6.983**	4.048*

Note: *$p < 0.05$; **$p < 0.01$

4   Conclusions

This research protocol presents that Depression is approximately at the same level between both groups, namely "School of Sciences" and "School of Humanities". In addition, the aforementioned study presented a high level of reliability ($\alpha = 0.881$) according to the calculation of Crombach's Alpha that was conducted. The high reliability of the Beck Inventory in its paper-and-pencil form has been verified by several studies [8–10] that show approximately the same level of reliability demonstrated by this study that used the Internet to deliver the aforementioned test.

Several research protocols examine depressive symptomatology among University students. The outcome of previous studies estimating symptoms of anxiety and depression among Medical Students and Humanities Students has shown a prevalence of 43% and 14% respectively among medical students of 52% and 12% respectively among the students of humanities [5]. Along the same research line, several studies concluded that medical students exhibit higher levels of depression and that the female students have higher levels of depression than male students [6].

References

1. Alansari, B.M.: Beck Depression Inventory (BDI-II) items characteristics among undergraduate students of nineteen Islamic countries. Soc. Behav. Personal. Int. J. **33**(7), 675–684 (2005)
2. Andrade, L., Gorenstein, C., Vieira Filho, A.H., Tung, T.C., Artes, R.: Psychometric properties of the Portuguese version of the State-Trait Anxiety Inventory applied to college students: factor analysis and relation to the Beck Depression Inventory. Braz. J. Med. Biol. Res. **34**(3), 367–374 (2001)
3. Beck, A.T., Steer, R.A., Brown, G.K.: Beck depression inventory-II. San Antonio **78**(2), 490–498 (1996)
4. Beck, A.T., Ward, C.H., Mendelson, M., Mock, J., Erbaugh, J.: An inventory for measuring. Arch. General Psychiatry **4**, 561–571 (1961)
5. Bunevicius, A., Katkute, A., Bunevicius, R.: Symptoms of anxiety and depression in medical students and in humanities students: relationship with big-five personality dimensions and vulnerability to stress. Int. J. Soc. Psychiatry **54**(6), 494–501 (2008)
6. Dahlin, M., Joneborg, N., Runeson, B.: Stress and depression among medical students: a cross-sectional study. Med. Educ. **39**(6), 594–604 (2005)
7. Furr, S.R., Westefeld, J.S., McConnell, G.N., Jenkins, J.M.: Suicide and depression among college students: a decade later. Prof. Psychol. Res. Pract. **32**(1), 97 (2001)
8. Fountoulakis, K.N., Iacovides, A., Kleanthous, S., Samolis, S., Gougoulias, K., St Kaprinis, G., Bech, P.: The Greek translation of the symptoms rating scale for depression and anxiety: preliminary results of the validation study. BMC Psychiatry **3**(1), 21 (2003)
9. Gallagher, D., Nies, G., Thompson, L.W.: Reliability of the Beck Depression Inventory with older adults. J. Consult. Clin. Psychol. **50**(1), 152 (1982)
10. Gomes-Oliveira, M.H., Gorenstein, C., Lotufo Neto, F., Andrade, L.H., Wang, Y.P.: Validation of the Brazilian Portuguese version of the Beck Depression Inventory-II in a community sample. Revista Brasileira de Psiquiatria **34**(4), 389–394 (2012)
11. Hammen, C.L., Cochran, S.D.: Cognitive correlates of life stress and depression in college students. J. Abnorm. Psychol. **90**(1), 23 (1981)

12. Hill, A.B., Kemp-Wheeler, S.M., Jones, S.A.: What does the Beck Depression Inventory measure in students? Personal. Individ. Differ. **7**(1), 39–47 (1986)
13. Jorm, A.F.: Sex and age differences in depression: a quantitative synthesis of published research. Aust. N. Z. J. Psychiatry **21**(1), 46–53 (1987)
14. Penley, J.A., Wiebe, J.S., Nwosu, A.: Psychometric properties of the Spanish Beck Depression Inventory-II in a medical sample. Psychol. Assess. **15**(4), 569 (2003)
15. Wang, Y.P., Lederman, L.P., Andrade, L.H., Gorenstein, C.: Symptomatic expression of depression among Jewish adolescents: Effects of gender and age. Soc. Psychiatry Psychiatr. Epidemiol. **43**(1), 79–86 (2008)

Engineering Education Using Professional Activity Simulators

Dorin Isoc[1]([⊠]) and Teodora Surubaru[2]

[1] Technical University of Cluj-Napoca, Cluj-Napoca, Romania
dorin.isoc@aut.utcluj.ro
[2] Group for Reform and University Alternative, Cluj-Napoca, Romania
office@graur.org

Abstract. The Engineering School is a set of broad-based activities designed to build an individual who, beyond knowledge, is also endowed with a particular way of thinking. In conventional school, knowledge and some technical skills take precedence. The way of thinking is build in a wider context, with well connected and justified activities, correctly connected to reality. A professional framework similar to the real-world professional environment, called the "*engineering office*" is introduced. In this framework, the technical simulator expands with a professional activities simulator. One describes the components of the professional activities simulator and their role in engineering thinking development. The major advantage of such a simulator is that it naturally integrates skills and knowledge essential through their frequency of use. So, the professional activity simulator put the student closer to structurally justifying details that the school dims and simplifies, sometimes not allowed for, for reasons of approach and methodology.

Keywords: Simulator · Regulation · Standard · Training ·
Engineering · Information · Active learning

1 Introduction

At this time, technical universities are concerned to provide theoretical knowledge to future specialists. The approach is natural and legitimate. It is less natural to consider that the formation of the future engineer would consist only of theory. Qualification of an engineer assumes, before knowledge, a certain way of thinking. Engineering manner to think is more than a habit: it is the second nature of the specialist.

The engineers thinking manner is not only a skill. This way of thinking is reflected in the way in which all training information is perceived, in the way they work with them. It is important to note that engineering thinking manner should be acquired in advance both in engineering and in work.

As principle, the engineering thinking learns at workplace. Such a solution reduces the effects of the controlled exercise under school conditions.

© Springer Nature Switzerland AG 2020
M. E. Auer and T. Tsiatsos (Eds.): ICL 2018, AISC 916, pp. 520–531, 2020.
https://doi.org/10.1007/978-3-030-11932-4_50

Forming engineering thinking is not a subject that acquires in books, as it is, just, as with certainty, its systematic knowledge of books is necessary.

The game or the simulator of phenomena or situations seems to be solutions for the development of such a skill. Literature directs applications as much as three directions.

The first direction is that of training by assisted simulation of human operators in complex technical systems [3,4,9]. It is easy to see that this research provides a framework for taking records of incidents logs, followed by their analysis and packing cases in simulator applications endowed with machine-machine interfaces. The human operator acquires the skill to track states and intervene operative by using the simulator. The domain is uniform and predictable if the information is available and the operators have a sufficient qualification captures in the expected patterns.

A second direction uses simulation by games [1,2]. The solution applies especially to the qualification of software specialists. The goal is direct and aims at developing applications on configurable environments. The most numerous applications find in the gaming area and it often one mentions that it would stimulate the creativity of the subjects. It is also noteworthy that the activity of the subjects is individual and particularly limited as application horizon.

Finally, the third direction of using simulators is to develop the skills to decide through specialized games [5,8]. In all situations, however, what searches is to stimulate the school activities by using a new challenge determined mainly by the interactive nature of applications.

This research aims to define and describe a simulator for engineering professional training. Although it includes a simulator implemented on a computer, the professional training simulator appears as a necessary assembly to assure the conditions of assimilation of working skills in engineering. A research like the application of simulators in veterinary medicine [10] expands here with units of an activity simulator. The aim of the research is to stimulate the development of a way to think of engineering.

The second chapter is dedicated to the relationship between training and simulation. It follows that conventional instruction, supported by disciplines that go through a certain order and which delayed integration and application for a long time, for the graduate. Simultaneously, simulation training discusses, where the first concern is an integrated approach that improves over time, respecting the specifics of the engineer profession.

The third chapter approaches the building of the professional training simulator. In its structure, there is necessarily a technical simulator and a professional activity simulator. Ensuring the functioning of the professional activity simulator develops through the organization specific to the school but with respect to the framework of an engineering office. It also describes a simulator structure of professional activity through its component units. One insists that such a simulator has, in particular, general components of engineer activity, with little specificity that provide by components resulting from conventional school activity.

The fourth chapter formulates the first interpretations and does not avoid the discussions that, such an approach, provoke even among the teaching staff.

2   About Relationships Between Training and Simulation

It is a fact that education represents, whatever its forms, a way of preparing an individual to enter, with a certain role, in society.

Despite the fact that only the training of the engineer will be further addressed, the treatment of any form of education, qualification, professionalism is similar.

The first observation, in essence, is that school disciplines, as they define and how classical schools perceive them, are entirely artificial and at most correspond to a much-simplified sketch of society.

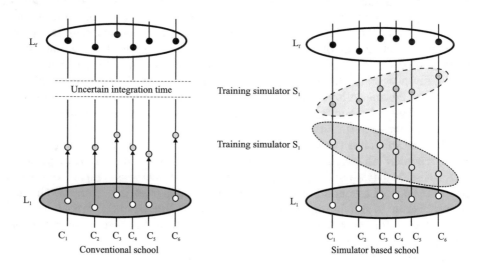

Fig. 1. Explanatory on how conventional and school-based simulation schools work.

As in Fig. 1, it assumes that the training process, with the goal of reaching the Li level of the student's initial state to the Lf level of qualification, sufficient to practice, requires a certain amount of time.

The training process can follow the way of a conventional school or of a school using training simulators. Following the way of the conventional school, each body of knowledge C_1, C_2, C_3, C_4, C_5, C_6 traces independently, and after a lasting integration of uncertainty, produces the true conversion of knowledge into skills.

On the contrary, in the school based on training simulators, learning involves several simulators, whether S_1, S_2. In these simulators, the same knowledge in C_1, C_2, C_3, C_4, C_5, C_6, integrates initially and further resumes in succession of professional activities until further training.

Improvement also occurs progressively through the consolidation of the assembly that leads to engineering thinking. The knowledge bodies discussed assimilates not necessarily to school subjects. They are rather assemblies of knowledge that lead to results or directly measurable effects, to name them generically deliverable. Following training, one expects that a professional inter-act with the environment. Through its professional and social role, the engineer can interact with the environment only through knowledge and skills that do not have a well-determined place of acquisition.

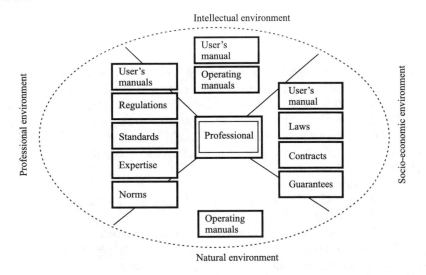

Fig. 2. Placing of a professional among the environmental components.

In Fig. 2 describes the main relationships that a professional, as an engineer, develops with the environment in which he/she lives. In order to emphasize the suggestion of the analysis, the environment divides into four types:

i. The professional environment, to which, the professional reports from the perspective of his/her basic qualification.
ii. The socio-economic environment, to which, the professional relates in the light of the results of his work and his/her partial products.
iii. The natural environment, to which, the professional relates by restrictions regarding the possible aggression of the activities he manages and the products of these activities.
iv. The intellectual environment, which implies all forms of non-material existence of the knowledge with which the professional operates.

One notes that specifically between the professional and the components of the environment there are knowledge organized in such a way that they are given, pre-exists before any individual professional initiative.

It also notes that any knowledge or skills that a professional possesses or accumulates will always be able to talk about restrictions or restrictive conditions.

The training of a professional, from the level of young people with general training, to the level of a professional who working in a job with a certain degree of qualification, is thus only possible by observing the knowledge of the professional restrictions. The training in question develops in engineering schools.

The questions that arise now seem to be: (a) What are the conditions in which engineer training is in itself an engineer? (b) Is the school a professional training simulator? (c) What is the structure of a professional training simulator?

3   Building the Professional Training Simulator

The education of an engineer is highly complex by the significant amount of specific information is imposes. This information, organized as knowledge, differs from the information used in the education of other professional categories.

All this information, which then builds on the skills, forms a complex educational background designed to build and stabilize specific manifestations such as professional responsibility, professional prudence, recognition and respect for the professional hierarchy.

In these conditions, it is a question of defining and operating a training structure able to assist the education at all stages and its forms of evolution in a school, either in engineering.

This training assistance structure is a common one for all school subjects. In principle, the structure intends to cover a set of activities without which the learning of knowledge and skills is devoid of consistency.

Without limiting or annihilating individual creativity, the professional work simulator offers the student's work a professional character, a character of stability. The simulator covers a number of ancillary activities, such as drafting professional documents, managing information, identifying and observing constraints or prohibitions, and, finally yet importantly, recognizing similar previous professional experience.

The professional work simulator builds as a collection of auxiliary and guiding components that is available to all engineering training modules.

At the time of research development, the component parts of the training simulator involve a series of regulations, rules, instructions, forms, fiches. In addition, a previous work experience portfolio consisting of works previously completed and completed in full or in the form of representative fragments was included in the simulator. A distinct segment of the simulator includes the standards and manuals for use and/or exploitation.

The entire documentation base is available in a fully-fledged format, for example in electronic format, available to all students and becomes mandatory for all their professional obligations.

The whole activity of the simulator subsumes into a specific functional unit. This unit does not need to identify the units or subunits specific to the school or, which the school does not even endeavor to assume.

If the analysis of basis of argumentation of the professional activity simulator, it is easy and without the fear of offending the followers of total creativity to recognize that this is a means of educating the routine. Professional routine is necessity and is the guarantee of professionalism. The routine does not contradict creativity and cannot counters against it. Thus, an engineering school builds the routine of the professionalism of its graduates and, in the same time, can stimulate inventiveness as a form of behavior face to the dynamics of the evolution of reality and its requirements.

3.1   Engineering Office

Developing a simulator implies in any case the identification of conditions that founded in a recognizable reality.

The diversity of jobs that a professional engineer can serve has prompted the researcher to appeal the *engineering office* as the most realistic hypothesis that an engineer can activate.

The arguments that have advocated for this hypothetical construction are multiple:

   i. The engineering office is a real job with complex and complete work.
   ii. The engineering office faces a significant number of beneficiaries with topics from a sufficiently extensive field.
   iii. Within the engineering office, acts the professional hierarchy of age and expertise accumulated simultaneously with horizontal collegiate relationships.
   iv. Within the engineering office, solving all technical problems involved by the technical projects in work coexists with organizational requirements in order to ensure the economic efficiency of the work.
   v. Within the engineering office, the quality of the works defines, organizes, evaluates, and certifies at each moment.
   vi. Within the engineering office, there are document feeds that are necessarily subject to predefined procedures.
   vii. Within the engineering office, manipulation and management of information requires compliance with rules to ensure efficiency.
   viii. The engineering office always requires the need to carry out individual or collective work with well-defined responsibilities for each professional.
   ix. The engineering office assumes planning and time management as a system assimilated to penalties generated by non-compliance or non-compliance with established deadlines.
   x. Any activity carried out in an engineering office puts the concept of a "deliverable" as a materialized unit that requires reception and delivery under preconditions to the beneficiary.
   xi. From the point of view of endowment, the engineering office includes equipment and technical means of general use, but totally harmonized with the specialized works.

Such a professional-economic unit one conceives as a environment for the professional training simulator.

As endowment, the engineering office includes equipment and technical means of general use. These means are totally harmonized with the specialized works.

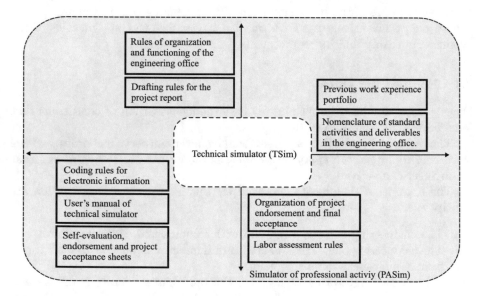

Fig. 3. Structure of the professional training simulator, an example.

3.2   Structure of the Professional Training Simulator

As a follow-up to those already anticipated in literature [6], the training simulator consists of a technical simulator and a professional activity simulator.

We resume here the discussion on the simulator topic - physical installation.

The subject has become more and more acute in recent years, especially since much of the engineering assimilates to computer science, and the realization of program sequences or computer applications to practical activities in engineering school. Divergence occurs when it comes to acquiring the ability to perform physical measurements and interventions.

Although this is not the subject of this research, we need to delineate some details.

First, the profession of engineer requires not only skills to make certain measurements but also the ability to organize and exploit technical measurements and interventions.

Secondly, a well-developed technical simulator is a means of economical training and avoids many other aspects that a physical installation implies.

Third, after training on a technical simulator, switching to a real-world operation whatever its nature and whatever the specificity of the intervention is always more effective than a direct entry to a real installation.

In Fig. 3 gives the proposed structure of the training simulator. The training simulator involves of a technical simulator, $TSim$ and a professional activity simulator, $PASim$.

The detailing in Fig. 3 of the professional activity simulator is just an example. However, it is necessary to specify that in $PASim$ the number of units change in order to meet appropriate requirements. In turn, existing units may have different degrees of detail, for example, for different levels of instruction.

3.3 Units of the Professional Activity Simulator

Generally, organizational elements, typically documents, will refer to as units of the professional activity simulator. It is obvious that these units are functional by the established working method and assured by the instructor's authority.

As a simulator, it is essential to have a role in the training process. The training simulator is a complex means of applying the activity to acquire skills specific to the profession of engineer.

The technical simulator, $TSim$, allows works that includes or views as projects. It is specific that each project has the characteristics of a similar engineering work. Essentially, the subject matter is that the projects aim to obtain deliverables using technical resources within a planned period and a fully regulated framework.

For the sake of exemplification, the technical simulator as in [6] is a platform for the design of control systems. The platform appears to users as a set of technical resources. These resources, together with a set of working and processing techniques, solve specific design themes for automatic control. Although it is a $MATLAB$ based platform, the student is not in the position to develop applications. The student has only functional function user functions according to a given user manual.

The simulator of professional activity, $PASim$, is a collection of regulatory and professional information documents. The emphasis is on procedural character, on instructional character, on normality.

For example, a set of such documents of a developed professional work simulator will be detailed:

Regulation of organization and operation of the engineering office. The document defines the engineering office and describes how it works. Each activity is supposed to carry out in the designated engineering office. It is necessary that the endowment, the access to resources, the flow of documents necessary to the professional obligations of the students must necessarily go through the specific flows of the engineering office.

Technical simulator user manual. The document builds to the highest standards and is so comprehensible that it covers all situations of operation and use of the technical simulator. Each resource of the $TSim$ technical simulator describes from the user's point of view and exemplifies in an application.

In this context, it should note that the technical simulator regards as a resource to which the user has access only if he/she authorizes. Authorization

15b.sofocl.01.150302.150317(15b.sofocl.010205(prosim.m, set_machine_0.m, set00_cc.m, set00_pa.m),
15b.sofocl.010205.mat,
15b.sofocl.01.pdf)

Fig. 4. The tree structure of a directory (folder) that includes the configuration files of an experiment and other associated information.

occurs when the student passes an authorization test. The authorization test is intended to confirm that the student has knowledge on the resource he/she will use, that he/she knows how to prepare the authorized resource, that he/she knows how to perform the operation, that he/she knows how to interpret the results and that he/she is able to draw up the working documentation.

Rules for drafting the technical reports. The drafting of technical reports regards as a current activity specific to the professional engineer. Drafting rules view as a document and they give through the contents and content of each sub-section of the technical reports associated with any activity carried out. Editorial rules present covering all the specific cases of the engineering office.

Coding rules for electronic resources. During information processing, each student arrives at files where applications or information are located. The external coding of these files imposes certain identification and information requirements. In turn, these files include in identifiable directories.

As an example, it will assumed that the files a, b, c are deposited in the $D1$ folder, the $D1$ folder places in the $D0$ folder along with the d and e files.

Such a tree structure describes as $D0$ $(D1(a, b, c), d, e)$.

Under standard conditions, filenames encoded to include a set of information that allows for effective targeting. In parallel, however, the technical simulator imposes a series of restrictions on the names of the configuration files used.

All the configuration files of a situation have exactly the name they appear in the simulator's pack. Personalization done only by the name of the directory that contains these files. In Fig. 4, the customization requires to: a: *prosim.m*, b: *set_machine_0.m*, c: *set00_cc.m*; D0: $15b.sofocl.01.150302.150317$; D1: $15b.sofocl.010205$; d: $15b.sofocl.010205.mat$; e: $15b.sofocl.01.pdf$.

Figure 4 shows the tree structure for the set of configuration files of the 05th experiment for the 02nd situation of the 01st work done in the second semester of the 2015 calendar year by the student 'em George Sofocle' who receives the conventional work name, '*sofocl*' consisting of six characters (letters). It also found that, the project with the number '01' of the same student had a teaching term in '150302' and was taught in '150317'. The results and report of the experiment are in the '$15b.sofocl.010205.mat$' and '$15b.sofocl.01.pdf$' files respectively. Obviously, the formation of filenames and directories is subject to unique rules.

Nomenclature of standard activities and deliverables in the engineering office. The designated engineering office has a certain field of activity. These defines and, for each of them, the related deliverables identified. Any developed project involves these activities, with the obligation for other new activities described and evaluated with similar information.

Labor assessment rules. Teamwork requires rules that need to be well known and properly applied. The intervention of the instructor in the engineering office environment is minimal and the regulations are those that operate as impersonal as possible.

Self-assessment and endorsement sheets. The peculiar importance of these professional documents lies in building peer-review mechanisms [7]. After completing a task, completing an activity, each student requires self-assessing and further, this self-evaluation can be subject to a hierarchical check by a colleague.

Previous work experience portfolio. The engineering profession is bounded by collaborations and projects. Even though thematic, these projects are different, as a way of approach, as a way of presenting, these papers can serve as examples for those who address them for the first time. An engineering office is bound to preserve the experiences and improve them.

Knowing the framework of the professional activity simulator helps the training process that can guide in terms of performance to strictly technical aspects.

The simulator of professional activity has as a model the engineering office and can supplement with its technical infrastructure, such as computers, computer networks and specific program applications, copied and scanned equipment, etc.

4   Results, Interpretations, Discussions

The definition and construction of the professional activity simulator appreciated and evaluated, by the quality of the technical reports achieved by the students.

The main effects of the existence of the work simulator are both quantitative and qualitative in nature:

  i. Effects exist if each student has a distinct theme or a distinct volume of work.
 ii. Carrying out the same professional papers, technical reports, allows identify an evolution that both the student and the instructor can control. Over time, there is no doubt a breakthrough based on a well-resumed replay.
iii. Correcting student progress greatly facilitates by the existence of a unit of model reporting.
 iv. Once the built-in form of the deliverables is set and tracked, the other elements of the simulator can act. It is about the association of terms, forms of peer review, the correct relationship - appreciation.
  v. The particularly important effect on the training process occurs when the previous professional experience portfolio is defined and materialized.

This research has been accomplished and experimented as an instructor's initiative. The results of the building of the engineering office are remarkable. However, it is worth noting the feeling of discomfort that many of the members of the trainers face over the novelty of the approach.

Arguments must at least present:

i. The relative increase in the workload required by students invokes. Experience shows, however, that increased workload is just an appearance. In fact, the unity of the provided activities leads to a phenomenon of stability of the students' work.

ii. One invokes the emergence of routine elements that would hinder the manifestation of the creative spirit, the development of imagination. It is true that the primary appearance highlights the presence of routine action flows. These flows, however, are specific to the engineering profession. First, the work of an engineer needs responsibility and this achieved only through thorough and completed experiences. In terms of creativity, it does not restrained, but it can educate.

5 Concluding remarks

Building a professional work simulator is an effort quite acceptable for any instructor and school. Once the built-in simulator, it is a professional issue for him to be accepted by the team of instructors.

The simulator of professional activity proves to be a necessary addition to the infrastructure of an engineering school. Whatever the specialty of the studies, the student must be prepared for the professional environment in which he/she will practice. A good simulator must be very close to the future professional future environment. By working within the simulator, extended to all disciplines and forms of training, the student, the future specialist will have the chance of an early integration of knowledge and skills undergoing assimilation.

The form and content of the simulator of professional activity is a specific problem for each school and possible professional associations. Professional associations can impose their own requirements that the school can provide with minimal effort.

Fundamental skills acquired from school by exercising on the professional work simulator will have a special effect on the uniformity of positive professional experiences.

A school is all the more powerful as its work can identifies as a good professional training simulator.

The absence of the above-mentioned features raises questions about the duration of the engineering school to bring its graduates to the professional standards of usability and social efficiency.

References

1. Baker, A., Navarro, E., Van Der Hoek, A.: An experimental card game for teaching software engineering processes. J. Syst. Softw. **75**(2), 3–16 (2005)
2. Cagiltay, N.: Teaching software engineering by means of computer-game development: challenges and opportunities. Br. J. Educ. Technol. **38**(3), 405–415 (2007)
3. Dozortsev, V.: Development of computer-based training simulator for industrial operators: main participants, their roles and communications. Autom. Remote Control **71**(7), 1476–1480 (2010)

4. Dozortsev, V.: Methods for computer-based operator training as a key element of training systems (present-day trends). Autom. Remote Control **74**(7), 1191–1200 (2013)
5. Ebner, M., Holzinger, A.: Successful implementation of user-centered game based learning in higher education: an example from civil engineering. Comput. Educ. **49**(3), 873–890 (2007)
6. Isoc, D.: Training using professional simulators in engineering education: a solution and a case study. In: M. Auer, D. Guralnick, I. Simonics (eds.) Teaching and Learning in a Digital World: Proceedings of the 20th International Conference on Interactive Collaborative Learning, pp. 208–218. Springer (2017)
7. Isoc, D., Isoc, T.: Practice of peer-review and the innovative engineering school. In: 2015 9th International Symposium on Advanced Topics in Electrical Engineering (ATEE), 7 May 2015, pp. 111–116. Bucharest (2015)
8. Kim, B., Park, H., Baek, Y.: Not just fun, but serious strategies: Using meta-cognitive strategies in game-based learning. Comput. Educ. **52**(4), 800–810 (2009)
9. Rogalski, T., Tomczyk, A., Kopecki, G.: Flight simulator as a tool for flight control system synthesis and handling qualities research. Solid State Phenom. **147**, 231–236 (2009)
10. Scalese, R., Issenberg, S.: Effective use of simulations for the teaching and acquisition of veterinary professional and clinical skills. J. Vet. Med. Educ. **32**(4), 461–677 (2005)

Experiential Learning Approaches in Automotive Engineering: Implementing Real World Experiences

Moein Mehrtash[✉] and Dan Centea

W Booth School of Engineering Practice and Technology, McMaster University,
Hamilton, ON, Canada
{mehrtam, centeadn}@mcmaster.ca

Abstract. This work aims to contribute to employing of the computer-based simulation in the high education in Automotive Engineering for experiential learning. In order to design a set of simulation laboratory activities, a pedagogical approach is presented on basis of Kolb's Experiential Learning Theory. The chosen content to be taught is vehicle dynamic with a focus on braking performance that is adapted and implemented in CarSim software. The pedagogical approach presented in this study can represent as a reference point for discussions in experiential learning environment for vehicle dynamics curriculum and automotive industrial standards, considering the use of the Kolb's theory as a model for development of teaching-learning process and computer-based simulations as a teaching tool. As part of pedagogical proposal, this study is also focused on development of real world experiment as concrete experiment in topics related to automotive industries. Finally, some suggestions are recommended in order to help future works.

Keywords: Vehicle dynamic curriculum · Kolb's theory ·
Experiential learning · Computer-based simulation laboratories ·
Automotive engineering education

1 Introduction

Recent advancements in computing have an important influence in the development of interactive and effective learning environments [1]. They complement traditional lecture-intensive courses with experiential learning approaches by employing computer simulation-based experiences. Dewey describes experiential learning as a methodology started when a person perceives an occurrence, practices former understanding to imitate it and then applies judgment to create a similar one in future [1]. Every experience used in learning environments may arouse students' interest and curiosity, strengthens initiative, and sets up desires and purposes from learnings [1, 2].

Effective experiential learning is seen when a person progresses through a cycle of four stages: (1) employing a concrete experience followed by (2) observation of that experience that precedes to (3) the development of intangible concepts and generalizations which are then (4) used to investigate premise in future [3–5]. Considering engineering as an applied science, computer simulations experiences can be used to

© Springer Nature Switzerland AG 2020
M. E. Auer and T. Tsiatsos (Eds.): ICL 2018, AISC 916, pp. 532–541, 2020.
https://doi.org/10.1007/978-3-030-11932-4_51

create all four stages of experiential learning, specifically in undergraduate engineering. This will also allow the students to work with real and complex situations in a controlled and safe environment. Thus, they are trained with hand-on experiences to understand abstract concepts or phenomena that are otherwise difficult to visualize [6–9].

Although computer simulation-based experiment has been growing to use in engineering curriculums, and employed as a tool for improving classroom learning, it can be partially faced with the uncertainties of developing, using, and incorporating computer simulation-based experiments effectively into educational environments [6, 10]. In order to contribute to reflect on the computer simulation use in teaching–learning process of vehicle dynamics in automotive engineering curriculum, in this paper, a pedagogical proposal is depicted to examine the practice of the Kolb's experiential learning theory, four-stage learning cycle [3–5], to originate a set of simulation laboratory activities.

The pedagogical methodology in this paper aims to develop a detailed class plan for lessons about major topics in road vehicle dynamics concepts: specifically bump-steer performance, brake proportioning and federal requirements, roll-over threshold determination and determining testing parameters, coast-down test for emission analysis, and hardware-in-the-loop for cruise control system. The "Vehicle Dynamic" course is a required course in a four-year bachelor of technology in Automotive Engineering stream. Experiential learning with focus on the practicing of real world problems in classroom is the major goal of this program.

In this course, the CarSim [11] and Adams/Car [12] software packages are used to model a detailed model of cars, environments, and drivers. Also, it allows the students to implement simulations based on exploration and discovery of in different scenarios that cannot safely be replicated in a real experiment, for example vehicle roll-over occurrence in cornering can hardly be implemented to be used in university labs considering the educational budget and student's safety factor.

This paper is organized as follows. First, it reviews the computer-based simulation effectiveness as an educational laboratory. Second, it includes the explanation of learning outcomes of special topics in vehicle dynamic according to automotive engineering curriculum. In Lab Experiment Strategy Section, we will then propose methodology using Kolb's experiential learning theorem. Finally, some overall conclusions along with our ongoing works are presented in Conclusions and Future Works Section.

2   Computer-Based Simulation as Educational Laboratory

Most modeling techniques and traditional simulation were developed based on an analytical approach. In this approach, the mathematical problem is solved by finding the exact solution of the equations that describe the model. However, analytical approach is often not able to deal with complex systems such as dynamic of vehicle that has so many moving parts [13]. The dynamic modeling of a vehicle as a complex system is not related to individual analysis of its elements, but in the interaction between of components. Currently, the modeling and simulation area is strengthened to systematically develop a complex simulation model. However, using of developed modeling and simulation packages needs extensive learning time. The selection of proper simulation package and approaches to visualize the main concept of the simulations further to learning simulation environment are the challenges that need to be addressed wisely. The computer-based simulation labs improve the teaching-learning with following main actions:

1. In computer simulation lab sessions, the students are required to be involved with practical understandings. It is vital skill to engineering students since Engineering is an applied science that requires hands-on capabilities.
2. The use of simulation allows students to perform the experiments in a shorter time. Thus, they acquire the knowledge in various situations and they attain the major concept about the main mechanisms that control the behavior of the analyzed model which are more difficult to develop in the traditional laboratory classes.
3. The students can be interactivity engaged in tasks to apply their gained knowledge to a real problem to advance skills in critical thinking and decision making. However, in traditional lab sessions, the students have only the chance to observe and analyze the data from a single experiment.
4. In simulated environments, the models can be tested to breakdown and failure, allowing the students to experience and understand failure mechanisms in a safe and controlled way. The students have the opportunity to gain knowledge about the testing procedure and standard requirements such as determining maximum roof-top-loading and vehicle roll-over during the turning.

The pedagogical proposal organizes hands-on simulation laboratory activities for teaching the vehicle dynamics, and it is composed by 2-hour biweekly lab sessions. In this paper, we will demonstrate the detail of simulation lab session for two concepts in vehicle dynamics. Two commercial softwares, CarSim and Adams/Car, that both have significant features to model vehicle dynamics are adopted to be used by compute-based lab sessions.

3 Learning Objective of Special Topics in Vehicle Dynamic

The pedagogical proposal organizes hands-on simulation laboratory activities for teaching the concepts of

1. Vehicle braking performance and application of Federal Motor Vehicle Safety Standard (FMVSS) 105 [14]
2. Vehicle rollover performance and application of fishhook test [15, 16]
3. Vehicle coast down testing to measure aerodynamic and tire resistant loads [17]
4. Vehicle handling performance to determine the stability and response time.

In all of these concepts, students need to know the analytical solution of the vehicle performance for concrete practical cases in automotive standards [14–17]. In this paper, we will discuss "Vehicle Brake Performance" experiential learning according to Kolb's theorem; however, similar approach has been developed for other concepts in vehicle dynamic course.

3.1   Vehicle Braking Performance

The braking decelerations achievable on a vehicle are simply the result of hydraulic application pressure and the brake gains up to a point where lockup will occur on one of the axles. Lockup will reduces the brake force on an axle, and result in some loss of ability to control the vehicle. It is well recognized that the preferred design is to bring both axle up to lockup point simultaneously, as shown in Fig. 1A. According to FMVSS-105 [14], the vehicle must meet the following braking performance requirements as:

- Dry pavement: Stopping distance ≤ 56 m from 100 km/h (0.7 g's)
- Wet pavement: Stopping distance ≤ 40 m from 50 km/h (0.25 g's).

Yet, this is not possible over the complete range of operating conditions to which a vehicle will be exposed. Balancing the brake outputs on both the front and rear axle is achieved by "Proportioning" the application pressure, Fig. 1B.

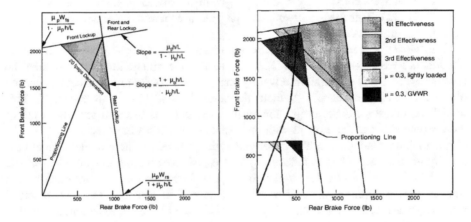

Fig. 1. (A) Brake proportioning [16] (B) Brake proportioning valve effect [16]

As shown in Fig. 1, the efficient proportioning line must be designed for braking system to use the maximum tire grip capacity. In addition, the braking force must provide certain deceleration rate according to federal requirement. The vehicle braking capacity is a function road surface (μ_p), axle loads (W_{fs}, W_{rs}), and height of the center of gravity of vehicle (h) [16]. Thus, the braking system is required to be designed to provide stopping distance in various road surface and vehicle loading conditions. This can be achieved with the design of brake proportioning.

4 The Teaching Strategy of Lab Sessions

In this section, a detailed planning of lab sessions employing four stages of Kolb's Experiential Learning Theory is presented. The detail of a single topic is considered for this study; however, six laboratory experiences are developed for the vehicle dynamic course. Each step of teaching strategy and Lab and classroom activities with the topic of "Braking Performance and Federal Requirements" will be presented.

4.1 Kolb's Concrete Experience: Automotive Braking Performance

First activity is Kolb's Concrete Experience, in order to address the subject to be developed, the instructor presents a comprehensive overview about the concept of braking performance [16] and provides a pre-designed template of a B-class sedan vehicle and road condition in CarSim software [11]. CarSim delivers the most accurate, detailed, and efficient methods for simulating the performance of passenger vehicles and light-duty trucks. CarSim is used extensively by 7 of the 10 largest automotive OEMs. OEM users consistently find close agreement between CarSim predictions and test results [11]. Thus, this software can provide a concrete experience in modeling braking performance.

The model of a real vehicle is provided to students in CarSim, shown in Fig. 2. CarSim has easy-to-use interface, Fig. 2B, to run the simulation and visualize the results thus Students can straightforwardly apply various brake pressure conditions and observe "stopping distance", "vehicle deceleration" and "tire grip capacity". This model has one-stage brake proportioning line and students can compare the result of the simulation-based experiment with the explained concepts in classroom.

In addition, students can also observe that the provided vehicle model cannot pass the requirements according to federal regulations [14] even though the provided vehicle is model of a real vehicle. This will actively provide the learning environment for the student to use automotive standard as well as critical thinking to solve practical problem, which can be the foundation of second stage of Kolb's experiential learning cycle.

4.2 Kolb's Reflective Observation: Automotive Braking Performance

Second activity is Kolb's Reflective Observation, students in a group of two should discuss and compare the braking performance, observed in the previous activity, based on key questions such that: "determine the maximum master cylinder pressure without the occurrence of wheel lockup during the braking", and "determine the minimum stopping distance of vehicle". This observation will help students to have better understand about braking performance according to tire capacity. With the increase of brake effort (increase the application pressure inside the master cylinder), the stopping distance becomes shorter; however, this will happen until the occurrence of wheel lockup, shown in Fig. 3. Students effectively observe the significant effect of wheel lockup during the brake. They can also compare the simulation result according to the theoretical concept of tractive/braking force generation in tire contact patch, shown in Fig. 3.

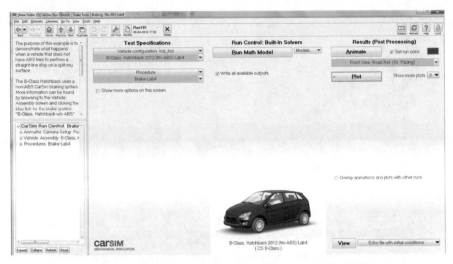

(A)   CarSim main window has easy-to-use interface to start simulation

(B)   CarSim graphics help to visualize the wheel lock-up and stopping distance

Fig. 2.   CarSim user-friendly simulation environment with animation interface

4.3   Kolb's Abstract Conceptualization: Automotive Braking Performance

Third activity is Kolb's Abstract Conceptualization, the professor reinforces the theory and the students are involved in thinking and forming a principle about the vehicle braking performance with use of industrial standard. The concept of "multiple-segment brake proportioning" will be applied to achieve a federal braking requirement. This stage will be a more general approach in braking system, since the students are needed to design a multiple-segment brake proportioning to pass the federal requirement in

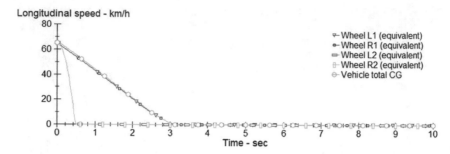

(A) The speed of each wheel is compared to vehicle speed to demonstrate the wheel lock-up occurrence (wheel velocity equals to zero)

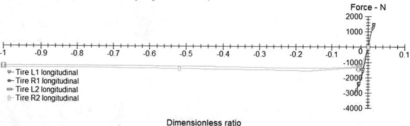

(B) The braking force generated in each tire contact patch according to dimensionless slip ratio

Fig. 3. Wheel lockup occurrence in braking, rear axle lock-up (wheel L2 and Wheel R2) occurs earlier than front axle due to longitudinal load transfer from rear to front

various vehicle loading and road surface condition, as explained in Fig. 1B. The designed multiple-segment brake proportioning system is designed in Microsoft excel using the theoretical concept of braking, shown in Fig. 4.

Students are asked to use the general information of the vehicle and road surface to determine and plot the axles lock-up regions, mathematical relations are derived during the lecture time, shown in Fig. 4. Students use FMVSS-105 standard to determine regions that provides required deceleration without occurrence axle lock-up during the vehicle braking, shown as green triangles in Fig. 4. Students can observe and compare the theoretical and simulation results. A single segment brake proportioning of the provided vehicle model cannot pass the requirement on the dry road surface. This result is obtained from the simulation using CarSim software in previous steps; however, this practice provides them the importance of designing a multiple-segment braking system, shown in Fig. 4. This approach during the lab session offers a concrete experience of industrial standard in real-world problems in combination with critical thinking in use of course principles.

4.4   Kolb's Active Experimentation: Automotive Braking Performance

Fourth activity is Kolb's Active Experimentation: It is the time that students implement the designed multiple-segment brake proportioning system for the vehicle in CarSim

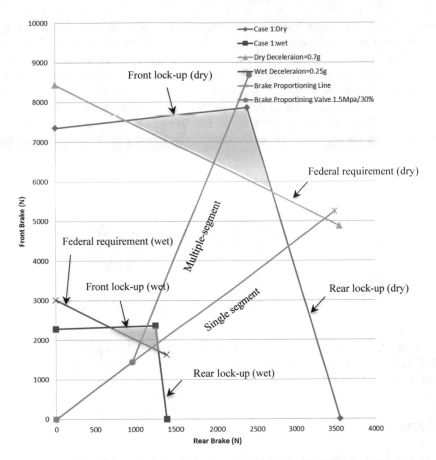

Fig. 4. Multiple-segment brake proportioning designed in Microsoft Excel

software. This experience is actively provided experience to test the designed braking system and check the performance with automotive standard FMVSS-105 [14]. As a lab homework, where each student is aimed to design a new braking system for a new provided vehicle, C-class vehicle, and demonstrate the performance according to standard. This experimentation of a real-world problem in vehicle dynamic can successfully be achieved with the use of CarSim as a simulation software.

5   Conclusion

The computer-based simulations are increasingly used as an approach to teaching Engineering Education. The pedagogical methodology in this paper contributes to re-design the use of computer simulation as a teaching tool in Automotive Engineering in the light of Kolb's experiential learning theory. As part of pedagogical proposal, this study is also focused on development of real world experiment as concrete experiment in topics related to automotive industries according to Kolb's learning cycles.

The students' satisfaction from the learning environment in four years in row is assessed with two questions: (1) how do you rate the value of this course compared with others you have taken at McMaster University and (2) independent critical judgment was encouraged. Table 1 demonstrates the result of student evaluation in reply to two mentioned questions, the scale for this question is from 1 to 5, 1: very poor and 5: excellent. In 2014, CarSim simulation software was use as a tool to visualize the result. From 2015, the real-world problem and use of industrial standards have been integrated with lab sessions. This new approach for experiential learning has relatively improved the value of the course and also advanced the critical thinking among students as shown in Table 1.

Table 1. Student evaluation from the course

Year	No. of students	Q1: Course relative value	Q2: Critical thinking
2014	32	3.65 (73%)	3.23 (64%)
2015	24	4.26 (85%)	4.30 (86%)
2016	34	4.82 (96%)	4.85 (97%)
2017	31	4.77 (95%)	4.81 (96%)

As an example of future works, the pedagogical approaches of this work need to be investigated in a series of real case studies, in order to evaluate this experiential and constructivist educational process.

References

1. Dewey, J.: Experience and education (Vol. original work published 1938) (1963)
2. Durkin, R.J.: Experiential learning in engineering technology: A case study on problem solving in project-based learning at the undergraduate level. J. Eng. Technol. **33**(1), 22 (2016)
3. Kolb, D.A.: Experiential learning: experience as the source of learning and development. FT Press (2014)
4. Kayes, A.B., Kayes, D.C., Kolb, D.A.: Experiential learning in teams. Simulation Gaming **36**(3), 330–354 (2005)
5. Healey, M., Jenkins, A.: Kolb's experiential learning theory and its application in geography in higher education. J. Geogr. **99**(5), 185–195 (2000)
6. Chen, W., Levinson, D.M.: Effectiveness of learning transportation network growth through simulation. J. Prof. Issues Eng. Educ. Pract. **132**(1), 29–41 (2006)
7. Balamuralithara, B., Woods, P.C.: An investigation on adoption of the engineering simulation lab exercise: a case study in Multimedia University Malaysia. Comput. Appl. Eng. Educ. **20**(2), 339–345 (2012)
8. Zavalani, O.: Computer-based simulation development of a design course project in electrical engineering. Comput. Appl. Eng. Educ. **23**(4), 587–595 (2015)
9. Ku, H., Fulcher, R., Xiang, W.: Using computer software packages to enhance the teaching of engineering management science: Part 1—Critical path networks. Comput. Appl. Eng. Educ. **19**(1), 26–39 (2011)

10. Botelho, W.T., Marietto, M.D.G.B., Ferreira, J.C.D.M., Pimentel, E.P.: Kolb's experiential learning theory and Belhot's learning cycle guiding the use of computer simulation in engineering education: a pedagogical proposal to shift toward an experiential pedagogy. Comput. Appl. Eng. Educ. **24**(1), 79–88 (2016)
11. CarSim. Mechanical simulation corporation. Ann Arbor, MI, 48013
12. Adams, M.S.C., Documentation, C.: Msc. Software Corporation (2005)
13. Blundell, M., Harty, D.: The Multibody Systems Approach to Vehicle Dynamics. Elsevier (2004)
14. Standard No. 105; Hydraulic Brake System, Code for Federal Regulations, Title 49, Part 571.105, Oct 1, pp. 199–215 (1990)
15. National Highway Traffic Safety Administration: Laboratory test procedure for dynamic rollover: the fishhook manoeuvre test procedure. Department of Transportation, Washington (DC), US (2013)
16. Gillespie, T.D.: Fundamentals of Vehicle Dynamics. Society of Automotive Engineers, Warrendale, PA (1992)
17. E/ECE/324 Regulation No. 83 (Uniform provisions concerning the approval of vehicles with regard to the emission of pollutants according to engine fuel requirements), p. 101, Annex 4, Appendix 3

Introducing of New Teaching Methods in Teaching Informatics

Informational Competences in Educating Teachers

István Szőköl[✉] and Kinga Horváth[✉]

J. Selye University, Komárno, Slovakia
{szokoli,horvathki}@ujs.sk

Abstract. Information and communication technologies play an important role in university studies. Researching and measuring the success of the modular system applied in education was carried out in the fall term in the framework of IT education at the Faculty of Education. In this paper we analyse the current stage of the problem, provide a definition of information competence, the standards of information competence are listed, emphasis is put on key competences as the concept of curricula, defining the modular system and subject system in IT education, explaining their advantages and disadvantages. The aim of the paper is to show the importance of the expansion of teacher's key competences, while also pointing out those other competences which are unavoidable in pedagogical work.

Keywords: Teaching methods · Teaching informatics · Modular system

1 Introduction

In most countries of the world, it is a trend to create expectations towards the results of school work, kind of standards, which can be controlled regularly. Most products are required to live up to predetermined standards and these standards or norms are strictly controlled. In schools, they do not do it. Nobody guarantees what kind of skills or knowledge the student of the school has.

The examiners or evaluators verify individual competencies in relation to the assessment criteria for general and professional qualifications, partial qualifications or full qualifications. Recognition of learning outcomes can be characterized as a set of standards or procedures to control the assessment of the learning process and learning outcomes, the validity check and the reliability of learning based on solid criteria.

Recognition of learning outcomes varies from country to country. The results of initial education in the Slovak Republic are recognized by successful completion of the final school leaving examination ("maturita"), the final exam, the graduation exam and the state final examination. There are no rules and regulations at European level.

Schools must keep pace with the rapid development of technology, research and social change. They must adopt a new form of education based on the results of research on how people learn and on the effective use of technologies and skills for the 21st century. Students coming to colleges and universities already have some skills and

© Springer Nature Switzerland AG 2020
M. E. Auer and T. Tsiatsos (Eds.): ICL 2018, AISC 916, pp. 542–551, 2020.
https://doi.org/10.1007/978-3-030-11932-4_52

knowledge from Information and Communication Technologies, but standards are different in all types of secondary schools. Universities must therefore adapt to new conditions and teach students what they do not know yet.

2 Key Competences, Concept, Definition

Nowadays, in the context of the changes in Slovak education, we are increasingly confronted with the concept of competence and key competencies. Key competences are usually considered as a new phenomenon in education. This concept arose in the 70 s of the 20th century in economics where it represented a set of specific requirements for job seekers. In the late 1990s, it went to the field of education to serve as a bridge between employers' demands on the labour market and the graduate profile. The term competence is used both in professional and common language; and ability, skill, capability, effectiveness, capacity, desired quality and others are used as synonyms for the group of terms. A person who has abilities and skills, motivation, knowledge, etc. to carry out tasks well in a particular field is considered competent. Competence is usually applied to individuals, social groups and institutions in case they successfully fulfil requirements and achieve goals set by their environment. The theory of key competences has not yet been completely formulated and a comprehensive and widely accepted definition does not exist. As Hrmo-Turek states in his publication *Key Competences*: "Competence is the behaviour (activity or set of activities), which characterizes excellent performance in a specific field. Key competences are the main competences of a set of competences. They are suitable for solving a wide range of mostly unforeseen problems which allows an individual to cope with rapid changes at work, personal and social life."

"Key competences are a set of interiorized, interconnected group of acquired knowledge, skills, abilities, attitudes and valuing approaches, which are important for the qualitative personal development of the individual, his/her active participation in society, application in employment and lifelong learning" [11].

Another definition states: "Having competence means having a complex equipment of personality, which allows the individual to successfully address challenges and situations in life, in which one is able to adequately orient, take appropriate actions and take a beneficial attitude. Key competences need to allow the individual to continuously refresh the skills and knowledge applicable in everyday life. For a person in training not all educational activities (cognitive, training, and educational) need to be beneficial, but especially those, which are useful in standard practice, provide quality education and correspond with company requirements in the labour market. Not only the attended educational process or certificate of the attended educational process are crucial, but also the learning outcomes.

Areas of key competences
Education in each field should be directed toward each individual creating the following key competences consistent with their levels and scholastic aptitude [4]:

– Informational
– Learning

- Cognitive
- Interpersonal
- Communicative
- Personal

2.1 Information Competences

Information technologies are key elements in building modern society based on knowledge. Information competences mainly include information literacy and computer literacy.

Information literacy may be explicitly defined as the ability to locate, evaluate and use information in a way that makes a person an independent, lifelong learning individual; as the ability to locate, evaluate, use and communicate information in various forms, such as the integration of written, computer, media and technological literacy, ethics, critical thinking and communication skills.

Computer literacy may be viewed as the ability to address problems, which means to educate and expand the following skills [3]:

- distinguish essential phenomena from non-essential,
- focus on and evaluate information,
- provide necessary information,
- choose (evaluate) and use appropriate methods, concatenate on or combine various methods to solve problems, or adapt or propose a new method, which solves a professional problem,
- express facts and their phenomena mathematically,
- carry out calculations,
- use outcomes – solve a problem.

Thus formulated computer literacy, respectively information literacy is not the only the content of a selected group of subjects which contain the expressions "computer technology and information technology" in their names, but all the subjects as a whole, the problems of which will be solved, while in addition to the mechanics of using computers, the emphasis is on the thinking process, evaluation, decision, optimizing and realization.

The model of securing IT competences – a study plan will be compiled for future teachers on the basis of the entrance test results – it will be recommended which modules they should attend, and it will set them the tasks to be carried out independently. Test results also determine which thematic areas (teaching units) should be repeated and how many times. Based on the requirements and self-evaluation, the student (future teacher) can develop his/her own, individual study plan. If he/she does not accept this responsibility, the teacher recommends an "optimal" study plan based on the result of the entrance test, which may or may not be respected. Tasks are defined in such a way so that by solving them, the learner should be able to get ready for passing the test successfully.

2.2   Learning, Cognitive and Interpersonal Competences

The development of learning competences supports mainly the knowledge of learning styles, which sum up preferred practices of teaching and learning in a particular period of the life of an individual, who develops, changes and improves from the innate basis. Learning competences include the readiness to learn how to learn, motivation to learn, in-depth access to learning, the overall process of learning.

Cognitive competences involve critical and creative thinking, and problem solving. Problem solving is closely linked to the ability of critical and creative thinking. Thanks to it, we can avoid many unnecessary errors while thinking and acting.

Interpersonal competence means the effective coexistence and cooperation, where you need [10]:

- the ability to work in a team (group) – cooperation (joint responsibility of planning, organizing, operating and evaluating a team; development of leadership and management skills),
- empathy (empathizing with the emotional state and situation of other people),
- solving conflicts in a non-violent way – assertiveness to enforce the rights, needs and interests (not being a passive and manipulative object),
- creating and maintaining harmonic and progressive interpersonal relationships (respecting the ethics of proper manners, responsibility and morality in terms of good relations with other people, mutual understanding and helping others),
- creating intercultural systems based on constructive negotiations, compromises, tolerance and pluralism (acceptance and recognition of differences in human views, opinions, values, faith, ethnic origins, various cultures, different areas of expertise, and thus the ability to live in an foreign environment – as a manifestation of ethical conduct),
- the development of a democratic civilian system, respect for human rights and fundamental freedoms, the maintenance of peace (an effort not to fight with others and not to rule others), a healthy environment.

2.3   Communicative and Personal Competences

The basis of communication competencies is to express a situation in writing and verbally, listen carefully, read with understanding. The European Union requires each citizen living in the European Union to know (except for the mother tongue) two foreign languages of the EU.

The personal qualities of individuals should foster the effective functioning of society; develop an authentic personal and professional life through co-operation and cooperation with the environment. Personal competences include self-awareness, self-control, self-motivation as well as being assertive.

3 The Competences of Teachers

There are several different views on the classification of key competences. These competences have been dealt with by many authors, like: Belz-Siegrist, Helz, Hrmo, Turek.

In addition to the already mentioned key competences, Hrmo and Turek outline the following competences of teachers [6]:

- Professional: the teacher as the guarantor of scientific basic, subjects of his/her own approbation,
- Psycho-didactic: the teacher as an individual, creating pleasant conditions for learning
- Communicative: the teacher as an individual, using an appropriate level of verbal and non-verbal communication with students, parents and colleagues,
- Diagnostic: the teacher is able to diagnose the problems of students,
- Planning and organizational: the teacher is able to plan his/her actions,
- Advisory and consultative: the teacher is able to help and advise the parents of his/her own students,
- Self-reflexive: the teacher is able to evaluate and modify his/her own educational activity.

In addition to the above-mentioned competences some countries also develop cultural awareness, occupational and entrepreneurial competences, and health education. Due to their poor representation in reviewed foreign systems, these systems are not going to be dealt with any closer in this paper.

4 Teaching Modules

They outline the learning outcomes in relation to the content of the lessons in the individual subjects. They express the forms, the way and the content of the activities that the student has to acquire in relation to the particular field. Teaching curricula/modules also include the time allocation for a given subject/module, learning objective, subject/module function, specific procedures, forms and organization of teaching, interdisciplinary relationships and learning resources.

Selection of learning content
The main criterion for selecting learning content is to determine those factors that most influence the achievement of the learning goal. As the goal varies according to social developments, the views on this choice have gradually changed.

Arrangement of learning content
In order to avoid unwanted empty spots in learning, where there is a lack of continuity between what we already know and what we should know, or unjustifiable duplication, the learning content needs to be properly structured and appropriately organized.

Under the **subject system of teaching**, we mean the teaching of the subject as a whole, while the layout of the subject is spiral (students do not lose their starting point and step by step, they extend their knowledge without interruption).

Under the **modular system of teaching**, we mean the teaching of a given subject divided into individual parts – modules, where the arrangement of the curriculum is modular (module building) [4].

4.1 Modular Arrangement of Curriculum

The term **module** is generally understood as a separate element (part, unit) that is complete, but which can be joined to other units and, together with them, can create a larger unit that serves to achieve wider goals or to solve more complex tasks. The module is therefore an independent part of a certain construction.

In education, the module refers to a separate part of the curriculum, and the modular arrangement of the curriculum of a certain type of school, course, etc. does not mean the division into lessons and thematic subjects and subject matters, but the curriculum is divided into modules. The time required to acquire the module's content is shorter than the time course of the subject, usually 15–40 lessons.

The **modular system of teaching** means the gradual acquisition of the curriculum of individual modules that make up a school programme, a course, a training course, etc. Certain education – obtaining a certificate, a diploma, etc. is understood to be a construction composed of a number of building blocks – modules. The successful acquisition of the content of this module is a permanent part of the construction – education and it is not necessary to acquire this content again after a certain period of time [3].

4.2 The Need to Implement the Modular System in Education

Quality is a measure of perfection, value, the usefulness of education and training, the fulfillment of requirements and expectations of school customers: pupils, students, parents, employers, citizens, and the state. The quality of education and training can be continually increased regardless of its current level.

Within the subject of Informatics at university, it is possible to achieve this only in a differentiated – modular way, so that each student learns only what he/she still does not know. Different types of secondary schools have different standards that students have to meet. There are big differences between students who apply for IT in terms of competencies and practical skills in the field. Although there are standards in this area that determine the level of information literacy of the secondary school graduate, in most cases the knowledge of students coming to the first year of university does not meet this standard. The causes of these shortcomings are different. One of the possible reasons is the type of secondary school attended, as well as the level of teaching Information and Communication Technologies (ICT). Another reason for the striking differences is that not every student has the same relationship with these disciplines and many lack proper internal motivation. The first two reasons can be included in the external conditions of education, which are closely related to the readiness of schools in terms of building the necessary material and technical background base and improving the knowledge and skills of students. The material-technical equipment of a school is a matter of finances and in a short time it may change from inappropriate to over-standard. The preparation of students it is not that simple. It seems that the various levels of knowledge at university will remain unresolved for a long time. That is why it would be possible to solve the problems of the subject of Informatics with a modular system – everybody learns what he/she does not know yet, on the basis of the written entrance test.

To provide information literacy at individual levels, the curriculum includes general and special subjects focused on computing, informatics and information and communication technologies.

4.2.1 The Educational System of Subjects in the Faculty of Education at the J. Selye University in Komarno

The subject Informatics is taught during two semesters in the Faculty of Education at JSU. The weekly range of classes is as follows: two lectures and two seminars per week, or 56 lessons of Informatics per semester. The winter semester is divided into 14 weeks, and each week has a specially defined content of lectures as well as exercises. This content of the subject must be completed by each student of the first year at FE JSU, taking into account their knowledge and skills acquired at secondary schools.

5   Proposing a Modular System of Education in the Faculty of Education at JSU in Komarno

In order to prevent redundancy in the teaching process, and to ensure that every student could upgrade his/her previous knowledge of Informatics, a modular system for teaching this subject has been developed, with the main emphasis being placed on the type of secondary school attended. Each type of secondary school has a standard from the subject of Informatics, which should be reached by the graduate of a given school.

Modules that serve as a proposal for teaching a modular learning system have specially defined instructions, introduction, module objectives, content and performance standards, teaching text, summary, self-test, supplementary literature, conclusion and bibliographic references.

The modular teaching system consists of five modules. Modules 1, 2 and 3 are taught in the first - winter semester at FE UJS and modules 4 and 5 in the second - summer semester. Each module consists of approximately 20 h of teaching - together 104 lessons per year. At the beginning of the first semester, the students take a modular entry test and, on the basis of the results, will apply for modules that they do not know yet.

Proposal of contents and objectives of individual modules

The study support module is divided and structured so that the learning and knowledge building among students is as effective as possible. Efficiency lies mainly in the fact that the learner can fully devote himself/herself to studying the educational content because he/she is not driven by the management of the study because the study text contains elements allowing for rapid and correct self-regulation. Students will receive study materials once they enroll in the course.

The input diagnostics of the module - through a diagnostic interview and in the form of an entrance test the lecturer determines the level of input and the ability of the participant to study. If required, the lecturer sets a separate diagnostic task (or even more tasks) that the student will solve and deliver electronically to the lecturer for assessment. This procedure will ensure a comparable level of input and ability of the student, thus creating a prerequisite for successful completion of the module.

5.1   Entrance and Final Test

One part of the education process is feedback from students. Due to the fact that students will be divided into groups based on the entrance test, tests cannot be anonymous. The individual questions have been selected on the basis of their practical usefulness for module evaluation and standards for four-year grammar schools developed by the State Pedagogical Institute.

At the beginning of the semester, students receive two types of tests. The first test deals with basic information on the type of secondary school attended, the number of computer science lessons per week as well as the basic knowledge of the subject. The second test is already divided into five modules, where students have to answer individual questions, where each question is worth one point. According to the results of this test, the students are provided an individual study plan according to which they attend individual lessons. At the end of the semester, students have to take the same test - the final test, which monitors the efficiency of teaching and the knowledge acquired during the semester, and compares them with the results of the entrance test.

5.2   Research Results

The total number of students taking the entrance test in the first semester was 181. We divided the students into two sets - A and B. There were 89 students in the A group – where Informatics was taught in the subject system, and in group B there were 92 students – where teaching was carried out according to the modular system. In the assessment, in each module, students had to reach at least 60% so to avoid having to take the module again.

Number of students in FE UJS in the first year (See Table 1).

Table 1. Results of the questionnaire in the first class

Modules	Don't have to take the module	Have to take the module	Total
Module 1	107	74	181
Module 2	99	82	181
Module 3	52	132	181
Module 4	20	161	181
Module 5	0	181	181

It is noteworthy that in the fifth module, no one answered the questions correctly, although the topics were included in the curriculum. This can be caused either by not teaching the curriculum, by the lack of interest of the students in the field of Informatics, or by forgetting the learned materials as a result of the lack of revision. As Informatics is taught in grammar schools only in the first year - as it is clear from the

Table 2. Number of required groups in each week of the first semester

Weeks	1–5	6–9	10–14
Need	74	82	132
Do not need	107	99	52
Number of groups per week	3	3	5

questionnaire - it is quite likely that the negative outcome of the fifth module is due to the fact that it is forgotten.

Since only modules 1, 2 and 3 are taught in the first semester, we divided the semester into three parts.

The number of contact hours according to the data from Table 2 will decrease by 38%. Initially, for 181 students it is necessary to create 6 study groups, i.e. the number of lessons in Informatics per semester would be 336. After the introduction of the modular system of teaching, we will reduce the number of contact hours per semester to 208. However, the exam from the subject must be completed by all students.

6 Conclusion

Given that the transformation of the subject system of teaching into the modular system of education is among the most up-to-date topics in higher education, we consider the following main benefits:

- The development of a modular system of teaching the subject of Informatics on a theoretical level as well as its implementation in practice.
- By the research, we found that by introducing a modular system of teaching it is possible to individualize the mass preparation of university students and rationalize the work of the teacher and achieve higher efficiency of the teaching process.
- The introduction of the modular structure of content and an adaptive method of teaching the subject of Informatics will save a number of contact hours. This results in less demand for special classrooms and equipment.
- The opportunity to use the "learning by doing" teaching method to compensate the completion of contact hours.
- The number of direct contact hours required is reduced by approximately 38%. Individual thematic units are likely to be differently represented in the curriculum.

When applying a modular system of teaching, it is necessary to deal with the process of input motivation, the process of assessment and the classification of individual modules, the process of the exposition of new learning content as well as the process of setting homework.

At the beginning of the semester, it is necessary to determine the level of knowledge of first year students at the Faculty of Education in the field of Informatics, and to detect their needs and their preferred learning styles.

References

1. Belz, H., Siegriest, M.: Klíčové kompetence a jejich rozvíjení, 375 pp. Portál, Praha (2001)
2. Bendíková, E.: Lifestyle, physical and sports education and health benefits of physical activity. In: European Researcher: International Multidisciplinary Journal. Academic publishing house Researcher, Sochi, vol. 69, no. 2–2, pp. 343–348 (2014). ISSN 2219-8229
3. Dostál, J.: Informační a počítačová gramotnosť – klíčové pojmy informační výchovy. In: INFOTECH 2007. PdF UP, Olomouc, pp. 60–65 (2007). ISBN 978-80-7220-301-7
4. Falus, I., et.al.: Bevezetés a pedagógia kutatás módszereibe. Műszaki Könyvkiadó, Budapest (2000). ISBN 963-16-2664-4
5. Horváthová, K.: Kontrola a hodnotenie v školskom manažmente 1. Vyd, 106 pp. Wolters Kluwer, Bratislava (2010). ISBN 978-80-8078-329-7
6. Hrmo, R., Turek, I.: Kľúčové kompetencie I. STU Bratislava, Bratislava (2003). ISBN 80-227-1881-5
7. Hrmo, R., Krištofiaková, L., Miština, J.: Building a quality system of technical and vocational education in Slovakia towards a European labour market. In: WEEF 2015, Proceedings of 2015 International Conference on Interactive Collaborative Learning (ICL), Florence, Italy (2015)
8. Hrmo, R., Miština, J., Krištofiaková, L.: Improving the quality of technical and vocational education in Slovakia for European labour market needs. Int. J. Eng. Pedagog. (iJEP) 6(2), 14–22 (2016). ISSN 2192-4880
9. Miština, J., Jurinová, J., Hrmo, R., Krištofiaková, L.: Design, development and implementation of e-learning course for secondary technical and vocational schools of electrical engineering in Slovakia. In: Auer, M., Guralnick, D., Simonics, I. (eds.) Teaching and Learning in a Digital World, ICL 2017, Advances in Intelligent Systems and Computing, vol. 715, pp. 915–925 (2017). ISBN print 978-3-319-73209-1. https://doi.org/10.1007/978-3-319-73210-7_104
10. Nagy, M., Makovický, P.: A bioritmus és a napirend kapcsolata a Selye János Egyetem hallgatói körében végzett felmérés tükrében. Actual problems of modern education in 21st century: Aktuálne problémy moderného vzdelávania v 21. storočí, pp. 131–137 (2012). ISBN 978-80-8122-065-4
11. Porubská, G.: Aktuálne problémy slovného hodnotenia na 2. stupni ZŠ. In: Slovné hodnotenie na druhom stupni základných škol: Zborník príspevkov z vedeckej konferencie Nitra: UKF, pp. 101–108 (2000). ISBN 80-8050-320-6
12. Puskás, A.: Assessing Young Learners in the English Language Classroom. 1. Vyd. 139 pp. Belvedere Meridionale, Szeged (2017). ISBN 978-615-5372-82-7
13. Szarka, K., Brestenská, B., Juhász, Gy.: Analýza aspektov hodnotenia autentických výstupov a komplexného monitorovania žiackych prác v chémii. In: Didaktika chemie a její kontexty: 24. Mezinárodní konference o výuce chemie. Masarykova univerzita, Brno, CD-ROM, pp. 200–208 (2015). ISBN 978-80-210-7954-0
14. Vass, V.: Creative school: renewing leadership for creativity. In: Marek, T., Karwowski, W., Frankowicz, M., Kantola, J., Zgaga, P. (eds.) Human Factors of a Global Society: A System of Systems Perspective. Education in Modern Society, pp. 969–974. CRC Press, Taylor and Francis Group, Florida, US (2015)

Game Based Learning

Gaming the Lecture Hall: Using Social Gamification to Enhance Student Motivation and Participation

Sebastian Mader[✉] and François Bry

Institute for Informatics, Ludwig Maximilian University of Munich,
Munich, Germany
sebastian.mader@ifi.lmu.de

Abstract. The traditional lecture is a teaching format that offers students few opportunities for interaction turning them into passive listeners of the lecturers' presentations what negatively impacts on their learning. With audience response systems, that is technology supporting classroom quizzes, breaks that re-activate the students can be introduced into the lecturers' presentations. This article reports on an audience response system coupled with a social gamification of quizzes based on teams: Each student is assigned to a team and the students' answers to quizzes contribute to their teams' success. An immediate overview of responses to quiz questions and the team standings motivates students to participate in the quizzes. The contribution of this article is threefold: First, a team-based social gamification of quizzes aimed at boosting participation in quizzes and attendance at lectures, second, original technological tools supporting the proposed team-based social gamification, and third, an evaluation of the approach demonstrating its effectiveness.

Keywords: Gamification · Audience Response Systems ·
Online Learning Environments

1 Introduction

The traditional lecture – that is, a lecturer addressing an audience of variable size – is a teaching format offering students few opportunities for interaction: Asking questions and receiving answers is often the only possible form of interaction in a traditional lecture. As lecture audiences grow, students tend to ask less questions because of a well-known inhibiting social barrier: The larger the audience, the greater the fear of speaking out [22]. A vicious circle sets in: The students' reduced activity results in their reduced involvement what in turn results in their reduced learning performances [27] contributing to further reduce their activity.

The reduced interactivity in traditional lectures can be addressed by slightly modifying the teaching format. A widespread approach consists in introducing

© Springer Nature Switzerland AG 2020
M. E. Auer and T. Tsiatsos (Eds.): ICL 2018, AISC 916, pp. 555–566, 2020.
https://doi.org/10.1007/978-3-030-11932-4_53

breaks in the lecturer's exposition that are used for re-activating the students, for example by posing questions to the audience, so called quizzes. While this is effective in lectures with small audiences of about 20, this turns out hardly practicable with audiences of a few ten and fully impossible with audiences of a few hundred students. Indeed, collecting and aggregating the answers of large audiences are time consuming tasks that disrupt the lecture. Audience response systems overcome the obstacle by handing these tasks to software. While such a use of audience response systems in lectures enhances participation and provides a number of students with "active breaks", there are in our experience still students who remain passive and do not answer the quizzes [19, p. 75], as well as students skipping lectures altogether. Indeed, regular or occasional absenteeism is another widespread problem of lectures [13]. The audience of a lecture generally drops by half during the semester and students that skip lectures try to discover, understand, learn, and practice lectures' contents one or two weeks or even only a few days before the lectures' examinations what mostly results in failure in the examination.

With the aim of improving both participation in quizzes and attendance at lectures, a gamification scheme with a social component, referred to in the following as "social gamification", has been devised. Social gamification is comparable to Nicholson's [18] engagement dimension. With the proposed social gamification scheme, students are assigned to teams that compete against each other in quizzes conducted during lectures. Each team member gives an individual answer, thus keeping the personal involvement necessary for an effective learning, but doing so contributes to their team's performance. After each quiz, updated team standings are shown to the audience. This article describes this social gamification of the lecture hall and reports on an experiment pointing to both the effectiveness of the approach and its positive reception by students.

The contribution of this article is threefold: First, a team concept aimed at boosting participation in lecture quizzes as well as attendance during lectures, second, original technological tools supporting the proposed team-based social gamification of quizzes, and third, an evaluation of the approach demonstrating its effectiveness.

This article is structured as follows: Sect. 1 is this introduction. Section 2 is devoted to related work. Section 3 gives a general overview of the concepts behind the team component and the interface elements supporting the team component. Section 4 introduces the lecture during which the team component has been evaluated and presents the results of the evaluation. Section 5 concludes the article and gives perspectives for future work.

2   Related Work

The social gamification of quizzes conducted during live lectures reported about in this article is a contribution to gamification and relates to audience response systems and peer discussion.

Gamification. Deterding et al. define gamification as "the use of game design elements in non-game contexts" [7, p. 10], a definition which leaves "game elements" open to interpretation. These authors cite Reeves [21] who identifies ten essential elements of games, among others feedback, "competition under rules that are explicit and enforced" [21, p. 80] and teams, and argue that these elements are not in themselves game elements and that it is their contextualization that turns them into game elements.

More support for teams and feedback as game elements or gamification mechanisms comes from Nicholson [18] who introduces various dimensions in which "meaningful gamification" can take place, one of them being "engagement" defined as "creating opportunities for participants to engage with others in meaningful ways" [18, p. 12]. According to Nicholson, leaderboards can be part of a gamification utilizing the engagement dimension for allowing comparison. Nicholson also argues that both cooperative and competitive elements can be implemented so as to achieve effects similar to those observed in sports: Cooperation within teams and competition between teams. Danelli [5] identifies competition and cooperation as drivers of engagement as well using various theories about games and play.

An example for a study in which teams were introduced into an education setting is the experiment conducted by Latulipe et al. [14] for a computer science class: Students in the class were randomly assigned to teams of about five. For the class, a flipped classroom design was adapted: Students studied the content before the lecture that was dedicated to running a set of quizzes; by answering the quizzes the students could earn points for their team. The authors report of a positive attitude among the students towards the team component deployed during the lectures.

Feedback – especially immediate and continuous feedback – is identified by McGonigal [16] as one of four defining traits of games, goal, rules, and voluntary participation being the other three. An area where gamified feedback is used are cars, e.g., for fostering safe or eco-friendly driving. In [24], a virtual passenger is simulated who reacts to the driver's driving style providing an immediate and tangible feedback to the driver. Eco-friendliness can be found in different implementations: From Chevrolet's system that requires to keep a green orb in a optimal position through eco-friendly driving to Ford's EcoGuide that makes the more leaves grow on a display, the eco-friendlier the driving [11]. In all cases actual optimization criteria are made tangible through a reification of the criterion instead of representing it, e.g., by an elusive numerical value (expressing, e.g., the fuel consumption).

While gamification often seems to produce positive results, there is criticism that most gamification systems are traditional "reward systems" motivating extrinsically (motivated by external rewards [23]) but not intrinsically (motivated by the task itself [23]) [18]. Nicholson argues that often the positive behaviour induced by rewards stops as soon as the rewards are no longer given if the user has no intrinsic motivation to further perform the task [18]. Indeed, Nicholson observes that there are situations where reward-based gamification can be positive, e.g., when used to teach a skill that has real-live applications because the learner most likely learns to the see value in the skill itself. While Hamari et al. [9] cite three studies coming to the conclusion that gamification has no long-term effects and that positive effects of gamification most likely may result from its novelty, they state that gamification generally does produce positive results.

Audience Response Systems. Audience response systems (ARS) allow lecturers to conduct quizzes during lectures and to provide immediate feedback to both, students and lecturer alike, on the correctness of quiz answers. In their survey of ARSs, Kay and LeSage [12] list various benefits of ARSs: Engagement, attendance, participation, and discussion among them. A meta-analysis conducted by Hunsu et al. [10] found that, among other, the use of ARSs had a positive effect on a number of learning outcomes such as knowledge transfer, but had no effect on other learning outcomes such as retention of the subject matter. ARSs come in form of clickers, that is, students are provided with a physical device that allows them to participate in quizzes, or in form of web-based platforms, which allow students to participate in quizzes using their own internet-capable devices. Examples of web-based ARSs are GoSoapBox [3] and Backstage [1].

Pedagogical Foundation. Peer instruction is a term coined, and a concept investigated, by Eric Mazur [15]. Peer instruction refers to posing questions to a lecture's audience, first asking the students to consider an answer on their own, then let the students discuss the value of their various answers, each student trying to convince their peers of their own answers [4]. The authors report on a positive effect of peer instruction on the learning performance of students. Byrd et al. [2] report on positive effects on academical results from using cooperative quizzes. It is worth stressing that peer instruction in the aforementioned form is hardly possible with lecture audiences of more than a few ten students and almost impossible with audiences of a few hundred.

An approach close to students' discussions of the subjects to learn is "learning by doing". This approach is traditionally deployed in science, technology, engineering, and mathematics (STEM) education. It is justified by Polya who states that humans "acquire any practical skill by imitation and practice" [20, p. 4]. The value for learning by imitation and practice is especially true for programming, a practical skill with deep mathematical foundations.

The importance of imitation and practice is clearly stressed by the emergence of "try-out" environments integrated in various educational software: Some MOOC platforms such as Codeacademy and freeCodeCamp make it possible to write and execute code directly from the learning material. Coding is thus closely integrated into the teaching material and therefore the learning activity. Stack Overflow, a question and answer platform for coding and software development, allows users to directly test answers in the browser if they are pieces of code in certain programming languages. Getting students to immediately try out code presented in a lecture is a highly desirable goal that can rarely be achieved in traditional lectures.

3   Team Competition

The team-based social gamification described in this article makes teams compete against each other in quizzes conducted during lectures. Participants earn points for their team by answering quizzes on their own. After each quiz, the points earned are computed and distributed and updated team standings are published.

3.1   Team Building and Reward System

The core component is a team building system that can be adjusted along three axes to fit the actual context it is deployed in:

- *Duration*: For what time span will the teams exist?
- *Formation*: Who forms the teams? Teams can be formed by lecturers, the software, or by students themselves.
- *Size*: How many students are there in a team?

The team building system does not offer one single team configuration expected to fit all situations. Indeed, a team configuration is heavily dependent on the context, and even for a same context different configurations may make sense.

Another component is the reward system that determines what, and under what circumstances, rewards are given to students' teams for their team members' answers to quizzes. After a quiz, one and only one of the following holds for every student:

1. The student participated and answered correctly.
2. The student participated and answered incorrectly.
3. The student was logged in and did not participate.
4. The student was not logged in.

For each of these four outcomes, an effect on the student's team score can be defined. In the experiment reported about below, both (1) and (2) were rewarded with points since the goal of the experiment was to improve participation. The points awarded for participation act as positive reinforcement (see [25]) for participation. Another possibility, but by no means to only one, would be to punish non-participating students with negative points what would act as positive punishment (see [25]). In the experiment reported about below, (3) and (4) had no effect on the student's team scores.

Subtracting points for non-participation could potentially have a stronger impact on participation because of loss aversion [26] but could also demotivate since it could be perceived as punishment to those team members who participated because through their team they would be punished for the actions of others.

3.2   System

This section introduces the underlying ARS and and its extension with the team-based social gamification described above, in the following referred to as the gamification component. The gamification was built as an extension to the existing ARS of the original teaching and learning platform Backstage.[1] Backstage's ARS provides the usual features: At any time during a lecture, a quiz can be conducted, students can answer using their laptop, the answers are aggregated, and the aggregated results are displayed in the lecture hall or on the students' screens.

Backstage was extended with an interface displaying in real-time the team participation, an overview of all teams' scores, and a dashboard displaying for each student the current score of their team. For quizzes, both multiple choice and fill-in-the-blank quizzes were used.

The display of a running quiz was extended with a real-time overview of the teams' participation. In this overview, each team was represented by a bar of a bar chart with each bar built from segments representing the team's members. Each segment is coloured in one of three colours: Light grey for students who are not logged in, dark grey for students who are logged in but have not yet answered, and blue for students who have answered. Representing each student as a unique segment serves as an immediate feedback and gives students a tangible representation of their contribution or of the absence thereof, respectively.

There are two reasons for including a live overview of team participation: First, a real-time overview of the teams' participation gives lecturers an opportunity to make quizzes more engaging by acting as moderator. As moderators, lecturers can for example call to participation members of teams with a low participation. Calling out specific teams should have a better impact on the participation than calling out to the whole audience because of "diffusion of responsibility", that is, the larger the group, the less responsible individuals feel [6]. While "diffusion of responsibility" is only researched for smaller group sizes and emergency situations, it is conceivable that the results are transferable to other settings, such as the education setting at hand. Latulipe et al. [14] included a live overview as well, but their implementation only contained the team standings, not a real-time overview of team participation.

Second, a real-time overview of the teams' participation directly influences all students to participate so as to contribute to their teams' score by providing both, *intra-group comparisons* (that is, how many people of my team have participated?) and *inter-group comparisons* (that is, how is the participation of the other teams compared to my team?). Social comparison theory [8] teaches that people generally aspire to reach the performance of people similar to themselves, and one could argue that team members and the rest of the audience fit this criterion. Mekler et al. [17] came the the conclusion that leaderboards – an element allowing comparison – tend to motivate people.

---

[1] https://backstage2.pms.ifi.lmu.de:8080/about/.

After a quiz has been conducted, updated team standings are displayed to all students what again provides both intra-group and inter-group comparisons. The updated team standings give information on participation and answer correctness both for each team and for the whole audience as well as possible changes in team placements.

On their own screen, each student is provided with an overview displaying the student's answers, the correct answers, and how many points the student has contributed to their team's score, again, providing a tangible and personalized feedback. Additionally, students are provided with an overview of the current team standings when logging into the ARS in form of a dashboard element.

4 Evaluation

A first evaluation on the effects of the team-based social gamification of quizzes on participation in quizzes and attendance at lectures has been conducted during a small-class lecture, the lecture part of a software development practical. In this practical, students are tasked with developing a browser-based game in the programming language JavaScript in groups of three to four students. The groups were created by the instructor and used as teams for the team-based social gamification of the quizzes run during the lecture.

In total, 24 students were partitioned into 7 teams: 3 teams consisted of 4 students and 4 teams consisted of 3 students. There were four lectures that introduced concepts required for the development of the software project. Lecture 1 had six, lecture 2 four, lecture 3 three, and lecture 4 four quizzes.

The configuration of the team-based social gamification (see Sect. 3.1) used in the evaluation was as follows: Lecturer-defined teams with a size of around 4 that lasted for the whole duration of the lectures were used. A correct answer was rewarded with 12 points, participation (regardless of the correctness of the answer) with 3 points. Every other action of a student had no effect on a student's team's score.

Methods. To evaluate the effects of the team-based social gamification on the participants of the course, two data sources were polled: the software tracking the students' actions and a survey conducted during the final lecture of the course. The survey covered questions of different types:

1. A first group of questions referred to the student's course of study, current semester, and gender.
2. A second block of questions aimed at measuring the motivation brought by the team-based social gamification.
3. A third block of questions aimed at measuring the engagement brought by the team-based social gamification.
4. A fourth block of questions collecting self-assessments of participation.
5. Two questions with free-text answers allowing students to give further feedback.

For (2), (3), and (4) the answers were given on a 4-point answer scale from *strong agree* to *strong disagree* with no neutral choice. Data collected directly from the system was quiz participation, team participation, and the history of team standings.

Results. The data acquired from the software refer 22 students for the first lecture and 24 students for all subsequent lectures. The reason for the difference in numbers is that two students had not registered to the software system by the time the first lecture took place. In the last lecture, a total of 19 students took part in the survey, 9 of them were female and 10 male. The students' current semester ranged from 2 to 8. With one exception, all participants enrolled in a computer science course of study.

Figure 1 shows how the team standings developed over time after each quiz. There seem to be two groups: A first group consisted of Team 7, Team 4, Team 1, and Team 2 – those teams participated in the competition for a place on the winner's podium. Team 5, Team 3, and Team 6 did not take part in the race for the first places: Team 6 took part at the beginning but starting with the third lecture the difference to the leading group grew steadily. Teams 5 and 3 show a nearly identical progression, the difference between the two being caused by the last quiz of the second lecture. The competition ended with four teams on the winner's podium, as the race between Team 1 and Team 4 ended in a draw.

Table 1 shows the teams' participations over the time-sorted lectures. The participation in quizzes increased for the first three lectures then dropped sharply. The average participation in quizzes was 75% with four of the teams beating the average and three teams performing below average, two of which were teams with three members.

Table 2 shows the survey results: The team competition seems to motivate the students to participate in quizzes with the real-time overview standing out as motivator (Block 2). The team-based social gamification of quizzes fostered engagement: Students discussed answer options with their team members before answering and tested the code the quizzes were referring to. The students expressed that the team-based social gamified quizzes make a lecture more engaging and more fun (Block 3). The majority of students would have taken part in the quizzes and brought a device even if there would have been no team competition. Only a minority of the students would have preferred to answer quizzes without the team-based social gamification (Block 4).

Discussion. The results indicate that the team-based social gamification had a positive effect on the participation in and on the engagement during the lectures. An experiment comparing participation in and on the engagement during the lectures with and without social gamification is outstanding and will be conducted in the forthcoming months.

The team standings given in Fig. 1 show that the team-based social gamification of quizzes introduces a competition: Indeed, there are changes in rankings, overtakes, and comebacks. While team standings alone are not an indicator of how students perceived the competition, the survey results clearly show that students perceived the competition introduced by team-based social gamification of quizzes and were motivated by that competition. The cooperation that took place within teams is evident: The majority of the students discussed their answers with their team before submitting them.

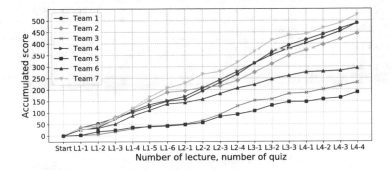

Fig. 1. Development of the accumulated team scores after each quiz (the x-axis represents the time-sorted quizzes, the y-axis represents the accumulated scores).

Table 1. Team participation in quizzes over the time-sorted lectures.

	Team 1	Team 2	Team 3	Team 4	Team 5	Team 6	Team 7	Average
Sizes	4	3	3	3	3	4	4	–
Lecture 1	0.78	1.00	0.33	1.00	0.67	0.67	0.88	0.76
Lecture 2	0.75	0.67	0.83	1.00	0.92	0.50	1.00	0.81
Lecture 3	1.00	1.00	0.67	1.00	0.67	0.50	1.00	0.83
Lecture 4	0.50	0.67	0.58	0.83	0.42	0.31	0.75	0.58
Average	0.76	0.84	0.60	0.96	0.67	0.5	0.91	0.75

The participation in the competition increased steadily during the first three lectures but dropped sharply from the fourth lecture onward. One could think that the positive effects of the social gamification are short-lived, but the authors presume another reason: The structure of the practical required students to actively start coding starting from the first week. Due to bank holidays, the fourth lecture took part in the sixth lecture week at which most teams were already deeply involved in the project and probably had already learned the techniques of lecture 4 on their own, therefore not seeing the point of attending the lecture. The effect on participation will be further investigated in further courses with larger audiences what should provide more conclusive answers. As of attendance to lectures, the same seems to be true: With the exception of the last lecture, no more than four students were skipping the lecture.

The survey shows that the competition had motivating effects on the students with each of the survey items in the category motivation (Block 2) ranging from "agree" to "strongly agree". According to the survey's results, one motivating

Table 2. Results for the questions of the survey grouped by the blocks described in Methods. Strong agree was assigned a value of 4, agree a value of 3, disagree a value of 2, and strong disagree a value of 1.

Statement	Mean	SD
Block 2: Motivating Components		
Motivated by the live overview of submitted responses	3.32	0.75
Motivated by competition with other teams	3.16	0.83
Motivated by the chance to contribute to team's score	3.21	0.71
Block 3: Engagement through Team Component		
Lecture became more engaging through the team component	3.16	0.69
Discussed answers with the team to get answer correct	3.21	0.63
Tried out quiz' code before answering to get answer correct	2.79	0.63
Competition was fun	3.16	0.60
Block 4: Participation without Team Component		
Would have participated without team component	3.16	0.83
Would have brought device without team component	3.16	0.83
Would prefer to solve on my own without points	1.90	0.57
Would prefer to solve on my own with points	1.84	0.69

factor for participation was the real-time overview. Thus, the results of the evaluation suggest the conclusion that the overview worked exactly as intended. Students favour the competition introduced by the teams over doing the quizzes by themselves and without being rewarded points. While the majority of students answered that they would have brought their device and participated in quizzes regardless of the team-based social gamification, 2 and 3 students respectively, answered that they would not have brought devices or participated without the team-based social gamification, which is no small proportion of the 19 students who took the survey.

The team component thus seems to introduce new layers of engagement: The majority of the students discussed their answers with their team mates before answering, and a smaller majority of the the students also tried out the code referred to in quizzes before answering. The team-based gamification of quizzes is perceived as introducing a fun element into lectures. The lecturers themselves had the feeling that the lecture audience was more lively during the quizzes than during the presentations.

A point of criticism expressed by several students was that the size of a team had an impact on the team's success since teams with four members always had the possibility to gain a 12 point advantage over teams of three per conducted quiz. The reward system will be corrected in the future so as to avoid such an effect.

5   Perspectives and Outlook

Learning from STEM lectures is often hindered by a low level of student activity what negatively impacts learning. To address this drawback of traditional lectures and make lectures with audiences of varying size more interactive, this article introduced an audience response system coupled with a team-based social gamification of quizzes. With this approach, teams compete in quizzes run during lectures, the individual participation of a student to a quiz contributing to their team's score. An evaluation conducted during a small-class lecture has shown the potential of the approach: It enhances interactivity and participation and adds new layers of engagement to lectures.

For future work, further game elements could be introduced to the team component like for example "power ups". Furthermore, the reward system must be adjusted so as to avoid giving larger teams an advantage over smaller teams.

Gamification injected directly in live, or face-to-face or traditional, lectures has the potential to make lectures more fun and more engaging for students – two factors of more positive learning outcomes. This article has demonstrated the positive effect of such an approach based on a team-based social gamification of quizzes.

Acknowledgements. The authors are thankful to Jacob Fürst for the implementation of the quizzes and the groundwork for the team-based social gamification described in the article which he did as part of his unpublished bachelor's thesis.

References

1. Bry, F., Pohl, A.Y.S.: Large class teaching with backstage. J. Appl. Res. High. Educ. **9**(1), 105–128 (2017)
2. Byrd, G.G., Coleman, S., Werneth, C.: Exploring the universe together: cooperative quizzes with and without a classroom performance system in astronomy 101. Astron. Educ. Rev. **3**(1), 26–30 (2004)
3. Carroll, J.A., Rodgers, J., Sankupellay, M., Newcomb, M., Cook, R.: Systematic evaluation of GoSoapBox in tertiary education: a student response system for improving learning experiences and outcomes. In: INTED2014 Proceedings (2014)
4. Crouch, C.H., Mazur, E.: Peer instruction: ten years of experience and results. Am. J. Phys. **69**(9), 970–977 (2001)
5. Danelli, F.: Implementing game design in gamification. In: Gamification in Education and Business, pp. 67–79. Springer (2015)
6. Darley, J.M., Latane, B.: Bystander intervention in emergencies: diffusion of responsibility. J. Pers. Soc. Psychol. **8**(4, Pt.1), 377 (1968)
7. Deterding, S., Dixon, D., Khaled, R., Nacke, L.: From game design elements to gamefulness: defining gamification. In: Proceedings of the 15th International Academic MindTrek Conference: Envisioning Future Media Environments, pp. 9–15. ACM (2011)
8. Festinger, L.: A theory of social comparison processes. Hum. Relat. **7**(2), 117–140 (1954)

9. Hamari, J., Koivisto, J., Sarsa, H.: Does gamification work?–a literature review of empirical studies on gamification. In: 2014 47th Hawaii International Conference on System Sciences (HICSS 2014), pp. 3025–3034. IEEE (2014)
10. Hunsu, N.J., Adesope, O., Bayly, D.J.: A meta-analysis of the effects of audience response systems (clicker-based technologies) on cognition and affect. Comput. Educ. **94**, 102–119 (2016)
11. Inbar, O., Tractinsky, N., Tsimhoni, O., Seder, T.: Driving the scoreboard: Motivating eco-driving through in-car gaming. In: Proceedings of the CHI 2011 Workshop Gamification: Using Game Design Elements in Non-Game Contexts, pp. 7–12 (2011)
12. Kay, R.H., LeSage, A.: Examining the benefits and challenges of using audience response systems: a review of the literature. Comput. Educ. **53**(3), 819–827 (2009)
13. Kelly, G.E.: Lecture attendance rates at university and related factors. J. Furth. High. Educ. **36**(1), 17–40 (2012)
14. Latulipe, C., Long, N.B., Seminario, C.E.: Structuring flipped classes with lightweight teams and gamification. In: Proceedings of the 46th ACM Technical Symposium on Computer Science Education, pp. 392–397. ACM (2015)
15. Mazur, E.: Peer Instruction. Springer (2017)
16. McGonigal, J.: Reality is Broken: Why Games Make Us Better and How They Can Change the World. Penguin (2011)
17. Mekler, E.D., Brühlmann, F., Opwis, K., Tuch, A.N.: Do points, levels and leaderboards harm intrinsic motivation?: an empirical analysis of common gamification elements. In: Proceedings of the First International Conference on Gameful Design, Research, and Applications, pp. 66–73. ACM (2013)
18. Nicholson, S.: A recipe for meaningful gamification. In: Gamification in Education and Business, pp. 1–20. Springer (2015)
19. Pohl, A.: Fostering Awareness and Collaboration in Large-Class Lectures. Doctoral thesis, Institute for Informatics, Ludwig Maximilian University of Munich (2015)
20. Polya, G.: How to Solve It: A New Aspect of Mathematical Method. Princeton University Press (2014)
21. Reeves, B., Read, J.L.: Total Engagement: How Games and Virtual Worlds Are Changing the Way People Work and Businesses Compete. Harvard Business Press (2009)
22. Russell, G., Shaw, S.: A study to investigate the prevalence of social anxiety in a sample of higher education students in the United Kingdom. J. Ment. Health **18**(3), 198–206 (2009)
23. Ryan, R.M., Deci, E.L.: Intrinsic and extrinsic motivations: classic definitions and new directions. Contemp. Educ. Psychol. **25**(1), 54–67 (2000)
24. Shi, C., Lee, H.J., Kurczak, J., Lee, A.: Driving infotainment app: gamification of performance driving. In: Adjunct Proceedings of the 4th International Conference on Automotive User Interfaces and Interactive Vehicular Applications, pp. 26–27 (2012)
25. Skinner, B.F.: About Behaviorism. Vintage (2011)
26. Tversky, A., Kahneman, D.: Loss aversion in riskless choice: a reference-dependent model. Q. J. Econ. **106**(4), 1039–1061 (1991)
27. Voelkl, K.E.: School warmth, student participation, and achievement. J. Exp. Educ. **63**(2), 127–138 (1995)

Gamification in an Augmented Reality Based Virtual Preparation Laboratory Training

Mesut Alptekin and Katrin Temmen[✉]

University Paderborn, Warburgerstr. 100, 33098 Paderborn, Germany
{mesut.alptekin,katrin.temmen}@upb.de

Abstract. Through technological progress during recent years, Augmented Reality (AR) technology can be used on ordinary smartphones with applications (Apps) in many formal and informal learning environments and educational institutions (e.g. [1, 2]). It is emerging as a suitable technology for teaching psychomotor skills. Simultaneously, gamification has become increasingly popular in the teaching field, providing famous examples, such as Duolingo (for the acquisition of foreign languages) or Codecademy (for learning programming languages) [3]. Many papers have already highlighted the beneficial aspects of gamification and AR for education and teaching (e.g. [1, 2, 4, 5]. While gamification is useful for improving students' motivation and engagement, AR can be applied to teach them operational skills without any time, costs and place constraints. Hence, this opens up numerous possibilities and forms to combine these two aspects (AR and gamification) for higher education teaching. However, there has been less research focusing on how gamification and AR can be combined in a useful manner to keep up students' initial motivation aroused through novelty effects of AR learning environments. Accordingly, this paper will present such a gamification concept for an AR based virtual preparation laboratory training to overcome the risk of demotivation, once AR will settle as a mainstream technology such as learning videos. The focus of the AR-App – presently being developed at the University of Paderborn – is to remedy the students' lack of practical skills when operating electro-technical laboratory equipment during their compulsory laboratory training.

Keywords: Augmented reality · Gamification · Electrical engineering education

1 Introduction

1.1 Definition and Categories

A quick review of the most popular Apps concerning the average ratings or down-loads (especially in the paid section) in the Google Play Store or Apple App-Store reveals that these are falling into the games or entertainment category. However, this initially (seemingly) surprising result is congruent with the fact that 66 percent of the population in Germany are active gamers. In the adolescence, this proportion is much greater yet with 97 percent for computer and video games [6].

© Springer Nature Switzerland AG 2020
M. E. Auer and T. Tsiatsos (Eds.): ICL 2018, AISC 916, pp. 567–578, 2020.
https://doi.org/10.1007/978-3-030-11932-4_54

For over one century, different concepts arise to combine game elements in aspects or systems other than entertainment or games, widely known under the term "gamification [7]. These non-game contexts can be found in different aspects of life with successful applications, for instance in economy or even science (e.g. Fold it), where game elements or -mechanisms are used to engage user participation for completing certain tasks [8]. These days it has become very popular also in the teaching aspect, with examples like Duolingo (for foreign language acquisition) or Codecademy (for learning programming languages) [3].

Using gamification does not only change the students' role from being passive to active but also their way of experiencing tasks differently. Being unsuccessful in completing a task does not mean a "failure" to them anymore, but rather some "approach" to optimal solution strategy.

There appear to be three types of gamification in the teaching concept: edutainment, game-based learning and serious gaming [3].

Generally, in edutainment attention and motivation are attracted by colorful animations. The basic idea of game-based learning is that the entertainment as well as the playful elements are paramount to the learners. Learning will then follow automatically and unconsciously, since skills and information for the solution of tasks are acquired playfully during the game [9]. One successful example is the game "Age of Empires", in the course of which the players are required to "go through" various human eras and to establish social structures.

In contrast, the focus of serious gaming is on the learning contents itself, extended by suitable game elements [10]. The gamification concept presented in this paper can be seen as one example thereof, since a feasible and prospective real life scenario is simulated.

1.2 Context and Background of the Research

The compulsory courses of the electrical engineering program at the University of Paderborn include three interdisciplinary laboratory practices to deepen theoretical lecture contents. One major problem is the operation of the electro-technical laboratory equipment. Both students and the supervising laboratory engineers criticize that the students have their first contact with the devices only within their practical work. To counteract this issue, a learning environment based on Augmented Reality (AR) is currently being developed, in which students will be able to acquire practical skills in handling the lab equipment irrespective of time or location constraints.

Although numerous studies have highlighted the potential of AR in different educational domains as well as its impact on learners regarding increased motivation and concentration on the topic [11, 12], it cannot be disregarded that these results are caused mainly by novelty effects of AR [13, 14]. Hence, this initial motivation will probably fade, once AR has settled as a mainstream technology, as in the case of learning videos.

To overcome the risk of demotivation and to improve learning outcomes, the AR learning App will be extended by game-based design elements through collaborative and competitive working.

2 Theoretical Background

2.1 Modern Learning Theories and Games

From findings in psychology three main learning theories can be derived, which explain the learning processes sensibly.

Behaviorism views the internal processes as a black box, where learning takes place through stimulus-response sequences [15]. Gamified approaches use rewards (e.g. scores) or penalties (e.g. loss of experience points) in order to seek a certain learning goal.

Cognitivism, however, considers learning as active information processing through amending and supplementing previous knowledge. In gamification this previous knowledge can be fully used through activating different levels [16]. Previous knowledge must be used to master more complex levels. In addition, according to the self-determination theory by Ryan and Deci, skills can be further enhanced by taking into account social integration (e.g. through ranking lists) [17].

Constructivism perceives learning as an individual and independent creation process [18]. Consequently, in learning games the autonomy and freedom of the individual learner must be as unrestricted as possible, e.g. by offering options to repeat levels according to the trial-and-error method with direct feedback. Hereby, learning will be enhanced by adjusting and optimizing solution strategies and leading to a so-called experience loop [19].

This quick overview shows that psychological findings and games have more in common than they appear to have initially. In the following section, the different gamification elements with their impact on learning and motivation will be further highlighted.

2.2 Gamification Elements

Direct User Interface (UI) Elements

Direct user interface elements are defined as game-design elements having a (direct) visual stimulus for action. The most common types are:

- *Narrative Elements (Backstory)*

Notwithstanding the fact that in gamification learning does not take place on account of the story itself, it is nonetheless the most important element [20]. Whether a gamified learning application is going to be successful is significantly dependent on the backstory which provides a framework for the application and thus creates a positive user environment [20, 21]. In addition, the story defines user targets to be reached whilst handling new tasks assigned. This can be included in the application through so-called onboarding, a process during which the user is introduced to the gaming mechanics of the App and the game's story [22].

- *Avatar (User-Profile)*

The Avatar enables the user to become more closely involved in the application, since, by virtue of the created character, he may take influence on the game's story'. By stronger identification with the Avatar, the user's interest and motivation can be increased consistently [23]. Furthermore, the increase of skills within the game can be displayed by way of transformation of the Avatars' appearance with items or clothes.

- *Quest and Missions (Tasks)*

Quest and missions are the (learning) tasks and (learning) objectives inside a story. A clear description of such and the expected rewards is a motivating incentive [24]. Being given the option of repeating Quests, the user may, pursuant to the experience loop, try out different approaches and thereby fortify the learning contents [20].

- *Points*

Points are satisfying the human need to collect things and help improve the users' motivation (for performing a certain action) [25]. There are different kinds of points to be distinguished for various aspects. Accordingly, experience points (XP) stand for increasing skills, while currency points indicate the level of successful task performance. The latter is actually a measure of how faultlessly or fast the task was being completed. On the other hand, reputation points visualize the team skills of the users.

- *Levels*

Levels may relate to the users visualizing their skill levels. They may also reflect the gaming progress and thus the difficulty level of the tasks. Generally, progress bars are used for visualization in both cases [22].

- *Badges*

Besides points, Badges are an excellent option of rewarding the user's activities. For thousands of years Badges have been of high symbolic character (e.g. in the form of furs and teeth from wild animals) [3] serving as the user's self-portrayal and for rewarding reached goals and outstanding accomplishments (e.g. accurate completion of tasks) [26]. All attainable Badges must be visible to the user right at the beginning to satisfy their need to collect and, furthermore, to depict the (learning) target. Furthermore, Badges may increase the students' awareness of their own studying habits [27].

- *Ranking lists*

Ranking lists are addressing the social demand to compare with others by appealing different needs, e.g. to compete [28]. A user ranking list may increase competition, whilst group ranking lists enhance commitment and collaboration between group members [21].

Indirect Elements

Indirect game-design elements are understood to be arising independently when combining the direct elements as well as the gaming structure and mechanics.

One feature of the indirect elements is the interlinking of learning application objectives with those of the students. The purpose here is to make the meaning and sense of the application recognizable by the user [29]. In order to increase the motivation according to the self-determination theory, the user's autonomy must be assured in spite of the numerous rules and restrictions of the game [5]. For this, a large scope of action must be provided, whereby various solution strategies may be tried [22].

The statement "Fun from games arises out of mastery. It arises out of comprehension." Reference ([30] p. 40), clearly shows that completing difficult tasks successfully and feeling an increase of skills already contributes to the motivation in the game. However, the application must always make this increase of skills visible and easily accessible [21]. This may be implemented by, e.g. changing the Avatar, a progress bar or activating new levels.

Further and very important indirect elements are feedback mechanisms. They should be constructive, and, according to Ryan and Deci, influence the performance as well as the user motivation, thus adding to the gameplay fun [7, 17]. Besides the typical "right" or "wrong" reply, also tips, hints and advice may be provided to let the users know what they must do to improve their results. The details of feedback elements are not addressed here, but further sources like [20, 31, 32] are referred to.

Creating competition and collaboration structures may also lead to a social bonding and thus to increased motivation. While competition can be created classically by ranking lists already, this is clearly more complex in collaborations [29]. According to McGonigal three issues must be taken into account here: The setting of common objectives (cooperating), the coordination of joint resources (coordinating) and a common result (cocreating). Further aspects, like random rewards, treasury rewards or so-called "easter eggs" and planned "special rewards" for completing a certain special task can be used to increase user motivation [33]. They rank among the surprise- or fortune-effects known in motivational psychology, according to which users are repeating tasks to get surprised or test their fortune and luck. These phenomena explain why many people participate in lotteries or other gambling activities.

3 Concept and Outcome

3.1 Methodological Approach

The App is developed and gamified in three phases on a modular basis as shown in Fig. 1.

Phase 1:

Presently, in the first phase the logical functions of laboratory devices, i.e. the multimeter, signal generator and oscilloscope are programmed. In this phase, an introduction will be included with explanatory descriptions of the devices. A quiz on the newly acquired theoretical operation instructions followed by a task required for handling the devices is also part of this phase. The development will take place in Unity3D using Wikitude SDK, allowing a marker-less application of AR on mobile devices and working device- and platform-independently.

Phase 2:

In the second phase, the first game-design elements are introduced, such as points and Badges. Additionally, the currently loose individual Unity modules are merged into

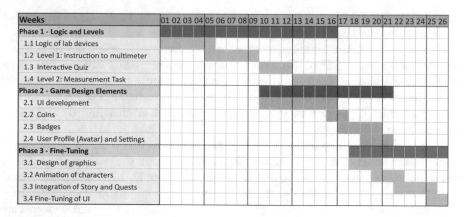

Weeks	01 02 03 04	05 06 07 08	09 10 11 12	13 14 15 16	17 18 19 20	21 22 23 24	25 26
Phase 1 - Logic and Levels							
1.1 Logic of lab devices							
1.2 Level 1: Instruction to multimeter							
1.3 Interactive Quiz							
1.4 Level 2: Measurement Task							
Phase 2 - Game Design Elements							
2.1 UI development							
2.2 Coins							
2.3 Badges							
2.4 User Profile (Avatar) and Settings							
Phase 3 - Fine-Tuning							
3.1 Design of graphics							
3.2 Animation of characters							
3.3 Integration of Story and Quests							
3.4 Fine-Tuning of UI							

Fig. 1. Phases in gamifying the App

one project and extended with a menu and a UI. A theoretical concept of the UI is presently being tested among students with the aim of identifying intuitive and comprehensible UI for an AR application.

Phase 3:

In the third phase, the graphics- and animation oriented development, the UI is embedded in the gamified story and equipped with Quests. Subsequently, the App is extended by cooperation opportunities.

The modular setup in each individual phase allows continuous tests with the students of electrical engineering and the involved laboratory engineers as well as gradual improvement of the App.

3.2   Gamification Elements

Story and Avatar

Through "Progressive Onboarding" and the personal "Tutor" by virtue of the humorous character of Einstein, the user is introduced to possible interaction opportunities of the App (technical environment) as well as to the story. The Einstein character symbolizes knowledge and new skills and is also popular from other learning applications (e.g. brain jogging). The main story of the AR-App is focused on the first employment of an engineer in a large electronic company. First of all, the user is requested to complete his profile by choosing a personal image of the Avatar and a user name. Subsequently, he will be introduced to the job and, to find his way in the App or the company respectively, he will learn about the location and purpose of different settings, e.g. where to switch between reading- and listening-mode for information.

In order not to overburden the user with excessive information and settings, he will first not be offered any further options, except for the profile image and the name, to personalize his work station. Further settings shall be made available when progressing on the character level and the experience gathered. At the same time, other customizable settings shall serve as rewards and motivate to use the App. The following Avatars are available (Fig. 2):

Fig. 2. Available Avatars in the final App

Quests and Levels

In line with the story, the Quests are given by the superior as project tasks on the laboratory devices. In view of the projects to be completed and the devices to be applied, their complexity keeps constantly increasing. Prior to each actual Quest, there is a brief story embedding the project work into the background story and outlining the purpose of the task. A small unit consisting of a story and the corresponding Quest form one Game-Level.j

The user is assumed to have no experience with AR applications at the time or in handling laboratory devices. He therefore commences with the lowest Level as a trainee. To increase motivation, however, a quick level rise is possible. Level 1 relates to the first working day, during which the trainee is introduced to technical working equipment available, such as the multimeter. The personal tutor will explain to the trainee the main functions of the different operating controls of the device (cf. Fig. 3 for the oscilloscope) and safety aspects.

Fig. 3. Personal Tutor explaining the oscilloscope

Since a number of students have previous experience in handling laboratory devices, the first working day may be skipped optionally. To determine whether previous experience is satisfying or the information imparted earlier "theoretically" was understood, the students will participate in a small mandatory and interactive quiz. The aim is to ensure that all students have the same level of knowledge before being able to start with their practical work.

Only then, the second level will be activated, where simple measurements must be performed at a virtual, preassembled workplace. After accomplishing this level, the user will be promoted from trainee to junior developer, and the third level will be activated along with a new multimeter. This device has a better measuring accuracy as well as additional functions to demonstrate the player why there is such an extreme price range (5 € - > 200 €) between the individual measurement devices. In doing so, the knowledge of multimeters is deepened by aspects such as correct measuring (e.g. correct voltage vs. correct current measurement).

The oscilloscope too is activated with an introduction and interactive quiz (Level 1), whereby the introduction is optional but the quiz mandatory. The user level rises to "Senior Developer" and is shown by increased salary (see section "Points and Badges").

In Level 4 and 5, analog to Level 2 and 3, detailed tasks on the operation and handling of the oscilloscope are posed. The tasks gradually increase in view of their difficulty level as well as the user's scope for action to solve these tasks. At a later date, the entire App may be supplemented by further levels to other devices with modular extensions. In these levels, the player will have been promoted by then to "Manager" in the company hierarchy. At the end of the App only, the player will level up to CTO (Chief Technology Officer) being permitted free access to the laboratory, inside of which all mastered (activated) devices are available to the user for planning and performing his own experiments.

Points and Badges

The Points in this App represent the process of a promotion, i.e. at the beginning of the project the user starts with a salary according to his position in the company. For the correct processing of a Quest he receives a salary increase as an interim result until the salary of the next hierarchy level is reached or the level was mastered. This way, the player is always made aware of the interim results and the aspired goals which also adds to the motivation and the learning success. Figure 4 is a prototypical representation of the points within the AR task for the oscilloscope.

Fig. 4. Prototype of the Progressbar with the virtual oscilloscope

Badges are offered to motivate the player to concentrate more intensely on a task or subject matter. They are classified into three categories: Time Management, Carefulness and Learning. To ensure that the user knows which Badges he may be rewarded for which merit, all collectable Badges are marked with a grey background and can be inspected together with those so far gained in the user profile (work table). Every category has a "Special Badge", a treasure chest in this game, to motivate the player by a surprising effect to work with the App for longer periods of time or more frequently. The table below shows the various types of Badges and the difficulty level required for obtaining these:

Leader Board

The ranking list is a weighted presentation of the Badges earned and of the actual income of all players. In a late version of the App, the multiplayer mode will be more extended, so that a department ranking list may also be implemented. While the latter has a stronger focus on competition among the players, the second ranking list reinforces the feeling of solidarity and cooperation. Since in real life the students must also perform their experiments in groups with other trainees, this mode is more closely inclined to real experiments and the future professional life (Table 1).

Table 1. Possible, obtainable Badges in the final App

Type	Name	Task	Difficulty	Icon
Time Management	Reasonable	Work 1 hour with no break	high	
	Nimble Fox	Solve Level 3 in less than 5 minutes	medium	
	Scrupulousf	Repeat Level 3 and 5 one day before the traineeship	low	
	Special Badge - Speed Run	Activate this Badge and solve Level 5 as quickly as possible	medium	
Carefulness	Unbeatable	Solve the quiz with 0 mistakes	low	
	Special Badge - Blacklist	Solve Level 2 and 4 without any help	high	
Learning	Teamplayer	Solve Level 4 and 5 in the multiplayer mode	medium	
	Special Badge - Staff of the Day	Solve all levels in one day	high	

To set up the ranking list and the multiplayer mode, a server instance is required with a data base and an overview of all active players. Also, the corresponding tasks must be adapted to the multiplayer mode. Therefore, these aspects shall only be realized subsequent to phase three and in a later version of the App.

4 Conclusion and Future Work

4.1 Conclusion

Learning playfully means fortifying knowledge stronger than this could usually be achieved by traditional methods. Thus, through gaming elements much learning theory knowledge from psychology can be ideally integrated into teaching, e.g. activating and blocking Levels depending on the previous knowledge of the students. Simultaneously, the increased use of AR technologies on ordinary smartphones opens diverse integration possibilities of gamification and didactics.

The concept introduced in this paper presents such a possibility of Game-Design-Elements in connection with the AR-App to prepare students for traineeships in electrical engineering. The essential design elements which can be featured through the UI in an App are: The Story, Avatar, Quests, Level, Points, Badges and Ranking lists.

The story in this App outlines the working day of an engineer in an electronics company. By personalizing the Avatar, the player may work actively in the company. In short stories he will be requested to solve quests in the shape of small to large project tasks. The unit consisting of a short story and quests form a level in the App that may be repeated indefinitely provided that it has been activated. Besides the Game-Level, also the Character-Level is increased. The latter is shown in the story by the promotion (from trainee to department manager). The Character-Level is furthermore expressed in shape of the salary through the points collected whilst processing a quest. Badges are also integrated into the collectable points, which might be awarded for special performances or efforts, e.g. solving a quest in a very short period of time or without any mistakes. On account of additionally required technical infrastructure, competitive and collaborative elements will be integrated into the App as ranking lists at a later date and in a subsequent version only.

4.2 Future Work

While previous works have predominantly been of conceptual nature, the next step in focus will be the concrete development of the application. Presently, the logic of the individual measurement devices is being implemented, while in future, the gamification concepts described in this paper will be programmed. As a conclusion of the first version of the application, a research study is planned to investigate the learning efficiency of the gamified AR-application among students of electrical engineering. Furthermore, the adaption and extension of the App for laboratory trainings of other disciplines, such as physics or chemistry, is planned.

References

1. Akçayır, M., Akçayır, G.: Advantages and challenges associated with augmented reality for education: a systematic review of the literature. Educ. Res. Rev. **20**, 1–11 (2017)
2. Saltan, F., Arslan, Ö.: The use of augmented reality in formal education: a scoping review. Eurasia J. Math. Sci. Technol. Educ. **13**(2), 503–520 (2016)

3. Knautz, K.: Gamification in der Hochschuldidaktik–Konzeption, Implementierung und Evaluation einer spielbasierten Lernumgebung. Verfügbar unter: https://docserv.uni-duesseldorf.de/servlets/DocumentServlet?id=36429. Accessed 19 April 2017
4. Johnson, L., Adams, S., Cummins, M., Estrada, V., Freeman, A., Hall, C.: NMC Horizon Report: 2016 Higher Education Edition. The New Media Consortium, Austin (2016)
5. Rigby, S., Ryan, R.M.: Glued to games: How video games draw us in and hold us spellbound: How video games draw us in and hold us spellbound. ABC-CLIO (2011)
6. Erenli, K.: The impact of gamification-recommending education scenarios. Int. J. Emerg. Technol. Learn. IJET 8(S1), 15–21 (2013)
7. Deterding, S., Dixon, D. Khaled, R., Nacke, L.: From game design elements to gamefulness: defining 'gamification'. In: Proceedings of the 15th International Academic MindTrek Conference: Envisioning Future Media Environments, New York, NY, USA, 2011, pp. 9–15 (2011)
8. Zichermann, G.: The purpose of gamification. A look at gamification's applications and limitations. Radar April, vol. 26 (2011)
9. Squire, K., Jenkins, H.: Harnessing the power of games in education. Insight 3(1), 5–33 (2003)
10. Michael, D.R., Chen, S.L.: Serious Games: Games that Educate, Train, and Inform. Muska & Lipman/Premier-Trade (2005)
11. Wu, H.-K., Lee, S.W.-Y., Chang, H.-Y., Liang, J.-C.: Current status, opportunities and challenges of augmented reality in education. Comput. Educ. 62, 41–49 (2013)
12. Bower, M., Howe, C., McCredie, N., Robinson, A., Grover, D.: Augmented Reality in education–cases, places and potentials. Educ. Media Int. 51(1), 1–15 (2014)
13. Di Serio, Á., Ibáñez, M.B., Kloos, C.D.: Impact of an augmented reality system on students' motivation for a visual art course. Comput. Educ. 68, 586–596 (2013)
14. N. Gavish et al.: Evaluating virtual reality and augmented reality training for industrial maintenance and assembly tasks. Interact. Learn. Environ. 23(6), 778–798 (2015)
15. Watson, J.B.: Psychology as the behaviorist views it. Psychol. Rev. 20(2), 158 (1913)
16. Piaget, J.: Biology and knowledge: an essay on the relations between organic regulations and cognitive processes (1971)
17. Deci, E.L., Ryan, R.M.: The "what" and "why" of goal pursuits: human needs and the self-determination of behavior. Psychol. Inq. 11(4), 227–268 (2000)
18. Vygotskij, L.S., Cole, M.: Mind in Society: The Development of Higher Psychological Processes. Nachdr. Harvard University Press, Cambridge, Mass (1981)
19. Kiili, K.: Digital game-based learning: towards an experiential gaming model. Internet High. Educ. 8(1), 13–24 (2005)
20. Kapp, K.M.: The Gamification of Learning and Instruction: Game-Based Methods and Strategies for Training and Education. Wiley (2012)
21. Sailer, M., Hense, J., Mandl, H., Klevers, M.: Psychological perspectives on motivation through gamification. IxD&A 19, 28–37 (2013)
22. Zichermann, G., Cunningham, C.: Gamification by Design: Implementing Game Mechanics in Web and Mobile Apps. O'Reilly Media, Inc. (2011)
23. Frery, A.C., Kelner, J., Moreira, J., Teichrieb, V.: User satisfaction through empathy and orientation in three-dimensional worlds. Cyberpsychol. Behav. 5(5), 451–459 (2002)
24. Barata, G., Gama, S., Jorge, J., Gonçalves, D.: Engaging engineering students with gamification. In: 2013 5th International Conference on Games and Virtual Worlds for Serious Applications (VS-GAMES), 2013, pp. 1–8 (2013)
25. Reiss, S.: Multifaceted nature of intrinsic motivation: the theory of 16 basic desires. Rev. Gen. Psychol. 8(3), 179 (2004)

26. Antin, J., Churchill, E.F.: Badges in social media: a social psychological perspective. In: CHI 2011 Gamification Workshop Proceedings, 2011, pp. 1–4 (2011)
27. Hakulinen, L., Auvinen, T., Korhonen, A.: The effect of achievement badges on students' behavior: an empirical study in a university-level computer science course. Int. J. Emerg. Technol. Learn. IJET **10**(1), 18 (2015)
28. Wood, J.V., Wilson, A.E.: How Important is Social Comparison? (2003)
29. Salen, K., Zimmerman, E.: Rules of Play: Game Design Fundamentals. MIT press (2004)
30. Koster, R.: Theory of fun for game design, p. 40. O'Reilly Media, Inc. (2013)
31. Schell, J.: The Art of Game Design: A Book of Lenses. AK Peters/CRC Press (2014)
32. Hunicke, R.: UX Week 2009/Wildflowers: The UX of Game/Play. (2009)
33. Chou, Y.: The six contextual types of rewards in gamification. Retrieved Dec. 15, 2014 (2013)

A Serious Game for Introducing Software Engineering Ethics to University Students

Michalis Xenos[✉] and Vasiliki Velli

Computer Engineering and Informatics Department, Patras University,
Patras, Greece
{xenos, velli}@ceid.upatras.gr

Abstract. This paper presents a game based on storytelling, in which the players are faced with ethical dilemmas related to software engineering specific issues. The players' choices have consequences on how the story unfolds and could lead to various alternative endings. This Ethics Game was used as a tool to mediate the learning activity and it was evaluated by 144 students during a Software Engineering Course on the 2017–2018 academic year. This evaluation was based on a within-subject pre-post design methodology and provided insights on the students learning gain (academic performance), as well as on the students' perceived educational experience. In addition, it provided the results of the students' usability evaluation of the Ethics Game. The results indicated that the students did improve their knowledge about software engineering ethics by playing this game. Also, they considered this game to be a useful educational tool and of high usability. Female students had statistically significant higher knowledge gain and higher evaluation scores than male students, while no statistically significant differences were measured in groups based on the year of study.

Keywords: Game-based learning · Computer engineering ethics · Usability evaluation

1 Introduction

While introducing students into software engineering ethics has been recognized as a necessity and many undergraduate computer science and computer engineering programs offer relative courses [1, 2], these courses are mostly based on lessons learned and theoretical essays [3, 4]. We argue that the use of game-based learning (or serious games [5]) into the area of software ethics could be a very helpful practice for the students. This practice allows the students to experience real life situations, during playing and while having fun. These situations are presented in scenarios that require moral judgement and solving ethical dilemmas. To the best of our knowledge, while there are serious games available for personal and social ethics [6], this is the first time a game is used for teaching software engineering ethics, that is based on students' choices on various scenarios that simulate real life situations.

The rest of the paper is structured as follows: In Sect. 2 we present the Ethics Game, the tool used to mediate the learning activity of software engineering ethics. In

© Springer Nature Switzerland AG 2020
M. E. Auer and T. Tsiatsos (Eds.): ICL 2018, AISC 916, pp. 579–588, 2020.
https://doi.org/10.1007/978-3-030-11932-4_55

Sect. 3 we present the setting and the materials of an evaluation study that involved 144 students and in Sect. 4 we present the results of this study. Finally, in Sect. 5 we discuss conclusions and future work.

2 The Ethics Game

A game based on a commercial storytelling platform[1] was developed, so to be available in most mobile devices. After installing the application on their mobile or tablet, anyone interested to play the Ethics Game have the option to either search for "*Ethical_Dilemmas*" (this was the title the game was registered) or to download the game using a direct link[2] offered by us. The game introduces the players to basic ethical dilemmas related to software engineering, based on the software engineering code of ethics [7], and offers them choices that will influence how the story will unfold. Some selections are straightforward and force the story to evolve in different paths, and in some cases in alternative endings, while other selections add up to internal scores (not revealed to the user) that also direct the path the story will follow.

The platform application is free to download and offers a variety of commercial stories/games in various genres (romance, drama, Hollywood, fantasy, mystery, comedy, action/adventure, and thriller/horror). Players selecting these stories need to purchase "diamonds" (the platform currency) to unfold the stories, but the rest of the stories/games created by users (as in our case) are free. Most of the commercial stories available on this platform seem to address female players (most of the main characters are females), so one of our research goals was to investigate if this was story-dependent, or is it apply in our Ethics Game as well.

To play each episode the player is using a "pass". Each user is given 3 passes at the first time they download the application and passes are frequently replenished, since the application adds 3 more passes every 4 h. A user can also purchase passes, but this is something optional and we advised students against it. Having the passes limitation in mind, we have created a story that unfolds in one week in work. During this week, Rose a software engineer working in a large firm, is facing with a lot of ethical dilemmas. The Ethics Game lasts for 6 days (corresponding to 6 episodes), therefore one needs only 6 passes to finish it. This ensures that every player can start playing the first 3 days of the game, then wait for 4 h to get more passes and then finish the game. Each one of the first 5 episodes represent one day at work (Monday to Friday), while the 6th episode presents one of the alternative endings, based on the player's choices. Figure 1 presents three instances from the Ethics Game; in the left and right images, the heroine Rose, is about to make a choice facing two alternative options, while in the image in the middle the game designer had zoomed on Rose to emphasise her comments.

---

[1] https://www.episodeinteractive.com/.

[2] https://www.episodeinteractive.com/s/6429627363229696.

Fig. 1. Screenshots from the Ethics Game

The author/creator of a story/game uses commands to change the background and to introduce animated backgrounds and sound. They can direct the story using commands such as (all the following commands are from the Ethics Game): @ROSE changes into work_outfit_4, @HAROLD stands screen right AND HAROLD faces left, @zoom on ROSE to 200% in 1.5, ROSE (talk_argue_defensive). The author can control the flow of the story (how the story will unfold) based on the choices the player selected, using a choice() command. Different choices lead to different branches of the story, which are controlled with if...then...else commands. Finally, the author can use variables to keep score of various elements related to the story. In our case we used choices and variables to evaluate the players' performance and to present them with alternative endings, based on their choices within the game.

Three alternative endings are available in the Ethics Game. Should a player manage to face all challenges successfully, the game ends with Rose being promoted. Players that made a lot of incorrect choices and repeatedly violated software engineering ethics they end the game by being fired from their position, while the rest of the players remain at their position, but at the end of the game they receive information about what they should have done better.

3   The Evaluation Study

This paper reports a within-subject pre-post study that investigates the learning effectiveness of the Ethics Game (post-test) compared to the lecture-based instruction (pre-test) in the context of campus-based higher education. In specific, these research questions were investigated by this study:

RQ1: Is there any effect of the Ethics Game activity on students' learning performance?

RQ2: Did students find the Ethics Game a useful educational tool?

RQ3: Did students find the Ethics Game a usable tool?

RQ4: Are there any differences in learning gain, perceived usability and perceived educational effectiveness related to gender?

RQ5: Are there any differences in learning gain, perceived usability and perceived educational effectiveness related to the year of studies?

The study took place in the context of campus-based classroom education and in specific in the course named "*CEID_Y232: Software Engineering*", during the academic year 2017–2018. This is a required course, offered to the students of the Computer Engineering and Informatics Department (CEID) at the University of Patras, during the second semester of their 4th year of studies (8th semester). CEID is a 5-year B.Sc. degree with an Integrated M.Sc., corresponding to 300 European Credit Transfer System (ECTS) units. The CEID_Y232 course includes 13 lectures, 4 compulsory assignments and 4 short elective assignments and offers 6 ECTS units to the students. All assignments are graded, the compulsory ones contribute to 30% of the final course grade, while a passing grade in all assignments is a prerequisite for participating in the final exams for the other 70% of the course grade. Participation in the short elective assignments is offered to aid students to improve the overall course grade. Students participating in CEID_Y232 course are introduced to basic ethical dilemmas related to software engineering during the 10th lecture and had the opportunity to play the game as part of the 4th elective assignment right after the lecture.

3.1 Participants

For the examined academic year 2017–2018, 378 students were registered for this course, but only 217 of them submitted all the required assignments so to be able to participate in the exams. Using the Ethics Game to mediate the learning activity was an elective assignment that contributed only 3% of their final grade, but nevertheless a total of 144 students have successfully completed it. This was the only one elective assignment having such a large participation, regardless of the small gain in the final grade, probably since playing a game was considered fun. The other three elective assignments had only: 32, 17 and 11 participants.

Before offering the Ethics Game to the students, a pilot test of the entire procedure performed by using 5 participants, whose responses were excluded from the results. Following the pilot test, 144 students participated in this study, 29 of them were female (20.1%) which is typical in computer engineering studies in Greece. From these students, 63 of them were at the 4th year of studies (current year), while 81 students belonged to higher years (students that had failed the CEID_Y232 course and repeating it). Most of the students used their smartphone to play ($N = 134$, 93.1%), while very few used a tablet ($N = 10$, 6.9%) or an emulator on their laptop ($N = 1$, 0.7%).

3.2   Materials

The students were asked to complete a knowledge test (pre-test) after the end of the lecture and before downloading the game. They had access to the game only after completing the knowledge test. The knowledge test included all issues related to software engineering ethics that were addressed in the game and comprised of 10 multiple-choice questions with four possible answers each. Then the students were asked to play all six episodes of the Ethics Game and report their result within the game, when they reach one of the possible endings. Finally, after finishing the game, the same test (post-test) was offered to them, and completing it was required to formally finalise this assignment.

Additionally, they were also asked to complete three additional scales as part of the post-test: (a) a 3-items scale rating their educational experience with the game from 1 to 5, (b) the standardized System Usability Scale [8], provided in participants' native language [9], and (c) the 7-point adjective rating question [10] with wordings from "worst-imaginable" to "best-imaginable". Finally, they had the option to comment, on an open question, on issues they feel we could improve in a future version of the game. This is a typical set of scales we have successfully used in similar usability evaluation studies [11–13]. The collected data were organized and pre-processed using Microsoft Excel 365 ProPlus and were analysed using IBM SPSS Statistics v20.0.

4   Results

First, reliability analysis was conducted for the questionnaires used in the study. To this end, the Cronbach's alpha measure of internal consistency was used [14]. The 10 questions knowledge test has marginal reliability, (Cronbach's alpha = 0.699, N = 10). We measured that removing the question number 9 could improve at (Cronbach's alpha = 0.715, N = 9), while removing any other question would resulted a Cronbach's alpha below the 0.700 threshold, but since the internal consistency was very close to the accepted limit for the 10 questions we decided to keep our original set of 10 questions for compliance with our educational model. SUS is a standardized scale [9, 15, 16] and had also adequate reliability for our dataset (Cronbach's alpha = 0.717, N = 10). Finally, the educational experience scale had also adequate internal consistency for our dataset (Cronbach's alpha = 0.746, N = 3).

Following the rationale reported in [17], we produced a composite variable for the normalized learning gain defined as the difference between post-test score and pre-test score ("observed gain") divided by the difference between the max possible score and the pre-test score ("amount of possible learning that could be achieved" [17]). Table 1 presents descriptive statistics of the dependent variables measured in this study (mean, median, standard deviation and 95% confidence interval).

4.1   Effect on Students' Learning Performance

For the 144 students, only 14 had negative ranks between the pre-test and the post-test, 27 had the exact same score and 103 had positive ranks. This is also reported from the

Table 1. Descriptive statistics of the depended variables of this study. Sample size N = 144.

Variable	M	Mdn	SD	95% CI
Pre-test score (0–100)	67.99	70.00	16.92	[65.20, 70.77]
Post-test score (0–100)	80.59	85.00	15.57	[78.03, 83.16]
Normalized learning gain (%)	35.04	33.00	46.01	[27.46, 42.62]
Educational experience rating (1–5)	3.87	4.00	0.80	[3.74, 4.00]
SUS score	83.14	85.00	8.06	[81.81, 84.47]
Usability adjective rating	4.84	5.00	0.86	[4.70, 4.98]

mean score of the composite variable "observed gain" which was high (+35.04%). Additionally, because the data for both test variables were skewed, a Wilcoxon Signed-Ranks test was run, and the output indicated that post-test scores were statistically significantly higher than pre-test scores ($Z = 8.321$, $p < .000$). Therefore, for RQ1, we can argue that the students did learn about software engineering ethics by playing the Ethics Game.

Table 2. Descriptive statistics of students' self-reported ratings of their educational experience

Question (1: strongly disagree; 5: strongly agree)	M	Mdn	SD	95% CI
Q1: I think that the Ethics Game is useful as an Educational tool	3.63	4.00	0.95	[3.47, 3.78]
Q2: I would recommend the Ethics Game to a colleague or friend who wants to learn about ethics in software engineering	3.80	4.00	1.11	[3.61, 3.98]
Q3: I would recommend the Ethics Game to a colleague or friend who wants to design something similar	4.19	4.00	0.93	[4.04, 4.35]
Overall scale (Cronbach's alpha = 0.746)	3.87	4.00	0.80	[3.74, 4.00]

4.2   The Ethics Game as a Useful Educational Tool

Participating students rated their learning experience with the Ethics Game in the post-test questionnaire. Table 2 presents descriptive statistics of these ratings per question and overall. So, regarding RQ2, the students self-reported ratings for their educational experience for the Ethics Game were relatively high (M = 3.87, SD = 0.8). In addition, students provided rather positive comments for their educational experience while playing the Ethics Game in the open-ended question of the post-test questionnaire.

4.3   The Ethics Game Perceived Usability

To assess the RQ3, after playing the Ethics Game, the participating students completed the SUS questionnaire and the adjective rating scale, both measures of a system's perceived usability. The Ethics Game received a mean SUS score of 83.14 (SD = 8.06). According to a dataset of nearly 1000 SUS surveys [10], this means that students found KLM-FA as "Good to Excellent" (SUS score from 71.4 to 85.5) in terms of perceived usability. Students' usability adjective ratings were also rather high (M = 4.84, SD = 0.86), indicating that the Ethics Game was perceived as "Good" (corresponds to 5).

Table 3. Descriptive statistics of the depended variables of this study, grouped by gender

Group	Variable	M	Mdn	SD	95% CI
Male students (N = 115)	Normalized learning gain (%)	20.42	28.50	53.83	[−0.05, 40.90]
	Educational experience rating (1–5)	3.37	3.33	0.98	[2.99, 3.74]
	SUS score	79.05	80.00	8.00	[76.01, 82.10]
	Usability adjective rating	4.24	4.00	0.99	[3.87, 4.61]
Female students (N = 29)	Normalized learning gain (%)	53.60	71.20	45.26	[36.38, 70.82]
	Educational experience rating (1–5)	4.24	4.33	0.51	[4.05, 4.44]
	SUS score	85.34	87.50	7.70	[82.42, 88.27]
	Usability adjective rating	5.34	5.00	0.72	[5.07, 5.62]

4.4 Gender Differences

As shown in Table 3, female students evaluated the Ethics Game higher than the male students in both SUS and usability adjective rating scores. Additionally, their self-reported ratings for their educational experience was also higher compared to male students and they also had higher normalized learning gain.

Since the assumption of normality was violated for all groups as measured by the Kolmogorov-Smirnov test, except for the SUS score for male students, which was also marginal (Sig = 0.200), a non-parametric test, the two-tailed Man-Whitney U test, was selected. The Normalized learning gain was higher for female students (M = 53.60, SD = 45.26) than male students (M = 20.42, SD = 53.83) and using a two-tailed Man-Whitney U test, we have found that the normalized learning gain was statistically significant higher for the female students (U = 1172, Z = −2.477, p = 0.013).

Regarding the SUS, although the female students evaluated the Ethics Game higher (M = 85.34, SD = 7.70) compared to male students (M = 79.05, SD = 8.00), the two-tailed Man-Whitney U test showed that these differences were not statistically significant (U = 1282.5, Z = −1.929, p = 0.054), although the result was marginal. For the usability adjective rating the female students evaluated the Ethics Game higher (M = 5.34, SD = 0.72) compared to male students (M = 4.24, SD = 0.99) and the two-tailed Man-Whitney U test revealed that the adjective evaluation was statistically significant higher for the female students (U = 1016.5, Z = −3.566, p = 0.000). Finally, for the self-reported ratings for their educational experience female students had also higher reported rankings (M = 4.24, SD = 0.51) compared to male students (M = 3.37, SD = 0.98) and the two-tailed Man-Whitney U test confirmed this (U = 1086, Z = −2.925, p = 0.003).

In conclude, for the RQ4, we can argue that female students enjoyed the Ethics Game more than male students, they found it a more valuable educational tool than the male students did, and by playing the game they improved their knowledge on the field significantly more than what male students did.

4.5 Differences Related to the Year of Studies

Table 4 presents the descriptive statistics for the groups of students that were at the current (4th) year or being in higher years. This could also be measured by grouping by age, since in our case was equivalent. Since the assumption of normality was violated for all groups as measured by the Kolmogorov-Smirnov test, a non-parametric test, the two-tailed Man-Whitney U test, was selected. Using this test, on the one hand, no statistically significant differences were found for these two groups for the normalized learning gain (U = 2470, Z = −0.329, p = 0.742) as well as for the self-reported educational experience (U = 2153.5, Z = −1.619, p = 0.106) and for the usability adjective rating (U = 2159.5, Z = −1.736, p = 0.083). On the other hand, the usability evaluation of the Ethics Game, based on the SUS, was statistically significantly higher for the students of the current year than this of students from higher years (U = 1932, Z = −2. 509, p = 0.012). Overall, regarding the RQ5, there are no statistically significant differences for students in different year of studies, with the exception that students on the current year evaluated significantly higher the Ethics Game usability based on the SUS.

Table 4. Descriptive statistics of the depended variables of this study, grouped by academic year

Group	Variable	M	Mdn	SD	95% CI
Current year students (N = 63)	Normalized learning gain (%)	34.94	33.10	43.66	[23.95, 45.94]
	Educational experience rating (1–5)	5.00	5.00	0.70	[4.82, 5.18]
	SUS score	85.16	85.00	6.67	[83.48, 86.84]
	Usability adjective rating	4.00	4.00	0.73	[3.82, 4.18]
Students from higher years (N = 81)	Normalized learning gain (%)	33.50	33.30	50.75	[20.72, 46.29]
	Educational experience rating (1–5)	3.66	3.67	0.87	[3.44, 3.88]
	SUS score	80.87	82.50	8.49	[78.74, 83.01]
	Usability adjective rating	4.60	5.00	0.96	[4.36, 4.84]

5 Conclusion and Future Goals

Although the game wasn't aimed to be particularly difficult, from the 144 students that participated, only 37 (25.7%) managed to reach the end where Rose gets promoted. At least, only 8 students (5.6%) reached the end where she gets fired. The rest of the 99 students (68.8%) reached the end where Rose remained at her position and they could read about which ethic related issues should be more careful at. This was one of the things that most students comment on, asking to remove it from an updated version. They mention that this was like spoilers that prevent them from playing again and they would prefer to play again and find out by themselves if they are able to perform better. This, among other issues, was improved in the version available today in the platform. Another issue that mentioned by the students is that they didn't like the waiting time for their passes to be refilled, so they proposed to combine days into three larger episodes, so they could play them all in once.

Since we have asked the students to fill-in the post-test questionnaire just after they had finished the game, we don't have any accurate measure of how many students kept playing the Ethics Game, even after fulfilling the requirement of the elective assignment. We can see that the Ethics Game had 921 reads, 14 days after the deadline of the corresponding assignment, which means that many of them probably did play it a few more times, but the platform does not report back on individual number of plays. This is one of the future goals to investigate further. Adding more episodes and expanding the game on other issues related to software engineering is another future goal.

Acknowledgment. The authors would like to thank the 144 students that participated in this study and helped us with their comments.

References

1. Quinn, M.J.: On teaching computer ethics within a computer science department. Sci. Eng. Ethics **12**(2), 335–343 (2006)
2. Larson, D.K., Miller, K.W.: Action ethics for a software development class. ACM Inroads **8**(1), 38–42 (2017)
3. Freedman, R.: Teaching computer ethics via current news articles. In Elleithy, K., Sobh, T. (eds.) Innovations and Advances in Computer, Information, Systems Sciences, and Engineering, pp. 1193–1204. Springer, New York (2013)
4. Heron, M.J., Belford, P.H.: A practitioner reflection on teaching computer ethics with case studies and psychology. eJournal Teach. Learn. (2015)
5. Michael, D.R., Chen, S.L.: Serious games: Games that educate, train, and inform: Muska & Lipman/Premier-Trade (2005)
6. Pereira, G., Brisson, A., Prada, R., Paiva, A., Bellotti, F., Kravcik, M., Klamma, R.: Serious games for personal and social learning & Ethics: status and trends. Procedia Comput. Sci. **15**, 53–65 (2012)
7. Gotterbarn, D., Miller, K., Rogerson, S.: Software engineering code of ethics. Commun. ACM **40**(11), 110–118 (1997)
8. Brooke, J.: SUS-A quick and dirty usability scale. In: Usability Evaluation in Industry, vol. 189, no. 194, pp. 4–7 (1996)

9. Katsanos, C., Tselios, N., Xenos, M.: Perceived usability evaluation of learning management systems: a first step towards standardization of the system usability scale in Greek. In: 16th Panhellenic Conference on Informatics, PCI2012, 2012, pp. 302–307 (2012)

10. Bangor, A., Kortum, P., Miller, J.: Determining what individual SUS scores mean: adding an adjective rating scale. J. Usability Stud. 4(3), 114–123 (2009)

11. Katsanos, C., Karousos, N., Tselios, N. Xenos, M., Avouris, N.: KLM form analyzer: automated evaluation of web form filling tasks using human performance models. In: 14th International Conference on Human-Computer Interaction (INTERACT), Cape Town, South Africa, 2013, pp. 530–537 (2013)

12. Tsironis, A., Katsanos, C., Xenos, M.: Comparative usability evaluation of three popular MOOC platforms. In: 2016 IEEE Global Engineering Education Conference (EDUCON), 2016, pp. 608–612 (2016)

13. Adamides, G., Katsanos, C., Parmet, Y., Christou, G., Xenos, M., Hadzilacos, T., Edan, Y.: HRI usability evaluation of interaction modes for a teleoperated agricultural robotic sprayer. Appl. Ergon. 62, 237–246 (2017)

14. Nunnally, J.C., Bernstein, I.H., Berge, J.M.T.: Psychometric theory. McGraw-Hill New York (1967)

15. Bangor, A., Kortum, P.T., Miller, J.T.: An empirical evaluation of the system usability scale. Int. J. Hum. Comput. Interact. 24(6), 574–594 (2008)

16. Kortum, P.T., Bangor, A.: Usability ratings for everyday products measured with the system usability scale. Int. J. Hum. Comput. Interact. 29(2), 67–76 (2013)

17. Nelson, L., Held, C., Pirolli, P., Hong, L., Schiano, D., Chi, E.H.: With a little help from my friends: examining the impact of social annotations in sensemaking tasks. In: Proceedings of the SIGCHI conference on human factors in computing systems, 2009, pp. 1795–1798 (2009)

The Challenge of Designing Interactive Scenarios to Train Nurses on Rostering Problems in a Virtual Clinical Unit

Catherine Pons Lelardeux[1]([⊠]), Herve Pingaud[2], Michel Galaup[3],
Arthur Ramolet[4], and Pierre Lagarrigue[5]

[1] IRIT, University of Toulouse (France), INU Champollion,
Serious Game Research Network, Toulouse, France
catherine.lelardeux@univ-jfc.fr
[2] LGC, University of Toulouse (France), INU Champollion,
Serious Game Research Network, Toulouse, France
herve.pingaud@univ-jfc.fr
[3] EFTS, University of Toulouse (France),
INU Champollion,
Serious Game Research Network, Toulouse, France
michel.galaup@univ-jfc.fr
[4] SGRL, University of Toulouse (France), INU Champollion,
Serious Game Research Network, Toulouse, France
arthur.ramolet@univ-jfc.fr
[5] ICA, University of Toulouse (France), INU Champollion,
Serious Game Research Network, Toulouse, France
pierre.lagarrigue@univ-jfc.fr

Abstract. The healthcare institutions in leading countries have undergone numerous changes to control their charges. These transformations generate an incredible impact on the work organization specially for caregiving staff. As a consequence, trainers in Nursing Schools need innovative tools to improve their courses. At the same time, there has been an increasing interest for immersive training environments which could represent with great fidelity a professional context. In this article, we present the method we set up to design educational interactive scenarios that take place in a socio-technical dynamic and complex system such as a clinical department. These scenarios aim to train nurses to plan their activity, deliver and organize cares for some fifteen inpatients. Using this method, we build a dozen of interactive non-linear scenarios. These scenario allow the trainee to freely act and make decision which could cause delay or equally bad and even far worse inadequate caregiving.

Keywords: Virtual Environment · Serious Game · Nurse Rostering · Nurse training

© Springer Nature Switzerland AG 2020
M. E. Auer and T. Tsiatsos (Eds.): ICL 2018, AISC 916, pp. 589–601, 2020.
https://doi.org/10.1007/978-3-030-11932-4_56

1   Context

All over the world and especially in healthcare institutions located in leading countries, the past decade has seen the rapid development of outpatient treatment in a day-care center. Most of hospitals offer ambulatory services that would enable them to pool resources. As a consequence, professionals have to face to an increasing number of patients in shorter caregiving periods. In spite of unpredictable situations and work interruptions, they must maintain the quality of service. This situation mainly impacts both the activity of healthcare teams who need to adapt and use a new workflow and the Health managers who need to solve scheduling and rostering problem. As a result, hospital staffs work with an increasing time pressure and daily workload.

Recently, healthcare professional experts and trainers point the importance to design educational environment and educational programs to reproduce with high fidelity the professional environment. It mainly relates to create experiential learning environment to help students to develop inter-professional skills [2].

Most of trainings in healthcare which take place in a digital world are designed for medical professionals and logically focused on technical skills and surgery [5]. A few ones are designed to train caregiving teams for better coordination and efficiency inside their operative unit [3,7]. Others have represented in a virtual operating room the inter-professional team activity [6,10].

Here, we plan to develop a virtual training environment to support teachers in the Nursing Schools to develop non-technical skills as work organization, decision making and situation awareness.

To that end, we need to define and use a structured method which enables us to create a library of non-linear interactive scenarios to train professionals who work in a clinical departement. A clinical department is a socio-technical and dynamic system where inter-professional teams need to cooperate and collaborate in order to deliver a service. Other well-known socio-technical systems are identified in nuclear industry, railroad and aviation industry. The method we present here can be used to build training scenario in others socio-technical and dynamic systems.

2   State of the Art

Writing an interactive scenario consists in combining knowledge and interactions in the virtual environment. According to the situational cognitive theory [8], a part of knowledge is in the virtual environment that provides the training context. Following the same idea, we refer to the affordance theory [4] that defines a possibility of action available in the environment. These concept has been developed by Norman [13] who focuses the design on users and iterative development cycle analyzing tasks, activity and users' needs. These two approaches: interaction design and explicit representation of knowledge are crucial to design an educational scenario that intends to represent a professional activity.

In the field of serious games, a scenario can be considered as a set of elements:(1) a briefing (mission): presentation of the current situation and expected

objectives to reach, (2) a virtual universe: objects, furniture, documents, characters... (3) a set of actions, pieces of information, documents, furniture... which
can be manipulated through the universe, (4) playful and educational locks (5)
educational skills to develop or acquire, (6) abstract or concrete concepts which
can be manipulated with interactivity through the environment: game play elements and educational concepts as programming, making decision... (7) levels
which compose the mission, (8) educational objectives to reach (9) a debriefing:
summary of outcomes with feedback that should help the player to succeed in
the future.

Firstly, a scenario proposes to the players a short storytelling of what is
the actual situation and what is the expected situation at the end. Secondly,
a scenario provides interactions that allow the players to achieve the mission
and locks (educational locks or playful locks) to prevent the player to succeed.
Finally, outcomes are compared to expected objectives and results are immediately displayed at the end of a game session.

This definition particularly suggests that interactive storytelling triggers
challenging opportunities in providing effective models for enforcing autonomous
behaviors for characters. In other words, players should have the possibility to
be wrong, fix their errors, succeed or fail.

Marfisi-Schottman and George [11] propose to support and guide teachers in
the design process to realize mobile collaborative games by providing three game
patterns that naturally integrate game mechanics and support specific learning
activities: Live action role-playing game, Mystery game and Treasure hunt.

Martens et al. [12] intend to provide tools to help game designers and experts
to generate interactive scenario and manage interactive storytelling. These tools
are based on linear logic that provides strong mathematical algorithms to calculate and deduce results.

In the case of training environment for high graduated students, the classical challenge to design an interactive scenario consists in (1) representing with
creativity but also with high fidelity the professional environment through the
virtual universe (2) providing opportunities to characters to choose their own
strategy, (3) providing interactions as part of the professional activity using
objects/equipment/furniture arranged in the virtual universe, (4) giving relative but controlled freedom to act in the universe in order to compare with the
expected behaviors.

Three kinds of scenario can be identified : (1) entirely controlled scenario: a
script defines every possible path to succeed or fail the mission, (2) controlled
scenario with a limited but real freedom: a large combination of paths is possible
to succeed or fail. As a consequence, none script defines all the possible combinations but algorithms can calculate them if necessary. (3) entirely free scenario :
machine learns from user's interactions and builds a statistically-realistic behavior. At the beginning, paths are unknown but machine learns from the user's
experience over the time.

Designing an entirely controlled scenario consists in determining in details
every available alternative and their consequences in the virtual world. Such a

scenario is called scripted or branching scenario and can be graphically represented by a tree-like structure. Using a tree-like structure to describe human activity is a widespread practice in virtual environment where an unique player is involved. Our research intends to design a set of controlled and non-linear educational interactive scenarios and involves teachers in the design process. All possible combinations are definitely not predetermined even if a list of constraints can be set to reduce the number of possible solutions considering the professional domain rules. The most important for us is to compare the behavior of the nurse with what is expected.

3   Purpose

Traditionally in Nurse Schools, teachers used to provide paper exercises working with small size case of studies (three or four patients) to challenge their students about the organization of a typical day. Nowadays, to the best of our knowledge, no digital educational environment proposes to train nurse to organize their daily work. We can propose some reasons why there are very few educational environments for care-giving teams: many constraints and difficulties restrict the development of this kind of training. The main constraints refer to the socio-technical environment itself, which is a complex and moving context. Consequently, recreating artificially the conditions of work in a clinical department where several dozens of patients are hospitalized is hardly impossible in a real educational place. Moreover, operational research has shown that the nurse rostering problem is an NP-hard combinatorial problem [1]. It is extremely difficult to efficiently solve real-life problems ex-ante, because of their size and complexity. Therefore, it should be easier to teach best practices in a controlled and well-defined system and to address the problem online in a secure and high fidelity context.

This article presents a method to design educational scenarios which aims to support trainers to teach nurses the best practices to organize their daily tasks facing to an increasing number of patients, to evaluate the importance of unpredictable events and all types of disturbances, to adjust their projected schedule, to face to a set of uncertainties relative to the tasks that have to be performed in an inter-professional and moving context. All the scenarios designed with this method will be available in this innovative digital and immersive environment. This research takes part of a global research that aims to provide a Real-time Digital Virtual Environment to train future nurses on scheduling their activity, providing and adapting their tasks to a moving context.

4   Approach

Our global research intends to design a real-time digital environment, which represents a clinical department in a virtual hospital, individual organization as well as interaction with their colleagues and includes embedded monitoring tools to control nurse's activity. It aims to transfer know-how from experienced nurses to junior by experiencing real-life situations in a virtual controlled and safe educational environment. Designing a virtual environment with a large library of well-known professional situations should support trainers to educate future nurses on work organization, situation awareness, leadership and decision-making.

The virtual environment should represent all the actors with whom the regular nurse is used to cooperate such as the care giving staff and their patients. The staff should be composed of a head nurse, a caretaker, a doctor, a kinesiologist, courier from blood analysis laboratory, hospital porter... The trainee should play the role of a regular nurse who has to organize their working day and take care about their patients. This digital environment should be designed combining game mechanisms and interactive features such as a scheduling system, a task shifting and a decision-making system.

The trainer should be able to choose an educational scenario from a library composed of various real-life ones. The nurse-student should play the role of a regular nurse and must manage the situation. The library of scenarios must be composed of regular situations as well as complex situations in which deficiencies or unpredictable events should occur or mistakes could be made and fix.

5   Method

Designing educational scenarios for scheduling training is particularly complex. It is extremely difficult to artificially reproduce a causal chain of events that foster students to change in the real-time their initial formulation of a rostering problem. It implies a large variety of contributing factors, such as human factors, technical failures, patients' pathologies... which are difficult to combine artificially. The method we propose aims to structure the design process to potentially build a large number of educational scenarios based on a large variety of real-life French professional situations which contains typical events and well-known hazards already experienced in the clinical departments.

The method is composed of three stages: the domain analysis, the human activity modeling and the scenario (see Fig. 1). The first row indicates these three stages. The second one indicates a set of steps which compose the stages. The last one lists the expected results and targets to be reached at each stage. The domain analysis stage allows to obtain information about the professional domain such as a typical clinical service organization, the average number of patients in a real one, inpatients, patient records, the composition of a medical staff, the care giving staff typical activities... It aims to highlight the concepts, the terminology, papers and documentation, the software and tools used... This step allows to collect elements to describe the virtual clinical service and a set of global professional activities which the care giving staff deals with in a real hospital.

The human activity modeling stage consists in structuring data to describe the universe and all elements that compose an interactive scenario.

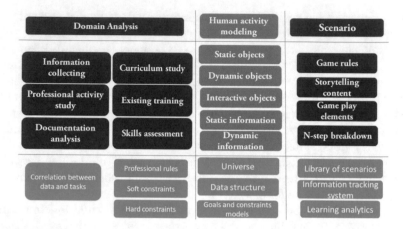

Fig. 1. The method is composed of three global stages: the domain analysis, the human activity modeling and the scenario scripting

5.1   Domain Analysis

This stage consists in interviewing experts, trainers and professionals to gather substantive information about the domain that will have implications for the assessment. Approximately some twenty semi-structured interviews have been conducted with a dozen of care experts and trainers who work either in Nursing Schools or in the Regional Healthcare Agency. Two dozens of experienced nurses have been interviewed and involved in the game design process. They were chosen because of their affiliation in eighteen different French hospitals. These criteria ensure to obtain a global consensus to represent a typical clinical service.

Real-life nurse's activity. The analysis of real situations related to the targeted skills enables us to identify tasks and patient records with a gradual complexity. For example, an educational scenario must deal with a shift type which is defined by a start and end time. It should begin either early: 6:30–14:30, or late: 14h30–21:30, or at night 21:30–6:30. The scenario must comply with a typical daily scheduler. The typical day is split into main stages into which similar tasks are performed for all patients. Some are hardly subject to constraints from the environment, some are not. For example, the delivery of breakfast is planned in the early morning just after the team shift from night to morning staff. The doctor in charge of the unit is making the visit of patient in the middle of the morning. This is considered as a hard constraint for two reasons. First, the doctor is often having a heavy workload including solicitations from emergencies or delay during surgical operations. It is likelihood that the visit will not be done at the expected time. Moreover, during the medical visit, doctors change the medical content of the Electronic Health Record (EHR) to adapt a therapeutic treatment or ask for new examinations. As the consequence, the nurse must revise their projected schedule for the remaining part of their service. This typical day description revealed both the existence of generic activities and a predefined time slicing that punctuates the work flow. All these substantive information allows to model a generic frame for a scenario and represent a typical clinical service.

Health records of real patients. The second step relates a detailed analysis of a set of real patient records and associated activities (nurse's activity, nurse assistant's activity, doctor's medical visit...). Fifteen health patient records of real patients from a dozen of different hospitals have been analyzed and compiled. This analyze has led to a standardized model of health patient record. A EHR is composed of three files: (1) an administrative file which contains administrative information about the patient, (2) a medical file which contains medical information about the health state and the health history, (3) a nurse file which contains care giving recommendations written by the previous nurses who gave health care to the patient.

Preliminary communications between experts lead rapidly to the conclusion of the diversity of representation of what health patient records are. It is a critical point as the nurse's tasks are daily reported and stored in files on the basis of the information delivered by the institution and by the doctors. The group of experts spent a long time to converge towards a unified and standardized model of a patient record. Differences concerned substantive issues on the semantics to use and the pieces of information to dispatch between the three files. If a rationalization of the EHR content has been reached, what we could call a convergence process of the expert views has really contributed to share a conceptual representation of the game design requirements and to strengthen the relationships between team members. As experts reached an agreement on a standardized patient record, they extracted valuable pieces of information from those 15records and lighted content about effective patients care needs as a key of educational purposes. At the end of the step, 15 virtual patient records have been designed (see Fig. 2).

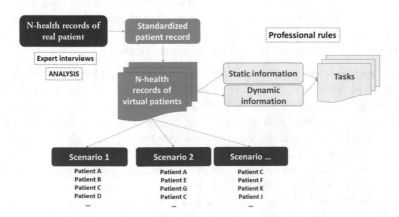

Fig. 2. A standardized health record results from the analysis of health records of real patients

Nurse's activity and constraints. This step consists in understanding the rules and the cognitive scheme used by the experienced nurses to organize their job. Nurses globally decide of their activity from the last pieces of information orally delivered by the night shift and the EHR. The starting point to understand how they proceed to organize their activity was to analyze how they schedule their activity with the 15 virtual patients studied in the last stage. Manual scheduling of 15 patients care plans is an expensive time consuming task, even for experienced nurses. However, it really allowed us to understand their way of thinking and the rules they used to comply with professional excellence. Analyzing their approach on self-scheduling facing this virtual workload of 15 patients, we noticed that they proceed on a step-by-step trial and error process. First, they mentally compare the specific cares a patient needs with a standard well-known scheduling driving what could be named a generic framework of daily nurse's activity. Secondly, they refresh this generic framework by task addition or/and task removal onto the timetable until they obtain a projected scheduling. Moreover, a feedback of our experience states that most of the time, nurses plan their activities using adjustments to a standard predefined planning after having performed a detailed analysis about the inpatient records.

An example of a task order practice is a breakfast postponing because the nurse must take a blood sample in fasting condition. Decision making is based on a kind of mimicry that is usually practiced in critical environments where service are delivered to people and affect their life.

The last step consists in balancing the workloads for the new set of tasks in a global perimeter with respect to all the rostering constraints: coverage constraints, time related constraints, hard constraints, soft constraints... Coverage constraints are constraints that refers to the number of professional roles needed in a working period. Time related constraints are constraints that refer to the scheduling, the right order of care delivered as well as other more personal activities. For example, personal lunch time, medication distribution time, patient lunch time, patient therapeutic care... Hard constraints are constraints that must be satisfied whatever event occurs. For example, give patient their medications. Soft constraints are constraints that are considered as recommended but they may relax if necessary to find a feasible solution.

A cross analysis of real-based situations and curriculum. At this step, we need to identify and classify the knowledge and skills which are concerned by the studied patient records. It implies a cross analysis on the curriculum of Nursing School and the EHR of the virtual patients. For Le Boterf [9] "the skill is the mobilization or activation of several knowledge, in a given situation and context". This cross analysis work was carried out in collaboration with trainers from Nursing Schools and didactic researcher. In the repository of activities and skills for a graduate nurse, we find that the university program involves ten skills. Among these ten skills, the skill number 9 entitled "Organize and coordinate care interventions" is the one we mainly target. We are going to propose an answer to the needs of trainers through the use of a serious game in initial training. But, how is it possible to ensure that the skill is mastered? In other words, we must provide to the designers the criteria to evaluate the students. The referenced French national book details the activities and skills which must be assessed to be a graduated nurse. It mainly points three criteria. The first one concerns the relevance of the intervention of the various participants. The second is about the Consistency in the continuity of care. And the third one relates to the reliability and relevance of the information provided. Each indicator is associated with another one which gives visible signs. Let's take for example the criterion named Consistency in the continuity of care. The indicators relating to this criterion are: (1) the links between the various professional interventions are identified and well defined, (2) the organization of the activities in order to optimize collaborative work is explained and argued, (3) the control of the assigned care is done, (4) continuity and traceability of care is ensured.

This collaborative work of skill identification makes the designers know what interactions must be monitored to assess the student at the end of a training session.

5.2   The Human Activity Modeling Stage

Representing the nurse's activity in a clinical unit implies to characterize a set of virtual patients, a set of non-player characters like the nurse assistant, the doctor, the courier, the hospital porter... who daily interacts with the nurse, a set of clinical rooms where the patients live, a nurse's room where the nurses cooperate, phone to other professionals and a room where nurses have a rest. We choose to define a virtual patient by all the pieces of information dispatched in their health record and all the pieces of information that can be delivered both by the patient itself and the other team members. The virtual patient model is exclusively based on the static information model and dynamic information model. The step of domain analysis showed that in average 40 pieces of information have been dispatched between the three files of the EHR and 40 tasks are associated to one virtual patient. The pieces of information are classified into two categories : a static information category and a dynamic information category.

This choice of model enables us to create as much as we need new pieces of information whose can concern both virtual patients or anything else like "Dr Vincent postpones the medical visit at 11:00".

A static information is an information which is set at the beginning of the scenario and will not change until the end whereas a dynamic piece of information is an information whose value can change and/or can only be delivered by a virtual character during a particular time period or after a task achievement. As an illustration, the patient's identity is a static piece of information that can be read in the EHR. The patient's blood pressure is an example of a dynamic piece of information which is delivered when the nurse measures the blood pressure. Its value may change over the time. A dynamic piece of information can be notified to the trainee at any time or can be delivered when the nurse achieves a task. The dynamic information system use the dynamic pieces of information to represent the moving context and enrich the narration. In previous part, we described events that are known as having an impact on the planning during a typical day. To inform the trainee that the current context is changing, the dynamic information system is used to notify an unexpected event. As an illustration, if the patient's health status is evolving either in a good way or in a bad way, the nurse must be informed and should be fostered to react very often. Another examples of unexpected events can illustrate the concept of dynamic information: a patient's call, a patient's family asks the nurse for an appointment, during the medical visit, a doctor changes their medical prescription or prescribes a new examination (radiography, MRI scanner, blood test...), the patient refuses to go with the hospital porter...

There are two categories of tasks, the generic tasks which can be achieved on any virtual patient such as 'distributing breakfast and meal', 'making the bed', 'measuring temperature', 'measuring pulse rate'... and the contextual tasks which are associated with pieces of information dispatched in the universe. Most of the tasks are freely available since the associated pieces of information have been read or previous tasks have been realized. These kind of constraints allow designers to make a task available at any time or only when the constraints have been fulfilled.

5.3   Scenario Stage

In terms of storytelling, designing educational scenarios based on real-life situations for professional training consists in two points. The first one consists in representing a perfect professional situation with competitive colleagues who realized all the prescribed tasks and do not postpone any care before the nurse begins their shift. The second one consists in representing a moving situation where medical prescriptions and unexpected events will occur around the clock and should foster the nurse to adapt their planning. If the nurse does not consider these new events in time, whatever their importance, problems should arise later or errors should be revealed as being part of the causal chain of events that could lead to an adverse event or a near miss. It should make the job more complex for the next Healthcare team. The model presented in Sect. 5.2 enables us to trigger new events that can disturb the nurse, enrich the narration and enforce the nurse to react. As an illustration, non quality of logistics services could be a root cause of disturbances. If a prescribed drug is not available, the unit is not supplied and it could be critical for a patient. A dynamic piece of information can be delivered to inform the student of a new drug prescription at 10:00 and another one can be delivered to inform the nurse that this drug is no longer available.

Each scenario is relative to the same clinical department with the option to choose the level of difficulty. The number of patients who need to obtain care is a parameter used to complicate a scenario. For example, a scenario addresses a group of 5 patients (A, B, C, D, E) while another one deals with 8 patients (A, C, F, G, H, I, J, H)and so on to build a scenario in a unit of 15 inpatients (see Fig. 2).

6   Results

This method has enabled us to identify the main features and interaction systems needed to represent the nurse professional activity in a virtual world designed to train future professional on organization skills. Using this method, it has been possible to design some ten scenarios which are all structured by five steps: (1) the briefing: to inform the students on their mission, (2) the communication step: to receive pieces of information from the previous staff and become aware of the situation when they shift at 6:30, (3) the scheduling step: to consult the patient's record and organize the activity of the day, (4) the activity itself: to provide care, organize medical examination, professional phone calls, patient discharge or arrival (5) the communication step: to inform the next staff on the current situation when they shift at 1:30pm.

During the domain analysis, we identified 6 features which would be part of game universe: (1) a virtual memory is attached to the nurse avatar. It should allow the trainees to use the pieces of information they read from the EHR or broadcast from other characters. It also stores a set of activities which can only be available since they become aware of particular information that are dispatched in the universe. This is the case of medical prescription for example. The task named 'Distributing per os treatment' is only available if they read the prescriptions on the medical patient record, (2) an information reading system to hide some pieces of information which could be relevant to use as argument during a decision making stage, (3) a scheduler feature which allows the player to place and organize their tasks on their own scheduler from a typical daily scheduler, (4) a managing and delegating feature to allow the nurse to assign some tasks to the nurse assistant, (5) an argumentation feature to enable the student to express and argue their decision when they need to adjust the scheduler, (6) a communication system to enable the student to obtain/broadcast information to the previous/next caregiving team.

7   Conclusion

This paper aimed to present a method to design interactive and non-linear educational scenarios for self-planning nursing tasks. The method used to build a library of a dozen of scenarios is composed of three stages: the domain analysis, the human activity modeling stage and the scenario stage. Each stage has been detailed and we showed how this method enabled us to build different scenarios combining a sub-group of virtual patients, all relevant for the training of nurses.

This research introduces a global work made to build a real-time digital environment (see Fig. 3) designed to train the nurses to schedule and manage their activity in a complex and dynamic inter-professional context. Future work aims to realize training using this digital environment that embedded these scenarios in a dozen of National French Nursing Schools.

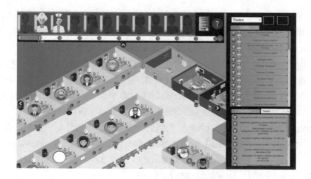

Fig. 3. The virtual environment graphically represents a virtual clinical department

Acknowledgments. These works are part of a global innovative IT program whose partners are University Champollion and the French Regional Healthcare Agency (Occitanie). The steering committee is composed of Ph.D. C. Pons Lelardeux, Ph.D.M. Galaup, Pr. H. Pingaud, Pr. P. Lagarrigue, C. Mercadier, V. Teilhol.

References

1. Burke, E.K., De Causmaecker, P., Berghe, G.V., Van Landeghem, H.: The state of the art of nurse rostering. J. Sched. **7**(6), 441–499 (2004)
2. Cornes, D.M.: Review of interprofessional education in the United Kingdom (1997–2013). J. Interprofessional Care **29**(1), 85–85 (2015)
3. Petit dit Dariel, O.J., Raby, T., Ravaut, F., Rothan-Tondeur, M.: Developing the Serious Games potential in nursing education. Nurse Educ. Today **33**(12), 1569–1575 (2013)
4. Gibson, J.J.: The ecological approach to the visual perception of pictures. Leonardo **11**(3), 227–235 (1978). http://www.jstor.org/stable/1574154
5. Graafland, M., Schraagen, J.M., Schijven, M.P.: Systematic review of serious games for medical education and surgical skills training. Br. J. Surg. **99**(10), 1322–1330 (2012)
6. Hu, J., Feijs, L.: A distributed multi-agent architecture in simulation based medical training. In: Pan, Z., Cheok, A.D., Mller, W., Chang, M. (eds.) Transactions on Edutainment III, pp. 105–115. No. 5940 in Lecture Notes in Computer Science. Springer Berlin Heidelberg (2009)
7. Kilmon, C.A., Brown, L., Ghosh, S., Mikitiuk, A.: Immersive virtual reality simulations in nursing education. Nurs. Educ. Perspect. **31**(5), 314–317 (2010)

8. Lave, J., Wenger, E.: Situated learning: legitimate peripheral participation. Cambridge University Press (1991)
9. Le Boterf, G.: Valuer les comptences. Quels jugements? Quels critres? Quelles instances. Educ. Perm. **135**(2), 143–151 (1998)
10. Lelardeux, C.P., Panzoli, D., Galaup, M., Minville, V., Lubrano, V., Lagarrigue, P., Jessel, J.P.: 3d real-time collaborative environment to learn teamwork and non-technical skills in the operating room. In: Interactive Collaborative Learning. pp. 143–157. Springer, Cham (2016)
11. Marfisi-Schottman, I., George, S.: Supporting teachers to design and use mobile collaborative learning games. In: International Conference on Mobile Learning, pp. 3–10 (2014)
12. Martens, C., Bosser, A.G., Ferreira, J.F., Cavazza, M.: Linear logic programming for narrative generation. In: International Conference on Logic Programming and Nonmonotonic Reasoning, pp. 427–432. Springer (2013)
13. Norman, D.A.: Cognitive artifacts. Designing interaction: Psychology at the human-computer interface, vol. 1, pp. 17–38 (1991)

Authoring Game-Based Programming Challenges to Improve Students' Motivation

José Carlos Paiva[1]([✉]), José Paulo Leal[2], and Ricardo Queirós[2]

[1] CRACS & INESC-Porto LA, Faculty of Sciences, University of Porto, Porto, Portugal
up201200272@fc.up.pt
[2] CRACS & INESC-Porto LA & ESMAD/IPP, Porto, Portugal
zp@dcc.fc.up.pt, ricardoqueiros@esmad.ipp.pt

Abstract. One of the great challenges in programming education is to keep students motivated while working on their programming assignments. Of the techniques proposed in the literature to engage students, gamification is arguably the most widely spread and effective method. Nevertheless, gamification is not a panacea and can be harmful to students. Challenges comprising intrinsic motivators of games, such as graphical feedback and game-thinking, are more prone to have longterm positive effects on students, but those are typically complex to create or adapt to slightly distinct contexts. This paper presents Asura, a game-based programming assessment environment providing means to minimize the hurdle of building game challenges. These challenges invite the student to code a Software Agent to solve a certain problem, in a way that can defeat every opponent. Moreover, the experiment conducted to assess the difficulty of authoring Asura challenges is described.

Keywords: Games · Gamification · Authoring · Learning · Programming · Competitive · Graphical feedback

1 Introduction

Motivation is what makes you try to do something [19]. One who feels unable to do an activity, or fails to value it and its outcome, is highly likable to lack motivation to engage in the activity, in which case he/she is said to be amotivated. In education, loss of motivation is one of the most outstanding problems. When students are amotivated, they tend to care less about educational activities and to stop striving to complete them. In programming courses, this results in an unsustainable lack of practice which is accompanied by recurring failures in assessments and later ends up in student dropout [1,6].

Several approaches have been proposed to mitigate this problem, such as problem-based learning [15,17], storytelling [8], pair programming [6],

© Springer Nature Switzerland AG 2020
M. E. Auer and T. Tsiatsos (Eds.): ICL 2018, AISC 916, pp. 602–613, 2020.
https://doi.org/10.1007/978-3-030-11932-4_57

competition-based learning [2,13], and gamification [7,21]. Undeniably, the most widespread approach is gamification, which consists of using game elements and mechanics to engage students. The most common gamification methods typically add extrinsic motivators, such as leaderboards, badges, or levels, to an existing learning environment. Even if these elements can increase students' engagement temporarily, they neither foster correct changes in attitudes and behaviors or longtime commitment. On the contrary, they undermine the intrinsic interest that one might have to perform a task. Completing the activity becomes a means to obtain the reward rather than to develop skills. Moreover, the best way to solve the task is the first to come to mind, taking risks or exploring are not options [10].

Gamification techniques with enduring effects typically rely on different game aspects, such as graphical feedback, game-thinking, collaboration, and in-game challenges. Another well-explored aspect of games in programming learning is competition. Despite the fact that it may have considerably harmful effects [9], it is also true that new graduates are increasingly facing programming contests after leaving universities, as a part of the recruitment process for top technology companies. Hence, a promising approach might be to combine these features. In fact, there are already some attempts to bring competitive games into learning, which challenge the student to code the Software Agent (SA) that controls the player and competes against other SAs. For instance, IBM CodeRally – a Java game-based car rally competition presented at the 2003 ACM International Collegiate Programming Contest (ICPC) World Finals – and Robocode – a Java-based virtual robot game – have been found to have a great potential to catch students and non-students attention [14,16]. Nevertheless, building these challenges involves a complex and time-consuming process which most times educators are not willing to perform.

This paper presents the authoring component of Asura, a game-based programming assessment environment that challenges students to code competitive SAs for a game. Asura is developed on top of Enki [18] – an existing pedagogical environment of Mooshak 2 [12] – and Mooshak 2 itself. The final goal of Asura challenges is to win a tournament among all submitted SAs, at the end of the submission time. The authoring component of Asura, named Asura Builder, is one of its key features. This component aims to provide a simple way of authoring new challenges. As a benchmark, the reference is the well-known ICPC problems whose automated assessment requires coding both a solution program and a test case generator. The goal is to keep the effort of authoring an Asura challenge similar to that of authoring an ICPC problem.

The development of new game-based challenges requires the teacher to extend an existing referee (i.e., manager) to implement the game rules and inquire SAs for their actions, implement the state interface that helps updating the game state, add a problem statement, and upload some image assets for the game. However, more complex challenges may demand the specification of wrappers for SAs submitted by students. Also, teachers can add their own SAs, so that students can do matches against them since the beginning. There is no limit on the number of SAs submitted by teachers.

The remainder of this paper is organized as follows. Section 2 reviews the state of the art on authoring tools for game-based challenges. Section 3 presents Asura, its concept and architecture. Section 4 details the component for authoring challenges. Section 5 describes the validation of Asura Builder, which aims to assess the increase in difficulty of building an Asura challenge when compared to that of creating an ICPC-like problem. Finally, Sect. 6 summarizes the main contributions of this paper and discloses the next steps on the presented work.

2   State-of-the-Art

Asura is a game-based assessment environment that aims to offer the teacher a way to motivate students to program and overcome their difficulties through practice, requiring a similar amount of effort to that of creating an ICPC-like problem. It engages students by challenging them to code an SA to play a game, supporting them with graphical game-like feedback to visualize how the SA performs against other SAs. The final goal of an Asura challenge is to win a tournament, like those found on traditional games and sports, among submitted SAs.

To the best of authors' knowledge, there is no tool to author challenges with these features. Therefore, the next subsections review the state-of-the-art on authoring tools for game-based challenges and present some environments similar to Asura, highlighting their extensibility for new challenges.

2.1   Authoring Tools for Game-Based Challenges

Even though tools to author game-based challenges are scarce, there is much research interest in such tools. Most of the attempts to create platforms for authoring game-based challenges focus on simple, yet attractive, characteristics of games, particularly the storyline, which are easy to adapt to heterogeneous contexts. StoryTec [5] is a digital storytelling platform for creating and experiencing interactive (i.e. non-linear) stories. It encompasses a story editor, an action set editor, a property editor and an asset manager. The story editor manages the structure of the story using an hierarchically organized graph consisting of scenes. The stage editor is a tool similar to a level editor, present in some video games, that enables scene creation using a drag-and-drop interface to insert objects from the assets manager. The action set editor is a visual environment, based on the UML activity diagram, for defining the possible actions and constraints of every scene. The asset manager is where the author can import various types of assets, such as cameras, lights, and models.

<e-Adventure> [20] is an authoring tool for story-driven educational games, that aims to introduce games in the learning process. It supports the creation of *third-person* and *first-person* adventure games by instructors with little or no programming background. Furthermore, it complies with standards and specifications of the e-learning field, allowing to export, modify, and reuse games as learning objects. The tool is an all-in-one game creator with game objects organized by types (e.g., scenes, items, conversations, etc.) which the author can add into the game.

Although these systems were designed for a pedagogical purpose, neither StoryTec or <e-Adventure> are specific for learning to code. Greenfoot system [11] is an interactive object world that provides a framework and an environment to create interactive 2D simulations. On the side of the solver, it features a full-fledged Integrated Development Environment (IDE), including code editing, compilation, object inspection, and debugging. Moreover, Greenfoot also offers graphical feedback to visualize the appearance, location, and rotation of the simulation objects as well as methods to directly interact with them.

2.2   Competitive Game-Based Programming Environments

SoGaCo [4] is a scalable web environment that evaluates competitive SAs, developed in several programming languages, that play simple mathematical board games. Its interactive GUI allows learners to see step-by-step how their programs play the game. Furthermore, it not only promotes competition among students but also collaboration, allowing them to share their SAs through a single bot address (URL).

The modular architecture of SoGaCo supports different games but there is no known framework or standard form to develop games for SoGaCo. Nevertheless, it already contains several board games, such as PrimeGame, Mancala and Othello.

CodinGame[1] is a web-based platform with several puzzles that learners can solve to practice their coding skills. Most of these puzzles require the user to develop an SA to control the behavior of a character in a game environment, and provide a 2D game-like graphical feedback. The SA programmed by the player must pass all test cases (public and hidden) to solve the puzzle. Players can choose one of the more than twenty programming languages available to write their SA, or even solve it in more than one language. Once the exercise is solved, players can access, rate, and vote on the best solutions.

The platform enables any user to contribute with programming challenges through a dedicated form. Nevertheless, it only allows them to create challenges based on input/output test cases without any game-like graphical feedback.

---

[1] www.codingame.com.

3   Asura

Asura is an environment for assessment of game-based programming challenges. The main goal of this environment is to minimize the hindrance of creating new game challenges, while allowing students to enjoy unique features of games, such as graphical feedback, game-thinking, and competitiveness. In Asura, students are challenged to develop an SA to play a game. This SA has the final objective of beating every other SAs in a tournament following a similar structure to those organized on traditional games and sports.

During the development of the SA, students can validate its effectiveness by executing matches against any previously submitted SAs that can be selected from a board in Enki. A match runs on the Asura Evaluator, which evaluates the code of the SA, starts a process with it, and leverages the rest of the evaluation on the game manager. The outcome of the match is a JSON object adhering to the JSON Schema[2] defined for a game movie. This object is given to the Asura Viewer by Enki, which transforms it into an adequate format to display to the learner. Figure 1a) presents the result of a validation of an SA against an opponent in a Bullseye Shooting game being displayed in the Asura Viewer integrated into Enki.

Tournaments can be organized by educators, once the time to code SAs ends. For that, they can use the wizard added to the administrator interface of Mooshak 2, shown in Fig. 1b). This wizard enables the educator to choose among the set of all accepted submissions, determine how much points are awarded per match, and define the structure of the tournament stages. After configuring the format of the tournament, the Asura Tournament Manager organizes and runs the matches of the tournament on the Asura Evaluator. A tournament produces JSON data adhering to the JSON Schema[3] defined for a tournament, which contains a reference to each match's movie, organized by stages and rounds, as well as partial and complete rankings of each phase. This data is presented in an interactive Graphic User Interface (GUI) to the students, allowing them to request the matches they want to see and check details of each phase. For instance, Fig. 1c) displays the interactive GUI of a tournament of Slalom Skying in a knockout phase, whereas Fig. 1d) presents the player's path after clicking in the player's name on the interactive GUI.

The architecture of Asura, depicted in Fig. 2, is composed of four distinct components, namely the Viewer, the Builder, the Evaluator, and the Tournament Manager. These components interact with tools already described in the literature, particularly Mooshak 2 and Enki. The Evaluator is a small package developed inside Mooshak 2, whose main class is the **AsuraAnalyzer**. This class is a specialization of the **ProgramAnalyzer**, the main analyzer of Mooshak 2 which is responsible for grading submissions to ICPC-like problems, for conducting the dynamic analysis using the provided JAR package. Both analyzers implement a common interface – **Analyzer** –, that allows evaluator consumers

---

[2] https://mooshak2.dcc.fc.up.pt/asura/static/match.schema.json.
[3] https://mooshak2.dcc.fc.up.pt/asura/static/tournament.schema.json.

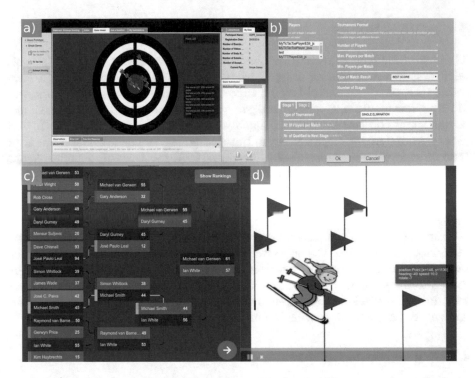

Fig. 1. Screenshots of the various components of Asura

to integrate seamlessly with any of them. In this case, the consumer is Enki. The Tournament Manager handles the set up and execution of the tournament, integrating with the Evaluator and Mooshak 2 administration GUI. The Viewer is an external Google Web Toolkit (GWT) widget that can integrate in any GWT environment, supplying it with an interface `TournamentMatchViewer` to enable them to display either tournaments or matches. Finally, the Builder is an independent component that produces the JAR package used by the Evaluator.

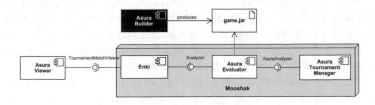

Fig. 2. Diagram of components of the architecture of Asura, highlighting Asura Builder

4 Asura Builder

The Asura Builder is an independent component composed of multiple tools dedicated to the authoring of game-based coding challenges, including a Java framework and a Command-Line Interface (CLI) tool. The Java framework provides a game movie builder, a general game manager, several utilities to exchange complex state objects between the manager and the SAs as JSON or XML, and general wrappers for players in several programming languages. The framework is accompanied by a Command-Line Interface (CLI) tool to easily generate Asura challenges and install specific features, such as support for a particular programming language, a default turn-based game manager, among others. Even though the authors are required to program the challenges in Java, players can use their preferred programming language to code their SAs.

Each of the next subsections describes a sub-component of Asura Builder. Subsection 4.1 describes the builder of graphical feedback. Subsection 4.2 details the referee and state management of the game. Subsection 4.3 presents the communication between the game manager and the players. Subsection 4.4 introduces the two kinds of wrappers that Asura supports to facilitate SA development. Subsection 4.5 provides an overview of the CLI tool.

4.1 Game Movie Builder

Most of the necessary effort for building video games is spent on graphics. They determine the players' first impression on the game and they provide the best feedback of the actions executed during the game. Asura games are not exceptions. In Asura, graphics are abstracted as a game movie, which consists of a set of frames, each of them containing a set of sprites annotated with information about their location and transformations. In this way, the representation of the game movie is very compact since each frame is just a collection of objects, completely described by four numbers. Besides that, a movie also includes metadata information, such as `title`, `background`, `width`, `height`, `fps` – number of frames to display per second –, `anchor_point` – sprite point relative to which coordinates are given –, the set of `players`, and the set of `sprites`.

This abstraction facilitates the construction of graphical feedback by defining a standard way to describe it, independently of the game. Furthermore, Asura Builder offers an interface (and an implementation) to easily build these game movies. The interface provides methods to manage metadata information, add frames, insert and transform items, include messages to players (e.g., logs of their SAs), set the observations and classification of a player, push and pop frame states from a stack, terminate the game movie indicating an error in the Builder component or an error in one of the players, among many others. Updates to the game movie are performed during the game state management.

There is no distinction between game movies built for validations or game movies created during the tournament, so the author does not need to change anything to execute tournaments.

4.2   Game Manager

Every game needs a controller (or referee) to ensure that the game rules are followed. The controller keeps the global state of the game, decides which player takes the next turn, declares a winner, among many other tasks. In the Builder component, these tasks are the responsibility of the `GameManager` who acts as the referee of the game.

The abstract manager provided by the framework connects to the input and output stream of the players' processes to receive their actions and update them with changes on the game state. Specialized managers determine the playing order and manage the game state accordingly. Some of these specialized managers, such as a turn-based game manager, are provided by the framework and can be easily integrated in a new challenge, using the CLI tool.

In order to manage the game state, which is unique to each game, the controller leverages on the `GameState` interface. This interface specifies methods to initialize the state before the game starts, update the game state according to the action of a concrete player, obtain the object that needs to be sent to a certain player to update it about changes to the game state, end the round when all players' commands were executed in that round, finish the game and declare a winner, and much more. Most of these methods receive a game movie builder as parameter, allowing the state object to manage the movie, reflecting any changes made to it.

As a referee, when a player breaks the rules of the game (e.g., takes too much time to play, does an invalid action, does not meet the communication protocol, etc), it must act. If the violation of the player prevents the game from continuing, the game ends marking the infraction of the SA in the game movie. Otherwise, the game proceeds but the faulty SA gets a "Wrong Answer" at the end.

4.3   Communication Manager-Players

The communication between the `GameManager` and the players can be done either through JSON or XML. The `GameManager` sends state updates to the players, containing a type, which identifies the state, and a comprehensive description of the current game state. The messages sent from the players to the `GameManager` contain the action (a command) that the player wants to execute as well as a list of messages for debugging purposes. This communication is handled without any action of the author or players, meaning that they are not aware of the type of messages being exchanged. Yet, the author knows that there is a channel that sends and receives objects. Figure 3 presents the structure of the data models that are exchanged during the game.

4.4   Wrappers

Wrappers are sets of functions, provided by both the framework and the author, that aim to give players an higher level of abstraction, so that they can focus

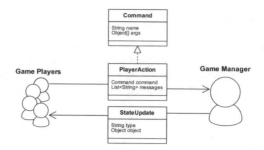

Fig. 3. Diagram of the communication between the manager and the players

on solving the real challenge instead of processing I/O. There are two types of wrappers: global and game-specific.

Global wrappers are defined by the Asura Builder framework. They implement functionality that is common to players of any game in a certain language. This includes methods to read and write JSON, log messages, and call the functions on the abstract and concrete players that implement the game/player-specific functionality.

Game-specific wrappers are provided by game authors and "extend" global wrappers with functionality related to a specific game. For instance, they can implement functions that process state updates, get and set values, or send actions. Besides that, they implement the player lifecycle, i.e., the player loop that reads updates or executes actions.

4.5 CLI Tool

The Command-Line Interface (CLI) tool is a command-line utility, based on cookiecutter,[4] one of the many existing CLI generators. A project with the codebase for an Asura game can be generated with a single command line `asura-cli --generate`. This command makes a series of queries to the user in order to obtain the required information to generate the project. The project generated by this utility is a Maven project containing a skeleton of a Game Manager and a Game State as well as the structure for SAs, wrappers, and skeletons.

Furthermore, the utility also imports pre-built game managers (e.g., `asura-cli --import-manager turn-based`) from a collection, including a turn-based game manager in which players act by turns, an all-at-once game manager in which players act all at the same time, among others. Support for new programming languages can also be managed through the CLI using the commands `asura-cli --add-language <language>` and `asura-cli --remove-language <language>`.

The deployment phase is also supported by the CLI tool. To package the game, the author can use the command `asura-cli --package`. A sample problem statement can also be generated using `asura-cli --add-statement`.

---

[4] https://github.com/audreyr/cookiecutter.

5   Validation

An experiment to validate the Asura Builder was conducted in an open environment with undergraduate students of the Department of Computer Science of the Faculty of Sciences of the University of Porto, enrolled either in the *First Degree in Computer Science* or the *Integrated Master of Science in Network Engineering and Information Systems* programs. This experiment aimed to assess the usefulness and ease of use of the Builder component on the authors' perspective only. However, students themselves played the role of authors. All students had an average background in Java acquired during the current semester in a Software Architecture course, and had no previous knowledge of Asura.

The experiment consisted on authoring both an ICPC problem and an Asura challenge, following one or more of the provided statements A,[5] B,[6] or C,[7] in increasing order of difficulty. These statements completely describe the challenges to be developed. Yet, they do not constraint the quality of the graphical feedback provided in the Asura game, which were left to the creativity of the authors. At the date of the experiment, the CLI tool was not finished yet. Thus, a Maven archetype was used instead to generate the project, requiring the authors to configure Maven before starting.

At the end of the experiment, students were asked to fill-in an online questionnaire to measure the user acceptance of the Builder. The questionnaire follows Davis' model [3] for evaluating perceived usefulness and ease of use of a system in a 7-value Likert scale, extended with questions about time spent in each type of problem, multiple-choice questions to compare the difficulty of authoring both types (in a 7-value Likert scale), and open text questions to identify weaknesses and strengths, and provide suggestions.

The global results indicate a perceived usefulness of 95.24% and a perceived ease of use of 76.19%. Regarding the comparison of developing an Asura challenge against creating and ICPC-like problem, a value of 73.81% was obtained (highest values are better). Nevertheless, statements regarding difficulty and time had a below average score, such as *I'm capable of developing an Asura challenge faster than an ICPC problem* (28.57%) and *It is easier to develop an Asura challenge than an ICPC problem* (71.42%). When asked to rate, in a Likert scale, the sentence *The additional hurdle of developing an Asura challenge is something that we can disregard considering the gains for students*, students agreed with 42.86%.

The free text answers highlighted the user acceptance of the Builder. For example, *The Framework for the development of challenges is very flexible and it's mechanics are easy to understand.*

---

[5] https://mooshak2.dcc.fc.up.pt/asura/static/asura-validation-problem-a.pdf.

[6] https://mooshak2.dcc.fc.up.pt/asura/static/asura-validation-problem-b.pdf.

[7] https://mooshak2.dcc.fc.up.pt/asura/static/asura-validation-problem-c.pdf.

6   Conclusion

Motivating students to learn in practice intensive courses, particularly in programming courses, is challenging. The time dedicated to solve exercises is typically inadequate for the amount of knowledge they have to acquire and techniques they must master. As students start getting bad results, their interest in the learning activities diminishes. To mitigate this problem, it is necessary to find new educational methods that promote coding practice outside of the classes. One of such methods is to wrap learning activities as game challenges.

Even though games are already widely used in programming education, there is a lack of tools for creating them. This paper presents the authoring component of Asura, an environment for assessment of game-based programming challenges. The goal of this component is to minimize the hurdle of creating these challenges, making its difficulty similar to that of creating an ICPC-like problem.

The conducted validation, even though with a very low number of participants due to the final exams, has demonstrated the usefulness of Asura Builder and its ease of use. Nevertheless, there is still a long way to go to achieve the fast creation of game-based programming exercises. The results highlighted a significant difference in terms of time, when comparing the two types of problems. It is expected that the CLI tool can make the generation, addition of features, and deployment phases faster, but not that much. More improvements and extra features are needed, particularly in the game movie builder. The collected comments have also revealed the need to support multiple programming languages in the creation of games, which was already planned in a future release.

The next step is to validate the effectiveness of Asura as an environment to keep students motivated while working on their programming assignments.

Acknowledgments. This work is partially funded by the ERDF – European Regional Development Fund – through the COMPETE 2020 Programme within project POCI-01-0145-FEDER-006961, and by National Funds through the FCT – Fundação para a Ciência e a Tecnologia (Portuguese Foundation for Science and Technology) – as part of project UID/EEA/50014/2013.

References

1. Bennedsen, J., Caspersen, M.E.: Failure rates in introductory programming. SIGCSE Bull. **39**(2), 32–36 (2007)
2. Dagiene, V., Skupiene, J.: Learning by competitions: olympiads in informatics as a tool for training high-grade skills in programming. In: ITRE 2004, 2nd International Conference Information Technology: Research and Education, pp. 79–83 (2004)
3. Davis, F.D.: Perceived usefulness, perceived ease of use, and user acceptance of information technology. MIS Q. 319–340 (1989)
4. Dietrich, J., Tandler, J., Sui, L., Meyer, M.: The primegame revolutions: a cloud-based collaborative environment for teaching introductory programming. In: Proceedings of the ASWEC 2015, 24th Australasian Software Engineering Conference, ASWEC 2015, vol. 2, pp. 8–12. ACM, New York, NY, USA (2015). http://doi.acm.org/10.1145/2811681.2811683

5. Göbel, S., Salvatore, L., Konrad, R.A., Mehm, F.: Storytec: a digital storytelling platform for the authoring and experiencing of interactive and non-linear stories. In: Spierling, U., Szilas, N. (eds.) Interactive Storytelling, pp. 325–328. Springer, Berlin (2008)
6. Han, K.W., Lee, E., Lee, Y.: The impact of a peer-learning agent based on pair programming in a programming course. IEEE Trans. Educ. **53**(2), 318–327 (2010)
7. Ibáñez, M.B., Di-Serio, A., Delgado-Kloos, C.: Gamification for engaging computer science students in learning activities: a case study. IEEE Trans. Learn. Technol. **7**(3), 291–301 (2014)
8. Kelleher, C., Pausch, R.F.: Using storytelling to motivate programming. Commun. ACM **50**, 58–64 (2007)
9. Kohn, A.: No Contest: The Case Against Competition. Houghton Mifflin Harcourt (1992)
10. Kohn, A.: Why Incentive Plans Cannot Work. Houghton Mifflin Company, Boston (1993)
11. Kölling, M., Henriksen, P.: Game programming in introductory courses with direct state manipulation. In: Proceedings of the 10th Annual SIGCSE Conference on Innovation and Technology in Computer Science Education, ITiCSE 2005, pp. 59–63. ACM, New York, NY, USA (2005). http://doi.acm.org/10.1145/1067445. 1067465
12. Leal, J.P., Silva, F.: Mooshak: a web-based multi-site programming contest system. Softw. Pract. Exp. **33**(6), 567–581 (2003)
13. Leal, J.P., Silva, F.: Using Mooshak as a competitive learning tool. In: The 2008 Competitive Learning Symposium (2008)
14. Liu, P.L.: Using open-source robocode as a java programming assignment. SIGCSE Bull. **40**(4), 63–67 (2008). http://doi.acm.org/10.1145/1473195.1473222
15. Lykke, M., Coto, M., Mora, S., Vandel, N., Jantzen, C.: Motivating programming students by problem based learning and lego robots. In: 2014 IEEE Global Engineering Education Conference (EDUCON), pp. 544–555 (2014)
16. Morris, C.L., Silberman, G.M.: Programming contests in academic environments. In: fie, pp. F1F7–7. IEEE (2003)
17. Nuutila, E., Törmä, S., Malmi, L.: PBL and computer programming-The seven steps method with adaptations. Comput. Sci. Educ. **15**(2), 123–142 (2005)
18. Paiva, J.C., Leal, J.P., Queirós, R.A.: Enki: a pedagogical services aggregator for learning programming languages. In: Proceedings of the 2016 ACM Conference on Innovation and Technology in Computer Science Education, pp. 332–337. ACM (2016)
19. Ryan, R.M., Deci, E.L.: Intrinsic and extrinsic motivations: classic definitions and new directions. Contemp. Educ. Psychol. **25**(1), 54–67 (2000)
20. Torrente, J., del Blanco, Á., Marchiori, E.J., Moreno-Ger, P., Fernndez-Manjn, B.: <e-adventure>: introducing educational games in the learning process. In: IEEE EDUCON 2010 Conference, pp. 1121–1126 (2010)
21. Utomo, A.Y., Amriani, A., Aji, A.F., Wahidah, F.R.N., Junus, K.M.: Gamified e-learning model based on community of inquiry. In: 2014 International Conference on Advanced Computer Science and Information System, pp. 474–480 (2014)

sCool - Game Based Learning in STEM Education: A Case Study in Secondary Education

Alexander Steinmaurer, Johanna Pirker, and Christian Gütl$^{(\boxtimes)}$

Graz University of Technology, Graz, Austria
a.steinmaurer@student.tugraz.at, {j.pirker,c.guetl}@tugraz.at

Abstract. The game sCool [10] is a game-based tool to support learners in STEM education. It is a multi-platform mobile learning game designed for school children supporting a flexible and easy integration of various subjects and different courses. Based on the tool sCool, we present a first integrated course, which encourages children to learn computational thinking and coding in Python in a playful way. The course consists of an exploration mode and a practical mode. Thus, first, the students learn different concepts in an exploratory way and then apply the acquired knowledge in a practical mode through small programming tasks. In this paper, we present a practical approach of the tool in a classroom experience. We conducted an experiment with a group of students and evaluated different dimensions, i.e. aspects of motivation, engagement, emotions, and gender-specific issues. In this first evaluation, we found that the learners were highly motivated and encouraged to learn more about programming. It showed that both male and female participants achieved about the same performance in the game. However, results also suggests that the participants had problems to transfer the learned concepts into similar problems.

Keywords: STEM · Game Based Learning · Computational thinking

1 Introduction

Despite the importance of STEM education, there is a lack of students' interest of STEM topics including computational thinking [8]. However, in today's school education, the ability of computational thinking plays a vital role. According to Wing, computational thinking *"involves solving problems, designing systems, and understanding human behavior, by drawing on the concepts fundamental to computer science"* [17].

There are many different approaches from organizations as well as politics, to increase the interest in STEM education. A leading organization in this field is code.org[1] which tries to encourage children to start coding through activities

---

[1] https://code.org/.

© Springer Nature Switzerland AG 2020
M. E. Auer and T. Tsiatsos (Eds.): ICL 2018, AISC 916, pp. 614–625, 2020.
https://doi.org/10.1007/978-3-030-11932-4_58

like the *Hour of Code*. The organization has famous supporters like Bill Gates, Mark Zuckerberg, or Barack Obama. There are also organizations in Europe that support STEM education like SCIENTIX[2] or STEM Alliance[3] that run different projects and campaigns. Furthermore, there are many national programs like the *Austrian Computer Society* that focuses on encouraging computer science and computational thinking; e.g they organize events like a coding contest (*"Biber der Informatik"*).[4] The need for an improved support for STEM learners has been identified and several programs as well as tools are implemented to support them. The current generation of learners, the "digital learners", however, tend to prefer digital tools, mobile tools, and wants to learn in a playful way [2]. In this paper, we present a tool to support this generation of learners and also support teachers in creating content for this tool.

In the remainder of this paper, we present the following contributions:

- Introduce the mobile game sCool [10] in an educational context and evaluate its practical usage.
- Conduct an evaluation with a group of 18 participants to find advantages and disadvantages of using sCool in STEM learning.
- Discuss the findings of the experiment to improve the tool in an educational way.

2   Related Work

The term "educational games" describes games that have the purpose to educate the player and impart the educational content in an explicit way [14]. The acceptance of video games in a classroom in an educational context is rising but there are also critical voices concerning the educational overvalue [4] and a gap between "relevant" and "irrelevant" knowledge [16]. Game Based Learning (GBL) is a combination of four aspects: curricular knowledge practice, pedagogical knowledge practice, scenario-based knowledge practice, and everyday knowledge practice [16]. The benefit of game-based learning is that students can learn in a personal and safe environment where they are also allowed to make mistakes and are provided with feedback [15]. GBL is also a fully student-centered possibility of learning [13]. The problem in the traditional approach is that students often face the problem that they don't understand theoretical concepts and don't know how to start [12]. In this case, game-based approaches can help to make these concepts more understandable.

The present generation of students in secondary school, know as *"digital natives"*, is growing up with various digital devices embedded in their everyday life. This generation is also connected to each other via mobile technology that plays an essential role in the daily routine. These technologies have a major role to unfold the possibilities in mobile learning [7]. The benefit of those technologies

---

[2] http://scientix.eu/.

[3] http://www.stemalliance.eu/.

[4] http://www.ocg.at/biber/.

is that learning isn't depending on a specific time or a certain location. In this way, an independent learning process can be increased and personal learning promoted [1]. Through mobile game-based learning the students receive feedback when they complete a task and thereby the motivation to reach higher levels will increase [5].

There are various game-based approaches to introduce programming and engage students in a playful way: CodeMonkey[5] is a game-based learning way where students can learn to code with CoffeeScript.[6] In different lessons they learn many concepts like conditions, loops, variables, or functions by helping a monkey on its journey. Educators are able to trace the learning progress of the students in a dashboard. Another web-based approach is Code Combat.[7] In this role-playing game the players can learn different concepts like objects, strings, functions, loops or variables in programming languages like Python[8] or JavaScript[9] by controlling a character and defeat enemies. It also provides a teacher dashboard to get a detailed progress over the students achievements. Lightbot[10] is a mobile game to learn programming concepts. Its approach is to use symbols instead of particular programming languages to teach coding. The players' goal is to navigate a robot through a maze by using different commands (walk, turn, jump) and concepts (loops, procedures).

While most of the described applications and tools only support game-based learning on PCs, sCool is mainly designed for mobile devices. This makes it in particular attractive and engaging for the new generation of digital learners. Additionally, most educational games only provide predefined learning content, which cannot be adapted to the learners or teachers needs. sCool is designed to be adaptive and flexible and allows teachers to add and edit learning content. The mobile game also allows the students to play anywhere and anytime - so it doesn't depend on a certain learning situation. In the next section, we describe the different parts of sCool in more detail.

3   sCool - A Practical Approach

sCool [10] - a project initiated in a collaboration between Graz University of Technology and Westminster University - is a pedagogical tool to make STEM education more engaging and is designed for students between 10–20 years old. The Unity-based mobile video game is a multi-platform learning game than can be used for different subjects and courses. In the current version, the children can learn computational thinking and coding with Python in a playful way. It is highly adaptive, so educators can create different courses and define content for

---

[5] https://www.playcodemonkey.com/.
[6] https://coffeescript.org/.
[7] https://codecombat.com/.
[8] https://www.python.org/.
[9] https://www.javascript.com/.
[10] http://lightbot.com/.

various subjects through a web platform. The web application is also responsible for game analytics so teachers have a clear overview of the students progress (Fig. 1).

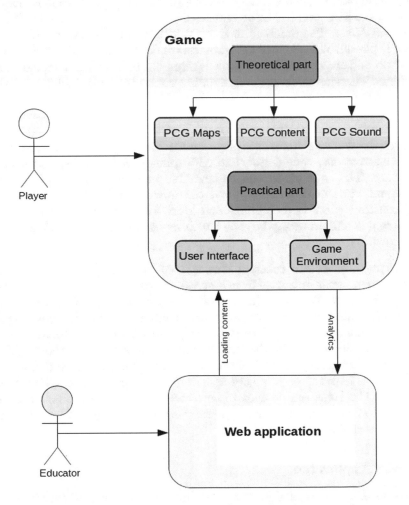

Fig. 1. Conceptual Architecture of sCool (Adapted after [11])

3.1 Mobile Game

The game typically consists of two different modes [10] (Figs. 2 and 3):

– *Exploration Mode*: In this mode students can explore a hostile planet, where they have to collect different pieces of information. The game maps and the playground are generated based on procedural content generation (PCG) algorithms, so that they look differently every time, which makes the game replayable. To enable an appropriate level of difficulty for the players' skills,

dynamic game balancing (DGB) techniques are used. According to the predefined difficulty level the players have to defeat the enemies and collect all disks.
After successfully finishing a level the specified learning content appears and
the players have to answer a single choice question. If the question is answered
wrong the whole level has to be repeated.

- *Practical Mode*: The practical mode provides a squared chessboard where a
  robot is placed. The aim is to navigate the robot over the playing field to reach
  the disks and avoid obstacles. The navigation has to be coded in Python by the
  students. Therefore different programming concepts like variables, conditions,
  and loops are needed to control the robot. The game environment provides a
  Python interpreter (IronPython[11]), where all commands can be executed. In
  case of syntax errors the players receive an error message from the interpreter.
  Besides the game environment (playground) the UI consists of three different
  tabs (instructions, code editor, and code panel), code blocks, and a virtual
  keyboard. The code blocks can be classified in four different sections: print
  command, variable declaration, move commands (left, right, down, and up),
  and control structures (conditions and loops). Players can control the robot
  via dragging and dropping the code blocks or writing the code with the virtual
  keyboard.

Both play modes are coherent through an underlying story. The players
received an emergency message from a space exploration team that has crashed
on a foreign planet. With the aid of a robot, the students should go on a rescue
mission and collect all lost disks, that are necessary to run the space shuttle
again. Since some disks are damaged, the players have to complete the code
on the disks and test it. In the exploration mode they learn about different
programming concepts that will help them to solve the coding examples.

After each task the players gain experience and for each achievement they
earn coins. The coins can be used to customize and extend the own avatar. It
is possible to buy different items, to make the robot stronger and more difficult
levels can be challenged.

3.2   Web Application

To provide new educational content, teachers can create new courses or edit
existing courses in the web application [10]. Each new skill or concept represents
a theoretical or practical task in the course. The content and the degree of
difficulty can be set in the web application and based on the input the map is
generated. On the web platform there is also the distinction between theoretical
and practical tasks. In the theoretical part educators can define learning content
and the corresponding single choice question for the players. Additionally, the
difficulty level for the exploration mode is set, so the map is generated according
to that value. For the practical tasks, educators can define the goal of the task

---

[11] http://ironpython.net/.

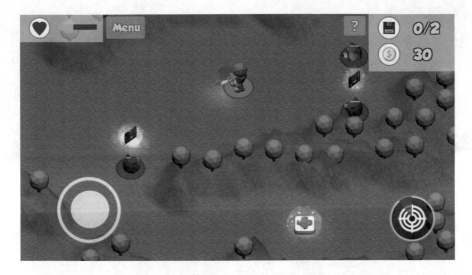

Fig. 2. Explorative Mode: Collect all pieces of information

Fig. 3. Practical Mode: The player has to navigate the robot over the field

and its reference output. This solution represents the expected output of the Python code to successfully finish the level. In this section it is also possible to define what kind of blocks can be used in the level.

The web platform also provides an assessment and analytics tool for educators to analyze the learning process of the students and get a detailed evaluation about the course. The mobile app communicates with the server via a JSON api. It loads the provided course contents and it also sends user-related data for analyzing purposes (Fig. 4).

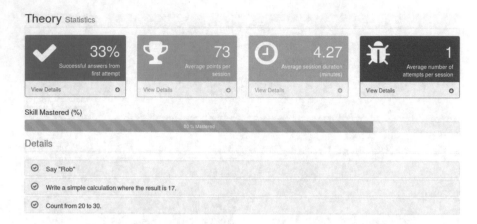

Fig. 4. Gamer Overview in the Web Platform

4   Learner Study in Secondary Education

The fundamental aim of the study was to evaluate the mobile game in its practical usage for classroom education. The main interest was to figure out the students motivation regarding the mobile game and STEM-learning, the students engagement, and also the learned concepts. Another considerable aspect was the girls points of view relating to the video game. The gained results provide the basis for an improved game version and improved learning experiences. Therefore, we tried to find out if the game can help to teach coding in a practical way with real world examples and how we should improve the courses respectively the game to design it better for learning. We focused on the motivation [6], the engagement [3], the emotions [9] and gender-specific issues. In groups of two students, they had 50 min to play the game. Therefore, a certain course was created and the children got additional learning materials. The purpose was that the students were able to solve a simple problem with basic programming concepts (variables and loops).

Participants. The participants were 18 students (11 girls and 7 boys) in the age between 12 and 14 years (M = 12.72; SD = 0.73) in an Austrian Secondary School. The class had no previous experience in computer science or programming.

Procedure. The whole experiment lasted 100 min (two lessons). At the beginning we introduced the mobile game. The next step was to build groups of two (each group should have an android device) and install the app via a provided link. Once the game was installed on each device, we handed out a paper with the first instructions: The students had to accomplish five given *Planet Exploration Missions* (theoretical part) and afterwards they had to solve three *Robot Missions*. Each level was based on the previous one and the whole course had a coherent story, where they had to help a robot collecting all peaces of information to escape from the foreign planet. The groups which finished within 50 min where handed out a worksheet with a simple problem that build upon the courses contents. The task was to print the three times table (3 to 30) within a single print command using loops. After 50 min the class had to fill out the evaluation forms. At the end of the lesson we asked the students about improvements and their personal opinion about the game.

Methodology and Instruments. After the practical part, the students were asked to fill-out a questionnaire about demographic information and their experience with mobile games and game-based learning. The questionnaire also included the Game Engagement Questionnaire (GEQ) [3], the Situational Motivation Scale (SIMS) [6], the Computer Emotional Scale (CES) [9], and some open-ended questions about the game and the game experience. A particular field of interest was the girls' play experience. Therefore the female participants got additional questions about the game mode. They also had a free text question for individual response.

Situational Motivation Scale. The Situational Motivation Scale [6] is a questionnaire with 16 questions and four parameters: intrinsic motivation, identified regulation, external regulation, and amotivation. Each question is assigned a different dimensions. They are rated on a Likert scale between 1 to 7.

Computer Emotional Scale. The Computer Emotional Scale [9] is a survey with 12 different items that are associated with four basic emotions (happiness, sadness, anxiety, and anger). On a Likert scale between 1 to 4 the participants have to report how they felt while playing the mobile game.

Game Engagement Questionnaire. The Game Engagement Questionnaire (GEQ) [3] consists of 19 items that can be assigned to the players engagement when playing the game. The questionnaire distinguishes between four different dimensions: immersion, presence, flow, and absorption. These states can be achieved by a Likert scale between 1 to 5.

Worksheet. To figure out if the students can assign the learned concepts to other given problems we prepared a worksheet. The requirement for this exercise was to complete all theoretical and practical missions in the mobile game, so all learned concepts are known. The worksheet was embedded in the whole game story where the students have to help the robot to escape from the planet. Therefore they had to fix the broken "calculation module" by writing a peace of code that counts from 3 to 30 in steps of 3. The objective of this exercise was to find out if the students can transfer the concepts of loops and variables from the domain of space mission to a mathematical field. They were provided with an additional hint about the concepts to be used (Fig. 5).

Results. Most of the students (66.67%) totally agreed to the statement *sCool encouraged me to learn more about programming*. Another 55.56% totally agreed that they *learned something while playing the game*. During the experiment the students were highly motivated and they worked very concentrated on the exercises. It also could be observed that they liked the theoretical part of the game most due to the exploratory character of the game.

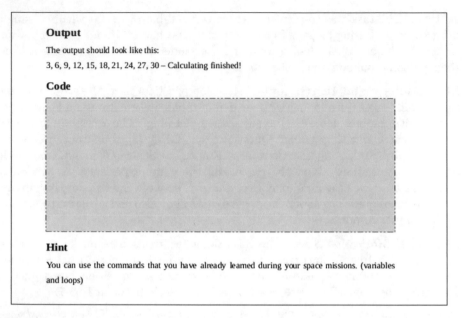

Output

The output should look like this:

3, 6, 9, 12, 15, 18, 21, 24, 27, 30 – Calculating finished!

Code

Hint

You can use the commands that you have already learned during your space missions. (variables and loops)

Fig. 5. Given Worksheet with Coding Task

After playing the game the students rated the immersion (M = 3.78; SD = 1.55) and the absorption (M = 3.78; SD = 1.47) quite high that means that they were highly engaged while playing sCool. The participants also rated the level of flow (M = 3.53; SD = 1.5) and presence (M = 3.04; SD = 1.32). Concerning motivation the group had a high level on intrinsic motivation (M = 5.72; SD = 1.27) and a low level of amotivation (M = 2.28; SD = 1.75). Considering that the experiment took place in a regular school lesson the dimension of external regulation is quite less (M = 2.86; SD = 2.11) compared to the intrinsic motivation. The game is also highly related with happiness (M = 3.39; SD = 0.63). There are very weak negative emotions about sCool: anxiety (M = 1.36; SD = 0.64), anger(M = 1.35; SD = 0.76) and sadness (M = 1.22; SD = 0.49).

All groups were able to pass the theoretical game tasks. The average duration for all groups to pass a theoretical level and answer the question right was 1.18 min. Just 33.39% of the theoretical tasks were passed the first time. The problem associated therewith was that the students did not read the learning concepts the first time and thus the questions could not be answered correct. This problem could be observed in each group at the beginning of the experiment. The working time on the practical examples was much higher, the average duration was 5.57 min for each task. 7 out of 9 groups were able to pass the second level (calculation with variables) and just 3 groups were able to complete all practical tasks within the given time. These three groups also tried to solve the worksheet with the programming task in about ten minutes. At the end of the working time no group had a correct solution, just one group had a broad idea how the task could be solved. The groups mentioned that they were not able to transform the concepts into other fields and they could not remember the commands. They also had some issues understanding loops.

The students stated that they liked the Planet Exploration best because the kind of game was more interesting for them. They also told that the Robot Missions were too hard and the user interface was to complicated: *"The steering should be easier. The programming examples in the Robot Missions were to complicated."* Only 27.78% totally agreed that the game control was easy. It could be observed that most groups had problems with dragging and dropping the commands into the editor.

Girls Perception. The girls were very interested in the game and 9 out of 11 totally agreed that they enjoyed playing the game. 72.72% totally agreed that they *liked the kind of game*. The girls also stated that they liked the theoretical part best. According on the average time for each the girls (M = 1.01 min) had a similar performance to the boys (M = 0.91 min) on the theoretical missions. In the practical missions, the girls average duration per level (M = 4.75 min; boys M = 4.58 min) was even when better compared with the theoretical missions.

5   Conclusions and Future Work

In this paper, we have introduced the educational mobile game sCool and presented a first experiment to introduce and evaluate the tool in a classroom experience. The game is designed as a client-server-based system and is highly adaptive. All learning contents are retrieved from the server. The client also sends user-based informations about the achievements to the API. In this way it is possible to gain necessary information about the students learning progress and habits.

The goal of the experiment was to figure out how the mobile learning game application *sCool* can be used in an educational context and how it improves the students' computational thinking skills. To evaluate sCool in a classroom experience, a school lesson was designed in a highly student-centered way so that children could play the game and solve the given tasks. Another essential point of view in the experiment was the girls' perception of the game.

We found that the students enjoyed playing the game. They were highly motivated and encouraged while playing. During the experiment we were able to show that the children liked the exploratory part of the game most. However, they focused more on the game-play instead of the learning content. Thus, in the end no group was able to solve the final work sheet. Therefore it is necessary to rethink the way new concepts are introduced and taught in both modes. It showed that the educators need to have a wide overview about the students learning progress. The advantage of sCool is, that it provides an overview of these progresses and its possible to adopt the learning concepts. Thus, courses can be created according to the needs of the students to reach a highly individualized learning experience.

As future work, we will adapt the lessons design according to our findings and rerun the study. In a second step, UI elements and game mechanics will be adapted according to the findings as well.

Acknowledgement. We would like to thank Aleksandar and Milos Kojic from Graz University of Technology as well as Daphne Economou and Markos Mentzelopoulos from Westminster University for their great effort in Designing and developing the basic version of sCool.

References

1. Bartel, A., Hagel, G.: Engaging students with a mobile game-based learning system in university education. In: 2014 IEEE Global Engineering Education Conference (EDUCON), pp. 957–960, April 2014
2. Bennett, S., Maton, K., Kervin, L.: The digital natives debate: a critical review of the evidence. Br. J. Educ. Technol. **39**(5), 775–786 (2008)
3. Brockmyer, J.H., Fox, C.M., Curtiss, K.A., McBroom, E., Burkhart, K.M., Pidruzny, J.N.: The development of the game engagement questionnaire: A measure of engagement in video game-playing. J. Exp. Soc. Psychol. **45**(4), 624–634 (2009)
4. Browne, K., Anand, C.: Gamification and serious game approaches for introductory computer science tablet software. In: Proceedings of the First International Conference on Gameful Design, Research, and Applications. Gamification 2013. pp. 50–57. ACM , New York, NY, USA (2013)
5. Cahyana, U., Paristiowati, M., Savitri, D.A., Hasyrin, S.N.: Developing and application of mobile game based learning (m-gbl) for high school students performance in chemistry. Eurasia J. Math. Sci. Technol. Educ. **13**(10), 7037–7047 (2017)
6. Guay, F., Vallerand, R.J., Blanchard, C.: On the assessment of situational intrinsic and extrinsic motivation: the situational motivation scale (sims). Motivation Emot. **24**(3), 175–213 (2000)
7. Huizenga, J., Admiraal, W., Akkerman, S., Dam, G.T.: Mobile gamebased learning in secondary education: engagement, motivation and learning in a mobile city game. J. Comput. Assist. Learni. **25**(4), 332–344 (2009)
8. Joyce, A.: Stimulating interest in stem careers among students in Europe: supporting career choice and giving a more realistic view of stem at work, Education and Employers Taskforce (2014)
9. Kay, R.H., Loverock, S.: Assessing emotions related to learning new software: the computer emotion scale. Comput. Hum. Behav. **24**(4), 1605–1623 (2008). Including the Special Issue: Integration of Human Factors in Networked Computing
10. Kojic, A., Kojic, M., Pirker, J., Gütl, C., Mentzelopoulos, M., Economou, D.: sCool – a mobile flexible learning environment. In: Online Proc. from 4th Immersive Learning Research Network Conference. Missoula, Montana (2018)
11. Kojic, M.: Procedural content generation in a multidisciplinary educational mobile game. Master's thesis, Graz University of Technology (2017)
12. Olsson, M., Mozelius, P.: Game-based learning and game construction as an e-learning strategy in programming education. In: XI International GUIDE Conference and IX International Edtech Ikasnabar Congress E-Learning 2016: New Strategies and Trends, Madrid, Spain, June 22–24, 2016. GUIDE Association. Global Universities in Distance Education

13. Shabalina, O., Malliarakis, C., Tomos, F., Mozelius, P.: Game-based learning for learning to program: from learning through play to learning through game development (10 2017)
14. Stege, L., Lankveld, G., Spronck, P.: Teaching electrical engineering with a serious game. In: Wiemeyer, J., Göbel, S. (eds.) Serious Games - Theory, Technology & Practice: Proceedings of the GameDays 2011, pp. 29–39. TU Darmstadt, Darmstadt, Germany (Presented at the GameDays 2011 conference) (2011)
15. Suglanto, N., Wiradinata, T.: An implementation of game-based learning using Alice programming environment. **20** 18.1 (02 2012)
16. Terracina, A., Mecella, M., Berta, R., Fabiani, F., Litardi, D.: Game@school. teaching through gaming and mobile-based tutoring systems. In: Poppe, R., Meyer, J.J., Veltkamp, R., Dastani, M. (eds.) Intelligent Technologies for Interactive Entertainment, pp. 34–44. Springer International Publishing, Cham (2017)
17. Wing, J.M.: Computational thinking. Commun. ACM **49**(3), 33–35 (2006)

Evaluation Methods for the Effective Assessment of Simulation Games

A Literature Review

Nilüfer Bas[1], Alexander Löffler[1]([⊠]), Robert Heininger[1],
Matthias Utesch[2], and Helmut Krcmar[1]

[1] Technical University of Munich, Munich, Germany
{nilufer.bas,alexander.loeffler,robert.heininger,
krcmar}@in.tum.de
[2] Staatliche Fachober- und Berufsoberschule Technik München, Munich,
Germany
utesch@in.tum.de

Abstract. Simulation games play an important role in the area of technology-based education. They allow the simulation of real-world problems in a risk-free environment and thereby intend to increase the learning experience of students. However, assessing the effectiveness of a simulation game is necessary to optimize elements of the game and increase the learning effect for students. For this, different evaluation methods exist, which do not always cover all phases when running a simulation game. In this study, we conduct a literature review to analyze evaluation methods for the pre-game, in-game, and post-game assessment of simulation games. In accordance with our inclusion criteria, we selected 31 peer-reviewed articles, and categorized them according to a didactic framework that describes four phases of running simulation games: Preparation, Introduction, Interaction and Conclusion phase. Based on the results, we provide a concrete evaluation strategy that can be used to assess simulation games throughout all phases. This study contributes to theory by providing an overview of evaluation methods for the assessment of simulation games within the different game phases. It contributes to practice by providing a concrete evaluation strategy that can be adapted and used to assess simulation games.

Keywords: Simulation games · Serious games · Game-based learning ·
Evaluation · Assessment

1 Introduction

For decades, simulations play an important role in the area of technology-based education. In 1956, the first largely known simulation game "Top Management Decision Simulation" was developed by the American Management Association for utilizing it in management seminars [1]. Simulation games not only allow replicating real-world problems, but also increasing the applicability of the acquired knowledge of real-world situations. One of the dimensions of these games is "Realism", which is defined as the game-users' perception on the degree of the simulation game's reflection of a life

© Springer Nature Switzerland AG 2020
M. E. Auer and T. Tsiatsos (Eds.): ICL 2018, AISC 916, pp. 626–637, 2020.
https://doi.org/10.1007/978-3-030-11932-4_59

situation. The more realistic a simulation game is, the higher is the degree that gamers learn from it [2]. Playful learning is a current trend in education which focuses on the hands-on practice of learning instead of on the sit-and-listen approach, spanning between free play and guided play [3]. In general, the term "playful learning" covers simulation games, business games, serious games and game-based learning. Among these, especially simulation games trigger experiential learning through engaging the learners or gamers in a dynamic experience in which problems are presented that have to be resolved or decisions have to be taken [4].

As one of the most important part of technology-based education, games and simulations are linked with constructivist pedagogy, which allows to get practical experience in content areas such as marketing, finance, management, or languages [5]. Especially, simulation games focusing on business management education has gained more importance, which guide student gamers to learn business skills while managing a company in teams [6]. In addition to empowering experience in these content areas, simulation games facilitate learning through discovery, experimentation, and practice with concrete examples in a risk-free environment [7]. Student's time management and team-work skills improves significantly through simulation games [11]. Moreover, student gamers learn how to make decisions, conserve past decisions critically, work in a team cooperatively, and manage time needed effectively. Overall, the mentioned skills become more and more important in a constantly changing global society. In their report, Hoberg et al. [8] surveyed companies about the required digital competencies. Only 17% of the companies agreed on the statement "*we have enough personnel with the skills necessary for the digital transformation of our company*", while 53% disagree. A competency model developed by Prifti et al. [9] for the future workforce in the digitized world lists eight important skills, which are "*leading and deciding, supporting and cooperating, interacting and presenting, analyzing and interpreting, creating and conceptualizing, organizing and executing, adapting and coping, enterprising and performing*". When the business world sounds so complex and requires a wide range of social and management skills, traditional learning methods such as reading materials, listening to lectures or notetaking are not enough to prepare students' for the modern business environment [10]. Technologically enhanced classroom teaching can prepare future employees best and equip them with the expected skills through simulation games. For instance, implementing a simulation game into a classroom and supporting learning with a scenario takes the learners beyond the traditional learning environment and makes them experience working with new digital technologies [11]. The purpose of simulation games are not only to be fun and entertaining, but also to be educational. That is why students prefer simulation games over other classroom activities [12].

Teaching the mentioned skills and competencies through simulation games as well as assessing their usefulness carry great importance to optimize elements of the game according to the student's learning style and pace, their performance, and motivational level to learn. Bellotti et al. [12] highlights the necessity to improve the design of simulation games as they lack accurate assessment. Furthermore, the assessment of students' performance is important because simulation games intend to increase the learning progress and learning outcomes and therefore need to be evaluated. Hence, the motivation of this contribution is to have a close examination of the literature on the assessment of simulation games and to bring to light which assessment types are used during the different phases of simulation games. Educational institutions, lecturers, and teachers need effective assessment methods and instruments to evaluate the usefulness of simulation games as well as their students' readiness, motivation, and learning outcomes. Aiming to analyze the assessment types used in simulation games as well as effective application of evaluation methods to these games, we pose the following research questions:

1. What are the existing evaluation methods aiming to assess simulation games and what are their success factors?
2. How can we apply evaluation methods to assess the success of simulation games effectively?

2 Related Work

Exploring from the point of view of each participant, what has occurred during the simulation games is fundamental for learning [4]. As highlighted above, simulation games lack accurate assessment methods and improving these games is a necessity [12]. Coming from the literature, which highlights the importance of evaluating students' learning processes as well as the games themselves, we describe the overview of the didactic framework of a simulation game process in the following section.

2.1 Didactic Framework

A didactic framework developed by Utesch [13] illustrates the flow of business games in four phases. The first phase called "Preparation" aims to manage organizational conditions needed to operate business games, during which participants are informed about the objectives of the course. This phase requires careful planning for creating a successful experiential learning atmosphere for the learners. Following is the "Introduction" phase, in which the students become familiar with the roles, management boards and the problems to solve in the game. In the "Interaction Phase", the participants face challenging tasks that they have to solve [13]. This phase consists of five sub-steps: analyzing the problem, developing a business strategy, implementing a business strategy, running the simulation and presenting the results [13]. Finally, there

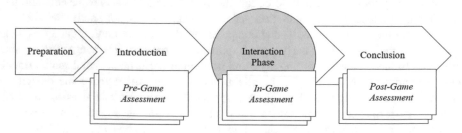

Fig. 1. Applied didactic framework for the assessment of simulation games (based on [13])

is the "Conclusion" phase, in which the achieved business objectives, the applied strategies and criterions to improve a company's success are summarized (Fig. 1).

This didactic framework inspired us to build our literature review of assessment types based on the different phases in simulation games. Therefore, we divided the assessment types into pre-game, in-game and post-game assessment consecutively. The identified assessment types from our literature review help us to answer the first research question as well as extend this didactic framework with evaluation methods for each phase. In the second research question, we will focus on the effective transfer and application process of the assessment types into a simulation game.

2.2 Assessment Types

In general, there are two types of assessments: summative and formative. Summative assessments measure and summarize students' learning and achievements by using certification for school completion [14]. Formative assessment refers to regular, interactive assessment to identify learning needs and adjust future teachings accordingly [14]. Formative assessment can help teachers to measure learning outcomes and make quick adjustments to improve them. However, assessments that are done after learning sessions in a game-based environment might miss important changes during the learning process, because an end of game assessment lacks immediate feedback during the game [15]. This type of in-game assessment is called "embedded" or "stealth assessment". Stealth assessment is an evidence-based process by which assessment can be integrated directly to the learning environments [16].

Ifenthaler et al. [15] divided types of assessments in game-based learning into three categories: game scoring, external assessment, and embedded assessment. Game scoring focuses on targets achieved, obstacles or time needed to complete a task or an iteration. Secondly, external assessment includes debriefing interviews, tests or surveys. Lastly, embedded or internal assessment gives information about the learner's behavior such as clickstreams or log files [15]. Later in this contribution, we will use the different types of rating to categorize valuation methods according to the four phases of the simulation games, as previously shown.

3 Methodology

In order to analyze the different assessment types for simulation games, we conducted a systematic literature review of empirical studies following the methodology of Webster and Watson [17] and vom Brocke et al. [18]. For our search, we chose the IEEE Xplore and the ERIC database, as they cover publications from Information Systems (IS), Economics, Computer Science, and Engineering, and furthermore have a strong focus on Education outlets. In order to analyze different kinds of assessment types for simulation games, we used the keywords "simulation games" or "business games" or "serious games" or "game-based learning", combined with "evaluation" or "assessment". Overall, as recommended by vom Brocke et al. [18], we conducted a representative literature review, meaning that we analyzed a broad number of articles in order to get a representative view on the different assessment types used to evaluate simulation games. Our search included all articles published in the mentioned databases until April 2018. We initially screened all hits based on the keywords, title and abstract, and filtered them according to the criteria whether they were related to simulation games and assessment types. Afterwards, we examined the remaining articles in detail, following the criteria whether they explained or applied a concrete assessment type or evaluation strategy on a simulation game from IS, Economics, Computer Science, or Engineering.

In total, we had 344 hits in the IEEE Xplore database, which lead to 38 articles after the first review and 24 in the final selection. In the ERIC database, we had 672 hits, which lead to 18 articles after the first review and seven in the final selection. In total, this resulted in 31 articles that were finally selected for a detailed analysis. For each paper, we analyzed whether they conducted any kind of pre-game, in-game, or post-game assessment types during simulation games. Afterwards, we categorized these results according to the three phases mentioned in the framework. According to Webster and Watson [17], we finally created a concept matrix that illustrates the results, which are explained in the following section.

4   Results

Based on 31 articles, a total of 37 assessment types could be identified. Table 1 illustrates all 37 types in clusters of pre-game, in-game, and post-game assessment types. Overall, our literature search lead to six types for pre-game assessment, 15 for in-game assessment, and 16 for post-game assessment.

In general, most of the authors focus on collecting data through questionnaires [12, 21, 28, 30–35]. The literature further highlights the importance of in-game assessment as only this type can provide instant feedback about the learning process [19]. For this, several authors provide concrete examples, such as counting implanted errors in the game [20], performance tracking [21], monitoring students' progress [22], Think-Aloud protocols [23, 24], field notes during classroom observations, heuristic participatory evaluation methodology, and storyboarding [25]. Besides the articles that only mention assessment types for one or two phases of a simulation game, seven of them use all three-assessment types in their studies.

Duin et al. [26] uses a serious game scenario of the production process of coffee machines to develop competencies in the domain of sustainable manufacturing. In pre-game and post-game assessments, questionnaires are used to measure the students' general and scenario related knowledge and perspectives. During the in-game phase, self-assessments are used to measure three competence performances of the students: "the ability to perform a life-cycle assessment, information gathering, and decision making".

The case study from Wilson et al. [27] aims to test the serious game "Macbeth", in which players in a fictional environment gather and analyze intelligence data to prevent simulated attacks. This study adopts an iterative evaluation methodology, which uses many qualitative assessment methodologies, such as focus groups, interviews, and one-on-one playtests. In the post-game phase, a follow-up test measures the learning outcomes of the participants.

Table 1. Assessment types for simulation games in the literature

Pre-game assessments	In-game assessments	Post-game assessments
Questionnaires [28–31]	Counting errors [20]	Questionnaires [28, 29, 31–36]
Analysis of cognitive improvements [30]	Assessment rules [37]	Data mining algorithms [38]
Demographic information [26, 39]	Performance tracking [21]	Tracking success of the teams [40]
Learner type information [39]	Questionnaire, semi-structured interviews [41]	Evaluation grids [42]
Multiple choice questionnaire (MCQ) [24]	Monitoring students' progress (Formative) [22]	Performance evaluation by observer [43]
Self-assessment of learning questionnaire (SAL) [24]	Self-evaluation by students [26]	Learning assessment tools (LATs) [44]
	Discussions, interviews [45]	Descriptive and causal analysis [46]
	Questionnaire, case study, testing [33]	Quantitative evaluation framework [47]
	Interviews, cognitive labs [23]	Survey, debriefing discussion [48]
	Unobtrusive observation [25]	Score evaluation (Summative) [22]
	Participatory heuristic evaluation methodology [25]	Questionnaire with exam-like questions [49]
	Storyboarding [25]	Online performance tasks [23]
	Think aloud protocols [23, 35]	Knowledge tests, aspect ratings, perceptions and learners preferences [39]
	Instant informative feedback to participants [50]	Self-assessment of learning questionnaire (SAL) [24]
	Field notes during classroom observation [36]	Checklist item assessing of learning outcomes [50]
		Teacher interviews, tests [36]

In Di Cerbo's [23] study on game-based learning, students complete a short pre-test composed of published or released items assessing area concepts. Then, cognitive labs and the Think Aloud methodology are used to assess the progress during the game. Think Aloud is a cognitive task analysis technique during which participants verbalize their experiences and reactions to the gaming experience. Finally, online performance tasks are conducted at the end of the game.

Tan et al. [25] aims to measure participants' social problem solving skills through the game named "Socialdrome". In three of the game phases, they use demographic questionnaires, unobtrusive observation, participatory heuristic evaluation methodology, and storyboarding respectively. During the unobtrusive observation, participants are observed by a research team during game playing and interaction happened only when a technical problem occurred. Participatory heuristic evaluation is a focus group discussion approach during which participants inspect the playability of the game in groups and their responses are recorded and prescribed. Storyboarding is a technique where students express their perspectives and ideas directly through their designs by using templates, paper, crayons, pencil colors, markers [25].

Cowley et al. [24] pilot a stakeholder management scenario in the game called "Target". The learning scenarios engage learners through employing interactive storytelling and evidence-based performance assessment. During simulated learning scenarios, learners take the role of a project manager, and have to show four competences: negotiation, communication, trust building, and taking risks and opportunities. As a pre-game and post-game assessment, Multiple Choice Questionnaires (MCQ) were distributed to the participants. As an in-game assessment, participants gave self-report feedback that aims to assess subjective mood of participants and filled out a Game Experience Questionnaire (GEQ) after every session. GEQ assesses participants subjective game experiences of feelings and thoughts while playing, and the measured attributes are competence, sensory & imaginative immersion, flow, tension, challenge, negative affect and positive affect [24, 51].

Widemann et al. [35] controlled a business simulation game named "Trivia" in their research. As a pre-game assessment, they used a demographic information questionnaire, which popped up in the students' computer. As an in-game assessment, a Think Aloud Protocol was used [35]. Clarity of the instructions and the usefulness of course materials were also measured via questionnaires after finishing the game.

Meerbaum-Salant et al. [36] used the educational visual programming language Scratch. They evaluated students and teachers through a pre-test, post-test, and interim-test. The interim-test is used to evaluate the learning progress of the students and to decide if they are capable of achieving a future rating through good performance [52]. Field notes during classroom observation, teacher interviews, questionnaires, and interim tests were the assessment tools during these three phases of the game.

5   An Evaluation Strategy for Simulation Games

In the previous chapter, we presented the results from analyzing several papers that implemented various qualitative and quantitative assessments into their game-based learning teaching environments. Moreover, we summarized seven papers that showed regularity by implementing assessments into all three phases. Based on these results, we propose an evaluation strategy that can be applied to evaluate simulation games successfully. An overview of the proposed strategy is shown in Fig. 2.

The implications of these results are manifold. First, the focus of research designs has representative characteristics with a variety of assessment types. For instance, during the pre-game phase, questionnaires are seen as the most frequently used efficient evaluation tool. This type of assessment allows feedback from a large number of students, as well as it provides respondents' opinions, attitudes, feelings, and perceptions about particular matter to the evaluator [53]. When educators want to inform

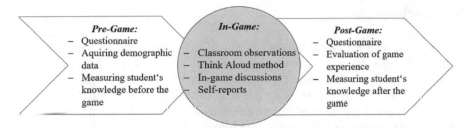

Fig. 2. Evaluation strategy for simulation games

participants about their learning progress, they can use retrospective questionnaires as a pre-game evaluation. This evaluation gives educators an opportunity to identify students' perceptions of their change in knowledge, skills, attitudes and behavior after their participation in an educational intervention [54]. Retrospective questionnaires help learners to look back in their learning journey and question themselves to make improvements with questions, such as "what went well? what have I learned? what still puzzles me?" [55]. Besides performance-based evaluation, lecturers can collect demographic data as well as information about learning styles of students through questionnaires.

As an in-game phase evaluation, classroom observations, Think Aloud protocols, and in-game group discussions can be used to assess the learning progress of players in the heat of the moment. *Classroom observation* or *unobtrusive observation* is a method which gives data to lecturers about the interaction happening in the classroom and helping the players only when a question or problem arises [25]. This method can be useful to get an instant picture about things happening in the classroom. A cognitive task analysis technique such as the *Think Aloud protocol* enables the players to verbalize their thinking processes during the game. Moreover, this technique helps to evaluate decision making processes of players as well as gives concrete explanations about the players' interaction with the user interface and other design elements [35]. Another term of in-game discussions is the participatory heuristic evaluation. Tan et al. [25] focused on measuring the playability of a game through *participatory heuristic evaluation* via asking questions, recording and transcribing answers. As a result, players experienced knowledge gain, skill improvement and fun through careful application of instructional system design and game principles [25]. Sharing the same aim with in-game discussions, self-reports can measure subjective mood of players as well as their experience, performance, and learning [24].

As a post-game assessment, students' knowledge gains and improvements can be measured with questionnaires. Simulation game should be evaluated considering its usefulness and effectiveness for use and questionnaire are the most common evaluation method to assess simulation games [20]. This assessment can also include evaluating the scenarios given in the game, both its quality and its degree of reflecting life situation [2]. Together with initial test in the pre-game phase, one can compare and analyze game players' knowledge gain with post-game *knowledge tests* [39].

6 Conclusion

In this paper, we conducted a literature review to analyze different assessment types for simulation games. We associated three phases of a simulation-game with pre-game, in-game, and post-game assessment types. We excluded the preparation phase because of the lack of enough evidence showing adequate association with pre-game assessment. Nevertheless, the preparation phase carries importance in playful learning environment because good preparation brings success in learning. To reach our goal, we analyzed 31 papers that focus on at least one of the three phases of assessment. Our results show that the majority of the papers only focus on one of the three phases in their assessment of simulation games. Overall, only seven papers mention a complete evaluation throughout all three phases. Furthermore, the literature showed that questionnaires are the most popular method to assess simulation games, as they are easy to conduct and allow evaluating a large number of students simultaneously. However, there exist further methods that provided good results, such as the Think Aloud methodology. Based on these findings, we provided an evaluation strategy for simulation games by presenting suitable assessment types for the pre-game, in-game, and post-game assessment of simulation games.

In conclusion, our findings provide a broad avenue for future research in the area of the development and evaluation of simulation games. On the one side, our literature review and the proposed evaluation strategy can be used for the assessment of future simulation games. Lecturers and researchers have the possibility to either directly apply the proposed assessment types or select different methods out of the overview we provided as part of the search results. On the other side, our work can be used as a basis to conduct more research on evaluation methods and assessment types from a theoretical perspective. For instance, the results of this literature review can be used to develop an evaluation model for simulation games, which includes further concepts of the assessment types considering different contexts, learner types, or environments. Moreover, assessing the efficiency and quality of the games will create a successful playful learning environment, which carries great importance to teach the skills and competencies required in a digitized working environment.

References

1. Hodgetts, R.: Management gaming for didactic purposes: a new look. Simul. Games **1**(1), 55–66 (1970)
2. Faria, A.J., Hutchinson, D., Wellington, W.J., Gold, S.: Developments in business gaming: a review of the past 40 years. Simul. Gaming **40**(4), 464–487 (2009)
3. Lillard, A.S.: Playful learning and Montessori education. Am. J. Play **5**(2), 157 (2013)
4. Thatcher, D.C.: Promoting learning through games and simulations. Simul. Gaming **21**(3), 262–273 (1990)
5. Marquardt, M.J., Kearsley, G.: Technology-Based Learning: Maximizing Human Performance and Corporate Success. CRC Press (1998)

6. Baume, M.: Computerunterstützte Planspiele für das Informationsmanagement: Realitäts-
   nahe und praxisorientierte Ausbildung in der universitären Lehre am Beispiel der CIO-
   Simulation. BoD–Books on Demand (2009)
7. Aldrich, C.: Clark Aldrich's Six Criteria of an Educational Simulation (2004)
8. Hoberg, P., Krcmar, H., Oswald, G., Welz, B.: Skills for Digital Transformation (2016)
9. Prifti, L., Knigge, M., Kienegger, H., Krcmar, H.: A competency model for Industrie 4.0
   employees. In: 13th International Conference on Wirtschaftsinformatik (2017)
10. Riedel, J.C., Hauge, J.B.: State of the art of serious games for business and industry. In: 17th
    International Conference on Concurrent Enterprising (ICE), pp. 1–8 (2011)
11. Löffler, A., Prifti, L., Levkovskyi, B., Utesch, M., Krcmar, H.: Simulation games for the
    digital transformation of business processes. In: IEEE Global Engineering Education
    Conference (EDUCON), Santa Cruz de Tenerife, Canary Islands, Spain (2018)
12. Bellotti, F., Kapralos, B., Lee, K., Moreno-Ger, P., Berta, R.: Assessment in and of serious
    games: an overview. Adv. Hum. Comput. Interact. **2013**, 1 (2013)
13. Utesch, M.C.: A successful approach to study skills: Go4C′ s projects strengthen teamwork.
    Int. J. Eng. Pedagog. (iJEP) **6**(1), 35–43 (2016)
14. OECD, CERI: Assessment for learning formative assessment. In: OECD/CERI International
    Conference on Learning in the 21st Century: Research, Innovation, Policy (2008)
15. Ifenthaler, D., Eseryel, D., Ge, X.: Assessment in Game-Based Learning, pp. 1–8. Springer
    (2012)
16. Shute, V.J.: Stealth assessment in computer-based games to support learning. Comput.
    Games Instr. **55**(2), 503–524 (2011)
17. Webster, J., Watson, R.T.: Analyzing the past to prepare for the future: writing a literature
    review. MIS Q. xiii–xxiii (2002)
18. Vom Brocke, J., Simons, A., Niehaves, B., Riemer, K., Plattfaut, R., Cleven, A.:
    Reconstructing the giant: on the importance of rigour in documenting the literature search
    process. In: European Conference on Information Systems (ECIS), vol. 9, pp. 2206–2217
    (2009)
19. Eseryel, D., Ifenthaler, D., Ge, X.: Alternative assessment strategies for complex problem
    solving in game-based learning environments. In: Multiple Perspectives on Problem Solving
    and Learning in the Digital Age, pp. 159–178 (2011)
20. Costantino, F., Di Gravio, G., Shaban, A., Tronci, M.: A simulation based game approach
    for teaching operations management topics. In: IEEE Winter Simulation Conference (WSC),
    pp. 1–12 (2012)
21. Smyrnaiou, Z., Petropoyloy, E., Menon, S., Zini, V.: From game to guidance: the innovative
    evaluation approach of the P4G simulation business game. In: Mathematics and Computers
    in Sciences and in Industry (MCSI), pp. 148–153 (2017)
22. Callaghan, M., Savin-Baden, M., McShane, N., Eguíluz, A.G.: Mapping learning and game
    mechanics for serious games analysis in engineering education. IEEE Trans. Emerg.
    Top. Comput. **5**(1), 77–83 (2017)
23. DiCerbo, K.E.: Building the evidentiary argument in game-based assessment. J. Appl. Test.
    Technol. **18**(S1), 7–18 (2017)
24. Cowley, B., Fantato, M., Jennett, C., Ruskov, M., Ravaja, N.: Learning when serious:
    psychophysiological evaluation of a technology-enhanced learning game. Educ. Technol.
    Soc. **17**(1), 3–16 (2014)
25. Tan, J.L., Goh, D.H.-L., Ang, R.P., Huan, V.S.: Participatory evaluation of an educational
    game for social skills acquisition. Comput. Educ. **64**, 70–80 (2013)
26. Duin, H., Pourabdollahian, B., Thoben, K.-D., Taisch, M.: On the effectiveness of teaching
    sustainable global manufacturing with serious gaming. In: Engineering, Technology and

Innovation (ICE) & IEEE International Technology Management Conference, pp. 1–8 (2013)

27. Wilson, D.W., et al.: Serious games: an evaluation framework and case study. In: 49th Hawaii International Conference on System Sciences (HICSS), pp. 638–647 (2016)

28. Utesch, M., Heininger, R., Krcmar, H.: Strengthening study skills by using ERPsim as a new tool within the pupils' academy of serious gaming. In: IEEE Global Engineering Education Conference (EDUCON), pp. 592–601 (2016)

29. Utesch, M., Heininger, R., Krcmar. H.: The pupils' academy of serious gaming: strengthening study skills with ERPsim. In: 13th International Conference on Remote Engineering and Virtual Instrumentation (REV), pp. 93–102 (2016)

30. Krassmann, A.L., Paschoal, L.N., Falcade, A., Medina, R.D.: Evaluation of game-based learning approaches through digital serious games in computer science higher education: a systematic mapping. In: 14th Brazilian Symposium on Computer Games and Digital Entertainment (SBGames), pp. 43–51 (2015)

31. Boyle, L., Hancock, F., Seeney, M., Allen, L.: The implementation of team based assessment in serious games. In: Games and Virtual Worlds for Serious Applications, pp. 28–35 (2009)

32. Bhardwaj, J.: Evaluation of the lasting impacts on employability of co-operative serious game-playing by first year computing students: an exploratory analysis. In: Frontiers in Education Conference (FIE), pp. 1–9 (2014)

33. Zolotaryova, I., Plokha, O.: Serious games: evaluation of the learning outcomes. In: 13th International Conference on Modern Problems of Radio Engineering. Telecommunications and Computer Science (TCSET), pp. 858–862 (2016)

34. Abdellatif, A.J., McCollum, B., McMullan, P.: Serious games: quality characteristics evaluation framework and case study. In: Integrated STEM Education Conference (ISEC), pp. 112–119 (2018)

35. Wideman, H.H., Owston, R.D., Brown, C., Kushniruk, A., Ho, F., Pitts, K.C.: Unpacking the potential of educational gaming: a new tool for gaming research. Simul. Gaming 38(1), 10–30 (2007)

36. Meerbaum-Salant, O., Armoni, M., Ben-Ari, M.: Learning computer science concepts with scratch. Comput. Sci. Educ. 23(3), 239–264 (2013)

37. Al-Smadi, M., Wesiak, G., Guetl, C.: Assessment in serious games: an enhanced approach for integrated assessment forms and feedback to support guided learning. In: 15th International Conference on in Interactive Collaborative Learning (ICL), pp. 1–6 (2012)

38. Zeng, L.Y.: An evaluation system of game-based learning based on data mining. In: International Conference on Computer Science and Network Technology (ICCSNT), pp. 1732–1736 (2012)

39. Hainey, T., Connolly, T.: Evaluating games-based learning (2010)

40. Merkuryev, Y., Bikovska, J.: Business simulation game development for education and training in supply chain management. In: Asia Modelling Symposium (AMS), pp. 179–184 (2012)

41. de Carvalho, C.V.: Is game-based learning suitable for engineering education? In: IEEE Global Engineering Education Conference (EDUCON), pp. 1–8 (2012)

42. Boughzala, I., Bououd, I., Michel, H.: Characterization and evaluation of serious games: a perspective of their use in higher education. In: 46th Hawaii International Conference on System Sciences (HICSS), pp. 844–852 (2013)

43. Michel, H.: Characterizing serious games implementation's strategies: is higher education the new playground of serious games? In: 49th Hawaii International Conference on System Sciences (HICSS), pp. 818–826 (2016)

44. Chatterjee, S., Mohanty, A., Bhattacharya, B.: Computer game-based learning and pedagogical contexts: initial findings from a field study. In: IEEE International Conference on Technology for Education (T4E), pp. 109–115 (2011)
45. Mettler, T., Pinto, R.: Serious games as a means for scientific knowledge transfer: a case from engineering management education. IEEE Trans. Eng. Manag. 62(2), 256–265 (2015)
46. Cleophas, C.: Designing serious games for revenue management training and strategy development. In: Proceedings of the Winter Simulation Conference, p. 140 (2012)
47. Escudeiro, P., Escudeiro, N.: Evaluation of serious games in mobile platforms with QEF: QEF (quantitative evaluation framework). In: IEEE 7th International Conference on Wireless, Mobile and Ubiquitous Technology in Education (WMUTE), pp. 268–271 (2012)
48. Tantan, O.C., Lang, D., Boughzala, I.: Learning business process management through serious games: feedbacks on the usage of INNOV8. In: 18th IEEE Conference on Business Informatics (CBI), vol. 1, pp. 248–254 (2016)
49. Yang, M.C., Xu, Z.T., Hsu, L.H.: On developing the learning game for graph theory: a new design model considering the learners' reflexiveness. In: 5th IIAI International Congress on Advanced Applied Informatics (IIAI-AAI), pp. 418–422 (2016)
50. Cutumisu, M., Blair, K.P., Chin, D.B., Schwartz, D.L.: Posterlet: a game-based assessment of children's choices to seek feedback and to revise. J. Learn. Anal. 2(1), 49–71 (2015)
51. IJsselsteijn, W., De Kort, Y., Poels, K.: The Game Experience Questionnaire (2008)
52. The Glossary of Education Reform: Interim Assessment. https://www.edglossary.org/interim-assessment/ (2018)
53. The University of Sheffield: Questionnaires. https://www.sheffield.ac.uk/lets/strategy/resources/evaluate/general/methods-collection/questionnaire (2018)
54. Davis, G.A.: Using a retrospective pre-post questionnaire to determine program impact (2002)
55. Waite, L., Lyons, C.: The 4 questions of a retrospective and why they work. https://www.infoq.com/articles/4-questions-retrospective (2013)

An Online Game for the Digital Electronics Course for Vocational Education and Training (VET) Students

Dimitrios Kotsifakos[✉], George Petrakis, Manthos Stavrou,
and Christos Douligeris

Department of Informatics, University of Piraeus, 50 Karaoli & Dimitriou St.,
18534 Piraeus, GR, Greece
{kotsifakos, cdoulig}@unipi.gr,
george.petrakis@yahoo.com, mstauroy@gmail.com

Abstract. The officially instituted curriculum of the Specialties of Vocational Education Training (VET) in Greece involves a wide set of various teaching models. The aim of these models is to introduce technical and technological education to the students of VET. The learning models implemented in VET as well as their respective teaching scenarios are of many types and are constantly being improved through a loosely defined curriculum development process. In our work, we propose a micro-activity (μ-activity) based on both the inductive image-word model and the learning based on an online game. As a case study, we have implemented a teaching addendum to the chapter Digital Electronics from the lesson "Principles of Electronics" that can be used as a first didactic approach to the subject of Digital Gates.

Keywords: Game-based learning · Digital electronics · M-activity · Teaching of electronics · Shanghai MahJong · Advanced web technologies

1 Introduction

1.1 Conceptual Framework

The Digital Gates subject is taught by the engineering teachers in Vocational Education and Training (VET), in the 1st class and in the 3rd class of the specialty of Electronic Engineers of the Greek Vocational Lyceum (EPAL). The books used for the lessons are "Principles of Electrical & Electronic Engineering" (Varzakas et al. reprint 2017) in the 1st class and "Digital Electronics" (Asimakis et al. reprint 2017) in the 3rd class. Recognizing that Digital Gates is a basic technical skill, our work aims at learning and correlating the relative symbolization - modeling through images in a direct and pleasant way. The basic knowledge of Digital Gates is also a fundamental knowledge for digital design in college Electrical and Computer Engineering (ECE) curricula.

The specific problem students face in this course, in general, and in this subject, in particular, is how to learn fast and with confidence the symbols of Digital Cates. That skill is very important for the next steps of the course and especially for their ability to design more complex digital circuits (Kleitz 2011).

© Springer Nature Switzerland AG 2020
M. E. Auer and T. Tsiatsos (Eds.): ICL 2018, AISC 916, pp. 638–649, 2020.
https://doi.org/10.1007/978-3-030-11932-4_60

For the course on the design of Electronic Circuits in VET, the teachers use specialized tools to describe the digital material and the description of electronic circuits and systems (Thomas and Moorby 2008). The electronic design is used for verification by simulation, timing analysis, test analysis (test analysis and error rating) and logic synthesis. Since digital design requires a great deal of knowledge of digital gates and its properties, we believe that with the application proposed in this paper we lay the foundations for the acquisition of this knowledge.

In this paper we exploit the potential of web technologies to organize a μ-activity through an online game. In this game, we place the relevant icons - symbols of the Digital Gates. In this way, we organize the pupils' upward development zone that contains at first the symbol recognition. For the online game the student must only recognize and combine the symbols of the Digital Gates without functions or the complicate combinations of complex Digital Circuits. The goal in this μ-activity is to learn and recognize the scheme of the basic Digital Gates in a simple and enjoyable way. In order to reach the students of VET in the state of recognition and understanding of the digital representation in a quick and pleasant way, the authors propose to abandon traditional teaching. With careful planning and the development of the appropriate support material, and with an aggressive structural design of a learning game, we strive for the student to move from simple patterns to more complex ones within a short period of time. In the closing section, our future work plans, future games features and extensions will also be delineated.

1.2 General View on Educational Issues

The teaching of digital design in Electrical and Computer Engineering (ECE) curricula is well-established (Amaral et al. 2005). However, the teaching of digital design to Computing Science (CS) students has not been discussed at length in the engineering education literature. Additional constraints in a CS program include

- the reduced number of hours dedicated to hardware-related subjects; and
- the fact that the incoming students' lack of background on switching theory, analogue circuits, electronics, and have a limited exposure to the concepts of concurrency, feedback, and timing.

With the proposed teaching methodology, we try to bridge the gap, to build a "zone of proximal development" and to solve some of the basic difficulties in teaching digital design and technology to the students of VET in Greece. As Vygotsky (Zavershneva and van der Veer 2018) states: "this zone of proximal development" corresponds to the distance between the actual developmental level as determined by the independent problem solving and the level of potential development as determined by the problem solving under the guidance of adults or in partnership with more capable peers. The theory of constructivism (Travers et al. 1993) states that it is true that only on the known knowledge we do "build" and enrich the unknown area. In the following sections, the technologies used in the development of a game will be presented. Indicative examples of the features of the platform will also be illustrated. We will express our personal expectations for this project along with conclusions and data we have gathered so far.

1.3   Picture – World Inductive Model, PWIM: A Teaching Approach

The teaching models related to the technical subjects of the EPAL curriculum are many and they are being constantly improved. For the specific design of the "DigiGame" (Game Based Learning, GBL) we propose the inductive Picture - World Inductive Model (PWIM). PWIM belongs to the general category of word-image teaching models and is specifically designed for VET. This game was evaluated by students in the school year 2017–2018 in a class of 25 students. As far as the curriculum is concerned, we are precisely in the first lessons (introduction phase) of the teaching of digital gateways. During the same year, laboratory developments will follow as well as more complex digital design systems and connection to other programming units (Morris Mano and Ciletti 2007). The students' access to the subject of digital gates at this stage is necessarily visual. The students see the charts, probably imagining what they mean, but they will understand them later through the exercises and applications.

The question is: what is the right balance between what students see and what they understand so that we can keep both of them in the students' consciousness in the next stages of teaching the material (Garces-Bacsal et al. 2018). For this phase, we propose the inductive word-image model. This learning model has to be implemented in an extremely fast pace. There is no time for a micro-activity, since the laboratory applications must follow. But this area of learning is of great importance for the students' later evolution towards the subject (Joyce et al. 2008). We could say that at this stage the foundations of construction and the delimitation of the field to be learned are laid. Thus, knowledge can be efficiently constructed in VET courses. According to PWIM, at the beginning of the school year 10–15 students will work for three lessons lasting two hours each. The initial actions will be for students to study the images (from the simplest to the more complex) and then to identify connections and functions.

The PWIM model helps students develop the knowledge of reading, recognition and design of the technical language and, in particular, of the technical symbols. The development of a meta-control of knowledge is of central importance. The students should develop capabilities within the process. The learning of the relevant technical language as well as the recognition of images (structural analysis) as the students develop skills of learning and organizing information through the curriculum is of central importance. A good technician (Electrician, Electronics or Informatics technician) must be able to develop the technical vocabulary of symbols, develop skills in their combinations and understand larger and more complex designs of representations of automation. In some respect, this is the ultimate constructivist model, because the general knowledge of the recognition of technical specifications is crucial to the knowledge related to the VET graduates' prospects (Shaman 2015).

1.4   Characteristics of Knowledge-Building Phases Based on the PWIM Model

The knowledge construction phases of the PIWM model (Novia 2015) follow certain steps:

- Students learn the relevant symbols through their basic categorizations, from the simplest to the most complex.

- Inductive thinking inherent in the human brain. Students, like all people, use sorting from birth as they try to understand the world. They are by nature creators of concepts.
- In technical language, things become even more demanding. No operation is possible without recognizing the meaning of each component separately but also the entire system as a whole. One can not organize automation without knowing what one is trying to use it for,
- Interaction with more mature technicians or classmates is the natural path to socialization. Interaction through the recognition of technical designs is an important part of the socialization of technical knowledge. Technical illiterates have a serious disadvantage in learning technical culture and are deprived of the pleasure that comes with learning as they interact with more mature technical specifications.
- Recognition, but also the smarter structure of technical systems, is a first step towards discoveries and patents. Without this, it is not possible to develop and optimize any industry's technical systems.

The links of the objects presented on the slides or in their book, and the actions or connections to be made enable students to physically go from the technical language to the actual application, inductively. First of all, students should learn to recognize them and then see all the possible adjustments and connections that these components may have. So, when they see the "Gate NOT", they recognize the "Gate NOT" image, they see the most comprehensive links and they "read" the gate functions (Fig. 6). The students see the image, combine it with the accessory, and monitor the logical flow of the system functions. A basic principle of the model is to enhance the recognition of the components and the way they are connected and to facilitate their technical and design recognition. Most students want to "make sense" of the universe that surrounds them and work diligently to solve the mysteries of the technical language they are studying (Gore et al. 2017). Consequently, this approach respects the students' technical development: the components they are learning and the ability to handle them properly through their connections are of great importance.

Memorization principles, especially the development of long-term memories, are clearly exploited as technical knowledge develops. What the students learn in a practical way they do not forget it. However, PWIM as a model deals with the evolution of the visual vocabulary as well as with the problem of maintaining the designed components and how we can transfer them to the long-term memory and make them available for studying and expanding the Digital Design the students will face either in the lab or during the real production later. The stage of project completion is important. After studying the individual components, the students should complete the communication circle between the virtual symbols and the completed drawings in their own words to describe the full meaning of a digital system as a whole. The students should also be able to distinguish an Integrated Digital Plan or the function of composite Gates (Multiplexers, Decoders, Registrars, or Counts) (Malvino and Leach 1986). Here, the role of the VET teacher is to guide metacognostic searches or to apply metacognostic control (Kelly 1996) based on which mechanism is likely to facilitate the process, which components are appropriate, which aren't and why (Tran and Le 2018). Thus, another principle that is applied in the inductive word-image model is that reading and writing are linked by nature and can be learned at the same time while in practice they are used at the same time. Therefore, knowledge in an EPAL which is rapidly and dynamically promoted through laboratory exercises finds here its largest application (Herrera 2018).

2 Technical Structure - Design and Implementation

The rules of the game follow the standards of Shanghai Mah-Jong (Kendall et al. 2008). The base level consists of several tiles. Each tile, when selected by the user, reveals a different selected digital portal image. There are only two identical gates across the level. The game ends when all the tiles are uncovered. The layout of the dashboard is created randomly from the computer. The part the player student sees first is the hidden part of the tiles that do not differ in shape, size and height. For the selection, the student has a specified amount of time to complete the level. Our main goal was that the platform would be developed combining three significant factors: cost-efficiency, resource-efficient implementation and cross-browser compatibility. The Uniform_Resource_Locator of the game is http://unipi.mstavrou.com (Fig. 1).

Fig. 1. Login screen

The game environment covers a part of the General Objectives of the curriculum for the Introduction to Electronics lessons, including the development of competences and skills related to:

(a) the use of symbolic data formats (symbols, images, digital gates portraits),
(b) the recognition and identification of the basic components of the technology in order to solve specific problems concerning technology and computational processing in the computer,

(c) the development of computational skills and complexity combinations to enhance automation and technology, and

(d) the preparation and expansion of hardware related to the upcoming development area around the knowledge and skills of digital design.

The μ-activity is approximately a fifteen minute micro-lesson on the subject which can then be developed in a full 45 min lesson, which is part of a wider thematic unit, a chapter, a book or a class. With this μ-activity, students acquire a stable resource for the development of their knowledge in Electronics. Having gained basic recognition skills, the students can now move on to complex combinations of gates, and pass on to the laboratory exercises by inductively identifying the elements or processing even more complex processes involving digital gates. Through continuous practice, the users are expected to manage to change their way of thinking and learning in Digital Electronics. This online μ-activity will take place in real time in a typical VET classroom within the allotted teaching time. Usually, the limited available time in a typical classroom does not allow such activities. With our implementation, beyond the fact that we succeeded in being consistent over time, we also found satisfaction from the students' side. From the evaluation we made in a real class, we have recorded several positive results.

At the μ-activity level, the game is about embedding an activity on mapping the symbols of the digital gates with their names and the functions they are programmed to perform. Embedding is done in a pleasant way that attracts pupils and raises their interest. Also, during the game, the pupils receive feedback as the teacher can supervise the results. Through the ability to record individual performance and to maintain individual scores, the student is encouraged to repeat the game at a more challenging level and thereby further improve the skills to which game based learning aims or to exercise recognition skills in more complex Electronic Circuits (Fig. 2).

Fig. 2. Gates and integrated circuits

Based on these pillars, the platform was designed using the following technologies: HTML5 (Hyper Text Mark-Up Language Standard, 5th version), CSS3 (Cascading Style Sheets, 3rd Version), PHP ((Personal HomePage Tools), Hypertext Preprocessor,

7.1 version), JavaScript, Bootstrap 3, JQuery, Ajax (Asynchronous JavaScript Xml), MySQL.

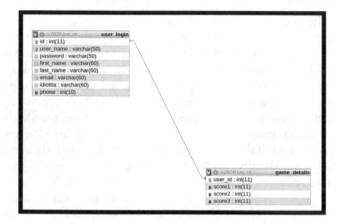

Fig. 3. Schema of the database used in the platform

To complete the construction, open source tools such as Gimp, Brackets, Atom and fileZilla were used. The game is based on a simple database. In the first table the pupil-user data is written and in the second the best scores scored in each level (Fig. 3).

3 Platform Access - Platform Roles – Segmentation. Platform Handling

When the students use the platform, they will encounter the platform's first screen, the Login Screen (Fig. 1). If a student doesn't have an account he must create one by entering his credentials (Fig. 4). It is the only way he can use the on line game.

Fig. 4. Registration form

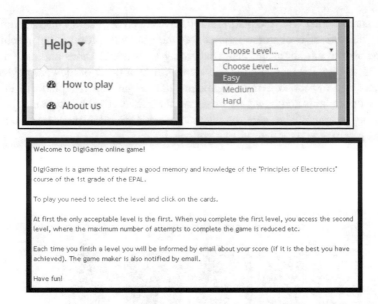

Fig. 5. Help menu and choose level

After creating an account, the student will be transferred to the course selection segment, where he can choose which course he wants to open. As mentioned above, the game is based on the standards of Shanghai Mah-Jong and only one matching is the right one. The web site has been built aiming at ease of use, attractive aesthetics, simple design and security. Thus, during user login (Fig. 1) or registration (Fig. 3), every possible error message (integrity check) is simple and comprehensible. During the game, the user-student can turn on or off the background music and select the desired level of difficulty: Easy, Medium or Hard (Fig. 5), depending on his performance which is stored in the database. At "how to play" section student can read instruction about the procedure. Noteworthy, the game is fully dynamic, so at each initialization the cards are either in different positions or in a different layout. At any time, the learner-user has access to the knowledge base of the game and can by its own choice obtain information about any digital gate he wants.

4   Expected Results - Level of Satisfaction

The game progress is simple. The student selects cards in a random order (Fig. 6). If the selection is correct (Fig. 7), the match stays stable and the student continues with the next choices. On the student-side, we expect students that have some previous knowledge on what a digital gate is to fully understand the transitions and properties of a digital gate: to know what exactly this digital gate is, how does it work, when one should use it to one's best interest and which are its most identifying characteristics. We expect to see students having enhanced their academic background in a way they haven't thought as a learning procedure. On the teaching staff side, we expect to see the teaching staff interact with the platform and have a better overall understanding of the students' theoretical background. Based on the students scoring, the teaching staff should be able to extract useful data concerning their initial status and how they have progressed, as they interact with the platform. From our point of view, as developers of this project, we expect to extract significant data regarding the platform interaction frequency on students and how they enhanced their theoretical background, judging by their scoring progress. The more the interaction with the platform increases, the better and more precise the data that will be collected.

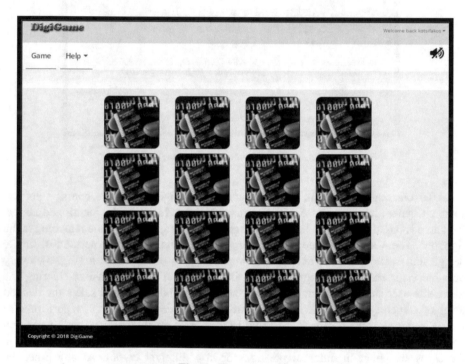

Fig. 6. Game progress

To evaluate the game, questionnaires were given to students. Their responses highlighted a positive feedback and a general acceptance. The questionnaires shared with the students were ranked on the Likert scale. We have measured the degree of agreement and acceptance among the students in six statements. The four (4) first statements concerned the degree of satisfaction with usability, accessibility, and acceptance of the process and the specific environment.

Acceptance rates of around 95% were recorded. The fifth statement concerned the degree of satisfaction with access to the knowledge base, the course book or the digital notes in the online classroom. This question had a low satisfaction rate of around 55%. The sixth question concerned the degree of willingness to extend the process to other learning objects or other lessons and accounted for 100% of the students' positive response. After counting the responses, we set requirements for the development of

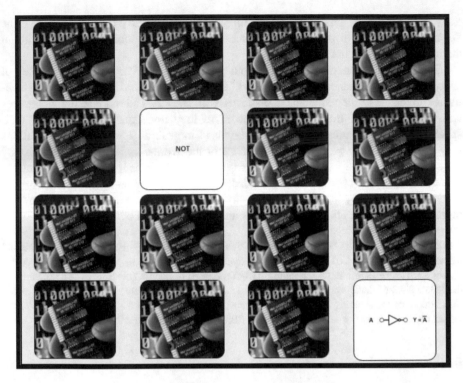

Fig. 7. Game progress

other games in course subjects or for collaborative science collaborations with other game developers in other courses.

Judging from the feedback, we believe that using DigiGame in the classroom as a teaching resource has achieved its educational goals and at the same time it has raised the students' interest. This is a dynamic, flexible and fully scalable environment. Prospectively, the platform on which DigiGame is set up can be a basis for extending to another knowledge base on a similar teaching subject that requires recognition skills of materials or technology symbols that may be relevant to Electrical Engineers, Informatics experts, Mechanical Engineers up to Nurses or Food Chemists (VET).

5   Conclusions and Future Work

Game-based learning applications motivate the students to learn by using a fun and interactive environment that encourages competition (Kotsifakos et al. 2018). Therefore, the player/student is focused and motivated to learn, instead of looking at learning as a chore and as a dull activity. By implementing games as a pedagogical tool, not only students are encouraged into studying, but their problem-solving skills, memory-retaining ability and computer fluency are also enhanced. Game-based learning is a proven and useful teaching method that should be used more often in education. Thus, there is a necessity to promote it by creating an easy-to-handle exercise. Therefore, the developed μ-activity provides a more accessible and easier tool for teachers without any expertise on the matter.

Beyond these general conclusions, our game aims specifically at the courses of Electronics. VET teachers "keep the keys" that ensure access and choice to the technical knowledge to their students (Kotsifakos et al. 2017a). The more technical features and elements enrich their repertoire, the more they understand the world of their specialty and the technical world that surrounds us in general. The more representations they have when they recognize and practice in the workshop, the more control they have on their material and the more choices they have in their lives inside and outside school, along with better access to knowledge and experience and greater self-learning ability. The more they understand how technical language works, the stronger they can be in their communication as technicians and as citizens (Kotsifakos et al. 2017b).

Acknowledgments. The work presented in this paper has been partially funded by National Matching Funds 2016–2017 of the Greek Government, and more specifically by the General Secretariat for Research and Technology (GSRT), related to EU project Sauron. The pictures of the "Gates" were designed by Chrysostomos Logaras, who is an undergraduate student at the Department of Informatics, University of Piraeus. We would like to thank all the students of the laboratory as well as the teachers' of the 3rd EPAL (Vocational Lyceum) of Piraeus.

References

Varzakas P., Paschos I., Tselekas P.: Elements of electronics. Diofantos. http://ebooks.edu.gr/new/books-pdf.php?course=DSEPAL-A123 (reprint 2017)

Asimakis A., Moustakas G., Papageorgas P.: Digital electronics, (Part A, Theory). Diofantos. http://www.pi-schools.gr/lessons/tee/electronic/biblia.php (reprint 2017)

Kleitz, W.: Digital Electronics: A Practical Approach. Prentice Hall. Greece Publication (1996); Kleitz, W.: Digital Electronics. Tziola Publications (2011). ISBN: 978-960-418-338-8

Thomas, D., Moorby, P.: The Verilog® Hardware Description Language. Springer Science & Business Media (2008)

Amaral, J.N., Berube, P., Mehta, P.: Teaching digital design to computing science students in a single academic term. IEEE Trans. Educ. **48**(1), 127–132 (2005)

Zavershneva, E., van der Veer, R.: Genres of writing. In: Vygotsky's Notebooks, pp. 51–55. Springer, Singapore (2018)

Travers, J.F., Elliott, S.N., & Kratochwill, T.R.: Educational Psychology: Effective Teaching, Effective Learning. Brown & Benchmark/Wm. C. Brown Publication, pp. 326–329 (1993)

Morris Mano, M., Cileti, M.D.: Digital Design, 4th edn. Pearson Education Inc., Prentice Hall. https://download-pdf-ebooks.org/files/download-pdf-ebooks.org-1519412449Jg0B2.PDF (2007). ISBN 0–13–198924–3

Garces Bacsal, R.M., Tupas, R., Kaur, S., Paculdar, A.M., Baja, E.S.: Reading for pleasure: whose job is it to build lifelong readers in the classroom? Literacy **52**(2). Special Issue: Reading for Pleasure: Supporting Reader Engagement, May 95–102 (2018)

Joyce, B., Calhoun, E., Hopkins, D.: Models of Learning, Tools for Teaching. McGraw-Hill Education, UK (2008)

Shaman, S.N.: Using the expanded keyword method to help K-12 students develop vocabulary knowledge. Ph.D. thesis, Washington State University, USA (2015)

Novia, F.: Promoting picture word inductive model (PWIM) to develop students' writing skill. Premise J. Engl. Educ **4**(1)

Gore, J., Ellis, H., Fray, L., Smith, M., Lloyd, A., Berrigan, C.,… Holmes, K.: Choosing VET: Investigating the VET Aspirations of School Students. NCVER Research Report. National Centre for Vocational Education Research (NCVER) (2017)

Malvino, A.P., Leach, D.P.: Digital Principles and Applications. McGraw-Hill (1986)

Kelly, L.: Challenging Minds· Thinking Skills and Enrichment Activities. Prufrock Press Inc (1996)

Tran, L.T., Le, T.T.T.: VET teachers' perceptions of their professional roles and responsibilities in international education. In: Teacher Professional Learning in International Education, pp. 29–50. Palgrave Macmillan, Cham (2018)

Herrera, L.M.: Cultural historical theory & VET–A contribution to broadening the theoretical grounds of research in VET. Nordic J. Vocat. Educ. Train. **7**(2), iii–vii (2018)

Kendall, G., Parkes, A.J., Spoerer, K.: A survey of NP-complete puzzles. ICGA J. **31**(1), 13–34. https://tinyurl.com/yaqth7ux (2008)

Kotsifakos, D., Zinoviou, X., Monachos, S., Douligeris, C.: A game-based learning platform for vocational education and training. In: Auer M., Tsiatsos T. (eds.) Interactive Mobile Communication Technologies and Learning. IMCL 2017. Advances in Intelligent Systems and Computing, vol 725. Springer, Cham (2018)

Kotsifakos, D., Kontaxis, A., Pangalos, S., Douligeris, C.: Introduction of technological programming in higher secondary education. In: 4th Scientific Conference of the Panhellenic Association of School Counsellors (PSSC) with Scientific Work, December 9–10, 2017, Agrinio. University of Ioannina (2017a)

Kotsifakos, D., Kasimati, A., Douligeris, C.: Educational policies and profiles of teachers and trainers for vocational education and training (VET) in the countries of the European Union. In: 3rd Scientific Conference of the Association of ASPETE, Marousi, 5–6 May (2017b)

Mobile Learning Design Using Gamification for Teaching and Learning in Algorithms and Programming Language

Vitri Tundjungsari[✉]

Faculty of Information Technology, YARSI University, Jl. Letjen Suprapto,
Jakarta, Indonesia
vitri.tundjungsari@yarsi.ac.id

Abstract. With the advent of mobile learning, educational systems are changing. Mobile learning, also known as m-learning, is an educational system which enable learning process through mobile devices, such as handheld and tablet computers, smart phones, and mobile phones. In this paper we design a mobile application using gamification to learn algorithms and programming, specifically in Java programming language. The objective of the application is to improve learning motivation of students in a private university majoring in computer science. The design of the application is applying a model for e-Learning, in order to enhance students' motivation and understanding for topic: introduction to Java programming. The application is developed because the need of fast-paced informatics competencies. In addition, Java programming is a subject which considered as attractive yet a difficult subject to learn. Therefore, approaches and methods to teach in this subject should be designed and delivered carefully. The usability testing result shows that the application considered as a usable application for each usability element. It is indicate that the application is potential to be used as learning alternatives in programming language. However there are some issues still remain related to the application's features which are very important to enhance students' motivation for learning algorithms and programming.

Keywords: Mobile learning · E-learning · Gamification · Usability · Programming

1 Introduction

Programming language subject for college and university students is considered as difficult subject to be learnt. Several authors mention that computer programs and algorithms are abstract and complex entities consist of concepts and processes, therefore it is not easy to teach and learn algorithms and programming [1–4].

Based on the problems encountered above, it is necessary to find alternative solutions to facilitate teaching and learning programming. We find out that mobile learning can be used as an alternative education tool for programming courses. Mobile learning or M-learning is educational technology and distance education using mobile

© Springer Nature Switzerland AG 2020
M. E. Auer and T. Tsiatsos (Eds.): ICL 2018, AISC 916, pp. 650–661, 2020.
https://doi.org/10.1007/978-3-030-11932-4_61

devices [5]. Mobile learning also enables students and teachers to use mobile devices in the education system in order to support teaching and learning [6].

However, mobile learning itself require good content in order to deliver the education materials to the students. In this research, we use gamification concept to be included in the application. We believe that games elements can improve students' motivation to use the application. In addition, the gamification concept is used in this application design because it can increase the level of student competition. We believe that competition is able to increase students' motivation because the need to get social recognition among students. Motivation to use the tool is very important, because the more often the tool is used it is expected that the knowledge and understanding of students is also increasing.

The purpose of this research is to design a usable mobile learning application for subject of Java programming language with specific topic of Introduction to Java language. This research is important because programming is one of compulsory subjects in computer science program in our university, a private university in Jakarta-Indonesia. The existing examination result for programming language subject show a tendency that is less good, which is about 30% of students fail in this subject each year. This shows the importance of new learning tools to increase the quantity and quality of learning, so as to increase the percentage of pass students in this subject.

Mobile learning and gamification can be combined to deliver better education's quality and quantity than traditional learning method. Literatur studies find out that the usage of gamification in computer programming course can enhance learning experience, students' motivation, recall ability, and performance [7]. We believe that by designing and implementing this mobile based learning application integrate with gamification concept can improve students' knowledge and understanding.

This paper is organized as follows. Section 1 explains the background and purpose of this research, Sect. 2 provides brief review of related works in the usage of mobile learning and gamification for teaching and learning programming. Section 3 presents the method used in this research and Sect. 4 explains the design of application. Finally in Sect. 5 concludes the research's summary and future work.

2 Related Works

There are several mobile learning applications using gamification concept that can be found in the literatures. Gamification is an application using game as representation of real life assignment which build to enhance users' motivation and engagement and also to stimulate users' learning behaviour [8]. Gamification also can be defined as the use of game elements in a nongame context to improve the engagement between humans and computers and to solve problems effectively [9, 10]. Author in [11] defined gamification as a method of game thinking and game mechanics in order to engages users and solves problems.

Alkhalaf et al. [12] study the usage of gamification elements for learning application. They find out that integration of game elements into learning application have potential to: (1) increase the level of fun and entertainment in learning environment; (2) motivate students to compete with each other; and (3) enhance gaming and learning skills. Game design element can be categorized into two types, i.e.: mechanics and dynamics [13]. Table 1 shows game-design elements.

Table 1. Game design element and motives [13]

Game design element		Motives
Game element: Mechanics	Game element: Dynamics	
Documentation of behavior	Exploration	Intellectual curiosity
Scoring systems, badges, trophies	Collection	Achievement
Rankings	Competition	Social recognition
Ranks, levels, reputation points	Acquisition of status	Social recognition
Group tasks	Collaboration	Social exchange
Time pressure, tasks, quests	Challenge	Cognitive stimulation
Avatars, virtual worlds, virtual trade	Development/organization	Self-determination

Zhang and Lu [14] have developed iPlayCode, a mobile based learning application using serious games for learning programming. This application is developed to provide programming learning for users with no programming skills. Using iPlayCode, users learn how to write correct syntax in C++ programming language. The application integrates game elements into learning content, as well as applies rewards, time constraints, the syntactic judgments and help system.

Swacha and Baszuro [15] propose a design of e-learning platform which is specifically used for programming teaching and learning. The e-learning platform employs gamification concepts in order to enhance students' commitment for learning. The gamification concept used is in form of reward for individual work, rivalry and team work. They found out that their e-learning platform is very promising to improve and stimulate students' motivation, participation, and engagement.

Tundjungsari [16] has introduced e-learning model to teach programming, as shown in Fig. 1. The model consists of three main components, i.e.:

1. Technical skills of e-Learning users. This component relates to computer literacy, including basic knowledge and ability to use and operate computer, word processor, and spread sheets.
2. Digital content. This component relates to programming language that chosen to be taught, curriculum and syllabus, resources availability. The digital content is design with gamification approach.
3. Tool and technology. This component consists of design and content of tool and technology used. There are several tool and technology can be considered.

Taking this into consideration, this research aims to fill the gap in our previous research by applying e-learning model for mobile learning development and integrating gamification concept in the learning content.

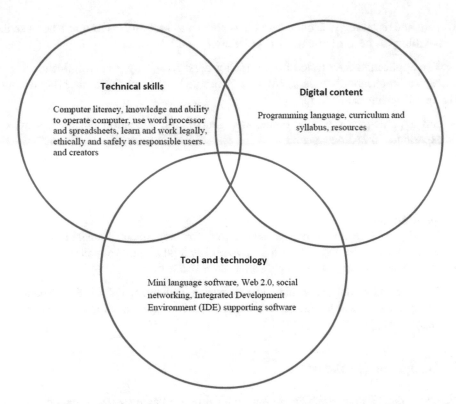

Fig. 1. Component used in our mobile based learning application [16]

3   Method

In this research, a mobile learning application is developed using the e-learning model proposed in our earlier research [16] combined with concept of gamification. The gamification concept is integrated into learning content, in order to produce fun learning environment yet motivating. Based on the model, we implement the mobile learning application, as follows:

1. Technical skills. The application is designed for our university's students as a supplementary learning tool, therefore all of the students have above the average of **technical skills**.
2. Digital content. The **content** of the application only focused for Introduction to Java Programming course, contain of these following material: computer structure and algorithm, the components of Java language, Loop, Array, Methods in Java, Recursion, Principles of Object-Oriented programming. The application is designed using game mechanics as game element, by implementing scoring systems, time pressure, task, and levels. The motive of using game element in the digital content design is to enhance motivation by achievement and social recognition.

3. Tool and technology. Mobile learning is chosen as **tool and technology** because its flexibility to be carried and used as learning tool.

The application then tested from usability perspectives by involving lecturers and students as application's testers. We have distributed questionnaire as usability testing. The questionnaire consists of:

- Part 1 participant's profile: name, age, gender, last education obtained, years of experience in last occupation.
- Part 2 information and computer literacy: skills in using computer, time in using the Internet, knowledge of basic Java programming.
- Part 3 experience of users while using the application during usability testing (how the application support their learning process, the benefits and the weaknesses of the application).
- Part 4 user's feedback of the application, for components: readability, content design, navigation and help, efficiency and flexibility, and error recovery.
- Part 5 user's opinion of the application after using it.

The questionnaires are written in Indonesian and evaluated by 30 respondents consist of students and lecturers. The output of the questionnaires is used to evaluate the design of application.

4 Design of Application

We apply the learning strategies in our application by having several features, i.e.:

- Lessons and tutorials. In this feature, we provide brief explanation of topic and also examples.
- Games. After learning the materials from lessons and tutorials feature, the user can start to do some tasks provided in three different levels. Level 1 is the easiest among others, while level 3 is the most difficult level compare to others. Whenever students finish for each level, they get the badge.
- Score. For each attempt, the user can see his/her score on the scoreboard. The scoreboard is very useful in order to motivate user.

The application is designed using game element, as follows:

1. Player: the target users are first year university students who take algorithm and programming course.
2. Scores: the value earned after completing the challenge.
3. Level: there are three gamification level available (level 1, 2, and 3).
4. Task: tasks are given to enhance students' skills related to algorithms and programming. Task should be complete in a limited given time.

Flow of the application:

1. User enter his/her name
2. User will be given lesson and tutorials related to algorithm and programming course

3. User will be asked to answer the game challenge, started from level 1. Every game challenged has full score of 100. Every time user has finished level 1 (having 100 score), he/she will get level-up to the next level (level 2). Figure 2 and 3 shows the interface of level 1.

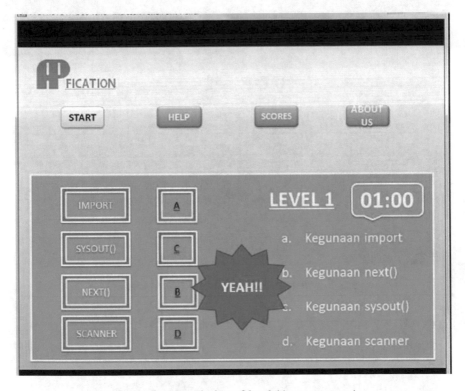

Fig. 2. Interface design of level 1(correct answer)

4. In level 2, users will be asked to answer the question by completing a word refers to algorithm and programming terms. Users will be given several questions until get full-score. When user get full score, he/she will play to the next level (level 3). Figure 4 shows the interface of level 2.
5. In level 3, user has to complete programming syntaxes in limited time. Figure 5 illustrates the interface of level 3.
6. As an addition, Help feature is provided to give users guidance to use the application (Fig. 6). Final score of each attempt can be seen in Score feature (Fig. 7).

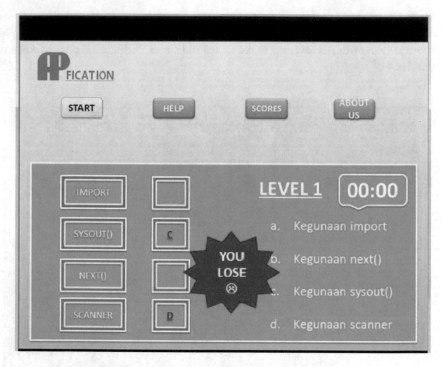

Fig. 3. Interface design of level 1 (wrong answer)

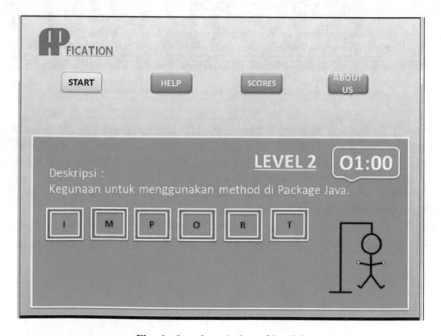

Fig. 4. Interface design of level 2

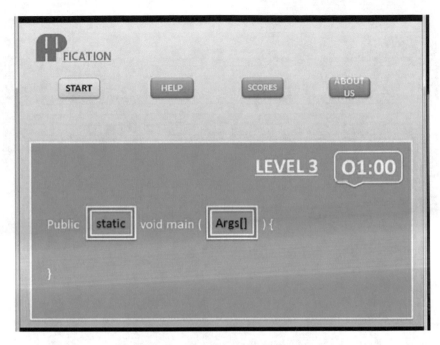

Fig. 5. Interface design of level 3

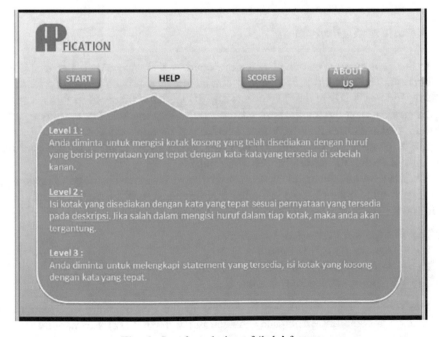

Fig. 6. Interface design of 'help' feature

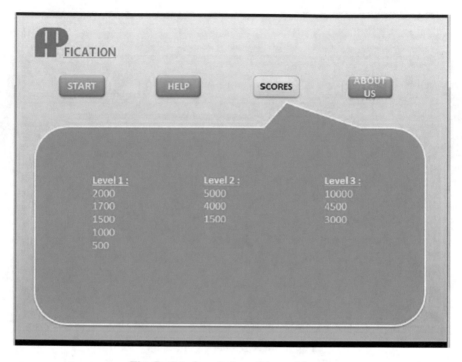

Fig. 7. Interface design of 'score' feature

5 Result and Discussion

After the prototype is designed and developed, we then evaluate the usability of application. The usability evaluation is conducted in a private university located in Jakarta, Indonesia. There are 30 students and 10 lecturers involved in this research to experience and test the application, related to usability matters. Usability is a quality attribute that measures how easy user interface of an application are to use (Nielsen nngroup). Usability is also one of important aspect to be considered for improving ease-of-use during the design process (nngroup).

We develop questionnaire as an instrument to measure usability. Questionnaire with Likert scale (range 1 to 4) is used as instruments to test the usability, with details as follows:

- The VG Answer (Very good) is rated 4
- The G Answer (Good) is rated 3
- The B Answer (Bad) is rated 2
- The VB Answer (Very bad) is rated 1

The usability components which we use in the questionnaire are: learnability, efficiency, memorability, errors, and satisfaction (Niel-sen 2012). Learnability refers to how easy the users can accomplish the basic tasks of the application at the first time they encounter the design. Efficiency relates to how quickly users can perform the task after they learned the design. Memorability is how easy users can perform their ability when they return using the application after a period of not using it. Errors relates to how many times users do errors and how easy they recover from the errors. Satisfaction refers to how pleasant users when interact with the application's design.

The questionnaire is given to the students after they experience the application in a week. The usability scenario then categorised into three groups, i.e.:

- The ability to complete the game at level 1
- The ability to complete the game at level 2
- The ability to complete the game at level 3.

From 30 students, it is found out that all of the students pass level 1 and level 2, while only about 57% students pass level 3. The results show that not all of students have ability to complete the task provided in the game. The reason of that, because the students have lack of motivation to self-learning and complete all level of the game.

The analysis of results of the evaluation shows, that all the usability elements has mean score above 3.50 (1 to 4 scale), which indicates that the mobile learning application usable. We find that learnability is the most important element of usability to determine students' motivation and therefore determines the success of application. We also find that lecturers as respondent tend to give high ratings than the students. The reason for the high ratings given by the lecturers is that they understand and appreciate better than students the complexity of the preparation of learning material which at the same time to be suitable for use on mobile devices.

The application is considered useful to students for subject Java Programming. Students said that the application is easy to use and usable in terms of readability, content design, navigation and help, efficiency and flexibility, and error recovery). From the test result it can be found out that learning Java programming become more easily understood, fun to learn and improve user ability in subject of Java Programming. However, the application needs to be improved in terms of materials of Lessons and Tutorials. The users also mention that the application should provide feedback for any error or mistake they make when answering questions.

We gather some opinions from the respondents to improve the application and conclude that the application should re-design by providing more game elements, such as:

- User profile. This game element provide brief information of user with picture.
- Level achievements. This game element shows global scores from all of users, provides badges to level-completers, and give stars to best score.
- Score and rewards. This game element shows leaderboard and top 10 scorers. Virtual gift is also given to best scorers.
- Status. This game element consists of result, report, dashboard, and progress bar.

Based on the result and discussion we can find out that we need to re-design the application, in order to enhance the application's usability based on users' feedback. Once the application's usability is improved, we believe that the motivation of users to use, engage, and participate in using application will be increase as well.

6 Conclusion

In this research, we have designed and developed a mobile learning application for teaching and learning algorithm and programming course as our first prototype. The result of the usability testing shows that the application has good usability (average for each usability element is above 3.50). It shows that our previous e-learning model can be applied in mobile learning environment. However, not all of the students able to complete level 3. Thus, it can be concluded that the application is usable but not sufficient to motivate the students to complete all levels.

In the future, we need to redesign the application by adding some more game elements and refine the quality of materials. We gather some answers from the respondents to improve the application and conclude that the application should re-design by providing more game elements, such as: user profile, achievements, reward schedule, and status. All those elements above are very important to enhance motivation by stimulating competition among users. More extensive usability assessment also required to be performed in our future research.

References

1. Olsson, M., Mozelius, P., Collin, J.: Visualisation and Gamification of e-Learning and programming education. Electron. J. e-Learn. **13**(6), pp. 441–454 (2015). www.ejel.org
2. Guzdial, M., Soloway, E.: Log on education: teaching the Nintendo generation to program. Commun. ACM **45**(4), 17–21 (2002)
3. Lahtinen, E., Ala-Mutka, K., Jarvinen, H.: A study of difficulties of novice programmers. Innov. Technol. Comput. Sci. Educ., 14–18 (2005)
4. Eckerdal, A.: Novice Programming Students' Learning of Concepts and Practice. Ph.D. Thesis Uppsala University (2009). http://www.avhandlingar.se/avhandling/6809751ebf/
5. Mehdipour, Y., Zerehkafi, H.: Mobile learning for education: benefits and challenges. Int. J. Comput. Eng. Res. **3**(6) (2013)
6. Alkhalaf, S., Amasha, M., Al-Jaarallah, A.: Using M-learning as an effective device in teaching and learning in higher education in Saudi Arabia. Int. J. Inf. Educ. Technol. **7**(6) (2017)
7. Fotaris, P., Mastoras, T., Leinfellner, R., Rosunally, Y.: Climbing up the leaderboard: an empirical study of applying gamification techniques to a computer programming class. Electron. J. e-Learn. **14**(2), 94–110 (2016) www.ejel.org
8. Marcewzski, A.: Gamification: A Simple Introduction and a Bit More. Amazon Digital Services, Seattle, WA (2012)
9. Deterding, S., Sicart, M., Nacke, L., O'Hara, K., Dixon, D.: Gamification. using game-design elements in non-gaming contexts. In: CHI 2011 Extended Abstracts on Human Factors in Computing Systems, pp. 2425–2428 (2011)
10. Deterding, S., Dixon, D., Khaled, R., Nacke, L.: From game design elements to gamefulness: defining gamification. In: Proceedings of the 15th International Academic MindTrek Conference: Envisioning Future Media Environments, pp. 9–15 (2011)
11. Zichermann, G., Cunningham, C.: Gamification by Design: Implementing Game Mechanics in Web and Mobile Apps. O'Reilly Media, Inc. (2011)

12. Alkhalaf, S., Amasha, M., Al-Jaarallah, A.: Using M-learning as an effective device in teaching and learning in higher education in Saudi Arabia. Int. J. Inf. Educ. Technol. **7**(6) (2017)
13. Blohm, I., Leimeister, J.M.: Design of IT-based enhancing services for motivational support and behavioral change. Bus. Inf. Syst. Eng. **5**(4), 275–278 (2013)
14. Zhang, J., Lu, J.: Using mobile serious games for learning programming. In: Proceedings of INFOCOMP 2014: The Fourth International Conference on Advances Communications and Computation (2014)
15. Swacha, J., Baszuro, P.: Gamification-based e-learning platform for computer programming education. In: Proceedings of X World Conference on Computers in Education, July 2–5 Toruń, Poland (2013)
16. Tundjungsari, V.: E-learning model for teaching programming language for secondary school students in Indonesia. In: Proceedings of the 13th Remote Engineering and Virtual Instrumentation (REV), 24–26 Feb, Madrid, Spain (2016)

Computer Based Business Simulation Application for Master Degree Students' Professional Education in Economics and Management Courses in Polytechnic University

Elena B. Gulk[1] and Dmitrii A. Gavrilov[2(✉)]

[1] Institute of Humanities, Department Engineering Pedagogy, Psychology and Applied Linguistics, Peter the Great St. Petersburg Polytechnic University, St. Petersburg, Russia
super.pedagog2012@yandex.ru
[2] Institute of Industrial Management, Economics and Trade, Department of Management and Business, Peter the Great St. Petersburg Polytechnic University, St. Petersburg, Russia
d.gavrilov@abc.org.ru

Abstract. Results of computer based business simulation application in educational process in Polytechnic University for economics and management educational component are described in this paper. An importance of active forms and methods of education, as well as educational and professional problems solving as a way to increase educational process efficiency for future engineers is demonstrated. Future engineers' capabilities for effective interpersonal communications, teamwork and leadership skills improvements are shown.

Keywords: Computer based business simulation · Active forms of learning · Professional and communicational tasks

1 Context

Modern polytechnic education should prepare a student to acquire both professional and universal competencies, such as communications skills, team work and leadership. An engineer should be able to solve technological, scientific, project and managerial issues. Hence an important high education didactic task is to develop these abilities and skills. It is especially important for educational programs which are dedicated to future managers, which would guide multinational and complex engineering teams. Moreover, professional education should create conditions for personal development of future professional, for individual capabilities growth, and for proactive social position as well as for capabilities of conscious and responsible professional and personal choice.

© Springer Nature Switzerland AG 2020
M. E. Auer and T. Tsiatsos (Eds.): ICL 2018, AISC 916, pp. 662–672, 2020.
https://doi.org/10.1007/978-3-030-11932-4_62

High education processes reforming followed professional education goals changing. Professional development model directed toward competences (i.e. integrative personal and actionable features) comes into play instead of adaptive and narrowly dedicated model. This model is based upon A.A. Leontiev's and S.L. Rubinstein's theory of human being in activity development [1, 2], as well as based upon the concept of developing and interactive learning which is represented in books and papers of Verbitskii, Vygotskii, Zaporozhets, Dyachenko, Kruglikov etc. [3–7] This change entails a necessity to look for new forms and methods of future engineer' professional training process setup; this setup have to be consistent with contemporary requirements.

As mentioned by V.N. Kruglikov, there are several definitions of interactive learning. It is considered from two standpoints in a literature: first of all, as an education, based on direct interaction, dialog, and communications between learner and teacher as well as between learners. Here "inter" (taken from "interaction") means "mutual", i.e. mutually active learning. This is learning when learner holds an active position, and learning process is a process of communication between learner and teacher, this is a "subject to subject relationships" process. Secondly, this is learning, based on communication, facilitated by contemporary communication media (computers, TV etc.). These two approaches sometimes considered as opposite to each other. We believe that computer business simulation allows, from one hand side, to use modern learning tools, and from another hand side, to provide high level of students' engagement, interpersonal communication, and decision making experience in an environment with gradually increasing complexity, which are very close to real life.

Studying the effectiveness of training graduate students in engineering University, we have identified theoretical and methodological approaches to the organization of the educational process using a personality-oriented approach to the construction of economic disciplines [8, 9].

The development of the educational process using business modeling involves: creating conditions for improving the primary communication between students; ensuring their interpersonal communication; careful understanding of the course content.

The implementation of these principles requires a new organization of education: training in small groups, the use of methods of associative dialogue aimed at intensive communication between students in the study of new material.

2 Purpose

The target of this research is to verify an effectiveness of computer based business simulation (based on The Fresh Connection and The Cool Connection business simulation software) for economics and management educational component for multinational teams of Master degree students in Polytechnic University.

Objectives:

1. To determine and check applicability and methodical efficiency of computer based business simulation for Polytechnic University students;

2. To explore a team work developments, self-assessment and students' communication skills in the business simulation environment when solving structurally complex tasks with gradually increasing level of complexity;
3. To explore students' attitude and readiness to implement computer based simulation into educational process in Polytechnic University.
4. To make a conclusions.

At the time of research these methods were applied: theoretical – an analyses of pedagogical and methodological textbooks, analyses of practical experience; empirical – surveys, testing, activities bottom line analyses, expert opinion; the results were processed with the help of mathematical statistics.

Based on the goal and objectives of research, these methods were selected:

1. T. Leary's method of interpersonal relationships investigation.
2. A.F. Fiedler's method of collective psychological atmosphere assessment.
3. Surveys which students went through.
4. An expert opinion about intra-team co-operation effectiveness and students' leadership.

3   Approach

The research was conducted during an academic year at the time of economics and management courses, which are devoted to Supply Chain Management, for multinational Master Degree students' teams of various educational programs (both engineering and management).

The educational process is based upon decision making in The Fresh Connection and The Cool Connection computer based online simulations which simulate four aspects of supply chain management (with roles of Sales, Operations, Supply Chain Management and Purchasing for The Fresh Connection, and roles for Sales, Operations, Purchasing and Finance for The Cool Connection).

Decisions in business simulations depended upon the role students played in and the type of simulation.

In The Fresh Connection business simulation Sales role is responsible for these decisions:

1. Customer agreements settings (customer service level, promised shelf life for products, payment terms, trade unit, promotional pressure level and promotional horizon, etc., and all these parameters has an influence on price company could charge for each product; it is worth to say that sensitivity for these settings is different for every customer);
2. Category management decisions – which products would be in portfolio for each customer, so it is possible to exclude particular products from portfolio or to cancel service for particular customer;
3. Customer priority – if particular product is of short supply, which customer is first in the line, which is second and which is third.

4. Forecast for each product, which is used to calculate material requirements and capacity requirements.

In The Fresh Connection business simulation Operations role is responsible for these decisions:

1. Production capacity (the set of equipment, and each unit of equipment is associated with particular features like productivity, cost of running, number of shifts, and a number of additional options for machinery usage optimization, as well as manufacturing personnel skills improvements like trainings and incentives);
2. Raw materials warehouse capacity and manning. In case of warehouse capacity and manpower shortage, overtime is automatically arranged, and if it is not enough to cover load, temporary personnel is hired and trained. If number of pallet locations is not big enough at any time to accommodate inventory, company use external warehouse paid for each pallet location per day.
3. Incoming inspection set up for raw materials and components;
4. Finished goods warehouse capacity and manning. In case of warehouse capacity and manpower shortage, like for raw materials warehouse, overtime is automatically arranged, and if it is not enough to cover load, temporary personnel is hired and trained. If number of pallet locations is not big enough at any time to accommodate inventory, company use external warehouse paid for each pallet location per day.

In The Fresh Connection business simulation Purchasing role is responsible for these decisions:

1. Supplier agreements settings (customer service level, payment terms, trade unit, etc., and all these parameters has an influence on price company should pay for each material; it is worth to say that sensitivity for these settings is different for every supplier);
2. Alternative suppliers from a list of suppliers on the market for each purchased material or component. Dual sourcing option could be applied instead of changing a supplier.

Supply Chain Management role probably the most complex one, because it is a balancing role which requires a deepest understanding of supply chain peculiarities and dynamics. This role in The Fresh Connection business simulation is responsible for these decisions:

1. Safety stock and lot size for purchased materials and components;
2. Frozen period for manufacturing, which means that it is not allowed to change production plan inside this period even if demand is changing; the longer the frozen period, the less is customer service level but the better for production plan adherence and for production costs;
3. Safety stock and production frequency for finished goods.

In The Cool Connection business simulation which was used as well Sales and Purchasing roles are identical for these roles in The Fresh Connection business simulation, Supply Chain role encompasses Operations and Supply Chain Management roles in The Fresh Connection business simulation. Finance role is introduced in The Cool Connection, and this role is responsible for these decisions:

1. Loans management (bank selection, loans in each bank, interest rate in each bank, additional terms and conditions for loans – for instance, Debt/Equity and Debt/EBITDA ratios);
2. Customer credit limits management;
3. Credit insurance;
4. Bank guarantees for suppliers etc.

Students were grouped in teams of 4 individuals, mostly of various nations, and they had to make decisions online which would improve ROI performance of a company.

Students selected teammates by their own; teacher didn't interfere in this choice.

Students played 6 rounds, and each round was more complex than previous one. The task of each team was to attain ROI (Return on Investment) as high as possible by making smart strategic and tactical supply chain management decisions, and save the company from disaster.

Students had one week on average to analyze and to make decision in each round. Teacher did business simulation orientation before simulation started. This orientation was devoted to goals and objectives of the simulation, as well as to review key decisions and settings.

Students mostly did analyses and run decision making meetings face-to-face and via Skype out of the class, because business simulation is an online tool. Based on student's estimation, they spent for each round from 2–3 up to 5–6 h on average, and some of them even more – up to 9–10 h. An experience showed that those teams who spent a little time and did superficial analyses in business simulation without proper co-ordination of decisions between team members, didn't prosper, being on average level at best. Those teams which spent much more time and efforts doing deep quantitative analyses in business simulation arrived to the end of simulation with better bottom line. Unfortunately, it was hard to assess more precisely the linkage between time spent and success of student's teams, this conclusion was done based on students responses to the teachers' questions and on expert opinion about amount of efforts they applied (how often students asked a teacher, what was the level of analysis, etc.).

The teacher had several goals for the business simulation:

1. To support theoretical material of Supply Chain Management course with practical activities in order to better comply with employers requirements for students skills and qualifications (this is an example when during student's interview in one of the biggest international FMCG companies knowledge and skills which student got during business simulation became a key point in getting a very attractive job);
2. To assess student's capabilities in Supply Chain Management decision making; business simulation was an important component of final ranking for the course – at least 30% of final ranking, sometimes up to 50%, depending upon master's degree program;
3. To give students a feeling what does it mean to make complex and interrelated decision in Supply Chain Management without hindering any existing company's

performance. This simulation is practical oriented so students have a chance to understand Supply Chain Management issues in 'like live' environment.

To reinforce student's understanding and skills, each team presented team's strategy and decisions to support a strategy by the end of business simulation. If presentation was persuasive, a team got bonus points, and final ranking became better.

4 Result

50 students of Peter the Great Saint-Petersburg Polytechnic University participated in this research, mostly of master degree level and a small number of bachelors. Business simulation was used as a component of Supply Chain Management course. It is important to mention that simulation was conducted in mostly multinational groups in both Russian and English.

Students were asked twice – in the middle (after 1st, 2nd or 3rd rounds) and at the very end of the game (after 6th round) – by a specially designed questionnaires. The questionnaires were designed to reveal students' attitude, team dynamics and self-assessment.

4.1 Results of Questioning

The questionnaire consisted of 34 questions in 9 sections. Questionnaire allowed us to analyze how students estimated teamwork, joint analysis engagement, focus on success, and leaders' performance. Besides it, students' attitude to business simulations as an educational technology was explored, as well as business simulation benefits and drawbacks. Additionally, students estimated the quality of the simulation, the relevance of the simulation, and the usefulness of business simulation for professional development.

Student's estimations are higher than average for all sections. Students believe that business simulation is thoughtful and contemporary for their professional development. Students noted a complexity of tasks suggested during simulation. However, an opportunity of discussions in teams, and their own standpoint comparison with teammates' standpoint really facilitated decision making. Students suggested improving an interface in this business simulation, for instance, they mentioned an option to assess changes of bottom line due to decisions that could be made. Business simulation as an educational technology was ranked high as well. Results we obtained showed clearly that students really value this kind of educational process organization for courses in economics and in supply chain management in particular. Average score in the beginning of semester was 20 points out of 25, and by the end of semester was 21 points. This tells us that this kind of educational technology is on of students' favor during simulation. Students believe that this form of learning is more effective when compared with conventional education. Students do think that business simulation experience is useful for their professional activity; they underline an importance of theoretical knowledge' practical application. The crucial point for students is an opportunity to solve educational and professional tasks by decision making in 'like real' environment.

Besides this, students were asked to estimate usefulness of this educational technology for teamwork training based on 5-point scale. Results are depicted on Fig. 1. We could see that rating really increased by the end of business simulation. Students mentioned an importance of this practice for teamwork even after first rounds, and this opinion is strengthened by the end of simulation.

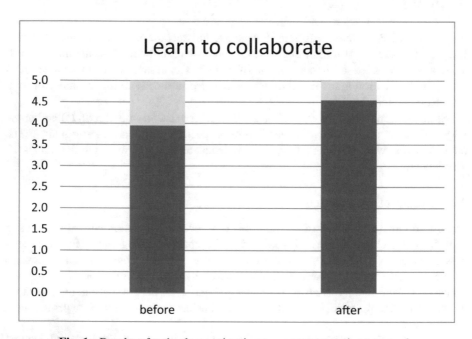

Fig. 1. Results of technology estimation as a way to practice teamwork.

Respondents were asked to assess a quality of teamwork from efficiency and cohesion standpoint, were they capable to discuss a task, to find a solution inside given sufficient timeframe, to balance risk taking, to take into account previous mistakes, were they focused on success. They were asked to assess team leader' performance, as well as and roles' fulfillment by all the team members. Maximal number of points was 50. Results of this estimation are depicted on Fig. 2.

Students mentioned team efficiency improvements during business simulation rounds, they learn to discuss issues and analyze they own performance. In general, students estimate their teams in a positive light – an average is 8,5 points (out of 10) in the beginning and 9 points in the end of their simulation journey. This conclusion is supported by expert's opinion; these experts were monitoring educational process.

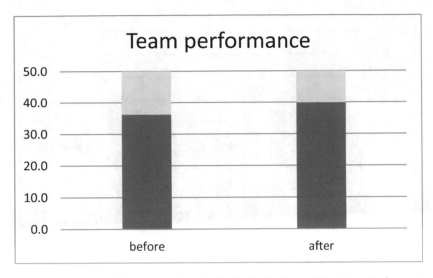

Fig. 2. Performance of the team estimation in the beginning and by the end of semester.

4.2   Results of A.F. Fiedler's Method Application

A.F. Fiedler's method was applied to assess psychological atmosphere in teams during joint decision making for tasks which simulate real business processes in competitive environment and under expert's supervision and estimation. Students were suggested to evaluate opposite pairs of words, which could be used to describe an atmosphere in team: friendliness – hostility; consensus – disagreement; contentment – discontentment; productivity – unproductiveness; warmth – coldness; cooperation – incoordination; mutual support – ill will; enthusiasm – indifference; entertaining – boredom; success – unsuccessfulness. This method is based upon semantic differential, when pairs are estimated with the help of 8 points scale, and the higher the point the more negative is the evaluation. Point the each pair is summarized. The total fluctuates between 10 (most positive evaluation) and 80 (most negative). Based on individual profiles an average profile is calculated, and this average profile is a sign of psychological atmosphere in a team. Average profile for teams before and after business simulation is represented on Fig. 3.

It is clear from the graph that evaluation done by students is positive, students assess relationships in teams as friendly, warm, cooperation focused, they mentioned mutual support and productive activities. Students are satisfied by their job.

An average score for teams after 2nd - 3rd round is 20,19 points, and after 6th round is 22,4 points. We believe that some minor changes after 6th round towards negative side is connected, first of all, with increased level of complexity of business simulation from round to round, and students should apply more and more efforts. Secondly, a student's self-assessment and peers assessment became more and more conscious. It is important to stress here that they evaluate productivity of their efforts higher than in the beginning of simulation. Thus, educational process organization with the support of business simulation allows creating working, emotionally positive

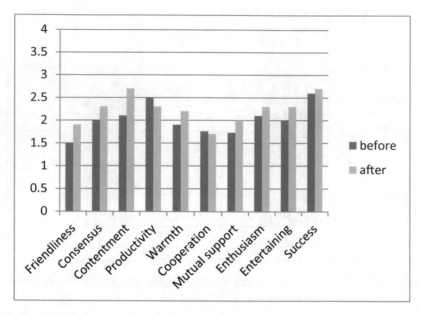

Fig. 3. Psychological atmosphere in teams before and after business simulation evaluation results.

cooperative atmosphere in a team, as a sample of departments in a company, even under conditions of estimation and competition. Students get an experience and develop skills of joint decision making.

4.3   Results of T. Leary's Method Application

T. Leary's method is directed towards a research of human being self-assessment, and towards revealing of dominate relationship type. Two components of this assessment are 'dominance – subordination' and 'friendliness – aggressiveness'. T. Leary differentiates 8 types of attitude to the people around: authoritarian, selfish, aggressive, suspicious, subordinated, dependent, friendly, and altruistic. The lower the score, the more adaptive interpersonal behavior is. Students were suggested to estimate themselves in a team they played business simulation.

Results obtained show that in the beginning of business simulation interpersonal behavior of 100% of students was adaptive by the majority of scales. Business simulation participation changed their self-assessment for 3 scales: authoritarian, subordinated, and dependent. Results are represented on Fig. 4.

In the beginning of semester 22% of students demonstrated overly authoritarian behavior, which was characterized by excessive insistence and aggressiveness in their own standpoint promoting without other's point of view taking into account. A number of such a students decreased to 12% after the game. Moreover, the number of students which are prone to be subordinated and dependent, decreased as well. We think that it is connected with an experience of effective interpersonal communication, joint

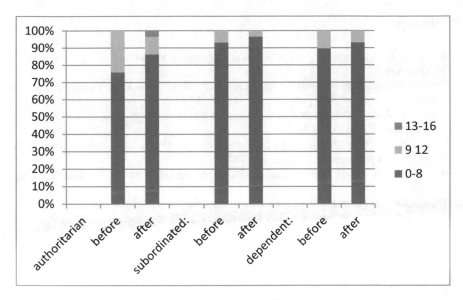

Fig. 4. T. Leary's method application results.

decision making capability development with regards to other team members potential and maximizing own potential application.

5 Conclusions

The conclusion is that an experience of business simulation in educational process in Polytechnic University for economics and management educational component revealed its effectiveness and relevance of this technology. Students tend to actively participate in educational process, and they assessed this form of learning higher than conventional forms of learning. They value that this form is practically oriented, and this form of learning is important to get teamwork experience and interpersonal communication skills.

An experiment demonstrated that business simulation application helped to increase students engagement in learning process, improved interpersonal communication in teams, created more favorable psychological atmosphere, as well as improved subject matter knowledge quality. Students developed abilities to analyze, prepare and make teams' decisions in structurally complex systems with gradually increasing level of complexity. Besides, this form of learning for future engineers facilitated communicative skills development, readiness to co-operate, teamwork, and self-assessment improvements. It allows us to recommend further implementation of business simulation learning technology to the educational process in polytechnic universities.

References

1. Leontiev, A.N.: Selected psychological works. Collected Works in 2 Volumes. Pedagogics, Moscow (1983)
2. Rubinshtein, S.L.: General psychology basics. Vol. 2S. L.Rubinshtein. Pedagogics, Moscow (1985)
3. Verbitskii, A.A.: Active learning in high education: contextual approach. A.A.Verbitskii, 207 p. High School, Moscow (1991)
4. Vygotskii, L.S.: General psychology problems. Collected Works in в 6 Volumes, vol. 2. V.V. Davydov, 504 p. Pedagogics, Moscow (1982)
5. Zaporozhets, A.V.: Selected psychological works/Collected Works in 2 Volumes. Pedagogics, Moscow (1986)
6. Dyatchenko, O.M.: Didactical games and exercises for intuitive modeling capabilities development. In: Dyatchenko, O.M., Govorova, R.I., Tsehanskaya, L.I. (eds.) Age-related features of preschoolers' cognitive capabilities development. L.A. Venger. Pedagogics, 224 p. Moscow (1986)
7. Kruglikov, V.N.: Interactive forms of professional learning: tutorial. In: Kruglikov, V.N., Olennikova, M.V. (eds.) 436 p. Saint-Petersburg: Publishing house of Polytechnic University (2015)
8. Arsentyeva, X.S.: Applying a collaborative learning technique in Ph.D. student groups with multinational structure during foreign language studying in technical university. In: Arsentyeva, X.S., Gulk, E.B., Kasyanik, P.M. (eds.) Proceedings of 19th International Conference on Interactive Collaborative Learning, ICL, pp. 117–130 (2016)
9. Gulk, E.B.: Educational process at the technical university through the eyes of its participants. In: Gulk, E.B., Kasyanik, P.M., Olennikova, M.V., Zakharov, K.P. Kruglikov, V.N. (eds.) Proceedings of 19th International Conference on Interactive Collaborative Learning, ICL, pp. 1041–1052 (2016)

The Effect of Tangible Augmented Reality Interfaces on Teaching Computational Thinking: A Preliminary Study

Anna Gardeli[(✉)] and Spyros Vosinakis

Department of Product and Systems Design Eng, University of the Aegean,
Mytilene, Greece
{agardeli, spyrosv}@aegean.gr

Abstract. Teaching introductory computer science is gradually shifting from learning a computer programming language towards acquiring more generic computational thinking skills. At the same time, more emphasis is placed nowadays in natural and playful approaches that can potentially increase student motivation and engagement. One promising such approach is the combination of tangible elements with augmented reality technology, where the instructions can be given in the real world by manipulating physical elements, and the output is presented in a digitally enhanced space. Despite its potential, this approach has not yet been evaluated in formal educational settings. In this paper we present the results from a preliminary study in an elementary school that compared a tangible AR game with the same game in an unplugged version, to examine the effect of the interface on student motivation, effectiveness and teaching practice. The results indicate that a tangible AR approach can improve the engagement and collaboration of students in classroom activities and affects the role of the instructor compared to unplugged activities. The paper concludes with a number of open issues that need to be further studied.

Keywords: Computational thinking · Physical programming ·
Augmented reality · Unplugged activities

1 Introduction

Contemporary education gradually leans towards more intrinsically motivated learning methods, wherein students themselves are the main driving force behind their own academic explorations and learning experiences. Thus, enjoyment and motivation are becoming increasingly more important learning factors in both formal and non-formal educational settings [1]. Within this scope, researchers and educators are in search of ways to increase engagement, attract students' interest and challenge them to take part in learning activities through active collaboration, social interaction, physical actions and game-like activities, based on new, more appropriate technology [2].

In regard to computer science, efforts have been made in the last decade to apply new tools and methods in the classroom and to assess their effect on student motivation and engagement. At the same time, the educational focus has shifted from learning how to program in a given language towards acquiring more generic problem solving and

© Springer Nature Switzerland AG 2020
M. E. Auer and T. Tsiatsos (Eds.): ICL 2018, AISC 916, pp. 673–684, 2020.
https://doi.org/10.1007/978-3-030-11932-4_63

algorithmic thinking skills, an approach termed as 'Computational Thinking' [3]. The first and still most popular category of efforts towards this end is visual programming, an approach which is based on the manipulation of visual elements to create a program instead of typing written code, and on the visualization of the code execution [4]. This approach helps students avoid syntactical errors and usually results to the development of more engaging programs compared to traditional programming environments. A more recent category that is gaining popularity are the unplugged activities, i.e. problems and algorithms that are solved in the real-world using everyday objects, such as pen and paper, toys, fruit, etc. [5]. In this case, students become familiar with algorithms and computational processes before even using any digital device, thus making it easier to disassociate computational thinking from computers. Finally, another interesting trend is physical programming, where students create programs through the manipulation of physical objects. They combine tangible interactive objects to construct their algorithms and they can observe the results of their instructions in the physical (e.g. through robots) or digital space [6]. Physical programming platforms have shown strong indicators of promoting engagement and motivation by combining the positive aspects of unplugged activities with the advantages of digital technology.

In the recent years, augmented reality (AR) technology has provided novel affordances for educational activities by presenting a digitally enhanced version of the physical world, thus creating further opportunities for physical computing. A number of researchers have recently combined camera-based AR systems with tangible artifacts to create physical programming environments with visual feedback. The aim was to introduce an engaging and motivating collaborative approach for children and help them learn abstract concepts of computational thinking [7, 8]. AR technology is nowadays easily accessible and affordable through mobile devices, and as a medium for physical programming platforms it shows great potential for integrating in formal settings, as (a) it can be more inexpensive, (b) it can be used collaboratively by groups of students, (c) it can be portable and (d) it makes learning more engaging.

Despite the fact that the combination of physical computing with mobile AR seems to be a promising direction for novel educational environments for computational thinking, the number of studies regarding their effectiveness and their impact on student motivation is limited. Furthermore, and to the best of the authors' knowledge, such systems have not yet been evaluated in formal educational settings.

The aim of our research is to study the effects of a tangible AR interface for teaching computational thinking in formal educational settings, regarding student motivation, learning effectiveness, collaboration and instructor support. We have designed and implemented a prototype tangible AR game for learning sequential instructions using mobile devices and 3D printed manipulatives, and performed a user study of its effectiveness during an educational intervention in a 3rd grade elementary school (8 to 9 years old). In the study we used two treatments: the experimental group played the tangible AR game, whilst the control group played the same game in an 'unplugged' version. The results of the study revealed some interesting findings regarding the suitability of the approach, its impact on student collaboration and on the role of the instructor, as well as some issues that need to be further studied.

2 Related Work

In recent years, a strong interest towards computational thinking (CT) is becoming apparent, regarding both education in the fields of STEM and computer science (CS), and students' overall development within modern society. Computational thinking refers to a specific way of understanding, analyzing and solving problems using known patterns, that can be directly and effectively visualized within the CS discipline but also apply to many different areas of science and everyday life. Instead of focusing on the syntax and semantics of computer programming, CT refers to the development of higher-level skills. However, it is a notion that has taken different dimensions over the years and the skills encompassing CT vary in the literature, as there is lack of a widely accepted definition. Critical thinking and problem solving have been reported as the most commonly related skills to CT [9]. Nevertheless, more skills have been correlated with CT in various studies, such as constructing algorithms, debugging, simulation and socializing [10].

Although many studies provide evidence that CT is more than a foundational skill in programming, implementing CT in typical education still faces many challenges. Some of those challenges have their roots in the core limitations of the current formal education settings, such as (a) the existing curriculum standards, (b) educators' lack of suitable skills and training and (c) limited institution resources which lack suitable equipment and infrastructures. As a result, much of the work in CT referring to K-12 education remains in out-of-school environments [11].

The most widely used approaches for teaching CT concepts are structured around students' interaction with computers, mainly based on block-based programming languages. A common approach in this direction is the use of visual tools, such as Scratch, where the program is written and its output is visualized in a graphical environment.

However, there is another approach gaining popularity in CT education, which includes activities with no use of digital devices: unplugged activities. Such activities are based on logic games using cards, strings or physical movements that represent computer science concepts such as algorithms or data transmission [12]. One of the main benefits of these activities is that they lowering the access barriers to the required equipment and infrastructures of educational tools [8], which results in easier implementation in field settings, consequently more concrete evidence concerning their use and effectiveness.

As technology evolves in exponentially increasing pace, block-based programming language in form of tangible interfaces have been gradually introduced in teaching CT, which enable programming by simply assembling physical blocks together. In this approach, each physical block stands for a respective command, and their combination creates a program. Students are programming on their desk or on the floor, which in the context of a classroom provides educators with the opportunity to more effectively determine and control the entire activity. Additionally, physical interfaces are exploiting the knowledge and skills that students already have through their own, non-computer related, experiences of everyday life, instead of forcing them to learn new skills, such as using commands of programming languages or non-physical interactions [13].

Physical programming platforms are following two approaches; the first include only physical parts and the second use the combination of digital and physical elements. One of the first examples of physical programming is the Programmable Bricks, created in the nineties at MIT [6, 14]. More recently, Google began its own research project within the field, with Project Blocks [14], which is based on Google visual programming language, Blockly.[1] This approach uses microcontrollers and microprocessors inside the blocks to communicate with the robot which increase the cost of construction. In the second approach, platforms like Tern [13], propose a solution where each physical block is read by an overhead camera, translated into commands, compiled by a computer and sent to a physical robot nearby. A more modern example is Osmo,[2] which uses the camera of an iPad device in order to execute the programming solution on screen with a digital character. This approach bears limitations concerning distance, light and color, but it uses widely available technology and reduce cost by combining the physical platform with a digital screen, gradually leading us to AR-based learning.

AR-based learning provides a new way to meet those challenges, concerning cost and complexity of the educational systems, while it could offer a natural way to present information based on reality [15], as it can enhance learning in terms of context visualization and learning interactivity [16]. AR allows augmented information for the surrounding real world to become interactive and manipulable [17], which could encourage students to have positive attitudes toward exploration, inquiry, logical thinking, and reasoning. In short, the two main affordances of tangible AR, i.e. visualization and interactivity [18], might have the potential to overcome some core difficulties that novices encounter in learning programming, such as abstraction and complexity. Furthermore, as a medium it may support the integration of modern technology in formal educational settings using widely available equipment.

A lack of studies is observed concerning concrete evidence on if, how and why tangible interfaces (TUIs) and more natural gestures create more effective learning experiences compared to other modern widely accepted approaches in formal education of computer science. Since now, there are cost and complexity issues, as the construction of tangible systems for programming is considered to be expensive, which provides an explanation for the lack of field evaluations [19]. Therefore, research in the field tends to work on how to make physical programming platforms more cost-effective, while enhancing interaction with widely available technology in formal education context [7, 8]. Mobile Augmented Reality (AR) technologies show great potential towards this direction.

---

[1] https://blockly-games.appspot.com/.

[2] https://www.playosmo.com.

3   A Tangible Augmented Reality Approach

We have designed and implemented "ALGO and his 'rithm", a collaborative game-based programming activity based on a widely accepted teaching model for CT concepts. The game-based learning approach is a shift from a traditional teacher-centered learning environment to a student-centered environment where students' participation and engagement are enhanced [20]. Our approach is based on mobile AR technology and is specifically focused in formal settings of primary education. This context of use significantly defined our design, as it creates a number of limitations and requirements. The core design decisions of our approach, in order to support learning and create an effective and controlled environment for both student and educators, are (a) the combination of tangible interaction with digital elements, (b) the collaborative context and (c) the low-cost and widely available equipment.

The educational goals of ALGO are to teach basic computational concepts, such as the development of algorithmic logic and sequential programming skills in a collaborative context. We address CT on the aspects of building algorithms and socializing within Computer Science discipline, as an introduction to programming for children. The main challenge introduced in the game is to create a programming solution – a sequence of steps – to assist a robot, named ALGO, collect a number of items and escape from a series of platforms (levels). "ALGO and his 'rithm" is following similar patterns regarding gameplay, challenges and educational goals to other games such as Light-Bot (2008)[3].

The game consists of two parts; the tangible part and the digital part. As every programming platform, there is the "programming language" (i.e. the commands) that is represented in tangible form and the execution that occurs on screen. It uses tangible blocks of commands in order to control ALGO on - screen, by taking advantage of marker – based AR using image targets. The user creates the programming solution of the given problem using the tangible blocks and then use the device to scan their solution and watch ALGO execute it on-screen (Fig. 1).

The prototype built for this study is limited to problems requiring sequential executions, which was the focus of our first study. The prototype is based on physical blocks of five different types, each of which represents a command that triggers ALGO's actions (to move in all four directions and to take an object). Respectively, the challenges introduced in the digital game are of the same complexity. In future implementation, we are planning to extend both the command set and the challenges to allow for conditionals and loops, and also to extend the gaming elements in the software.

The prototype has been implemented in the Unity game engine and uses Vuforia SDK which integrates AR capabilities in the game, in order to create a real-time interaction. The application runs on Android devices, tablet and smartphone. In our approach we have taken advantage of 3D printing as construction method for the tangible elements of the system, as they have been specially designed to snap together. However, the system is flexible regarding the physical elements, so it can support the

---

[3] http://lightbot.com/.

Fig. 1. The tangible AR version of "ALGO" game.

use of DIY tangible blocks using other materials like paper or cardboard and only the markers are required in order to play the game. That way our system supports an important aspect of this approach, as mentioned above, the requirement of widely available equipment and low-cost materials.

Before using our prototype in the main study, we have conducted a small lab session to evaluate its usability and improve any critical issues that might occur. For our evaluation we used six (6) students, age 8 to 9, forming two groups, where they had to complete three goals of the game. Our preliminary results showed that the interaction with the game was intriguing and challenging, as the participants did not have prior experience with AR, and indicated high levels of motivation and engagement for the kids. However, some technical and usability issues surfaced related to marker detection. The main problem was that the preliminary prototype supported linear "scanning" of the blocks (i.e. all blocks should be arranged horizontally), which made it difficult for the AR camera to detect them all simultaneously. This issue was resolved in the latest version of ALGO, by allowing for more than one lines of commands. Further minor usability issues that were observed during the tests, such as the size of specific virtual elements and the lack of audio or visual feedback for some actions, have also been improved.

4   User Study

This study is driven by the following research motivations: (a) what is the perceived value and effectiveness of using a tangible interface in an AR-based programming activity compared to an unplugged activity, in terms of programming performance and motivation, then (b) how the role of the educator is differentiated in class settings concerning those two approaches, in terms of involvement and engagement and finally, (c) does the AR-based approach support the educational goals, while addressing the limitations of the current formal education settings, such as educators' skills and training, and institution resources? What are the factors and the characteristics of the system that contribute to that?

We have performed a controlled experiment for addressing the above questions using the tangible AR approach as the experimental condition and an 'unplugged' version of the same game as the control condition. We used the same challenges (three tasks to be solved) in both treatments. In the experimental condition users interacted with the tangible AR prototype described in Sect. 3, whilst in the control condition they played the same game in an 'unplugged' version, by interacting only with physical elements. Specifically, the unplugged version had a physical tableau, a plastic cup and some colored tokens for the execution of the algorithm, the commands were given in pen and paper, and the algorithm was 'executed' by the human instructor. Figure 2

Fig. 2. Photos during students' interaction with ALGO in the AR (left) and unplugged (right) version.

shows students interacting with the game in both versions.

The study adopted a between-subjects approach to avoid the learning effect of the successive use of both conditions.

4.1   Setting and Participants

The study was held during regular school hours in an elementary public school in Syros, Greece as a special activity, in collaboration with their educator, who attended and participated in the process. The role of the educator in the process is critical, whereas there is a turning point from a traditional instructor to a facilitator in modern learning settings. As previous studies have highlighted that educators' understanding and engagement can be limited, in this study we purposefully involved the educator in all sessions. The authors had the role of the instructors during the activity, supported by the educator.

The experiment took place in the context of a classroom. The classroom was divided in two; for half workspaces a smartphone and the physical blocks was laying on the table and for the other half it was the unplugged prototype. Students were sitting around a table in groups of 3 to 4. Then, they were given the instructions by the authors. We described the goal, the commands and the execution process. The experimental group watched a short demonstration, as it was the first time for the student to interact with AR technology.

The participants consisted of a third-grade class of eighteen (18) students gender balanced (9 girls and 9 boys) 8 to 9 years old. As expected, most of them were familiar with mobile devices, such as smartphones and tablets. Although they were novices in computer science concepts, they had some basic prior experience with block-based programming tools.

4.2   Procedure

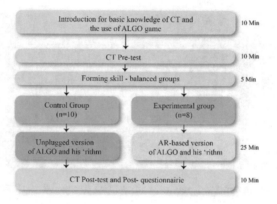

Fig. 3. The study procedure

The process during each session was the following (Fig. 3):

- *Introduction to algorithms and presentation of the activity*: a short introduction to the concepts of CT was held. Then the goal, the commands and the execution process in order to complete the activity was described.
- *CT pre-test*: students had to complete a short pre-test regarding their knowledge on CT concepts.
- *Formation of groups and placement*: students were divided in groups of 3 to 4 in order to work in a collaborative context. There was an existing grouping formed by the educator for every group to be skill-balanced in CS concepts. We respected the existing grouping for the assignment of the experimental and control conditions. Because of the formation of the teams we had two teams of three students and one team of four students (10 in total) in control group and two teams of four students (8 in total) in the experimental group.
- *Game-based learning activity in both conditions (experimental and control)*: the control group interacted with the unplugged prototype, while the experimental group interacted with the AR-based prototype. Each session took place in formal settings in a separated space, during regular school hours. The duration of each

session last around 60 min. During the activity students had to complete three levels i.e. create three programming solutions in both groups.

- *CT post-test and questionnaires*: students had to complete a short post-test and a questionnaire concerning their feelings towards the game-based activity.
- *Wrap up discussion*: a discussion followed the whole process with both students and the educator.

4.3 Measures

The data collection methods adopted during the experiment were observation, think aloud, questionnaires and open discussion, in order to collect both qualitative and quantitative data. The questionnaire aimed to provide feedback about student perception concerning enjoyment and ease of use. Each item of the questionnaire was scaled on the 5 - point Likert scale adapted for children, i.e. using simple language and graphic representations (smileys) instead of numbers. Combined with the questionnaire, we adopted a simplified adaptation of the CT test in order to access the level of CT in students participated in our research, before and after the experiment [12]. Our version of CT test was adapted to the visualization of ALGO's interface (i.e. images of our prototype, game's character etc.) and it focuses on sequences, as a computation concept. At the end of each session an open discussion with the educator was carried out. The performance on programming was evaluated per group, based on the observation of both the research team and the educator. However, the contribution of each student in the collaboration aspect was considered.

5   Results and Discussion

As an educational intervention, the study was quite successful. Both groups participated actively in the process, and most of them managed to solve the challenges. Notably, some of the groups that finished earlier asked for further, more challenging problems, which the instructors provided, at least in the unplugged version.

Regarding the enjoyment factor, the majority of the students of both sessions enjoyed participating in the experiment, as it was shown from the questionnaires where all answers were above neutral (i.e. above 3). The minimum average per question is 3.5, while most of them is around 4.5. The enjoyment was also apparent through observation of students' reactions and comments. For instance, there was a student of the experimental group who asked if they can play it at home. Despite the fact that the AR- based system was an early functioning prototype and engagement factors commonly used in game environments was limited (i.e. only basic animations and sounds was implemented), the new way of interaction provided made the activity challenging and intriguing.

Being the first time for most students to interact with AR technology, it created reactions of surprise and excitement, but also concerns. The most important feedback we received from the students was through the think aloud protocol. We gathered comments like "Wow, it can see the blocks!", referring to the smartphone, but also "Did it get it now?", expressing concern weather the device records the commands, misconceiving the indicators of the action's feedback. Although the students were not familiar with AR technology, which raised concerns and required time to adapt, they were familiar and feeling at ease with the tangible part. Many students described the physical blocks like a puzzle and it was easy for them to understand how they were connected together, which considerably helped to avoid misinterpretations of the semantics - which was crucial as we will explain later below.

Students' prior experience with introductory programming had a mixed effect on the activity. It has been made evident from the CT tests, as within the control group 70% of participants completed correct all answers in pre-test (80% in post-test), while within the experimental group 75% completed correct all answers in pre-test (87% in post-test), and from the discussion with the teacher that they already had an understanding of fundamental concepts through other introductory activities, which made it easier for them to understand what they were supposed to do. However, it also led to some misinterpretations, as they had already created a mental connection of semantics to specific actions. Specifically, in the context of our game the action of collecting an item is successful only if the robot is already on the same block with it. The majority of the students were inclined to use it when ALGO was in a neighboring block. For thirteen (13) students out of eighteen (18), around 68% of all participants, that misconception was obvious in the pre-test. Throughout the activity most of the students of the experimental group understood the difference of semantics through a trial and error procedure. They took the initiative to try out the meaning of the blocks/commands, before they start working towards the goals. On the other end, the control group did not have that opportunity, as they had to wait for the instructor to check their solution. In this respect, the tangible AR activity provides the opportunity for more initiative actions, personalized features and direct feedback, while the unplugged version is solely dependent on the instructor for feedback.

Concerning collaboration, we observed different types of organization and management of the tasks and roles inside every group. We did not give any directions regarding collaboration patterns and there was no specified role assignment inside the group, i.e. someone to control the device or someone else to issue the commands. Therefore, every group created their own pattern of collaboration. For instance, in the experimental group one team decided that one member would solely handle the device and the other decided to take turns. The AR-based version facilitated more structured and clear collaboration patterns, as it has clear sequence of actions guided by the system. On the other end, the unplugged version does not define clear steps by itself, as they face all at once the "execution tableau", the problem and the means to create the solution, which created distractions and made the role of educator crucial. Indeed, the most incidents of disagreement occurred during the control group session, which considerably affected the total time spend on every goal, prolonging the session. That indicates two things, the AR-based version (a) facilitates control by the instructors, which means that during the experimental session, instructors could maintain control and guidance easier as they had less involvement in the activity itself and (b) supports more effective and structured ways of collaboration.

Finally, through our open discussion with the educator another challenge occurred for long term use of the AR-based version of ALGO in formal settings, concerning the need of educators of keeping track of students' performance. The game is keeping the score temporarily but it does not provide the educator with the feedback they require in long term. Therefore, they have to keep track for themselves, which increases the level of involvement. During this study there was some limitations in order to provide more concrete evidences, such as our early functioning prototype i.e. only a few levels were implemented, while more complex computational concepts are planned for future iterations. It is of significant importance to test our approach with a full functioning prototype of ALGO and his 'rithm in long term use, although open discussion with the educator gave us important feedback concerning that matter.

6 Conclusions

In this paper we present the results from a preliminary study in an elementary school that compared a tangible AR game with the same game in an unplugged version in teaching computational thinking, to examine the effect of the interface on student motivation, effectiveness and teaching practice. The results were encouraging, in terms of improving engagement and collaboration, while they provided interesting insights towards teaching practice. Both approaches show positive outcomes on students' enjoyment. The AR-based version seems to prevail in terms of (a) collaboration, (b) educator's control, (c) excitement; emerge from interacting with technology combined with physical object, (d) direct feedback, while both approaches share (d) the benefit of low cost or widely available equipment.

Field evaluations are very meaningful, as real context requirements and limitations surface. More field sessions focusing on qualitative and quantitative results concerning educational goals in long term use of our approach are needed. Some interesting directions for this approach arose through the study. It will be interesting to create a more challenging context, outside the game, i.e. by expanding the interaction between the groups in term of collaboration or competition in kind of a multiplayer game.

A section of free play is a plausible solution for future work and research, in order for students to try out and understand the meaning of semantics before they start playing, without being discouraged, as a trial and error way tend to make children feel disenchanted. Finally, as an educational tool intended for classroom settings, an educator version that will keep track of students' performance, in long term, may be an interesting direction.

References

1. Walker, C.O., Greene, B.A., Mansell, R.A.: Identification with academics, intrinsic/extrinsic motivation, and self-efficacy as predictors of cognitive engagement. Learn. Individ. Differ. **16**(1), 1–12 (2006)
2. Papastergiou, M.: Digital game-based learning in high school computer science education: impact on educational effectiveness and student motivation. Comput. Educ. **52**(1), 1–12 (2009)
3. Guzdial, M.: Education Paving the way for computational thinking. Commun. ACM **51**(8), 25–27 (2008)

4. Myers, B.A.: Taxonomies of visual programming and program visualization. J. Vis. Lang. Comput. **1**(1), 97–123 (1990)
5. Bell, T., Alexander, J., Freeman, I., Grimley, M.: Computer science unplugged: school students doing real computing without computers. N. Z. J. Appl. Comput. Inf. Technol. **13** (1), 20–29 (2009)
6. McNerney, T.S.: From turtles to tangible programming bricks: explorations in physical language design. Pers. Ubiquitous Comput. **8**(5), 326–337 (2004)
7. Goyal, S., Vijay, R.S., Monga, C., Kalita, P.: Code Bits: an inexpensive tangible computational thinking toolkit For K-12 curriculum. In: Proceedings of the TEI 2016: Tenth International Conference on Tangible, Embedded, and Embodied Interaction, pp. 441–447. ACM (2016)
8. Klopfenstein, L., Fedosyeyev, A., Bogliolo, A.: Bringing an unplugged coding card game to augmented reality. INTED Proceedings, pp. 9800–9805 (2017)
9. Kranov, A. A., Bryant, R., Orr, G., Wallace, S. A., Zhang, M.: Developing a community definition and teaching modules for computational thinking: accomplishments and challenges (2010)
10. Kazimoglu, C., Kiernan, M., Bacon, L., MacKinnon, L.: Learning programming at the computational thinking level via digital game-play. Procedia Comput. Sci. **9**, 522–531 (2012)
11. Lee, I., et al.: (2011). Computational thinking for youth in practice. ACM Inroads **2**(1), 32–37
12. Brackmann, C.P., et al: Development of computational thinking skills through unplugged activities in primary school. In: Proceedings of the 12th Workshop on Primary and Secondary Computing Education- WiPSCE 2017, pp. 65–72. ACM Press, New York, New York, USA (2017)
13. Horn, M.S., Jacob, R.J.K.: Designing tangible programming languages for classroom use. In: Proceedings of the 1st International Conference on Tangible and Embedded Interaction (TEI 2007), (February), pp. 159–162 (2007)
14. Paulo Blikstein (Stanford University, USA), Arnan Sipitakiat (Chiang Mai University, Thailand), Jayme Goldstein (Google), João Wilbert (Google), Maggie Johnson (Google), S. V. (Google), & Zebedee Pedersen (Google), W. C. (IDEO). (2016). Project Bloks : designing a development platform for tangible programming for children
15. Carmigniani, J., Furht, B.: Augmented reality: an overview. In: Handbook of Augmented Reality, pp. 3–46. Springer, New York (2011)
16. Teng, C.-H., Chen, J.-Y., Chen, Z.-H.: Impact of augmented reality on programming language learning. J. Educ. Comput. Res., 073563311770610 (2017)
17. Santos, M.E.C., Chen, A., Taketomi, T., Yamamoto, G., Miyazaki, J., Kato, H.: Augmented reality learning experiences: survey of prototype design and evaluation. IEEE Trans. Learn. Technol. **7**(1), 38–56 (2014)
18. Azuma, R.T.: A survey of augmented reality. Presence Teleoperators Virtual Environ. **6**(4), 355–385 (1997)
19. Sapounidis, T., Demetriadis, S., Stamelos, I.: Evaluating children performance with graphical and tangible robot programming tools. Personal Ubiquitous Comput. **19**(1), 225–237 (2015)
20. Sung, H.Y., Hwang, G.J.: A collaborative game-based learning approach to improving students' learning performance in science courses. Comput. Educ. **63**, 43–51 (2013)

Using Gamification in Teaching Public Relations Students

Elizaveta Osipovskaya[1] and Svetlana Miakotnikova[2(✉)]

[1] RUDN University, Miklukho-Maklaya St. 10/2, Moscow, Russia
e.osipovskaya@gmail.com
[2] Perm National Research Polytechnic University,
Komsomolsky Av. 29, Perm, Russia
myakotnikova@yandex.ru

Abstract. The emergence of modern technologies and their successful application to teaching and learning has been modifying the educational system at higher schools. A new millennial generation, so-called digital natives, learn and process information differently. They are digitally savvy, are keen on games and express themselves openly by posting comments and sharing posts on social media. New trends in education technologies have to adopt these peculiarities and urge educators to apply them to teaching. The article covers aspects regarding the application of Gamification strategy to teaching of Public Relations students at higher school. Game-based learning is used as a part of instruction in two different courses (Graphic Design and Business English). The paper investigates the implementation of game elements into e-Learning environments, namely the Moodle platform. It is assumed that their application to solving business problems at the lessons enhances students' motivation and creates the atmosphere of a real Public Relations department. All together it generates and develops various competencies of future specialists in terms of developing professional, social and personal values.

Keywords: e-Learning · Gamification · Higher education · Public Relations

1 Introduction

Gamification is the use of game mechanics and game design techniques in non-game contexts. It uses natural desire for competition, achievement, status or collaboration. Educational institutions are interested in educating students in a gamified way, because it improves their learning outcomes. Given the potential benefits of the gamification in education, we are interested in identifying gamification components: (1) objectives – a behavioral mechanic type encouraging a user to take action for the reward; (2) progression – moving a user through the content; (3) feedback – informing a user of their status.

Game elements include interactivity, instant feedback, progress indicators (motivational factors that show students what they need to do to finish a particular level), time limits (digital learners enjoy competing against the time), repetition, scoreboards (leaderboards), badges and awards, and social interaction.

© Springer Nature Switzerland AG 2020
M. E. Auer and T. Tsiatsos (Eds.): ICL 2018, AISC 916, pp. 685–696, 2020.
https://doi.org/10.1007/978-3-030-11932-4_64

This article presents the results of the experiment involving gamification conducted for 3rd year students in a Graphic Design course at the Russian University of People's Friendship (RUDN University), Moscow and for 2nd year students in the Business English course at Perm National Research Polytechnic University (PNRPU), Perm in the winter semester of 2017/2018 academic year. Two researchers ran classes with two different groups of students who study Public Relations as their major. Each researcher chose one group to be an experimental group, while another one was a control group. Gamification method was introduced in the experimental group whereas in the control group classes were given in the usual way. Students in all 4 groups were approximately of the same age: 18–20 years old; gender was not taken into account.

2   Literature Review

2.1   Conceptualizing Gamification

Games are the oldest forms of everyday human social interaction. Human beings are defined as homo ludens, a concept proposed by Huizinga [1]. He describes game concept as a free activity standing quite consciously outside of «ordinary» life, as being «not serious» , but at the same time intensely absorbing. It is a discharge of superabundant vital energy, by others as the satisfaction of some «imitative instinct».

Gamification is worthy of special mention when talk about a game. Nick Pelling [2] defines it as the application of game-like accelerated user interface design to make electronic transactions both enjoyable and fast. The term has changed a lot since then, embracing various aspects, presenting learning content and using game-like elements to boost student engagement. Apparently, the most widespread definition of gamification is «the use of game design elements in non-game contexts», proposed by Zichermann [3]. He also defined core game mechanics, such as points, badges, levels, challenges, and leaderboards.

Game-based learning is the use of games as a part of instruction. Games can be used in a course that has been gamified; on the other hand, using games does not mean that you have a gamified course. A game is a system in which players are engaged in an artificial conflict defined by rules that results in a quantifiable outcome. A game is a structured experience with rules, goals and progress that is fun. A game is the use of techniques to guide and motivate a player throughout a journey.

A review of the literature on gamification presents its various elements in educational and learning contexts. Researchers incorporate levels, leaderboards, challenges and badges into teaching. Barata [4] noticed that a gamified approach led to greater students' engagement and increased the number of lecture downloads; it boosted attendance. Unfortunately, it did not significantly improve grades. Brewer [5] emphasized that game elements enhanced learners' motivation. This is consistent with Eleftheria [6] who confirmed that gamification made the learning experience more engaging, enjoyable and productive. Kumar and Khurana [7] said that students did not express much interest in the traditional way of learning. However, they were interested in a gamified approach. According to Kapp [8] who used storytelling and feedback as game design elements, frequency, intensity, and immediacy of feedback are important for sustaining engagement throughout the learning process.

2.2 Moodle as an e-Learning Tool

There has always been a strong relationship between education and technology. Tools and technologies in their broadest sense are important drivers of education. Writing, one of the most important tools in the development of human civilization, was not invented for education but for commerce. Books were initially used to spread the word of religion, not to educate. Education adopted both but had little effect on driving the development of either. Even slide presentation tools were invented for the business community. We have to acknowledge that, typically, education does not drive technological invention. Instead, it appropriates the useful inventions of business and leisure industries. In the age of rampant technological invention this becomes a critical issue [9].

Attitudes concerning e-learning range from neutral to positive. On the one hand, it is described as an effective traditional educational process [10] and it is stated that there are no major differences in academic performance between these modes of instruction [11]. On the other hand, many researchers argue against the positive effect of e-learning [12].

There is a lot of evidence for the claims that teaching delivery styles have different degrees of success. Concerning online teaching, some studies indicate that it has a positive impact on performance, for example, Smith and Hardaker [10]. Other studies however, find that e-learning, on the contrary, has a negative impact on performance [13].

In the field of our study many definitions of e-learning can be found. As mentioned by Alonso [14], this term encompasses electronic delivery of teaching and learning as well as theoretical dimensions of cognition to ensure the effective use of technological tools. Others highlight that it is the use of new multimedia technologies and the Internet to improve the quality of learning by facilitating access to resources and services; it is remote exchange and collaboration [15].

For this study, online learning is defined as learning that takes place partially or entirely on the Internet. This definition includes online platforms and applications that have a significant Web-based component. By e-learning we mean combined or blended (sometimes called hybrid) learning with face-to-face instruction.

For the past several years, many Web 2.0 technologies have been developed. These services have been immediately utilized by students. It heralded a shakeup of the classic lecture model. One such service is Learning Management Systems (LMS) that is freeware and open source. An excellent case is Moodle, an acronym for the Modular Object-Oriented Dynamic Learning Environment. The first public version of this software was released in 2001 by Martin Dougiamas. It is one of the most promising Learning Management Systems which has been translated into twenty-seven languages. Moodle is a free online platform enabling educators to create private websites filled with dynamic courses that extend learning. It has a modern, responsive and accessible design; it is easy to navigate on both desktop and mobile devices. Materials can be accessed at any time and the coursework is self-directed.

A virtual classroom creates engaging and interactive learning environments. Moodle integrates such features as a whiteboard, chat, audio, video, comment box, and embedded PowerPoint presentations.

The Moodle platform has several useful and engaging features. One of the advantages of Moodle is that it gives a teacher a lot of freedom. They can add texts, links, images, videos by clicking on an appropriate icon. There is an option to upload additional support documents, to set a date when a teacher wants learners to send in their works with due dates or cutoff dates. Additionally, a teacher can choose a submission type: to upload one or several files, to specify the type of a file. With a Moodle text a teacher-editor can make an assignment to write an essay with a set word limit. By pressing the feedback button a teacher can set a comment line that allows them to type directly on learners' works just as they would do when grading a paper.

Moodle facilitates students' learning. Such a feature as a gradebook gives them access to their course grades and real-time updates. Grade history allows them to see different grades when they are modified. "A completion tracking" option allows students to see their progress throughout the course using check or tick boxes on the sides of activities. The boxes with dotted lines are ticked automatically when students meet certain criteria.

There are Moodle features which make learning more engaging, collaborative and fun. Emoticons help students to express emotions and humor and strengthen messages. Learners can collaborate in online groups and then deliver presentations in front of a class. An entire class can also watch their project on a group's whiteboard.

Finally, a Moodle platform has wiki features. Wikipedia is a collaborative platform which allows to edit pages by adding information. Likewise, Moodle has a wiki too. A teacher can set up such an option for leaners to do a group project. It also has "History" that permits people to see changes on a wiki and to revert them.

As regards game elements that can be used in Moodle courses, they are the following:

1. Crossword - a game takes words from glossary and generates a random crossword puzzle. It gives students the sense of feedback, a teacher allows student to take it as many times as they need to master terms.
2. Millionaire game – a game takes words from multiple choice quiz questions and creates a "Who wants to be a Millionaire" style game complete with the three lifelines. It interfaces with quizzes. Student have "50 × 50", "Phone your friend", "Ask the audience". There are a lot of repetitions in this game which is a great way to study and master basic terms.
3. Quizventure – an adventure game which includes questions from a question bank. Possible answers come down the screen as spaceships and students have to shoot the correct one to gain points. Students use space bar to shoot the aliens ships in the sky and arrow keys to navigate back and force.
4. Badges – an online recognition of students' achievements and skills. It is a great motivation and engagement tool to award learners for the best assignments.
5. Certificate of achievement – it shows that a student has accomplished a specified level of achievement. It is based upon requirements that a teacher has set.

6. Scoreboards. A teacher displays scores to encourage students and motivate them to proceed.
7. Progress bar. This item is a good visual indicator for students who monitor their progress. It can produce reports for a teacher too.

That covers the description of key concepts and characteristics of Moodle as the e-learning platform and principals of interaction in digital environment between a teacher and a student.

The aim of this paper was to broaden current knowledge of using gamification in higher education and look into how Public Relations students can be taught their professional skills by application of the new approach.

3 Methodology

It was decided that the best methods for the current research would be quantitative and qualitative methods: observation, a survey in the control and experimental groups and a questionnaire.

The researchers were interested in observing the changes in students' engagement, participation, motivation, productive learning levels before and after the courses (Graphic Design Course at RUDN and Business English Course at PNRPU).

The control groups provided a baseline for the effectiveness of game implementation. The control groups allowed to draw a meaningful comparison against the experimental groups in the form of a survey about class performance, home assignments, class attendance, test results and final grades.

The questionnaire was used as a tool of receiving favorable feedback about attitudes to gamification from respondents. It included close-ended questions, as it is believed that they are time-efficient and responses are easy to code and interpret. At the same time there was a place to express opinions in a free manner.

4 Incorporation of Gamification Elements into Graphic Design Course

Gamification was conducted in the form of an experiment in a Graphic Design class (GD) for 3rd year students of Public Relations at RUDN University. The researcher randomly chose two groups: an experimental one in which gamification methods were introduced and a control group in which classes were conducted in the same way as in previous years. Each group of students was of the same size (approx. 18 people). All activities and topics were taught in the same manner, using the same training documentation.

The curriculum content introduced in the experimental group was a quest-based game. Using a platform called Readymag, the teacher designed content as series of quests for students to complete; students were rewarded with points and badges. The content was provided in a multimedia format and included teacher-created and external public information.

Students assumed the roles of designers and PR-managers who apply for a tender to organize a Science Forum. They wrote a comprehensive guide that embraced a theme, speakers, branding (logo, a visual design of print and digital publications, an event website) and a promotional plan.

In the control group all classes were conducted in a classic manner. The students were taught how to create icons, infographics in a vector-based program Adobe Illustrator; they expanded skills in a pattern design; learned how to manipulate font into an interesting lettering piece and played with the texture.

After the experiment each group of students was provided with a questionnaire to assess the content of the class.

Gamification mechanisms used in the course included badges of honor, leaderboards, two levels of competitions, time bars, and immediate feedback.

Badges of Honor. The best PR-team was awarded the highest grade. Students earned badges for progression through the quest levels. For example, the first level of the quest was centered around presentation of a bid (a main idea, a Forum topic). Students had to take part in a presentation battle and prepare a pitch about their idea. At the end of the contest the teacher chose the best presentation. The second level involved development of a marketing plan to raise awareness and attract attention to the event. The third level connected with the creation of a backgrounder (interactive PDF) that contained information about the Forum, facts and stories related to key speakers. On the forth level students got an assignment to create square-shaped promo videos for Instagram and Facebook.

Leaderboards. On Moodle students could see a thumb-nail image of their team on a real-time leaderboard. On the one hand, it displayed achievements of other students but, on the other hand, it motivated them to improve their performance.

Two Levels of Competition. The first level evaluated students' individual participation, while the second showed the results of the teamwork and its synergy.

Time Bars. Sections and colors of the bars varied depending on the expected completion date. For example, green indicated an activity that was accomplished, red indicated that team did not meet the deadline, and blue showed that the expected completion date had not been met yet.

Immediate Feedback. The teacher noticed that students who were given immediate feedback showed a significantly larger increase in performance, than those who had received delayed feedback.

A survey was conducted on the results of a class performance in both control and experimental groups. The results are presented in Table 1.

5   Incorporating Gamification Elements into Business English Course

Business English course was conducted for 2nd year students of Public Relations at Perm National Research Polytechnic University. Two groups of students (control and experimental) were approximately of the same size (13–15 people in each group) and of the same age – 18–19 years old. The researcher used the same educational materials in each group. They included topical texts, vocabulary exercises, videos and case studies. The difference was that for the experimental group the resources were uploaded on the Moodle platform. The Moodle platform facilitates gamification of the learning process. Muntean [16] and Henrick [17] describe Moodle gamification capabilities as using avatars; possibility to track students' progress, to display quiz results and to give immediate feedback.

The Business English course syllabus is made up of 3 modules: Management, Advertising and Marketing, and Social Responsibility. Both experimental and control groups started the course with the Management module, followed by Advertising and Marketing and ending with Social Responsibility.

Students in the control group had lessons in a traditional manner: reading and discussing topical texts, learning a target vocabulary, doing vocabulary exercises, watching videos and giving summaries, making presentations, doing case studies. The other differences included the absence of a consistent scenario, the absence of regular feedback after each lesson and the absence of gamification elements in delivering the course for this group.

Students in the experimental group studied in the form of a game. They were asked to establish their dream company. Their goals were to create an innovative product, to develop an advertising and marketing campaigns for this product and to think of socially responsible activities connected with the product which their company can offer to society. The home assignments included: drawing (or making) a new product and describing its specifications; presenting a management structure of their company, the activities of the departments; developing an advertising campaign for the product and creating a video commercial; working out a social responsibility policy for their company and connecting it with the product.

Gamification mechanisms used in the course included crosswords, badges of honor, leaderboards, two levels of competitions, time bars, and immediate feedback.

Crosswords. To encourage students to study the topic vocabulary the teacher placed a link to "Quizlet" on the university Moodle platform. Quizlet is a digital tool that allows to create digital flashcards with words and their definition or translation. Students can listen to the pronunciation of a word or/and see its image. Different activities with flashcards include a spelling activity, a timed matching task and a space gravity game. Lindsay Warwick believes that these activities help learners move the vocabulary into long-term memory [18]. Quizlet Live feature allows students to compete against each other in teams. "Class Progress" feature on Quizlet's teacher account enables to monitor students' progress.

Crosswords were used as a part of a lesson or a home assignment.

Badges of Honor. The best company was awarded the highest grade. Throughout the game students were awarded additional badges of honor for the best completed current assignments.

Two Levels of Competition. Zichermann considers that competition increases students' engagement [3]. The first level evaluated students' individual performance, while the second presented the results of the established companies.

Leaderboards. On Moodle (at home or in the classroom) students could see the images of created products, watch commercials and compare them with their own creations. They could better understand why other groups were assigned that number of points/those badges and what place in leaderboard they occupied. Leaderboards, on the one hand, allow students to view their achievements compared to other students, but on the other hand, they create the sense of belonging to a similar minded group [19].

Time Bars. Students always had time restrictions which prevented them from procrastination of assignments.

Immediate Feedback. Kapp states that giving feedback is critical in learning. It also helps track individual and group progress [8]. Students received immediate feedback after each lesson in face-to-face meetings or on the Moodle platform after completion of their home assignments.

Throughout the experiment the researcher observed students' work and collected data on students' attendance, participation at the lessons, and homework assignments. Students sat for the vocabulary test and did the case study. The results are presented in Table 2.

6 Results and Discussion

The results of the experiment included both quantitative and qualitative indicators in Graphic Design (GD) and Business English (BE) classes. Tables 1 and 2 present quantitative data which took into account the results of the project work (GD)/vocabulary test (BE), workshop (GD)/case study (BE), and doing homework. Additional indicators included presence or absence at the lessons, final grade students got at the exam and the number of students who passed the test.

Table 1. The quantitative results of the class performance in Graphic Design course (GD)

	Project (out of 5 points)	Workshop (out of 5 points)	Homework (out of 5 points)	Attendance (%)	Final grade	Exam passed
Control group (18 students)	4.2	4.1	3.8	79%	4.0	17 students
Experimental group (18 students)	4.7	4.9	4.6	96%	4.8	18 students

Table 2. The quantitative results of the class work in Business English course (BE)

	Test (out of 5 points)	Case study (out of 5 points)	Homework (out of 5 points)	Attendance (%)	Final grade	Exam passed
Control group (13 students)	4.1	4.0	3.9	80%	4.0	12 students
Experimental group (15 students)	4.6	4.8	4.7	94%	4.7	15 students

On average the results confirm that the students in both control and experimental groups at both courses were rather successful in their achievements. However, the Business English course experimental group grades were higher by 13,5% and 100% of the students passed the exam, compared to the control group where 2 students failed. As for Graphic Design experimental group results, students' grades were higher by 14% and also 100% of learners passed the exam, compared to the control group where 1 student failed.

Qualitative research consisted of in-class observations and questionnaires filled out by students as part of the course feedback.

The in-class observations revealed the differences in students' attitudes to the course in general and to the given assignments in particular. Having received their assignments, students in the control groups regarded them as routine assignments. In contrast, students in the experimental groups got excited and enthusiastic from the very beginning. They took their time to invent products (in the Business English course) at the initial stages of the game and continued being engaged throughout the game. This corresponds with the findings of W. Hsin-Yean who states that gamification affects students' behavior, commitment and motivation, which in its turn can lead to improvement of knowledge and skills [20].

The students of both control and experimental groups at RUDN and PNRPU were provided with the same questionnaire to evaluate their attitudes to the courses and gamification elements employed at the studied subjects (Graphic Design and Business English) The questionnaire included the following questions:

1. Did you like the class?
2. Did the class help you to achieve your learning goals?
3. Did the class engage and motivate you?
4. Did the class help you to do home assignments?
5. Did the class facilitate your preparation for the final exam?

The researchers used the same questionnaire in two different subject courses to find out if the subject can have any effect on students' attitudes to gamification. Table 3 presents the results.

Among the best engaging gamification elements in the Business English course, Quizlet was mentioned for its simplicity to train vocabulary and availability of leaderboards which encouraged a competitive spirit. For the Graphic Design course,

Table 3. The questionnaire on students' attitudes to the Graphic Design (GD) and Business English (BE) course

	I liked the class	The class helped me to achieve my learning goals	The class engaged and motivated Me	The class helped me to do home assignments	The class facilitated my preparation for the final exam
GD control group (18)	18	16	15	15	17
GD experimental group (18)	18	18	18	18	18
BE control group (13)	12	12	11	12	13
BE experimental group (15)	15	15	15	15	15

badges of honor appeared to be the most enjoyable element. Students also liked the presentation battle game that mastered their skills in public speaking.

To sum up, "My dream company" (the Business English course at PNRPU) game and the quest (Graphic Design course at RUDN University) have served its purpose of raising engagement by using game elements. Unfortunately, a classic approach does not produce such incredible learning outcomes, and even sometimes leads to students' dissatisfaction and lack of emotional connection with a teacher.

For both Graphic Design and Business English courses the researchers have chosen a blended type of learning that combines in-class face-to-face instruction with learning via an online platform. The researchers believe that such a teaching strategy will make in-class interaction more active and collaborative and will minimize or even eliminate a passive transmission of information. Therefore, students will acquire information at home (online), read additional content beyond their university curriculum, browse through a designer's portfolio that creates cutting-edge projects, watch supplementary tutorials and focus more on practical assignments in-class.

Moodle-based courses are expected to create the following effects on Public Relation students:

- Improve learning skills. Students can access courses at any time through 24/7 accessibility and are able to learn from any topic they wish. Moreover, learners find it enjoyable to connect to other students online and exchange views, participate in discussions and engage in forums.
- Allow students to acquire information in different formats. Information can be presented in multiple formats (text, audio, video and graphic). It enriches the experience of students and enhances their capacity to absorb information.

- Develop self-efficacy. Students have to respond individually to questions posted in blogs and forums. So, they think and act for themselves. They learn to set forth arguments, while both defending their own position while simultaneously cutting down someone's arguments.
- Build self-discipline. They learn to manage multiple tasks and do assignments before the deadline This is a crucial competence, because a PR professional constantly operates on tight deadlines and needs to be able to organize their workload in order to meet strict deadlines.
- Improve communication. Online communication requires students to organize ideas in a way that allows readers to understand the nuances of a message as well as its content. Discussion boards help all class members, especially those who might be more reticent to contribute to discussions, to participate equally and receive quality individual feedback from the faculty members and peers. That is undoubtedly true, that it does help to establish mutually respectful relationships.
- Activate prior knowledge. Repetitive teaching techniques are extremely useful with homeschool tutoring. A teacher can include prior knowledge on subjects in assignments for better learning.
- Provide feedback. Meaningful feedback through forums and comment boxes greatly enhances learning and improves student achievement.

To sum up, e-education improves student learning more than traditional lectures where students may remain passive and refuse to make contributions.

7 Conclusion

The researchers are at the beginning of introducing game elements into their courses. Plenty of further research is to be done in the field of gamification to collect enough empirical data to support the theory and to make universal recommendations.

At the same time some preliminary conclusions state that if educational institutions are looking for the ways to improve teaching effectiveness, they should consider aiming at developing e-teaching strategies. The incorporation of game elements into courses leads to greater student engagement, motivates them to complete assignments, fosters collaboration, activates a competitive spirit and enhances their digital competencies. However, gamification is not a universal panacea, some elements may work better than others, some may even fail. Gamification projects have to be carefully designed to address the real challenges of universities. The publication has been prepared with the support of the «RUDN University Program 5-100».

References

1. Huizinga, J.: Homo Ludens: A Study of the Play Element in Culture. Trans. RFC Hull, pp. 26–31. Boston, Beacon (1955)
2. Pelling, N.: The (short) Prehistory of Gamification. Funding Startups, pp. 36–41. Haettu (2011)
3. Zichermann, G., Cunningham, C.: Gamification by Design. O'Reilly Media Inc, Sebastopol (2011)

4. Barata, G., Gama, S., Jorge, J., Goncalves, D.: Engaging engineering students with gamification. In: 5th International Conference on Games and Virtual Worlds for Serious Applications, pp. 1–8 (2013)
5. Brewer, R., Anthony, L., Brown, Q., Irwin, G., Nias, J., Tate, B.: Using gamification to motivate children to complete empirical studies in lab environments. In: 12th International Conference on Interaction Design and Children, pp. 388–391 (2013)
6. Eleftheria, C.A., Charikleia, P., Iason, C.G., Athanasios, T., Dimitrios, T.: An innovative augmented reality educational platform using gamification to enhance lifelong learning and cultural education. In: 4th International Conference on Information, Intelligence, Systems and Applications, pp. 1–5 (2013)
7. Kumar, B., Khurana, P.: Gamification in education–learn computer programming with fun. Int. J. Comput. Distrib. Syst. 2(1), 46–53 (2012)
8. Kapp, K.M.: Games, gamification, and the quest for learner engagement. Train. Dev. 66(6), 64–68 (2012)
9. Laurillard, D.: Teaching as a Design Science: Building Pedagogical Patterns for Learning and Technology, p. 5. Taylor and Francis, UK (2017)
10. Smith, D., Hardaker, G.: E-learning innovation through the implementation of an internet supported learning environment. Educ. Technol. Soc. 3, 1–16 (2000)
11. Cavanaugh, C.S.: The effectiveness of interactive distance education technologies in K-12 learning: a meta-analysis. Int. J. Educ. Telecommun. 7(1), 73–88 (2001)
12. Maier, R.: Knowledge Management Systems, 3rd edn. No. 8/9, pp. 506–516. Springer, Heidelberg, Perth, Western Australia (2007)
13. Johnson, G.M.: Student alienation, academic achievement, and WebCT use. Educ. Technol. Soc. 8, 179–189 (2005)
14. Alonso, F., López, G., Manrique, D., Viñes, J.M.: An instructional model for web-based e-learning education with a blended learning process approach. Br. J. Educ. Technol. 36(2), 217–235 (2005). UK
15. Boss, S.: PBL for 21st Century Success: Teaching Critical Thinking, Collaboration, Communication, and Creativity, pp. 5–7. Buck Institute for Education, California, USA (2013)
16. Muntean, C.: Raising engagement in e-learning through gamification. In: 6th International Conference on Virtual Learning ICVL, pp. 323–329 (2011)
17. Henrick, G.: Gamification - What is it and What it is in Moodle. Slide Share (2013). http://www.slideshare.net/ghenrick/gamification-what-is-it-and-what-it-is-in-moodle
18. Warwick, L.: Quizlet. Efficient, engaging, flashcard tool. The Digital Teacher (2017). https://thedigitalteacher.com/reviews/quizlet
19. O'Donovan, S., Gain, J., Marais, P.: A case study in the gamification of a university-level games development course. In: Machanick, P., Tsietsi, M. (eds.) Proceedings of the South African Institute for Computer Scientists and Information Technologists Conference (SAICSIT 2013), pp. 242–251. ACM, New York (2013)
20. Huang, W.H.-Y., Soman, D.: Gamification of Education. Toronto: University of Toronto. Inside Rotman (2013). http://inside.rotman.utoronto.ca/behaviouraleconomicsinaction/files/2013/09/GuideGamificationEducationDec2013.pdf

Creating and Testing a Game-Based Entrepreneurship Education Approach

Ines Krajger, Wolfgang Lattacher$^{(\boxtimes)}$, and Erich J. Schwarz

Alpen-Adria University, Klagenfurt, Austria
{ines.krajger, wolfgang.lattacher,
erich.schwarz}@aau.at

Abstract. Entrepreneurial games are a promising tool for entrepreneurship education. Students may gain a range of competences and – if feeling satisfied with the gaming episode – develop increased interest in the field. Despite some progress over the last decade, game development and accompanying research are still in exploratory stages. Thus, further best practice examples and insights in game mechanisms are essential. We support progress in two ways. First, we present a self-developed entrepreneurial game, called *inspire! build your business*. Second, based on data from three *inspire!* game events, we explore factors influencing gamer's satisfaction via a quantitative approach. We find that the relationship between teacher and students, the structure of the game session and the perceived gains in competency significantly explain variance in satisfaction. These insights along with the best practice example of the *inspire!* game may contribute to game development and teaching.

Keywords: Business game · Satisfaction · Gamification ·
Entrepreneurship education

1 Introduction

Over the last decade, gamification has become a key teaching approach in entrepreneurship education. Universities along with other education institutions and game producers seek for game settings that foster students' competences and satisfaction. Recent research reveals that only a few games focus on the front-end of the entrepreneurial process [1]. Consequently, the field of this type of entrepreneurial games still finds itself in an early stage. Further progress will depend on the creation and better understanding of these games. This requires creativity and rigorous empirical research [2].

What is known so far, is that business games help to build up a variety of competences [3]. In addition, research ascribes entrepreneurship education – and in particular games – the potential to influence attitudes and intentions [4–6]. Academia attributes students' satisfaction as important explanatory variable for subsequent behavior [7]. It thus can be assumed that students who feel satisfied with an entrepreneurial gaming experience may build intentions to further deepen their entrepreneurial knowledge and/or start entrepreneurial endeavors.

From a learner's perspective, satisfaction can be defined as an outcome of an experience that is associated with positive emotions [8, 9]. When analyzing the full

© Springer Nature Switzerland AG 2020
M. E. Auer and T. Tsiatsos (Eds.): ICL 2018, AISC 916, pp. 697–709, 2020.
https://doi.org/10.1007/978-3-030-11932-4_65

scope of educational offerings from a satisfaction perspective, academia utilizes a range of different theoretical approaches (e.g. customer satisfaction theory, technology acceptance theories as well as theory of planned behavior). Empirical evidence on antecedents of satisfaction is intensively discussed in studies on e-learning. As far as the context of entrepreneurship education is concerned, empirical evidence is, however, rather scarce. A comprehensive study on measuring the performance of a business game is provided by [3]. She theorizes and tests overall gaming performance of "TOPSIM General Management II". Given the broad range of measured game and context variables, this study provides a good basis for further quantitative endeavors. Yet, accounting for the key role satisfaction plays in entrepreneurship education, it appears promising to focus research interest on satisfaction as important outcome.

The present study aims to contribute to research and practice in two ways. First, we present the entrepreneurial game "inspire! build your business", which focuses on business model development. This game was developed at the Department of Innovation Management and Entrepreneurship, University of Klagenfurt, to offer participants a context for gaining both, competences and impressions associated with entrepreneurial challenges. Second, considering the necessity to supplement game creation with rigorous empirical testing, we provide initial empirical results. In particular, we evaluate the importance of a variety of variables for student satisfaction.

Our research aim requires a two-step methodological approach. In fact, we will qualitatively introduce the entrepreneurship game "inspire build your business!" and subsequently we will quantitatively evaluate which game and context characteristics influence the satisfaction of the participants. In this way, we hope to contribute to future business game development and teaching.

2   Research Context

2.1   Game-Based Approaches in Entrepreneurship Education

Games can be viewed as structured processes with clearly defined starting and ending points. Within this process, players are exposed to challenges that have to be overcome to reach specific targets. The degree of aim fulfillment determines whether participants are perceived as winners or losers.

Unlike traditional approaches in education, games offer features that lead to superior outcomes beyond mere knowledge transfer. First, games are found to intrinsically create joy, particularly among young learners [10]. [11] highlights the importance of game mechanics which let players immerse themselves in a flow-condition. This fosters engagement and thereby sets the basis for rich learning. As games often require students to form teams, they train a range of social skills. Confronted with challenges, team members eventually may face conflicts with colleagues that have to be solved. Further, games allow participants to experiment within a riskless, yet realistic setting.

2.2   Satisfaction with Educational Games

Satisfaction is a complex construct that can be analyzed from a range of different perspectives [9]. Student satisfaction is defined as the student's assessment of services offered by universities and colleges [12]. Thus, satisfaction with educational games is the participant's assessment related to this specific educational service.

Focusing the research on satisfaction is valuable due to at least three reasons. First, research found a link between satisfaction and subsequent behavior [7]. Thus, it can be argued that satisfied students may build intentions to further deepen their entrepreneurial knowledge and/or start an entrepreneurial endeavor of their own. In this way, satisfaction becomes a key variable for one central aim of entrepreneurship education: fostering interest in this field. Second, satisfaction thereby shapes students' intention to continue learning, fosters students' retention and therefore positively influences learning outcomes. Third, satisfaction has become a key indicator for quality evaluation within and among educational institutions. It is of interest for various stakeholders (students, quality instructors, state) and provides the potential for improving education quality [7, 13].

From both – research and practical perspective – the question arises which factors influence students' satisfaction in an educational context. Looking for relevant literature, it becomes apparent that there are only a few studies that discuss this question in an educational gaming context. This is particularly true for entrepreneurial games. We therefore decided to broaden our search field and include studies that evaluate satisfaction determinants with educational courses (offline and online). As Table 1 illustrates, there is a large heterogeneity in results, which can partly be attributed to the diverse contexts these studies stem from. Yet, by grouping findings, one can see that teaching competence, course determinants as well as student and technology characteristics have been found to play an essential role in the formation of satisfaction.

While this collection of influencing factors is informative, one has to keep in mind that these studies have been carried out in different contexts or settings (course evaluations, online learning, business games). One study that is settled within the entrepreneurial game context and at the same time discusses a broad variety of influencing factors was conducted by [3]. She has built a model of overall game success, determined by a complex interrelation of student satisfaction, competences and motivation. This model is based on 28 variables, of which most are measured via four items or more. Given the study focus and the broad range of variables investigated, we perceive this study as an adequate basis for our own investigation. Yet, it is necessary to adapt single constructs, given our specific focus (details are provided in the methodology section).

3   Inspire! Build Your Business-Approach[1]

As games are still a rather recent phenomenon in entrepreneurship education, creating and sharing ideas about how to build innovative game solutions is an important issue. We therefore contribute by presenting the game "inspire! build your business", which has been created at the Department of Innovation Management and Entrepreneurship, University of Klagenfurt. The main motivation for creating a new entrepreneurship game was to combine theory and practice in an innovative way and to increase

---

[1] Game description has been taken from Krajger, I. and E. Schwarz (2018). Inspire! build your business - a Game-Based Approach in Entrepreneurship Education. ECSB 3E Conference, Enschede, the Netherlands. pp 50.

Table 1. Factors influencing satisfaction

Categories of factors found to be influencing student satisfaction		Teacher		Course		Social interactions among students	Evaluation	Gains		Student characteristics	Technology characteristics
		Social competence	Teaching competence	Design	Materials			Competency expansion	Relevance for further life		
Example for category		e.g. attitude of teacher towards students	e.g. teacher's presentation style	e.g. high quality course structure	e.g. easy course materials	e.g. group dynamics	e.g. exam style	e.g. recalled performance	e.g. adequacy in light of students' long-term goals	e.g. age	e.g. flexibility of access
Authors	Context										
[14] Arbaugh (2002)	e-learn.	x	x			x					x
[15] Arbaugh and Duray (2002)	e-learn.			x							x
[16] Baena-Extremera et al. (2015)	course	x	x							x	
[17] Bekele (2010)	e-learn.			x					x	x	x
[18] Canuana et al. (2016)	marketing game							x		x	x
[19] Hong et al. (2011)	e-learn.	x			x	x				x	x
[7] Li et al. (2016)	e-learn.				x		x		x	x	x
[20] Mikulic et al. (2015)	course	x	x	x			x			x	
[21] Sun et al. (2008)	e-learn.	x	x	x						x	x
[22] Thurmond et al. (2002)	e-learn.		x			x					x
[3] Trautwein (2010)	entrepr. game	x	x	x	x	x		x	x	x	

"x" indicates that the referring study found at least one factor within the related category to significantly explain variance in satisfaction

students' awareness, motivation and skills for entrepreneurship. Additionally, it should lead to a better understanding of the entrepreneurial role and ecosystem.

3.1 Principles of the Game

Our game-based approach follows the learning process developed by [23]. In his didactical model, experience or situational learning are key elements. The learner has to undergo a four-step process. The process starts with concrete experiences, which serve as the basis for observation and reflection. According to [23], knowledge can only be created by the transformation of experience. An interactive setting rather than one-way communication facilitates learning [24, 25].

When gamifying our business development process we considered the following elements of a game to be essential.

- Focus on experiential learning and learning by doing
- Challenges should address real world tasks
- Time-pressured tasks with uncertain outcomes
- Experience sharing by bringing different stakeholders together
- Interactive setting and teamwork
- Combination of different game activities (role-play, matching…..)
- Feedback loops and reflection
- Competition and rewards

3.2 The Process Behind the Game

inspire! build your business focuses on the pre-stage of venture creation, deals with the different steps of business modeling and depicts different components of the entrepreneurial ecosystem. The business model development process starts with the identification and evaluation of the opportunity and ends with the development of a prototype and a business plan. Step by step, the business idea becomes more concrete and enhanced. The whole process is designed as an open innovation process, so the identification of important stakeholders in the entrepreneurial ecosystem is critical for business model development. Furthermore, this process combines a classical stage gate process with the Lean Start-up approach. Figure 1 portrays our four stage gate business model development process.

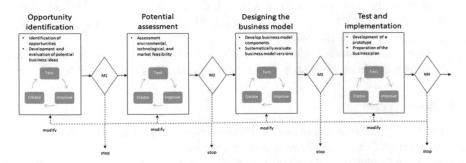

Fig. 1. Business model development process based on [26]

Gamers start with the identification of a raw idea/opportunity. Subsequently, this idea is assessed for its potential. Stage 3 aims at the development of a business model based on the idea. In stage 4 the business plan and necessary prototypes are developed.

The whole process is agile and iterative. In each stage a circular track consisting of three steps – create, test, improve – is implemented. In the first stage, "opportunity identification", the step "create" embraces a problem analysis, the identification of potential business ideas and the development of hypotheses. The second step, "test", includes first interviews (with e.g. customers, experts) and critical reviews in order to prove these assumptions. In the last step, "improve", the evaluation of interviews and learning from feedback can lead to a modification or radical change of the business idea. In each stage, the players must move around the circular track. Another key element of the proposed process model are milestones, which separate each stage. Milestones are decision points and should help to determine if the idea should be pursued, modified or rejected. The business model development process is supported by selected methods and instruments. Furthermore, questions for every milestone provide a good tool for the assessment and evaluation of the business idea.

3.3   Game Board and Components

As a graphical underpinning we developed a game board, which illustrates the business model development process. During the game, players move their playing pieces around a track which is divided into the four stages. To progress from one stage to the next, participants have to fulfill different tasks and pitch their idea. Traffic lights symbolize the gates. The game includes rewards (points), a dice and game cards.

The game cards can be described as follows:
- Risk and opportunity cards: These cards introduce the risks and opportunities of business model development derived from the entrepreneurial ecosystem. By collecting these cards, points can be generated or lost.
- Challenge cards: Participants have to fulfill different tasks such as explaining entrepreneurship terms (in the inspire! basic version for high-school students), interviewing customers and experts (in the advanced version for bachelor students and professionals). If they complete the task successfully, points are collected.
- Joker cards: In every stage the team with the highest score gets a joker. These cards allow participants to double points, exchange challenge cards, …
- Tool cards: In each stage teams get tool cards, which support the business model development process (e.g. problem-solution fit, personas, business model canvas) and help to progress the business idea.
- Learning and reflection cards: These cards help students to reflect on the executed challenges and to integrate learnings in the next step of business model development.

3.4   Game Versions

inspire! build your business is available in different versions which are tailored for a set of target groups. This study focuses on two case-based versions of the game, inspire! basic (for high-school students) and inspire! advanced (for bachelor students and professionals). While the game mechanisms described above are similar for all versions, there are some specific differences. In the basic version students can recall and test knowledge with the help of a so-called "activity zone", where players have to describe, draw or act a given term in the field of entrepreneurship and team members have to find out what it should be. In the advanced version, an additional role is introduced. Some participants can become an investor who has to analyze and evaluate all business ideas in the different stages of the game. A structured feedback and reflection section at the end of the game is designed to support the transformation of individual experience into knowledge.

4   Methodology

We employ a quantitative methodology using basic descriptive analyses (means, standard deviation, correlation) and multiple linear regression with manual variable selection as well as penalized linear regression and automated variable selection approaches (i.e. elastic net and stepwise backward regression).

4.1   Data Collection

Data was collected over the course of three game events, which took place between January and May 2018. To ensure coherent and comparable statistical information all sessions were instructed by the same teacher. 37 high-school students (12th grade) played the inspire! basic version, 12 MBA students played the advanced version.

Consequently, our sample consisted of 49 persons. During regression analysis, we controlled for these two groups. In line with the findings of [7], we found no significant group-related influence on satisfaction.

Data was collected via a questionnaire, which was distributed to each study participant at the end of the gaming event. Given the discussed adequacy of Trautwein's research, we decided to utilize her questionnaire design. For the present study, 13 variables, measured via 44 items were found to be relevant. All of these were measured using a six-point Likert scale from (1) "does not apply" to (6) "completely applies". This approach offered an adequate range of evaluation and prevented centralization tendency of responses [3]. The variables found to be relevant for our study context are:

Dependent variable:

- Satisfaction: is formed by four items measuring students' satisfaction perceptions. Two items address the overall impression (game, gaming event), one the willingness to recommend and one the positive remembrance of the game.

Independent variables:
- Competency gain: is formed by four items measuring the perceived increase in competences.
- Working relationship in group: is formed by six items, referring to a set of characteristics of effective group work, e.g. constructive working relationship, effective task solving and decision taking.

- Atmosphere in group: addresses a group's social climate. It consists of four items, measuring perceptions of belongingness and appreciation of social interaction among the group members.
- Task distribution in group: three items measuring whether each group member knows her/his function as well as range of tasks within the group.
- Teacher expertise: consists of five items, which measure the instructor's technical knowledge, her/his expertise with regard to the game as well as teaching skills.
- Structure of session: measured via four items referring to the structural quality of the game experience, i.e. clarity of episodes and timelines.
- Teacher engagement: measured by six items, e.g. the degree of the instructor's (positive) involvement in the game, her/his preparation and availability for students. The variable is formed using six items.
- Teacher participant relationship: consists of three items, e.g. whether students felt a good social relationship with the instructor.
- Traceability: two items measuring if students perceive game results as plausible.
- Game instruction clarity: indicates on a single item basis the degree to which the game material (e.g. manual and game cards) is easy to understand.
- Adequacy of room: one item addressing the students' perception of room facilities.
- Participant characteristics: refer to the classification of a gamer participant – either being a high-school or MBA student. This item acts as a dummy variable controlling for the two different groups within our sample.

4.2 Data Analysis

We used SPSS 22 for data analysis. The reliability assessment based on Cronbach's α demonstrated satisfying item fits within each construct (see Table 2). Due to the low number of participants, factor analysis could not be realized in a reliable way.

Table 2. Cronbach α values (values > 0.6–0.7 are acceptable; [27])

Satisfaction	Competency gain	Working relationship in group	Atmosphere in group	Task distribution in group	Teacher expertise	Structure of session	Teacher engagement	Teacher participant relationship	Traceability
.87	.82	.87	.81	.84	.83	.75	.80	.79	.88

Descriptive analysis across all variables was performed. Table 3 presents the means and standard deviations of each variable as well as the correlations between variables.

The descriptive statistics reveal high average means of most variables. Most constructs do not show normal distribution of responses, but a left-skewed distribution. Given the specific shape of response distribution (highest response rate at the far right), a power transformation did not solve this issue. Thus, conducting regression analysis we could only find the best *linear* unbiased estimates but not the best unbiased estimator. Further, some independent variables demonstrate moderate correlations (e.g. teacher engagement and teacher expertise r = 0.670**).

We conducted a multiple linear regression to identify the predictors for satisfaction. Considering the moderate collinearities among independent variables, we used an elastic net approach for initial variable selection. This approach, combining lasso-based

Table 3. Descriptive statistics

	Means	SD	Correlations											
			(1)	(2)	(3)	(4)	(5)	(6)	(7)	(8)	(9)	(10)	(11)	(12)
(1) Satisfaction	5.429	0.582												
(2) Competency_gain	4.888	0.903	.475**											
(3) Working_relationship_in_group	5.215	0.734	.218	.318*										
(4) Atmosphere_in_group	5.417	0.604	.388**	.267	.519**									
(5) Task_distribution_in_group	5.042	1.003	.222	.302*	.570**	.237								
(6) Teacher_expertise	5.580	0.453	.457**	.221	.273	.330*	.302*							
(7) Structure_of_session	5.423	0.495	.595**	.373**	.370**	.390**	.341*	.554**						
(8) Teacher_engagement	5.578	0.440	.531**	.287	.343*	.543**	.273	.670**	.630**					
(9) Teacher_participant_relationship	5.571	0.509	.557**	.176	.244	.259	.164	.437**	.549**	.613**				
(10) Traceability	5.255	0.902	.050	.007	.116	.164	.133	.089	.120	.036	-.037			
(11) Game_instruction_clarity	5.327	0.689	.254	.303*	.292*	.319*	.432**	.462**	.380**	.327*	.308*	.282*		
(12) Adequacy_of_room	5.878	0.331	.143	-.064	.054	.316*	.058	.288*	.291*	.472**	.465**	.142	.088	
(13) Participant_characteristics	1.245	0.434	-.032	-.114	.089	.304*	.094	.111	.041	.225	-.049	.156	.075	.213

Table 4. Model overview

Model	R	R Square	Adjusted R Square	Std. Error of the Estimate	Change statistics				
					R Square Change	F Change	df1	df2	Sig. F Change
1	.714[a]	.510	.477	.42064	.510	15.613	3	45	.000

[a] Predictors: (Constant), Competency_gain, Teacher_participant_relationship, Structure_of_session

variable selection and ridge regularization to overcome possible multicollinearity, resulted in a set of two independent variables forming the optimal model: Teacher participant relationship and structure of sessions were found to be best in terms of the elastic net optimization criteria. According to the optimal elastic net lasso paths, the next best variables to include are competency gain and atmosphere in group. Based on this insight, we conducted a linear regression with stepwise addition of the variables found to be "next best". In the initial version, both variables (relationship and structure) have a significant influence on satisfaction with an R^2adj of 0.405. Adding competencies as a third predictor leads to an increase of R^2adj to 0.477, with all explanatory variables being significant. The variable "atmosphere in group", however, is insignificant. We thus find the three predictor model comprising teacher participant relationship, structure of sessions and competencies to be an optimal solution (Tables 4 and 5).

To confirm this finding, and to find an optimal model with unbiased parameter estimates, we conducted a stepwise backward multiple regression (probability of F with an entry value of 0.05 and a removal rule at 0.10). This approach led to the same model characteristics. We thus feel convinced that our three-predictor model offers the best explanatory power of satisfaction.

5   Results and Discussion

One central aim of our study was to explore determinants of student satisfaction in the context of entrepreneurial games. Following a multiple linear regression approach we found three factors with significant explanatory power for satisfaction: The relationship between teacher and students ($\beta = 0.342$, $p < 0.01$), the competency gain perceived by students ($\beta = 0.305$, $p < 0.01$) and the structure of the game events ($\beta = 0.294$, $p < 0.05$) explain 47.7% of variance in satisfaction.

Our findings demonstrate that at least three spheres influence student satisfaction: teacher, game creator and game participant. From a teacher's perspective, building up a positive relationship with participants is essential. Further, the teacher may try to (positively) influence the structural clarity of gaming sessions by providing relevant information to students (e.g. timelines). A second actor that may have an important influence on structure is the game creator. She/he provides the frame the teacher acts in. As [3] argued, a logical course structure supports students' knowledge gains and motivation. The third relevant actor is the participant and his/her perception of gaining competences through playing the game.

Table 5. Coefficients

Coefficients[a]

Model		Unstandardized coefficients		Standardized coefficients	t	Sig.	Collinearity statistics	
		B	Std. Error	Beta			Tolerance	VIF
1	(Constant)	.421	.764		.551	.584		
	Structure_of_session	.345	.156	.294	2.215	.032	.620	1.613
	Teacher_participant_relationship	.391	.143	.342	2.736	.009	.698	1.433
	Competency_gain	.197	.073	.305	2.712	.009	.859	1.164

[a] Dependent Variable: Satisfaction

6   Conclusion

With the present study we contribute to the field of entrepreneurial games in two respects. First, we provide a best practice example of an entrepreneurial game, which students perceived in a very positive manner (see high means referring to satisfaction, competence gains, structural clarity). Second, conducting a multiple regression analysis we explore factors influencing satisfaction. Our findings reveal that at least three actors (teacher, game creator, game participant) have a substantial impact on the perceived overall game satisfaction. From the perspective of a university, which aims to create entrepreneurial games, it appears that student satisfaction can be substantially influenced via game structure and teaching quality (particularly with respect to relationship). Based on the theorized link to future intentions, schools and universities – via game and teaching style – create a satisfying experience and in this way contribute to an increased interest in entrepreneurial activities.

References

1. Krajger, I., Schwarz, E.: inspire! build your business. In: ECSB 3E Conference, Enschede, The Netherlands, p. 50 (2018)
2. Kriz, W.C.: Systemkompetenz als Zieldimension komplexer Simulationen. Berufs- und Wirtschaftspädagogik online **10**, 1–26 (2006)
3. Trautwein, C.: Unternehmensplanspiele im industriebetrieblichen Hochschulstudium, vol. 1, p. 307. Gabler Verlag, Wiesbaden (2010)
4. Garcia, J.C.S., et al.: Entrepreneurship education: state of the art. Propositos Y Representaciones **5**(2), 401–473 (2017)
5. Kriz, W.C., Auchter, E.: 10 years of evaluation research into gaming simulation for german entrepreneurship and a new study on its long-term effects. Simul. Gaming **47**(2), 179–205 (2016)
6. Rauch, A., Hulsink, W.: Putting entrepreneurship education where the intention to act lies: an investigation into the impact of entrepreneurship education on entrepreneurial behavior. Acad. Manage. Learn. Educ. **14**(2), 187–204 (2015)
7. Li, N., Marsh, V., Rienties, B.: Modelling and managing learner satisfaction: use of learner feedback to enhance blended and online learning experience. Decis. Sci. J. Innovative Educ. **14**(2), 216–242 (2016)
8. Hennig-Thurau, T., Klee, A.: The impact of customer satisfaction and relationship quality on customer retention: a critical reassessment and model development. Psychol. Mark. **14**(8), 737–764 (1997)
9. Virtanen, M.A., et al.: The comparison of students' satisfaction between ubiquitous and web-based learning environments. Educ. Inf. Technol. **22**(5), 2565–2581 (2017)
10. Huang, Y.M., Huang, S.H., Wu, T.T.: Embedding diagnostic mechanisms in a digital game for learning mathematics. Etr&D-Educational Technol. Res. Dev. **62**(2), 187–207 (2014)
11. Csikszentmihalyi, M.: Flow: Das Geheimnis des Glücks, 15th edn. Klett-Cotta, Stuttgart (2010)
12. Wiers-Jenssen, J., Stensaker, B.R., Grøgaard, J.B.: Student satisfaction: towards an empirical deconstruction of the concept. Qual. High. Educ. **8**(2), 183–195 (2002)

13. Ozgungor, S.: Identifying dimensions of students' ratings that best predict students' self efficacy, course value and satisfaction. Egitim Arastirmalari-Eurasian J. Educ. Res. **10**(38), 146–163 (2010)

14. Arbaugh, J.B.: Managing the on-line classroom: A study of technological and behavioral characteristics of web-based MBA courses. J. High Technol. Manage. Res. **13**(2), 203–223 (2002)

15. Arbaugh, J.B., Duray, R.: Technological and structural characteristics, student learning and satisfaction with web-based courses: an exploratory study of two on-line MBA programs. Manage. Learn. **33**(3), 331–347 (2002)

16. Baena-Extremera, A., et al.: Predicting satisfaction in physical education from motivational climate and self-determined motivation. J. Teach. Phys. Educ. **34**(2), 210–224 (2015)

17. Bckclc, T.A.: Motivation and Satisfaction in Internet-Supported Learning Environments: A Review. Educational Technology & Society **13**(2), 116–127 (2010)

18. Caruana, A., La Rocca, A., Snehota, I.: Learner satisfaction in marketing simulation games: antecedents and influencers. J. Mark. Educ. **38**(2), 107–118 (2016)

19. Hong, W., et al.: User acceptance of agile information systems: a model and empirical test. J. Manage. Inf. Syst. **28**(1), 235–272 (2011)

20. Mikulic, J., Duzevic, I., Bakovic, T.: Exploring drivers of student satisfaction and dissatisfaction: an assessment of impact-asymmetry and impact-range. Total Qual. Manage. Bus. Excellence **26**(11–12), 1213–1225 (2015)

21. Sun, P.-C., et al.: What drives a successful e-Learning? An empirical investigation of the critical factors influencing learner satisfaction. Comput. Educ. **50**(4), 1183–1202 (2008)

22. Thurmond, V., et al.: Evaluation of student satisfaction: determining the impact of a web-based environment by controlling for student characteristics, vol. 16, pp. 169–190 (2002)

23. Kolb, D.: Individual learning styles and the learning process. Working Paper # 535–71. Sloan School of Management (1971)

24. Kapp, K.M., Blair, L., Mesch, R.: The Gamification of Learning and Instruction Fieldbook. Wiley, San Francisco (2014)

25. Kolb, A.Y., Kolb, D.A.: The learning way: meta-cognitive aspects of experiential learning. Simul. Gaming **40**(3), 297–327 (2009)

26. Schwarz, E., Krajger, I., Holzmann, P.: Prozessmodell zur systematischen Geschäftsmodellinnovation. Geschäftsmodellinnovationen, pp. 65–77. Springer, Wiesbaden (2016)

27. Hair, J., et al.: Multivariate Data Analysis, 7th edn. Pearson, Upper Saddle River (2010)

Enhanced Virtual Learning Spaces Using Applied Gaming

Panagiotis Migkotzidis[1], Dimitrios Ververidis[1(✉)], Eleftherios Anastasovitis[1],
Spiros Nikolopoulos[1], Ioannis Kompatsiaris[1], Georgios Mavromanolakis[2],
Line Ebdrup Thomsen[3], Marc Müller[4], and Fabian Hadiji[4]

[1] Centre of Research and Technology, Thessaloniki, Greece
ververid@iti.gr
[2] Ellinogermaniki Agogi, Athens, Greece
[3] Aalborg University, Aalborg, Denmark
[4] goedle.io Gmbh, Cologne, Germany

Abstract. Online virtual labs, emulate real laboratories where students
can accomplish a number of learning tasks and receive courses that otherwise would be difficult if not infeasible to be offered. In reaching this challenging goal, we developed two virtual labs enhancing students knowledge by providing an entertaining and engaging experience. Ensuring
that the educational requirements of educators and students will not be
overlooked, extensive evaluations of the virtual labs were deployed. Both
virtual labs were accompanied with game analytics enabling detailed
tracking of learner behavioral data. Tracking and understanding behavioral data facilitated decision-making at the design level of a lab, but
also allowed for adapting learning content to the personal needs of the
students.

1 Introduction

As computer technologies become an integral part of modern life, school administrators, teachers, and researchers strive to incorporate the technologies into
classrooms to improve student learning outcomes [3]. The concept of immersive education has been applied to all aspects of education, including formal-
institutional education, informal massive education, and professional training
in companies [9]. One important type of such educational games are those for
Science, Technology, Engineering, and Mathematics (STEM) education [10].

Virtual Laboratories are a useful and efficient educational tool for acquiring
knowledge and enhancing learning. Video games with the advances of graphic
hardware have fostered the creation of realistic virtual labs [4]. A rapid and
drastic fall in prices, a huge leap in the computer processing power, the prolif-
eration of World Wide Web and the prevalence of broadband connections have
aggravated the use of desktop VR [6]. Augmented Reality and Mixed Reality
are also attractive sections of the game industry offering immersive experiences.
Serious games can exist in the form of mobile applications, simple web-based

© Springer Nature Switzerland AG 2020
M. E. Auer and T. Tsiatsos (Eds.): ICL 2018, AISC 916, pp. 710–721, 2020.
https://doi.org/10.1007/978-3-030-11932-4_66

solutions, and combinations of social software applications or in the shape of "grown-up" computer games. Employing modern games technologies to create virtual worlds for interactive experiences may include socially based interactions, as well as mixed reality games that combine real and virtual interactions, all of which can be used in virtual labs applications. The main strengths of serious gaming applications may be generalized as being in the areas of communication, visual expression of information, collaboration mechanisms, interactivity and entertainment [1]. In our work, we exploit the recent advances of 3D gaming in order to improve the look and feel of two 2D labs namely one for Chemistry and one for Wind Energy, originally developed in the GoLabs project.[1] Our 3D and the old 2D labs can be played freely in our project web page.[2]

The purpose of our work is to offer a solution towards optimizing the learning process in virtual labs and therefore maximize their impact in education. In reaching this challenging goal, we managed to migrate knowledge from the neighboring domain of digital games, where the capture and analysis of detailed, high-frequency behavioral data has reached mature levels in recent years. In digital games, Game Analytics (GA) is used to profile users, predict their behavior, provide insights into the design of games and adapt games to users. Tracking and understanding behavioral data can facilitate decision-making at the design level of a lab, but also can allow for adapting learning content to the personal needs and requirements of students.

The rest of the paper is organized as follows. Section 2 reviews related work. Section 3 provides the details of gamification and game design for the two case studies, namely the *Chemistry Lab*, and the *Wind Energy Lab*. In Sect. 4, the results of the conducted user evaluation for our Virtual labs are presented. Section 5 concludes our work and proposes future developments.

2   Background and Related Work

Several studies test the hypothesis that video games have a positive effect on supporting educational goals. These studies on the application of videogames in school curricula concentrate on the impact of the material in the game to learning goals [2]. In [5], the pedagological foundations of modern educational computer video games, that were developed between 2000 and 2007 strictly for educational purposes to facilitate achievement of specified learning objectives are studied. A ten-year critical review (1999–2009) of empirical research on the educational applications of VR was presented in [7]. Results show that although the majority of the 53 reviewed articles refer to science and mathematics, researchers from social sciences also seem to appreciate the educational value of VR and incorporate their learning goals in Educational Virtual Environments (EVEs).

In [9], a summary of the state of the art in virtual laboratories and virtual worlds in the fields of science, technology, and engineering can be found. The main research activity in these fields is discussed, but special emphasis is put on

---

[1] https://www.golabz.eu/.
[2] http://www.envisage-h2020.eu/virtual-labs/.

the field of robotics due to the maturity of this area within the virtual-education community. In most cases they are specific to an educational context, but do not offer possibilities for generalization to a platform applicable to a wider class of engineering disciplines. These laboratories have different levels of technical complexity. The most recent publications include the *Captivate* project [10]; a mobile game for STEM in higher education. In [8], a research study on the effectiveness of a new interactive educational 3D video game *Final Frontier* can be found. The game supports delivery of scientific knowledge on the Solar system to primary school students.

Relatively recently, Artificial Intelligence (AI) is also used to analyze games, and model players' profiles as game developers need to create games that appeal diverse audiences. Facebook games such as *FarmVille* were among the first to benefit from continuous data collection, AI-supported analysis of the data and semi-automatic adaptation of the game. Nowadays, games such as *Nevermind* can track the emotional changes of the player and adapt the game accordingly [12].

In our work, the cardinal objective is to optimize the design and functionality of virtual learning labs from the perspective of both the teacher (i.e., the designer of a lab) and the student (i.e., the user of a lab) by applying technologies developed in the gaming industry for authoring games, and for collecting game analytics encoding the activity of learners.

3   Methodology

In order to achieve a correct balance between entertainment and education, we collaborated with educators from *Ellinogermaniki Agogi* school.[3] that helped us with the initial requirements of the games. Next, the Labs were designed and implemented using Unity3D game engine[4] and Blender design software.[5] The Labs are compiled for WebGL format in order to be easily accessible by the students. Moreover, Analytics tracking was also implemented inside the virtual labs for monitoring players' activities and modeling their learning behavior. Evaluations by questionnaires were filled in by the educators from *Ellinogermaniki Agogi* in two cycles providing valuable feedback. Pilots with students from the same school were conducted, collecting a broad amount of player's data, as well as feedback for improving the general design of the games. Lastly, we managed to equip virtual labs with tools that perform Dynamic Difficulty Adjustment (DDA) and semi-automatic adaptation of the learning parameters according to personal requirements of the learners. In the following an extensive description of the two Virtual Labs is presented.

---

[3] http://www.ea.gr/ea/index.asp?lag=en.

[4] https://www.unity3d.com.

[5] https://www.blender.org.

3.1   Virtual Labs Development

The Chemistry Lab: The Chemistry Laboratory aims to educate students about the science of chemistry through an engaging and visually appealing experience. The general concept of the game has been designed to suit the needs of chemistry curriculum in primary and secondary schools. The game focuses on molecules names, formulas and 3D structures. During the development of the game, two pilots took place inside the classrooms of two schools. The initial pilot was performed from students and teachers of *Ellinogermaniki Agogi* school. Another pilot run took place in a Greek vocational training institute with a different audience (3D graphics students) that also suggested changes that helped making the game more appealing and entertaining.

The Lab immerses the player to feel like he or she is really working in a realistic chemistry lab environment. The main and first interactive scene of the game, is a 3D environment as shown in Fig. 1. The look and feel of the lab environment has been designed based on real chemistry labs in order to immerse the player in the virtual world, but also display the rational appearance of a lab working environment. Discrete signs were placed inside the lab environment to inform the player about potential hazards and safety practices.

Regarding the main gameplay elements, the player learns about chemistry molecules through two mini-game quizzes. The first one is called molecule Naming and tests player's knowledge on molecules formulas and names. The second one, is called Molecule construction and gives the ability to the player to construct the 3D shape of a molecule from its given formula. The player can be transferred in these mini-games by interacting with two different virtual laptops that can be found inside the lab environment.

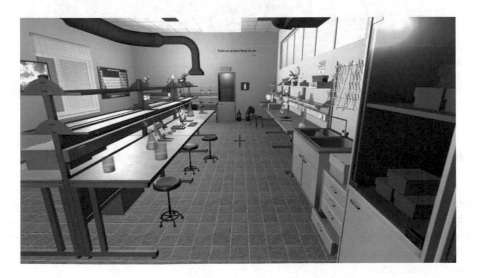

Fig. 1. The chemistry lab game environment.

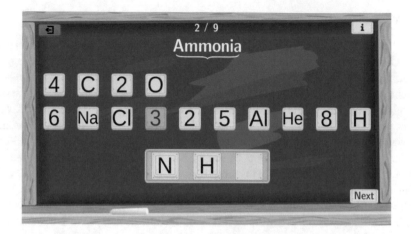

Fig. 2. The formula naming 2D mini-game.

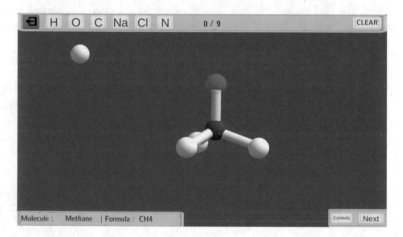

Fig. 3. The molecule construction 3D mini-game.

More specifically, Molecule Naming as shown in Fig. 2, is a 2D mini-game where the player is given a molecule name and must choose from a list of given atoms, the ones needed for assembling the molecular formula type and place them in the correct order. Molecule Construction mini-game on the other hand has a more complex interface where the player can create specific elements and try to place them in a 3D dimensional space in order to create the molecules 3D structure as shown in Fig. 3. The learner can rotate the 3D molecule model and drag and drop elements onto the structure by moving them in different dimensions. A score is displayed that validates his overall performance during the two mini-games.

The Wind Energy Lab: In the Wind energy lab the player controls a wind farm in order to provide electrical energy to a small town. The player understands how random changes in wind speed and power requirements of the town affect the use of this natural energy resource. It has been integrated inside the educational curriculum of the *Ellinogermaniki Agogi* school, where two pilot runs were conducted from three different classes participating and providing feedback.

The lab has three phases namely (a) the construction, (b) the simulation, and (c) the answering to some questions regarding strategic future decisions on the farm. Concerning the construction process, when the game starts a sheet is displayed to the learner that presents specific requirements and energy needs. Subsequently, based on the requirements, the learner must make a choice about the installation location of his wind energy farm. There are three available choices such as Mountains, Fields, or Seashore as shown in Fig. 4, where each one has its own pros and cons regarding installation cost and wind speed. After the learner chooses a main region to install the wind farm, he/she is transferred to the 3D environment of the region in order to inspect the game terrain and choose a specific area in the region to install the park (Fig. 5). However, each of these areas have different characteristics and risks. For example, some areas contain natural resources or archaeological places that induce a score penalty. In the final phase of the installation, the player can choose the type of wind turbine to use among different types having different efficiency and size.

In the second phase, the learner is transferred to the specified 3D simulation environment in order to control the wind farm (Fig. 6). The goal is to produce a balance between the amount of energy produced with the one required. To achieve this, the learner can activate or deactivate some wind turbines. The wind speed is changing rapidly so as to force the learner to change his/her plans and make new actions. In the final phase, when the simulation ends, a multiple

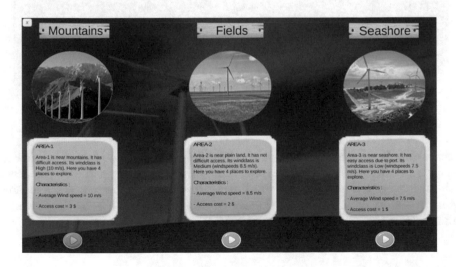

Fig. 4. The main region selection scene.

Fig. 5. The installation area selection stage.

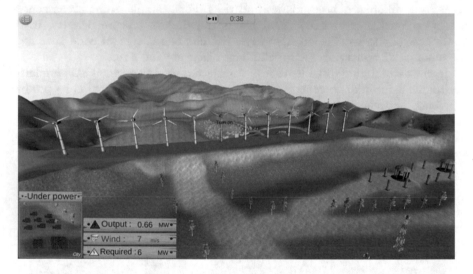

Fig. 6. The simulation scene with the installed wind farm.

choice quiz is presented where the learner should answer questions related to future scenarios such as demographic changes in the area. In the end, a score tab is displayed that evaluates the learner based on his/her choices and how well the initial requirements of the exercise were covered, e.g. by displaying a summary of energy balance efficiency (Fig. 7).

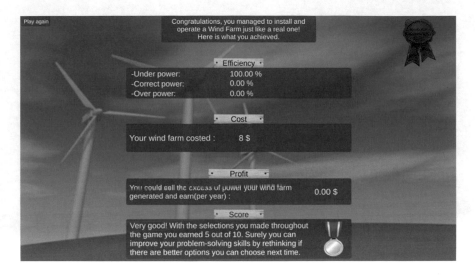

Fig. 7. The final scene evaluating players performance.

4   User Evaluation

4.1   Evaluation Protocol

In order to evaluate the labs we have conducted some surveys. We have let the learners play the labs and then 18 questions-statements were posed to them as shown in Table 1. The questionnaire is used to evaluate the learning content, the usability, the adoption, and the engagement with the virtual labs. A Likert-type scale was used to collect the data; each of eighteen questions was rated on a five-point scale ranking: (i) strongly agree, (ii) agree, (iii) neutral, (iv) disagree, and (v) strongly disagree.

The *Chemistry Lab* was first tested, followed by the *Wind Energy Lab*. The first serious game was started with the entertainment instruction ("MISSION: Congratulations! Today is your first day in the Chemistry Laboratory. Lets produce some chemical compounds, by using the Labs equipment. Pay attention to your assistants advices."), while the second was started with the mission instruction ("Your team has undertaken to study the construction of a wind farm. The wind park should meet the energy needs of a village with the following characteristics: Population: 4.000, Households: 1.800, Requested power: 6 MW".). Afterwards, the students filled out the questionnaire. There was no time restriction for playing and answering the questionnaires.

4.2   Results

The results presented here regard the survey on 40 individuals (19 to 48-years old, average: 23, 5% female, 95% male) from a Greek vocational training institute with specialization in multimedia applications. These results regard the latest

Table 1. The statements on questionnaires.

Q1:	The content presented in the virtual lab is correct and well balanced
Q2:	The virtual lab fits well with the curricula
Q3:	The virtual lab presents the learning content in a relevant manner for the students
Q4:	The quality of the learning content did not meet my expectations
Q5:	It is difficult to integrate the virtual lab into a learning content
Q6:	The learning material is presented in structure and complexity that suits the students' competencies
Q7:	The user interface of the virtual lab (menus, buttons etc.) was easy to understand
Q8:	I believe that the virtual lab will give the students a better understanding of the topic
Q9:	The learning goals for the virtual lab are clear
Q10:	I believe it will be hard for me to evaluate the student's performance in the virtual lab
Q11:	I have a good sense of how the students will work with the virtual lab
Q12:	The virtual lab supports differentiated learning
Q13:	I would use the virtual lab in my teaching
Q14:	I would like to change part of the virtual lab to better support my teaching
Q15:	I believe students will find the virtual lab engaging
Q16:	I believe students will find the virtual lab challenging
Q17:	I believe students will enjoy using the virtual lab
Q18:	I believe the virtual lab will stimulate the students' interest or curiosity in the subject

versions the labs. Their participation was volunteered and they all informed for a bonus grade in final exams, as a reward. The majority of participants declared great familiarity with desktop and web-based 3D games, but limited experience with virtual learning. Each student played the game alone following some general instructions that were displayed on a projector. The average playing time for each game was 40 min with an intermediate break of 15 min.

The statistics of the survey are shown in Table 2. Regarding the content and cognitive value of the virtual labs there is a general acceptance. In Q1, 65.0% of participants found the presented content in the virtual labs correct and well-balanced, while only one tester (2.5%) expressed a negative opinion. In addition, 62.5% of the volunteers were satisfied with the presentation of the learning material. More specific, the structure and the complexity suit the students competencies (Q6). There is no negative feedback in Q2 and Q3. Specifically, 67.5% thought that the virtual labs fit well with the curricula. The same percentage evaluated that the virtual labs present the learning content in a relevant manner

Table 2. The aggregated results.

Statement	Strongly agree	Agree	Neutral	Disagree	Strongly disagree
Q1	22.5%	42.5%	32.5%	2.5%	0.0%
Q2	12.5%	50.0%	37.5%	0.0%	0.0%
Q3	10.0%	52.5%	37.5%	0.0%	0.0%
Q4	7.5%	20.0%	32.5%	37.5%	2.5%
Q5	10.0%	25.0%	32.5%	30.0%	2.5%
Q6	25.0%	37.5%	35.0%	2.5%	0.0%
Q7	42.5%	32.5%	22.5%	0.0%	2.5%
Q8	32.5%	47.5%	7.5%	10.0%	2.5%
Q9	27.5%	47.5%	20.0%	5.0%	0.0%
Q10	5.0%	12.5%	47.5%	25.0%	10.0%
Q11	15.0%	55.0%	20.0%	7.5%	2.5%
Q12	7.5%	55.0%	30.0%	5.0%	2.5%
Q13	25.0%	45.0%	20.0%	5.0%	5.0%
Q14	12.5%	42.5%	35.0%	10.0%	0.0%
Q15	17.5%	45.0%	32.5%	5.0%	0.0%
Q16	10.0%	35.0%	45.0%	7.5%	2.5%
Q17	17.5%	45.0%	27.5%	7.5%	2.5%
Q18	30.0%	47.5%	15.0%	5.0%	2.5%

for the students. The responses in Q5 and Q4 statements were balanced, which refer to users' expectations and virtual labs integration with the learning content.

From usability perspective, there is a positive evaluation. More specific, in Q7, 75.0% of participants expressed a positive opinion about the user interface of the virtual labs, while only 2.5% was against this statement. In Q8, 80.0% thought that the virtual labs will give the students a better understanding of the topics, while 12.5% disagreed. Moreover, 75.0% of participants supposed that the learning goals for the virtual labs were clear, in contrast with 5.0% that disagreed (Q9). The balanced results in Q10, risk any assessment compared with the contribution of the virtual labs in students' evaluation.

Regarding the adoption of the virtual labs by participants in the learning process, their general acceptance is detected. The percentage of dissidents is limited to 10.0%. Particularly, 70.0% of testers said in Q13, that they would use the virtual labs in their future teaching. The same percentage agreed with Q11, having a good sense of how the students will work with the virtual labs. In Q12, 62.5% of participants recognized the support in differentiated learning, by using the virtual labs. Moreover, 55.0% referred in Q14 that they would like to change part of the virtual lab to better support their teaching.

From engagement's perspective (Q18), a massive percentage of 77.5% admitted that the virtual labs will stimulate the students' interest or curiosity in the subject, while only 7.5% doubted. Regarding students' engagement with the virtual labs, in Q15, 62.5% was positively expressed, in contrast with 5.0%. In Q17, 62.5% believe that students will enjoy using the virtual lab, and 45.0% consider that these are challenging in Q16, while a 10.0% of participants are opposite to these statements.

5   Conclusions and Future Work

In this paper we presented the design and development of two virtual laboratories that support the educational process in STEM lessons. In addition, we have displayed the results of a quantitative user evaluation with 40 participants - specialists in 3D games. As concerns the content and cognitive value of the *Chemistry Lab*, and the *Wind Energy Lab*, it was rated positively, well balanced, and met the expectations. Regarding the usability, the navigation into the virtual environments was easy, and the graphical user interfaces were reliable. The virtual laboratories were tagged as engaging, challenging, and entertaining even with their mid-quality 3D graphics, due to their web-based orientation. Moreover, the adoption of these educational tools in the learning process was clearly expressed. In addition to the positive evaluation of these educational serious games, an important element that has emerged is that users would like to create and customize virtual laboratories themselves.

In ENVISAGE project have also developed a web-based authoring tool, taking advantage of recent 3D graphic technologies such as three.js.[6] The instructor is able to create the virtual environment, and design the learning context. The virtual lab can compiled in WebGL or standalone-desktop format through Unity3D game engine, and disseminated to students. However, the important thing is that the teacher will be able to adapt the educational lab appropriately, using the analysis of pupils' behavior [11].

Acknowledgement. The research leading to these results has received funding from the European Union H2020 Horizon Programme (2014-2020) under grant agreement 731900, project ENVISAGE (Enhance virtual learning spaces using applied gaming in education).

References

1. Anderson, E.F., McLoughlin, L., Liarokapis, F., Peters, C., Petridis, P., De Freitas, S.: Developing serious games for cultural heritage: a state-of-the-art review. Virtual Real. **14**(4), 255–275 (2010)
2. Gros, B.: Digital games in education: the design of games-based learning environments. J. Res. Technol. Educ. **40**(1), 23–38 (2007)
3. Hodges, G.W., Wang, L., Lee, J., Cohen, A., Jang, Y.: An exploratory study of blending the virtual world and the laboratory experience in secondary chemistry classrooms. Comput. Educ. **122**, 179–193 (2018)

---

[6] http://threejs.org.

4. Ibáñez, L.A.H., Naya, V.B.: An enhanced navigation kit for virtual heritage exploration using a game engine. Digi. Herit. **2**, 755–756. IEEE (2015)
5. Kebritchi, M., et al.: Examining the pedagogical foundations of modern educational computer games. Comput. Educ. **51**(4), 1729–1743 (2008)
6. Lee, E.A.L., Wong, K.W., Fung, C.C.: How does desktop virtual reality enhance learning outcomes? A structural equation modeling approach. Comput. Educ. **55**(4), 1424–1442 (2010)
7. Mikropoulos, T.A., Natsis, A.: Educational virtual environments: a ten-year review of empirical research (1999–2009). Comput. Educ. **56**(3), 769–780 (2011)
8. Muntean, C.H., Andrews, J., Muntean, G.M.: Final frontier: an educational game on solar system concepts acquisition for primary schools. In: 2017 IEEE 17th International Conference on Advanced Learning Technologies (ICALT), pp. 335–337. IEEE (2017)
9. Potkonjak, V., Gardner, M., Callaghan, V., Mattila, P., Guetl, C., Petrović, V.M., Jovanović, K.: Virtual laboratories for education in science, technology, and engineering: a review. Comput. Educ. **95**, 309–327 (2016)
10. Smith, K., Shull, J., Shen, Y., Dean, A., Michaeli, J.: Overcoming challenges in educational stem game design and development. In: Simulation Conference (WSC), 2017 Winter. pp. 849–859. IEEE (2017)
11. Ververidis, D., Chantas, G., Migkotzidis, P., Anastasovitis, E., Papazoglou-Chalikias, A., Nikolaidis, E., Nikolopoulos, S., Kompatsiaris, Y., Mavromanolakis, G., Ebdrup-Thomsen, L., Liapis, A., Yannakakis, G., Mller, M., Hadiji, F.: An authoring tool for educators to make virtual labs. In: Proceedings International Conference on Interactive Collaborative Learning (2018)
12. Yannakakis, G.N., Togelius, J.: Artificial Intelligence and Games. Springer (2018). http://gameaibook.org

A Game for Entrepreneurship Training Supporting Dual-Career Paths

Ian Dunwell[✉] and Petros Lameras

School of Computing, Electronics, and Mathematics, Coventry University,
Coventry, UK
i.dunwell@cad.coventry.ac.uk

Abstract. This paper presents the early-stage user workshop findings and subsequent design of a game-based learning approach to entrepreneurship. The specific context is to address the dual-career training needs of athletes, as part of a large-scale European online course (MOOC). An interactive card-based activity was used within a small scale focus group, with the purpose of enabling dialogue between end-users (n = 11), as to elicit their experiences of using games for understanding athletes' dual career training needs. Initial findings from this exercise was a suggestion that quantifiable performance indicators (scores, points, achievements) could be preferable to less quantifiable measures (e.g. narrative progression, or multiple scenario outcomes). Establishing meaning, sense-of-purpose, and identity within the game were also highlighted as desirable features by the focus group, when compared to other options as detailed in this paper. The subsequent prototype of the game is presented, with reference to these findings.

1 Introduction

Challenges experienced by athletes in dual-career planning have been widely identified. These include difficulty in balancing the time needs of a demanding training regime against academic study or employment, difficulty engaging with courses or activities whose content and purpose may be far-removed from their sporting career [1]. Furthermore, athletes may tend to overestimate how long their athletic career will continue, with the most difficult transitions often related to unanticipated events such as injury or shifts in funding support beyond their control. Significant diversity also exists amongst athletes in terms of their engagement with dual career planning, which has been suggested to be strongly affected by their social context [2]. Few athletes will generate enough income during their professional career to avoid the need for a career transition [3]. Hence, such transitions often face the (former) athlete with a considerable psychological strain, potentially including simultaneous needs to cope with injury, financial worries, and loss of identity.

Numerous approaches to easing this transition have been developed, typically with a principal focus on providing access to education and training during an athlete's career to prepare. These range from policy-level approaches [4], through to specific educational resources. In the scope of this paper, the focus is on game development within the Gamified and Online Learning to Support Dual-Careers of Athletes (GOAL)

© Springer Nature Switzerland AG 2020
M. E. Auer and T. Tsiatsos (Eds.): ICL 2018, AISC 916, pp. 722–732, 2020.
https://doi.org/10.1007/978-3-030-11932-4_67

project [5]. A primary rationale for selecting games as a medium for dual-career training is their potential capacity to mirror many common aspects of professional sports, for example by implementing clear and achievable objectives, or mechanics for competition or collaboration. Through this, they may be hoped to enhance engagement and motivation with dual-career planning. Focusing upon these gaming aspects of dual-career training, the background of this paper (Sect. 2) considers research related to game design, and how it may be mapped to the needs of athlete's dual career planning. This provides an outline consideration of how learning typologies could suggest certain mechanics or features which may best suit athletes. However, noting the fluidity of, and controversy surrounding, such learner typologies as a basis for design, a pragmatic workshop was designed (Sect. 3), with results providing direct insight into athletes' opinions on what game-based learning could bring to their dual-career planning. Results (Sect. 3.2) identified discrete mechanics which are mapped into the prototype game (Sect. 4). As work-in-progress, this game will be subsequently used and evaluated by a cohort of European amateur and professional athletes, with this and other considerations for future work presented in Sect. 5.

2   Background

Serious games and gamification techniques are frequently deployed in online educational platforms, such as Moodle [6]. The rationale for their inclusion can broadly be categorised as to promote engagement, scaffold competition or collaboration, or to present content and address learning objectives in novel ways. To-date, no large-scale, validated examples exist which directly aim to support dual-career development amongst athletes. However, a limited number of gaming or gamification examples exist for promoting related skills, such as coaching, or career development and planning, as detailed in this section. Principally, the question asked in this section is whether specific themes, considerations, or design paradigms are common to the areas of games for career development, or when designing for an audience of athletes.

Focusing on the use of serious games targeting an audience of athletes, consideration of the use of serious games for athletes and coaching similarly does not yield a large number of significant studies. It should be noted that this is scoped in terms of digital games relevant to GOAL; indeed, a professional athlete may view their sport in itself as a "serious" game. The use of games for career planning and development has also been explored in a range of games and associated studies. Work on the game *MeTycoon* demonstrated the ability of the game to significantly increase engagement times with existing online career training materials, such as videos [7]. Such use of existing educational content as a basis for a serious game that relates more widely to gamification strategies, which can fit around content to increase engagement rather than replace existing material. Other examples of the use of games or game-like tools for career guidance include the use of AI to facilitate dialogues on career choices [8], or simulate job interviews [9]. In both cases, evidence of efficacy is limited, given the significant technical undertaking required to accurately and effectively simulate or replace the role of a human interviewer or career counsellor. Despite these examples, there still exists limited presence of games as applied to career guidance. A report from the EU's Comenius Networks *CareerGUIDE* project demonstrated this paucity in 2007 with only a handful of examples across Europe [10]. However, games developed in subsequent LLPs, such as the *youth@work* game [11], have sought to address career guidance for school-age audiences across Europe. These have demonstrated the need to focus on engagement as well as education: in the case of the *youth@work* game, a range of mini-games addressing learning objectives are scaffolded within an overall fictional narrative to stimulate this engagement. Whilst the more recent examples above evidence the relevance and increased interest in games in this area, there remains little empirical evidence on the ideal content types and design patterns which might yield the greatest effectiveness.

Combining games and gamification within GOAL seeks to provide dual benefits: serious games allow salient learning objectives to be addressed through novel gaming experiences, whilst gamification can stimulate engagement with the wider course materials and objectives. Therefore whilst the section has thus far noted examples, and the lack thereof, for using serious games, gamification is an addtional consideration for GOAL. We distinguish here between a serious game as bespoke software, designed to address a learning objective, whilst gamification seeks to take an existing process and insert game elements towards improved outcomes [12]. Gamification is often construed at a simple level as adding elements such as points, badges, and leaderboards to a process, though these in isolation have been shown to not have a predictable or universal positive effect [13]. Rather, consideration needs to be given to underlying behavioural theory and individual contexts, and how layered, game-based measures, may influence these. For example, collaboration and competition can both be scaffolded at group and individual levels through the addition of leaderboards and badges; whether these have positive effect, however, depends on the nature and motivation users at both the group and individual level. Introducing leaderboards can encourage high-performers to compete, yet it can also disengage or discourage low performers, and indeed past studies have suggested an even greater degree of complexity exists than this simple paradigm [14].

An interesting design-level link in existing literature is between studies relating personality type to gaming preferences [15], and analysing personality type relationships to athletes. Large-scale studies of personality type have shown links to particular traits and physical activity [16], as well as sports leadership [17]. Athletes, as a subset of highly physically active individuals, are demonstrably more likely to score highly on extraversion and conscientiousness. In the study of Peever et al. [15], this is shown to also equate to positive inclination towards, under Peever et al. classification, "Party, Music, and Casual Games", and a marginally negative relationship to "role-playing games, MMORPGs, action role-playing games, turn-based strategy, and real-time strategy games". Conscientiousness, which is also shown as a significantly stronger trait amongst professional athletes, is linked to preferences for, perhaps unsurprisingly, "sport [games]", as well as "racing, … simulation, and fighting games". Whilst the classification and findings here are rather broad, they reflect a significant statistical effect at $p < 0.05$, with the smallest power in the above studies being $n = 418$. A further consideration of the use of personality types to predict gaming preferences is the added influence of different cultures on preferences, as the studies cited [15, 16, 18] used sizable but not homogenous samples. These studies are also not specifically focused upon serious games, which may need to make further affordances in their content and mechanics to address learning objectives. Hence, the most preferred game by end-users may not necessarily by the most effective in terms of learning outcomes. This initial link, however, could be summarised as suggesting it more likely that casual, social, or simulation aspects would be more likely to appeal to professional athletes than role playing or strategy, a consideration which will be adopted in game design for GOAL.

Relating this also to prior work on links between gamification, rather than gaming preferences and personality type allows a degree of anticipation of the types of gamification that may be best received by athletes. A recent study showed in comparison of competitive, collaborative, and control version of a gamified platform that extraversion positively moderates performance under competition [19]. Therefore it might be anticipated that a competitive, rather than collaborative approach to gamification would be most effective, given the personality traits commonly reflected by athletes. In essence, these conclusions may appear self-evident, in that they reinforce an obvious assumption that athletes are likely to respond more positively to a competitive environment; however, where this consideration adds value is in that this response appears likely to occur beyond the confines of their professional career. The above studies utilise the five-factor model of personality type, a widely validated model, however one with limitations with respect to predictive value and generalistic tendency. Personality type is not necessarily static, and difficult to predictively measure; moreover type theory itself is open to critique, as classifying and categorising people, in a broad sense, is a subject open to debate in terms of both epistemological and ethical viability. Yet, in the absence of an empiricially-validated best practice for the design of games and gamification strategies for dual-career athletes, they allow formation of a set of early-stage considerations which are summarised below:

- Athletes are statistically more likely to exhibit personality types which have been linked to preferences for social, casual, simulation/sports, and competitive games [15, 16];
- Athletes are statistically more likely to exhibit personality types which have been linked to preferences for competitive, rather than collaborative, gamification paradigms [16, 19];
- A need exists to gauge the requirement for engagement, as well as education, for game and gamification components at an early development and user requirements elicitation stage [11];
- Games can significantly boost consumption (in terms of usage time) of career guidance resources, and it may be beneficial to consider how games integrate and blend with other course materials [19].

These initial considerations must be taken cautiously, however. The notion of using learner styles or typologies has attracted much attention recently, with a strong critique emerging around the consequences of its misuse; which can include 'pigeonholing' learners, by failing to appreciate the fluidity of learner preferences - simply because a learner prefers one mode of learning with a certain subject matter and context, does not mean this will necessarily transfer to other subjects and contexts. Whilst it remains unquestionably appealing to the designer or developer of an online resource to be able to categorise learners into a finite number of groups, and react by serving content accordingly, there is no clear evidence showing persistent 'learning styles' [20]. Thus, to react to this limitation, the following sections describe a pragmatic early-stage design workshop intended to elicit and analyse primary data from potential end-users regarding their perspectives on game design.

3   Design Workshop

3.1   Method

A co-creation methodology was used to underpin the design of the workshop, drawing upon aspects of best-practice relating to co-creative ideation exercises. Principally, these included an initial introduction to objectives, 'hands-on' activities that encouraged participation, and the presence of a moderator acting to encourage equal discussion, whilst keeping ideation on-topic and seeking clarity and consensus, or noting observation of the lack thereof, with participants' views. The workshop was pragmatically designed to elicit the opinions of a small sample, with awareness of the risk in seeking to generalise these views; however, in the absence of a concrete existing template or mapping to drive design, this provided an opportunity to gain a degree of insight despite these limitations.

Recruitment and Sampling
Participants self-selected from an email sent to staff and students of a UK university. There was no incentive for participation beyond the opportunity to feed-in to early stage game design work. The selection criteria was broad, including amateur (student) athletes with an interest in developing a professional (dual) career, academic staff with expertise in sports science, and professional or ex-professional athletes. 11 participants were recruited via this method. This small-scale sample was therefore not expected to be representative of the diverse range of cultures and careers targeted by the project, however, it was a pragmatic approach given time and resource constraints.

Activities
As an initial activity following an overview of the project and warm-up, participants were asked to list issues on post-it notes based on their own prior experiences and understanding. These were translated to a flipchart, creating a Venn diagram categorising these issues. Categories within the diagram, agreed amongst participants, were issues around learning provision (including e-learning); issues around professional career pressure and overlap; and issues around links to prospective employers and identification of transferrable skills.

The second activity used group ideation, based on the challenges identified against the first question, to suggest best-practices in content, context, and structured for an e-Learning module supporting dual-careers in athletes. Individual group's comments and suggestions were presented and fed back on by the whole cohort. Several of the issues noted in reference to the first question persisted: primarily that participants externalised the target audience. There was an agreed consensus to move away from "top of game think" (i.e. considering primarily professionals competing at national or international level); and to attempt to focus instead of semi-professional or amateur athletes.

The third and final activity challenged participants to identify aspects of game design which were most relevant to the objectives and GOAL, and it's e-Learning module. Given that participants were not game designers, this was scaffolded by using a deck of playing cards, split into 4 suites representing game mechanics, social mechanics, victory conditions, and motivators. Each card has a simple written description of a feature, allowing cards to be rapidly sifted and aligned to creating a game design template, without requiring a detailed knowledge of game design theory. Since, as noted, participants were not game designers, and moreover the curriculum for GOAL had not been designed at the point of the workshop, there is a noteworthy risk of the results elicited diverging from the future curriculum needs. However, this appeared a worthwhile early stage task, allowing participants to be engaged with the concepts of games and gamification, and discuss high level strategies for its implementation. Participants were split into groups and each worked through a suite of the deck of cards, followed by a roundtable discussion of each groups' choices, and how the components might interact to form the basis for games. The cards used were developed by PlayGen, as part of their addingplay toolkit[1].

3.2   Results

A list of points elicited from participants in the first activity is presented below:
For the athletes:

- "Motivation to take the course; identifying need for [e]ffecting change"
- "(1) Time (2) Education (3) Knowing the world of work (4) If high income drop in income (5) Family structures"
- "What types of problem do the[y] experience? – Subject-content – Career orientation – Psychology"
- "Culture sports mentality –> business mentality – CV – what do they have? Support – clubs, management, sports organisation"
- "Getting over their ego"
- "Not being aware of the possibilities; What the need if they take one possibility; Need [exposure] to people from other fields/industries"
- "What sciences do they like? Can they link a number of those sciences? Are there jobs in that common [area]?"

For prospective employers:

- "Acquiring transferable skills"
- "What skills employers desire from athletes"
- "Lack of skills; flexibility in [part-time] working; disability of injured"
- "Mapping identifiable skills – time [management] as an athlete to time [management] in workplace"
- "Lack of appropriate qualifications"

---

[1] http://gamification.playgen.com/.

In terms of discussion and synthesis, there were several additional observations as a faciliator:

- Participants tended to externalise the target audience, even though they themselves were amateur or semi-professional athletes. This is evidenced by the frequent use of third-person in the above notes, and direct comments (e.g. "Getting over their ego"). This is not uncommon in intervention design workshops – participants often find it easier to talk about the needs of others rather than themselves. Yet it risks design grounded in perception rather than reality.
- In the UK, high profile cases of top sporting professionals who have publically struggled with mental health issues after a career as a top athlete have been widely reported by UK media. These were raised by participants as examples of the 'problem', though this is not necessarily representative of the typical problem the majority of the GOAL target audience is likely to face. Avoiding lengthy discussion of these cases required some steering, and it seems worthwhile to note again the perception of the target audience of GOAL is prone to straying from the group which, in demographic terms, represents the most sizable and likely to benefit.
- Participants had received a presentation on GOAL prior to the task to set the context for the workshop, and therefore had been somewhat "primed" as the materials covered several key issues.

Outcomes from the second activity centred around two themes:

- **Diversity.** The wide audience, and near-infinite range of potential careers was identified as the foremost barrier to tangible course design. In addition to previously noted variance in professional or amateur status, culture, age, and sporting or athletic discipline, some further potential differences between athletes raised interesting questions. Could, for example, athletes from team-based sports be expected to prefer a collaborative learning environment, whereas athletes from individual-based sports expected to prefer a competitive one? There is, as was noted by participants, no clear "one size fits all" solution – followed by assertion of the "need to define [the] audience in more specific terms".
- **(External) Goal Setting.** One the one hand, it was argued athletes are used to external goals, whether these be incremental, for example increase performance over a month, or competitive, long terms goals, for example Olympic success. Yet on the other hand, amateur or semi-professional athletes may, in absence of substantial coaching, be equally used to defining their own goals related to their current performance and perceptions. "Used to a structured framework" was a well-discussed assumption and assertion: this may translate to their expectations and preferences in an online course. However, on discussing this, the point was also raised that there may be a need to "[facilitate] movement [towards] less structured learning". Meant by this was the need to reflect upon GOAL as a component of a dual-career solution, rather than a comprehensive solution to the problem. If athletes are used to structured, frequent goal-setting and feedback, rather than catering for this need, to allow transitions to other careers and learning programmes, challenging and changing their learning styles may be a necessary path to take.

In the third exercise, the following game attributes were selected from the deck of ∼ 70 cards:

- **Score/Points.** The rationale given by participants here for their inclusion was enabling immediate and highly-structured feedback. Potential implementations were discussed as including point-scoring mini-games, or overall platform gamification giving points for productive actions (e.g. post counts, modules completed). Scoring was also included; a broadly understood distinction was scoring as relative, whilst points were absolute; i.e. "you can gain 100 points for an action, and a score of 100 points is more than…".

- **Variable Challenge.** This was noted and argued as essential for a diverse audience. One concept raised and discussed was whether this variable challenge should be self-selected (i.e. easy/medium/hard) or more fluid (i.e. a single main goal, and additional "stretch" goals). This linked strongly to discussion of goals and objectives, widely perceived as highly relevant to athletes.

- **Goals/Objectives.** Linking to the previous two discussions, the need to frame goals clearly and in a way relatable to athletes was viewed as desirable. In the context of variable challenge, suggestions were raised in terms of personal goal-setting, though participants struggled to define concrete examples.

- **Race.** Perhaps selected due to its arguable linguistic appeal to athletes, the overriding concept here was competition. A discussion followed here of the tendency for traditional classroom pedagogy to avoid placing students in direct competition, and its obvious contrast with a sporting or athletics context.

- **Identity.** The notion of "playing with" identity was suggested as a potential avenue for games within goal. What if, perhaps, the player was placed as a coach managing the career of an athlete after injury, with the player learning skills and challenges by proxy. This was discussed as a potential "good fit" to the tendency of individuals, participants included, to view the dual career problem as one relating to others rather than themselves.

- **Meaning.** This was identified as a key barrier. Without meaning, engagement and motivation are unlikely to follow: whilst the overall benefits of sound dual career planning can be readily identified and communicated, participants were cautious that atheletes may see them as 'things to do in the future', rather than 'things to do right now'. Games could potentially impart immediate meaning, a discussed example being getting a high score; this was also noted as requiring careful implementation to avoid it diverging from the learning objectives.

- **Communal Collaboration.** A collaborative effort as a "GOAL community' was also suggested. This links to identity, in that the GOAL platform could be seen as an "online team" (or range of competing teams), though this brings the need to tightly scaffold collaboration to foster inclusion and address risks of exclusion.

- **Achievement.** A final proposed key element was a sense of meaningful achievement. Discussed here was whether this could come in the form of the direct achievement of developing dual-career skills, or identifying a dual career, or indirectly, for example being the best team on the GOAL platform.

4 Game Design

A recognised tension exists in educational game design between pedagogical goals, and engaging game mechanics. In designing the game thus far, attempt has been made to achieve a successful balance between the two by identifying learning requirements, then approaching, from an entertainment perspective, genres and mechanics that might best fit the needs formed from these objectives. As such, the features identified in the previous section drove both the genre selection, and the bespoke design of individual features and components. Overall, the game adopts a 'business sim' approach, allowing the user to build a company be selecting and staffing various "areas". By linking each area to a learning objective, an expansible, component-based approach is implemented. Current areas cover ideation ("creative space"), translating idea to product, manufacturing, stock management, and recruitment, with current and future work planned to explore concepts such as service-based industry and servitisation, staff development and training, and finance. The current prototype is illustrated in Fig. 1.

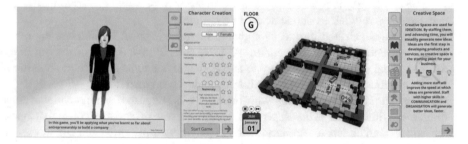

Fig. 1. Screenshots from the prototype game

The game itself was developed using Unity, which enabled rapid prototyping under a agile methodology. In advancing and iterating the design, several features were added and refined, including the ability for the player to create their own character and assign "skill points" allowing them to reflect, or experiment with, specific strengths and weaknesses. These are also applied to recruited staff, adding a layer of strategy to the game in recruiting and deploying staff in the most effective way possible. A short (5–10 min) interactive tutorial was also implemented, seeking to negate the need for any additional documentation beyond the game, and ease new players in to the overall mechanics of creating spaces, deploying staff, managing budget, and developing products. With reference to the workshop findings in Sect. 3, we sought to explore identity through the character creation process; variable challenge through a design which can scale between a "sandbox" mode, with an infinite budget, or challenging scenarios which involve either difficult financial starting constraints, or the player entering an existing business to solve problems. Achievements, which will drive goals and objectives, are currently being implemented as part of the iterative development approach.

5 Conclusions and Future Work

In terms of design workshop outcomes, a general finding relevant to the creation of educational games is that it appears important to steer discussion of curriculum and objectives towards a clearly-defined, and scoped demographic. The tendency of participants to externalise the problem, suggested they themselves either did not percieve it as a personal challenge, or tended to relate to extremes rather than norms. Furthermore, as evidenced by focus on e.g. "science" above, there appeared also a tendency to view certain careers as 'exemplar' of dual-careers, but these are not realistic solutions for *all* athletes. In particular coaching appeared a popular career consideration, a finding supported by literature, though not every athlete is likely to succeed as a future coach given the ratio of coaches to atheletes. Participants also discussed that, particularly for athletes who entered a professional career at a young age, material covering how to learn effectively, as well as the content itself, may likely be beneficial. It was noted that in cases where a higher or further education qualification had not been sought prior to, or during, a professional career, an athlete may have experienced a single learning style (linked to their coaching environment), and may struggle to adopt other pedagogical approaches used in mainstream further, higher, and distance education. Awareness of possibilities appears a necessary first-step prior to any specific training.

Given the diversity of athletes, and the diversity of the global job market, it would seem very difficult and exclusive to provide courses linked directly to specific future jobs or roles. Indeed, this may disengage quickly athletes who don't see a 'list' of 'potential professions' as mapping to their needs and motivation. Yet links, and exposure to, other fields and industries was agreed as an important factor. The challenge, therefore, is how to raise awareness of possibilities, without restrictively listing a set of 'ideal' dual-careers. In terms of achieving this through the game itself, iteration, user-feedback, and testing are typically essential components of the design and development process for any game. Current work is evaluating the game in terms of usability and pedagogical content, with subsequent deployment within the MOOC enabling examination of athletes' reponses and usage patterns. Subsequent findings are expected to provide concrete insight into the reception to the game's various features, enabling further iteration and refinement.

Acknowledgements. This work has been part-supported by the European Commission's Erasmus + programme, under Grant No. 2016-3258/001-001 "Gamified and Online Activities for Learning to Support Dual Careers of Athletes" (GOAL).

References

1. Li, M., Sum, R.K.W.: A meta-synthesis of elite athletes' experiences in dual career development. Asia Pac. J. Sport Soc. Sci. **6**, 99–117 (2017)
2. Lupo, C., Guidotti, F., Goncalves, C.E., Moreira, L., Doupona Topic, M., Bellardini, H., Tonkonogi, M., Colin, A., Capranica, L.: Motivation towards dual career of European student-athletes. Eur. J. Sport Sci. **15**, 151–160 (2015)

3. Aquilina, D.: A study of the relationship between elite athletes' educational development and sporting performance. Int. J. Hist. Sport **30**, 374–392 (2013)

4. EC: EU Guidelines on Dual Careers of Athletes: Recommended Policy Actions in Support of Dual Careers in High-Performance Sport. Publications Office of the European Union, Luxembourg (2013)

5. Tsiatsos, T., Douka, S., Politopoulos, N., Panagiotis, S., Ziagkas, E., Zilidou, V.: Gamified and Online Activities for Learning to Support Dual Career of Athletes (GOAL) (2018)

6. Daloukas, V., Dai, V., Alikanioti, E., Sirmakessis, S.: The design of open source educational games for secondary schools. Presented at the Proceedings of the 1st International Conference on PErvasive Technologies Related to Assistive Environments, Athens, Greece (2008)

7. Dunwell, I., Lameras, P., de Freitas, S., Petridis, P., Star, K., Hendrix, M., Arnab, S.: MeTycoon: a game-based approach to career guidance. In: 2013 5th International Conference on Games and Virtual Worlds for Serious Applications (VS-GAMES), pp. 1–6 (2013)

8. Srivathsan, G., Garg, P., Bharambe, A., Varshney, H., Bhaskaran, R.: A dialogue system for career counseling. Presented at the Proceedings of the International Conference; Workshop on Emerging Trends in Technology, Mumbai, Maharashtra, India (2011)

9. Chang, B., Lee, J.T., Chen, Y.Y., Yu, F.Y.: Applying role reversal strategy to conduct the virtual job interview: a practice in second life immersive environment. In: 2012 IEEE Fourth International Conference on Digital Game And Intelligent Toy Enhanced Learning, pp. 177–181 (2012)

10. CareerGuide: An overview of the Tools, Systems and Games used to support Career Guidance in Europe INSEAD, Ellinogermaniki Agogi, Orientum, Alba, Technical University of Dresden, University of Louis Pasteur, Technical University of Sofia, IPA S.A., Alpha-Omega Communications, Newman College of Higher Education (2007)

11. Boyle, E., Allan, G., Moffett, J., Connolly, T., Oudi, H., Badii, A., Einarsdóttir, S., Hummel, H., Graur, A.: Theoretical foundations of the Youth@Work game. Presented at the EduLearn16 (2016)

12. Deterding, S.: Gamification: designing for motivation. Interactions **19**, 14–17 (2012)

13. Mekler, E.D., Brühlmann, F., Opwis, K., Tuch, A.N.: Do points, levels and leaderboards harm intrinsic motivation? An empirical analysis of common gamification elements. Presented at the Proceedings of the First International Conference on Gameful Design, Research, and Applications, Toronto, Ontario, Canada (2013)

14. Sun, E., Jones, B., Traca, S., Bos, M.W.: Leaderboard position psychology: counterfactual thinking. Presented at the Proceedings of the 33rd Annual ACM Conference Extended Abstracts on Human Factors in Computing Systems, Seoul, Republic of Korea, (2015)

15. Peever, N., Johnson, D., Gardner, J.: Personality & video game genre preferences. Presented at the Proceedings of The 8th Australasian Conference on Interactive Entertainment: Playing the System, Auckland, New Zealand (2012)

16. Rhodes, R.E., Smith, N.E.I.: Personality correlates of physical activity: a review and meta-analysis. Br. J. Sports Med. **40**, 958–965 (2006)

17. Judge, T.A., Bono, J.E., Ilies, R., Gerhardt, M.W.: Personality and leadership: a qualitative and quantitative review. J. Appl. Psychol. **87**, 765–780 (2002)

18. Star, K.: Gamification, Interdependence, and the Moderating Effect of Personality on Performance. Ph.D., Serious Games Institute, Coventry University, Coventry (2016)

19. Dunwell, I., Dixon, R., Bul, K.C., Hendrix, M., Kato, P.M., Ascolese, A.: Translating open data to educational minigames. In: 2016 11th International Workshop on Semantic and Social Media Adaptation and Personalization (SMAP), pp. 145–150 (2016)

20. Newton, P.M., Miah, M.: Evidence-based higher education – is the learning styles 'myth' important? Frontiers Psychol. **8**, 444 (2017)

A Serious Game Approach in Mitigating Performance Enhancement Culture in Youth (GAME Project)

Vasileios Barikoukis[1], Thrasyvoulos Tsiatsos[1],
Nikolaos Politopoulos[1], Panagiotis Stylianidis[1(✉)],
Efthymios Ziagkas[1], Andreas Loukovitis[1], Lazuras Lambros[2],
and Ypsilanti Antonia[2]

[1] Aristotle University of Thessaloniki, Thessaloniki, Greece
{bark,eziagkas}@phed.auth.gr,
{tsiatsos,npolitop,pastylia}@csd.auth.gr,
louko-vitis@hotmail.com
[2] Sheffield Hallam University, Sheffield, UK
{L.Lazuras,A.ypsilanti}@shu.ac.uk

Abstract. Anti-doping education is currently at an early stage and there are several needs that must be addressed with respect to the design, implementation, and evaluation of anti-doping education programmes. Firstly, anti-doping education should be based on state-of-art learning pedagogies that will enable effective engagement, learning and retention of the learned material. Secondly, currently there is a lack of such a systematic approach for evaluating the behavior change outcomes of anti-doping educational interventions. Thirdly, there is a lack of other known anti-doping educational interventions that promote a positive approach to doping prevention. Finally, anti-doping educational interventions should incorporate the learning process in the context of new learning technologies that can also facilitate behavior change outcomes.

The aim of this study is to present highlight the importance of anti-doping, conduct a state of the art about serious games design, present an initial design of a serious game for anti-doping education the project, and summarize GAME's future work.

Keywords: Serious games · Anti-doping · Sports

1 Introduction

Doping use is defined as the use of Performance and Appearance Enhancement Substances (PAES) that are prohibited by the World Anti-Doping Agency (WADA), such as synthetic forms of human growth hormone, testosterone and related derivatives, masking agents, stimulants and other drugs that were originally designed to treat diseases in humans and/or animals, as well as designer synthetic drugs that have been developed to improve athletic performance (Baron et al. 2007; Lazuras and Barkoukis 2015). In elite sports doping is an ongoing issue and the recent scandals involving elite

© Springer Nature Switzerland AG 2020
M. E. Auer and T. Tsiatsos (Eds.): ICL 2018, AISC 916, pp. 733–742, 2020.
https://doi.org/10.1007/978-3-030-11932-4_68

athletes (e.g., use of meldonium by Sharapova) and the Russian anti-doping agency in the Rio Olympics cast a pall on sports and shake athletes' and the public's confidence about the rules of fair play and clean sports (Barkoukis et al. under preparation). According to studies that used indirect measures and anonymous and confidential self-reported surveys, doping use in competitive and elite sports ranges from 15% to 39% (de Hon et al. 2015; Laure 1997). Recent studies also show that doping use is fast becoming a crisis in amateur and grassroots sports too. In particular, a study funded by Erasmus + Sport in 2016 showed that doping use can be initiated among amateur athletes as young as 10 years old (Nicholls et al. 2017). Accordingly, Project SAFE YOU, that was funded by Erasmus + Sport in 2015, showed that, on average, 1 out of 5 young amateur athletes and exercisers aged between 16–25 years, have used doping substances at least once in their lifetime, with higher prevalence rates being reported in South-East European countries like Greece (27.6%) and Cyprus (28.9%), and lower prevalence rates in Germany (17%) and the UK (14.6%; Lazuras et al. in press).

The use of doping substances not only contradicts the spirit of sports and fair play rules, it also poses a range of health risks to users, including sexual dysfunction and hormonal imbalance, mood fluctuations, anxiety and aggressive behaviour, as well as potentially lethal heart and kidney dysfunction, and even sudden death, especially among younger users (Christou et al. 2017; Darke et al. 2014; Hartgens and Kuipers 2004; Frati et al. 2015). Therefore, the uncontrolled use of doping substances represents an emerging public health concern that may inflict a significant proportion of young people (at least 20% of amateur athletes and exercisers in the EU according to Lazuras et al. in press) unless preventive action is taken.

Following a systematic review of 51 studies, Nicholls et al. (2017) identified nine key risk factors for doping use among young athletes aged between 10–21 years: age, gender, participation in sports, sport type, beliefs/behaviours of coaches and athlete's entourage, as well as psychological variables, and use of nutritional supplements. Clearly, not all of these variables are amenable to interventions against doping use, but there are a lot of psychological and social aspects of doping use that can be directly targeted by tailor-made educational interventions. In this respect, another meta-analysis of 63 independent studies on doping behaviour in adolescent and adult athletes showed that doping behaviour is better understood as a goal-directed, intentional process, and that variables such as attitudes, self-efficacy, and perceived social norms (e.g., social approval from referent others such as fellow athletes and coaches; perceived prevalence of doping among referent others) directly predicted athletes' intentions to use doping substances in the near future (Ntoumanis et al. 2014).

Project SAFE YOU also showed that an urgency to seek for immediate performance and appearance benefits, and to recover quickly from heavy trainings or injuries during training were among the top five reasons for doping use in young amateur athletes and exercisers (Lazuras et al. in press).

Another line of research has highlighted the psychological and social factors that act protectively against doping use, that is the factors that can be targeted by educational interventions in order to strengthen attitudes against doping use, and empower athletes to "stay clean" even in the face of internal (e.g., performance anxiety and stress) or external situational pressures and temptations (e.g., peer pressure, coach pressure).

These protective factors include health beliefs and awareness of the adverse health consequences of doping use; factual knowledge about the actual and alleged effects of doping use on athletic performance (and on physical appearance where exercisers are concerned); self-regulation and resilience to social pressures; and a "self-determined" approach to exercise and sport participation, whereby athletes are motivated to participate in sports for the sake of participation and intrinsic motivation and not for external rewards and the need to outperform others (Chan et al. 2015a; Chan et al. 2015b; Erikson et al. 2015; Lazuras et al. in press; Mohamed et al. 2013).

Taken together, these findings indicate the reasons and motivations that would "push" athletes into the dark side of performance enhancement (doping use), as well as the factors that would act protectively to prevent doping use.

This evidence can be utilized to inform, design and evaluate tailored anti-doping educational interventions.

The main aim of this paper is to propose an innovative anti-doping educational intervention that incorporates the learning process in the context of new learning technologies and especially serious games.

This paper is structured as follows. The next section presents the anti-doping education needs and problems along with an overview of the Erasmus + Sport project titled "GAME". This work is presented in the context of GAME project. The fourth section presents the proposed technological solution along with a first scenario. The last section presents our concluding remarks along with our vision about the next steps.

2 Anti-doping Education Needs and the GAME Project

GAME project is motivated by the need to advance anti-doping education intervention targeting competitive and recreational athletes. The project's consortium recognizes the need to move forward and transform the way anti-doping education is designed, delivered and evaluated and aims to utilize: (a) updated research from the social and behavioural sciences on doping use; (b) state-of-art learning pedagogies; and (c) cutting-edge serious gaming design and technology in order to deliver an innovative and impactful anti-doping educational intervention.

Updated evidence from the social and the behavioural sciences will help in designing problem-based learning scenarios that can realistically depict risk-conducive situations and contingencies in a serious game environment. This can maximize the relevance, engagement and impact of the envisaged anti-doping serious game.

Furthermore, blending learning pedagogies with serious gaming technology will result in an effective learning approach that facilitates and stimulates independent and active learning, better retention of information, and greater likelihood for persuasion, attitude and behaviour change.

Finally, applying behaviour change indicators (e.g., changes in beliefs, intentions, and actual behaviour) will allow the consortium's experts to validate the effectiveness and impact of the envisaged anti-doping serious game, and also to recommend specific policy actions that are needed to optimize the effectiveness of anti-doping education. More specifically, behaviour change can be seen as the dynamic interaction between capabilities (e.g., role modelling, environmental barriers or facilitators of the target behaviour), opportunities for behaviour change, motivation, and actual behaviour (Michie et al. 2014). This approach helps the consortium's experts to identify the key intervention areas as well as different policy-making categories that are relevant to optimizing anti-doping education and intervention outcomes.

On the whole, it is the first time an anti-doping serious game with these features is proposed, and the aims of "The Game" Project are to:

- Utilize cutting-edge behavioural science research about the risk and protective factors against doping use in amateur and grassroots sports to inform the development of an anti-doping serious game.
- Use an "open innovation" framework to co-design the anti-doping serious game, through the active collaboration of SG designers and young people engaged in amateur and grassroots sports.
- Apply and evaluate the effectiveness of the doping prevention SG in changing young people's learning, motivation, beliefs and behaviour towards the use of PAES, and in promoting a more positive mentality about drug-free and health-enhancing physical activity and sports.
- Train the trainers on how to promote the serious game into several the project's target groups, namely adolescents and young competitive and recreational athletes.
- Develop research agenda and policy recommendations for the wider application of SG technologies for the prevention of doping and the promotion of health-enhancing physical activity and sports in young people.

GAME consists of a consortium of various organizations with the participation of people with expertise in appropriate fields such as sport policy and practice (training, competitions, coaching, etc.), interactive ICT-based tools, serious game designers, sports marketing and entrepreneurship with academic expertise as well as their ability to reach out wider audiences. More specifically the consortium includes Aristotle University of Thessaloniki (Greece), Sheffield Hallam University (UK), National University of Physical Education and Sports Bucharest (Romania), Lithuanian Sports University (Lithuania); The Cyprus Sports Organization (Cyprus), Young Men's Christian Association of Thessaloniki (Greece) and The European Network of Academic Sports Services (The Netherlands).

The innovation and novelty of project GAME is evaluated against previous Erasmus + projects in anti-doping as well as other doping-related projects funded under other European Commission funds. In this context, GAME is innovative in the following respects. Firstly, the envisaged GAME training will be developed as a serious mobile game. Although there have been attempts to develop web-based anti-doping education tools (e.g., the ALPHA program by the World Anti-Doping Agency) there have not been similar attempts at a European level, and there has been no mobile anti-doping video games as yet even at a global level. This is an important novelty of GAME because smart phone applications have become widespread for a wide range of uses and functions in everyday life, from monitoring household bills and energy expenditure, to keeping track of physical activity and exercise goals (Falaki et al. 2010; Lane et al. 2011). Having a serious game for anti-doping education on a smart phone application is expected to further improve the impact and reach of GAME in the target population of young exercisers and athletes. Studies showed that by the time a teenager begins his/her professional career he has already played over 10,000 h (Prensky 2001, European Summary Report 2012) or the equivalent to 5 years of fulltime employment. In recent years, the interest on gaming has led to a rapid growth of the game industry, and in particular commercial entertainment games. Video games allow individuals to reach high levels of motivation and engagement and they have proven to be more successful than schools in attracting interest from young people (Caperton 2007; Boyle et al. 2016).

Secondly, GAME will utilize findings, data and resources from previous projects in order to create educational scenarios that will then be the basis of the serious mobile game. This way material that has been tested and evaluated can be made more accessible and presented in a way that raises motivation and awareness to the subject (Pomales and Trevino 2014).

Thirdly, the envisaged GAME training will be developed on the basis of contemporary approaches in learning pedagogies (i.e., game-based learning). It is noteworthy that, although previous projects have considered the development of anti-doping education and training, this is the first time game-based learning will be used to inform the ways and the training/education material that will be deployed and presented to the intended target groups.

3   Towards a Technological Solution to Support Anti-doping Education

There are many ways digital technology could be exploited to support education related with sports. Examples are Massive Open Online Course (MOOCs) in C4BIPS and GOAL Erasmus + Sport projects (Tsiatsos et al. 2018a). Digital games to support educational activities (i.e. serious games) have been also implemented or planned in the context of Erasmus + Sport projects (e.g. GOAL Erasmus + Sport project) (Tsiatsos et al. 2018b).

The movement towards the use of serious games as training and learning is proliferated by the perceived ability of such games to create a memorable and engaging learning experience. Various commentators and practitioners alike argue that serious games may develop and reinforce 21st century skills such as collaboration, problem solving and communication. While in the past, practitioners and trainers have been reluctant in using serious games for improving skills and competencies there is an increasing interest, to explore how serious games could be used to improve specific skills and competencies. The overarching assumption made is that serious games are built on sound learning principles encompassing teaching and training approaches that support the design of authentic and situated learning activities in an engaging and immersive way.

Developing serious games based on activity-centered pedagogies that enable trainees to engage actively with questions and problems associated with sport activity and dual careers is an empowering approach with benefits for subject learning as well as for developing a wide range of important high-order intellectual attributes including the notion of 'transferability – that is being able to situate specific skills in different settings and contexts, a competence much needed.

The serious games and gamification mechanics of GAME will be based on real life scenarios that offer a range of outcomes based on the users' choices throughout the serious game. The reason for this is to let the user experience personal dilemmas and also the consequences of his actions, thus creating a multi-faceted view of the subject that promotes in-depth understanding.

In order to develop personal skills and based on the statistical analysis, it was decided to create adventure games and especially interactive movies.

An interactive movie contains pre-filmed full-motion cartoons or live-action sequences, such as dialogues, where the player controls some of the moves of the main character or his/her answers. For example, when in danger, the player decides which move, action, or combination to choose. In these games, the only activity the player has is to choose or guess the move the designers intend him to make. Elements of inter-active movies have been adapted for game cut scenes, in the form of Quick Time Events, to keep the player alert.

In order to evaluate the existing game engines suitable to create these types of games, an evaluation table was created and importance points were assigned to features that are needed on the platform. Importance scale is 1 to 3, 1 slightly important, 2 important and 3 very important. If a game engine qualifies for a feature it was assigned an X on the table. The values are summarized at the bottom of the Table 1.

Table 1. Game engines state of the art

Characteristic	Degree of significance	Unity + Fungus	GameMaker	Ren'Py	AGS	eAdventure
Dialogue tools	3	X	X	X	X	X
Graphic tools	3	X	X	X	X	X
Sound tools	2	X	X	X	X	X
Variables	3	X	X	X ·	X	X
Free export to WEB	3	X	–	–	–	–
Free export to MOBILE	3	X	–	–	–	–
Scenario editing	3	X	X	X	X	X
VLE incorporation	1	–	–	–	–	X
Cross-platform – Free export	3	X	–	–	–	–
Total		23	14	14	14	15

As, can be seen from the evaluation, the game engine that most suits the needs of the project is Unity (https://unity3d.com/). A concept design of these games was created (Fig. 5) to demonstrate the main structural idea behind them. Every player will be involved in every day scenarios that assess and improve his personal skills, such as communication skills and allow him to self-regulate.

Fig. 1. Interactive movie game prototype design

Figure 1 presents the main components of the GAME project technological proposal. The game simulates pedagogical scenarios satysfing specific needs and exploiting pedagogical material form Open Educational Resources (Fig. 2).

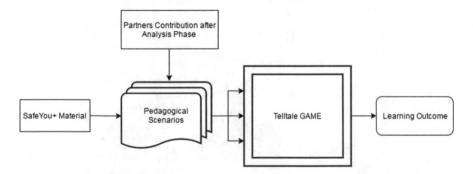

Fig. 2. Main components of GAME project technological proposal

A challenge in the game mechanics design is the co-operation among e-learning experts, game developers, psychologists and anti-doping education experts. The development team exploits flowcharts (Fig. 3) and mock-up (Fig. 4) tools to create a common understanding among its members. More specifically the psychologists and anti-doping education experts are presenting their pedagogical scenarios using flowcharts in order to decide the gameflow, the user feedback, the awards, badges and the scoring system of the game.

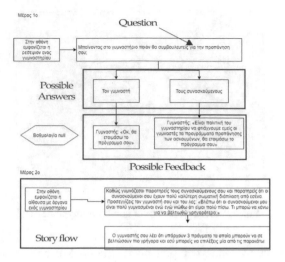

Fig. 3. Sample scenario flowchart

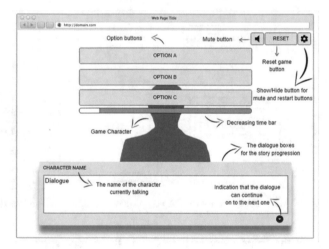

Fig. 4. Game Interface Mockup

4 Conclusion—Future Work

This paper presents a proposal to satisfy the need to advance anti-doping education intervention targeting competitive and recreational athletes.

GAME consists of a consortium of various organizations with the participation of people with expertise in appropriate fields such as sport policy and practice (training, competitions, coaching, etc.), interactive ICT-based tools, serious game designers, sports marketing and entrepreneurship with academic expertise as well as their ability to reach out wider audiences.

Having a serious game for anti-doping education on a smart phone application is expected to further improve the impact and reach of GAME in the target population of young exercisers and athletes. The next step, is to deploy the game trough the project website and the google play store. At the end of this stage, a pilot study will be conducted to measure learning outcome and user satisfaction.

Acknowledgement. This project has been funded with support from the European Commission. This publication reflects the views only of the author, and the Commission cannot be held responsible for any use which may be made of the information contained therein.

The authors of this research would like to thank GAME team who generously shared their time, experience, and materials for the purposes of this project.

References

De Hon, O., Kuipers, H., van Bottenburg, M.: Prevalence of doping use in elite sports: a review of numbers and methods. Sports Med. **45**(1), 57–69 (2015)

Laure, P.: Epidemiologic approach of doping in sport. A review. J. Sports Med. Phys. Fitness **37** (3), 218–224 (1997)

Baron, D.A., Martin, D.M., Abol Magd, S.: Doping in sports and its spread to at-risk populations: an international review. World Psychiatry Off J. World Psychiatr. Assoc. WPA **6**, 118–123 (2007)

Lazuras, L., Barkoukis, V., Tsorbatzoudis, H.: Toward an integrative model of doping use: an empirical study with adolescent athletes. J. Sport. Psych. **37**, 37–50 (2015). https://doi.org/10.1123/jsep.2013-0232

Nicholls, A.R., Cope, E., Bailey, R., Koenen, K., Dumon, D., Theodorou, N.C., Chanal, B., Saint, Laurent D., Muller, D., Andres, M.P., Kristensen, A.H., Thompson, M.A., Bau-mann, W., Laurent, J.F.: Children's first experience of taking anabolic-an-drogenic steroids can occur before their 10th birthday: a systematic review identifying 9 factors that predicted doping among young people. Frontiers Psychology **8**, 1015 (2017)

Christou, M.A., Christou, P.A., Markozannes, G., Tsatsoulis, A., Mastorakos, G., Tigas, S.: Effects of anabolic an-drogenic steroids on the reproductive system of athletes and recreational users: a systematic review and meta-analysis. Sports Med. 47 (2017). https://doi.org/10.1007/s40279-017-0709-z

Darke, S., Torok, M., Duflou, J.: Sudden or unnatural deaths involving ana-bolic-androgenic steroids. J. Forensic Sci. **59**, 1025–1028 (2014). https://doi.org/10.1111/1556-4029.12424

Hartgens, F., Kuipers, H.: Effects of androgenic-anabolic steroids in athletes. Sports Med. **34**(8), 513–554 (2004)

Frati, P., Busardò, F.P., Cipolloni, L., Dominicis, E.D., Fineschi, V.: Anabolic an-drogenic steroid (AAS) related deaths: autoptic, histopathological and toxicological findings. Curr. Neuropharmacol. **2015**(13), 146–159 (2015)

Ntoumanis, N., Ng, J.Y., Barkoukis, V., Backhouse, S.: Personal and psychosocial predictors of doping use in physical activity settings: a meta-analysis. Sports Med. (Auckland, N.Z.) (2014). https://doi.org/10.1007/s40279-014-0240-4

Chan, D.K., Lentillon-Kaestner, V., Dimmock, J.A., Donovan, R.J., Keatley, D.A., Hardcastle, S.J., Hagger, M.S.: Self-control, self-regulation, and doping in sport: a test of the strength-energy model. J. Sport Exerc. Psychol. **37**, 199–206 (2015). https://doi.org/10.1123/jsep.2014-0250

Erickson, K., McKenna, J., Backhouse, S.H.: A qualitative analysis of the factors that protect athletes against doping in sport. Psychol. Sport Exerc. **16**, 149–155 (2015)

Michie, S., West, R., Campbell, R., Brown, J., Gainforth, H.L.: ABC of Behaviour Change Theories. Silverback Publishing, London, UK (2014)

Falaki, H., Mahajan, R., Kandula, S., Lymberopoulos, D., Govindan, R., Estrin, D.: Diversity in smartphone usage. In: MobiSys 2010, June 2010

Lane, W., Manner, C.: The impact of personality traits on smartphone-ownership and use. Int. J. Bus. Soc. Sci. **2**(17), 22–28 (2011)

Caperton, I.: Video Games and Education. OECD Background Paper for OECD-ENLACES Expert Meeting (2007)

Treviño-Guzmán, N., Pomales-García, C.: How can a serious game impact student motivation and learning? In: Industrial and Systems Engineering Research Conference, Montreal, IIE, Norcross (2014)

Tsiatsos, T., Douka, S., Politopoulos, N., Stylianidis, P.: Massive open online course for basketball injury prevention strategies (BIPS). In: Auer, M., Tsiatsos, T. (eds.) Interactive Mobile Communication Technologies and Learning, IMCL 2017. Advances in Intelligent Systems and Computing, vol. 725. Springer, Cham (2018a)

Tsiatsos, T., Douka, S., Politopoulos, N., Stylianidis, P., Ziagkas, E., Zilidou, V.: Gamified and online activities for learning to support dual career of athletes (GOAL). In: Auer M., Tsiatsos T. (eds) Interactive Mobile Communication Technologies and Learning, IMCL 2017. Advances in Intelligent Systems and Computing, vol. 725. Springer, Cham (2018b)

Design, Creation and Evaluation of TEAM, A Serious Game for Teamwork Development

Martha Sereti, Angeliki Mavropoulou, Panagiotis Stylianidis$^{(\boxtimes)}$,
Nikolaos Politopoulos, Thrasyvoulos Tsiatsos, and Stella Douka

Aristotle University of Thessaloniki, Thessaloniki, Greece
{mpsereti, pastylia, npolitop, tsiatsos}@csd.auth.gr,
{angelikim, sdouka}@phed.auth.gr

Abstract. The aim of this study is to create an educational game which, through the simulation of realistic situations, enhances the development, practice and improvement of skills, which are necessary for the people's professional progress. In particular, this game will aim at developing people's teamwork skills, as they need to work closely with others in order to achieve their common goals.

Initially, a detailed description of the necessary theoretical concepts, namely game-based learning and serious games, is displayed. In addition, the term soft skills, is presented, while the chapter concludes with presentation of teamwork skills. This analysis of the theoretical concepts explains the importance of soft skills and in particular of teamwork skills, while at the same time it explains why the creation of an educational game has been chosen in order to strengthen them. Then, after comparing the available technologies, it emerged that the most appropriate tool was the Unity 3D platform along with the Fungus package. Afterwards there was a comparison in order to find the appropriate questionnaire for using to create the game's script. The steps followed for designing and implementing the game are described below. After the game's implementation an evaluation activity was conducted. The results were encouraging concerning the ease of use, the enjoyment and the usefulness of the game. This article ends with suggestions for future research.

Keywords: Game-based learning · Soft skills · Teamwork skills · Unity 3D

1 Introduction

According to researches which conducted in the 21st Century (Glenn 2008; Mitchell et al. 2010), employers want to recruit candidates who also have strong interpersonal skills (soft skills) in addition to having hard skills. Unfortunately, it seems that young graduates do not meet their expectations. Although companies regard such skills as important, they pay little attention to the training of their employees and often the superior complains because soft skills are essential to the success of the organization (Klaus 2010). For this reason, educational institutions should equip their trainees with the necessary soft skills. Particular emphasis should be placed on teamwork skills as

© Springer Nature Switzerland AG 2020
M. E. Auer and T. Tsiatsos (Eds.): ICL 2018, AISC 916, pp. 743–754, 2020.
https://doi.org/10.1007/978-3-030-11932-4_69

they will cooperate with their colleagues in the workplace as future employees (Evenson 1999; Robles 2012).

In recent years, with the rapid development of technology, educational computer games have attracted attention, researchers have indicated that they can effectively help to gain knowledge through an educational environment that provokes student interest (Cagiltay 2007; Papastergiou 2009) cited in Sung and Hwang (2013). So-called serious games are a form of electronic games designed to support the learning process for learning and are designed in a way that balances the entertainment with education (Zyda 2005; cited in Romero et al. (2012).

So, according to all of the above, this research addresses the implementation of an educational game, aimed at acquiring, practicing and improving their teamwork skills. In the following sections, the terms game-based learning, serious games, soft skills and teamwork skills are described. Then the analysis of the available technologies is conducted in order to select the best option for the game's implementation, the choice of the appropriate questionnaire in order to create the game's scenario, the game design and the implementation process and finally an evaluation of the game user interface satisfaction.

The main focus on this research is to examine the students' experience of using the educational game as a learning tool. Thus, this article is exploring user's perceives of usefulness, ease of use, ease of learning, and satisfaction after an evaluation activity where the students played the game. Moreover this paper is also exploring whether the game offers enjoyment to the players.

2 Theoretical Background

2.1 Game Based Learning

Nowadays, as technology is increasingly integrated into everyday life and culture, education tends to focus on gaming-based learning, namely the application of learning-based learning methods and the way it is supported by the use of various innovative digital technologies (Demetriadis 2014). The approach of gaming-based learning involves the design of educational games that combine balanced learning with entertainment while good design satisfies students in their use. In this way, bridging between formal and informal learning is achieved, always aiming that the experience of the game is being a learning experience at the same time.

The benefits expected from this approach are enough. Initially, there is a strong motivation for students to engage in the game, as they want to experience the game's experience with what it has to offer. Thus, the objective of the game is through the increased engagement and interaction of the students, the activation of the appropriate cognitive processes, in order to achieve deep knowledge in the cognitive subject. Moreover, according to Warschauer and Matuchniak (2010), well-designed games can help players to develop skills such as abstract thinking, experimentation, collaboration and critical thinking, skills which are necessary in all areas for the individual's professional career.

The game designers should give special attention, since they have to take into account that good design encourages students to continue playing, while the good scenario makes the game even more attractive. In addition, the script is a basic feature of game design and perhaps the most important is the relationship between the script and the learning mechanism. The scenario should incorporate the learning mechanism in such a way that the learner learns during the game without interrupting its development. Attention should also be given to the level of challenges as they are a factor, which influences students' attitudes towards the game as it gives them a sense of success within the game's competition (Chen and Yang 2013).

2.2 Serious Games

Serious games are designed for a particular purpose beyond pure entertainment. Most of these are real-world simulations or processes through everyday reality that are designed to solve a problem. Their basic role is educator, but most of them involve entertainment, the main feature of the games. Serious games provide students with learning ways through which they can gain practical knowledge, professional skills and innovative thinking. However, the condition for achieving is the good design of the game, the constant review of the educational material and the orientation of the game to the needs, abilities and interests of the target audience. These games have been demonstrated to provoke active learner involvement through exploration, experimentation, competition and co-operation.

Storytelling games are games with minimal game play. Typically the majority of player interaction is limited to repetitive mouse clicks for keeping text, graphics and audio in motion, while making narrative choices along the way.

2.3 Soft Skills

The employers are seeking mostly for workers, who beyond the technical skills and knowledge of work they also possess some other personal qualities, called soft skills, which are necessary to succeed in their workplace. Soft skills are a combination of people skills, social skills, communication skills, character traits, attitudes, career attributes, social intelligence and emotional intelligence quotients among others that enable people to navigate their environment, work efficiently with others, perform well, and achieve their goals with complementing hard skills.

It is very important for the future employee, in addition to hard skills, to be equipped with the necessary soft skills in order to succeed, since in today's competitive labor market, they are the characteristics that make the difference in recruiting a person (Evenson 1999). The lack of soft skills could be a major obstacle to the professional development of a person who possesses the necessary hard skills and professional experience but doesn't have the necessary interpersonal skills (soft skills) (Klaus 2010). According to Sutton (2002), soft skills are an important criterion for hiring future employees in several professions, since an organization whose employees have soft skills becomes highly competitive (Glenn 2008). So these skills are considered to be essential for working in high-performance environments and to be productive (Wilhelm 2002).

A Washington's Protocol School survey, which was applied to three universities, showed that 15% of a person's workforce is technical skills and knowledge, while the 85% of the success is due to the soft skills of the individual. So it is the university's responsibility to fully equip its graduates so as to be able to find a job, and this means that the development of soft skills should be integrated into the curriculum of universities combined with hard skills (Hairuzila 2009). Therefore, for a better training of soft skills, it is proposed to integrate them in the training of hard skills. In this way the lesson will become more attractive and the success rate will increase, so graduates will be more appropriately equipped for their workplace.

2.4   Teamwork Skills

The changes in the industry's field brought changes in the field of education as well and thus changes in the traditional way of teaching, which didn't promote the idea of human interaction (Frank et al. 2003). Due to these changes passive learning is no longer sufficient as graduates have high technical but limited team and communication skills (Hoyt et al. 2003 refers to Karamudin et al. 2012).

So universities have to create graduates who can work and adapt to different languages, cultures and ethics so as they can work with other people. But to do so, they need to be educated in teamwork. The team consists of a group of people who work together, coordinate roles and responsibilities, and share the same goal. Due to the development of technology the communication and collaboration are facilitated (Duarte and Snyder 2001; Gatlin-Watts et al. 2007), so as individuals can work together without obstructing spatial and temporal distances. Therefore, graduates of universities should be ready to work in global organizations where they have to work effectively and solve problems as a team member. In order for a team to work harmoniously and effectively, its members must have some skills. So a fully equipped team member should:

1. Listen to the ideas of the other and supporting his comments
2. Ask questions to trigger interaction through the discussion
3. Respect the opinions of others, support the ideas and efforts of others
4. Persuade or be convinced about the exchange of ideas
5. Help other members of the team
6. Share his ideas and report the conclusions
7. Participates in the work of the team.

3   Technological Analysis

3.1   Game Development Tool

The requirements for the game's implementation must be met by the chosen technology which will be used in order to create the game. In particular, the choice of technology must be made in such a way as to cover the following features:

1. Dialogue tools in order to help manage the screenplay dialogues
2. Graphics tools in order to help in editing and interacting with game graphics (such as characters, background, and user interface elements)
3. Audio tools, which are used to add sound effects and music tracks to the game
4. Scenario processing tools that allow the scripting texts to be edited by people who are not familiar with programming
5. Variables, which will be influenced by the player's choices and decisions
6. Free export to mobile, which allows you to export files to iOS, Android and Windows Phone
7. Free online export, which allows the export of HTML or WebGL files
8. Integration into Virtual Learning Environments (VLE), which enables the game to integrate into an environment such as Moodle.

After an overview of the available technologies, the game engines are compared in the table below (Table 1). The comparison is based on the features needed to implement the game so the most appropriate game engine will be selected. The table has been completed by following the rating X when a feature is fully supported. For each feature, a Degree of Significance (BS) has been assigned:

- "1": Desirable feature - Low Significance
- "2": Useful feature - Medium Significance
- "3": Essential feature - High Significance

Table 1. Comparison of game development engines based on implementation features

Eatures	DF	Visual studio	GM	Ren'Py	eAdventure	AGS	NSc-ripter	Unity + Fungus
Dialogs tools	3	X	X	X	X	X	X	X
Graphic tools	3	X	X	X	X	X	X	X
Audio tools	2	X	X	X	X	X	X	X
Variables	3	X	X	X	X	X	X	X
Free export to WEB	3	–	–	–	–	–	–	X
Free export to MOBILE	3	–	–	–	–	–	–	X
Script editor	3	X	X	X	X	X	X	X
VLE integration	1				X			
Cross-platform – Free export	3	X	X	X	–	X	–	X
Total		17	17	17	15	17	14	23

According to the results of the above table (Table 1), comparing the available development game engines the most suitable is considered Unity 3D in combination with the Fungus package.

The Unity 3D platform is a platform that is ideal for game development. Fungus is a package that allows easy character insertion and the creation of "story-telling" games. It allows storytelling features to be added to games created through "Unity". It is ideal for creating "visual romances" and interactive games.

3.2   The Questionnaire's Choice

For the game's implementation, the appropriate questionnaire needs to be selected, whose questions will be converted into the game's scenario. For this reason, a survey was carried out to find the most appropriate questionnaires which are based on teamwork skills.

After comparing and evaluating the questionnaires from the research the questionnaire from the University of Kent has collected the highest score, which appears to be the most suitable questionnaire. This questionnaire is suitable for this study because it fulfills the following basic criteria:

1. Sufficient number of questions, appropriate to make a complete scenario
2. It is divided into 7 categories, thus calculating 7 different types of teamwork skills
3. The answers to the questionnaire are given on a 4-level Likert scale, which helps to convert the questions into game's scenario
4. The questionnaire gives results which helps to display results in the game as well.

4   Game Design and Development

The purpose of this study is to create a storytelling game. This game offers the user a profile based on the questionnaire's seven types of teamwork skills. Players through the use of the game have the opportunity to develop, improve and practice their teamwork skills. Creation of the game requires the following requirements to be met:

- To integrate a realistic and interesting scenario
- To create a database for the storage of users and their results
- To recognize the user and customize the game
- To provide a graphical presentation of the results for displaying the progress.

The first step is to create the script, which is based on the chosen questionnaire. The user undertakes the role of a member of a group of traditional dances, which must organize a sports event. Based on the player's selections specific values based on a grading model will be assigned to the questions in the similar scale with the questionnaire. After the player has completed the game for the first time, the game creates a user's profile based on the seven kinds of teamwork skills. A basic requirement for the user is to create a game account so the game will be capable to store the user's data and results.

4.1   Scenario and Grading Model

The game is based on the chosen questionnaire, which consists of 28 questions. For each question on the questionnaire a script was created, in which the user is asked to make a decision. After the script was created, it was reviewed by experts in order to see if the situations described in it were realistic. The grading model of the game should capture the four different values of the questionnaire's scale for each one of the user's choices. The user will have to choose between two bad choices and two good ones. The model for each question is shown in the figure below (Fig. 1).

2 BAD   2 CHOICES   2 GOOD

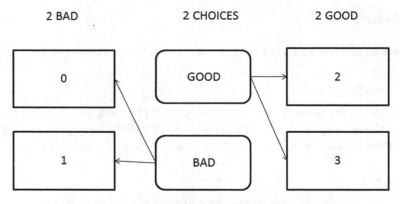

Fig. 1. The grading model

4.2 Main Game Screen

The user is the one who controls the flow of the game and interacts with the game by choosing among the available options in each situation. On the main game's screen the script's characters appear in the center of the screen and below them the dialog box that contains the script's dialogs. When the user has to choose between the four possible options, the options are displayed in the center of the screen. If the user has played the game before, seven bars are displayed at the left of the screen, showing the results of his previous attempts. For each type of team skill, there is a bar, which increases dynamically according to the choices made by the user in his current effort. The main game screen is displayed in Fig. 2(a).

4.3 Results Screen

The last screen of the game is about the display of the results. After the user completes his/her attempt, the users' profile is displayed based on the seven types of teamwork skills. The results screen is displayed in Fig. 2(b).

Fig. 2. (a) Main screen (b) results screen

5 Method

This section presents the methodology followed in order to evaluate our research work.

5.1 Participants

In order to test the research questions of this article 42 elementary school students, 19 boys and 23 girls, aged 12 years old participated. These students spend 2.24 days per week playing at least 30 min a video game. Each participant played the game 3 times.

5.2 Instruments

In order to evaluate game's usability, players were asked to complete an online questionnaire after the activity was concluded. The particular questionnaire was based on Lund's "USE" questionnaire (Lund 2001), which is a short questionnaire designed to measure effectively the most important aspects of a product's usability. It consists of 32 questions grouped in four dimensions: (1) Usefulness, (2) Ease of use, (3) Ease of Learning, and (4) Satisfaction. The questions' type is a 7-point Likert rating scale with the following anchors: 1 strongly disagree, 2 disagree, 3 slightly disagree, 4 neutral, 5 slightly agree, 6 agree and 7 strongly agree.

In order to explore whether the game offers enjoyment to the players, the participants were asked to complete a second online questionnaire, the EGameFlow questionnaire (Fu et al. 2009). It consists of 56 items grouped in eight dimensions (1) Concentration, (2) Clear Goal, (3) Feedback, (4) Challenge, (5) Autonomy, (6) Immersion, (7) Social Interaction, (8) Knowledge Improvement. The questions' type is a 7-point Likert rating scale respectively representing the lowest and the highest degree to which respondents agree with the items.

6 Data Analysis

6.1 Descriptive Analysis of Responses to Use Questionnaire

The statistical analysis was performed using Excel. For each one of the four dimensions (usefulness, ease of use, ease of learning and satisfaction) descriptive measures of central tendency, such as mean, were estimated for each Likert-type item from one to seven. The mean values of the four dimensions are being presented in Table 2.

Table 2. Mean values of the four dimensions

	Usefulness	Ease of use	Ease of learning	Satisfaction
Mean	4.47	4.15	4.48	4.44

As indicated by the results of the analysis, the players' opinion was positive, since the mean value (M) for the "Usefulness" dimension is 4.47, whereas M is 4.40, for the "Satisfaction" dimension. Moreover, it could be deduced that the players did not encounter any major difficulties in comprehending and using the tool, since the mean value for the "Ease of Use" dimension is equal to 4.15, while M = 4.48, for the dimension "Ease of Learning". The frequency diagram presented the user interface satisfaction is being presented in Fig. 3.

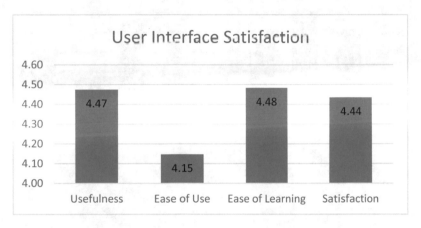

Fig. 3. Frequency diagram of user interface satisfaction

6.2 Descriptive Analysis of Responses to EGameFlow Questionnaire

For each one of the eight dimensions (concentration, clear goal, feedback, challenge, autonomy, immersion, social interaction and knowledge improvement) descriptive measures of central tendency, such as mean, were estimated for each Likert-type item from one to seven. The mean values of the eight dimensions are being presented in Table 3.

Table 3. Mean values of the eight dimensions

	Concentration	Clear goal	Feed-back	Challenge	Autonomy	Immersion	Social interaction	Knowledge improvement
Mean	4.71	4.80	4.66	4.52	4.64	4.74	4.46	4.62

As indicated by the results of the analysis, the players' opinion was positive, since the mean value (M) for the "Concentration" dimension is 4.71, whereas M is 4.80, for the "Clear Goal" dimension and 4.66 for the "Feedback" dimension. Moreover, the M for the "Challenge" dimension is 4.52 while the M for the "Autonomy" dimension is 4.64. It could be deduced that the game offers social interaction and immersion since the means values are 4.46 and 4.74 respectively. Furthermore the players believe that the game improves their knowledge since the M is 4.62. The frequency diagram presented the gamer enjoyment is being presented in Fig. 4.

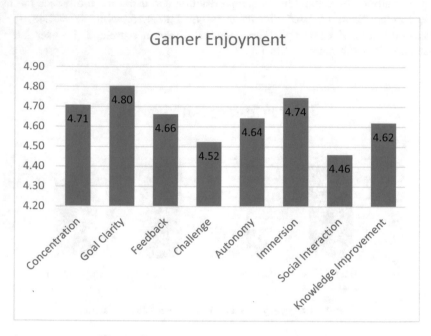

Fig. 4. Frequency diagram of gamer enjoyment

7 Conclusions

It is a fact that other educational institutions and learners attach greater importance to the acquisition of hard skills. However, because of the competitiveness of the labor market in order to succeed, they must also be equipped with the necessary soft skills. Due to the changes that have been made in the industry field, it is necessary to change education so that graduates are fully equipped with the necessary soft skills. Particular emphasis should be placed on acquiring teamwork skills as, as future's employees, they will need to work with other colleagues to fulfill a common purpose. Most educational games don't focus on human interaction with other people. For this reason, a story-telling game has been created in this study which, through the real situations that it simulates, assists in the acquisition, practice and improvement of teamwork skills in an environment which isn't stressful for the user. The evaluation results are encouraging since the student's opinion was positive. They found the game easy to use, easy to learn and satisfactory. Moreover the results indicate that the game offers enjoyment to the players.

7.1   Future Studies

Future studies should evaluate the game in order to ascertain if its results are the ones we're expecting and if there is in fact an improvement at user's teamwork skills.

Acknowledgement. This project has been funded with support from the European Commission. This publication reflects the views only of the author, and the Commission cannot be held responsible for any use which may be made of the information contained therein.

The authors of this research would like to thank GOAL team who generously shared their time, experience, and materials for the purposes of this project.

References

Cagiltay, N.E.: Teaching software engineering by means of computer-game development: challenges and opportunities. Br. J. Educ. Technol. **38**(3), 405–415 (2007)

Chen, H.H., Yang, T.C.: The impact of adventure video games on foreign language learning and the perceptions of learners. Interact. Learn. Environ. **21**(2), 129–141 (2013). http://dx.doi.org/10.1080/10494820.2012.705851

Duarte, D.L., Snyder, N.T.: Mastering Virtual Teams, 2nd edn. Jossey-Bass, San Francisco (2001)

Evenson, R.: Soft skills, hard sell [Electronic version]. Tech. Mak. Educ. Career Connect. **74**(3), 29–31 (1999)

Frank, M., Lavy, I., Elata, D.: Implementing the project- based learning aproach in an academic engineering course. Int. J. Technol. Design Educ. **13**, 273–288 (2003)

Fu, F.L., Su, R.C., Yu, S.C.: EGameFlow: a scale to measure learners' enjoyment of e-learning games. Comput. Educ. **52**(1), 101–112 (2009)

Gatlin-Watts, R., Carson, M., Horton, J., Maxwell, L., Maltby, N.: A guide to global virtual teaming. Team Perform. Manag. **13**, 47–52 (2007)

Glenn, J.L.: The "new" customer service model: customer advocate, company ambassador. Bus. Educ. Forum **62**(4), 7–13 (2008)

Hairuzila, I.: Challenges in the integration of soft skills in teaching technical courses: Lecturers' perspectives. Asian J. Univ. Educ. **5**(2), 67–81 (2009)

Hoyt, B., Prince, M., Shooter, S., Hanyak, M.: Engineering education a conceptual framework for supporting faculty in adopting collaborative learning. In: Proceedings of the 2003 American Society for Engineering Education Annual Conference & Exposition (2003)

Kamarudin, S.K., Abdullah, S.R.S., Kofli, N.T., Rahman, N.A., Tasirin, S.M., Jahim, J., Rahman, R.A.: Communication and teamwork skills in student learning process in the university. Procedia Soc. Behav. Sci. **60** 472–478 (2012)

Klaus, P.: Communication breakdown. California Job Journal **28**, 1–9 (2010)

Lund A.M.: Measuring usability with the USE questionnaire. STC Usability Interface **8**(2) (2001). http://hcibib.org/search:quest=U.lund.2001

Mitchell, G.W., Skinner, L.B., White, B.J.: Essential soft skills for success in the twenty-first century workforce as perceived by business educators. Delta Pi Epsilon J. **52**, 43–53 (2010)

Papastergiou, M.: Digital game-based learning in high school computer science education: Impact on educational effectiveness and student motivation. Comput. Educ. **52**(1), 1–12 (2009)

Robles, M.M.: Executive perceptions of the top 10 soft skills needed in today's workplace. Bus. Commun. Q. **75**(4), 453–465 (2012)

Romero, M., Usart, M., Ott, M., Earp, J.: Learning through playing for or against each other? Promoting collaborative learning in digital game based learning. In: *E*CIS 2012 Proceedings, Paper 93 (2012)

Sung, H.Y., Hwang, G.J.: A collaborative game-based learning approach to improving students' learning performance in science courses. Comput. Educ. **63**, 43–51 (2013)

Sutton, N.: Why can't we all just get along? Comput. Canada **28**(16), 20 (2002)

Warschauer, M., Matuchniak, T.: New technology and digital worlds: analyzing evidence of equity in access, use, and outcomes. Rev. Res. Educ. **34**(1), 179–225 (2010)

Wilhelm, W.J.: Research on workplace skills employers want. Meeting the demand: Teaching "soft" skills, pp. 12–33. Delta Pi Epsilon Society, Little Rock, AR (2002)

Zyda, M.: From visual simulation to virtual reality to games. IEEE Comput. **38**(9) (2005)

Δημητριάδης, Ν.Σ.: Θεωρίες Μάθησης & Εκπαιδευτικό Λογισμικό, κεφάλαιο 6, 8. Ελληνικά Ακαδημαϊκά Ηλεκτρονικά Συγγράμματα και Βοηθήματα (2014)

K-12 and Pre-college Programs

Cyber and Internet Module Using Python in Junior-High School

Doron Zohar, Tamar Benaya[✉], and Ela Zur

Computer Science Department, The Open University of Israel, Haifa, Israel
{doron.zohar, tamar, ela}@openu.ac.il

Abstract. Nowadays there has been considerable activity surrounding Computer Science (CS) education on all levels especially in junior- high school. In this paper we present the CS and Robotics curriculum for junior-high schools in Israel which has been implemented since 2011. In addition we will focus on the Cyber and Internet module which recently introduced Python as its programming language. We then display the results of a survey aimed to evaluate the teachers' attitudes towards the new Cyber and Internet module with Python. We sum up with a discussion regarding the new module.

Keywords: Computer science curriculum · Junior-High school

1   Background

In the last few decades, there has been a considerable amount of effort in developing the computer science (CS) curricula, first in the university level (e.g., [1–3]), and later-on in the high-school level (e.g., [4–6]). The goal of the K-12 designers in the US and Israel was to create a curriculum that could be widely disseminated and delivered to high-school students. The primary objective of this curriculum [5, 6] is that every CS student will learn and understand the nature of the field and the place of CS in the modern world. Recently there has been a growing amount of interest in developing CS curricula for junior-high schools. Although research concerning junior-high school computing education is a young field there are some conferences dedicated to this area. One of them is the Workshop in Primary and Secondary Computing Education (WiPSCE). Duncan and Bell [7] present in their paper a comparison of computing curricula for late primary school years in three countries: U.S.A [8], England [9] and Australia [10].

Israel's CS and Robotics junior-high school curriculum was designed by a committee appointed by the Ministry of Education. The program was first implemented, using corresponding learning materials, in 2011 [11, 12]. The Israeli curriculum includes similar components to those found in American, British and Australian curricula mentioned above. A detailed description of the Israeli program is given in [13].

The following section describes the CS and Robotics junior-high Curriculum.

© Springer Nature Switzerland AG 2020
M. E. Auer and T. Tsiatsos (Eds.): ICL 2018, AISC 916, pp. 757–767, 2020.
https://doi.org/10.1007/978-3-030-11932-4_70

2   CS and Robotics Curriculum

The CS and Robotics curriculum is divided into three modules:

- Introduction to CS - taught in 7th grade;
- A choice between two modules: Client Side Programming or Introduction to Robotics – taught in the 8th grade;
- Cyber and Internet – taught in 9th grade.

2.1   Module 1: Introduction to CS

The purpose of the module is to expose the students to fundamental concepts of algorithms. The module is implemented using Scratch. It is intended for 7th grade and it requires 60 school hours. The topics included in this module are:

- Introduction to Algorithm
- Variables, I/O and Expressions
- Multiple Sprites and Messages
- Repetition Statements
- Conditional Statements
- Advanced Algorithms (such as counters, accumulators.....)
- Final Project

2.2   Module 2: Client Side Programming or Introduction to Robotics

This module includes a choice between two subjects: Client Side Programming or Introduction to Robotics. The module is intended for 8th grade and it requires 60 school hours.

2.2.1   Client Side Programming

The purpose of the module is to deepen and widen the students' algorithmic knowledge particularly with conditions and loops using a different technology than the one taught in the first module. This module focuses on providing knowledge of innovative technologies in web development, abstract thinking and modular problem solving.

At the end of the study the students submit a project based on client-side development which implements a game or an application or a web-site based on one of the topics taught. The module is taught in two weekly hours in a laboratory. The students perform predefined tasks and are given free time for self-study and experimentation.

The topics included in this module are:

- HTML
- Introduction to JavaScript
- Variables, I/O and Expressions
- Conditional Statements
- Repetition Statements
- Advanced Algorithms (such as counters, accumulators.....)
- Functions using on Click

- Graphics using Canvas
- Animation using Canvas
- Final Project.

2.2.2   Introduction to Robotics
The purpose of this module:

- To develop integrated multi-disciplinary logical engineering thinking for developing, analyzing and building robotic solutions.
- To instill basic concepts and principles of robotics.
- To expose students to mechatronic technologies (including electronics, mechanics, software and control).
- To encourage self-study and to promote independent coping with challenges.
- To provide a pleasant learning environment.

  The topics included in this module are:

- Introduction to robotics
- Review of robot components
- Movement – electrical energy, mechanical energy and basic transmission
- Sensors – object detection and measurement of physical quantities by the robot
- Open-Loop control
- Closed-Loop control
- System analysis and design
- Artificial intelligence and learning systems.

2.3   Module 3: Cyber and Internet

The purpose of the module is to develop students' investigative capabilities and expose them to various subjects in computer science using real life problems. The students are exposed to the importance of solving real-life problems; they investigate different solutions, choose a solution and implement it.

The module is taught in a laboratory. The students perform predefined tasks and are given free time for self-study and experimentation. The module is intended for 9th grade and it requires 40 school hours.

The topics included in this module are:

- Introduction to Digital Design
- Data communication
- Encrypted data transmission
- Cyber Protection.

This module has been taught for the past several years. Several complaints have been received from students and teachers regarding this module. Some claimed that the module was too theoretical and did not include hands-on experience. Following this feedback, the ministry of education decided to develop an experimental module with the same goal which will be taught in Python. Python was selected for the following reasons:

- The committee thought that adding Python in 9th grade completes the junior-high school CS curriculum which starts with Scratch for basic programming concepts, followed by HTML and JavaScript for Client Side Programming.
- Python has become popular as a programming language used in the introductory course [14].
- Python has become popular as a programming language used in the high-tech industry.
- Python is taught in Israeli high schools as part of a new Cyber Protection program and therefore it is advantageous to introduce it already in junior-high school.

The following section describes the syllabus of the experimental module which was first introduced in the school year 2015–2016.

3 Cyber and Internet Module Using Python

In addition to the goals of this module as mentioned in the previous section, it was decided to include basic programming skills including condition and repetition statements and introduction to graphics. The committee thought that it is important for students to be exposed to a fun programming experience so that it will encourage them to select CS in high school.

Table 1 shows the detailed syllabus for the Cyber and Internet Module using Python.

Table 1. Cyber and internet module using Python syllabus

Chapter	Topics	Class hours	Lab hours	Total hours
1	Introduction to cyber - Cyber space - Cyber components: Internet, technology and human - Cyber threats - Python in the cyber space	1	1	2
2	Introduction to Python - PyCharm IDE: file, opening and saving a projects - Simple and complex I/O instructions - Programing conventions: documentation, indentation and file names	1	1	2
3	Variables and operators - Variable definition - Input (raw_input()) - Assignment (Natural numbers, Real numbers, String and Boolean) - Arithmetic operators - Division and remainder operators - Variable naming conventions	1	1	2

(continued)

Table 1. (*continued*)

Chapter	Topics	Class hours	Lab hours	Total hours
4	Introduction to functions - The main function (main()) - Defining and calling functions - Functions with no parameters and functions with no return values	1	1	2
5	Introduction to graphics - Tkinter module - User interface components: Button, TextBox, Label, ComboBox and Menu - Event handling	2	4	6
6	Numeral systems and base conversion - Numeral Systems: Decimal, Binary, Octal, Hexadecimal - Conversion from source base to decimal - Conversion from decimal to target base - Relationship between bases (binary, octal and hexadecimal)	3	3	6
7	Algorithm representation - Flow diagrams - Pseudo code	1	1	2
8	Conditional execution - Logical expressions - Conditional statements - Compound conditions (and, or, not)	2	2	4
9	Repetitive execution - Predefined repetition - Conditional repetition - Counters, accumulators - Nested repetition	3	3	6
10	Handing in and presenting a Base conversion calculator exercise	0	2	2
11	Introduction to cryptography - History of cryptography and the need for data encryption - Problem-based learning: encryption methods (Transposition ciphers and substitution ciphers)	2	2	4
12	Encryption using Python - Encrypting data using Caesar cipher - Encrypting data using modular multiplication	2	2	4
13	Strings - String operators (+, *) - String slicing ([], [:]) - String methods (count, find, isalpha, isnumeric, len, replace) - Problem-based learning: method investigation	2	2	4

(*continued*)

Table 1. (*continued*)

Chapter	Topics	Class hours	Lab hours	Total hours
14	Functions - Functions with parameters - Functions with return values	2	2	4
15	Final encryption project - Project proposal - Project folder - Project implementation - Project presentation	2	4	6
16	Data communication - Basic concepts: IP address, MAC address, DNS, Router, Switch, Hub, DHCP, Default Gateway, ping, tracert - Problem-based learning - Warriors of the Net: https://www.youtube.com/watch?v=PBWhzz_Gn10	1	3	4
Total		26	34	60

4   Teachers' Views and Attitudes Towards the New Cyber and Internet Module Using Python

The experimental "Cyber and Internet Module using Python" was first taught in six classes in 2015–2016. We conducted a preliminary study in order to examine Teachers' Views and Attitudes towards the new Cyber and Internet Module. We first briefly describe the study's methodology in Sect. 4.1, and then display the results in Sect. 4.2.

4.1   Methodology

4.1.1   Tools

We designed a nine-question questionnaire that was aimed at examining teachers' views and attitudes. The questionnaire focused on Python as a programming language regarding its suitability for 9th grade students and for the subject matter taught in the "Cyber and Internet module".

4.1.2   Population

The questionnaire was given to the six junior-high school teachers who taught the new Cyber and Internet module using Python.

4.2   Results

We present below a summary of the teachers' answers to the questionnaire:

What is your CS education?

All of the respondents have at least an undergraduate degree. Three of them have an undergraduate degree and a teaching certificate in CS. Four of them have a graduate degree.

For how many years have you been teaching Computer Science and Robotics in junior-high school?

- 4 respondents have been teaching CS for 3 to 6 years.
- One respondent has been teaching CS for more than 6 years and one of them for less than 3 years.

Which CS subjects have you taught or are you teaching in junior-high school?

Table 2 shows the number of respondents teaching the different subjects in the CS junior-high school curriculum.

Table 2. Number of respondents teaching each of the CS subjects

CS subjects	Number of respondents
Introduction to computer science	5
Client side programming	5
Introduction to robotics	3
Cyber and internet (with Python)	6

Three of the teachers teach all four subjects. Two of the teachers teach all subjects expect for Introduction to Robotics and one of the teachers teaches only the Cyber and Internet module.

Do you also teach CS in high school?

All of the teachers, but one, teach CS also in high school.

What in your opinion were the advantages and disadvantages of the Cyber and Internet original module (which was taught before the new module using python)?

The advantages mentioned were:

- The students are introduced to new topics such as: Internet communications, counting methods and encryption (3 teachers);
- The material is very well organized and has many examples.

The disadvantages mentioned were:

- Not enough lab hours, too theoretical (4 teachers);
- Too much material;
- This module disrupts the programming continuity which started in the previous modules taught with Scratch and JavaScript;
- This module lowers the students' motivation and may discourage them from selecting CS in high school;
- The module was boring and it had too many unrelated topics.

To what extent do you agree with the following statements regarding the Cyber and Internet module in its experimental version? (Strongly agree, agree, partially agree, disagree and strongly disagree).

The following list shows the responses of the respondents to each of the statements:

- Teaching the module using Python makes the module more interesting for the students: All of the teachers agreed or strongly agreed with this statement.
- The students were interested in the learning material of the Cyber and Internet module: All of the teachers but two agreed or strongly agreed with this statement.
- Python is suitable for teaching in 9th grade: All of the teachers agreed or strongly agreed with this statement.
- Enough hours are allocated for teaching this module: Only two teachers agreed or strongly agreed with this statement, three teachers partially agreed and one teacher strongly disagreed.
- Python is suitable for teaching programming after the students have learnt Scratch and JavaScript: All of the teachers strongly agreed with this statement.
- It is important to teach the topic of Numeral systems and base conversion within the module of Cyber and Internet: All of the teachers agreed or strongly agreed with this statement.
- It is important to teach the topic of Data communications within the module of Cyber and Internet: Three teachers agreed or strongly agreed with this statement, two teachers partially agreed and one teacher strongly disagreed.
- It is important to teach the topic of Cryptography within the module of Cyber and Internet: All of the teachers but one agreed or strongly agreed with this statement. One teacher partially agreed.
- The Cyber and Internet module encourages students to select to study CS in junior-high school: Four of the teachers strongly agreed with this statement. Two teachers partially agreed.
- The Cyber and Internet module encourages students to select to study CS in high school: All of the teachers but one agreed or strongly agreed with this statement. One teacher partially agreed.
- The Cyber and Internet module is too difficult for the 9th grade and it should be taught in high school: All of the teachers but one disagreed or strongly disagreed with this statement. One teacher agreed.

Which programming language in your opinion is most appropriate for teaching in the 9th grade (Java, C#, Python, JavaScript, Scratch, other)?
All teachers supported Python as the most appropriate programming language for 9th grade. One of the teachers included also Scratch and JavaScript, one of the teachers included also Assembly and one of them included Java and C#.

The teachers' comments were as follows:

- Python makes the module more interesting for the students. Adding graphics would make it more interesting for the girls.
- Python is excellent for students who will continue with the Cyber Protection program in high school. Python is also excellent for introducing algorithmic thinking.

- Python, Scratch and JavaScript are friendly languages and provide a good foundation for studying CS in high school.
- Python is very intuitive, it is easy to understand and it has a simple environment.
- Java, C# and Python are full fledged programing languages and therefore more appropriate for 9th grade than JavaScript or scratch.

Do you feel that you have enough professional support from other CS teachers or workshops etc.?

Four teachers felt that they had enough professional support. They pointed out that they received support from their peer teachers who organized a support group and shared material and exercises.

Two teachers felt that they did not have enough professional support. One teacher pointed out that she could not attend the support group which met at an inconvenient time.

What in your opinion were the advantages and disadvantages of the new experimental Cyber and Internet module (in addition to those mentioned in the questions above)?

All teachers felt that the new module had advantages and three teachers felt that the new module also had some disadvantages.

The advantages mentioned were:

- This module is more interesting for girls;
- This module has more Lab hours;
- This module is a good introduction to CS in high school (3 teachers);
- Python is a natural programming language for implementing Cyber topics;
- Python is a natural continuation to Scratch and client side programming. For those who selected the Robotics module (and not the Client Side Programming module), Python brings them back to algorithmic thinking;
- This module is much more practical as opposed to the previous version of the module which was too theoretical;
- This module is a good exposure to programming (3 teachers);
- This module is more interesting for students.

The disadvantages mentioned were:

- It is difficult for some students to simultaneously grasp new concepts of programing together with theoretical material such as: encryption and Numeral systems and base conversion;
- There was not enough time to thoroughly teach all the topics of the module;
- It is a shame that Packet Tracer was removed from the module because it was a good tool for illustrating network communication.

5 Discussion and Conclusion

In the last few decades, there has been a considerable amount of effort in developing the CS curricula, first in the university level, later-on in the high-school level and recently there has been a growing amount of interest in developing CS curriculum in junior-high school. Israel's CS and Robotics junior-high school curriculum was designed by a committee appointed by the Ministry of Education. The program was first implemented, using corresponding learning materials, in 2011. In this paper we described the CS and Robotics junior-high Curriculum which includes three modules: Introduction to CS using Scratch, Client Side Programming or Introduction to Robotics and The Cyber and Internet module. Following negative feedback from the Cyber and Internet module, a new version of the module was introduced using Python. Following the first year of its implementation we conducted a survey among the teachers who taught the module. All of the teachers, but one, are experienced teachers who also teach CS in high school.

The main insights from the survey are:

- Most of the teachers found that the original version of the Cyber and Internet module was too theoretical and boring.
- All teachers found that using Python makes the module more interesting for the students, that Python is suitable for teaching in 9th grade and that it is suitable for teaching programming after the students have learnt Scratch and JavaScript.
- Most of the teachers complained that not enough hours are allocated for teaching this module.
- Most of the teachers claimed that the Cyber and Internet module encourages students to select to study CS in junior-high school and in high school.
- All teachers supported Python as the most appropriate programming language for 9th grade. They pointed out that Python is a full fledged programing language and it will be good for students who will continue with the Cyber Protection program in high school.

Following positive feedback from the teachers who participated in teaching the new version of the module with Python, the ministry of education decided to adopt the new version of the module with few minor modifications. Some of the main modifications include:

- The topics of Conditional and Repetitive Execution were moved forward and will be presented before Introduction to Functions;
- The topics of Functions with Parameters and Functions with Return Values will be taught after Introduction to Functions;
- The topic of Strings will be taught before Introduction to Cryptography;
- The topic of Algorithm representation was removed.
- Since some of the teachers felt that they did not have enough professional support and we invested more effort in teacher training in the summer so that they will be better prepared for the following year.

In the year 2016–2017 more schools started teaching the new version of the module. From this research we can conclude that the teachers were very satisfied with the Cyber and Internet module using Python. We expect that the minor modifications which were made in this module will solve some of issues raised by the teachers who taught the experimental program. In addition, the teacher training will help support the teachers with the new materials of the module particularly with Python.

References

1. ACM Curriculum Committee on Computer Science: Curriculum '68, recommendations for academic programs in computer science. Commun. ACM **11**(3), 151–197 (1968)
2. Tucker, A., et al.: Computing curricula 1991, a summary of the ACM/IEEE-CS joint curriculum task force report. Commun. ACM **34**(6), 69–84 (1991)
3. Joint IEEE Computing Society/ACM Task Force on Computing Curricula: Computing Curricula 2001 Final Report (2001). http://www.acm.org/education/curric_vols/cc2001.pdf
4. Merritt, S., et al.: ACM model high school computer science curriculum. Commun. ACM **36** (5), 87–90 (1993)
5. Tucker, A., et al.: A model curriculum for K–12 computer science: final report of the ACM K-12 task force curriculum committee (2003). http://csta.acm.org/Curriculum/sub/k12final1022.pdf
6. Gal-Ezer, J., et al.: A high school program in computer science. Computer **28**(10), 73–80 (1995)
7. Duncan, C., Bell, T.: A pilot computer science and programming course for primary school students. In: Proceedings of the 10th Workshop in Primary and Secondary Computing Education, pp. 39–48 (2015)
8. Seehorn, D., Carey, S., Fuschetto, B., Lee, I., Moix, D., O'Grady-Cunniff, B., Owens, B., Stephenson, C., Verno, A.: CSTA K-12 Computer Science Standards; Revised 2011, Technical Report, New-York, NY, USA (2011)
9. Brown, N.C.C., Sentence, S., Crick, T., Humphreys, S.: Restart: the resurgence of computer science in UK schools. ACM Trans. Comput. Educ. (TOCE) **14**(2), 9:1–9:22 (2014). Special Issue on Computing Education in (K-12) Schools
10. Falkner, K., Vivian, R., Falkner, N.: The Australian digital technologies curriculum: challenge and opportunity. In: Proceedings of 16th Australian Computing Education Conference (ACE 2014), pp. 3–12 (2014)
11. Zur Bargury, I.: A new curriculum for junior-high in computer science. In: The 17th Annual Conference on Innovation and Technology in Computer Science Education ITiCSE 2012, Haifa, Israel, pp. 204–208 (2012)
12. Zur Bargury, I., et al.: Implementing a New Computer Science Curriculum for Middle School in Israel. In: Proceedings of Frontiers in Education FIE 2012, Seattle, Washington, pp. 886–891 (2012). http://fie2012.fie-conference.org/sites/fie2012.fie-conference.org/files/FIE2012_Proceedings.pdf
13. CS and Robotics Junior-High School Curriculum (in Hebrew). http://cms.education.gov.il/EducationCMS/Units/MadaTech/csit/TochnitLimudim/chativa/tohnit_nisuit_hativa.htm. Accessed 03 Apr 2017
14. Shein, E.: Python for beginners. Commun. ACM **58**(3), 19–21 (2015)

The Problem of Dropout from Technical Universities: Early Professional Orientation Approach

Alexander Solovyev[✉], Ekaterina Makarenko, and Larisa Petrova

Moscow Automobile and Road Construction State Technical University
(MADI), Moscow, Russia
soloviev@pre-admission.madi.ru,
{makarenko_madi,petrova_madi}@mail.ru

Abstract. The fact is that the quantity of enrolled students in technical universities does not equal to the number of graduates. According to the Russian statistics, about $1/5^{th}$ of universities' students interrupt their studies and do not get diplomas. The reasons of the withdrawal are the following: (1) the accidental reasons such as health problems or changed life circumstances, (2) the personal motives or decisions, (3) the administrative consequences of poor progress of a student. The dropouts from higher education institutions become a big problem for the university, for the society and for the national economy as a whole. The administrative dropouts of students by the reason of their unsatisfactory evaluation are the most often at the first or at the second years of study. The reasons of such dropouts are the subject for the discussion in this paper. The goal of the paper consists in the elaborating of a possible way for the situation improvement by application of the early professional orientation at the stage of high school education. We interviewed 80 students of the 1st, 2nd, and 3rd years of study trained on the different engineering specialties in the MADI University. The average assessment of progress of each student added to completed questionnaire provides the possibility to compare his/her professional attitude with academic performance and with the probability of dropout.

Keywords: Technical university · Dropout · Professional orientation · Sociological survey

1 Introduction

The opinion about the great value of higher education for the successful future of the younger generation still dominates in the Russian society. The evidence of this fact is supported by the results of the sociological survey covering 9,000 families across Russia [1]. About 90% of respondents believe that a higher education is of the great importance for their children. Furthermore, the impact of the education level and the professional status of parents, location of the family residence, the cultural status of the family did not affect this statistic. Most of lads and lasses (supported by their parents) after leaving school want to enter a higher education institution no matter whether it is free or fee-paid. Those who have a higher Unified State Exams rating apply for the

© Springer Nature Switzerland AG 2020
M. E. Auer and T. Tsiatsos (Eds.): ICL 2018, AISC 916, pp. 768–777, 2020.
https://doi.org/10.1007/978-3-030-11932-4_71

State funding. The fact is that the quantity of enrolled students in universities is less than the number of graduates. According to the statistics, about 1/5th of students of Russian technical universities interrupt their studies and do not get diplomas, and the similar situation is in some other countries [2]. The reasons of the withdrawal are the following: (1) the accidental reasons, such as health problems or personal circumstances, (2) the personal motives or decisions, (3) the administrative reasons as consequences of poor progress of a student. The dropouts from higher education institutions become a big problem for a student and his/her family, for the university, and for the society and national economy as a whole. Therefore, the problem of dropouts from higher educational institutions is at the forefront of many families in Russia.

The aim of the paper is to investigate why a good number of the enrolled students does not transform into a high number of graduates. The administrative dropouts result from ineffective learning are the most frequent at the first and the second year of study. The reasons of such dropouts are the subject for the discussion in this paper. The goal of the paper is to study all possible ways to organize professional orientation at high and secondary school.

2 State of the Art

There are many different ways to reduce the number of students expelled from universities. Some of them have an international dimension and others are specific to the Russian system of higher education. Below is the review of the possible solutions to reduce the dropout rate from a university during the pre-university phase of education. There is no universal understanding of the challenge "What should be done?" among the authors of the reviewed papers.

Let us begin with the point of view of N.D. Bas and the colleagues [2]. "Learning skills" is the key word in the paper, and the conclusion is that a student's dropout depends on her/his low learning skills. According to the authors, there are five categories of the learning skills: communication, collaboration with course mates, problem solving, self-learning and evaluation, IT literacy. The authors suggest the model to make the learning skills more effective. They suppose this could decrease the dropout rate of the university students. However, to realize this approach it may be necessary both to invite the additional academic staff and to include in the timetable additional hours. It is clear that a class schedule is usual overloaded and it is problematic to increase the number of study hours.

In the paper [3] the relation between the Unified State Exams (USE) results and the higher education performance is studied. This applies to the Russian system of admission to the universities: all school leavers are taking part in the competition for admission to a University based on the USE results. First, entrants of a technical university have to pass the so-called STEM exams (Scientific, Techno, Engineering and Math). The authors found out that student's performance depends on their USE points at the first year of study only. Student's progress at the second and the subsequent years depends on the USE points indirectly; it relates just to the performance of the first year. The conclusion of the paper is that according to a schoolchild's USE results, we can judge the future performance of the student. Therefore, it is logical to take these estimates as a criterion for admission to the University on competition. Therefore, we can suggest including some extra hours in the school schedule for the STEM subjects.

In which grade the additional STEM hours should be introduced in the schedule? We found the answer in the paper of D. Popov and his colleagues [4]. They see the connection of the personal educational achievements of the secondary school graduates (9th grade in Russia) with their future educational intentions and with the USE points after graduating a high school. Therefore, the extra STEM hours should start at the earliest possible school grade.

The paper [5] presents the results of the multiple surveys on the priority choice of the career and the life path of the school leavers. About 80% of respondents want to make a professional career. Only 25% of boys and less than 10% of girls say they want to be engineers. As a result, technical universities face the problem of the insufficient number of the schoolchildren applying for the admission to engineering specialties. The low percentage of students wishing to become engineers is due not just to insufficient information about engineering professions, but the low level of STEM subjects training either. The thorough investigation of the problem is one of starting points of our present paper.

In our previous study [6, 7] we considered the facts that many university freshmen cannot pass exams in STEM disciplines. This is the reason of the dropouts from the university after the first year of study. We think the fact has two feasible explanations: (1) the freshmen have no realistic educational and social goals; (2) they cannot asses their STEM knowledge themselves. The situation could be rectified by two possible ways: early professional orientation of the school leavers and additional training in mathematics and natural sciences. By means of sociological survey among MADI students and graduates we tried to clarify: what influenced their choice of the university during last years at school, and did they have any additional training in this period? The results of the first survey of 2005 (about 350 respondents) work to update the system of additional training at MADI: professional advice sessions for the schoolchildren by university's staff were introduced. We have found that parents and other close people play the most important role in choosing the University to enter. Therefore, professional advice was also given to the schoolchildren's parents. The comparison with the later similar survey results (of 2014) has shown the effectiveness of the approach: the number of dropouts from the university lowered to about 10% less. The results of the new survey of 2017-2018 we shall describe and discuss below.

3   Method

The task of our study is to discover the link between career orientation and pre-admission preparation of schoolchildren on one hand and the results of their training at the University on the other. We carried out the sociological survey among the students of Moscow Automobile and Road Construction State Technical University (MADI). The successful comparison could not be carried out by the anonymous survey, so we used the method of semi-standardized sociological interview of the students. Semi-standardized interview is conducted according to the general plan and is based on a predetermined set of questions to respondent. The interviewer can change a form of questions and their sequence in a course of conversation, and the respondent is free to choose any form of answers. The possibility to get straight and clear, complete and meaningful answers and use grounded theory is an evident advantage of the method.

We used the principle of cluster sample when analyzing the answers - the whole groups of students were interviewed. Students of the 1st, 2nd, and 3rd year took part in the survey. Among them, there were 18 freshmen of the Management faculty, 37 2nd year students of the Mechanical faculty, and 25 students of the Road and Bridges Construction faculty. We supposed that these three clusters are homogeneous. The sample is representative because different year students and specialties were involved.

It is necessary to take into account the peculiarity of the Russian higher education system [8]. It includes not only three Bologna levels (Bachelor, Master and PhD) but also the so-called "Specialist" level. The last one is inherited from the Soviet system of higher education, when students studied five or six years and graduated from the university with the qualification of an "Engineer". The traditional system is popular among those who choose engineering as the future occupation. Management students study Bachelor's degree program and all others students study Specialists programs.

Questionnaires prepared by the sociologist were distributed in the clusters (groups), and after answers to the questions, the experts further interviewed students. Average assessment of respondents' educational progress was made in parallel using the data from the dean's offices.

There were seven questions in the questionnaire:

1. What forms of the professional orientation did you meet at school (if any)?
2. What determined the choice of your future occupation and of the university?
3. Why did you decide to choose physics in the list of the Unified State Exams?
4. Is the engineer career a tradition in your family?
5. Do you think that adaptation to the University life after leaving high school is a problem?
6. What is your idea of employment after graduation?
7. Do you think that knowledge acquired at school is of importance for the profession of engineer?

Now some comments and explanations concerning the questions of the questionnaire are discussed. The first question uses the term "professional orientation" which is specific for the Russian education. The terms "Vocational guidance" or "Professional counselling" are equivalents (but not synonyms) of those in English and American papers. The third question linked to the Russian system of admission to a university using the results of USE assessment, and requires the following comment. Each school leaver has to pass two obligatory exams: Basic mathematics and the Russian language. A person gets the School Leaving Certificate if passes both exams with satisfactory results. However, it is not enough for the admission to a university. For example, an entrant of a technical university must show positive results in Special mathematics and in physics. Consequently, he/she have to choose these exams at school.

The official Russian statistics of 2017 shows that about 703 000 persons passed at least one of USE; among them about 391000 chose Special mathematics and only about 155 000 chose physics. Taking into account that not all of them passed the exams successfully we can estimate the number of potential school leavers entering the technical universities.

4 Results

Below we give the analysis of the received answers in the same order, as in the questionnaire.

The forms of the professional orientation.

During the interview we identified forms of the professional orientation or kinds of career guidance (we did not arranged them in the order of frequency of mentioning).

A. Psychological testing **at schools**.
B. **"Open Day"** in universities.
C. **Additional preparation courses** for passing USE in technical universities (including MADI).
D. He/she graduated the **school**, which **cooperates with MADI**.
E. **Private tutors'** advice.
F. **Family**.

The results show the influence of cluster type upon the answers. It turned out that the most pro-oriented students studied at the Mechanical faculty. This faculty is the most complicated to study at MADI: the percent of administrative dropouts due to unsatisfactory students' learning results is very high. Therefore, only persons who are highly motivated by the profession choice can successfully study at this faculty. However, we suppose that ambitions to become a mechanical engineer cannot compensate for the lack of school STEM knowledge.

The survey shows that children usually receive the best professional orientation in the family that is understandable. If parents see the future of their child in engineering, they try to guide him from the early childhood. Often, one or both of the parents are engineers in such families. There are some examples of answers on the family role: "I choose my profession following the example of my mother"; "My daddy is a Builder. From the age of 16 I wanted to enter MADI and to build bridges"; "For me the career guidance is the dialogue with my relatives"; "In my native city my parents sent me to a radio club since my childhood, and I decided to enter the technical faculty". The last one shows that the earlier the discussion on the further career starts in the family the better are the results. Such conversations in the family are natural, and children do not perceive it as a deliberate professional advice.

The so-called **"Open Days"** of universities play a significant role in professional orientation. This indicates that universities are properly prepared for such traditional kind of events.

Positive feedback received from the respondents about the role of **preparatory courses for schoolchildren** at universities (and in MADI, in particular). The respondents noted that the university not only helped to better preparation for exams (we mean USE), but also gave the opportunity to learn more about a profession and about the university.

Cooperation of Moscow universities with schools in professional guidance is a long-term good tradition. Such form of **University-School cooperation** is a well-proven form of organizing the career guidance. MADI has the agreements on pedagogical **cooperation** with several Moscow high schools; therefore, the choice of the institution is predetermined for the schoolchildren. One of the respondents said: "My school cooperates with MADI, so all my schoolfellows are focused on technical professions and especially on MADI".

The analyses brought to light the lack of the career guidance **at school:** "We had career guidance lessons at school, but they were not interesting"; "The school played no role in my profession choice". About 50% of respondents (35 people) indicated the absence of career guidance at school. Many students mentioned testing by a school psychologist who explained them the predisposition for certain activities. Only six students noted that this helped them but the others called it profanation and formality. This students' opinion corresponds with the thesis of Dr. J. Raven [9].

About 40% of respondents (30 people) had additional preparation for the USE by **private tutors:** they were trained in physics (16) and in math (14). We do not think that it is a particular form of career guidance because parents chose the additional tutoring and pay for it.

Factors determined the choice of the future profession and of the university.

Generally, students assessed their future profession as socially significant and prestigious. Here are some answers of the Mechanical faculty students: "There is no State without engineers", "It is interesting to be engaged in engineering activities", "I wanted to enter technical university from my childhood", "I am interested in weapon and military vehicles…" etc. The typical answer of the Road and Bridges Construction Faculty's students is "Further employment by profession".

Families influence on the profession choice is significant: "My father wanted me to become an engineer, so he sent me to physical & math Lyceum". Family traditions are strong in MADI: almost half of the respondents said that their parents or relatives/friends have graduated MADI University.

However, not everyone is going to work by profession after graduating. Not many people chose MADI because of its specific educational programs. We got answers showing rather utilitarian motives of the university choice: "Convenient location, near to the metro station", "To hang around with friends", "The idea came spontaneously", "I did not win competition in the Architectural University and came to the Road and Bridges Construction faculty in MADI". We are not sure that students without powerful motivation will go on studying until graduation.

The choice of physics as an additional exam in the USE system.

We got many "politically correct" responses on physics exam choice and that was expected: "Interest in this science", "Awareness of the importance of its study", "Everyone should know physics; this is a basis of our life", "Important science which is used in different spheres".

Many students noted that it was easier for them to study physics and mathematics than the Humanities: "At school I slept at history classes but physics is interesting for me", "For me physics comes easy". We should note that among the respondents, there were students graduated physical and mathematical classes; for them it was logical to pass physics exam. Some students had pragmatic motives: "There is less competition in such universities where physics exam is accepted", or "I chose physics as an additional exam on the off-chance".

Some answers are related to the influence of a school teacher on professional choice of a learner: "I had a wonderful teacher of physics at school and I fell in love with the subject". Therefore, the role of teachers is very important for future life of a person; this emphasizes the fact that the love of science and the vision of person's abilities begin at school.

Adaptation to study in the University

For the majority of students an adaptation to educational process is more difficult than adaptation to other sides of university's life. Study in the university gives more autonomy than at school, but it requires more responsibility of a student. After the first exam session and seven months of studying, students say: "Studying at the university requires greater organization and discipline", "Teachers don't look after you as it was at school, and they lecture quickly". Indeed, freshmen have no necessary skills and discipline for successful adjustment to the university's organization. It will take several months to recognize the relations between students and teachers another way than at school: "Here you can leave the lecture and no one will say anything", "Teachers don't explain in details, they suppose that student must learn or go away". At the second year of study: "The disciplines became 10 times more difficult", "It is necessary to work hard and those who doesn't must leave the university".

Communication is another important aspect of adaptation. Freshmen will have to integrate to the students' community, to build relationships with fellows, to get new friends. This problem is extremely important for students arrived from other regions and living in campus: "At my school everyone knew each other but here there is more isolation even in my group". The students who lives at home are faced other difficulties. For example, the longer distance from the place of residence to the University compared to the school situated nearby.

Of course, 3rd year students are already adapted to the university's life, and they perceive problems more easily. Being fully adapted they do not remember past difficulties at all.

Further employment in the profession.

The responses on further professional employment differ in all three interviewed clusters. Most of the students of technical faculties are more or less optimistic in their future employment perspectives. The students of the Road and Bridges Construction faculty (3rd year) consider: "Builders are always needed", "I shall get the degree and I am confident that it will work", "I will work at my father's company to continue the family business", "My future profession is a very highly paid one", "I have no friends in this area but I'm sure I will get a job". The forecasts of the Mechanical faculty students (2nd year) are more cautious: "I hope to find a job in a design engineering bureau", "I'd like to work according my specialty but I'm not sure that I will find such a job". The freshmen of the Management faculty expressed the greatest uncertainty: "Somewhere I'll get a job", "We will see...", "I do not want to work by the profession but I just want to get a higher education", "I don't want to go in the profession, may be only at first".

5 Discussion

According to the results of our study, we conclude on the personal and administrative reasons for the dropout of students. Among personal reasons we see the frustration with the profession due to the lack of school career guidance and the uncertainty of future employment that increases during University studies. Administrative reasons are related to the low results of examination sessions, which in turn is a consequence of the lack of school training in the STEM subjects.

Disappointment in the profession is a complex gradual and multifactorial process. For some students it occurs during the internship when he/she found that the reality was not the same as he/she had imagined. For others it comes from the inability to study adequately as we found among the student answers. N.D. Bas et al. in [2] call this situation "absence of study skills". We agree that the study skills concept is a very broad one; these skills are based on individual psychological specifics and are formed from the infancy.

The cluster type influences the factor of the uncertainty about the future employment. The analysis of the responses received from the three different faculties has shown that only students of the Road and Bridges Construction faculty do expect to get a job by profession. Therefore, the uncertainty factor cannot be for them the reason of leaving the University. What is wrong with the two other faculties? The explanations are as follows.

The Management faculty gets students who are not professionally oriented; they think that the "Manager" is a profession applied to many kinds of activities. Therefore, a perspective of unemployment is not the main reason of their leaving the University. The entrants of this faculty know that they will not study such complicated disciplines as the engineering students. For them it was the main reason to choose a faculty. We can summarize that students of this faculty are not motivated enough to study and this is the reason of the administrative dropouts. We think that use of the term "absence of motivation" instead of "absence of study skills" is more correct.

The situation at the Mechanical faculty is quite different. The students consciously chose the profession, so they were professionally oriented indeed; and most of them hope to get employment by specialty in the future. The main reason of dropouts at this faculty is an administrative one, because they are not good prepared in the STEM disciplines.

Poor students' results of exam sessions means by the university regulations, that a student who did not received the positive marks during the exam session must be expelled from the university. We call this "the administrative dropout"; and most part of expelled students (90%) have this reason. We have indicated some reasons of unsatisfactory assessment. The main of them are low motivation (or low study skills), low STEM knowledge (skills got at school), and difficulties of adaptation to the university life. We suppose that to overcome the last problem is a duty of the university staff working with students. The special university department working with the entrants should resolve the two first problems.

6 Conclusions

The pre-university educational stage is very important to diminish dropout rate. One of the mainstreams is a modification of the system of professional advice. It is obvious that the engineering students' dropout rate falls inversely proportional with raising Unified State Exams' points. This depends on STEM subjects teaching at school. The problem of dropouts directly relates to early professional orientation that must be accompanied by STEM preparation at the earliest grades at school. For schoolchildren, it is necessary to choose the school with wide program of STEM subjects or join any additional preparation course. Characteristics of an individual formed in the early childhood effect his/her study skills and potentially the academic performance in the STEM disciplines.

We conducted our study under the assumption that the role of engineering in the modern society and the demand of engineers at the labor market must attract youth to the corresponding professional choice. The survey brought to light important aspects of personal specifics of professional orientation, the reasons of career selection and of a university choice. The role of family traditions, specifics of "school – university" adaptation, future employment expectations, social importance of an engineer are of crucial importance. The average grade of progress of interviewed students gives a chance to compare their professional attitude with their academic performance. Namely, low academic performance is a consequence of low motivation. Therefore, this study may be useful for the high school administration while recruiting students for senior grades, and for pre-admission departments of universities.

The comparison of the nowadays students' answers with the results of the similar 7-year ago survey gives us the chance to see changes of opinion of students of different generation on their career and on the profession of an engineer. Now more students recognize the importance of the engineering activities. However, for most of them study in the technical university is very complicated, and the main reason of this is the insufficient preparation in the STEM disciplines at high school.

Our survey confirms the statement that family plays the essential role in the kids' early professional orientation. Parents directing their child to the carefully selected high school increase the chances to enter and successfully graduate a university in the future.

References

1. Petrenko, E., Galitskaya, E., Shmerlina, I.: The value of higher education. Educ. Stud. Moscow **4**, 187–206 (2010). (In Russian)
2. Bas, N.D., Heininger, R., Utesch, M., Kremar, H.: Influence of study skills on the dropout rate of universities: results from a literature study. In: 20th International Conference on Interactive Collaborative Learning, ICL 2017, 27–29 September 2017, Budapest, Hungary, pp. 297–308 (2017)
3. Khavenson, T., Solovyova, A.: Studying the relation between the unified state exam points and higher education performance. Educ. Stud. Moscow **1**, 176–199 (2014). (In Russ., abstract in Eng.)

4. Popov, D., Tyumeneva, Y., Larina, G.: Life after 9th grade: how do personal achievements of students and their family resources influence life trajectories. Educ. Stud. Moscow **4**, 310–334 (2013). (In Russ., abstract in Eng.)
5. Selivanova, Z. K.: Life goals and professional preferences of older adolescents. Sociologicheskie issledovaniya [Sociological Studies] **5**, 51–55 (2017). (In Russian)
6. Solovyev, A., Petrova, L., Prikhodko,V., Makarenko, E.: Pre-admission education for better adapting freshmen, pp. 26,1239.1–26 1239.9 (2015). https://pccr.asce.org/24576
7. Solovyev, A., Petrova, L., Prikhodko,V., Makarenko, E.: Improving the quality of engineering education by controlling the incoming contingent. In: Proceedings of 2015 International Conference on Interactive Collaborative Learning (ICL), 20–24 Sept 2015, IEEE, Florence, Italy (2015). 978-1-4799-8706-1/15 ©2015
8. Narbut, N.P., Puzanova, J.V., Larina, T.I.: Life of a student in the European dimension. Sociologicheskie issledovaniya [Sociological Studies] **5**, 47–50 (2017) (In Russ., abstract in Eng.)
9. Raven, J.: The Tragic Illusion: Educational Testing. Oxford Psychologists Press

Enhanced Assessment Approaches in Immersive Environments

Meeting Competency-Oriented Requirements in the Classroom

Joachim Maderer[1]($\boxtimes$), Johanna Pirker[1], and Christian Gütl[1,2]

[1] Graz University of Technology, Graz, Austria
joachim.maderer@student.tugraz.at,
{johanna.pirker,c.guetl}@tugraz.at
[2] Curtin University, Perth, Australia

Abstract. The broader availability of high-tech devices enables the research and application of enhanced and novel teaching technologies in STEM education, such as 3D learning environments. Also, automated assessment and guidance mechanisms should be considered that meet the requirements of modern competency-oriented curricula and support teachers in the classroom. In this paper, we investigate the current adoption of digital assessment tools and simulations, how teachers and research experts assess students in competency-oriented learning settings and present an extended conceptual architecture for flexible automated assessment in immersive environments. First results indicate that basic e-assessment techniques are still not fully prevalent in secondary education, but most teachers have expressed interest in simulations with automated assessment features.

Keywords: Assessment · Immersive environments · Physics education · Competences · STEM education

1 Introduction

The improvement of e-learning techniques has been an interest of academics and practitioners for several decades now. However, the realization of digital learning activities in the classroom (especially secondary education) has not yet been widely adopted. The broader availability of high-tech devices like smartphones or tablets enable the research and application of enhanced and novel teaching technologies in STEM education, such as 3D learning environments [1].

But with these new technologies, also assessment paradigms are subject to change as tracking of user behavior is a trivial task in such environments, which eventually enables the assessment of competencies [2, 3]. As with any learning activity, immersive environments should consider effective and efficient assessment as a critical component (*cf.* [4]), especially in terms of formative feedback and guidance [4, 5]. With regard to national differences, competency-oriented curricula roughly describe what learners should be able to do in terms of learning outcomes [6]. For instance, the Austrian

© Springer Nature Switzerland AG 2020
M. E. Auer and T. Tsiatsos (Eds.): ICL 2018, AISC 916, pp. 778–789, 2020.
https://doi.org/10.1007/978-3-030-11932-4_72

competence model for science education is organized into the three dimensions of *acting* ("observing & describing", "investigating & interpreting", and "assessing & judging"), subject-dependent *content* and *levels of complexity*, which are jointly assigned to a task [7]. A further aspect that might be considered is the ethics involved in permanent tracking of learners [3]. However, in the Austrian school system, teachers are currently entitled (in several subjects) to give final grades based on the observation of active cooperation alone; whereas formal assessments (such as tests and oral exams) are strongly limited and regulated in terms of available time and quantity. Hence, the inclusion of automated assessment technology to support teachers seems reasonable, although the balance between the needs of students and teachers is certainly open for discussion (*cf.* [3]).

In this paper, we investigate how practitioners and pedagogical researchers actually assess – or intend to assess – students' learning progress in competency-oriented learning settings. The results are used as a foundation for an improved conceptual architecture, extending the work described in [8, 9], in order to develop flexible systems for competency-oriented formative and summative assessment strategies in 3D virtual learning environments. Thus, the paper focuses on the following main objectives in a selected STEM subject:

- identifying the actual adoption of digital assessment technologies in the physics classroom as selected application domain, collecting feature requests from experienced teachers and aligning our research approaches with the current classroom situation;
- applying the findings to evaluate and improve an exemplary student assignment that can be designed and implemented in a 3D virtual learning environment, such as "Maroon" [1];
- extending our existing conceptual architecture in order to include competency-oriented learner profiles and provide a more detailed strategy to store and manage the progress of learners in the context of science education.

The remaining paper is structured as follows: the next section summarizes related work on assessment approaches in immersive environments; Sect. 3 introduces the enhanced conceptual architecture including the description of the exemplary student assignment; Sect. 4 reports about the findings of a teacher survey and an expert evaluation used to improve the architecture; whereas Sect. 5 finally discusses the results and future work.

2 Related Work

Immersive environments have been in the focus of educational research for some time, but only few attempts have been made to integrate automated e-assessment approaches. For instance, the Sloodle^{TM} project combines the Moodle^{TM} platform with Second Life^{TM}, thus serving traditional e-assessment items such as quizzes as well as using other features like the grade book; or quizHUD^{TM} which associates in-world locations with assessment activities [3]. Regarding the assessment of competencies, the authors in [10] use a "stealth assessment" approach to collect evidence for the development of competencies. It is based on evidence-centered design (ECD) as promoted by Mislevy et al., relying on the completion of specific tasks to be used as indicator for progress in certain competence areas. These competence areas are modelled as Bayesian network – a directed graph of conditional probabilities – where task achievements update probabilities of minor competence areas which again update major competence areas. Promising results have been reported in three exemplary game-based settings. In contrast, the "SAVE framework" as introduced in [11, 12] describes 3D objects and their associated interactions with semantic ontologies. The sequence of actions a user performs during exercises is compared to a solution graph by a reasoner in order to estimate the correctness of the solution and to produce generic feedback messages, such as hints regarding the order of actions or allowed objects to be used with these actions. The solution graph is first created out of recorded actions carried out by a domain expert and afterwards annotated with further information, e.g. to declare optional actions or allow an arbitrary order of certain actions. However, the approach is targeting the training of procedural skills and does not yet include interfaces to competence models or learner preferences, neither is it possible to provide custom or more complex feedback messages. It is further mentioned that the reasoner does neither consider the current state of other 3D objects for assessment, nor past events that might influence the assessment objective associated with current actions. Considering these last shortcomings, the approach presented in [9] had already used an algorithm that matches a sequence of actions under consideration of dynamic state changes of independent nearby objects to estimate the correctness of a measurement procedure. Subsequent custom feedback messages can depend on previously detected patterns. Another related aspect is a concept named "micro-adaptivity" that is also grounded in probabilistic models and adapts the game-environment based on estimated competence levels by instrumenting pedagogical agents represented as non-player-characters (NPC) [13].

3   Enhanced Conceptual Approach

Combining the aforementioned aspects and extending the work described in [8, 9], this section presents an enhanced conceptual architecture that supports different virtual- and augmented reality environments, enables teachers to implement assignments with integrated assessment measurements – such as behavior patterns and traditional e-assessment items – and interconnects them with corresponding learner models to enable adaptive feedback (*cf.* [4]) and tracking of learners' progress. The approach is further applied to a conceptual show case in the domain of physics education; whereas domain specific aspects have been revised based on an expert feedback (refer to Sect. 4.2).

3.1   Conceptual Architecture

The conceptual architecture is depicted in Fig. 1. Students can initiate a learning session either by starting an immersive task from the learning management system (**virtual reality** or simple **augmented reality**) or by activating special equipment with their

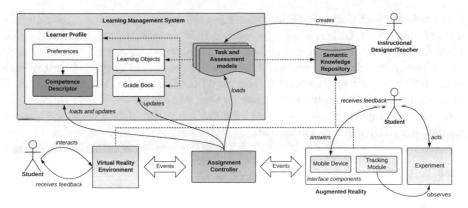

Fig. 1. Enhanced conceptual architecture for flexible assessment in immersive environments.

mobile devices, e.g. using QR-codes or NFC tags (on premise infrastructure, **augmented reality**). The virtual reality environment or tracking system establishes a connection to the **assignment controller** which in turn loads the corresponding **task and assessment model** that describes student assignments with embedded assessment and guidance measurements. These measurements are based on behavior patterns – considering sequences of learner actions and state changes of nearby objects – as well as traditional e-learning content and assessment items that might be referenced from learning object repositories. Assignment authors further select *competence descriptors* as well as levels of educational requirement to enable the reasoner to adapt the intensity, style and richness of feedback based on learners' preferences and abilities (*cf.* [4]). Depending on the available infrastructure feedback messages might trigger special actors such as pedagogical agents to deliver feedback. The **grade book** can optionally be connected to record summative assessment aspects. The communication between the different components is based on event messages that share a common set of semantic ontologies which are stored in the **semantic knowledge repository**. The **learner profile** utilizes a hierarchical structure to manage multiple competence models possibly realized as Bayesian networks, whereas descriptors might be assigned to different dimensions to support a wider range of competence models. Data updates are based on probability values that are connected to one or more descriptors, depending on the dimensionality of the model, and carry a timestamp.

3.2 Show Case

In our exemplary assignment, the students are supposed to explore the oscillation behavior as well as other physical properties of a spring pendulum. The laboratory setup is supposed to be rather similar to the PhET simulation available in [14] and the equipment should therefore be reused for other related experiments in the future. The assignment consists of several tasks which are assigned to descriptors of the Austrian competence model for science education, a description of the expected outcomes, several possible problems as well as feedback mechanisms and interventions that might be triggered by the assessment system.

Preconditions and Learning Objectives. It is assumed that the students have already learned about the related example of a simple (mathematical) pendulum and participated in a demonstration that explored the oscillation behavior of such a system based on different parameters, e.g. length of thread or different mass. They have further been instructed on the concept of a cycle duration and how to calculate the frequency. In this exercise, the students are supposed to transfer their knowledge and skills to explore the behavior of the spring pendulum by varying the setup with different springs, masses and deflections. In addition, the behavior of such a system should be understood from a kinematic and energetic perspective. Thus, the assignment is separated into two tasks.

Task 1. The description states *"Explore the effects of different masses and coil springs on the oscillation behavior of a spring pendulum by comparing the resulting frequencies. Record your measurements in the provided table"*. It is expected that the students use a stop watch to measure the time of several cycles and calculate the frequency, further to repeat these steps systematically using different masses and coil springs, and to enter the results into tables, selecting appropriate headers and units.

Anticipated issues: First, it is supposed that the results of the measurements are not well in line with the simulation. Table 1 exemplarily summarizes possible causes of the problem and proposes different feedback measures based on the estimated competence levels of the learner. Further aspects that might be discussed and demonstrated in virtual dialogs could include the following points:

- mass pieces or springs are not changed in a systematic way, or the values are recorded incorrectly;
- Students do not explore if different deflections influence the frequency but use arbitrary deflections while changing other parameters.

Table 1. Problems with inaccurate measurement results and possible feedback.

Problem	Indicators	Feedback measurement	
		Unexperienced	Experienced
The concept of a cycle duration was not understood	The stop watch is not stopped after full cycles	The actual problem is pointed out directly and a pedagogical agent explains and demonstrates how to measure a cycle duration	The student is asked to study the idea of a cycle duration once more
The formula was not applied correctly	The measurement is correct, but result is significantly off the expected value	The formula is explained and the value is calculated step by step with the student	The student is asked to check for mistakes in their calculations
The result is close to the expected value, but exceeds the required accuracy	The difference between measured and expected value is too high. Very few cycles were used to measure the cycle duration	Considering the relation between exercise time and progress, this step might be optional	Based on the remaining exercise time the student is possibly advised to develop a better strategy (e.g. using multiple cycles)

Consequences: Depending on the feedback that was necessary, the task is only recorded as partially fulfilled, in some cases even as not fulfilled. The probability values of the associated competence areas are increased or decreased based on the fulfillment of the task. In that regard, feedback is supposed to be intensified (to lower competence levels) if the problems last.

Task 2: The description states *"Describe the speed and acceleration behavior as well as the composition of different forms of energy at different deflection points"*. The learners are supposed to create "statues" at different deflections and describe the relevant *kinematic*, respectively *energetic* values in terms of discrete categories (such as *zero, medium, max*). A missing conceptual understanding is expected as primary issue. Feedback measurements could include general advices as well as additional visualizations of the concepts through arrows or energy bars (using a slow-motion feature). The consequences are similar as in Task 1, however, this task might be classified as advanced achievement, when it comes to the formation of final grades.

4 Surveys

In order to improve and evaluate our conceptual approach two independent studies have been conducted. For the first survey, a general questionnaire was distributed among Styrian (Austrian province) physics teachers in order to investigate the current application of digital learning tools, particularly simulations and assessment tools, as well as the experience with the Austrian competence models for science education, the relevance of experimental tasks for assessment and the general attitude towards simulations with automated assessment systems. The second survey was supposed to target dedicated research experts in the field of physics didactics to evaluate and improve the correctness of two competency-oriented show cases for learning and assessment to be implemented in 3D immersive environments as future work. It further contained a proposal for managing competency profiles and obtaining grades from the collected data.

4.1 Teacher Questionnaire

The rather small group of teachers (N = 25; 11f, 13 m, 1una) that participated in the survey is rather equally distributed among the age groups (*under 30*: 4, *30–39*: 5, *40–49*: 6, *50–59*: 6, *60 and more:* 4). The participants have an average experience as physics teachers of 18.1 years (SD = 12.5).

Use of Digital Learning Tools. First, the participants were asked how often several types of digital learning tools are used in the classroom, based on Likert scale between 1 (never) and 5 (very often). As shown in Table 2, the overall usage of digital tools is rather average, especially with assessment tools being the second lowest category.

The category of other digital materials did not reveal unexpected tools, as some of them anyway fit into the provided categories. The mentioned tools include – but are not limited to – the *Moodle* platform, *Geogebra*, videos, presentations, web pages, apps, educational trails and self-created or scanned materials such as worksheets; although individual teachers even reported about self-programmed simulations or applets.

Table 2. Usage of digital learning tools in the physics classroom.

Category	Mean	SD
Simulations (virtual experiments)	3.28	1.00
Assessment tools (e.g. online quiz tools)	**2.40**	**1.20**
Apps for recording and analyzing measurement data (e.g. gravity sensors, motion analysis, etc.)	2.24	0.81
Office-based applications (spreadsheets etc.)	2.64	0.97
Digital schoolbooks	2.60	1.55
Other digital materials	3.60	1.33

Subsequently, the participants should name up to five tools (apps or websites) which they use for assessment purpose. As expected answers included popular tools such as the *Moodle* platform for quizzes and assignment submissions (4), *Kahoot* (4), *Socrative* (3), and *HotPotatoes* (3). However, many of the participants mentioned different simulation applets or web pages at this point. It remains inconclusive if the participants simply misunderstood the question or are actually using simulations as part of their assessment procedures. If that is the case, they probably only use it as support for oral exams or exercises, as the mentioned tools – especially the PhET simulations – do not feature any explicit assessment capabilities. As we were particularly interested in how simulations are currently used in the physics classroom, we asked about the usage frequency in three possible settings as shown in Table 3, also using a Likert scale between 1 (never) and 5 (very often). The most important simulation tools used by the participants include *PhET* (13) (https://phet.colorado.edu), simulations on "LEIFI Physik" (13) (https://www.leifiphysik.de), *Geogebra* (3), as well as simulations by Walter Fendt (3) (http://www.walter-fendt.de) and Paul Falstad (2) (http://www.falstad.com).

Table 3. Frequency of simulation usage in different learning settings.

Setting	Mean	SD
I use simulations to explain concepts in physics (teacher-centered)	3.44	1.13
Students are supposed to use certain simulations on their own/dedicated devices to perform tasks in groups or individually	2.52	1.17
Students use simulations at home	1.96	1.00

If the participants use simulations for student assignments, assessment information is either not collected (selected 15 times), obtained from direct observations (3), or the tools generate assessment reports automatically (2). Other methods (5) included filling in spreadsheets and uploading them, writing a report, or having to work out an assignment or worksheet. Finally, we were interested in the perceived quality and quantity of eventually automated assessment features, as well as pedagogical guiding systems, that might be implemented into the used simulations tools. The participants confirm the availability of generic progress information such as the number of completed tasks (selected 9 times), numeric values describing the overall success (3) and detailed information about conceptual understanding and achievements (3). These findings suggest that only a few tools feature qualitative assessment reports at all; and even the participants who claim to have experienced sophisticated assessment algorithms did hardly mention appropriate simulation tools. Indeed, if we connect these answers to the mentioned simulation tools it might even be debatable if the participants understood what is meant by an automated assessment feature. The last question was how many simulations feature pedagogical guiding mechanisms. If we leave out all participants who did not answer or could not give an answer, the remaining 15 answered with an average of 2.13 (SD = 0.88), on a Likert scale between 1 (none) and 4 (almost all).

Familiarity and Application of the Competence Model. In this section of the survey we investigated how familiar the teachers are with the relatively new Austrian Competence Model for Science Education and how the definitions and descriptors of this model influence the class management in terms of experimental exercises and grading systems. Thus, we asked the participants to evaluate a series of statements regarding the competence model, using a Liker scale between 1 (does not apply) and 5 (applies). The results are presented in Table 4. While most participants state to be familiar with the competence model the practical relevance of experimental tasks as well as the application of the model for designing and grading experimental tasks appears rather average. The progress in such competence-oriented experiments is either observed directly (selected 7 times) or determined through written protocols afterwards (11); no other methods have been mentioned.

Acceptance for Automated Assessment Approaches. The final section of the

Table 4. Statements regarding the Austrian Competence Model for Science education.

Statement	Mean	SD
I am familiar with the Austrian competence model for science education	4.12	1.03
In my grading system, experimental exercises have a great impact on students' final grades	2.88	1.21
In my grading system, active cooperation in the classroom has a great impact on students' final grades	3.96	1.08
My exercises (especially experimental tasks) and my grading system is aligned with the Austrian competence model for science education	3.08	1.13
Incorporating the competence model into exercises/assignments for the classroom requires too much effort	3.48	1.10

questionnaire was concerned with the acceptance, respectively interest in automated assessment approaches for summative assessment aspects as well as guiding and formative feedback. While nearly every participant is open to simulations with automated guiding and feedback mechanisms (Mean = 4.40, SD = 0.89), there remain some reservations about automated approaches for summative assessment aspects (Mean = 4.00, SD = 1.44). Particularly one participant was entirely against the automated (summative) evaluation of students and concerned with assessment regulations; although confirmed acceptance for simulated learning settings if not applicable or too cost intensive in real world settings. Other participants, however, expressed several wishes including reusable, explorative learning setting (regarding materials and measurement devices), compatibility with all devices (based on HTML5, explicitly excluding Java and Flash solutions), and interfaces to other software (spreadsheet applications, learn management systems).

4.2   Expert Feedback

The expert feedback on our conceptual approach was collected by sending a description of our assessment and feedback mechanisms applied to experimental assignments (without technical details about the architecture itself), combined with an anonymous online questionnaire. This draft included an analysis and description of how we intended to store competency-related achievements as well as determine (partial) grades based on this information. The task descriptions of the experiments explained the desired outcomes, possible feedback measurements and the recording of related achievements in regard to the competence model. The questionnaire is based on a Likert scale between 1 (does not apply) and 5 (applies), as well as free text answers. The group of possible experts was determined by contacting known research experts and educators in subject didactics, completed by a list of participants recommended by a federal government department. Only these responses have been considered for evaluation that confirmed a profession as didactical expert or teacher trainer, resulting in a group that is very familiar with didactical research (Mean = 4.57, SD = 0.73) and the Austrian competence model (Mean = 4.86, SD = 0.35).

Table 5. Evaluation of the learner profile and grading concept.

Statement	Mean	SD
The initial situation on the informative content of the competence model and the assessment regulations have been identified correctly	3.71	0.70
The presented data structure enables a sufficiently detailed recording of the competencies	3.14	1.36
The competence model has been interpreted correctly with regards to the definition of the degree of fulfillment and an individual decrease of the complexity level that may arises	3.00	1.20
The difference between "unfulfilled", "partially fulfilled", "completely fulfilled" is detailed enough to describe one level of complexity	3.29	1.67
The calculation of a recommendation of a (partial) grade is in accordance with the assessment regulations	2.57	1.29

Learner Profile and Grading. Although the overall idea was received rather positively, particularly the concept of managing competency-related achievements and determining grades as presented in Table 5 has been received critically. Several experts pointed out that the Austrian competence model has not been empirically tested.

Especially the complexity levels are supposed to be incompatible with the grading system. Considering these concerns in conjunction with the preference of teachers on the guiding and feedback features (see Sect. 4.1), we decided to keep the competence models separated from the grade book and use the competence model only for an estimation of the appropriate feedback mechanisms for the time being.

Experiment. Regarding the proposed experiments, the two tasks of the spring pendulum have received higher approval (refer to Table 6) than an alternative experiment targeting the concept of inertia. The alternative is not described in more details, due to the limited scope of this paper. In addition, other competence models were suggested which would fit experimental tasks better that the rather generic Austrian model. This aspect is anyway considered in our architecture as multiple competence models can be mapped into the learner profile.

Table 6. Evaluation of the spring pendulum assignment.

Statement	Task 1		Task 2	
	Mean	SD	Mean	SD
The presented (experimental) assignment is suitable for physics lessons	**4.14**	**0.64**	**3.43**	**1.29**
The assigned descriptors of the competence model correspond to the expected solution	3.86	1.12	3.86	0.99
The major problems that may arise during the conduction of the experiment have been identified	3.86	0.99	3.71	0.45
The outlined feedback measurements are reasonable	**3.86**	**0.35**	**3.57**	**0.49**
The described consequences and protocolling measurements (assessment) are reasonable.	3.14	0.64	3.00	0.76

Further Aspects. There have been several individual comments that are worth discussing: First, one expert explicitly praised the described feedback system but is rather looking forward for experiments that are hardly possible in real settings. The latter is certainly true; however, a simple experiment is currently more likely to support subsequent studies regarding the comparability to traditional settings. Second, although the draft referred to automatically guided learning activities, some experts expressed the requirement to clearly separate between (performance) assessment and learning settings. That is surprising, as the Austrian grading system implies to also obtain summative assessment data from student's active cooperation in the classroom. Finally, we have been advised to leave room for creative solutions, test our concepts empirically in the field and examine related work of didactical researchers targeting the measurement of competencies in virtual laboratories. Nevertheless, the need for efficient exploratory but strongly guided learning settings has been confirmed.

5 Conclusion

In this paper, we presented an enhanced conceptual architecture for flexible assessment in immersive environments, extending our approach in terms of supported environments, authoring concepts and particularly the integration of competence models and adaptive feedback. Based on this architecture we described an exemplary show case in a selected STEM subject. We further investigated how practical teachers use digital assessment tools and simulations today, take competence models into account and explored their acceptance and expectations regarding automated assessment and feedback strategies in simulated environments. By combining these results with an additional expert evaluation, we further refined the approach.

The findings suggest that basic e-assessment techniques are still not fully prevalent in secondary education, simulations are rather used by teachers to explain concepts instead of instructing students to experiment on their own and homework tasks involving simulations are hardly relevant. Particularly the answers to questions regarding automated assessment and feedback approaches were rather contradictory compared to the actual features of mentioned simulation tools. Besides and existing familiarity with the Austrian competence model for science education, its application to experimental tasks combined with summative assessment aspects was rated considerably lower. Nevertheless, automated guidance and feedback mechanisms are highly welcome, followed by little less interest in summative assessment features, although with demands for flexibility and compatibility of the simulation tools. The expert feedback on our conceptual approach most importantly revealed that the competence model was not intended to be used for summative assessment as it is currently rather incompatible with legal requirements. Thus, recommendations included to focus only on the dimensions of content and acting and leave out the complexity levels entirely, or to even use different competence models for pure experimental tasks. Besides minor didactical details, the show case of a spring pendulum was preferred among alternatives and thus selected for future research activities.

In contrast to the extensive research already conducted in immersive education, we conclude that such approaches will be new to most secondary school educators; even more such systems that would yield usable assessment results automatically. We believe that the full potential is unlikely to be exploited unless digital teaching exceeds the limits of standard software, simple standalone quiz tools and simple simulations. By providing a flexible assessment framework for 3D virtual learning environments, we believe to contribute towards effective and efficient blended classroom settings that meet the requirements of modern competency-oriented curricula. Thus, we are planning to test new prototypes that are currently being implemented based on our conceptual architecture in real classroom settings, as well as the relevant authoring tools on domain experts and practitioners.

References

1. Pirker, J., Lesjak, I., Gütl, C.: An educational physics laboratory in mobile versus room scale virtual reality–a comparative study. Int. J. Online Eng. (iJOE) **13**, 106 (2017). https://doi.org/10.3991/ijoe.v13i08.7371
2. Ibáñez, M.B., Crespo, R.M., Kloos, C.D.: Assessment of knowledge and competencies in 3d virtual worlds: a proposal. In: Reynolds, N., Turcsányi Szabó, M. (eds.) Key Competencies in the Knowledge Society, pp. 165–176. Springer, Heidelberg (2010). https://doi.org/10.1007/978-3-642-15378-5_16
3. Crisp, G.: E-assessment that enhances graduateness and employability skills. In: Coetzee, M., Botha, J.A., Eccles, N., Nienaber, H., Holzhausen, N. (eds.) Developing Student Graduateness and Employability, pp. 529–548 (2012)
4. Spector, J.M.: Conceptualizing the emerging field of smart learning environments. Smart Learn. Environ. **1**(1), 1–10 (2014). http://www.slejournal.com/content/pdf/s40561-014-0002-7.pdf
5. Nicol, D., Milligan, C.: Rethinking technology-supported assessment practices in relation to the seven principles of good feedback practice. Innov. Assess. High. Educ. 64–77 (2006)
6. Méhaut, P., Winch, C.: The European qualification framework: skills, competences or knowledge? Eur. Educ. Res. J. **11**(3), 369–381 (2012). https://doi.org/10.2304/eerj.2012.11.3.369
7. Haagen, C., Hopf, M.: Standardization in physics–first steps in the Austrian educational system. In: E-Book Proceedings of the ESERA 2011 Conference: Science learning and Citizenship (2012). https://www.esera.org/publications/esera-conference-proceedings/esera-2011
8. Maderer, J., Gütl, C., Al-Smadi, M.: Formative assessment in immersive environments: a semantic approach to automated evaluation of user behavior in open wonderland. In: Proceedings of Immersive Education (iED) Summit (2013)
9. Maderer, J., Gütl, C. (2013). Flexible automated assessment in 3D learning environments: Technical improvements and expert feedback. In E-iED 2013 Proceedings of the 3rd European Immersive Education Summit (pp. 100–110)
10. Shute, V.J., Wang, L., Greiff, S., Zhao, W., Moore, G.: Measuring problem solving skills via stealth assessment in an engaging video game. Comput. Hum. Behav. **63**, 106–117 (2016)
11. Greuel, C., Denker, G., Myers, K.: Semantic Instrumentation of Virtual Environments for Training (2017). https://pdfs.semanticscholar.org/4758/97c345d490c35d86359d4051da551c71b5d0.pdf
12. Greuel, C., Myers, K., Denker, G., Gervasio, M.: Assessment and content authoring in semantic virtual environments. In: Proceedings of the Interservice/Industry Training, Simulation and Education Conference (I/ITSEC) (2016)
13. Kickmeier-Rust, M.D., Albert, D.: Micro-adaptivity: protecting immersion in didactically adaptive digital educational games. J. Comput. Assist. Learn. **26**(2), 95–105 (2010)
14. Masses and Springs: PhET Interactive Simulations. https://phet.colorado.edu/sims/mass-spring-lab/mass-spring-lab_en.html

Children's Reflection-in-Action During Collaborative Design-Based Learning

Zhongya Zhang[1](✉), Tilde Bekker[1], Panos Markopoulos[1],
and Perry den Brok[2]

[1] Department of Industrial Design, Eindhoven University of Technology,
Eindhoven, Netherlands
{Z.Zhang,M.M.Bekker,P.Markopoulos}@tue.nl
[2] Department of Social Science, Wageningen University & Research,
Wageningen, Netherlands
perry.denbrok@wur.nl

Abstract. Supporting children's reflection-in-action during Design-based Learning (DBL) processes can help them to make sense of their design activities and optimize their action on the spot. Earlier work has not examined the process in a way to inform the public on detailed design decisions for scaffolding. In addition, the social character of reflection-in-action is underappreciated in previous work. In this paper, we begin with building a conceptual framework. We identify four elements that define reflection-in-action: surprising event, knowing-in-action, improvisation to respond to surprise, effects on ongoing action. We describe a qualitative study that examined how these elements manifested themselves during a collaborative DBL workshop with 9 children. Our study uncovered six types of *reflective discourses* and shows how these on-the-spot reflections can affect subsequent group behaviors. Based on our results, we discuss the social process of reflection-in-action, the problems revealed and the requirements to design technological reflection scaffolds in a collaborative DBL context.

Keywords: Reflection-in-action · Design-based learning · Collaboration · Reflection scaffolds

1 Introduction

Design-Based Learning (DBL) is a pedagogical approach that engages students in solving real-world problems through design process with authentic, hands-on activities in a classroom setting [1]. DBL is promising for fostering 21st-century skills and for eliminating the divide between knowledge gained in class and that gained by children experiencing the 'real-world'. Within a DBL context, a design thinking process is employed to help students define their problems and develop a solution. The student takes an active role in knowledge construction [2]. However, a risk in DBL is that students focus on solving the problem at hand by creating designs, but take less time for the learning process in the context.

© Springer Nature Switzerland AG 2020
M. E. Auer and T. Tsiatsos (Eds.): ICL 2018, AISC 916, pp. 790–800, 2020.
https://doi.org/10.1007/978-3-030-11932-4_73

To counter the risk of falling short of learning, it is crucial to promote students' reflection during and after DBL activities via so-called scaffolds. In the domain of learning, the concept of scaffolding refers to temporary support provided for the completion of a task that learners might otherwise not be able to complete [3]. We note that many studies that focused on supporting children's reflection during authentic learning processes assumed that by providing enough information for children to reflect on, they can reflect automatically. However, we employ the concept of scaffolding to address the advanced and systematic support for children's reflection.

Schön's theory of reflection, which addresses reflection-in-action, is helpful in understanding the learning through doing process of DBL. Schön illustrates reflection-in-action as the process of knowledge acquisition and restructuring when people reflect in the midst of action [4]. According to Schön's elaboration, it is an individual, cognitive process. In our context, where children engage in design activities with a coordinated effort, the knowledge is shared and constructed. Reflection-in-action can happen during this collaborative practice. Given our aim to design technological tools to scaffold children's reflection-in-action during DBL activities, it is important to understand how reflection-in-action occurs in a collaborative DBL setting.

To address this challenge,

- We review literature to develop a framework of the reflection-in-action process.
- Subsequently, we set the criteria functioning as analytic scheme for identifying children's reflection-in-action.
- We propose that when children engage in DBL activities, they will apply their tacit knowledge related to the activity. During the DBL process, children may become aware that something unexpected happens in their action. When children pause and respond to the unexpected situation, it is a moment when children' reflection-in-action will occur.
- We discuss the problems of children's DBL activities uncovered by analysing children's reflection-in-action process and generate design requirements for designing reflection scaffolds.

2   Conceptual Framework

Understanding reflection-in-action processes: from epistemological to phenomenological perspective

Schön's theory of reflection-in-action emphasised knowledge acquisition and restructuring of knowledge acquisition through reflection in the midst of doing. Originally, Schön invented the concept of reflection-in-action to complement the notion of rational thinking. Reflection-in-action occurs when the activity at hand involves some puzzling, troubling, or interesting phenomenon with which the individual is trying to deal, as he tries to make sense of it, he reflects on the understandings which have been tacit in the action (namely, knowing-in-action). He further identifies a sequence of moments to elaborate the process: (1) a situation to which people bring spontaneous, routinized responses. (2) The routine response produces a surprise. (3) The surprise leads to people's thoughts turning back to the surprising phenomenon. (4) This gives rise to an on-the-spot experiment [4].

Arias et al. [5] describe reflection-in-action as a process in which humans act in a situation, that when they experience breakdowns, they reflect upon their activities, instead of studying large information spaces, they explore information spaces associated with the activity. Arias et al. address that the different contexts will lead to context-related exploration of information, i.e. the context will determine the knowing-in-action. Yanow and Tsoukas [6] further developed Schön's theory taking a phenomenological perspective. They identified three types of surprise: *malfunction*, *temporary breakdown*, and *total breakdown* according to the amount of effort required for addressing the surprising event. However, they also illuminated that realizing that a surprise event occurs is an intuitive response. Different types of surprise will lead to different types of improvisation of reflection-in-action. Improvisation is one's own cognitive invention, which is made up and tested on the spot. Examining instances of breakdowns and contextual information exploration is a promising approach for identifying reflection-in-action.

Schön emphasised that what distinguishes reflection-in-action from other kinds of reflection is its immediate significance for action [4]. The result of reflection-in-action is created through constant improvisation. In summary, we identify four elements of reflection-in-action that could help us to define reflection-in-action. These elements are: (1) a surprise leading to a breakdown, (2) knowing-in-action referring to the contextual information used for reflecting, (3) improvisation responding to a surprise and (4) effects on on-going action.

Reflection-in-action in a Collaborative DBL context: reflective discourse

In the context of children's collaborative DBL, Schön has elaborated the reciprocal reflection-in-action in a coach-student context. Levina [7] introduced the collective reflection-in-action regarding to peers' reflection-in-action. The researcher addressed the significance of participant's explicit expression and sharing their views through audible and visual artefacts.

Gourlet et al. [8] designed a "research diary" to offer children a public place to take pictures of their design artefacts. They intended to create interaction moments with peers. They reported the increase of the frequency of reflection-in-action by the indicator that the pupils modify their products on the spot, or discuss new possibilities of their projects within groups.

Similarly, in our DBL context, collaboration is a coordinated, synchronous activity that is the result of a continued attempt to construct and maintain a shared conception of a problem. The articulation plays a crucial role. Because articulation refers to a process that an individual makes the 'invisible thinking' explicit in a way that enables sharing. It is the premise for collaboration in any kind of activities [5]. We assume that as a child studies individually, he may have a silent reflection-in-action, while when children work collaboratively, either personal action or group action can be a possible source for the emergence of surprising events. In group work, taking action responsive to the surprise, they explicitly express their reflection through audible or visual artifacts, which can also trigger other group members' reflection-in-action. In this way a personal surprise becomes a group surprising event. Meanwhile, they explore information spaces associated to the activity which will lead to immediate effect on their ongoing action. Hence we suggest a framework to analyse the *group action* during a DBL activity. In this exploratory study we define this kind of reflection-in-action as a *reflective discourse* which consists of reciprocal reflection-in-action among peers. To summarize, in this study we examine the reflective discourse of a group of children based on several *observable* characteristics:

- A surprising event, which refers to the emerging topic of the group's reflective discourse.
- Shared Knowing-in-action to reflect on the contextual content that collaborators shared in the group to appreciate the surprising event.
- Improvisation responding to a surprising event through the audible or visual artefacts they created.
- Effects on the on-going action.

3   Method

DBL Workshop Design
We designed a workshop for children to confront a real-world and ill-defined design challenge. The process was constrained in time, to favour creative and synthetic thinking rather than critical and analytical thinking. Students work on a design challenge of 'Designing for a new student to make him/her feel at ease'. The whole DBL process was divided into several design phases: Understanding, defining, ideation, prototyping and testing. A collaborative workbook was designed comprising all design phases and guidance for the activities. More details are presented in Fig. 1.

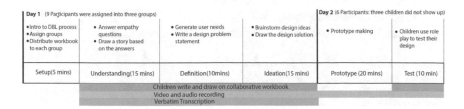

Fig. 1. Overview of the DBL activities

Nine children (5 boys and 4 girls, age 11–13) from a Dutch secondary school participated in the workshop. We conducted our workshop in their design classroom. Children were asked to divide themselves into three groups.

Two types of data were collected. First, we video and audio taped group activities and transcribed verbatim. Second, we collected the collaborative workbooks of all the groups as a reference to analyse their group interactions. In this exploratory study, we focus on the process of children's generation of design solutions. We transcribed the videos of three group processes of generating design solutions. First, we excluded the off-topic content. Through an initial analysis, we separated the transcripts into 47 continuous episodes (20 for group 1, 12 for group 2, 15 for group 3). To examine the quality of the coding approach, two coders judged whether a reflective component was visible in the episodes. Two trained coders used our framework to determine whether an episode contained a reflective discourse or not. Discussing the codes helped to improve the description of the coding approach.

4   Results

After the analysis process, 20 reflective discourses were coded (see Fig. 2) from their classroom activities. We distinguished six categories of reflective discourses (RD) from the 20 reflective discourses. Table 1 gives an overview of the number of reflective discourses per group. Below, we provide an example of a typical reflective discussion for each category, and elaboration on how our framework applied in scenario. Group 1 created a story in which a new student named Dave, suffering ADHD (Attention Deficit Hyperactivity Disorder) who disturbs classes on the first day at school, and as a consequence, his teacher and his classmates think he is annoying. They designed a new fidget spinner toy. Group 2 made a design for Lisa, who is a shy girl, but can sing very well. They assumed that Lisa is too shy to make friends and needs to enhance confidence. They generated a solution that offered an opportunity for Lisa to sing in public to enhance her confidence and make friends. Group 3 chose to design for "Dave". They thought that due to Dave's extensive enthusiasm about making new friends it is very likely that other students would not like him. However, Group 3 failed to draw the story and make a design solution.

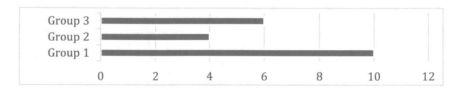

Fig. 2. The number of reflective discussions of each group

RD type 1: the role of debate
Type 1 reflective discourse (RD) is an episode that the group experienced a moment of debate triggered by one student's articulation of his or her improvisation to his/her awareness of the division about a question. This RD type is displayed by two teams.

At the start, student T from Group 3 thinks the new student 'Dave' is nervous (see Table 2, line 1.1). G responds rapidly with framing a question (line 1.1) and rejects T's idea (line 1.3), even though at that moment he cannot say why he does not agree with T. His utterance creates an uncertain moment for G. G improvises with framing a question for the group (line 1.3). Student B begins to review T's idea as well and improvises with his support to G's utterance (line 1.4). As the group is trying to appreciate the situation that is created by the debate, G shares his knowing gained from

Table 1. Distribution of reflective discourses

Type of reflective discourse	Gr. 1	Gr. 2	Gr. 3	Total
The role of debate	1	0	1	2
Awareness of skills deficiencies for design tasks	1	1	0	2
Rediscovering design task requirement	3	0	5	8
Team equity	2	2	0	4
Documentation trouble	1	1	0	2
Awareness of doing better	2	0	0	2

the persona card for backing up why 'Dave' could not be nervous (line 1.5). The information provided by the workbook is becoming shared knowing-in-action for the group to understand the situation formed by their group action. Finally, T revisits his idea.

Table 2. Typical type 1 reflective discourse

Group 3		
1.1	G:	Yes, those things. How did you draw nerves again?
1.2	T:	Yeah, how would I know
1.3	G:	But he's not nervous, he's just not nervous
1.4	B:	Really not, no!
1.5	G:	Yeah, he can't wait to make new friends.
1.6	T:	So (Dave is) Happy

RD type 2: becoming aware of skill deficiencies for certain design tasks

Type 2 reflective discourse is a moment of a group being aware of skill deficiencies for a certain design task. All three groups experience skill deficiency when they tackled drawing their stories. They finally developed strategies to compensate for the deficiencies. For example, group 1 started drawing together, while group 2 began to combine drawing with writing. When Group 2 began to draw their story, student M was aware that the design task required drawing. Her utterance shows (see Table 3, line 2.3), that she realized her deficiency of drawing skills. This led her to reframe her current situation (line 2.3), which is the improvisation of M's reflection-in-action process. Her utterance creates a moment for other students to be aware of this unexpected event; it triggers student A to reflect on this situation and respond with a solution (line 2.4) They then decide very quickly to combine drawing with writing to better express their story.

RD type 3: being aware of the deviations from design task requirements

Type 3 RD is a moment when children notice that what they are doing is not appropriate for the task. As a consequence, they review the requirement of the design task. This RD type emerged in sessions of Group 1 and 3. When students of Group 1 were answering the questions, student R notices that what they are doing needs to be

Table 3. Typical type 2 reflective discourse

Group 2		
2.1	M:	So, we have to draw a story about Lisa.
2.2	A:	Yes
2.3	M:	I think we have to draw a couch over there and where no one is looking at her. But I can't draw.
2.4	A:	Yes, we can also just draw a story together with writing. Right?
2.5	M:	Yes, okay we will draw a couch

visualized later (see Table 4). For student R it is a surprise: her improvisation to this surprising event is to stop the group action and share her understanding of the task requirement with the group members (line 3.3). Subsequently, this triggered L to respond with changing his way of expressing 'annoying' immediately into a more vivid way in sharing his knowledge about "annoying boy" (line 3.4).

Table 4. Typical type 3 reflective discourse

Group 1		
3.1	L:	He is annoying
3.2	D:	Yeah, write it down!
3.3	R:	No, stop, you should also visualize it here.
3.4	L:	Hey, you do see how someone is throwing a paper pellet at you!
3.5	R:	Hahaha
3.6	D:	He is annoying because he is too energetic.
3.7	R:	I have got it

RD type 4: paying attention to deviations from team equity

Some students may get distracted during a DBL activity. When another student realizes the distraction, he stops and asks the distracted student to participate in the work.

Student L from group 2, was playing with sticky notes, while he barely participated in the group activity. Only the other two members kept working on the discussion and writing the workbook. Suddenly, M realized that L was distracted from their teamwork (see Table 5). For M, her awareness of L's distraction is the surprising event: her improvisation is pointing out L's problem (line 4.1). Student M's utterance made L reflect on his occasional contribution and said that he thought a lot (line 4.2, line 4.4). The interaction between M and L became a surprise for student A. At that moment, A reframed the problem that M and L were arguing about and responded with a suggestion about what is the appropriate way to contribute to teamwork (line 4.6).

Table 5. Typical type 4 reflective discourse

Group 2		
4.1	M:	You are doing nothing at all. Except for playing with this (sticky notes).
4.2	L:	I thought a lot.
4.3	M:	Like what?
4.4	L:	Books, family, just use it
4.5	A:	No, you have to do it on here. Write on the right site, right?
4.6	L:	OK, what is the idea?

RD type 5: documentation trouble

When a group is engaged in collaborative work, they always have a dialogue containing a lot of ideas and information while they are writing or drawing. This can lead to a confusing situation and led to one child being aware of this situation.

When Group 1 was empathizing with their user 'Dave', they started to discuss the problem 'Dave' would run into. They were engaged in active discussion (see Table 6). When the time came to make a decision about what to write in the workbook, L did not know what to write, because too much information had appeared in an unorganized way (line 5.1–5.4). He could not summarise the main idea. When L met this type of unexpected situation, he framed and articulated the problem that emerged from the discussion (line 5.5). At last, D took over the documentation work (line 5.6).

Table 6. Typical type 5 reflective discourse

Group 1		
5.1	D:	He is energetic and he is annoying, like this (doing some gestures)
5.2	L:	And what's your opinion?
5.3	R:	He is hyperactive
5.4	D:	Yes and sick(they mean sick in his head), energetic
5.5	L:	Hey, what do I have to write down?
5.6	D:	Ok, I am going to draw

RD type 6: becoming aware of doing better than expected

This type of DR occurs during children's teamwork, when students find that they could cope with the design task very well and create moments to stop and praise their work or compliment each other.

When group 1 was drawing their story about 'Dave', at first they found a deficiency in their drawing skills. However, while they continued drawing and everyone

Table 7. Typical type 6 reflective discourse

Group 1		
6.1	D:	That's better
6.2	L:	Beautiful
6.3	D:	The best drawing ever, I'm going to put it on the wall
6.4	R:	We're really good at this, we're good team
6.5	D:	Oh, I can do that really fast! Pay attention.

contributed their ideas on the drawing, they realised that they were drawing better than they expected. Then, there was a moment that they stopped and reviewed what they had drawn, and had comments on their work (see Table 7). The effect related to this reflection was engaging team members to immediately feel positive about their work.

The different discourse types indicate the various ways in which the group aspect of the reflection contributes to the group process. Referring to what Schön describes as "immediate significance for action", we showed that these reflective discourses can amount to a group repair action (type 1, type 3 and type 4), an improvement action that children explore in the context to find a way to optimize their way of doing DBL tasks (type 2, type 5) or an increase of positive emotions in the group (type 6).

5   Discussion and Conclusion

In this exploratory study, we provide an extension of the reflection-in-action concept in a peer-to-peer collaborative DBL context. We have defined four elements to be used as an initial version of an analytic scheme to detect collaborative reflection-in-action from transcripts. Hereby the findings can develop our understanding of children's reflection while they are engaging in DBL activities. We can see explicitly how "reciprocal reflection-in-action" [4] occurs in our elaboration of the reflective discourse. Initially, reflection-in-action begins with the individual reflection-in-action, the surprise is tacit. However, as a collaborator, a student is likely to verbalise his improvisation responding to the surprise that he noticed. A verbalisation can also be the surprise that triggers another group member to engage in reflection-in-action. As more team members are engaged in reflection-in-action, this creates a very short stop moment for collaborative reflection focusing on one theme.

Zooming in on the structure of the reflective discourse, the first student of a group who explicitly expresses his reflection-in-action falls into two models:

(1) If the surprise comes from the student's own action, he is likely to improvise as framing a question with clearly illustration of the situation. In that case, a question reframing action is easy to be observed.
(2) If the surprise comes from the group's action, he is likely to stop the ongoing activity and point out the problem existing with supplement evidence to help others to re-appreciate the situation. In that case, an explicit pause is easy to observe.

Reflective discourses we found in this study show that children try to understand the uncertain situation, and share their appreciation with the other group members. To deal with the uncertain situation, children will modify their action on-the-spot and develop new understanding through group effort.

However, the classification of reflective discourses reveals that children lack abilities to recognize and appreciate desirable or undesirable qualities of design artefacts and tools. Schön argues that the process of reflection-in-action depends on a repertoire of experience gained from similar situations [4]. In our case, children began to design even though they do not yet know what design thinking means. Children acted based on their experience of doing collaborative schoolwork. Therefore, the reflective discourses that we have identified mainly refer to strategies to finish the DBL tasks in a way to fulfil the requirements written in the workbook. We rarely identified any reflective discourses relevant to the quality of their process productions (story-telling for instance) and design solutions.

Insights for designing reflection scaffolds

The utterances of children who started the reflective discourse always had a complex structure. In addition, children's visualization skills are varied. As the verbal or visual product plays a role for improvisation in reflection-in-action, it is crucial to scaffold children's expression to provide a child-friendly DBL environment.

Our results reveal that children were unable to make sense of their design activities through reflection-in-action [4], mainly due to their lacking appreciation on the design thinking mind-sets and design quality while doing DBL activities. It poses a challenge in engaging children in DBL appropriately through scaffolding reflection-in-action: how to integrate design thinking in a way beyond merely guiding them to go through DBL processes and to finish tasks. Meanwhile, enhancing children's awareness of unexpected situations related to design quality can engage children in reflection-in-action.

In summary, we propose two design requirements for designing reflection scaffolds in collaborative DBL contexts which can support children making sense of their DBL activities:

- Designing representation [9] aligned to children's cognitive level and previous experience that (1) reveals the learning elements (e.g. knowledge gaining) of a design thinking process, (2) and enables sharing understanding among children with different expression (verbal and visual) abilities during their collaboration.
- Along with the DBL process unfolding, a "talk back" mechanism is required which can enhance children's awareness of unexpected situations, and scaffold them appreciate the unexpected situations based on the knowledge gained during the design action.

In this exploratory study, we only examine the peer interaction during DBL with limited data. Future work will apply a research through design method to understand children's reflection practices in a collaborative DBL context based on more data.

Acknowledgments. We thank all the teachers and the children from the Heerbeeck College in Eindhoven. We thank Rong-Rong and Luc for helping with the transcripts and the data analysis. The first author gratefully acknowledge the grant given by the China Scholarship Council (CSC).

References

1. Doppelt, Y., Mehalik, M.M., Schunn, C.D., Silk, E., Krysinski, D.: Engagement and achievements: a case study of design-based learning in a science context. J. Technol. Educ. **19**, 22–39 (2008)
2. Scheer, A., Noweski, C., Meinel, C.: Transforming constructivist learning into action: design thinking in education. Des. Technol. Educ. **17**, 8–19 (2012)
3. van de Pol, J., Volman, M., Beishuizen, J.: Scaffolding in teacher-student interaction: a decade of research. Educ. Psychol. Rev. **22**, 271–296 (2010)

4. Schön, D.A.: Educating the Reflective Practitioner: Toward a New Design for Teaching and Learning in the Professions. Jossey-Bass (1987)
5. Arias, E., Eden, H., Fischer, G., Gorman, A., Scharff, E.: Transcending the individual human mind—creating shared understanding through collaborative design. ACM Trans. Comput. Interact. **7**, 84–113 (2000)
6. Yanow, D., Tsoukas, H.: What is reflection-in-action? A phenomenological account. J. Manag. Stud. **46**, 1339–1364 (2009)
7. Levina, N.: Collaborating on multiparty information systems development projects: a collective reflection-in-action view. Inf. Syst. Res. **16**, 109–130 (2005)
8. Gourlet, P., Eveillard, L., Dervieux, F.: The Research Diary, supporting pupils' reflective thinking during design activities. In: Proceedings of IDC 2016 - The 15th International Conference on Interaction Design and Children, pp. 206–217 (2016)
9. Fiore, S.M., Wiltshire, T.J.: Technology as teammate: examining the role of external cognition in support of team cognitive processes. Front. Psychol. **7**, 1531 (2016)

ICT in STEM Education in Bulgaria

Valentina Terzieva, Elena Paunova-Hubenova[(✉)],
Stanislav Dimitrov, and Yordanka Boneva

Institute of Information and Communication Technologies of Bulgarian
Academy of Science, Sofia 1113, Bulgaria
valia@isdip.bas.bg, {eli.paunova, boneva.yordanka}
@gmail.com, sdimitrov85@abv.bg

Abstract. Contemporary information and communication technology (ICT) and its applications are not only becoming an integral part of the everyday life of our society but also standard in most of the schools across the globe. Furthermore, the technology integration offers a qualitative transformation of all components of educational process. As innovative tools are getting an important requisite for all kind of educational institutions, the researchers should be aware of their influence on teaching and learning processes. This paper presents parts of the outcomes from a massive survey aimed at investigation of ICT integration in Bulgarian schools. In order to assess the acceptance of technology resources and their institution-wide adoption, it is important to understand teachers and students' perceptions on the technology-enhanced teaching. It is essential to find out to what extent the teachers' practice is in line with the expectations of the students. The research of dynamics of this issue helps to better understand the processes and provide valuable guidance for the effective use of innovative tools in a learning context. Here we focus especially on the teaching practice in science, technology, engineering, and mathematics (STEM) subjects in secondary education, where many different innovative tools are widely implemented.

Keywords: Technology-supported teaching · STEM education · Survey

1 Introduction

In today's digital world, STEM subjects are of vital importance for the development of the economy in a global aspect. STEM concern everything around the world, as science, technology and engineering (together with built-in mathematics) is the base of innovations and progress. Hence, it is substantially the education in these subjects to be at a relevant level. Furthermore, the contemporary society and jobs demand not only an intelligent, well-educated workforce but also technologically well-skilled persons. Since ICT has become a driving force for the whole economy, so is inevitable to be integrated into education to ensure its quality and effectiveness. Technologies provide powerful and flexible tools for teaching, which despite meeting the challenges of the digital era, inspire and facilitate students to learn more efficiently, which is especially beneficial for the STEM subjects. Extensive research explores the essential factors that

© Springer Nature Switzerland AG 2020
M. E. Auer and T. Tsiatsos (Eds.): ICL 2018, AISC 916, pp. 801–812, 2020.
https://doi.org/10.1007/978-3-030-11932-4_74

support effective STEM teaching approaches and identifies both the classroom environment and school characteristics that contribute to engaging students in these subjects [8]. It asserts that the use of ICT in teaching is of great importance. Therefore, the pedagogy starts to change by embracing technology. Many surveys demonstrate the essential advantages of computer-supported virtual laboratories and technology-rich learning environments in comparison to the traditional ones – students' understanding improves [4]. It is also underlined that except the appropriate technology and link with experimental work, a good teacher's attitude and proper pedagogy is necessary [1, 6].

In fact, the process of technology integration in education was started decades ago all around the world and Bulgaria is not left behind. Years ago, the government launched a strategy that outlined the fundamental issues in this process at all levels of academic institutions [9]. As a result, ICT tools became supportive resources in schools. However, recently the government program "ICT in the system of pre-school and school education" [7] has started that aims to develop supporting networks in educational infrastructure, providing modern means for access to educational and training resources in schools, links with innovative data centres and cloud services. Following government and sponsorship funding, latterly few centres of natural sciences have opened mostly in schools in the capital. They provide new digitized classrooms for STEM subjects (e.g. chemistry, physics, biology, mathematics, robotics), where some of them equipped with laboratories. In such an environment, besides computers, interactive boards, and scientific equipment, students will be able to use their tablets and smartphones as part of the learning process. Students can learn, as they prefer - by doing, watching and even immersing.

Applying different approaches is not only a necessity in modern education but also a way for the teacher to get personal job satisfaction. Interactive technologies help overcome existing attitudes and reluctance to learn so that lessons can be absorbed more easily and quickly. The technology-enhanced teaching should be applied not in general but by the appropriate tools according to the profile and individual abilities of students. In the previous years, even educational game-resources that adapt their level of difficulty to learner's style of play are created [2].

Recently, almost all industries have become increasingly technology-oriented, so the implementation of ICT especially in STEM education is of increasing importance. The goal of utilising technological resources and tools in pedagogical contexts is to make the teaching process more active, motivating and flexible. This transformation requires ongoing research into the effects and benefits of educational technology on learning outcomes.

Responding to the need to find out what is the actual use of technology in teaching practice, the researchers have started work on the study "Data Analysis of the Integration of ICT resources in Bulgarian schools". It aims at gathering reliable, extensive information about the use of ICT in classrooms through collection and analysis of data concerning the implementation aspects of new technologies in schools that can be quantified and evaluated. To gain a better notion, authors go further and analyse it according to many parameters to reveal the significant relations and tendencies, so that to outline changes in the attitude of teachers and the expectations of students. Teaching and learning are two sides of an educational process which deep understanding can help analysing and interpreting information about the dynamics of changes in participants' opinion. The findings will contribute to an evidence-based assessment of the impact of technology-based pedagogical strategies, improvement of the effectiveness of training methods, support of adaptive teaching and learning approaches, etc. Thus, the outcomes of the survey can help the assessment of the current state of technology integration in education, regarding the government's strategies for effective implementation of ICTs in education and science [7, 9]. According to them, the national budget will finance only school infrastructure, essential software, wireless networks, while the support of so-called soft measures - learning content and teachers' training will be from other sources.

The conducted survey explores how Bulgarian schools integrate ICT in teaching to address global technological challenges of the 21st century. The research examines the adoption and integration of ICT in school education across the country, as well as the factors that support or obstruct pedagogical staff in this process. Furthermore, the survey investigates the perceptions and attitudes of teachers and students to the use of innovative technologies considering many issues (e.g. types of used resources and tools, the frequency of their use, how they are used to support students' scientific knowledge acquisition).

In this research the motivation to pay particular attention to technology integration into STEM teaching in the secondary schools is due to the great importance of these subjects for the contemporary digital society – they are the engine to development of economics and innovations. According to many previous studies, ICT infrastructure supports pedagogical activities that correspond with teachers' beliefs regarding the advantage associated with the integration of technology-supported teaching methods into STEM-related practices [5, 6]. Teachers' perceptions of the advantages of these pedagogical approaches in assisting students' cognitive and academic performances also affect the usability assessment of ICT resource and tools. In addition to teachers and students' opinions, there are interrelations between many factors that affect their attitudes regarding ICT infrastructure and advantage associated with its use in education. Therefore the conducted research aims to understand teachers' and students' perceptions of the technology-enhanced teaching and to find out to what extent the teachers' practice is in line with the expectations of the students.

2 Methodology

The subject of this research is the touch between ICT and different aspects of their application in school education. In the study, the authors use a quantitative research methodology – in particular, design a survey to measure the implementations of the ICT resources and tools towards the study parameters in the STEM educational context. The approach used is an anonymous structured online questionnaire. It covers the following issues:

- Profiling of the respondents (for more detailed analysis);
- Conditions and obstacles for applying ICT in schools (e.g. technology infrastructure in schools, teachers competence);
- Implementation of ICT in teaching activities (e.g. frequency of usage and usefulness of resources, teachers' role, used e-learning platforms);
- Influence of e-resources on students (education-related parameters, social, cognitive and personal skills).

Researchers have developed four different surveys [10] aimed at teachers and students in the three school stages. The number of questions for teachers is 21, while for the students is 16. The questionnaires are designed in Google Forms with two classic types of questions – closed and open. The first ones include predefined choice (single and multiple) and Likert-scale type questions. Of particular importance also are open questions and comments as they give extra information when analysing data from closed ones. These options allow creating flexible and comprehensible questions, which offer an opportunity to easy process and classify the findings. The developed detailed questionnaires indicate a high probability of differentiating the variety of perceptions and the practices of the integration of ICT into the traditional classroom teaching. The successful implementation of ICT in the school education depends on many factors, among which the positive attitude and matching of expectations of teachers and students are essential.

As the goal is national representative research, study population covers teachers and students from all stages of the school education in Bulgaria. The appeal for teachers was to fill in the questionnaire, to forward it to colleagues, and to ask the students to participate in the surveys specially designed for them. Methods of dissemination are as follows: sending e-mails to regional education authorities to spread the studies among all districts schools, direct contact with teachers during thematic events, personal visits in some schools in different regions of Bulgaria and announcements in pedagogical professional networks and forums.

The study sample includes more than 1600 teachers and above 7000 students from secondary schools (2577 lower and 4681 high secondary students) of all regions in Bulgaria. Data analysis is performed employing standard statistical procedures, so to focus on the individual attitudes themselves and their relation to some investigated criteria: professional experience, computer literacy, and technology provision. The authors interpret and analyse the survey data in depth using quantitative research statistical methods, as well as calculate some indicators, e.g. frequencies, means, and standard deviations.

From the total number of respondent teachers, the number of STEM teachers is 534. For technology subjects, the authors consider the information technologies and informatics and further in the paper they are mentioned as IT subjects or teachers. The distribution of respondents is presented at Fig. 1, as the most widely represented are the

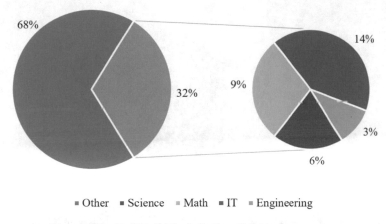

= Other = Science = Math = IT = Engineering

Fig. 1. Distribution of respondent teachers

IT (222 persons), following by math (153) and science teachers (104). The smallest one is the share of engineering teachers (55). In this paper, the authors do not consider the answers by the primary school teachers (about a third of all). Thus, the STEM teachers are approximately half of those in secondary schools.

3   Outcomes

The authors analyse the students and teachers' attitudes of the implementation of ICT in teaching STEM subjects and compare the data gathered in both surveys. The parallel is twofold – first, it is explored students' perceptions of learning by technology-enhanced activities in comparison to teachers' perceptions on this issue. Second, researchers aim to juxtapose the assessment of the usefulness of the diverse types of used technology resources and tools from both groups of respondents. Additionally, the authors analyse the findings according to the research hypotheses and study parameters.

3.1   Prerequisites to ICT-Based Teaching in STEM School Education

As digital competence is crucial for integration of ICT in school practice, it is necessary to explore STEM teacher's self-assessment on this issue (see Fig. 2). For the more accurate measurement, a Likert five-point scale from 1 "no competence" to 5 "excellent competence" is used. In total 60%, declare "excellent" (33%) and "very good" (27%) digital competence that is higher than average values for all respondents. This finding shows that the majority of STEM teachers are ready to teach in the technology-rich environment and to implement a variety of innovative tools (see Fig. 6). Teachers with no or poor competence are in total 20%, which is relatively high. It is clear that they use hardly any technology tools, despite the tendency for integration of ICT in a learning context. Most of them find an excuse for that in lack or not appropriate enough ICT qualification courses. In fact, most respondents comment that training courses should be more practical oriented.

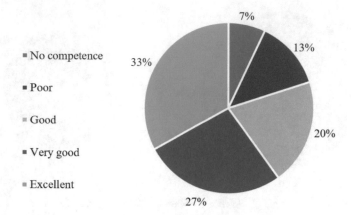

Fig. 2. STEM teachers' self-assessment of competence to use ICT for education

Other important factors that contribute to the wide spreading of new technologies in STEM teaching is the availability of adequate technology-enhanced school environment. The survey explore these conditions and Table 1 shows findings. Above the half teachers assert that the schools have necessary technical equipment, special software, digital learning resources and broadband or optical Internet connection. It seems that most of the school authorities are at least able to provide STEM teachers with appropriate ICT tools and resources. Although, according to more than third of respondents, all the necessary elements for contemporary teaching process are only partly available and between 10% and 14% even there are not any conditions for that. The need for modern learning resources, specialized software and computer technique relevant to the STEM subject is evident, as well as they should be available at every school. As a whole, the findings regarding STEM education do not vary significantly compared to secondary education in general. The straightforward reason is that usually, all the teachers in one school use the same computer classrooms, while only a few schools have specially designed STEM classrooms. All mentioned factors are addressed in the government strategy [9].

Table 1. Usage of technical equipment and e-resources in the STEM subjects

Condition	Availability		
	Yes (%)	No (%)	Partly (%)
Technical equipment	56	10	34
Specialised software	53	12	35
E-learning resources	50	14	36
Fast Internet	55	13	32

3.2 Usage Frequency of Technical Equipment and E-Resources in STEM Subjects

Figures 3 and 4 show the frequency of ICT usage in STEM subjects for low and high secondary school, according to the students' answers. The graphics also present the average values, where 1 means "never use" and 4 is "everyday use". The results for the capital city and the smaller towns are very close, and for both stages, they practically coincide. As the surveys explore students' opinions only about general education subjects, the following results exclude engineering subjects.

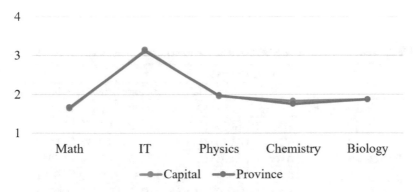

Fig. 3. The frequency of ICT usage in STEM subjects in the capital city and the province – high secondary school students

As a whole, the ICTs are almost equal frequently used in both school stages. It is not surprising that the new technologies most often are used in IT teaching – 3.5 (between every day and once a week) for the low secondary schools and 3 (i.e. weekly) for the high ones. For the science subjects, the average frequency of using ICT is nearly once per month; in math classes, it is similar, while in high secondary schools it is even smaller. The survey findings clearly show that in the lower stages of schools the technology resources are used more often in comparison to the higher ones.

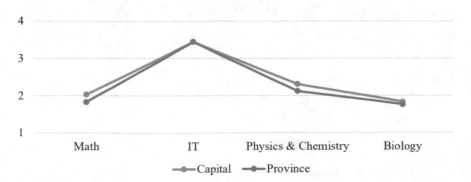

Fig. 4. The frequency of ICT usage in STEM subjects in the capital city and the province – low secondary school students

Table 2 presents the statistical mean (M), standard deviation (SD), and standard error (SE) about the frequency of use of technical equipment and e-resources in the STEM subjects according to the students. The standard deviations are significant, which means that the answers are heterogeneous. There are substantial differences in the use of innovative tools between the schools and teachers' practices. Thus, some students have the opportunity to use new technologies for learning, but a big part of them does not.

Although the teachers find the ICTs useful in different situations, they do not apply them so often. The reasons are mainly the lack of technical equipment in the classrooms, lack of appropriate resources and educational games, as well as not satisfactory technological competence of the teachers.

Table 2. The frequency of use of technical equipment and e-resources in STEM subjects

	Statistics	Subject				
		Math	IT	Biology	Physics	Chemistry
Technical equipment	Average	1.6651	3.1108	1.8773	1.9634	1.8091
	SD	1.0133	1.1553	1.0693	1.1261	1.0761
	SE	0.0209	0.0238	0.0221	0.0232	0.0222
E-resources	Average	1.6025	2.5841	1.7235	1.7754	1.7068
	SD	0.9833	1.2894	1.033	1.0741	1.0491
	SE	0.0203	0.0266	0.0213	0.0222	0.0217

3.3   Appropriate Activities for Application of ICTs

The teachers express their opinion about the relevance of ICTs for applying in different activities in the teaching process (see Fig. 5).

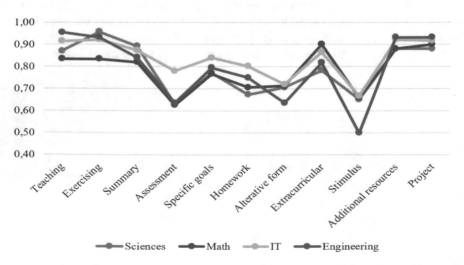

Fig. 5. Appropriateness of ICTs for application in teaching activities

The average values are presented where 0 is "not appropriate", 0.5—"partially appropriate" and 1 means "appropriate" for the situations. As a whole, teachers consider new technologies as useful for most of the activities especially for teaching, exercising, extracurricular classes, providing additional resources and project work. To the least extent, but still useful, the respondents find ICTs suitable for assessment and stimulus or reward. There are no significant discrepancies between the opinions of the four groups of STEM teachers. It is not a surprise that the IT teachers find new technologies more appropriate than the others do (average 0.84 comparing to 0.79 for each of the rest groups).

3.4   Used Types of ICT Resources

The researchers asked teachers what types of ICT equipment and resources they use in their practices (Fig. 6). The most used technical means are computers (between 60% and 72%) and projectors (between 62% and 85%), while the most used e-resources are educational websites (between 27% and 62%) and e-textbooks (between 18% and 57%). The smallest usage rates for all e-resources give the answers of the engineering subject teachers. The most probable reason is that there are not enough sufficient e-tools for these subjects. The opposite: the less used are interactive boards and educational games. The new technologies and resources are mostly used in IT classes (49%) and less used in engineering classes (32%). The average values for math and science teachers are respectively 38% and 41%.

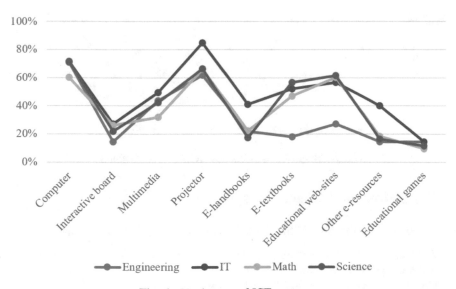

Fig. 6. Used types of ICT resources

Figure 7 presents the usage frequency of different types of ICT resources (1 means "not use" and 4 – "everyday use"). As a whole, specialised software is used most frequently in IT classes, while virtual laboratories and process simulations are rarely used, predominantly in science subjects. This fact is not a surprise, because such resources are very appropriate to these subjects for clear presentation of the learning matter. Teachers find e-resources very useful for the learning process, but it makes an impression that they do not apply them as frequently. Respondents state that the reasons for this are the lack of technical equipment in classrooms and insufficient resources. This fact should be a sign for the authorities to have in mind the need for such types of resources when creating the future policies in the field. The ICT resources are used most frequently by IT teachers (2.58) followed by science teachers (2.25) and are less used in mathematics (2.05) and engineering (2).

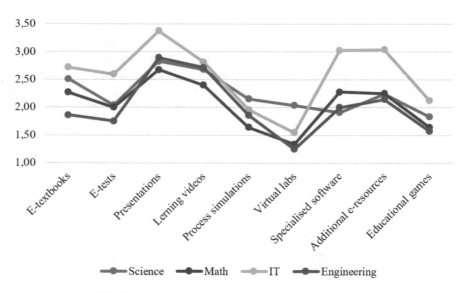

Fig. 7. Frequency of usage the different ICT resources

4   Discussions and Conclusions

Before conducting the survey, the researchers build the following hypotheses:

(1) Adequate ICT infrastructure positively influences the implementation of technology-based teaching activities;
(2) Competence of teachers positive influences the attitude to technology resources;
(3) Appropriate ICT tools and resources positive influence their frequency of use;
(4) ICT resources positive influence knowledge acquisition in STEM school subjects.

When analysing survey findings, it became clear that only about half of Bulgarian schools provide conditions enabling the technology-enhanced teaching process, adequate to necessities of today's digital society. These include modern ICT infrastructure and resources, as well as opportunities for teachers to improve their ICT competence.

Recently, gain popularity substantial changes in educational methods that include extended use of project-based, inquiry-based, problem solving, and exploration approaches. Many studies show that these innovations appear to be efficient and lead to enhanced learning outcomes. According to [3] the essential characteristic of all these approaches is that they are student-focused, but the authors recognise them also as enabled by ICT. To practice innovative learning activities in a technology-rich environment, teachers need a proper qualification both in ICT and in pedagogy. Survey findings show that the majority of STEM teachers self-assess as enough qualified to benefit from the modern digitalised classroom. The relatively high frequency of use and variety of used ICT resources in teaching also proves the research hypotheses in general.

Today the question is not if to integrate ICTs into the teaching process, but how to implement them. The aim is to ensure an efficient education that complies with the requirements of the digital era. The survey indicates that in a considerable part of Bulgarian schools, technology resources and tools like computers, projectors, multimedia, and the internet have already been usual for classroom infrastructure but not always are adequate in quantity or technical characteristics. Some other statistics show that only less than a third of schools have enough technology equipment bought in the last five years. Therefore, over the past four years, a re-equipment of computer rooms with contemporary solutions has begun. Thus, the students can perceive the science subjects in a much more attractive way. A teacher from the respondents said that "… ICTs enable doing everything, in fact. Digitization allows working innovatively and interdisciplinary on several subjects at the same time." A student shares: "Because of the new equipment it is much easier to understand things as in fact, we see them, which is much better for me."

The future development of society depends on the natural sciences, but so far, there have been no optimal conditions for teaching in these subjects in schools. This research presents a detailed view of the use of technology in the STEM teaching. Data shows that the use of innovative tools is widening, and they have reached a significant level of integration. More and more teachers already are going beyond the usual ICT applications and try to employ specific ones. Students are those who mainly will take advantage of technology resources, while teachers are those who should enable the access to contemporary teaching approaches. However, the learning applications designed particularly for STEM education are still insufficient, except those for mathematics and informatics.

Acknowledgment. This work has been supported by project M02/1 of the Bulgarian National Science fund: "Learning data analytics for ICT resource integration in Bulgarian schools", contract DM 02/1, 13.12.2016.

References

1. Barak, M: Closing the gap between attitudes and perceptions about ICT-enhanced learning among pre-service STEM teachers. J. Sci. Educ. Technol. **23**(1), 1–14. https://doi.org/10.1007/s10956-013-9446-8 (2014). Last accessed 02 May 2018
2. Bontchev., B., Georgieva, O.: Playing style recognition through an adaptive video game. Comput. Hum. Behav. **82**, 136–147 (2018). https://doi.org/10.1016/j.chb.2017.12.040
3. Connor, A.M., Karmokar, S., Whittington, C.: From STEM to STEAM: strategies for enhancing engineering & technology education. Int. J. Eng. Pedagog. **5**(2), 37–47. http://online-journals.org/index.php/i-jep/article/view/4458 (2015). Last accessed 02 May 2018
4. Devlin, T.J., Feldhaus, C.R., Bentrem, K.M.: The evolving classroom: a study of traditional and technology-based instruction in a STEM classroom. J. Technol. Educ. **25**(1), 34–54. https://scholar.lib.vt.edu/ejournals/JTE/v25n1/pdf/devlin.pdf (2013). Last accessed 02 May 2018
5. European Schoolnet: ICT in STEM Education - Impacts and Challenges: On Students. A STEM Alliance Literature Review. Brussels, Belgium. http://www.stemalliance.eu/ict-paper-2-on-students (2017). Last accessed 02 May 2018
6. European Schoolnet: ICT in STEM Education - Impacts and Challenges: On Teachers. A STEM Alliance Literature Review. Brussels, Belgium. http://www.stemalliance.eu/ict-paper-3-on-teachers (2017). Last accessed 02 May 2018
7. National Program: Provision of Modern Educational Environment. https://mon.bg/bg/1375. Last accessed 02 May 2018
8. National Research Council: Successful K-12 STEM Education: Identifying Effective Approaches in Science. Technology. Engineering and Mathematics. The National Academies Press, Washington DC. http://www.nap.edu/ (2011). Last accessed 02 May 2018
9. Strategy for Effective Implementation of Information and Communication Technologies in Education and Science in the Republic of Bulgaria (2014–2020). http://www.strategy.bg/StrategicDocuments/View.aspx?Id=904. Last accessed 07 Feb 2018
10. http://hsi.iccs.bas.bg/projects/MPIKT

Internet Addiction and Anxiety Among Greek Adolescents: An Online Survey

Theano Yfanti[1], Nikolaos C. Zygouris[1(✉)], Ioannis Chondropoulos[1], and Georgios I. Stamoulis[2]

[1] Computer Science Department, University of Thessaly, Lamia, Greece
{tyfanti, nzygouris, ichondrop}@uth.gr
[2] Electrical and Computer Engineering, University of Thessaly, Volos, Greece
georges@uth.gr

Abstract. The relationship between Internet addiction and anxiety symptomatology is a scientific field of particular research interest. Internet addiction creates emotional, psychological and social dysfunctions in several aspects of an individual's life such as at home, at school and/or at work. During the sensitive period of adolescence these dysfunctions may be determinants for the development in adulthood. The main purpose of the present study was to identify the Internet addicted students aged from 13 to 15-years-old and assess the coexistence of anxiety psychopathological symptomatology. Another aim of the present study was to use an online application in order to deliver the Internet Addiction Diagnostic Questionnaire of Young and the Revised Children's Manifest Anxiety Scale. The results of the present study revealed that 15.3% of participants were found to be Internet addicted. These adolescents presented higher anxiety symptomatology compared with their average peers in the aforementioned psychopathological scale. Evolution of research in problematic use of Internet and the subsequent emerge of effects on adolescence psychology requires innovation of highly sophisticated measurement and assessment tools.

Keywords: Internet Addiction · Anxiety · Adolescents · Online Addiction

1 Introduction

Technological breakthroughs and the widespread use of the Internet have been crucial components of today's digitalized world. These technologies had received a lot of criticism during the last two decades, despite their undeniable contribution to prosperity. Significant emphasis has been placed on pathological online behavior which can lead to Internet Addiction (IA). Undoubtedly this is an interdisciplinary field of research in recent years. Computer science, psychology, sociology, medicine and law have dealt with the study of the IA phenomenon following different approaches.

Over two decades research and debate on IA, has shown that there is no agreement about what IA really is. The first definition for IA described an impulse control disorder, without substance use, which is referred to as a behavioral addiction. The first research work in the field used an adapted version of the pathological gambling criteria from DSM-IV [1]. Another hypothesis argued that the Internet is a vector for

© Springer Nature Switzerland AG 2020
M. E. Auer and T. Tsiatsos (Eds.): ICL 2018, AISC 916, pp. 813–823, 2020.
https://doi.org/10.1007/978-3-030-11932-4_75

expressing other addictions and not the cause of the addiction [2] that is based on six factors: projection, change of mood, tolerance, withdrawal, conflict and relapse [3].

The attempt to approach this phenomenon has created several definitions and terms such as: Pathological Use [4, 5], Problematic Internet Use [6, 7], Generalize Problematic Use of Internet [8], Problematic or Maladaptive Use [9], Addiction in Cyberspace [10]. More general and comprehensive terms such as "Pathological Use of Electronic Media" [11] have been proposed recently, which are not solely restricted to Internet use.

Despite the disagreements, the dominant view about IA characterizes a situation of excessive use with negative effects in many aspects of an adolescent's life such as personal relationships, health, psychological well-being, employment and education [12]. This classifies IA as a behavioral addiction, in the intermediate region of impulse control disorders and substance use disorders [13]. The development of addictive behaviors is facilitated by the particular features of the Internet such as an easy access to reward - pleasure stimuli and as a modern anonymous way of communication [14]. Over and above, IA has been described as a primary disorder but also as a secondary disorder that develops as a means of addressing the emotional discomfort created by a wide range of mental disorders [14].

Due to the lack of consensus about diagnostic criteria, the prevalence of adolescent IA in the international literature ranges from 0.8% in Italy to 26.7% in Hong Kong [15]. Surveys in Greece, conducted to determine the prevalence of IA, have found rates ranging from 11% to 22%. Research by the Greek Society for the Study of Internet Addiction Disorder in 2008 found a prevalence of 11% to 16%, while the city of Volos has the highest rate recorded. One year later, a survey in Chios identified a 15% prevalence of IA, while three years later in the school population of Kos it registered at 16.1%. A recent survey in secondary school students in the city of Volos found the prevalence of IA at 22%, the highest ever recorded in Greece [16].

Several studies have repeatedly demonstrated the associations between most behavioral addictions with anxiety symptomatology and their bidirectional relationship to initiation, relapse or contribution. Anxiety disorders are among the most frequent mental health problems in adolescence (31.9%), followed by behaviour disorders (19.1%) [17]. Anxiety is experienced as a negative emotional state [18]. Young [1] first found that 34% of those who met the IA criteria had a history of anxiety disorder. The root cause attributed to this observation was that the Internet is an attractive and socially acceptable medium that is used as a means of escaping from anxiety. As stated in [4] the Internet is being used by people with mood regulation deficits as a means of regulating this condition. Failures are experienced as personal inadequacies while their self-image is disturbed. Internet excessive use easily leads to relief from painful feelings.

Past experience of peer-to-peer interaction can act as a defining factor between rejection and depression. During the middle childhood, the quality of friendship combined with childhood popularity can be important loneliness precursors. Many studies suggest that children who have been rejected by their peers, face difficulties in social adaptation and present high levels of loneliness [19].

The association of IA with the emergence of social phobias has been found in similar studies in China and Taiwan. Since Internet use can provide social support, adolescents with social phobia can benefit by avoiding exposure to face-to-face stress with others [20, 21]. As long as their social difficulties in the real world are not improving, adolescents with social phobia can only seek and accept social support through the Internet. The above facts reduce their motivation to meet others in the real world and increases the risk of IA. Along this line, social anxiety is associated with Problematic Internet Use expressed by preference for online social interactions [22]. This preference leads to negative outcomes, creates new problems at home, at school and/or at work. The most important predictor factor is loneliness, since lonely people have a negative perception of social competence and communication skills. Moreover, psychosocial health predicted levels of preference for online social interaction [8].

According to Davis [5], Pathological Internet Use is an abnormal behaviour resulting from individual´s problematic cognitions. The cognitive-behavioral model of PIU considers the maladaptive cognitions of an individual's thoughts about the self and the world. The existing psychopathology (depression, social anxiety) consists the predisposition. An anxiety life event is the promotor for the emerge of PIU process. Low self-esteem, feelings of consciousness and ruminative thinking of Internet use are cognitive distortions that maintain or amplify the PIU while at the same time inhibit action and problem solving.

A recent hypothesis based on Davis´ theory, argues that psychopathological symptoms and dysfunctional aspects of personality (such as low self-esteem, shyness-hesitancy, stress sensitivity and tendency to procrastination) can cause Generalized Internet Addiction (GIA). In addition, social isolation, lack of social support, predis-posing characteristics, cognitions, Internet use expectancies, coping patterns and self-regulation are related to GIA [23]. It has repeatedly demonstrated that addictive behaviors and technological addictions, in particular [14], are a relief from anxiety symptoms by providing immediate gratification and thus functioning as maladaptive anxiety mechanisms [14, 18, 24]. Anxiety-related factors, such as avoidance and sensitivity to stress, have been associated with addiction severity. The most common anxiety factors are loss or threat of loss, separation, a school failure and a surgery operation without prior psychological preparation [25]. In addition, anxiety has been associated with IA, especially with online social networking [14] as it has been determined that different psychopathology characterizes different online behaviors.

A large number of studies on adolescence IA and the coexistence with other mental disorders have been conducted in countries with extensive Internet use. South Korea and China conduct research and also provide education and treatment for the problem. Correlation studies in those countries show that the prevalence of IA ranged from 1.8% in a sample of 1,573 adolescents in Korea [26] to 26.6% in a sample of 8,941 male adolescents in Taiwan [27].

In a two-year prospective study in Taiwan, on a population of 2,162 students, 10.8% were classified as having IA. Risk factors for IA were found to be Attention Deficit Hyperactivity Disorder, depression and social phobia. The most significant predictor of IA among male adolescents was hostility and among females Attention Deficit Hyperactivity Disorder [28]. Other studies carried out on individuals with IA determined comorbidity with anxiety disorders up to 50% [13] and furthermore up to 71.7% [29].

The association of IA is well established with social phobia and phobic anxiety [20, 28], antisocial and aggressive behavior [30], family dysfunction [20, 31], reduced school performance [31–33], negative relationship with teachers [31], hostility [21, 30], suicidal ideation [26], low self-esteem and low life satisfaction [34].

The main purpose of the present study was to identify the Internet addicted students aged from 13 to 15-years-old and assess the coexistent of anxiety symptomatology. Another aim of the present study was to use an online application in order to deliver to the participants Young's scale of Internet Addiction [1] and the Revised Children's Manifest Anxiety Scale [35]. In every case, the null hypothesis (H_0) was that addicted children will not present anxiety symptomatology.

2 Method

2.1 Participants

Participants in the study were 131 High School adolescent students, 66 males (age range 13–15 years-old, M = 13.86, SD = 0.762) and 65 females (age range 13–15 years-old, M = 13.82, SD = 0.659). It is worth to notice that 25 out of 131 children did not complete all the answers of the RCMAS questionnaire, so they were excluded from the group of participants. However, they completed the IA questionnaire, so their answers were included in the aforementioned procedure. Children that completed the aforementioned questionnaires were 106, 48 males (M = 13.92 years-old, SD = 0.71) and 58 females (M = 13.86 years-old, SD = 0.712). The High School is located at the municipality of Lamia in central Greece. The completion of the questionnaires took place online, during the class of Informatics.

The first step was a detailed briefing to teachers and parents on the research protocol, the scales filling, the pure scientific purpose of the survey, the assurance of the anonymity and the compliance with the ethical rules which relate to the use of psychometric tools. Also, clearly described the process, the stages of data collection and data management. It was particularly emphasized that the questionnaires were self-filled which ensured the anonymity of the participants and the absence of personal interface between researcher and student.

We did not characterize questions as compulsory so we had optional scale completion by students. Participation was voluntary and participants could independently decide to withdraw from the study at any point. Additionally, all children that participated in the present study did not have a history of major medical illness, psychiatric disorder, developmental disorder, significant visual or auditory impairments according to their medical reports available at their school.

2.2 Materials and Procedure

The research was conducted through two weighted questionnaires and demographic information was requested about age and gender.

The Internet Addiction Diagnostic Questionnaire of Young (DQY), was used to detect IA. The DQY consists of eight questions which refer to: (1) persistence of

Internet use, (2) the need to use Internet in an ascending order, (3) the attempts to control, restrict or interrupt Internet engagement, (4) feelings resulting from the attempt to stop using Internet, (5) the time spent online, (6) lie and conceal the degree of engagement, (7) risk of relationships, school and/or work, and (8) use as a means of escape. We considered as Internet addicted a threshold of five positive responses, for the last six months [1].

For the detection of levels and the nature of anxiety the Revised Children's Manifest Anxiety Scale (RCMAS) was used. The RCMAS is a questionnaire with three anxiety-related subfactors: (1) physiological manifestations of anxiety (e.g. sleeping difficulties, nausea and fatigue), (2) worry and oversensitivity (e.g. worries about many things, fears, or emotional isolation) and (3) social concerns and concentration (e.g. questions that include distractions, fears of social or interpersonal nature). The RCMAS contains a number of mood items and items that have to do with attentional, impulsivity and peer interaction problems. The scale raw scores were measured from one to three points. It must be noted that lower scores assign higher probability for the presence of anxiety symptomatology [35].

All questionnaires delivered were self-reported, as children had to select the answers that better described their association with their personal computer and their emotional and anxiety condition.

2.3 Implementation

The online survey was applied through the use of a Google Forms application, a tool that allows collecting information from users via survey. The information is collected and automatically connected to a spreadsheet - Google Sheets. The spreadsheet is populated with the survey responses. We had select to use software that is very easy to access. Also, the above mentioned web applications are available for use at no monetary cost.

For the research purposes, a website was created that was named "web-science", a title representative of our study. The multiple-choice questionnaires and an online interface environment with relevant photos and images that would serve the purpose of the research were created. The Google Forms application was embeded to the website.

All the answers were given during the time of informatics class. The teacher made a small briefing about the procedure of the ongoing scientific project, helped the students to type correctly the URL and then stayed away from the students' screen in order to ensure the privacy of the participants and the anonymity of the answers. The website was displayed on the screen and then the user-participant answered the questionnaires and registered the responses while the data were automatically stored in the Google Forms. The export of the data was done with the use of Google Sheets. Initially, a pilot application project with a small-scale sample was conducted to determine the adequacy of the webpage and the data manipulation procedure.

3 Results

Statistical analysis took place as follows: a computation was first performed on all test answers of both questionnaires in order to find the mean scores of answers and classify each child to Internet addicted or control group. Also, a descriptive statistical analysis was performed in order to present the percentage of school children that fell within the addiction criteria according to Young's test. The percentages are presented in the Table 1. It is worth to keep in mind that the present test was completed by 131 children.

Table 1. Percentage of children that fell within the addiction criteria according to Young's questionnaire

Group	Frequency	Percent
Addicted	20	15.3%
Control group	111	84.7%

According to the statistical analysis 15.3% of the students that participated in the present protocol were found to be addicted to Internet.

One-way Anova was performed by comparing the answers of addicted children and the control group in the RSMAS questionnaire. It is worth to note that all addicted children (N = 20) completed all answers to the questionnaires in comparison to the control group where 25 children did not complete the second questionnaire, so they were excluded from the statistical analysis. Table 2 presents the mean scores of all answers of Internet addicted children and control group to the RCMAS questionnaire. It must be noted that higher raw scores in RCMAS indicate less likeability.

Table 2. Mean scores of addicted children and control group in RCMAS questionnaire

Tests	Addicted		Control group		
	Mean	SD	Mean	SD	F
RCMAS	1.95	0.36	2.36	0.29	28.73***

Note 1. *** p < 0.01

The evidence suggests that Internet addicted children presented lower scores in RCMAS questionnaire.

In order to present the questions of the RCMAS test, to which Internet addicted children presented lower scores in comparison to the control group, a one-way ANOVA analysis was conducted. Mean scores of RCMAS analysis are presented in Table 3.

As revealed by statistical analysis, children that scored higher in Young's Internet Addiction Questionnaire presented statistical significant (p < 0.05/p < 0.01) lower scores in 21 items of RCMAS.

A correlation analysis was performed in order to estimate the differences between ages and gender of addicted children. There was no statistical significance (p > 0.05) between genders. However, it was found that Internet addicted children's age was correlated with their answers of the RCMAS (p < 0.01) and, as it is demonstrated by the mean scores of answers, 14-year-old children scored lower in the aforementioned tests.

Table 3. Mean scores of questions of RCMAS that Internet addicted children presented lower scores in comparison to control group children

Items of RCMAS	Addicted		Control Group		
	M	SD	M	SD	F
Q1	2.05	0.51	2.50	0.55	11.22***
Q2	2.05	0.76	2.45	0.62	6.20***
Q3	2.05	0.67	2.37	0.60	4.48*
Q4	2.00	0.79	2.40	0.62	5.92*
Q5	2.15	0.87	2.60	0.56	7.14***
Q6	2.10	0.85	2.74	0.58	16.61***
Q7	1.35	0.49	2.13	0.71	21.21***
Q9	2.10	0.64	2.40	0.58	4.35*
Q13	1.90	0.64	2.40	0.62	10.26***
Q14	2.30	0.86	2.88	0.36	22.97***
Q15	2.05	0.69	2.45	0.66	5.94*
Q16	2.45	0.69	2.93	0.30	23.54***
Q18	2.10	0.31	2.60	0.52	16.70***
Q19	1.95	0.67	2.43	0.64	8.84***
Q21	1.70	0.80	2.51	0.61	25.44***
Q22	2.15	0.74	2.50	0.63	4.69*
Q23	2.20	0.89	2.75	0.48	14.86***
Q24	1.65	0.67	2.30	0.72	12.25***
Q26	2.05	0.66	2.58	0.66	9.55***
Q27	1.50	0.61	1.82	0.65	4.12*
Q28	1.80	0.72	2.23	0.73	9.29***

Note 1. *p < 0.05 ***p < 0.01

Lastly, Cronbach's alpha was calculated in order to estimate the reliability of RCMAS questionnaire, as to the best of our knowledge, was never before delivered via a web application. The results suggested that the use of RCMAS ($\alpha = 0.90$) was reliable.

4 Discussion

The results of the present protocol suggested that 15.3% of children were found to score highly in Young's Internet Addiction Questionnaire, a score that follows the data presented by other studies [e.g.15]. In the international literature there is a wider variation in prevalence of IA. In studies of adolescent populations using the D.Q.Y. the prevalence of IA prevailed from 1.6 in a sample of 1,573 teenagers in Korea [26] to 26.6% in a sample of 8,941 adolescents in Taiwan [21]. The present study reveals that Internet addicted children score lower in RCMAS test. The data suggests that Internet addicted children present anxiety symptomatology. A significant relationship between anxiety, addictive behaviors and technological addictions, in particular, has been repeatedly demonstrated. Moreover, the correlation analysis has shown that 14-years-old Internet addicted children presented lower anxiety and mean scores of answers in comparison to children aged 13 and 15 years-old.

As it is deduced from the results of the present protocol, children that score high in Young's scale of IA also present more than normal positive items of RCMAS which describe an anxiety symptomatology. Furthermore, in RCMAS addicted children scored statistically significantly lower (more likeability) compared with not addicted children in 21 items of the questionnaire. These results are in agreement with the main line of studies that reveal more or less comparable results. From the items it can be inferred that Internet addicted children are more likely to face problems and difficulties in interaction with parents and friends [21], problems or negative evaluation of their academic achievement [32, 33, 36] worries and fears about the future suggests that adolescents feel somewhat concerned that they are unable to meet the expectations of other important individuals in their lives and are inadequate or unable to concentrate on their duties. This is in consensus with previous research indicating that the personality of the anxious children is characterized by lack of spontaneity and self-confidence, introversion and increased anxiety around rewarding and enjoyment. There is an increased difficulty in separating and emotional dependence, which is often presented as an inappropriate or diminished claim [25].

Also, in this study, there was a statistically significant difference between addicted and non-addicted adolescents on sleeping problems [37] (including nightmares, feeling alone) and physical symptoms of stress (such as stomach aches and breathing difficulties).

As the results are in consensus with an extensive body of literature which suggests that addictive behaviors provide immediate gratification and relief from anxiety and distress [3, 14, 18, 24], we can also support the demonstrated position that IA may reveal a psychopathological way of coping with the primary states of anxiety or that it constitutes a consequence of anxiety disorders [13]. Therefore, early IA diagnosis in puberty can contribute to early diagnosis and treatment of anxiety disorders within a broader context.

5 Conclusions

Rapid development of technology provokes psychosocial changes. An impact study requires evolution of applied measurement and assessment tools. The present research method combines the best elements of pure via on line and pure pen & paper classic research methods.

Several advantages of the method had emerged during the implementation of the project. We had good acceptance of the participants and even or better level of results reliability compare with other relative projects. We choose an agile survey process that demands minimal research resources such as time, money, personnel, hardware, software, infrastructures, capacity building.

The main limitation of the present study is the relatively small sample size but it is still sufficient to confirm expected results. The method demands a very low budget and is capable for a significantly larger number of participants. It would be feasible for a future survey to deal with a much bigger sample size studying other parameters related with IA.

Measurement, evaluation and assessment of the IA and anxiety symptoms through the implementation of research projects like the present can be crucial in the design of health education programs for: adolescent students (mental health promotion and illness prevention), parents and citizens (information, sensitization and good practices training), teachers (strengthen their major role to educate young people in healthy Internet use).

References

1. Young, K.S.: Internet addiction: the emergence of a new clinical disorder. Cyber Psychol. Behav. **1**(3), 237–244 (1998)
2. Griffiths, M.: Internet addiction-time to be taken seriously? Addict. Res. Theor. **8**, 413–418 (2000)
3. Griffiths, M.D.: A components model of addiction within a biopsychosocial framework. J. Subst. Use **10**(4), 191–197 (2005)
4. Morahan-Martin, J., Schumacher, P.: Loneliness and social uses of the internet. Comput. Hum. Behav. **19**, 659–671 (2003)
5. Davis, R.A.: A cognitive-behavioral model of pathological internet use. Comput. Hum. Behav. **17**(2), 187–195 (2001)
6. Shapira, N., Lessing, M., Goldsmith, T., Szabo, S., Lazoritz, M., Gold, M., Stein, D.: Problematic internet use: proposed classification and diagnostic criteria. Depress. Anxiety **17**, 207–230 (2003)
7. Aboujaude, E.: Problematic internet use: an overview. World Psychiatry **9**, 85–90 (2010)
8. Caplan, S.: Preference for online social interaction: a theory of problematic Internet use and psychosocial well-being. Commun. Res. **30**, 625–648 (2003)
9. Beard, K.W., Wolf, E.M.: Modification in the proposed diagnostic criteria for internet addiction. Cyberpsychology Behav. **4**(3), 377–383 (2001)
10. Suler, J.: Cyberspace as psychological space. R. Suler Psychol. Cyberspace (1999). http://www.rider.edu/suler/psycyber/basicfeat.html
11. Pies, R.: Should DSM-V designate "Internet Addiction" as a mental disorder? Psychiatry (Edgemont) **6**(2), 31–37 (2009)
12. Widyanto, L., Griffiths, M.D.: An empirical study of problematic internet use and self-esteem. Int. J. Cyber Behav. Psychol. Learn. **1**(1), 13–24 (2011)
13. Starcevic, V., Khazaal, Y.: Relationships between behavioral addictions and psychiatric disorders: what is Known and what is yet to be learned? Front. Psychiatry **8**, 53 (2017). https://doi.org/10.3389/fpsyt.2017.00053
14. Starcevic, V., Billieux, J.: Does the construct of internet addiction reflect a single entity or a spectrum of disorders? Clin. Neuropsychiatr. **14**(1), 5–10 (2017)

15. Kuss, D.J., Griffiths, M.D., Binder, J.F.: Internet addiction in students: prevalence and risk factors. Comput. Hum. Behav. **29**(3), 959–966 (2013)
16. Karapetsas, A., Fotis, A., Zygouris, N.: Adolescents and internet addiction: a research study of the occurrence. Encephalos **49**, 67–72 (2012)
17. Merikangas, K., et al.: Life time prevalence of mental disorders in US adolescents: results from the national comorbidity study-adolescent supplement (NCS-A). J. Am. Acad. Child Adolesc. Psychiatry **49**(10), 980–989 (2010)
18. Akin, A., Iskender, M.: Internet addiction and depression, anxiety and stress. Int. J. Educ. Sci. **3**(1), 138–148 (2011)
19. Zygouris, N., Karapetsas, A.: Loneliness. Current scientific data. Encephalos **52**, 1–3 (2015)
20. Ha, J.H., Yoo, H.J., Cho, I.H., Chin, B., Shin, D., Kim, J.H.: Psychiatric comorbidity assessed in Korean children and adolescents who screen positive for internet addiction. J. Clin. Psychiatry **67**(5), 821–826 (2006)
21. Yen, J.Y., Ko, C.H., Yen, C.F., Wu, H.Y., Yang, M.J.: The comorbid psychiatric symptoms of Internet addiction: attention deficit and hyperactivity disorder (ADHD), depression, social phobia, and hostility. J. Adolescent Health **41**(1), 93–98 (2007)
22. Caplan, S.: Problematic internet use and psychological well-being: development of a theory-based cognitive-behavioral measurement instrument. Commun. Res. **18**(5), 553–575 (2002)
23. Brand, M., Young, K., Laier, C.: Prefrontal control and internet addiction: a theoretical model and review of neuropsychological and neuroimaging findings. Front. Hum. Neurosci. **8**(375), 1–13 (2014)
24. Douglas, A., Mills, J.E., Niang, M., Stepchenkova, S., Byun, S., Ruffini, C., Lee, S.K., Loutfi, J., Lee, J.-K., Atallah, M., Blanton, M.: Internet addiction: meta-synthesis of qualitative research for the decade 1996–2006. Comput. Hum. Behav. **24**(6), 3027–3044 (2008)
25. Klein, R.G.: Anxiety disorders. J. Child Psychol. Psychiatry **50**, 153–162 (2009)
26. Kim, K., Ryu, E., Chon, M.Y., Yeun, E.J., Choi, S.Y., Seo, J.S., Nam, B.W.: Internet addiction in Korean adolescents and its relation to depression and suicidal ideation: a questionnaire survey. Int. J. Nurs. Stud. **43**(2), 185–192 (2006)
27. Yen, C.F., Ko, C.H., Yen, J.Y., Chang, Y.P., Cheng, C.P.: Multi-dimensional discriminative factors for Internet addiction among adolescents regarding gender and age. Psychiatry Clin. Neurosci. **63**(3), 357–364 (2009)
28. Ko, C.H., Yen, J.Y., Chen, C.S., Yeh, Y.C., Yen, C.F.: Predictive values of psychiatric symptoms for Internet Addiction in adolescents: a 2-year prospective study. Archives Pediatrics Adolescent Med. **163**(10), 937–943 (2009)
29. Bozkurt, H., Coskun, M., Ayaydin, H., Adak, I., Zoroglu, D.: Prevalence and patterns of psychiatric disorders in refered adolescents with Internet addiction. Psychiatry Clin. Neurosci. **67**, 352–359 (2013)
30. Ko, C.H., Yen, J.Y., Liu, S.C., Huang, C.F., Yen, C.F.: The associations between aggressive behaviors and Internet Addiction and online activities in adolescents. J. Adolesc. Health **44**(6), 598–605 (2009)
31. Xina, M., Xingb, J., Pengfeia, W., Hourua, L., Mengchenga, W., Honga, Z.: Online activities, prevalence of internet Addiction and risk factors related to family and school among adolescents in China. Addictive Behav. Reports **7**, 14–18 (2018)
32. Mythily, S., Qiu, S., Winslow, M.: Prevalence and correlates of excessive internet use among youth in Singapore. Ann. Acad. Med. Singapore **37**(1), 9–14 (2008)
33. Ko, C.H., Yen, J.Y., Yen, C.F., Chen, C.S., Wang, S.Y.: The association between internet addiction and belief of frustration intolerance: the gender difference. Cyber Psychol. Behav. **11**(3), 273–278 (2008)

34. Wang, L., Luo, J., Bai, Y., Kong, J., Gao, W., Sun, X.: Internet addiction of adolescents in China: prevalence, predictors, and association with well-being. Addiction Res. Theor. **21**(1), 62–69 (2013)
35. Reynolds, C.R., Richmond, B.O.: What I think and feel: a revised measure of children's manifest anxiety. J. Abnorm. Child Psychol. **25**(1), 15–20 (1997)
36. Young, K.: Psychology of computer use: XL. addictive use of the internet: a case that breaks the stereotype. Psychol. Reports **79**, 899–902 (1996)
37. Rikkers, W., Lawrence, D., Hafekost, J., Zubrick, R.: Internet use and gaming by children and adolescents with emotional and behavioral problem in Australia-results from the second Child and adolescent Survey of Mental Health and Wellbeing. BMC Public Health **16**, 399–415 (2016)

Intelligent Robotics in High School: An Educational Paradigm for the Industry 4.0 Era

Igor Verner[1(✉)], Dan Cuperman[1], Tal Romm[1,2], Michael Reitman[2], Shi Kai Chong[3], and Zoe Gong[3]

[1] Technion – Israel Institute of Technology, Haifa, Israel
ttrigor@technion.ac.il, dancup@ed.technion.ac.il,
ty.romm@gmail.com
[2] PTC Corp., Haifa, Israel
reit@ptc.com
[3] Massachusetts Institute of Technology, Boston, MA, USA
{cshikai,zoegong}@mit.edu

Abstract. This paper presents an outreach program aimed to introduce high school students to the basic concepts of intelligent robotics, internet of things (IoT), digital simulation, and programming in Robot Operating System (ROS). The challenge of teaching the emerging technologies to school students was addressed through collaboration of engineering educators from Technion and MIT and professional engineers from PTC Corp. The participants were 15 high school students majoring in computer science. The program included three parts studied at the Technion. In the robotics course, the students constructed and programmed mobile robotic arms to pick and place objects of unknown weight and communicate via the IoT platform ThingWorx. In the second course, the school students programmed a TurtleBot Waffle robot equipped by LIDAR sensor to navigate in an environment by means of simultaneous localization and mapping. In the mini project the students programmed two mobile robotic arms and the Waffle to implement a warehouse scenario, in which the robots jointly carried out loading, transportation and warehousing of a cargo. Assessment results indicated that the students acquired initial understanding of the studied concepts. The program raised their interest in intelligent robotics and awareness about technological and social changes associated with Industry 4.0.

1 Introduction

New technologies and concepts that are revolutionizing the industry bring drastic changes in engineering processes, digitalization of factories and transformation of entire industry domains. This general transformation of the industry is widely referred as Industry 4.0 [1, 2]. The other side of the fourth industrial revolution is that it entails a significant change in the requirements for science and technology education. By blurring the borders between science and technology, between software and hardware engineering, between local and remote, between physical and digital – it urges

© Springer Nature Switzerland AG 2020
M. E. Auer and T. Tsiatsos (Eds.): ICL 2018, AISC 916, pp. 824–832, 2020.
https://doi.org/10.1007/978-3-030-11932-4_76

educators to rethink basic approaches to engineering education and develop new efficient models appropriate for the new era [3].

The development of the abilities to learn in an ambiguous and fast paced environment, to combine knowledge from different areas of science and technology, and creatively apply it in the field of complex engineering systems – becomes the greatest challenge for the future engineers, and correspondingly the challenge for educators and teachers. The coming disruption and transformation of education driven by Industry 4.0 requirements is so significant that it is often referred as Education 4.0 [4]. The need for accommodation of Education 4.0 in high school is critical since, on one hand, it is the most important age group to develop motivation for science and engineering careers, and, on the other hand, the lack of knowledge of technology among school graduates makes it difficult for them to study technological disciplines at tertiary levels of education.

Neither high school, nor university can cope with Education 4.0 challenges on their own. Close cooperation between teachers, educators, engineering researchers, and industry experts is needed to develop efficient solutions for new pedagogical models having the ultimate goal for developing the next generation of scientists and engineers for the Industry 4.0 era.

2 Cooperation Framework

This section describes the cooperation between the Technion Center for Robotics and Digital Technology Education (CRDTE) and other centers and programs for development and implementation of the high school Intelligent Robotics course. The cooperation is based on the common goals and complementary expertise of the partners, as presented below.

CRDTE is dedicated for development, implementation, and evaluation of innovative programs in the targeted areas of engineering education. CRDTE fulfills this mission through cooperation with academic institutions, schools, and hi-tech industry. The program "Learning with learning robots", discussed in this paper, engages students in setting up experiments in which robots acquire new behaviors through learning algorithms [5]. Development of the program involves the challenges of creating an innovative robotic environment and a strategy for teaching advance engineering concepts to school students, and their evaluation through practical implementation in class. To cope with these challenges, we established and developed a novel collaboration model including several partners in academy, industry and high school jointly unified by shared goals and complementary expertise. This collaboration model was developed, validated and fine-tuned through multi-year joint research having ultimate objective to address challenges of bringing the teaching of new advanced concepts and technologies to the high schools. The main partners of CRDTE in this collaboration framework have been MIT (Beaver Works Center, MISTI program), PTC Corporation, and Hugim high school (Haifa, Israel).

Beaver Works Center is chartered by the MIT School of Engineering and MIT Lincoln Laboratory to facilitate partnership projects in research and innovative engineering education. One of its central educational programs, the Beaver Works Summer

Institute (BWSI), is a rigorous world-class program in the field of robotics and autonomous systems, addressed to talented high-school students. The program that started in 2016 as a U.S. national program is rapidly growing with the aspiration to build a network of institutes to collectively improve engineering education worldwide (Karaman et al. 2017). A possible channel suggested by BWSI for international collaboration is through MIT International Science and Technology Initiatives (MISTI) program.

MISTI creates hands-on, international research and educational experiences for MIT students with the aspiration to broaden their academic, professional, and personal horizons. The two MISTI student programs relevant for our collaboration are Internships and Global Teaching Labs (GTL). In GTL MIT students practice teaching science, technology, engineering and mathematics (STEM) in high schools abroad, while using online learning materials developed at MIT. The Internships program allows students to perform R&D projects in leading universities and companies around the world on their summer vacations. In our case, the internships and teaching practice were hosted by our Technion Center for Robotics and Digital Technology Education (CRDTE) through partnership with PTC Corporation and Hugim High School Haifa.

PTC is a global software company which creates software products and platforms for design, manufacturing, and management of smart connected products and engineering systems. These platforms implement wide range of the modern concepts and technologies related to Industry 4.0, such as internet of things (IoT), digital twin, design for additive manufacturing, and augmented reality (AR). PTC promotes and strongly supports learning these concepts and technologies in universities, colleges and schools, including development of e-learning tools and their implementation through academic, school outreach, and teacher training programs.

Hugim is a comprehensive secondary school in Haifa which has a program for high achieving students who study advanced level mathematics, physics and computer science. The program promotes accelerated learning by encouraging the students to take basic university courses and participate in advanced outreach programs.

The established collaboration model allowed combining the rich BWSI online courseware for teaching advanced topics in software engineering, robotics and AI with professional tools and industry concepts developed by PTC – and developing a novel pedagogical model that was practically implemented and evaluated in high school in the framework of the intelligent robotics course described in detail in the next section.

3   The Program

The outreach program was delivered to a group of 15 high school students majoring in computer science and interested to learn advanced topics in robotics and software development in an academic environment. The program consisted of three parts:

1. The robotics and IoT course was developed through collaboration with PTC Corp. and delivered by instructors of CRDTE. In parallel, the students got lessons on Python programming taught by the school teacher.

2. The robot operating system (ROS) course was developed using the instructional materials of the BWSI program and given by two MIT students participated in the MISTI framework.
3. The final project was conducted under joint guidance of the instructors from CRDTE and the MIT students.

3.1 Robotics and IoT Course

The students that participated in the course came with a good computer science background but had no knowledge in robotics or robot programing. To deal with this situation, and close the knowledge gap, we started with a weekly, 2 h outreach course that was delivered for more than two months. The 22 h course consisted of the following subjects:

- Robotics basics: The students learned about robots as technological systems, got to know robot subsystems and the role of programing behind robots' autonomous behaviors. Those subjects were convoyed by basic hands on activities with BIO-LOID educational robotic kits.
- Robot construction and programing: The students constructed basic wheeled robots using detailed assembly instruction. The robots were programed by the students using the robot control language RoboPlus, to demonstrate basic functions, such as locate an object and move to it and stop at a specified distance.
- Advanced construction and programing: The students constructed a gripping arm and attached it to the robot, so that turned out to be a mobile robotic arm. The students then programmed this robot to pick up and place objects.
- Robot calibration: The students performed a scientific experiment to find a way to estimate the weight of the object lifted by the mobile robotic arm, even though the robot doesn't have a sensor designated to measure weight. The students found that if turning off the torque to the gripping arm for a known period, the arm will descend proportionally to the weight. The students then calibrated the measurement using known weights, enabling the robot to calculate the weights' mass in grams.
- Robot reinforcement learning: The students programed the robots to perform trial and error experiments aimed to determine stable postures for manipulating objects of different weight (Fig. 1). In each discrete experiment, the robot lifted an object,

Fig. 1. Robot reinforcement learning to lift a load.

estimated its weight, and elongated the gripping arm until toppling over. The robot recorded the parameters of weight and point of robot critical stability.

- IoT basics: the students learned about smart connected products and got acquainted with the IoT platform ThingWorx. Then the students practice in using ThingWorx to create a Thing object and an IoT user interface.

3.2   Robot Operating System (ROS) Course

The three weeks 14 h course was delivered by two MIT students who visited Technion in the framework of the MISTI GTL program. In the course the high school students learned basic principles of control and programming robots in ROS and performed practical exercises in programming the mobile autonomous robot TurtleBot Waffle. The subjects learned were as follows:

- Introduction to ROS. The students learned about modular design of ROS, publisher and subscriber, topics and messages. They performed lab experiments with ROS and 3D robot simulator Gazebo. The students participated in a roleplaying game where they act as nodes and topics to improve understanding of the ROS computation graph.
- Open- and closed loop control. The students learned about limitations of open loops and the need for feedback. They learned the 3 basic types of feedback control: proportional–integral–derivative (PID). They practiced in programming ROS nodes in Python and in publishing messages to move the robot (in the simulation mode) along a predetermined path.
- PID control. The students learned about bang-bang, P, PD, and PID controllers, and on how to read data from the LIDAR sensor of the TurtleBot Waffle robot (Fig. 2a). They implemented the controllers and tested them in the robot simulation mode.
- Robot navigation and mapping. The students learned how to determine position and orientation of the robot and how it can be programmed to navigate in an unknown environment by means of simultaneous localization and mapping (Fig. 2b). In the laboratory practice the students tested robot navigation in ROS, so that the constructed map was displayed using the Gazebo simulator, while commands were teleoperated to the robot controller using a Python script.

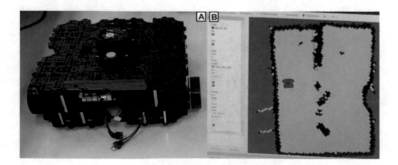

Fig. 2.   a. Delivery robot Waffle; b. Robot location and map displayed in Gazebo.

3.3   Mini Project

The mini project allowed the students to apply the knowledge and skills they acquired during the two courses. The assignment was to implement an automated warehouse scenario involving three robots. Accordingly, the Loader (mobile robotic arm #1) detected the load, picked it, estimated the weight, and put it on the Transporter (TurtleBot Waffle) that delivered the load to the destination location, by following the wall and avoiding the obstacles. Upon its arrival, the Unloader (mobile robotic arm #2) picked the load from the delivery-robot, estimated the weight, and put the load on the floor (Fig. 3a).

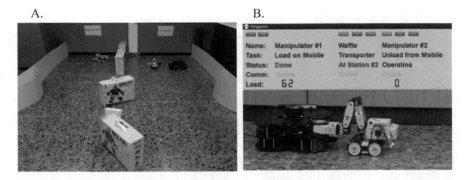

A.

B.

Fig. 3. a. The automated warehouse assignment; Unloader picks the load from the Transporter.

To work on the project, the students were divided into three teams, so that each team was responsible for programming one of the robots. Each robot operated autonomously and communicated with other robots by transmitting data over the internet to the cloud platform ThingWorx. To provide data communication, the students connected a Bluetooth module to each manipulator, and activated RaspberryPi computers, that served as mediators transmitting the data to ThingWorx via WiFi. In the mini project the students got RaspberryPi units pre-programmed and ready to work. The Transporter communicated with ThingWorx directly using its built-in WiFi. The students remotely monitored the robots' performance using a Mashup that served as the dashboard. The Mashup was given to them pre-programmed and ready to work (Fig. 3b).

4   Educational Study

The purpose of our case study was to evaluate the possibility of using the proposed approach to teach the core concepts of intelligent robotics technology to high school students. The question was: Whether and how the course influenced the development of students' understanding of the studied concepts and prepared them for further study of the subject? In the study, we used knowledge questionnaires and personal interviews. The questionnaires were administered at the beginning and at the end of the program.

Among the 15 school students, participated in the program, 11 were tenth graders and 4 twelfth graders. 53% of them did not study robotics before the program, 33% had some little experience, and only two students (14%) participated in outreach robotics courses. In the questionnaires before and after the program we asked the students to explain the meaning of several concepts and evaluated their answers, as presented in Table 1.

Table 1. Gain in understanding concepts of information technology

Concepts to explain	Students' answers (%)			
	Pre-course		Post-course	
	Partial	Correct	Partial	Correct
Programmable controller	60	–	67	13
Electronic sensor	54	13	67	20
Analog and digital signals	20	–	54	13
Interactive website and user interface	34	–	27	13
Internet of Things	20	7	54	13
Smart connected product	–	–	73	13

The table indicates that at the beginning of the program the majority of the students could not explain the concepts related to computer hardware and communication. Some of the students gave partial explanations and only few answers were full and correct. A possible reason for this lack of knowledge is that the school computer science curriculum concentrates on the study of programming languages, while the study of information technology does not get sufficient attention. Our program contributed to closing this gap. The absolute majority of the students acquired basic understanding of the concepts. Some students gave full and correct answers and explained how the concepts were practically used in the program.

The outreach program was intended to impart only basic knowledge of robotics intelligence technology. As robotics is not part of the school curriculum, the only way for students to continue studying the subject was to carry out the project. Therefore, we were interested to know to what extent the program prepared the students to perform a project which will require them to more deeply study and practically apply the subject. In the post-program questionnaire, we asked the students to what extent they are confident that they can self-dependently learn and apply the technologies studied in the program. There answers are summarized in Table 2.

From the data presented in the table, the absolute majority of the students acquired confidence in their ability to learn and apply the noted skills. Part of the students rated their confidence as high, especially for programming robot learning procedures and design of Gazebo simulations.

Table 2. Confidence in the ability to learn and apply the robotics technologies

Subjects for further study and application	Confidence to succeed (%)	
	Medium	High
Programming a robot in ROS	66	34
Design of simulations of dynamic systems using Gazebo	40	53
Design of connected systems where the components share data through the cloud	53	27
Programming trial & error procedures for robot learning	7	93

5 Conclusion

Intelligent robotics technology is becoming the key factor in the development of robotics and automation, and of technological progress in general. It is important to attract the best young minds to this emerging field and prepare them for its interdisciplinary challenges. The purpose of our ongoing research is to develop a methodology for high school outreach in the area of intelligent robotics. The concepts and methods of intelligent robotics are currently studied in senior-level university courses, far beyond the school curriculum, and can be introduced to school students only through academic outreach programs. Beaver Works Summer Institute that has been operating at MIT since 2016 presents the flagship example of such an outreach program. BWSI calls to build an international network of institutes to collectively advance engineering education in schools.

Our intelligent robotics outreach program developed at the Technion Center for Robotics and Digital Technology Education is a model of partnership between two academic institutes (MIT and Technion) and a hi-tech company (PTC Corp.) for the advancement of engineering education in schools. The program imparted to high school students majoring in computer science basic knowledge and skills in robotics intelligence technology and prepared them for further study of the subject. We continue to develop the program towards a deeper exploration of its methodology and broader participation of school students.

References

1. Porter, M.E., Heppelmann, J.E.: How smart, connected products are transforming competition. Harv. Bus. Rev. 92(11), 64–88 (2014)
2. Liao, Y., Deschamps, F., de Freitas Rocha Loures, E., Felipe Pierin Ramos, L.: Past, present and future of Industry 4.0 - a systematic literature review and research agenda proposal. Int. J. Prod. Res. 55(12), 3609–3629 (2017)
3. Richert, A., Shehadeh, M., Plumanns, L., Grob, K., Schuster, K., Jeshke, S.: Educating engineers for industry 4.0: virtual worlds and human-robot-teams. Empirical studies towards a new educational age. In: Proceedings of EDUCON, pp. 142–149 (2016)

4. Motyl, B., Baronio, G., Uberti, S., Speranza, D., Filippi, S.: How will change the future engineers' skills in the industry 4.0 framework? Questionnaire Surv. Procedia Manuf. **11**, 1501–1509 (2017)
5. Verner, I., Cuperman, D., Fang, A., Reitman, M., Romm, T., Balikin, G.: Robot online learning through digital twin experiments: a weightlifting project. Online Engineering & Internet of Things, pp. 307–314. Springer, Cham (2018)

Design and Use of Digitally Controlled Electric Motors for Purpose of Engineering Education

Milan Matijevic[1](✉) and Milos S. Nedeljkovic[2]

[1] Faculty of Engineering at University of Kragujevac, Kragujevac, Serbia
matijevic@kg.ac.rs
[2] Faculty of Mechanical Engineering, University of Belgrade, Belgrade, Serbia
mnedeljkovic@mas.bg.ac.rs

Abstract. This paper describes some pedagogical approaches of using electrical motors within engineering study programs. We are discussing the education of students who are without any electrical engineering background, and who only aim to use motors within different engineering applications like robots, hydro-mechanical or control engineering application. Laboratory support is typically expensive, especially for countries like Serbia, but laboratories are necessary for engineering education. The paper promotes problem-based learning approach which is based on affordable or as cheap as possible laboratory experiments.

Keywords: Problem based learning · DC motor · Laboratory models

1 Introduction

Laboratory support is a necessary part of engineering education. According to criteria for accreditation of engineering study programs by National Accreditation Body in Serbia (up to now - Commission for Accreditation and Quality Assurance (CAQA)), laboratory part of engineering education should be organized within groups of up to eight students in master study programs or up to twenty students in undergraduate study programs. Trained laboratory staff and 8 or 20 laboratory setups are basic infrastructure for laboratory work. Laboratory infrastructure is typically very expensive. In reality, this limitation causes different kinds of limitations in quality of the educational process: less capacity and quality of laboratory work, hands-on laboratory work is substituted by demonstrative laboratory work, and so on. In [1–5] some potential concepts and solutions are described for better quality of engineering education, despite limitations of educational resources.

This paper deals with possible improvements of engineering education in Serbia. The paper promotes the design of low-cost experimental setups, transfer of good practices from EU to Serbia, and illustrates some pedagogical and problem-based learning (PBL) approaches in engineering education in Serbia. The subject of this paper is educational aspects using electric motors in mechanical engineering applications. In the first part of this paper a case study of the design of DC servomotor at Department for Applied Mechanics and Automatic Control at University of Kragujevac is described. The second part of the paper is dedicated to laboratory models with electric

© Springer Nature Switzerland AG 2020
M. E. Auer and T. Tsiatsos (Eds.): ICL 2018, AISC 916, pp. 833–844, 2020.
https://doi.org/10.1007/978-3-030-11932-4_77

motors. The third part of the paper illustrates the possibility of using low-cost laboratory models for advanced educational purposes.

2 Problem Based Learning and Laboratory Models

Problem based learning in engineering education is typically based on the scale models produced by firms with long time experience and reputation in production of education equipment like: Feedback Instruments Limited; Armfield Limited; G.U.N.T. Gerätebau GmbH; Quanser Consulting Inc. etc. [5]. These laboratory models are typically flexible and versatile learning platforms to provide students with a practical introduction to a wide range of topics and issues in different fields of engineering. The close interlinking of practical skills with theoretical/analytical aspects is promoted thorough learning. Students can learn systematically categorized learning content or just combine complex material into integrated project work as well. However, these laboratory models are usually high-quality products that are expensive, with specialized user interface and limited educational potential. The software tools for the use with these products are mainly developed with expensive software packages like Matlab/Simulink or LabVIEW. Users get tutorial for using of already developed software with recommended educational methodology [5]. An alternative approach is own independent construction of laboratory models based on open source software. The power of this approach is consisting of an overall improvement of institutional resources and processes, including project-oriented education within a lot of interconnected educational subjects. For example, in Fig. 1b an experimental setup, produced through a problem-oriented educational approach at University of Kragujevac, is depicted, which has the same purpose in engineering education as commercial products depicted in Fig. 1a Produced laboratory model in Fig. 1b is much cheaper than commercial laboratory model AMIRA DR300. But, laboratory model AMIRA DR300 is a serial product with guaranteed performances, accompanied by standardized tutorials and laboratory practicums. Also, standardized laboratory models can be used, with more efficiency, like benchmark experimental setups for verification of results in research and education. For narrow educational and research purposes, and also, because of safety reasons, commercial laboratory models are more comfortable for using.

Servo DR300 – AMIRA (see Fig. 1a) is a servomechanism consisting of two identical DC motors, which are connected by a mechanical clutch. The first motor is used for control of the rotation speed or the shaft angle. The second one, in the following called as a generator, is used for an emulation of load torque [6]. The whole system consists of three components. The first part is an I/O card MF614 (Humusoft (http://www.humusoft.cz/)) with analogue and digital inputs and outputs. The second, power part, contains sources, current sensors and amplifiers. The third part contains two motors and sensors of the rotation speed and the shaft angle. The plant is represented by a permanently excited DC-motor of which the input signal (armature current) is provided by a current control loop. The sensors for the output signal (speed) are a tachogenerator and an incremental encoder. The free end of the motor shaft is fixedly coupled to the shaft of a second identical motor which is used as a load.

Fig. 1. a, AMIRA DR300 laboratory model [6], b. Servo300 laboratory model – produced at University of Kragujevac in 2010 [7]

The system is controlled by the Real Time Toolbox of MATLAB. Note that the servomechanism is relatively small and its behaviour is nonlinear for low values of the input voltage; its mathematical description by the linear control theory is very inaccurate. Therefore, we control the rotation speed in the linear area only [6].

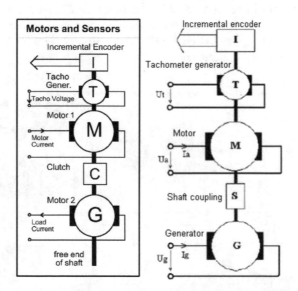

Fig. 2. The concept of Servo300 laboratory model (Amira http://www.ict.com.tw/AI/Amira/amira/dr300en.htm)

Servo300 Laboratory Model (see Fig. 1b) has the same concept as Servo DR300 – AMIRA (see Fig. 2) but it is much cheaper. It is based on second-hand DC motors ESCAP 28D2R and flexible possibilities for controller implementation via PLC, PC or microcontroller. Our aim was to cover a larger range of engineering problems in the context of problem-based learning applications, including the design, fabrication, and use of laboratory models. Namely, our goal is teaching and learning overall aspects in design, manufacturing, and verification of digitally controlled DC servo drive with a variable load. In this manner, our education approach can be more effective and useful for students if there are available high competent and organized teaching staff.

Within different educational subjects, students can realize different phases of design, manufacturing and use of the laboratory model for testing different approaches of signal processing, model identification, and different control algorithms implementations. Problem based learning approach under overall scale process model development, as an engineering product, can contribute to students to better understanding all aspects of engineering product, and use of different theoretical concepts.

The the first step, applicable for non-commercial laboratory models, is principal design and fabrication.

3 Design of Digitally Controlled DC Servo Drives

The subjects "Sensors and Actuators" and "Industrial Computer Systems" at University of Kragujevac offer to students a problem based learning project where students should design electronic interface for signal adjustment from incremental encoder and motor drive implementation in order to enable the motor shaft control related to desired angular position and velocity by use of control computer, Arduino or PLC. For design of DC servomotor, brushless DC motor ESCAP 219P (12 V) with following characteristics is used (Figs. 3, 4, 5 and 6):

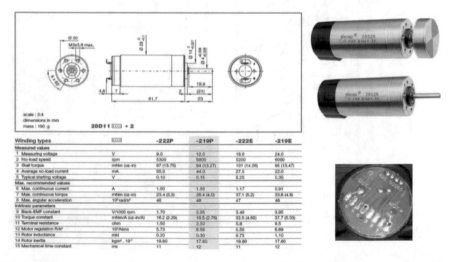

Fig. 3. a. Data sheet of the DC motor ESCAP 28D2R [8], b. MotorESCAP 28D2R with integrated incremental encoder which generate 144 impulses per rotation (2 pulses dislocated at 90°)

Students should develop electronics an interface between the DC motor and the computer support, as well as mathematical and simulation model of the DC motor based on the data of the motor [8],

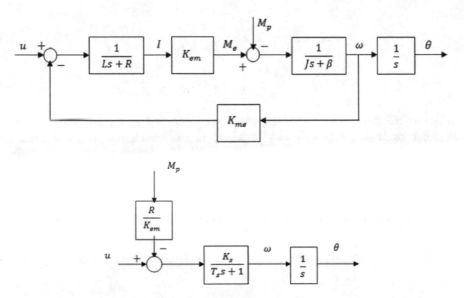

Fig. 4. Modeling of DC motor

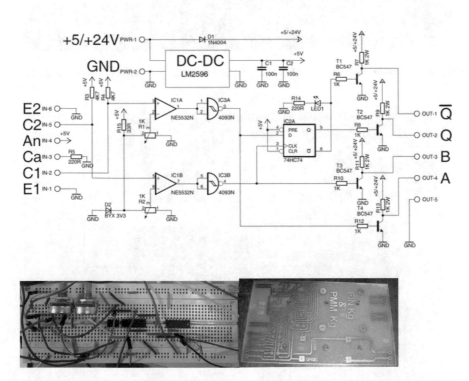

Fig. 5. Electronic interface for DC ESCAP 219P motor: (a) Electrical scheme, (b) Implementation on breadboard and printed board

$$\omega(s) = \frac{K_s}{T_s s + 1} U(s) - \frac{R}{K_{em} N} \frac{K_s}{T_s s + 1} M_p(s) \tag{1}$$

$$K_s = \frac{K_{em} N}{R\beta + K_{em} K_{me} N^2} \quad T_s = \frac{JR}{R\beta + K_{em} K_{me} N^2} \tag{2}$$

and compare step responses of experimental setup and simulation model. Power amplifier dynamics, which is not included in the simulation model, should be commented.

Other typical tasks for students are: identification of the static characteristic and its comparison with the theoretical one from model, identification of dynamical characteristics, design of speed servo (PI controller implementation) and design of positional servo (P control implementation) – simulation and experimental verification.

Fig. 6. Produced exp. setup: DC ESCAP 219P motor with Arduino and electronic interface

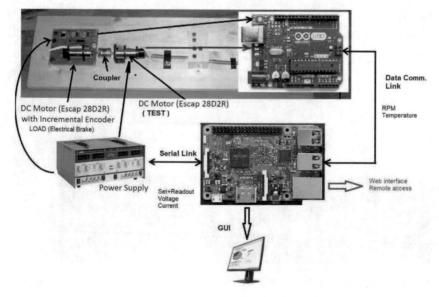

Fig. 7. Low-cost laboratory model based on two DC motors (ESCAP 28D2R with encoder) for remote experimentation [12]

In [12] an experimental setup in Fig. 7 is proposed, for conducting of remote experiments according to proposed experimental protocol. By laboratory work, students obtain important practical knowledge, about the real behavior and performance of the tested equipment. In order to evaluate an electrical motor, certain parameters are needed, such as energy consumption, torque and speed in idle and load state of the system, oscillations (vibrations), temperature drifting and noise exposure [12].

4 Complex Laboratory Models Based on Digitally Controlled Electric Drives

The next, more advanced step, is coupling two or several DC motors in order to build a more complex mechatronics system. For example, the structure from Fig. 2 can be realized with fixed or flexible coupling between the motor and variable load. Also, there is a possibility to solve more complex problems in motion control concerning synchronization of several motors. Students have an opportunity to get to know the hardware, theoretical and software aspects in design of DC servo motors. Besides mechatronics systems, digitally controlled electrical drives can be used within process control systems (see Fig. 8).

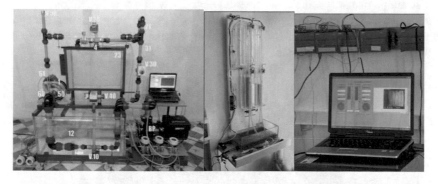

Fig. 8. Digitally controlled electrical drives in process control educational systems

Laboratory models in Fig. 8 are produced at University of Belgrade and University of Kragujevac, respectively, and both of them are much cheaper than commercial ones. Hydraulic pumps are included in both laboratory models, and they are generally used in nearly all industries including, construction, steel milling, mining, manufacturing, machining, heavy equipment, etc. Sizing electric motors correctly for hydraulic power units can save a sizable amount of money over the life of the equipment. At University of Kragujevac, laboratory model of four coupled tanks is used with DC driven pumps. DC motors have been used as pump drive motors due to their variable speed control ability, especially at low speeds, simple control system, high starting torque and good transient response. Students use this laboratory model within educational subjects

"Industrial Computer Systems" and "Design of Control Systems" for programming SCADA applications and testing different control and diagnostic algorithms.

At University of Belgrade, Faculty of Mechanical Engineering, Hydraulic Machinery and Energy Systems Department the laboratory model for demonstration of several pump operation modes and determination of its full performance curves is produced, with additional possibility of flow transducer calibration and measurements, and learning about cavitational phenomenon. Detailed presentation of this setup has been described in [10]. The booster pump is frequency regulated. The LabVIEW application is designed for measurement and control.

Laboratory models in Fig. 8 illustrate importance in design and use of digitally controlled electric motors for engineering education purposes.

5   Low-Cost Laboratory Models and Implementation of Complex Control Algorithms

Good example that complex theoretical concepts can be illustrated and verified on low-cost experiments is laboratory model in Fig. 9. The experimental setup is extremely cheap and simple. Function of incremental encoder is based on pulse wave produced by photo sensor and "windmill" attached on DC motor shaft. DC motor is practically a toy with accompanied very simple incremental encoder with a low resolution. Laboratory model is easy for students to understand as well as the programming function of incremental encoder. Speed of DC motor's shaft is calculated by Arduino UNO based on the number of pulses in time.

Fig. 9. Laboratory model of DC servomotor: DC motor with attached "incremental encoder" and Arduino UNO [11]

Regardless of simplicity, this laboratory model is used in MAS Group, at Ruhr-Universität Bochum, Germany for implementation of adaptive control algorithms for education purpose of international students [11] (Fig. 10).

Namely, adaptive controller is a controller that deals with a complex system having unpredictable parameter deviations and uncertainties. The basic objective of adaptive control application is to maintain consistent performance of a system in the presence of uncertainty and variations in plant parameters. An adaptive controller differs from an

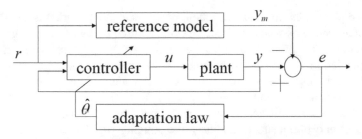

Fig. 10. Model reference adaptive control (MRAC) method [11]

ordinary controller in that the controller parameters are variable, and there is a mechanism for adjusting these parameters online based on the signals in the system. There are two main approaches for constructing an adaptive controller, one is a model reference adaptive control (MRAC) method and the other is a self-tuning method. In [11], the general idea behind the MRAC is to create a closed loop controller with parameters that can be updated to change the response of the system. The output of the system is compared to a desired response from a reference model. The control parameters are updated based on this error. The goal is for the parameters to converge to ideal values that cause the plant response to match the response of the reference model [11]. Simulation schemes in Matlab/Simulink and comparison of experimental and simulation results is given in [11] (Fig. 11).

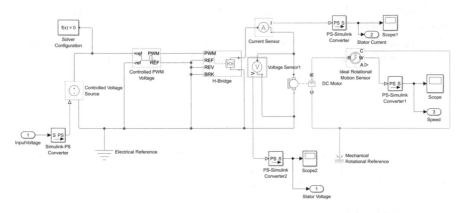

Fig. 11. Simulink model of a DC motor used for reference model and plant simulation in [11]

6 Talking About Teaching

According to our experience, students like to accept any effort if they believe that is beneficial for them. They are truly very interested for knowledge outcomes when they believe that these knowledge and skills are applicable in their future engineering practice. However, students are under high pressure of own environment, existential reality, information from media, local ethics norms, and beliefs of their relatives, friends, and professors. In such a situation, when local labor market is mostly looking for mediocrity students who can accept specific job requirement (low wages, frequent travels, trainings for different positions, extended working hours, etc.) - majority of student population is focused to complete their studies, as soon as possible, with minimum efforts. Opposite of that, a part of the student generation plans migration from local environment, and wants to get the best education. Of course, there are students who are not motivated by future plans, but they are dedicated to studies with a great interest and optimism. In addition, the funding of both state and private universities is based on the number of students, so that there is pressure from the management of universities to maintain financial stability. Under that conditions, the quality of the teaching process is very vulnerable, and very much depends on enthusiasm and pedagogical abilities of teachers. In countries with the aforementioned problems, with teaching resources which are decreasing, variable motivation and initial knowledge of students, contemporary engineering education is a very difficult task.

Our experience in teaching is always linked to the student generation, teaching resources and requirements of faculty management to achieve the appropriate statistical parameters under any limitation. Because of that, laboratory work is a flexible category that always provides opportunity to students to reach more knowledge and skills through compulsory laboratory exercises, student research work and their BSc and MSc theses.

The laboratory work must be precisely guided and well prepared, so that at least 80% of students can do all tasks without much effort. For example, at beginning, students feel helpless in relation how to do an overall complex laboratory task, but if they are well guided through all phases, this concern turns in great satisfaction with the achieved outcomes of knowledge.

Depending on the specific conditions, the teaching process changes from year to year. Unfortunately, these changes are not always the better. Because of that, this paper deals with some limitations in engineering education and possibilities of their removal.

7   Conclusion

Still more than two centuries ago, famous Benjamin Franklin (1706–1790) has pointed out the value of education approach which today we call problem oriented education or problem based learning, by the sentence

"Tell me, and I forget.
Teach me, and I remember,
Involve me, and I learn."

– Benjamin Franklin

Aim of this paper is to promote problem based learning approach in engineering education, based on low-cost laboratory models. As a very general study case in engineering education, for our illustrative example, we have chosen design of digitally controlled electric motors and its use in engineering education.

This paper describes some experiences from different educational environments (University of Kragujevac, University of Belgrade and RUB) about possibilities to organize low-cost laboratory models in order to apply different kinds of problem oriented engineering education. Producing and using low-cost of laboratory models is a good opportunity for schools which are dedicated to engineering education with limited financial support. This approach can save money for expensive equipment but not for high competent and trained teaching staff. Overall motivation and success of students depend on content and organization of educational process, proper guidance and explanation of laboratory part education, and high competent and motivated teaching staff is necessary. Because of that engineering education can not be cheap. Our pedagogical approach of using low cost laboratory models for educational purposes can be useful for enthusiastic teachers to overcome temporary financial limitations.

Acknowledgment. This work has been partly funded by the SCOPES project IZ74Z0_160454/1 "Enabling Web-based Remote Laboratory Community and Infrastructure" of Swiss National Science Foundation and partly by projects TR 35046 and TR33047 Ministry of Education, Science and Technological Development Republic of Serbia, what is gratefully acknowledged. Courtesy of the Department for Mechanics of Adaptive Systems at Ruhr-Universität Bochum in Germany and its Chair, Prof. Tamara Nestorović is gratefully acknowledged for providing the access to laboratory facilities and insight in experimental student education activities through collaboration supported by the German Academic Exchange Service (DAAD).

References

1. Heradio, R., et al.: Virtual and remote labs in control education: a survey. Ann. Rev. Control **42**, 1–10 (2016)
2. Kalúz, M., et al.: ArPi lab: a low-cost remote laboratory for control education. In: 19th World Congress The International Federation of Automatic Control, August 24–29, 2014, Cape Town, South Africa, pp. 9057–9062 (2014)
3. de Graaff, E., Kolmos, A.: History of problem based and project based learning. Manag. Change. Sense Publisher. pp. 1–8 (2007)
4. Mills, J.E., Treagust, D.F.: Engineering education—Is problem-based or project-based learning the answer. Australas. J. Eng. Educ. **3**(2), 2–16 (2003)
5. Matijevic, M., Nedeljkovic, M., Cantrak, Dj., Jovic, N.: Remote labs and problem oriented engineering education. In: IEEE Global Engineering Education Conference (EDUCON), April 25–28, 2017, Athens, Greece, pp. 1390–1395 (2017)
6. Roubal, J., Augusta, P., Havlena, V., Fuka, J.: Control Design for Servo AMIRA-DR300, 15th International Conference on Process Control 2005, Slovak University of Technology, Bratislava, June 7–10, 2005
7. Babajic, N., Jokovic, V., Matijevic, M., Stefanovic M.: Speed servo system controlled via Internet. In: 54thETRAN Conference 2010, June 7–11, Donji Milanovac, Serbia (2010)
8. Rojko, A., Hercog, D.: Control of Nonlinear Mechanism – Documentation for Students, e-publishing, University of Maribor, Maribor (2008)
9. Radojevic, S.: Control of water level in two tanks. MSc Thesis, FE at University of Kragujevac (2013)
10. Nedeljkovic, M., et al.: Engineering education lab setup ready for remote operation – pump system hydraulic performance. In: IEEE Global Engineering Education Conference (EDUCON), April 17–20, 2018, Canary Islands, Spain, pp. 1175–1182

11. Tanaka, S.: Internship Report. Mechanics of Adaptive Systems Research Group, Ruhr-Universität Bochum, Germany (2017)
12. Baltayan, S., Kreiter, C., Pester, A.: An online DC-motor test bench for engineering education. In: IEEE Global Engineering Education Conference (EDUCON), April 17–20, 2018, Canary Islands, Spain, pp. 1490–1494
13. Vukosavic, S.N.: Digital Control of Electrical Drives. Springer. Series: Power Electronics and Power Systems (2007)
14. Vukosavic, S.N.: Electrical machines. Springer. Series: Power Electronics and Power Systems (2012)
15. de Silva, C.W.: Sensors and Actuators – Control System Instrumentation. CRC Press (2007)

High School – University Collaborative Working in Melotherapy

Mirela Sabau[1], Doru Ursutiu[2(✉)], and Cornel Samoila[3]

[1] "Dr. IoanMesota" High School Brasov -Romania, Braşov, Romania
sabaumirela@yahoo.com
[2] Transylvania University of Braşov - Romanian Academy of Scientists AOSR,
Braşov, Romania
udoru@unitbv.ro
[3] Transylvania University of Braşov - Technical Sciences Academy ASTR,
Braşov, Romania
csam@unitbv.ro

Abstract. The brain determines our emotional states, our perception and our reactions to the world around us. The way we respond to events directly affects the strength of each system in our body; strengthening or weakening the ability to repair, adjust and resist the disease. This is why so many physical and emotional conditions originated in the mind and brain. The last century of research in the field of neurosciences and technological advances allowed new methods of approaching information related to the human brain to be found. By using non-invasive methods, such as placing a sensor on the scalp, there may be patterns of electrical signals in the brain. By interpreting the results we can identify new aspects regarding the brain's response to divergence and music (music).

Keywords: Melotherapy · MindWave · Sensors

1 Introduction

The Creativity Lab of the Department of Electronics and Computers - The IESC Faculty, Transylvania University of Brasov, is accessible to college and university students as well as to high school students from Brasov. The Creativity Lab involves pupils in activities that provide a good experience through the exchange of ideas, experiences, opinions and beliefs. In this framework, they are familiarized with graphics programming languages, such as LabVIEW from National Instruments [1], or with the latest electronic devices and the MindWave Headset [2] is the subject of this research.

The root of all our thoughts, emotions and behavior is the communication and interaction of the neurons in our brain. Neurons inter-communicate through electrical impulses and these pulses produce cerebral waves. Brain outlets can be detected using scalp sensors. They are classified according to the wavelength/frequency, forming a continuous spectrum of knowledge; slow, strong and functional at the beginning and

© Springer Nature Switzerland AG 2020
M. E. Auer and T. Tsiatsos (Eds.): ICL 2018, AISC 916, pp. 845–854, 2020.
https://doi.org/10.1007/978-3-030-11932-4_78

then fast, subtle and complex. In order to study the brain waves, under the musical notes, we have used the MindWave sensor (headset).

2 Description of the Work Method

The study was lead upon 7 participants, volunteer highschool students: 5 girls and 2 boys, within the age of 17–18. In order to raise their interest, they were asked to take part in two games: a focus game (Fig. 1) and a meditation one (Fig. 2) using the MindWave Headset. For each subject, the level of attention and meditation was measured, making the subjects to virtually explode a barrel (Fig. 1) and lift a ball with the power of their mind (Fig. 2). In the first game, fulfilling the task consisted in observing the level of their attention, which was reflected in the level of instant burning of the barrel. It was perceived as a game of competition and students were encouraged to obtain a time record.

In the second game, the pupils showed initially a certain amount of mistrust, as they were challenged to close their eyes to foster meditation, but they were curious about the way the ball moves, as it could not be controlled with their eyes open.

Furthermore, a set of 10 songs[1] were used from various genres: classical, rock, Latino, partying songs or disco, song that are not really familiar to the subjects participating in the study. And after each song, the level of oxygenation and the pulse (using the oximeter), the tension and the pulse (using the Blood Pressure Tensiometer) and the temperature were measured.

Using the MindWave device, we determined the dominated brain waves throughout the melody and we also watched the variation in the level of attention and meditation. After each song, slight variations in temperature, pulse and oxygenation were observed, without exceeding the normal values. The results were centralized for each measured parameter.

The study was completed by resuming the two games, but this time in the "presence" of the songs that stimulate attention /meditation, stimuli identified during the measurements. The conclusions of the study were made on the basis of the graphics and the recordings made during the proposed games.

3 The Influence of Music on Medical Parameters

Parameters and physical values were recorded in tables using the Excel application. The same application was used for graphical representation of physical parameters / dimensions as well as for graphical representation of their variation trend in situations where significant variations were observed. All parameters were represented in a single graph for each subject taken into consideration. The order of listening songs represents the variable on the Ox axis. The instantaneous representation of the recorded parameters may be confusing as variations have been recorded from one song to another,

---

[1] Ursuțiu D., Samoilă C., Drăgulin S., Constantin F.A., Investigation of Music and Colours Influences on the Levels of Emotion and Concentration, pg. 6–7.

Fig. 1. Game using meditation

Fig. 2. Game using attention

therefore, we have chosen to trace the trend of evolution during the hearing session for the represented measures.

In order to achieve a comparative approach of the medical parameters for each parameter, a table was compiled with the parameter values for each subject as well as for each part of the chosen repertoire. By realizing a graphic mediation, the tendency of variation of the medical parameter was made. Through this mediation, a comparative study can be made between the subjects regarding their time evolution during the song listening.

Digital resting pulse oximetry offers information on the ability of the lungs to provide the body with enough oxygen in the absence of effort (94–98%). Decreasing this parameter indicates a manifest respiratory failure (below 94%).

In the graphical representation (Fig. 3) it can be noticed that 71.5% (5 subjects) have a decrease in oxygenation and only 28.5% (2 subjects) have registered a tendency to increase oxygenation.

By analyzing the variations in oxygenation trends for each subject, we can notice that:

- for the subjects with a trend of growth, the slope of the representation is of low value

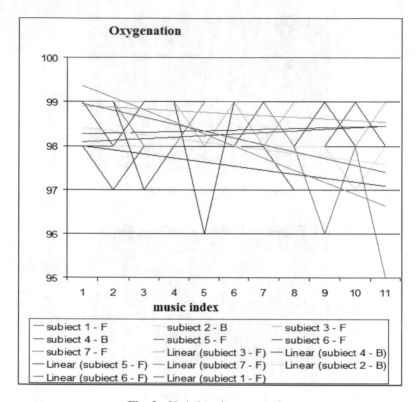

Fig. 3. Variations in oxygenation

- for the subjects with a downward trend, the slope of the representation is of greater value.

It should be noted that the range of variation of this parameter is between 99% and 95%, which reflects an optimal saturation of oxygen in the blood. During the passive music therapy session no oxygenation values were recorded reflecting a significant decrease.

Temperature is a parameter that indicates health or a possible change to it. For a healthy teenager, body temperature is between 36 and 37.8 °C.

Trends in body temperature variation (Fig. 4) measured during the study imply that:

- temperature variations are no more than 1.5°
- in a proportion of 71.5%, the overall trend was a decrease temperature
- for one subject, were recorded temperatures that exceeded the range specified in medical standards, but during the session the temperature stabilized rapidly.

Normal pulse has values between 60 and 100 beats per minute in rest. Pulse was measured with both the pulse oximeter and the Blood Pressure Tensiometer. The chosen values are those measured with the Blood Pressure Tensiometer. During each song, the pulse oximeter tracked the pulse variation and it was observed that the

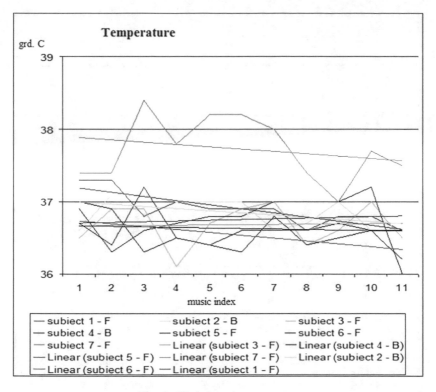

Fig. 4. Variations in temperature

instantaneous variations followed the melody rhythm. During the musical listening, it was noticed that in a proportion of 71.5% the general trend was pulse increase and only 28.5% had a trend of pulse decrease (Fig. 5). Two values exceeded the maximum, one value for two different subjects.

4   The Influence of Music on Attention and Meditation

During the musical auditions we followed the attention and meditation indicators and we marked the average values for these parameters. After recording the values, we noticed that for the same subject, we recorded identical values for more than one song. In this case, the choice of the song was determined by the subject, as he/she was asked what song she/he preferred to listen among the ones with a maximum value for attention /meditation.

The recorded values for attention are presented in Table 1.

The time "gain"/record were calculated for each subject and it was observed that each subject recorded better times while listening the songs. The best effect was recorded for subject number 5 (Fig. 6).

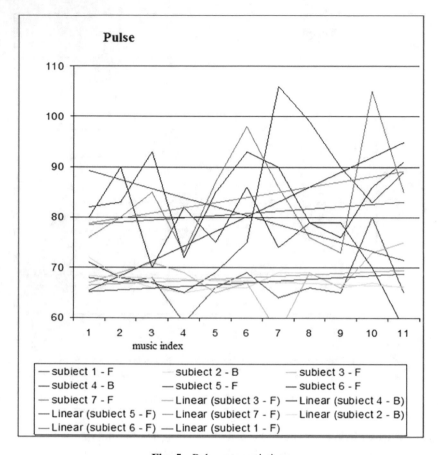

Fig. 5. Pulse rate variations

Table 1. Recorded variations of attention

Subject	Attention			
	Music index	t 1 (s) – time to explode barrel in the absence of music	t 2 (s) – time to explode barrel in the presence of music	$\Delta t = t2 - t1$
1	4	39.85	23.31	−16.54
2	9	32.61	19	−13.61
3	10	23.77	22.33	−1.44
4	7	30	19	−11
5	2	128	41.77	−86.23
6	3	23	19	−4
7	3	47.71	41.1	−6.61

Subject	Music index	Attention		$\Delta t = t2 - t1$
		t 1 (s) – time to explode barrel in the absence of music	t 2 (s) – time to explode barrel in the presence of music	
1	4	39.85	23.31	-16.54
2	9	32.61	19	-13.61
3	10	23.77	22.33	-1.44
4	7	30	19	-11
5	2	128	41.77	-86.23
6	3	23	19	-4
7	3	47.71	41.1	-6.61

Fig. 6. Recorded variations of attention

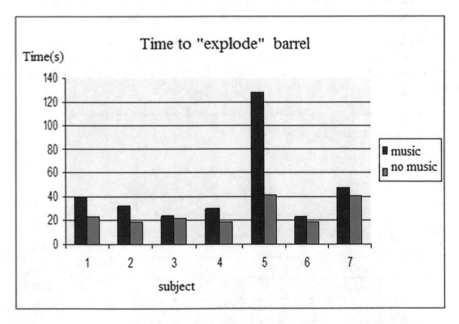

Fig. 7. Time to "explode" barrel

Table 2 indicates the recorded data for each subject on meditation. Indices 1 refers to parameters recorded in the absence of songs, and indices 2 refers to the same parameters but while listening the songs.

Table 2. Subject recorded data on meditation

Subject	Meditation						
	Music index	t_1 (s)	h_1 (m)	v_1 (m/s)	t_2(s)	h_2(m)	v_2 (m/s)
1	7	57.24	4.65	0.081237	258.46	28.65	0.110849
2	5	56.31	8.21	0.1458	46.95	38.42	0.818317
3	1	14.36	2.3	0.160167	18.11	3.22	0.177802
4	9	12.79	1.28	0.100078	13.69	3.67	0.268079
5	6	149.89	16.67	0.111215	260.18	39.89	0.153317
6	5	17.75	2.33	0.131268	25.48	4.68	0.183673
7	4	32.94	4.03	0.122344	161.41	11.3	0.070008

Symbols:

- $t1$ – time lifting the ball in the absence of the song
- $h1$ - the maximum height at which the ball was lifted in the absence of the song
- $v1$ - the average velocity of moving the ball in the absence of the song
- $t2$ – time lifting the ball in the presence of the song
- $h2$ - the maximum height at which the ball was raised in the presence of the song
- $v2$ - the average velocity of the ball in the presence of the song

85.71%, the subjects in the study obtained higher scores in the presence of music. Significant increases were recorded for subject 2 (velocity increased 5.6 times in the presence of music) and for subject 4 (velocity increased 2.68 times in the presence of music) (Fig. 8).

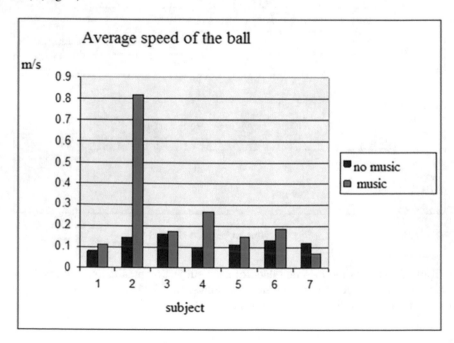

Fig. 8. Average velocity of the ball

Only one subject, representing 14.29% of the sample surveyed, experienced a decrease in velocity of 1,745 times of the baseline Fig. 7.

5 Conclusions

The study above shows that it is possible to "train" the brain by monitoring brain waves under various stimuli, and this workout is possible with a well-chosen musical stimulus.

Attention can be generally controlled if focused on visual level. In order to work with this parameter, it is recommended to keep a series of thoughts aimed at achieving a single process or imagining the expected outcome. In the case of meditation, it is generally desirable to relax. This can be achieved by reaching a state of calm in which the mind is liberated of thoughts and actions that distract attention. To reach a meditative state, the experiment is recommended to be done with eyes closed. Patience is an essential factor in achieving the expected results.

We are all different, especially when it comes to the distribution of brain waves. Stimulating a brainwave can be good for a person and emotionally uncomfortable for another. One and the same song for the same-age subjects with similar schooling experience, produces different "effects" in terms of stimulated brain waves, but also different medical parameters as a value or time evolution.

Our brain profile and our daily experience are inseparable. When our brain waves are out of balance, there will be consequently problems in our emotional or neurophysical health. The research has identified brain patterns associated with all sorts of emotional and neurological disorders.

The brain mediates our perception: every wave of emotion, every thought, every sensation we have in our brain activity. The brain drives our ability to pay attention, to our emotional balance, to the central nervous system tone, to autoimmune function and many more.

It is extraordinary to observe that the human brain is capable of an extraordinary and miraculous change by using well-chosen music but also by eliminating uncomfortable patterns and restoring the system to balance, the optimal function can be restored, and the secondary symptoms naturally fade away.

In this study, we observed how the brain wave is stimulated by musical tracks as well as the influence of music on some medical parameters. The games made the subjects perform actions only with their mind power and "explode a barrel" or "raise a ball" using a Bluetooth headset from NeuroSky called MindWave that was connected to a laptop running a simple application.

The technology that was featured in SF movies has become reality. The helmets used in the study transmitted information (electrical pulses produced at neural level and interpreted as brain waves) to a laptop that, based on special algorithms, could recognize when electrical signals for certain brain activities appeared. People are different and, by consequence, the results obtained are different from one person to another.

References

1. LabVIEW training course. http://sine.ni.com/tacs/app/fp/p/ap/ov/lang/ro/pg/1/sn/n8:28
2. MindWave review. https://www.laptopmag.com/reviews/accessories/neurosky-mindwave-mobile

Software Platform for the Secondary Technical School E-Learning Course

Roman Hrmo[1], Juraj Mistina[2(✉)], Jana Jurinova[2],
and Lucia Kristofiakova[1]

[1] DTI University in Dubnica Nad Vahom, Dubnica Nad Váhom, Slovak
Republic
{hrmo, kristofiakova}@dti.sk
[2] University of SS. Cyril and Methodius in Trnava, Trnava, Slovak Republic
{juraj.mistina, jana.jurinova}@ucm.sk

Abstract. There is a wide array of the e-learning tool platforms to choose from
and a lengthy list features to take into consideration. Authors created a comparison of the top eLearning platforms on the market today. Articulate Storyline
is one of the most popular e-learning authoring tools. The courses can be
adapted to tablet and smartphone screens, providing an optimal view of your
course on every device. It is a standalone tool with great interactivity including
simulations, screen recordings, drag-and-drop interactions, click-and-reveal
activities, quizzes, and assessments. The authors present their experience in the
development of the English e-learning course for presentation skills and professional competences development, using the Articulate software. The paper
provides the readers with practical examples and illustrations, as well as recommendations concerning methodology of the course design.

Keywords: E-learning course · Secondary technical schools ·
Quality of education · Global labour market ·
Communication and presentation skills

1 Introduction

In a professional environment, we encounter different interpretations of processes,
phenomena, methods and forms of e-learning ant terminology closely connected to it.
The rapid development of educational technology and technologies at all has led not
only to a confusing variety of terms in the field of distance education - distance
learning, e-learning, online learning, virtual education, blended learning, flexible
learning, m-learning, etc. It has brought a confusion in understanding and interpreting
the terms even among professionals from educational area. Since our project deals with
practical aspects of the development of e-learning course, first, we would like to
explain some of the terminology.

It is understood that, compared to [1, 2], which defined e-learning at the end of the
1990s, current definitions reflect technology developments and are therefore much
more complex [3–7]. However, even at present, education with the support of electronic media is not understood and defined unambiguously. For example, [8] defines e-

© Springer Nature Switzerland AG 2020
M. E. Auer and T. Tsiatsos (Eds.): ICL 2018, AISC 916, pp. 855–865, 2020.
https://doi.org/10.1007/978-3-030-11932-4_79

learning as an umbrella term that describes education using electronic devices and digital media. It encompasses everything from traditional classrooms that incorporate basic technology to online universities. At the same time, it points to the traditional perception of e-learning as the one that may include educational films and PowerPoint presentations. These types of media can provide students with content that is more dynamic and engaging than textbooks and a whiteboard. Edutainment, or content that is designed to be educational and entertaining, may be used to keep students' attention while providing knowledge about a particular topic, drawing attention to the digital learning environment. On the other hand, [9] provides a simplified definition that is close to our understanding of e-learning. It is understood as a learning utilizing electronic technologies to access educational curriculum outside of a traditional classroom. In most cases, it refers to a course, programme, or degree delivered completely online. As an added value, we also see the possibility of working outside the classroom with CD or DVD-ROM media, or with type Dropbox, Google Drive, Microsoft OneDrive, etc.

There are several other aspects we consider important for the intended e-learning course development. They are: multimedia, interactivity, technical and skill-based content, as defined in [10–14].

2 Course Development

The research conducted at 37 secondary technical schools in Slovakia and in the target environment of companies that are potential employers of graduates of these schools showed the priority interest of employers in professional communication and presentation skills, and confirmed insufficient training of students of these schools in the field of presentation techniques and skills. This was the very reason why we worked out the project "Improving the quality of graduates into practice at secondary technical and vocational schools" supported by the National Grant Agency of the Slovak Ministry of Education. The project deals with the creation of a methodology aimed at improving professional communication skills, including presentation techniques and skills. Presentation skills belong to transferable competencies of graduates of vocational education, what makes them to be a benefit in the selection process at the labour market. The project included a rigorous analysis of students' needs, which demonstrated the need to provide students with new, interesting training forms and methods appropriate to the societal technological progress [15].

2.1 Goal of the Project

The goal of the Project team was to develop an e-learning course that would have the attributes of interactivity, multimedia character, and an interesting animation environment. The objectives of the project are to create a multi-module e-learning course composed of three modules:

- presentation structure and language,
- non-verbal communication,
- PowerPoint slides and graphics.

Our aim was to integrate a fully-fledged LMS with a cloud-based authoring tool, meaning that users can work with e-learning content and deliver it from the same dashboard. The authoring tool component is designed to be simple enough to be used by individuals with basic design knowledge, negating the need for design experts, but powerful enough to create strong content.

2.2 Approach to the Course Development

Firstly, we analysed the market with e-learning software to set up the e-learning environment according to the priorities. There is a wide array of the e-learning tool platforms to choose from and lengthy list features to take into consideration. Therefore the software platform we have chosen was Articulate Storyline Software, that lets you publish courses to a wide range of formats, including Flash, HTML5, and the Articulate Mobile Player app. Host published courses on the web, a learning management system (LMS), or local media (CD, DVD, etc.). You can even create handouts or a transcript by publishing to Microsoft Word.

In the development of the course, we organised our work in the following strategy:

Part I – Introduction

1. Getting started
2. What is needed to develop an e-learning course?

Part II – Designing an e-learning course

3. Identifying and organizing course content
4. Defining instructional, media, evaluation and delivery strategies

Part III – Creating interactive content

5. Preparing content
6. Creating storyboards
7. Courseware development

Part IV – Managing and evaluating learning activities

8. Course delivery and evaluation
9. Learning platforms

So far, we have achieved points 1–7 and we are in the phase to deliver, test and evaluate the course in educational practice of selected secondary technical schools.

Because, apart from animations, we needed to reinforce multimodality with audio recordings, we utilized free online Text To Speech (TTS) service with natural sounding voices: <http://www.fromtexttospeech.com/> that enabled us to design sounding male and female avatar speaking British and American English using presentation phraseology throughout the presentation structure. The advantage is that students can use the course both online and offline on their own pace.

3   Articulate Storyline - the Software Platform

In this paragraph, we would like to explain, why we chose the Articulate Storyline as a software platform for designing the intended e-learning course. As it has been mentioned above, there are multiple e-learning tool platforms to choose from and a lengthy list features to take into consideration. In the preparatory phase, we used several strategies to select a software platform for the needs of our course. First, we studied reviews in prestigious media. According to the following criteria, we chose seven of the most popular software according to the preferred features intended to apply in the e-learning course. The criteria were: *import from Microsoft PowerPoint, preview presentations, image library, embeddable webpages, webcam recording, survey tool, free form quizzes, customizable activities, screen recording, publish as scorm, to mobile devices (iOS, Android), publish to CD, drag and drop, video incorporation, save as online and local files, course/quiz template.* The two other criteria limited our selection – the price/performance ratio and availability of a trial version for testing. Therefore, we download, tested and compared the trial versions of the six following e-learning authoring tools: *Articulate, Adobe Captivate, Lectora Inspire, Elucidat, iSpring, gomo learning.* Besides the others, the five final criteria helped us to decide for Articulate Storyline software – user friendliness, animation advancements, updated image library, interactivity for user and last but not least, the availability to publish courses to a wide range of formats (Fig. 1) as the precondition for the course as the result of the learner needs analysis.

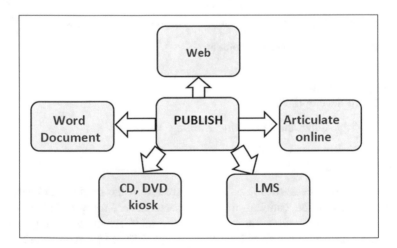

Fig. 1. Articulate storyline publishing platforms

3.1   Story Size

When you create a new Articulate Storyline project, the slide size defaults to 720 pixels wide by 540 pixels high, but you can change it to any size you would like. We used a default option as a suitable option for the course graphics. The overall size of the published output was slightly larger than your slide dimensions. That is because the player (the interface around the perimeter of the slides) added some width and height. The player added up to 260 pixels to the width and up to 118 pixels to the height, depending on the chosen player features (see Table 1 for details).

Table 1. Player features of the slides.

Player feature	Width	Height
Player frame (with or without payer features)	+20 pixels	+20 pixels
Title		+23 pixels
One or more topbar tabs		+24 pixels
Volume controller, seekbar or navigation buttons		+51 pixels
Sidebar	+240 pixels	

3.2   Using Story View

Articulate Storyline uses a hierarchical structure of scenes, slides, and layers to organize content. Scenes are the largest organizational units. Each scene contains one or more slides. Each slide can have multiple layers (Fig. 2).

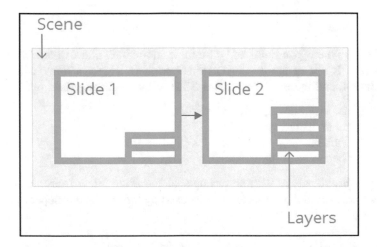

Fig. 2. Articulate Storyline multiple layers used in the course

Story View gives a big-picture view of the entire project. It is where we managed scenes and the layout of the course. Every Storyline project has at least one scene. Our project has 30 scenes (green boxes in Fig. 4). Story View is where you manage scenes, including adding scenes, deleting scenes, and rearranging the slides that appear in those scenes. The starting scene is the first scene learners see when they launch your course. In Story View, the starting scene has a small red flag next to its title (Fig. 3).

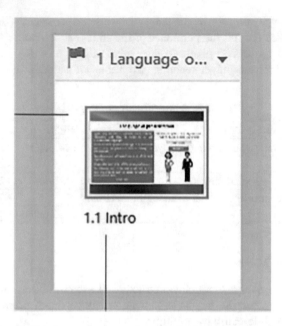

Fig. 3. Starting scene as can be seen by the course developer

3.3 Slide View

Slide View is where you build and customize individual slides in an Articulate Storyline course. It allows the developer to add content, trigger actions, adjust timings, manage layers, and more. Therefore, the Slide View is characterised by:

– building and customizing individual slides,
– adding content, trigger actions,
– adjusting timings,
– managing layers, and more.

Figure 5 provides you with an example of the Slide View from the course. It is divided into sectors and each sector is to be used by the course developer to provide an action.

3.4 Storyline Technical Specification and Terminology - Interactivity for User

In Articulate Storyline, the term "player" refers to the interface around the perimeter of your slides. It can include several optional features, such as navigation buttons, a seekbar, a clickable menu, slide notes, a glossary, a list of supplemental resources, and

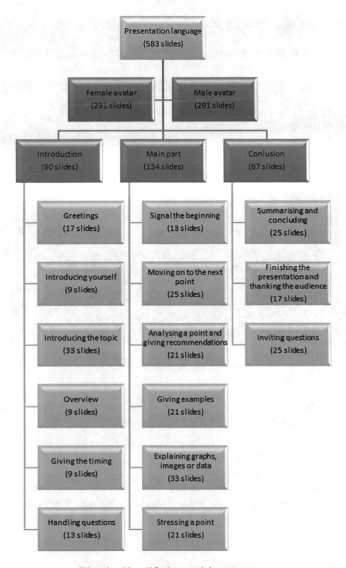

Fig. 4. Simplified organising structure

more. You can even customize the player with specific colours and fonts. For a completely chromeless look, you can make the player invisible, which is a nice option if you plan to build your own custom navigation buttons (Fig. 6).

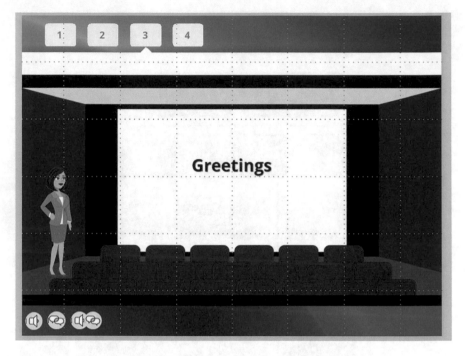

Fig. 5. The course story view

A. navigation tools - transition between components is free, i.e. depends on user + repeatedly can return to individual components:

- A1 complete presentation menu,
- A2 menu of individual parts – variants,
- A3 ways of audio playback (Soundtrack, Text, Soundtrack + Text),
- A4 fluent motion throughout the presentation (NEXT and PREV button).

B. user feedback:

- B1 highlighting the used parts in the menu - grey text colouring,
- B2 highlighting the used individual slides - green stripe above the slide number button,
- B3 highlighting the used parts of audio and text recordings - green "swoosh" at the button.

C. search box - based on metadata, not the slide content.
D. Volume control (on/off/mute).
E. About authors (contact).
F. Zoom regions (Fig. 7).

Fig. 6. Preview of navigation buttons

Fig. 7. Example of zoom regions showing zooming the text and the avatar to attract the user attention

4   Conclusion

The course development has been done purposefully, target group oriented, focused on the target topic unit supported by partial modules to improve students' professional communication in the global labour market. The paper will provide you with technical as well as methodological details of the animated multimedia interactive e-learning course with illustrative figures related to the course. Because, apart from animations, we needed to reinforce multimodality with audio recordings, we utilized free online

Text To Speech (TTS) service with natural sounding voices: <http://www.fromtexttospeech.com/> that enabled us to design sounding male and female avatar speaking British and American English using presentation phraseology throughout the presentation structure. The advantage is that students can use the course both online and offline on their own pace. The following figure provides an illustrative example of the course content and interactivity.

Last but not least, we must not forget to make lessons interesting not only for the lower age categories. We plan to incorporate the elements of the game into the course. We inspired by the he process of game-based learning in education described in [16].

The course is a unique and original outcome of a systematic study of an experienced project team of teachers, methodology experts and technology designers and software specialists. We believe that presenting the course in an international scientific forum can provide a significant exchange of views and experiences that can be mutually beneficial. We did not copy the courses we know from the learning environment, but we tried to use our own expertise to design a course that can be both useful and interesting for the target group of students towards their professional development.

Acknowledgment. The paper is supported by the KEGA National Grant Agency of the Slovak Ministry of Education as an output of the project "Implementation of the project for improving the quality of graduates into practice at secondary technical and vocational schools".

References

1. Kirsh, D.: Interactivity and multimedia interfaces. Instr. Sci. **25**(2), 79–96 (1997)
2. Purchase, H.: Defining multimedia. IEEE Multimed. **5**(1), 8–15 (1998)
3. Sangra, A., Vlachopoulos, D., Cabrera, N.: Building an inclusive definition of e-learning: an approach to the conceptual framework. Int. Rev. Res. Open Distrib. Learn. [S.l.] **13**(2), 145–159. http://www.irrodl.org/index.php/irrodl/article/view/1161/2146 (2012). Last accessed 29 May 2018. ISSN 1492-3831
4. Vališová, A., Šubrt, J., Andres, P.: e-learning from the technical university students point of view. In: Proceedings of the 10th IASTED International Conference on Computers and Advanced Technology in Education, pp. 185–189. ACTA Press (2007)
5. Mironovová, E., Chmelíková, G., Fedic, D.: Developing e-Learning skills within English language training in the Slovak University of Technology. In: International Conference on e-Education, e-Business, e-Management, and e-Learning, IC4E 2010, pp. 203–206. IEEE (2010)
6. Poulova, P., Simonova, I.: e-Learning reflected in research studies in Czech republic: comparative analyses. Procedia-Soc. Behav. Sci. **116**, 1298–1304 (2014)
7. Simonova, I., Poulova, P., Kriz, P.: Reflection of intelligent e-learning/tutoring-the flexible learning model in LMS Blackboard. In: Transactions on Computational Collective Intelligence XVIII, pp. 20–43. Springer, Berlin (2015)
8. Christensson, P.: E-learning definition. https://techterms.com (2015). Last accessed 10 May 2018
9. http://www.elearningnc.gov/about_elearning/what_is_elearning/ © Copyright, eLearningNC.gov (2018). Last accessed 04 Apr 2018

10. Berková, K., Krejčová, K., Králová, A., Krpálek, P., Krelová, K.K., Kolářová, D.: The conceptual four-sector model of development of the cognitive process dimensions in abstract visual thinking. Probl. Educ. 21st Century **76**(2) (2018)
11. Andres, P., Dobrovská, D.: Managing interaction skills in the engineering pedagogy programme. In: 2014 International Conference on Interactive Collaborative Learning (ICL), pp. 112–116. IEEE (2014)
12. Andres, P., Svoboda, P.: Multimedia as a modern didactic tool–windows EDU proof of concept project at Czech Technical University in Prague. In: International Conference on Interactive Collaborative Learning, pp. 29–40. Springer, Cham (2016)
13. Chmelíková, G., Hurajová, Ľ.: How to foster students´ study activity via ICT and real projects. J. Teach. Engl. Specif. Acad. Purp. **5**(2), 173–177 (2017)
14. Szőköl, I.: Educational Evaluation in Contemporary Schools, pp. 159. Belvedere Meridionale, Szeged, Austria (2016)
15. Smékalová, L., Němejc, K.: Transferable competencies of graduates of vocational education: a retrospective survey 2007–2014. In: Proceedings of the International Scientific Conference Rural Environment, Education, Personality (REEP), vol. 9, pp. 106–113 (2016)
16. Herout, L.: Application of gamification and game-based learning in education. In: EDULEARN2016: 8th International Conference on Education and New Learning Technologies, pp. 978–984 (2016). ISBN

Entrepreneurship Education by Youth Start - Entrepreneurial Challenge-Based Learning

Johannes Lindner$^{(\boxtimes)}$

Center for Entrepreneurship Education & Value-Based Business Education,
University Teacher College & Eesi-Center of the Austrian Federal Ministry of
Education, Science and Research & Initiative for Teaching Entrepreneurship,
Vienna, Austria
johannes.lindner@kphvie.ac.at

Abstract. The development of our entrepreneurial mindset does not begin with
the start of our professional lives but is initiated in earlier phases of our
socialization. Entrepreneurial mindset has to be learned – in fact from generation
to generation anew. Entrepreneurship Education for kids and youth can trigger
and support this process. This paper provides an introduction to Entrepreneurship Education, with a focus on the Youth Start Program with the Entrepreneurial Challenge-Based Learning.

Entrepreneurial Challenge-Based Learning takes place in an environment
with competent role-models; if we consciously perceive a situation as a challenge; if we confront the challenge in a deliberate and active manner and successfully master it; if we reflect on ourselves, the development of our ideas and
our self-evaluation – metacognition. The Youth Start Program uses challenges
from various areas as learning opportunities.

Keywords: Entrepreneur · Entrepreneurship education ·
Entrepreneurial challenge-based learning ·
TRIO model for entrepreneurship education · Youth start program ·
www.youthstart.eu

1 Introduction

Don't say 'begin', but 'take part' when you want to achieve something. A responsible
market economy needs self-confident entrepreneurs and mature citizens who help
shape their own and their society's future offensively. Without the dreams of vision-
aries and people who put ideas actively into practice we would live in another world
nowadays. There would be no fine arts and no schools, no cars and no medication, no
state of law and also no consumer protection if people did not take a stand for ideas and
did not change social rules with civil courage.

Entrepreneurial Mindset has to be learned – in fact from generation to generation
anew. It is wrong to believe that economy and democracy can be inherited. Each
generation is once again challenged to develop these competences, ideas and values
which are important for people's lives and society.

© Springer Nature Switzerland AG 2020
M. E. Auer and T. Tsiatsos (Eds.): ICL 2018, AISC 916, pp. 866–875, 2020.
https://doi.org/10.1007/978-3-030-11932-4_80

I am strongly convinced that education within the family and educational training play a decisive role for a responsible market economy. All future entrepreneurs and employees are at school now. The way they are currently educated and trained will leave a mark on their social and economic understanding. Mature citizens don't fall into a society's lap, they need fundamental competences to reflect the execution of their ideas. Education is, therefore, never neutral, it is either an instrument of liberation or it is an instrument of adaptation. As a consequence we should promote entrepreneurship education which connects and fosters creativity, self-responsibility, decision-making ability, knowledge acquisition and independence.

We need entrepreneurial thinking everywhere, the more there is within a society, the more spirited and sustainable it will be [7]. Entrepreneurship is a mental mindset, where people participate actively in society in such a way that they firstly identify what has to be done to be able to work out a solution afterwards.

This article will answer the following questions:

- What do the terms entrepreneur, entrepreneurship and entrepreneurship education mean? – Definition
- Which competences does promoting entrepreneurship convey? – Strengthening entrepreneurship competences
- What are key activities of an Entrepreneurial Challenge-Based learning?

2   Definitions

The terms "entrepreneur" and "entrepreneurship" are derived from the French word "entreprendre" ("to undertake"/"to launch"). In their current meaning, they were introduced by Joseph Schumpeter [14] and are now frequently used in English-speaking countries.

Entrepreneur: Entrepreneurs are independent protagonists. According to Schumpeter they are the key drivers of economic and social dynamics. Schumpeter [14] emphasized their skills and abilities in the independent development and implementation of ideas and pointed out their innovative power, which encompasses the creation of new products, production processes, organizational structures or alternative distribution channels. Entrepreneurs play a relevant role in all subsystems of our society: from business, religion, sciences and politics to education and sports.

Generally speaking, anybody can become active as an entrepreneur. The terms *intrapreneurs* and *co-entrepreneurs* describe entrepreneurs who are not self-employed but decide to become active within an existing organization. *Social entrepreneurs* or *change-makers* are individuals who combine entrepreneurial and social initiatives to bring about a positive change in society. They implement ideas in areas such as education, environmental protection or the creation of jobs for persons with disabilities.

Entrepreneurship: While the term "entrepreneur" refers to a person, "entrepreneurship" describes the process of developing an idea, identifying an opportunity and implementing the idea as a team [8]. The term *social entrepreneurship* is used for initiatives that focus on solving certain problems of society; it partly overlaps with the term *social business*.

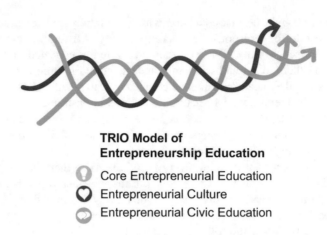

**TRIO Model of
Entrepreneurship Education**

Core Entrepreneurial Education

Entrepreneurial Culture

Entrepreneurial Civic Education

Fig. 1. TRIO Model of Entrepreneurship Education, author's graphic.

Entrepreneurship Education: The term *Entrepreneurship Education* [12] refers to the development of independent ideas and the acquisition of the respective competences that are necessary to implement these ideas (see Sect. 3). Emancipatory approaches to entrepreneurship education emphasize its social and pedagogical relevance for society. Entrepreneurship Education thus pervades areas from the entrepreneurial sphere itself to the personal qualities required for socially proactive citizenship. The *TRIO Model of Entrepreneurship Education* [1] offers a good overview of the most important interlinked segments of Entrepreneurship Education (see Fig. 1): Segment I *Core Entrepreneurship Education* teaches core competences that foster entrepreneurial and professional independence and support independent decisions for one's private life. Segment II, *Entrepreneurial Culture*, focuses on the promotion of a culture of independence, passion, inspiration, open-mindedness, empathy and sustainability that encourages relationships and communication. Segment III, Entrepreneurial Civic Education, aims at promoting a culture of autonomy and responsibility to face social challenges. This is achieved through the development and argumentation of ideas for social initiatives and a personal commitment to their implementation.

Entrepreneurial Learning [2] takes place in an environment with competent role-models; if we consciously perceive a situation as a challenge; if we confront the challenge in a deliberate and active manner and successfully master it [13]; if we reflect on ourselves, the development of our ideas and our self-evaluation – metacognition. This kind of Learning is also called Entrepreneurial Challenge-Based Learning, for details on the You[th] Start Entrepreneurial Challenges see Sect. 4.

3   Strengthening Entrepreneurship Competences

Entrepreneurship competences [5] are the individual's ability to implement ideas. It requires creativity, innovation and the willingness to take risks as well as the capacity to plan and implement projects in order to reach certain objectives. It enhances the individual's everyday personal and social life and enables employees to consciously perceive their working environment and grasp opportunities. It is the basis on which entrepreneurs build their initiatives in a social or business context.

The **Framework of Reference for Entrepreneurship Competences** (see Fig. 2) provides orientation for the learning process in the following competence areas: "developing ideas", "implementing ideas" and "sustainable thinking" (intended to inspire value-oriented thinking). The competences of learners are expressed in *'Can Do'* *statements*, describing the specific attitudes, abilities and skills that learners are supposed to develop. The Framework of Reference for Entrepreneurship Competences is meant to serve as an aid for curriculum development teams and an inspiration for the development of teaching-learning arrangements (see Sect. 4). The *Framework* [11] was elaborated on the basis of the *Common European Framework of Reference for Languages*, CEFR. Entrepreneurship competences develop over time, in a process that starts long before individuals embark on their professional careers, which is why the framework of reference for entrepreneurship competences includes several competence levels (see Fig. 3).

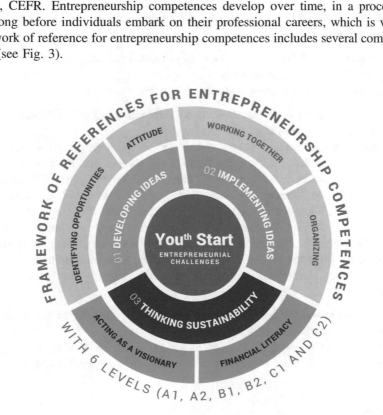

Fig. 2. Structure of the framework of reference for entrepreneurship competences [12].

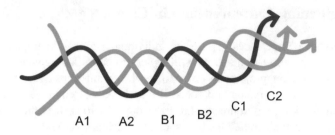

**Different Competence Levels
of the Framework of
Entrepreneurship References**
A1, A2 ... Beginners
B1, B2 ... independent and
C1, C2 ... competent learners

Fig. 3. Different Competence Levels [12].

These levels (see Fig. 3) correspond to the conventional division into primary, secondary and tertiary levels:

- competence level A (Beginners) refers to primary or elementary,
- competence level B (independent) to secondary,
- competence level C (competent learners) to tertiary applications.

The individual competence levels are subdivided into a higher and lower level each, creating a total of six competence levels.

This article primarily discusses the awareness and development phase of education through entrepreneurship = *learning to become entrepreneurial*: the main objective is to strengthen the "entrepreneurial mindset" (capacities), i.e. the general disposition that enables someone to "launch an enterprise"; with its potential teaching-learning arrangements and adopts a transdisciplinary interpretation. Working with the target group in question, the aim is to foster their disposition to show entrepreneurial initiative and strengthen their entrepreneurial orientation. Only for a small segment of the target group the actual intention to launch a business will be a realistic immediate goal. The promotion of an entrepreneurial mindset in this target group comprises three components:

- **professional initiative**: developing and implementing ideas up to the foundation of an enterprise or within a company;
- **personal initiative**: implementing plans and ideas for one's own life;
- **social initiative:** acting independently as empowered and responsible citizens.

4 Youth Start - Entrepreneurial Challenge

The *Youth Start - Entrepreneurial-Challenge* is based on the TRIO Model (see Fig. 1) and the Framework of Reference for Entrepreneurship Competences (see Fig. 2). *Entrepreneurial Challenge-Cased Learning* is modelled on the *learning cycle* [10] (see Fig. 4): "Challenge - Feedback - Reflection" [10] and adopts various learning methods (see Fig. 5). The *Youth Start Program* uses challenges from various areas as learning opportunities (see Fig. 6).

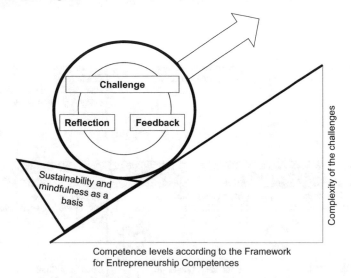

Fig. 4. Entrepreneurial Challenge-Based Learning, author's graphic.

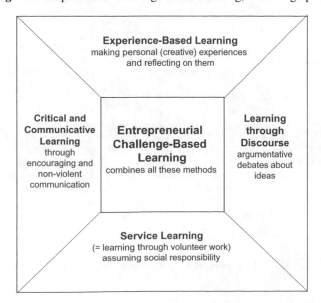

Fig. 5. Entrepreneurial Challenge-Based Learning combines various learning methods [12].

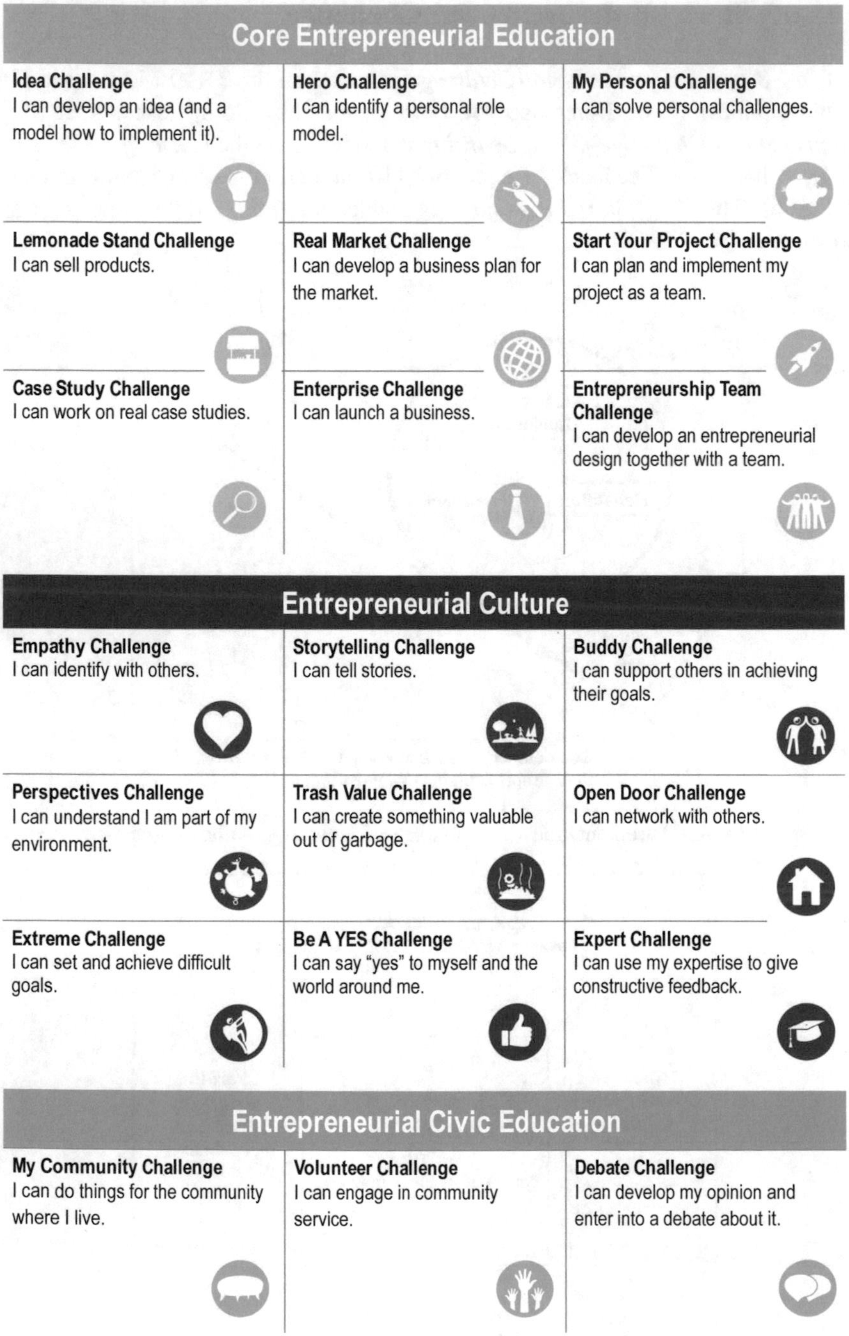

Core Entrepreneurial Education

Idea Challenge
I can develop an idea (and a model how to implement it).

Hero Challenge
I can identify a personal role model.

My Personal Challenge
I can solve personal challenges.

Lemonade Stand Challenge
I can sell products.

Real Market Challenge
I can develop a business plan for the market.

Start Your Project Challenge
I can plan and implement my project as a team.

Case Study Challenge
I can work on real case studies.

Enterprise Challenge
I can launch a business.

Entrepreneurship Team Challenge
I can develop an entrepreneurial design together with a team.

Entrepreneurial Culture

Empathy Challenge
I can identify with others.

Storytelling Challenge
I can tell stories.

Buddy Challenge
I can support others in achieving their goals.

Perspectives Challenge
I can understand I am part of my environment.

Trash Value Challenge
I can create something valuable out of garbage.

Open Door Challenge
I can network with others.

Extreme Challenge
I can set and achieve difficult goals.

Be A YES Challenge
I can say "yes" to myself and the world around me.

Expert Challenge
I can use my expertise to give constructive feedback.

Entrepreneurial Civic Education

My Community Challenge
I can do things for the community where I live.

Volunteer Challenge
I can engage in community service.

Debate Challenge
I can develop my opinion and enter into a debate about it.

Fig. 6. "You[th] Start – Entrepreneurial Challenges" Program. www.youthstart.eu [9].

The learning cycle (see Fig. 4) comprises "challenge – feedback – reflection":

- A challenge is defined as a demanding and complex task that is tailored to the target group and reflects their daily reality. Learners are challenged to develop and implement ideas (often in collaboration with others) for the specific situation in question. Children and adolescents should learn from early on that they can develop their own ideas and tackle challenges themselves.
- Adequate feedback uses the "backwards design model". The competences that are to be achieved are made clear at the beginning of the learning cycle. Following a phase of independent and creative work, feedback is provided in the form of "teacher assessment", "peer assessment" (students) or "self-assessment".
- Reflection offers learners an opportunity to process their personal experience in the challenge and the feedback phases [3] and to develop internal structures or attitudes. The strengthening of attitudes has to do with a person's self-efficacy. It is precisely this internal conviction that determines whether we succeed at what we are doing or not.

Entrepreneurship education centres on the development of new ideas and their creative and conceptual implementation [8]. *Entrepreneurial Challenge-Based Learning* is based on a mix of different learning methods (see Fig. 5).

- Advocates of **experience-based learning** [4] argue that competences are best acquired if they are experienced in real situations. Entrepreneurship education promotes an experimental process of inquisitive learning in close correlation between theory and practice, combined with creativity and teamwork [6].
- **Critical and communicative learning** in entrepreneurship education is realised through communication between teachers and students in the form of dialogue that is characterised by respect, empathy and encouragement. The focus is on the development of respectful relationship.
- **Service learning** means **learning through commitment** to an existing idea. This method introduces students to the possibility of performing practical community *service* that is related to content they learn in school. "Service" and "learning" will thus benefit from each other. Assuming social responsibility is a key element in entrepreneurship education. Volunteer work (see *My Volunteer Challenge*) opens up an opportunity to develop ideas to solve social problems (as social entrepreneurs, see *My Community Challenge*).
- **Learning through discourse** encourages learners to engage in debates, thus offering insight into the diversity of positions and interests with regard to questions of society. In the context of entrepreneurship education, debate clubs are a learning arrangement (see Debate Challenge).

The *"Youth Start - Entrepreneurial-Challenge Based Learning"* is a program, which was funded within the framework of an EU Policy experimentation project in cooperation with the Ministries of Education from Luxembourg, Portugal, Austria and Slovenia. The methodological development work was carried out and coordinated by IFTE.at. The implementation will be accompanied by randomized field research over a period of three years. The Danish Entrepreneurship Foundation and a team of scientists in all participating countries are responsible for coordinating the scientific support. The overall coordination is carried out by PEEP.

Each of these categories features (see Fig. 6) individual variants for the different Competence Levels of the Framework of Reference for Entrepreneurship Education. An example: The *Idea Challenge* at Level A1 comprises a workshop called "Get your ideas moving". Level A2 introduces various examples of *Design Thinking*, with different, age-appropriate starting points. At level B1 students develop a sustainable entrepreneurial design, using *Entrepreneurial Design*. Level B2 is based on *The Business Model Canvas*.

As a *summary*, some recommendations based on the experiences with the "You[th] Start – Entrepreneurial Challenges" for the implementation of Entrepreneurship Education for kids and youth.

I recommend …

- to start with Entrepreneurship Education at school at primary level and continue until the upper secondary level. EACH child should be empowered to work self-determined on different entrepreneurial challenges several times through its school career. This approach corresponds to a structured care of the next generation of entrepreneurs. The You[th] Start Entrepreneurial Challenges Program is developed as an open source program – open to any user. It refers to the You[th] Start Reference Framework of Entrepreneurship Competences and is mapped to EntreComp, which makes it usable on each school level.
- to use a holistic approach of Entrepreneurship like the TRIO model of Entrepreneurship Education, which has been used for the You[th] Start Entrepreneurial Challenges Program. The TRIO model is covering three areas Core Entrepreneurial Education, Entrepreneurial Culture and Entrepreneurial Civic Education.
- to learn entrepreneurship trough "Entrepreneurial Challenge based Learning". This is defined as a demanding and complex task that is tailored to the target group and reflects their daily reality. Learners are challenged to develop and implement ideas (often in collaboration with others) for the specific situation in question. The You[th] Start Entrepreneurial Challenges Program offers a good basis for developing and strengthening entrepreneurial competencies and mindset. This learning method combines an action with a reflection-oriented pedagogical approach.
- to use a modular Entrepreneurship Program, which allows to integrate entrepreneurship in different subjects or as standalone subject Entrepreneurship or even shaping the future up to an *Entrepreneurship School*.
- to implement regularly entrepreneurial learning tools from elementary to high school. We recommend to show clearly to the students, that each single module is part of a whole program with the aim to foster their entrepreneurial competences. It turned out to be a good idea to design a poster for the overview and to add photos of the implementation.
- to accompany the implementation of Entrepreneurship Education at schools with the following 3 measures to make it successful: Learning Materials, like www.youthstart.eu, Personnel Development (Training of teachers and headmasters) and School development (involving teachers, headmasters and stakeholders).

References

1. Aff, J., Lindner, J.: Entrepreneurship education zwischen "small and big ideas" – Markierungen einer Entrepreneurship Education an wirtschaftsberuflichen Vollzeitschulen. In: Aff, J., Hahn, A. (eds.) Entrepreneurship-Erziehung und Begabungsförderung an wirtschaftsberuflichen Vollzeitschulen, pp. 83–138. Innsbruck (2005)
2. Bandura, A.: Self-Efficacy. The Exercise of Control. Worth Publishers, New York (1997)
3. Boud, D., Keogh, R., Walker, D.: Reflection: Turning Experience into Learning. Routledge Falmer, London (1985)
4. Dewey, J.: How We Think: A Restatement of the Relation of Reflective Thinking to the Educative Process, 2nd edn. D.C. Heath & Co., Lexington (1933)
5. European Commission: Recommendations of the European parliament and the council on key competences for life-long learning. KOM 548, Brussels. Retrieved from https://eur-lex.europa.eu/legal-content/EN/TXT/HTML/?uri=LEGISSUM:c11090&from=EN (2005). Last accessed 07 July 2018
6. European Commission/EACEA/Eurydice: Entrepreneurship Education at School in Europe, Luxembourg. Retrieved from https://webgate.ec.europa.eu/fpfis/mwikis/eurydice/images/4/45/195EN.pdf (2016). Last accessed 05 May 2018
7. European Council: Conclusions on entrepreneurship in education and training. Retrieved from http://eur-lex.europa.eu/legal-content/EN/TXT/PDF/?uri=CELEX:52015XG0120(01)&from=EN (2014). Last accessed 07 July 2018
8. Faltin, G.: Brains Versus Capital - Entrepreneurship for Everyone: Lean, Smart, Simple. Entrepreneurship Foundation, Berlin (2013)
9. Jambor, E., Lindner, J.: Youth start entrepreneurial challenges, materials for teachers and students. Retrieved from www.youthstart.eu (2018)
10. Kolb, D.A.: Experiential Learning: Experience as the Source of Learning and Development. Prentice Hall, New Jersey (1983)
11. Lindner, J.: Reference framework for entrepreneurship competences, Version 15. In: EESI Austrian Federal Ministry of Education/IFTE (eds.) Vienna. Retrieved from http://www.youthstart.eu/en/approach (2014). Last accessed 07 July 2018
12. Lindner, J.: Entrepreneurship Education, In: Faltin, G. (eds.) Handbuch Entrepreneurship, pp. 407–423 (2018)
13. Neck, H., Greene, P., Brush, C.: Teaching Entrepreneurship, a Practice-Based-Approach. Edward Elgar, Cheltenham/Northampton (2014)
14. Schumpeter, J.A.: [1911] The Theory of Economic Development: An Inquiry into Profits, Capital, Credit, Interest and the Business Cycle, translated from the German by Redvers Opie. Transaction Publishers, New Brunswick, U.S.A and London, U.K. (2008)
15. Sternad, D., Buchner, F.: Wissenschaftliche Grundlagen des Lernens durch Herausforderung. In: Lernen durch Herausforderung, pp. 41–46. Wiesbaden (2016)

Defining Higher Order Learning Objectives for Software Development that Align with Employability Requirements

Veronika Thurner(✉) and Axel Böttcher

Department of Computer Science and Mathematics,
Munich University of Applied Sciences, Lothstraße 64, 80335 Munich, Germany
{veronika.thurner,axel.boettcher}@hm.edu
http://www.cs.hm.edu/

Abstract. Experience shows that many students of STEM-subjects have difficulties in acquiring skills on higher levels of Bloom's revised taxonomy for learning objectives [1]. However, it is crucial to foster these skills in our students, as they will be required from our graduates in professional life, where it is not sufficient to just "know" about things, but necessary to actively use the acquired skills, both systematically and creatively, to solve problems that were hitherto unknown. As a basis for devising teaching and learning methods that systematically develop these high level skills, we define learning objectives for some of those competences that will be required of our graduates. These learning objectives do not only address technical competences. Rather, they also comprise those *non-technical* key competences that are essential for developing the required *technical* competences in the first place, and which later on are necessary for applying and enhancing these *technical* competences professionally throughout one's working life.

Keywords: Software development · Employability ·
Learning objectives

1 Motivation

Nowadays, good software engineers must be able to perform in a highly professional way in complex surroundings and within a variety of different contexts. To cope with these demands, they must be technically highly qualified and, in addition, possess a variety of non-technical competences as well [4]. For example, they must be able to communicate appropriately with both technical and non-technical people, e.g. to ensure that the system to be developed really meets customer requirements. As well, they must be able to critically reflect on both their own and their team's skill, working process and on the achieved results, to ensure product quality and process effectiveness.

For being able to meet the technical skill requirements of their domain, students need to develop high cognitive skills that go far beyond memorizing fact

© Springer Nature Switzerland AG 2020
M. E. Auer and T. Tsiatsos (Eds.): ICL 2018, AISC 916, pp. 876–887, 2020.
https://doi.org/10.1007/978-3-030-11932-4_81

knowledge and reproducing it for an exam [9]. As these skills do not develop magically on their own, lecturers must devise methods of teaching and learning that help to systematically foster these high level technical competences. The basis for devising these methods is a detailed definition of the desired learning outcomes that we aspire to foster in our students. These learning objectives specify what kind of behaviour constitutes a certain skill on a specific skill level.

In technical domains, these learning objectives mainly focus on technical competences. However, a variety of non-technical key competences is necessary as well, both for developing the technical competences in the first place, and for being able to act professionally in everyday working life. Therefore, it is necessary to specify learning objectives not only for the technical competences, but for selected, highly relevant non-technical competences as well.

2   Goal of This Contribution

To prepare our students appropriately for their working life later on, we aim at developing their technical competences up towards the higher cognitive levels of the revised taxonomy of Bloom [1]. As a first step, we define learning objectives across the different levels of the taxonomy, for those technical skills that we aspire to foster in our students within the introductory classes on software development, which are mandatory in the first and second semester of our Bachelor curricula on computer science and on information systems and management.

Furthermore, we define learning objectives for selected key competences, which are essential not only for developing the high level technical competences, but also for applying and enhancing them in working life. These learning objectives build the foundation for devising teaching and learning methods that systematically foster the development of these competences in our students.

3   State of the Art

Following [10], we define *competences* as those characteristics and skills that are necessary for executing a certain kind and amount of tasks in an appropriate way. In accordance with [5,9], we distinguish four areas of competences, i.e. personal, methodical, social and technical competences. Adhering to [12], by *core competences* we denote those competences that supplement technical competences in a way that enables a person to fulfill their own needs, live together with others and to pursue a useful work that provides a sufficient income.

A well known approach for defining learning objectives in the cognitive area is Bloom's taxonomy of educational objectives [2], which defines competences across different skill levels. The notion of high level competences of this approach was refined later on by Anderson et al. [1]. In the following, we adhere to this revised version, with the competence levels *1: Remember, 2: Understand, 3: Apply, 4: Analyse, 5: Evaluate* and *6: Create*.

In our definitions of competence oriented learning objectives that we provide in this work, we combine a specific content with the aspired competence level and specify in prose text the observable behaviour that constitutes the corresponding skill for this content and on this level.

4   Learning Objectives for Technical Competences

The tables introduced in this section define learning objectives for the technical content that is focused on in our introductory classes on software development. Each table focusses on a specific technical content and the related competenes, with rows denoting the different competence levels. Within the definitions on the different levels, we deliberately avoid using those verbs which label the competence levels of the taxonomy. Instead, we use verbs that describe an action or a result that can be observed from the outside. Thus, we ensure that it is possible to assess whether a person has, or has not, mastered a certain competence.

For each technical content that we include in the learning objectives for our students, we the lecturers must have a clear understanding of whether or not this content will be relevant for our students in their later professional life, and if so, why. As well, to realize which level of the respective competences we must aspire to develop in our students, we must have a notion of the settings and tasks in whose context our professionals-to-be will be required to use this technical content.

In their professional life, software developers might be faced with a multitude of different tasks, each requiring a different skill level:

- Understanding existing code in order to find bugs, fix these bugs, or extend the existing code—which are three related, but different, contexts. In each of these cases, the focus is on understanding programming concepts in the way they occur within a given source code.
- Writing new code. This requires the application of programming concepts, and decisions on which language concepts are best to be used in order to solve a problem.
- Transferring programming concepts to a different programming language, thus abstracting from a specific language and concretizing to another programming language that is more or less unfamiliar yet.
- Designing and creating new software, especially larger systems, in teams. This requires a multitude of technical, methodical and communication skills.

Note that learning objectives for different programming concepts, but on the same competence level, tend to be rather similar. Therefore, to reduce redundancy in our definition of learning objectives, in Table 1 we define learning objectives for programming concepts and software development in a generic way, as a template that can be concretized for specific programming concepts as needed.

As an example, Table 2 concretizes this template for the programming concept *control structures*. Splitting the flow of control into alternative branches, based on the result of a certain decision, is a central basis of all programming languages. So is the concept of loops, which is the basis for automating repetitive tasks.

Table 1. Generic learning objectives for programming concepts and software development

L	Students ...
1	... define the basic concepts of the respective programming concept
	... correctly identify and name the concepts/elements that occur within a given artefact (i.e. a requirements specification, test case, source code,...)
	... use the correct syntax and adhere to syntax conventions
2	... explain in their own words the meaning of the basic concepts of the respective programming constructs
	... describe in their own words the differences between programming constructs
	... explain which programming construct to use in which conctext, and why
	... elaborate why software development comprises more than just implementation
3	... implement an existing design specification (provided either as text, or graphically) in a specific programming language. The design specifies the required classes, including their attributes and methods. For the methods, the algorithm is already roughly sketched as well. The resulting source code fulfills fundamental quality criteria (i.e. readability, testability, correctness)
	... identify for a given implementation and given input values the generated result
4	... state for a given implementation what it does in principle, abstracting from specific input values
	... derive from an informal problem statement the underlying requirements, and document them adequately (as text or graphically)
	... identify within a given artefact its underlying structure and interdependencies, and document them adequately (as text or graphically)
5	... reason in a systematic way which technical concept or programming construct is best suited for realizing a specific requirement.
	... identify strenghts and weaknesses in a given artefact (e.g. requirements specification, test case definition, design, source code, ...)
	... critically assess a self-created artefact with respect to established quality criteria
6	... develop for a simple problem, and based on a given requirements specification, a design that specifies both the overall structure of the solution as well as the required algorithms. The design adheres to established quality criteria
	(By "simple problem", we denote tasks that can be solved in an object oriented way with ten or less classes. More complex problems will be addressed in more advanced classes later on in the curriculum.)

The provided template can be used analogously to specify learning objectives for other programming concepts that are covered in this class, e.g. classes, data types, variables, visibility, packages, algorithms, recursion, strings, arrays, lists, inheritance, generics, collections and exception handling. Since all of these learning objectives are highly redundant in principle, specifying them explicitly does not provide a significant benefit as compared to the generic definition.

Table 2. Generic learning objectives concretized for control structures

L	Students ...
1	... name the different kinds of control structures
	... define the semantics of each of these control structures
	.... correctly identify and name the control structures that occur in a given artefact
	.... use the syntax of control structures correctly and adhere to syntax conventions
2	... explain in their own words the meaning of the different control structures
	... describe in their own words the differences between different kinds of loops
	... describe the differences of different statements for distinguishing cases
	... explain which control structure should be used in which context, and why
3	... implement an existing specification of an algorithm in a specific programming language, while making use of the appropriate control structures. The resulting source code fulfills fundamental quality criteria
	... correctly evaluate for a given implementation the contained control structures for specific given input values, and derive the generated result
4	... state for a given combination of control structures what it does in principle, abstracting from specific input values
	... derive from a given requirements specification the control structures that are appropriate for solving the task at hand
5	... reason in a systematic way which control structure is best suited for realizing a specific requirement
	... identify strenghts and weaknesses of the usage of control structures in a given artefact (e.g. algorithm specification, source code)
	... critically assess a self-created artefact with respect to established quality criteria
6	... develop for a simple problem, and based on a given requirements specification, an algorithm or source code that uses the appropriate control structures

However, for some other technical content, the corresponding competences differ in principle from the ones that we defined above in a generic way. Thus, we handle them separately in individual definitions of learning objectives.

Quality criteria for software are closely related to creating understandable source code, as it is much easier to understand another developer's code when it adheres to well established programming conventions. Furthermore, adhering to quality criteria is a crucial precondition for writing maintainable code. Therefore, in Table 3 we define learning objectives for software quality criteria. Note that for the introductory studies on software development, levels 1–4 (*remember, understand, apply, analyze*) usually would suffice. In professional life, however, the higher levels must be mastered as well, especially level 5, *evaluate*.

Nowadays, professional software development comprises systematic testing as a matter of course. Testing adds value to a product, in that it increases the trust

Table 3. Learning objectives for software quality criteria

L	Students ...
1	... name fundamental quality criteria for software, e.g. readability, testability, correctness, efficiency
	... define the meaning of these quality criteria
2	... explain in their own words the meaning of the different quality criteria
	... describe in their own words why these quality criteria are relevant, and the consequences that will arise if they are violated
3	... adhere to the specified quality criteria in software development
4	... analyse to what extent a given artefact adheres to the specified quality criteria
	... identify violations of these quality criteria
5	... evaluate critically whether the specified quality criteria are adequate and sufficient for a specific usage scenario
6	... create new quality criteria for new, complex usage scenarios

into the correctness and reliability of the software system. Therefore, higher order testing skills are essential for professional developers, and for professionals-to-be.

However, most text books for introductory programming classes do not cover testing at all, or only towards the end of the book or curriculum. This insinuates that testing might not be essential for learning how to develop software. As a consequence, testing is still under-rated, and under-estimated, by many of our students.

In principle, software development is a highly constructive process that (hopefully) creates a working software product—and thus some kind of resulting artefact. In contrast to this, testing is a somewhat destructive activity, as it aims at destabilizing the system in order to identify faults and weaknesses. (Of course, the overall purpose of testing is to make the resulting system better and more stable.) Due to this principal difference in the approach, testing requires different kinds of competences than the generic software development ones. We define those in Table 4. Here again, professionals should reach at least level 5, *evaluate*.

Another essential skill for professional developers is debugging, i.e. the systematic search for error causes. Here, too, the focus is not on constructing some kind of virtual artefact. Instead, debugging comprises the usage of tools and strategies for identifying and removing error causes.

From an economic point of view, debugging activities usually do not add value to a software product, at least not in a direct way. However, good debugging skills are indispensable in a professional programmer, since it is almost impossible to avoid creating bugs when writing complex software systems. Thus, in a professional software project, the effort spent on debugging should be kept as small as possible, yet as high as necessary.

As with testing, the competences required for debugging differ in principle from the generic software development ones. Therefore, we define them explic-

Table 4. Learning objectives for testing

L	Students ...
1	... define the meaning of unit testing
	... name different assert-methods that can be used in JUnit-tests
	... name different annotations for JUnit-tests
	... use the syntax for test classes/methods correctly and adhere to conventions
2	... explain in their own words the necessity for automated testing
	... explain the relevance and contribution of a complete test coverage
3	... implement basic test cases in a schematic way
	... systematically use tools that measure the test coverage
	... use appropriate tools for running tests automatically
4	... examine which typical cases occur within the method to be tested, cluster them into appropriate equivalence classes and define suitable test cases as typical members, to appropriately cover each equivalence class
5	... evaluate whether the amount and kind of specified tests is sufficient to ensure the quality of the software
6	... develop test strategies for complex, new problem situations
	... develop a new test infrastructure or a new testing framework

itly in Table 5. Similarly to the competences on quality criteria and on testing, professionals should reach at least level 5, *evaluate*.

5   Learning Objectives for Key Competences

Many students of computer science related subjects experience some difficulties when aspiring to develop technical expertise. Experience shows that this phenomenon does not necessarily relate to a certain specific technical content. Rather, students find it difficult to reach a certain skill *level* at all, regardless of the specific technical content. This leads to the assumption that it is not mainly a lack of technical skill that inhibits students in acquiring a higher competence level for a specific technical content. Rather, some essential key competences need to be developed sufficiently, before a higher level of Bloom's revised taxonomy is reachable at all.

To support this assumption, we analyzed a large variety of typical errors that occur during the process of learning how to program and develop software. We based this analysis both on literature (e. g. [6–8,11,13]) and on our own observations. After collecting these errors, we categorized them and mapped them onto the different competence levels of Bloom's revised taxonomy. On this basis, we analyzed and categorized the underlying causes of these errors. As a result, we realized that some competence levels require that certain key competences are sufficiently developed in the first place [14].

Table 5. Learning objectives for debugging

L	Students ...
1	... define the meaning of debugging
	... name different approaches and strategies for debugging
	... enumerate the basic functionality of debugging tools
2	... explain in their own words the advantages of using a debugging tool over print-debugging
3	... use the debugger schematically, to visualize program behaviour
	... compare the behaviour and values displayed by the debugger to the own mental expectations, in order to identify any discrepancys between intended and actual system behaviour
4	... draw appropriate conclusions from the gaps between intended and actual system behaviour, as detected by using the debugger
	... narrow in the possible causes of the underlying errors, to identify the position in the source code that should be corrected
5	... evaluate whether an applied or observed debugging strategy is appropriate and efficient
6	... develop new strategies or tools to search for software errors and finally locate their causes

5.1 Methodical Competence: Abstract Thinking

Many subdomains of computer science deal with concepts that are physically intangible, and thus require a high amount of abstract thinking. For example, automating a certain behaviour by software requires to abstract from different concrete scenarios, in order to identify the underlying behavior patterns.

Therefore, especially in the context of computer science, the ability of *abstract thinking* is a highly relevant precondition for all competence levels except the first one. Level 2, *understand*, requires abstraction as a basis for identifying and applying rules and patterns of thought that are necessary to understand the underlying mechanism, instead of learning a multitude of single instances by heart (e.g. understanding the mechanism of adding natural numbers, rather than memorizing that $2 + 3 = 5$, $4 + 1 = 5$ and so on). In the context of computer science, level 3, *apply*, often involves choosing a proper strategy for abstraction. Abstract thinking is also required for assessing the appropriateness of structural interdependencies (level 4, *analyze*) and for rating the relevance of possible critical issues (level 5, *evaluate*). Finally, level 6, *create*, involves abstract thinking when new kinds of rules and principles are developed.

As abstract thinking is a highly relevant methodical competence in computer science, in Table 6 we define desired learning outcomes for abstract thinking on the different levels of Bloom's revised taxonomy. To achieve a level 4, *analyze*, or even better level 5, *evaluate*, in our students would be highly desirable. However, many students find it hard even to schematically *apply* (level 3) abstract thinking

Table 6. Learning objectives for abstract thinking

L	Students ...
1	... define the meaning of abstraction and concretion
	... define the competence of abstract thinking
	... name the basic concepts of abstraction and define their meaning
2	... explain the role of abstraction in the domain of computer science
3	... derive concrete examples or conclusions from a given simple abstraction or a simple set of rules
	... extract from a given set of simple examples the underlying set of simple rules
	... apply a given guideline to extract from a set of nontrivial examples the underlying set of rules
	... derive from a given meta model relevant conclusions for the underlying type level
4	... find an appropriate approach for deriving from a given set of examples an underlying set of rules, while using an adequate strategy
	... use a meta strategy in order to identify an adequate abstraction strategy that is suitable for solving a specific abstraction problem
5	... assess whether an abstract model or set of rules is an appropriate abstraction for representing a given set of concrete examples
	... critically evaluate whether the pursued abstraction approach is adequate and efficient in the current context
6	... devise a new meta model for a certain context
	... invent new strategies for abstraction

appropriately, e.g. when deriving a set of rules, a model or a class structure from a set of concrete examples.

5.2   Social Competence: Ability to Handle Criticism

Nowadays, software is developed as a team effort, involving intensive communication both within the development team as well as with the customer or potential future users. Especially within agile development, feedback from the customer to the development team takes place on a regular basis. As well, the team itself regularly reflects both on the achieved result and on the process that led to it. Even if everybody involved works as best they can, situations will arise which leave room for improvement. These will be addressed on a regular basis, e.g. during reviews with the customer or during retrospective team meetings.

Therefore, it is necessary for software developers to be able to handle criticism—both as a recipient and as a provider of feedback. Table 7 reflects the required competences. On their way to graduation, students should at least reach level 4, *analyze*, or even better level 5, *evaluate*.

Table 7. Learning objectives for the ability to handle criticism

L	Students ...
1	... define the meaning of the ability to handle criticism
	... name the two roles involved in the context of criticism
	... name the basic rules for giving and receiving feedback
2	... elaborate the value of receiving feedback for one's own personal development
	... explain why the basic rules for giving and receiving feedback make sense
3	... voice their feedback in accordance with the feedback rules
	... adhere to the feedback rules when receiving feedback
4	... identify a goal that they want to pursue when giving/receiving feedback
	... think about the manner in which they should voice feedback, so that it becomes acceptable for the recipient
	... mull over the feedback they received and try to understand it
5	... reflect, as feedback provider, which of the identified items of possible criticism are relevant for the goal of the current feedback situation
	... reflect, as feedback recipient, which of the received items of criticism are feasible
6	... devise a guideline and/or a scenario that helps others to practise giving/receiving feedback

Table 8. Learning objectives for self reflection

L	Students ...
1	... define the meaning of self reflection
2	... explain in their own words the role of self reflection for their own personal development
3	... follow a given guideline to reflect on their own situation, process or a self created artefact, and document their findings on the level of observed symptoms (rather than the underlying causes)
4	... identify a goal for their self reflection, i.e. a deeper insight that they aim at gaining by self reflection
	... identify the causes for the observed symptoms
	... identify specific measures that are suitable for avoiding negative symptoms in the future, or for recreating positive symptoms again, as lessons learned
5	... assess whether a given approach for self reflection is appropriate for reaching the desired goal
	... observe their own reflection process and evaluate to which extent it really takes place and whether it is suitable for reaching the desired goal
6	... devise a guideline and/or a scenario that helps others to practise self reflection

5.3   Self Competence: Self Reflection

For achieving level 5, *evaluate*, of Bloom's revised taxonomy for any kind of skill, the ability for critical reflection is highly relevant. Moreover, when evaluating self created artefacts or self executed processes, the related (but even more difficult ability) for self reflection is called for.

Furthermore, being aware of one's strengths and weaknesses is a prerequisite skill for any efficient learning process. Therefore, the ability for self reflection is highly relevant not only for a successful study process, but for life long learning as well, which is of increasing relevance in our quickly evolving world.

Table 8 specifies the competence of self reflection on the different levels. Here again, reaching level 5, *evaluate*, would be highly desirable for students, to smoothen both their academic study process and their ongoing intellectual growth in their professional life later on.

6   Conclusions and Future Work

The paper specifies learning outcomes across the different levels of Bloom's revised taxonomy, both for technical and for selected methodical, social and self competences that are relevant for achieving and employing competences in software development on a professional level.

For each competence in focus, we elaborated why it is highly relevant in professional software developers. As well, we specified in detail what kind of specific skills constitute the competence in focus on the different skill levels of Bloom's revised taxonomy, both for technical and non-technical core competences. Thus, we clarified not only the technical expertise we try to foster in our students, but also which abilities are relevant for acquiring this technical expertise in the first place, and for applying it successfully in a professional context.

We developed the initial version of these learning objectives in 2014, and revise and extend them as necessary. Since then, we communicate them explicitly in class to our students, to clarify expectations and deliverables. As well, they are the basis for the blueprint design of the formative assessments of our software development classes.

Furthermore, based on these specifications of learning objectives, we can now develop and refine approaches for teaching and learning that focus in a dedicated way on fostering both these core and the technical competences in our students (e.g. [3] for teaching abstraction).

References

1. Anderson, L.W., Krathwohl, D.R., Airasian, P.W., Cruikshank, K.A., Mayer, R.E., Pintrich, P.R., Raths, J., Wittrock, M.C.: A Taxonomy for Learning, Teaching, and Assessing: A Revision of Bloom's Taxonomy of Educational Objectives, 1st edn. Longman, New York (2001)

2. Bloom, B.S., Engelhart, M.B., Furst, E.J., Hill, W.H., Krathwohl, D.R.: Taxonomy of Educational Objectives: The Classification of Educational Goals. David McKay Company, New York (1956)

3. Böttcher, A., Schlierkamp, K., Thurner, V., Zehetmeier, D.: Teaching abstraction. In: 2nd International Conference on Higher Education Advances (HEAd 2016), pp. 357–364 (2016)

4. Böttcher, A., Thurner, V., Müller, G.: Kompetenzorientierte Lehre im software engineering. In: Software Engineering im Unterricht der Hochschulen (SEUH), pp. 33–39 (2011)

5. Chur, D.: Schlüsselkompetenzen – Herausforderung für die (Aus-)Bildungsqualität an Hochschulen. In: Stifterverband für die Wissenschaft (ed.) Schlüsselkompetenzen und Beschäftigungsfähigkeit – Konzepte für die Vermittlung überfachlicher Qualifikationen an Hochschulen, pp. 16–19. Essen, Juni (2004)

6. Hristova, M., Misra, A., Rutter, M., Mercuri, R.: Identifying and correcting java programming errors for introductory computer science students. In: Proceedings of the 34th SIGCSE Technical Symposium on Computer Science Education, SIGCSE 2003, pp. 153–156. ACM, New York (2003)

7. Humbert, L.: Didaktik der Informatik mit praxiserprobtem Unterrichtsmaterial. B.G. Teubner, 2 edn. (2006)

8. Kaczmarczyk, L.C., Petrick, E.R., East, J.P., Herman, G.L.: Identifying student misconceptions of programming. In: Proceedings of the 41st ACM Technical Symposium on Computer Science Education, SIGCSE 2010, pp. 107–111. ACM, New York (2010)

9. Schaeper, H., Briedis, K.: Kompetenzen von Hochschulabsolventinnen und Hochschulabsolventen, berufliche Anforderungen und Folgerungen für die Hochschulreform. HIS-Kurzinformation (2004)

10. Schott, F., Ghanbari, S.A.: Modellierung, Vermittlung und Diagnostik der Kompetenz kompetenzorientiert zu unterrichten - wissenschaftliche Herausforderung und ein praktischer Lösungsversuch. Lehrerbildung auf dem Prüfstand 2(1), 10–27 (2009)

11. Sirkiä, T., Sorva, J.: Exploring programming misconceptions: An analysis of student mistakes in visual program simulation exercises. In: Proceedings of the 12th Koli Calling International Conference on Computing Education Research, Koli Calling 2012, pp. 19–28. ACM, New York (2012)

12. In der Smitten, S., Jaeger, M.: Kompetenzerwerb von Studierenden und Profilbildung an Hochschulen. In: HIS-Tagung 2009 – Studentischer Kompetenzerwerb im Kontext von Hochschulsteuerung und Profilbildung, pp. 1–26. HIS, Hannover (2009)

13. Sorva, J.: The same but different – students' understandings of primitive and object variables. In: Proceedings of the 8th International Conference on Computing Education Research, Koli 2008, pp. 5–15. ACM, New York (2008)

14. Zehetmeier, D., Böttcher, A., Brüggemann-Klein, A., Thurner, V.: Development of a classification scheme for errors observed in the process of computer programming education. In: International Conference on Higher Education Advances (HEAd), pp. 475–484 (2015)

Supporting the Co-design of Games for Privacy Awareness

Erlend Bergen, Dag F. Solberg, Torjus H. Sæthre,
and Monica Divitini(✉)

Norwegian University of Science and Technology, Trondheim, Norway
divitini@ntnu.no

Abstract. Privacy is a well-known concern connected to teenagers' usage of e.g., social media, mobile apps, and wearables. However, providing proper learning in this area is challenging. Games have recently been proposed as a tool to increase awareness of privacy concerns. It is important that these games are relevant and engaging. In this paper, we present a workshop to involve teenagers in the co-design of games to promote privacy awareness, describing the workshop process together with the cards and the board that support the process. We evaluated the workshop together with students between 15–17 years of age divided in groups of 3–4 participants. Results show that all the groups were able to generate interesting game ideas and the workshop was perceived as entertaining.

Keywords: Co-design · Serious games · Privacy

1 Introduction

Privacy is an ever-growing concern. With the technological development and increase in use of connected devices, data is being collected everywhere. Terms of service are complicated, leaving people unaware of what type of data they share, with whom and what it is used for [17]. The new General Data Protection Regulation (GDPR) in Europe, in effect May 2018, addresses some of these concerns, but individuals still have to be aware of privacy issues and act accordingly in a rather complex context [4].

Teenagers are a user group for which concerns are higher. They are heavy users of digital services and might lack knowledge about data sharing and underestimate the risks. For example, a study conducted by NorSIS [14] shows that only 28,4% of Norwegian youth received training in information security in the last two years.

Serious games have recently emerged as a way for children to learn about sharing of personal data and privacy in an engaging and evoking way. Just to mention a few examples of privacy related serious games (hereafter simply games):

– *Friend Inspector*, described in [3], is a game that aims to raise the privacy awareness of Social Network Sites (SNS) users, like Facebook. The conceptual design of the game focuses on the discrepancies between perceived and actual visibility of shared items. It is a memory-like game where the player is asked to

© Springer Nature Switzerland AG 2020
M. E. Auer and T. Tsiatsos (Eds.): ICL 2018, AISC 916, pp. 888–899, 2020.
https://doi.org/10.1007/978-3-030-11932-4_82

guess the visibility of an item. To give the user a relevant context, the frame story is based around items shared on the user's own profile.

– *Master F.I.N.D.*, described in [16], also focuses on awareness about privacy risks in SNSs. The game is a fake SNS and is developed to be played individually by teenagers. A player takes the role of a web detective and attempts to solve missions through searching for information on profiles on the fake SNS. An example mission is to try to locate a person at a certain moment.
– *Google's Interland,*[1] aims at educating children in four areas of internet security: Cyber bullying, phishing, password creation and sharing awareness. The player controls a character through different games, scoring points for completing tasks, while learning about safe Internet behavior at the same time.

Most of the existing games are addressing a limited number of risks, mainly focusing on sharing of information on social media, neglecting for example emerging risks connected to data collected through sensors. The aim of our research is to investigate how to foster human-centered design of novel games for promoting awareness about privacy by providing tools to engage teenagers in idea generation. Focusing on the recognized importance of the ideation phase in any design method [6], this paper presents a card-based ideation workshop, i.e. a tool supporting the collaborative formulation of initial game concepts. The workshop, called *Privacy Game Co-design Workshop*, is intended for non-experts, i.e. users without previous knowledge on the field of privacy or formal training in design techniques, with focus on teenagers as the main target group. The proposed workshop is an adaptation of the Triadic Game Design workshop [8]. It provides: (1) a structured process to guide ideation; (2) a board to focus the contribution of the players; and (3) a set of cards to focus on different aspects of the games.

The design of the workshop was an iterative process. We evaluated its usefulness in informing and guiding idea generation during two pilots and a final evaluation with 32 participants divided in 9 groups. Data was collected through observations, questionnaires, artifact analysis, and, for the pilots, a final group interview. All material is available on request under a Creative Common License.

2 Related Work and Background

The involvement of children in the co-design of privacy games and learning material has been lately recognized as important in, for example, [1]. In this paper the authors propose the use of Collaborative Inquiry method. The work presented in this paper is instead positioned in the research that aims at using card-based approaches to promote idea generation and playful user involvement in co-design [15]. As examples, in [12] the authors propose a set of cards and a structured workshop to promote co-design of IoT systems. Similar approaches are also used in game design, as e.g., in the work connected to tangible interfaces for learning games [5], for exertion games [13]; and to

---

[1] Interland - Be Internet Awesome. Retrieved October 1, 2017 from https://beinternetawesome.withgoogle.com/

design for playfulness [11]. Cards are an effective vehicle to convert theoretical frameworks to guidelines that can be manipulated by designers [5], keeping users at the center of the design process [10, 11] and facilitating creative dialogue and shared understanding. Cards can be a source of inspiration [11], facilitate collaborative and divergent thinking by providing a medium for conversation between stakeholders and designer [2, 7], and providing a common ground [1]. As summarized in [12], card-based tools are "...(i) informative: helping to describe complex concepts to non-experts, (ii) inspirational: helping trigger and guide brainstorming and idea generation, (iii) collaborative: engaging users by helping collaboration and creative dialogue..." However, cards should not be seen as stand alone, but rather complemented by clear guidance on how to use them [13], possibly in the context of a structured workshop process. In this context, we chose the Triadic Game Design [8] workshop as a foundation for our Privacy Game Co-design Workshop. The Triadic Game Design is intended to support the design of serious games by pushing the designer to address in turn three core perspectives:

1. *Play*: how to make a game entertaining. Only considering this element would be the same as designing a regular game with no learning goals.
2. *Meaning*: how to make the game education. The game designed should provide a value beyond play itself like educating or raising awareness.
3. *Reality*: to ground the game in a specific real-world context.

In order to make a successful serious game, these three perspectives must be balanced, and they can complement each other or be conflicting. The proposed workshop is intended to have a flexible format and to adapt to different needs. In the original version of the workshop, participants are divided in groups of 3–4, and after an ice-breaking activity, they go through different assignments, the first three focusing in turn on each of the three core perspectives listed above plus a last one to bring the three elements together. For each assignment, a deck of cards is provided, identifying possible choices for the participants. In addition, a set of worksheets is used to provide questions that guide the creation of the game as well as space for recording design choices.

The Triadic game design workshop focuses on the creation of concepts rather than graphics. This is the main reason it has been chosen as starting point for the proposed approach. However, it has been adapted to target privacy and suit better to teenagers.

3   The Co-design Workshop

The Privacy Game Co-design Workshop aims to include the target group as participants in a workshop to help generate ideas for serious games focused on privacy awareness. The goal is to be able to run the workshop in a classroom-setting with groups of 3–6 people and therefore generate multiple ideas (Fig. 1). The design of the workshop has been an iterative process. The authors used the Triadic Game Design workshop as a core, and made changes to adjust the workshop time scope, audience and altered the focus from "any" problem to privacy. The resulting workshop includes (1) a structured process to guide ideation; (2) a board to focus the contribution of the

Fig. 1. Students during one of the workshops

players; and (3) a set of 30 cards helping participants to focus on different aspects of the games they are conceiving. The 30 cards are divided into 7 Reality cards, 1 Meaning card, 14 Play cards and 8 Technology cards.

Each phase should take approximately 30 min. It is difficult to set a firm time-limit on each step within the 30 min, as they are fluid and often overlap, though Step (iii) should take the most time, as it is where groups generate their ideas.

Rather than an initial ice-breaker activity like in the original workshop, the workshop includes an initial introduction to privacy. Though this initial part might be tailored, we have developed a Kahoot! quiz (https://kahoot.com/welcomeback/) and a short lecture about: What is privacy? What is online privacy? Risks of sharing personal information with other people/friends, and Risks of sharing personal information with companies or organizations through usage of services.

3.1 The Board

The original workshop provides detailed worksheet templates to document design choices. Since we aim at a shorter activity and at the involvement of teenagers, in our adaptation we decided to substitute the worksheets with a board. The board is used: (i) to scaffold the process, (ii) to collect ideas and notes during the process, and (iii) to support cooperation and interaction within the group. Because of its size (A2 format), the board enables 3–4 people to easily work around it.

At the beginning of the workshop, each group receives a board that they can write on. The board is divided in 4 areas, one for each of the workshop phases (Fig. 1, right). The areas are covered, and the groups have to discover the areas only during the related workshop phase. This is intended to help them focus. When an area is open, there are two sheets supporting the discussion. As an example, Fig. 2 shows the two sheets for the Reality phase. On one side there is a short description of the phase and the steps that have to be followed. On the other, there are some questions that are intended to trigger the discussion within the groups and an area to annotate the discussion and ideas. In the sheet they can also select if they want to address challenges connected to the private sector or related to the use of personal data by companies.

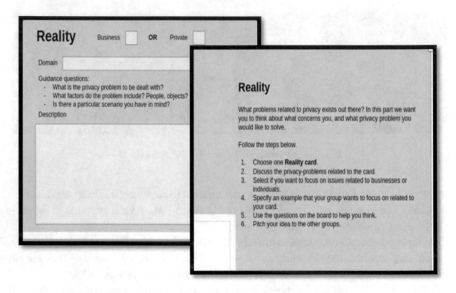

Fig. 2. Board components of *Reality*

3.2 Cards

The Privacy Game Co-design Workshop uses four sets of cards, one for each phase of the workshop.

– *Reality*. While the original workshop is open to any domain, in our workshop we focus on privacy and all the cards for reality are on privacy, each representing a different privacy scenario that can be addressed in the game. The reality cards are: Location Sharing; Smart Cities (example in Fig. 3, left); Health Devices; Activity Trackers; Social Media; Mobile App Permissions; Loyalty programs. The scenarios have been defined by analyzing cases reported in the media. The list of privacy problems is not exhaustive and can be extended to address other scenarios.

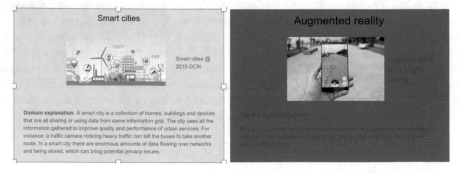

Fig. 3. Example of a reality card (left) and a technology card (right)

The description of the scenarios is, by choice, broad enough to be interpreted in different directions, but still specific enough to provide focus on privacy.

– *Meaning*. The original workshop includes a number of cards for promoting creativity around meaning. However, since the game that we aim at designing are connected to increasing awareness of privacy, we limit to the most relevant card, "Awareness and Attitude", i.e., the developed games will all focus on increasing awareness or change attitude towards data sharing.

– *Play*. The cards to support participants in thinking about different types of game are the same than in the Triadic workshop, but text has been simplified to fit better to the target group and the game examples have been updated.

– *Technology*. This deck of cards does not exist in the original workshop, but we have introduced it to promote the development of games that use a broader spectrum of technologies. Technology cards specify what kind of technology the serious game will be utilizing. Having a specific technology to design the game for may help the participants to move away from traditional PC games and promote creativity. The technology cards are: Augmented Reality (example in Fig. 3, right); Virtual Reality; Mobile; Computer; Console; Interactive Surfaces; Interactive Devices.

4   User Studies

The workshop has been evaluated through two small pilots, mainly intended to fine-tune the workshop, and then a larger evaluation. Data was collected through: a questionnaire using a 1–5 Likert-scale and focusing on fun and perceived difficulty level; artifact analysis, i.e. the annotated boards; and observations by three of the co-authors who also acted as facilitators, with individual observations discussed in the team after the workshop. For the two pilots, the study also included a semi-structured audio recorded group interview with all participants [4]. For the final evaluation, no final interview was conducted because being in a school there were more time constraints.

The participants to the studies were all teenagers in upper secondary schools. The first pilot was conducted with 3 participants who were spending two weeks at the university as part of their vocational education in ICT (Information and Communication Technology) and service design. The second pilot was conducted with 6 participants that were working at the university as part of a national program for which students in secondary schools can work one day in companies to collect money for a charity. The first group was therefore not compensated, whereas the second group received indirect compensation, circa 50 euro each, to charity. The final evaluation was conducted with two classes of a school with specialization in ICT, with a total of 32 students divided in 9 groups. The pilots were conducted at the university premises, while the final evaluation was conducted at the school. Participation of girls was very low, with only two girls attending the second pilot and 1 the final evaluation. We therefore do not perform any analysis of gender issues. The first pilot was conducted with an earlier version of the workshop. The workshop was then revised based on the results. The workshop as described in the previous section is the one resulting from this revision and it is the version that is evaluated in the second pilot and in the final study.

4.1   Results from the Pilots

During the first pilot, the 3 students were put into one group. Participants were given a first version of the board, the Privacy cards as described above, all the Play and Meaning cards in the original Triadic workshop (updated and simplified), and the Technology cards. The group was able to conceive a relevant and interesting game idea, but they did get stuck on several occasions, needing help to get back on track. They also struggled to detach their ideas from the game examples in the cards. However, the questionnaire results show that the participants enjoyed the workshop. Their answers suggest that Part 1 (Reality) was the most boring, with a fun rating of 3.33, and the most difficult to combine with the other elements. The group discussion after the workshop confirmed the observations. The main concern of the participants was the difficulty to put together all the previous steps in the final game, especially the scenario from the Reality phase. As stated by one of the participants: *"The difficult part is to make the privacy an essential part of the game while still keeping it interesting"*. Discussing the Meaning cards after the workshop, there was also a general consensus that many of the cards in the deck are difficult to understand, and that "Attitude" is the card best related to privacy risks. Many of the meaning cards wouldn't actually make sense in the given context.

As a result of the evaluation, the following changes were made:

- Participants are able to choose the Reality card (privacy scenario) they want to work with, but all the cards are presented at the beginning of the process. Combining all the elements proved too difficult, and Reality the most difficult one to incorporate. By letting the participants choose reality card it will be something they understand.
- All the Meaning cards are removed from the deck, except for the "Attitude and Awareness" to focus on the fact that the games that have to be designed are aimed at changing attitudes and increase awareness, not developing any generic skill.
- Redesign of the board to use better the available space, but also to help participants to concentrate more on the task at hand.

The participants of the second pilot were divided in two groups. Both groups were able to generate a relevant game. The process was smoother, with less breakdowns. The results from the questionnaires confirm the observations. The participants appreciated the presentation of each reality card before they selected one, as opposed to Pilot 1 where they drew a card blindly. As one participant stated:

"It was nice to be able to choose [reality card]. It made it easier to come up with interesting angles for the game. The Play part was more difficult since the genres were untraditional and we had to think outside the box."

Facilitator: *"Is that a bad thing?"* *"No, creating yet another Call of Duty2* [a successful first-person shooter game] *would have been boring. It was fun but challenging."*

In the second pilot there was no evidence that Phase 4 (working on technology and combining all previous parts) was hard. The fun-rating of part 4 was also higher than in Pilot 1. The groups felt they had sufficient time for each task, supporting the results from the first iteration. As a result of the second pilot only minor changes to the text on some cards and on the board were introduced.

4.2   Results from the Main Evaluation

The participants seemed to enjoy the workshop and worked well with the tasks, though they had to be reminded frequently to write down their ideas in the board. The different phases received increasing higher score in the questionnaire, with the last phase receiving the highest score, over 4 on average. The workshop seems to hit an appropriate difficulty level, with 23 out of 32 participants reporting the workshop to be neither easy nor hard, and only 2 experiencing it as difficult. Most of the participants also felt that they had enough time for the workshop (26 out of 32).

A general positive attitude was also observed during the pitches, during which students seemed to enjoy presenting their ideas and listening to what the other groups had done. It is however worth to note that some of the pitches were very effective in presenting the ideas, while others were harder to follow, with poorer explanation of the context. Questions had to be asked to facilitate the pitching and clarify details.

The proposed game ideas were evaluated by the three facilitators when the groups performed their final pitch. The average of these scores can be seen in Fig. 4. The facilitators independently rated the ideas based on:

- Privacy Scenario, how well defined the problem statement/scenario was. Did they think of the different roles, why it is a problem, provide an example.
- Raising Awareness, did the participants find a problem to promote awareness for? Did they find a game, and did they modify it in a meaningful way?
- Entertainment Value, did they define goals, rules, and story for the game? Did it seem like a fun game to play?
- Innovative, did the group come up with a creative new game concept? Did they combine existing concepts in an interesting way?
- Overall Impression, the subjective overall impression.

The maximum possible score was 50 points, the highest given score 39 and the lowest just above 28. Most of the scores were in the mid 30's range. Most of the groups scored high on innovative thinking, with 7 of 9 groups with a score of more than 7 out of 10.

Table 1 provides an overview of the game ideas generated during the final evaluation, specifying which cards have been used, the game concept, and the score.

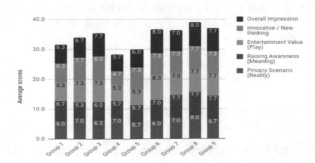

Fig. 4. The average scores for each group in the five aspects their ideas were rated.

Table 1. Table showing selected cards, game concept, and total score evaluating the game.

ID	Reality	Play	Tech.	Game concept	Score
1	Social media (Business)	Strategy	Aug. reality	The player explores the real world and using his phone with AR can hack the information of virtual companies. The information can be traded for money and other goods	31,7
2	Social media (Private)	Shooter	Virtual reality	Your job is to explore the world and detect fake profiles on Tinder. By using a shotgun you exterminate the fake users one by one	34,0
3	Social media (P)	RPG + Adventure[1]	Virtual reality	In a VR world the player takes pictures of objects and post them to social media. This can give the player fame, or have grave consequences if wrong picture is posted	35,3
4	Smart cities (B)	Survival Horror	Console	The player must survive in a smart city using stealth to not be detected by the government or hacked	28,3
5	App permissions (P)	Survival Horror	Computer	A puzzle game where the player give permission to all his personal information. If he doesn't finish the puzzle everything is posted to social media	30,0
6	Smart cities (P)	Adventure, Survival Horror	Console	A game where the state has gathered a lot of personal data about the player in a post-apocalyptic setting, and the player must prevent them from abusing it	36,7
7	Health devices (P)	Platform	Computer	Open world game, player is prompted to share private information. Can interact with other people to learn from mistakes	36,3
8	Social media + Mobile App (P)	Adventure	Computer	The player discovers that an SNS uses private information illegally and must decide what to do in a decision-based game	39
9	Smart cities (P)	Action	Computer	First person stealth game, where the player attempts to infiltrate and take down an "evil" organization that abuses personal data without giving away personal data	37

5   Discussion

The Privacy Game Co-design Workshop proved successful in supporting the co-design of serious games for privacy awareness. The results show that, in a limited amount of time, the participants were able to:

- Select and elaborate a privacy-related scenario
- Give a meaning to an existing game, i.e. turning an existing game into a game with a learning purpose
- Come up with a fun new game in a specific genre
- Reflect and combine the elements into one serious game for privacy awareness.

On the overall, the changes made to the original workshop are evaluated positively for the intended purpose. The workshop was perceived by students as an engaging activity and all the groups managed to come up with relevant ideas. As shown in Table 1, the groups produced ideas for different scenarios. It is interesting to underline that only 4 out of 9 ideas are related to social media, that is what normally students get information about. Also, 5 ideas do not use the computer as underlying technology, again increasing the potential innovativeness of the game.

Having a structured *process* proved to support generation of creative ideas. Through the different phases participants focus on different perspective of serious games and advance their design. In the pilot tests we experimented with letting the participants choose all their cards, as opposed to draw them, but feedback showed that this only lead to confusion. The participants were often excited to include different cards that did not seem to fit together, i.e., Social Media, Virtual Reality and Role-Playing Games. The resulting game idea was often very innovative and successful. That creativity permeates the entire process is also visible in the results, with 7 out of the 9 final game ideas receiving high scores on innovation.

The *cards* played their expected role of informing participants about different options, triggering discussion and idea generation, and promoting cooperation providing specific concepts for focus on.

The *board* provided a focal point for group interaction and scaffolding of the process, by providing different hints about the process as well as triggers to help the group to focus. The evaluation revealed however that the participants did not use the board as much as intended, often forgetting about the guidance questions meant to help their creative process. This might result in games that are less elaborated as well as in a more frustrating process. It is also important to note that the boards are an important outcome of the co-design workshop and are essential for designers who want to take the games further. It is therefore important that the workshop facilitator makes sure to give clear instructions and reminds participants about the proper use of the board.

Several of the games designed by the participants could be promising tools to raise privacy awareness. A challenge with advancing the ideas to game development is that they are often very complex as well as costly and difficult to realize. However, asking the participants to only create simple games is very likely to hinder their creative process and affect the final ideas. It is also important to note that the facilitators of the workshop are not necessarily looking for a final concept to implement, but rather ideas that can be combined or used as inspiration for creation of relevant serious games.

A recurring theme in the games from the workshop is to raise awareness by having in-game actions result in consequences. This applies to both negative actions, such as over-sharing of information, and positive actions, such as making good decisions. A drawback of using consequences of all actions as a mechanism to teach privacy awareness is that it requires a lot of resources in development to foresee and design all possible outcomes in the serious game.

The proposed workshop is intended to last between 2–3 h to provide an activity that can easily be integrated into a busy school day. However, the evaluation shows that an extension of the activity might be beneficial. In particular, if there is time, the facilitator might consider using more time to provide: a more extensive introduction to privacy; more time for discussion after the pitches to generate knowledge exchange among the groups; starting a class discussion among the ideas.

6 Conclusions

In this paper we presented a workshop to promote co-design of games aimed at promoting awareness of privacy among teenagers, with focus on the early idea generation. The workshop includes a structured process to be used together with a board and cards. The workshop is an adaptation of the Triadic Game Design Workshop previously proposed in the literature. In addition to a general update of the cards proposed in the original workshop, the main proposed changes include a focus on privacy through the introduction of a deck of cards capturing different privacy scenarios; the introduction of a technology perspective and related cards, to promote the design of games adopting novel interaction approaches; the introduction of a board to scaffold the process and promote cooperation. The workshop has successfully been evaluated with 32 students.

The participants of the main evaluation were all ICT students aged 15–17, with only one girl. The workshop needs therefore to be evaluated with a more diverse population. As part of our future work, we also aim at studying how the workshop can be used not only as a co-design tool, but also as a tool to promote learning of privacy in schools.

Acknowledgements. The research is co-funded by the NFR IKTPLUSS project ALerT, #270969 and by the EU project 'UMI-Sci-Ed (H2020-SEAC-2015-1). We thank the students and teachers who joined the workshop.

References

1. Brandt, E., Messeter, J.: Facilitating collaboration through design games. In: Proceedings of the Eighth Conference on Participatory Design: Artful Integration: Interweaving Media, Materials and Practices, vol. 1, pp. 121–131 (2004)
2. Carneiro, G., Li, Z.: i|o cards: a tool to support collaborative design of interactive objects. Proc. DESIRE, 357–2 (2011)
3. Cetto, A., Netter, M., Pernul, G., Richthammer, C., Riesner, M., Roth, C., Sänger, J.: Friend Inspector: A Serious Game to Enhance Privacy Awareness in Social Networks (2014)
4. Crabtree, Andy, Tolmie, Peter, Knight, Will: Repacking 'Privacy' for a networked world. Comput. Support. Coop. Work. (CSCW) **26**(4–6), 453–488 (2017). https://doi.org/10.1007/s10606-017-9276-y
5. Deng, Y., Antle, A.N., Neustaedter, C.: Tango cards: a card-based design tool for informing the design of tangible learning games. In: Proceedings of the 2014 Conference on Designing Interactive Systems, pp. 695–704 (2014)
6. Fowles, Robert A.: Design methods in UK schools of architecture. Des. Stud. **1**(1), 15–16 (1979)
7. Halskov, K., Dalsgård, P.: Inspiration card workshops. In: Proceedings of the 6th Conference on Designing Interactive Systems, pp. 2–11 (2006)
8. Harteveld, C., Van de Bergh, R.: Serious game design workshop (2009). http://resolver.tudelft.nl/uuid:df34341f-5430-4792-a9d0-c19d41a979a3. Accessed 3 Oct 2017
9. Hornecker, E.: Creative idea exploration within the structure of a guiding framework: the card brainstorming game. In: Proceedings of TEI, pp. 101–108 (2010)

10. IDEO: IDEO method cards: 51 ways to inspire design. William Stout (2003). https://www. ideo.com/post/method-cards
11. Lucero, A., Arrasvuori, J.: PLEX cards: a source of inspiration when designing for playfulness. In: Proceedings of the 3rd International Conference on Fun and Games, pp. 28–37 (2010)
12. Mora, S., Gianni, F., Divitini, M.: Tiles: a card-based ideation toolkit for the internet of things. In: Proceedings of the 2017 Conference on Designing Interactive Systems, pp. 587–598 (2017)
13. Mueller, F., Gibbs, M.R., Vetere, F., Edge, D.: . Supporting the creative game design process with exertion cards. In: Proceedings of the 32nd Annual ACM Conference on Human Factors in Computing Systems, pp. 2211–2220 (2014)
14. NorSIS: Ungdom og digital sikkerhetskultur (2017). https://norsis.no/ungdom-digital-sikkerhetskultur/. Accessed 26 Sep 2017
15. Vaajakallio, Kirsikka, Mattelmäki, Tuuli: Design games in codesign: as a tool, a mindset and a structure. CoDesign **10**(1), 63–77 (2014)
16. Vanderhoven, Ellen, Schellens, Tammy, Valcke, Martin: Educating teens about the risks on social network sites. Huelva **22**(43), 123–131 (2014)
17. Click to agree with what? https://www.theguardian.com/technology/2017/mar/03/terms-of-service-online-contracts-fine-print. Accessed 27 Nov 2017

Mobile Learning Environments Applications

Mobile Phone Usage Among Senior High and Technical School Students in Ghana and Its Impact on Academic Outcomes – A Case Study

Christiana Selorm Aggor[1], E. T. Tchao[1(✉)], Eliel Keelson[1], and Kwasi Diawuo[2]

[1] Department of Computer Engineering, Kwame Nkrumah University Science and Technology, Kumasi, Ghana
ettchao.coe@knust.edu.gh
[2] Department of Computer and Electronics Engineering, University of Energy and Natural Resources, Sunyani, Ghana

Abstract. There are currently over 7.6 billion mobile phone connections in the world, which is approximately equivalent to the world's population; as if to say that everyone in the world is hooked up to mobile telephony. By the next decade, it is estimated that the number of mobile phone subscribers would increase by a billion with 90% of this growth coming from developing countries. This exponential growth coupled with its ubiquity highlights its essentiality in all spheres of human endeavour. Despite the fact that many sectors in developing countries worldwide are embracing its usage as an effective tool, the Ghana Education Service (GES) has placed a ban on the use of mobile phones in senior high schools, technical and vocational institutions. Bearing in mind that most of these institutions provide boarding facilities for their students for long periods of time – spanning over several months; the teaching and learning experience is denied access to mobile phones over this long period. This study administered a survey to determine whether this denial has any impact on educational outcomes and what the impact would be if done otherwise. Using variations of selective and random sampling methods on a sample size of 150 respondents from the Nkoranza municipality, results from the study showed that the use of mobile phones in the Ghana educational system will have a positive impact on the teaching and learning process. Teachers and students would have easy access to up-to-date educational materials. However, the study recommends that for this to be effective and sustainable, proper usage supervision must be provided especially for students in order to avoid negative influence that stem from some social media.

Keywords: Mobile technologies · Mobile learning · Mobile phone usage · Educational institutions

© Springer Nature Switzerland AG 2020
M. E. Auer and T. Tsiatsos (Eds.): ICL 2018, AISC 916, pp. 903–913, 2020.
https://doi.org/10.1007/978-3-030-11932-4_83

1   Introduction

At the central core of development of almost all sectors in the world lies Information and Communication Technology (ICT). By providing effective processing and delivery of information via reliable communication channels, ICT has become an integral part of modern societies; changing the lives of many. In its evolution, there has been the introduction of varying gadgets that harness the power of different communication technologies to suffice its users with flexible access to up-to-date information. Most common among these gadgets are desktop computers, laptops, tablets, personal digital assistants (PDAs), smart feature phones and smartphones. Mobile phones (smart feature phones and smartphones) penetration in the world tops among the listed gadgets. Currently there are over 7.6 billion mobile phone connections worldwide; almost an equivalent of the world's population, if not more. It is estimated that by 2020 there would be an additional 1 billion mobile subscribers, 90% of which would stem from developing countries (GSMA 2016). Factors such as the standardization of mobile communication technologies like Global Systems for Mobile Communication (GSM), improved spectrum efficiency and the provision of international roaming services have contributed to this high penetration rate.

For Africa and most developing countries, mobile phone usage has dominated and is still on a constant surge over the use of other ICT gadgets like laptops mainly due to the low cost of mobile handsets, long battery life, mobility and the drastic reduction in the cost of communication via these phones as a result of keen competition among Mobile Network Operators (MNOs). These factors have made mobile phones the primary and most essential information access tool in almost all sectors of these developing countries. Aker et al. (2010) highlighted on how the provision of this service via the mobile phone has contributed immensely to Africa's economic development.

Despite these unclouded empirical evidences of its essence, some authorities and governing bodies of institutions in African developing countries have ban its use. A typical example is the Ghana Education Service – the main governing body overseeing primary, secondary, technical and vocational institutions in Ghana, has placed a ban on the use of mobile phones in these institutions. As a result of this ban the learning and teaching experience is devoid of the usage of these mobile phones, for as long as school is in session, which for most of these institutions spans over several months. However, there are advocates who strongly believe that allowing mobile phone usage in these institutions could positively affect teaching and learning outcomes and enhance academic work. This study investigates this hypothesis by reviewing literature on research conducted by various authors on issues related to this as well as carrying out a survey in the Nkoranza Municipality of Ghana. This survey solicits the views of a selected sample of affected teachers and students.

2   Literature Review

The advent of technology is changing our way of life. In Ghana, there have been strides to leverage the opportunities of technology in improving areas of health (Tchao et al. 2017a), transportation (World Bank 2018) and Governance (Tchao et al. 2017b). In the educational sector, technology has provided improved access to educational resources and the internet has played an important role in the teaching and learning process (Ofosu 2015). The Internet contains a wealth of knowledge that is available instantly upon any search. Because of this, the Internet has superseded libraries as a source for information gathering and research. Many teachers now ask students to visit specific websites to study from home, and online encyclopedias provide masses of knowledge on almost every topic imaginable. The variety of sources allows students to pursue subjects in much greater detail rather than being limited to whatever the teacher sends home.

It used to be that students that forgot work, missed a lecture or couldn't remember an assignment were out of luck until talking face to face with a teacher or a classmate. However, the Internet allows instantaneous connection to their classmates and teachers. Improving communication between students and teachers allows teachers to assist students without having to stay after class. It also allows for students to have greater efficiency when working on projects with their peers when everyone cannot attend or asking for clarification when something is unclear.

A number of universities, such as Harvard, Yale and Stanford, have opened up free courses on a variety of subjects that are accessible to anyone for free. These typically come in the form of lectures on video, but some also have notes attached. This means there is easy access to plenty of free lectures without emptying your bank account to pay tuition. The Internet also makes education accessible to impoverished communities. The "Granny Cloud," for example, made use of Skype as a number of volunteers, mostly retired teachers, read stories aloud over Skype to children in India to teach them how to read.

In Ghana, many people do not have access to wired internet access. Many internet subscribers rely on mobile telephony operators for access. Many Ghanaian students in universities use smartphones and other mobile digital devices to access the internet (Ofosu 2015). The situation is entirely different when it comes to internet access in high schools and vocational institutions as internet facilities are almost not existent. The very few institutions that have access, restrict the usage of internet by students whiles on campus. In most of these Senior high school institutions, the usage of mobile phones is banned mainly because the Ghana Education Service (GES) is of the view that most students in these institutions would use their mobile devices for recreational activities such as playing games, listening to music or just building up their social media presence. Advocates of the ban are firm in their conviction that the improper interaction between tutors and student on the usage of mobile phones, coupled with the failure of some educators to take into consideration the distinct learning styles of students could have a negative impact on academic performance.

Over the last couple of decades, several researchers have carried out a number of studies on the effect of mobile phone usage in education. Some notable ones would be reviewed in this section. Ryan De Kock et al. (2016) in a research on mobile device usage in educational institutions in South Africa, highlighted some of key reasons that have contributed to some educational institutions allowing students and teachers to bring their own mobile devices to school. Among these reasons included the fact that these personally owned mobile devices complemented the institutions' efforts to provide ICTs for teaching and learning especially where these institutions cannot provide all the necessary state of the art technology for all its students and teachers. He also added that these devices (phones) provide a means of getting access to online educational resources via the internet.

Morphitou (2014) also carried out a survey in Cyprus using a descriptive quantitative methodology in determining the effects of the use of smartphones among students on their education and social life. Results from 124 respondents on what they used their mobile phones for was skewed in suggesting that in the coming years the smartphone was going to replace the laptop and desktop computer on many grounds. This goes to buttress the earlier mentioned point raised by Ryan De Kock on institutions allowing mobile phones as a measure to complement the efforts of ICTs especially in developing countries where the provision of these computing devices for the teaching and learning process is often found to be insufficient. Being convinced by the results of her study on the great potential mobile phones hold in promoting the success of the educational institutions, Ria suggested that teachers and instructors encourage their students to leverage these benefits.

Faradina et al. (2016) having realized this potential, as shared by the previous author, developed a mobile phone application by name Mpianatra which used efficient pedagogical techniques to transmit knowledge to students while sustaining their interest. This application provided functionalities to improve student-lecturer interaction even in crowded classroom environments or situations where students felt disposed to avoid notice. Mpianatra also provided features to increase student participation and collaboration. It also allowed teachers and instructors to easily disseminate questions and auto-grade them after they have been answered by students with little or no effort. By using this application, the teaching and learning process continued even after the class schedule was over. A survey conducted on the usage of Mpianatra from 100 respondents evidently showed that it was massively accepted by both teachers/instructors and students. Pinar Kirci et al.' (2015) research on game-based education with smartphones produced similar results. It was realized that the learning process was more of fun and devoid of stress and bashfulness when this technique of gamification was used to relay and assess understanding of theoretical concepts.

Despite these benefits delivered by mobile devices to educational institutions, there have been concerns raised on the proclivity of students and/or instructors to misuse these handsets for improper activities during school time; thus, making the mobile phone more of a tool for distraction than of education. This is worth considering, especially when these smart devices are capable of accessing any resource on the internet; be it constructive or destructive. This is probably one of the major reasons why GES has banned the use of mobile phones in educational institutions in Ghana. Some advocates propose that heads of schools could forge partnership with the mobile phone service providers to enable them to monitor how the students use the phones while on campus.

Mahesh G. et al. (2016) proposed a solution which deals effectively with this area of concern. Their research on a smartphone integrated smart classroom proposed a learning environment where students' usage of mobile phone applications is controlled and monitored. Their solution incorporates the use of an android application installed on the smartphone which allows the student to only access whitelisted applications as well as puts the phone into airplane mode; blocking all external connections. Where there is the need to access resources online, a connection is provided to the institution's network which is configured to allow access to only whitelisted online resources. Kadry et al. (2017) result was in the same area of providing an innovative way of controlling how effectively students use their mobile phones while in the classroom. It is also worth mentioning that some mobile device manufacturers such as Apple and Kindle over the past few years have taken pragmatic steps to introduce applications and operating systems which allow teachers and instructors to audit and monitor

In the light of these proposed solutions Quist and Quarshie (2016) suggest that the Ghana Education Services' restrictions on the use of mobile phones in schools should be reviewed to enable students to take advantage of the technology. The authors further entreat instructors, lab technicians and teachers in these educational institutions to keep themselves updated on current technologies so that they can effectively update the implemented control measures. Based on findings from these reviewed literatures the study went further to carry out a survey to test the initial stated hypothesis.

3   Research Methodology

3.1   Study Area and Sample Population

The study area for the study was the Nkoranza South municipality. The population of Nkoranza South is currently 147,301. The selected institutions for the study were Nkoranza south Senior High and Technical Schools in Nkoranza municipality. The sampled schools are accredited by the Ghana Education Service (GES). The study was a descriptive study since it portrayed the accurate profile of the sampled respondents.

For the purpose of this study, the population consisted of teachers and students of Nkoranza South Senior High and Technical Schools in the Nkoranza municipality. The target populations consisted of all teachers and students who were available at the time of questionnaire administration and interview. A simple random and convenience sampling was used to ensure the possible representation of all section of the target population.

3.2   Sampling Size Determination

For the purpose of having more precise estimate proportion of the target population, a total of one hundred and fifty (150) respondents from the target population was selected for the study. This sample size was chosen based on the expense of data collection, and the need to have sufficient statistical power. The sampled population comprised of municipality one hundred (100) students and fifty (50) teachers. The selection of the target population was randomly done to give a fair representation of the general view of on perception of mobile devices usage.

3.3 Sampling Technique

Probability and non-probability sampling were selected as the target respondents' identity were undefined and therefore there was no sampling frame of target respondent. The one hundred (100) students were sampled using simple random sampling method which belongs to non-probability sampling design. Also, selective and purposive sampling which belongs to probability sampling was used in sampling the teachers who have in-depth knowledge on this subject matter.

3.4 Research Design and Data Collection Instrument

It was acknowledged that direct observation was a plausible social research technique that could be used for data collection in this research, however due to the constraints of time and the need to avoid subjectivity, this approach was not considered. The approaches used include document analysis where related literature on the subject matter was reviewed as seen in the literature review section above. It also includes the use of face-to-face conducted interviews and administered questionnaires. The questionnaires were exclusively used to solicit the views of the respondents on the research hypothesis. The questionnaires were given out to the teachers and students of Nkoranza Senior High and Technical School. The questionnaires captured both closed and open-ended questions. The research design could therefore be said to be mainly quantitative in nature and objective. In addition, the study critically evaluated the cost-benefit analysis of the entire study by conducting a feasibility analysis within the study area.

3.5 Sources of Data

In administering the questionnaires and interviews various questioning techniques were used to produce primary data from respondents and interviewees. In addition to the above reviewed literature, other secondary data sources consulted were guidelines and published materials which gave the research team information on the impact of labour turnover on the productivity of Nkoranza South municipality. Data from the field was processed and interpreted appropriately to make an objective analysis. Where necessary, editing was done to correct errors, check for non-responses, accuracy and correct answers. Interpretation and data coding was done to facilitate data entering and a comprehensive analysis. This research used descriptive statistics as the main means of analyzing the datasets.

4   Results

Most of the teachers who participated in the study were males. The male respondents constitute 74% of the total respondents. In addition, 67% of the students were males while 33% were females. From the results collected on the academic qualification of the teachers, it was observed that majority of the teachers were degree holders, thus 29 of the teachers, representing 58% are degree holders whiles 21 teachers representing 42% are diploma holders. The age bracket of the students was also looked into. For the students, their age bracket was categorized into 15–16 years, 17–18 years, 19–20 years and those above 20 years. 60% of the students had obtained the ages of 17–18 years, 24% had also attained the ages of 19–20 years, 10% had obtained ages above 20 years while those who had obtained ages between 15–16 years form the remaining 6%.

Teachers were asked to indicate whether they think mobile phone usage influenced students' academic performance. 78% of the teachers indicated that the usage of mobile phones affects students' academic performance while 22% indicated that mobile phones usage does not affect students' performance. The results indicate that, majority of teachers believe that the usage of mobile phones does influence students' academic performance. In addition, 76.9% of the 78% of the teachers who agreed that mobile phone usage influence students' academic performance indicated that mobile phone usage by students negatively affect their performance while 23.1% think otherwise. Results from the study summarized in Table 1 show that about 50% of the teachers agreed that the use of mobile phones makes them more innovative and updated in the current educational trend. A further 42% of teachers strongly agreed to this perception. On the other hand, only 2% of the teachers strongly disagreed with the statement whereas 6% of them disagreed. Majority of the teachers believed that the usage of mobile phones and other connected devices could positively affects their teaching since it assists them to undertake research so that they can be abreast with the current trend of education.

Table 1. Teachers' responses on the usage of mobile phones on students' academic performance

Statement	SA	A	NEU	D	SD
It makes teachers more innovative and updated in the current educational trend	42% (21)	50% (25)	–	6% (3)	2% (1)
Reduces the problem of text book shortage	10% (5)	20% (10)	8% (4)	30% (15)	32% (16)
Can be a burden to parents	40% (20)	30% (15)	–	16% (8)	14% (7)
Will be more biased towards the poor and those in deprive needy	48% (24)	20% (10)	2% (1)	20% (10)	10% (5)

Majority of the teachers (32%) strongly disagreed with the statement that students' use of mobile devices reduces the problem of text book shortage. An additionally 30% further disagreed with the statement while 8% remained neutral. 20% of the teachers however agreed the adoption of mobile devices could reduce the problem of text book shortage while a further 10% strongly agreed to it. When asked to indicate whether students' use of mobile phones could be a financial burden to parents, majority of the teachers either strongly agreed (40%) or agreed (30%). The study had 16% of the teachers disagreeing while 14% strongly disagreed.

The study was also set to determine the type of technology, content and mobile devices that can be used for Senior High school students. All the respondents who participated in this study were given a number of questions. To begin with, the students were asked to indicate the type of mobile device they would prefer to use in the classroom. This included personal computers, laptops, tablets, PDAs and notebooks. Figure 1 summarizes the results. From the results obtained, 27.5% of the students would prefer to use Laptops in the classroom while 24.3% preferred tablets, 23.9% of them preferred mobile phones while the remaining 15.5% preferred notebooks. Additionally, students were asked to indicate the extent to which they agreed or disagreed to the proposal of delivering educational content through mobile devices. Table 2 summarizes the results. On whether they would prefer learning with animations, majority of the students indicated their likeliness for this proposal by indicating their strong agreement (16%) or agreement (54%) with the proposal. However, slightly greater than a quarter of them (26%) disagreed with the proposal whiles 2% strongly disagreed.

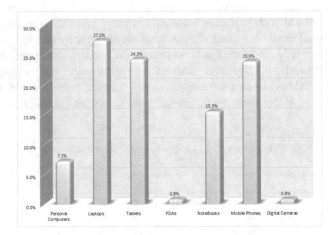

Fig. 1. Type of mobile devices students prefer to use in class

Table 2. Students' responses on mobile devices used in school

Statement	SA	A	NEU	D	SD
I like learning with animations	16% (16)	54% (54)	2% (2)	26% (26)	2% (2)
I like teaching delivered through audio devices	13% (13)	27% (27)	7% (7)	43% (43)	10% (10)
Learning is best understood when it is delivered in video mode	51% (51)	23% (23)	7% (7)	17% (17)	2% (2)
I prefer to learn from interactive whiteboard	54% (54)	31% (31)	9% (9)	6% (6)	–
I always want to access information from the internet through e-learning	54% (54)	41% (41)	3% (3)	2% (2)	–

The study further sort to find out if students would like the option of their course content being delivered through audio recording devices. Majority of the students (43%) disagreed with the statement that they like teaching delivered through audio devices. An additional 10% further strongly disagreed with the statement. Meanwhile, 27% agreed to the statement whiles 13% strongly agreed to it. When asked to indicate whether learning is best understood when it is delivered in video mode, majority of the students either strongly agreed (51%) or agreed (23%). The study had 17% of the students disagreeing while 2% strongly disagreed. On the other hand, when students were asked whether they prefer to learn from interactive whiteboard, 54% strongly agreed to it, 31% agreed, 9% stayed neutral while 6% disagreed. On whether they would prefer to access information from the internet through e-learning, 54% strongly agreed, 41% agreed, 3% remained neutral while 2% disagreed

In order to evaluate the interest of the teachers to deliver lessons through a particular mode, teachers were asked to indicate the extent to which they preferred to deliver the content of their lessons. The results of this evaluation are summarized in Table 3. A majority of the teachers (68%) agreed that they will like to deliver lessons with animations. 28% did not like the idea of delivering lessons with animations whereas 2% of them remained neutral.

Table 3. Teachers responses on how they prefer contents to be delivered

Statement	SA	A	NEU	D	SD
I will like to deliver lessons with animations	28% (14)	68% (34)	4% (2)	–	–
I like teaching delivered through audio devices	24% (12)	44% (22)	20% (10)	12% (6)	–
Learning is best understood when it is delivered in video mode	72% (36)	24% (12)	–	4% (2)	–
I prefer to teach from interactive whiteboard	20% (10)	68% (34)	4% (4)	8% (4)	–
I always want to access information from the internet to assist my students	52% (26)	40% (20)	4% (2)	4% (2)	–

When the teachers were asked about their preference for delivering their lectures through audio recording devices, 68% of the respondents indicated that delivering lessons through audio recording and playback devices would benefit students and they would prefer this mode of lesson delivery. On the contrary, 12% disagreed that they would not like to deliver lessons in audio form. 20% of the teachers however remained neutral. When it comes to delivering lessons in video mode 72% of the teachers indicated their preparedness to incorporate video in their classroom activities whiles the remaining teachers indicated their unwillingness to incorporate videos in their classroom activities. Teachers also expressed their views on the usage of interactive whiteboard, 88% of the respondents would prefer using it to teach by strongly agreeing to it adoption while 8% of the teachers would not like to use it.

On course information access, majority of the teachers (52%) strongly agreed that they always want to access information from the internet to assist their students. 40% also agreed that accessing information from the internet is vital and they always wish to do so while 4% remained neutral. Only 4% of the respondents disagreed on this assertion.

4.1   Implication of Study Results

Majority of the teachers who agreed that mobile phone usage influences students' academic performance indicated that mobile phone usage by students negatively affect their performance. This finding conflicts with that of (Goundar, 2011) who recognized that mobile phone use has become a pervasive communication tool among youth culture and created recommendations and guidelines on using this communication technology to support a sustainable learning culture with students. He further stated that mobile phone Short Message Service (SMS) prompting was very successful in both enhancing student participation and motivating them to meet deadlines for assessment. Both of these contribute to improved learning.

From our research, half of the teachers agreed that the use of mobile phones makes teachers more innovative and updated in the current educational trend. This they attributed to the fact that it assists them to undertake research so that they can be abreast with the current trend of education. However, teachers strongly disagreed with the statement that students' use of mobile phones reduces the problem of text book shortage while accepting the claim that the use of mobile phones could bring financial burden to parents since most of the parents in the survey municipality are subsistence farmers and as such lived under less than $4 a day.

4.2   Conclusions and Recommendations

The internet is vital resource of education and digital education by extension depends mainly on internet. A country's economic development depends of their digital system. From the literature review section, it has been established that if any country or academic institutions should use digital education for their educational system, that country will be on course in achieving a digital system. From the analysis of this study, it was realized that the use of ICTs in the educational system could positively help in the teaching and learning process. For the teachers, mobile devices could help them in gaining access to research material and updated information to be able to deliver accurate information to students when teaching. However, for the students, if not properly supervised, it could go a long way to affect them negatively in studying. This is a result of the many distractions available on mobile devices. Access to the social media platforms and other websites via the internet and the numerous apps available can reduce the amount of time they spend researching on educational material and studying with the devices.

The study however recommends provision of ICT gadgets purely for educational purposes to the Senior High Schools and Technical institutions in Ghana to enable students be abreast with the current technological trends. These gadgets should be restricted in the websites they can access and the applications that can be installed on them to ensure they do not provide unnecessary distractions to students.

References

Aker, J.C., Ksoll, C., Lybbert, T. (2010). ABC, 123: can you text me now? The impact of a mobile phone literacy program on educational outcomes. https://www.cgdev.org/sites/default/files/1424423_file_Aker_Mobile_Phone_Literacy_FINAL.pdf

De Kock, R., Futcher, L.A.: Mobile device usage in higher education institutions in South Africa, 2016 Information Security for South Africa (ISSA). Johannesburg **2016**, 27–34 (2016). https://doi.org/10.1109/ISSA.2016.7802925

Faradina, F.H., Rabevohitra, F.H.: Mpianatra, a mobile phone application to improve the quality of college education, 2016 International Conference on Advances in Human Machine Interaction (HMI), Doddaballapur, pp. 1–10 (2016). https://doi.org/10.1109/hmi.2016.7449186

Goundar, S.: What is the potential impact of using mobile devices in education? In: Proceedings of SIG GlobDev Fourth Annual Workshop, Shanghai, China, 3 Dec 2011, pp. 6 (2011)

GSMA: statistics on 1 billion new mobile subscribers and currently 7.6 billion connections (2016). https://www.gsma.com/mobileeconomy/archive/GSMA_ME_2016.pdf

Kadry, S., Roufayel, R.: How to use effectively smartphone in the classroom. In: 2017 IEEE Global Engineering Education Conference (EDUCON), Athens, 2017, pp. 441–447 (2017). https://doi.org/10.1109/educon.2017.7942884

Kirci, P., Kahraman, M.O.: Game based education with android mobile devices. In: 2015 6th International Conference on Modeling, Simulation, and Applied Optimization (ICMSAO), Istanbul, pp. 1–4 (2015). https://doi.org/10.1109/icmsao.2015.7152220

Mahesh, G., Bijlani, K.: A smart phone integrated smart classroom. In: 2016 10th International Conference on Next Generation Mobile Applications, Security and Technologies (NGMAST), Cardiff, pp. 88–93 (2016). https://doi.org/10.1109/ngmast.2016.31

Morphitou, R.N.: The use of smartphones among students in relation to their education and social life. In: 2014 International Conference on Interactive Mobile Communication Technologies and Learning (IMCL2014), Thessaloniki, 2014, pp. 315–319 (2014). https://doi.org/10.1109/imctl.2014.7011155

Ofosu, W.K., Tchao, E.T.: From the environment to the classroom: A Sub-Saharan African scenario. In: Proceedings of IEEE International Conference on Teaching, Assessment and Learning for Engineering: Learning for the Future Now, TALE 20147062560, pp. 341–345 (2015)

Quist, S.C., Quarshie, H.O.: The use of mobile phones among undergraduate students-a case in Ghana. S. Am. J. Acad. Res. 1–7 (2016) (Special edition May 2016)

Tchao, E.T., Diawuo, K., Ofosu, W.K.: Mobile telemedicine implementation with WiMAX Technology: a case study of Ghana. J. Med. Syst. **41**(1), 17 (2017a)

Tchao, E.T., Keelson, E., Aggor, C., Amankwa, G.A.M.: e-government services in Ghana: current state and future perspectives. In: Proceedings of Computational Science and Computational Intelligence, Las Vegas, 14th Dec 2017 (2017b)

World Bank (2018). White Paper-Toolkit on Intelligent Transport Systems for Urban Transport, Case Study–Ghana. https://www.ssatp.org/sites/ssatp/files/publications/Toolkits/ITS%20Toolkit%20content/case-studies/accra-ghana.html. Accessed May 2018

Teaching and Learning Microbiology for Engineers in a Digital World: The Case of the FIT Courses at the Tecnologico de Monterrey, Mexico

Josefina Castillo-Reyna[1], Rebeca M. García-García[1],
Alicia Ramírez-Medrano[1], Maribell Reyes-Millán[2],
Blanca R. Benavente-Vázquez[2], Claudia D. Chamorro-Urroz[2],
and Jorge Membrillo-Hernández[1(✉)]

[1] Escuela de Ingeniería y Ciencias, Tecnológico de Monterrey, Monterrey,
Mexico
jmembrillo@tec.mx
[2] Dirección de Educación Digital, Tecnológico de Monterrey, Monterrey,
Mexico

Abstract. The availability of online courses has increased dramatically; however, the effectiveness of online teaching remains uncertain. The Tecnologico de Monterrey has implemented the Tec21 educational model that includes online teaching with the modality of the FIT courses (Flexibility, Interaction and Technology), a type of well-designed courses using the CANVAS web-based platform and using ZOOM as a tool for communication with students. To compare the objective learning outcomes between a FIT course and a face-to-face course (Microbiology for Engineering), an analysis was made between these two types of courses, with the same or different instructor. Our results clearly indicate that the grades of the students enrolled in the FIT groups were higher than those of the classroom courses. These results were confirmed when the historical record files of both types of teaching were examined. Satisfaction surveys however revealed that students prefer face-to-face courses to online courses. Our data implies that it is necessary to work more on the tools of the distance-learning courses to make them extremely attractive and interesting for students.

Keywords: Learning · Online teaching · Traditional classes · FIT courses

1 Introduction

The advent of the Internet and the design and implementation of new (and increasingly easy to use) digital communication tools, have transformed the way in which information is transmitted, stored and shared, eliminating the difficulties of time and location [1]. This has impacted the world of academia and many universities have designed new courses, taking into account approaches that lower barriers and increase the access of many potential students [1]. Accordingly, the Instituto Tecnológico de Estudios

© Springer Nature Switzerland AG 2020
M. E. Auer and T. Tsiatsos (Eds.): ICL 2018, AISC 916, pp. 914–922, 2020.
https://doi.org/10.1007/978-3-030-11932-4_84

Superiores de Monterrey (ITESM) has launched the *Tec21* Educational Model that bases its success on improving competitiveness by boosting abilities and developing the skills required in different professional fields. The *Tec21* educational model is based on 4 components that allow the formation of leaders capable of successfully facing the challenges of the 21st century: (1) *Challenge-based learning*: It exposes the student to real problems, allowing the development of transforming leadership skills, making it more competitive in today's world [3], (2) *Inspiring teachers*: Every teacher is interested in the student, challenging, guiding and empowering his development, (3) *Memorable experience*: Enforcing the students involvement in a global, diverse and multicultural learning community, (4) *Flexibility*: Offering students options on what, how, when and where of their professional training process can take place. Part of this last component are the newly designed on-line Flexible-Internet-Interactive with synchronic Technological communication courses (FIT courses).

The flexibility and convenience in time and location for both instructor and student are valuable only if the courses facilitate student learning [4]. According to previous studies there is no significant difference between online and face to face courses [3–11], however most of the studies were carried out by comparing courses that were not identical or the evaluation instrument differed [5, 11, 13]. In addition, Siemens et al. [12], reported the current status of distance, blended and online learning, concluding that well-designed courses are crucial for greater adoption of online learning. While these results are positively reflected in online and blended learning in general, they do not examine combined and online learning outcomes in science courses, specifically those at the undergraduate level [2, 14]. Here we report on a study on teaching an undergraduate-level Microbiology course for Engineers using a new platform designed at ITESM: The FIT courses. In this comparison we analyzed the very same Microbiology course given in two modalities: The traditional *In person*, face-to-face (hereafter "classroom") and the new FIT course.

2   Methods

2.1   Study Purpose

The purpose of this study was to compare both didactic strategies for the Microbiology course taught by the same teacher (Teacher A, TA). Additional data were provided by a different teacher (Teacher B, TB) giving the classroom course. All courses were taught in the same semester, the topics were reviewed in parallel in the same time frame. Courses were measured by: (1) Percentage of student withdrawals, (2) Average of exam grades, (3) Average of research exercises scores, and (4) Student satisfaction with the instructor, the course and self-perception of obtaining knowledge.

2.2   Academic Settings and Study Population

The Microbiology class consists of two weekly sessions of 90 min each for 16 weeks. This course belongs to the first third of engineering professional careers (mainly biotechnology engineering). The age range of students taking this class is 18–21 years.

The classroom courses had 23 students (TA) and 26 (TB) while for the FIT course 23 students were enrolled (Total n = 72). The students registered in the FIT course belonged to 4 different ITESM campuses in Mexico.

The contents of the course include a historical perspective of Microbiology, the properties of different microbes, identification and classification techniques, medical and biotechnological perspectives, cases of bioremediation, the role of microorganisms in human health and microbial ecology. Both, classroom and FIT courses used the same text books, course notes, learning objectives, similar evaluation criteria, same core activity (creation of a microbiological consultancy that solved 3 real-life cases), and identical exercises and presentations. All contents, directions and strategies of the course were uploaded into the CANVAS platform (http://canvas.instructure.com), a teacher-friendly environment to keep everything in order. It is very important to note that the previously designed CANVAS-based calendar of sessions was strictly respected in both systems. Both courses had three exams, two partial exams (Master Quizzes) and a final exam. The exams were 95% multiple choice and 5% short answer. Generally 20% of the questions requires a higher level of synthesis and application of knowledge and 10% of the questions come from text readings.

2.3   Similarities and Differences Between Courses

Classroom students received lessons with the teacher present using exactly the same PowerPoint presentations, exercises, assignments, and assessments as the FIT course where the students used the ZOOM tool (https://zoom.us) for communication. We have to point out that the FIT courses are synchronous, which implies that the length of the sessions for both teaching strategies was similar. Previous studies have made this a key issue on the interpretation of the outcomes [14]. In addition, the students of both types of courses had access to off-the-calendar advisory sessions. A key difference between the courses was the fact that each class of the FIT course was recorded and immediately uploaded on the CANVAS platform and available for review by the students. The classroom classes were not recorded. It is important to point out that both the classroom course and the FIT course used the REMIND tool (http://www.remind.com) for communication between the teacher and the students. This application is important to send short messages to all or to a particular student, who requires it. Should collaborative work was required (break-out rooms and team-assignment) a specific period of time was assigned. Table 1 shows the similarities and differences between both courses.

2.4   Background and Training of the Course Instructors

The instructors of the courses were two Ph.D. in Biotechnology, with more than 20 years of teaching experience. Before starting up the course, both teachers received specific training in an immersion week with the goal of designing contents for the FIT course that were exactly the same for the classroom course. This training is a collegial work using different templates to guarantee the quality of the courses, likewise, different strategies are carried out to achieve the construction of all of the contents of the

Table 1. Instructional elements of the FIT (online) and classroom microbiology courses

	FIT	Classroom	Similarities/Differences
Exams	2 Partial 1 Final	Same	The questions are identical between courses
Quizzes	4 Academic 2 On the use of technological tools	4 Academic	Quizzes have the goal to reinforce every module of teaching
Homework	Weekly homework	Same	The opening of every class was the review of the homework
Learning objectives	Defined in the FIT course training session	Same	All quizzes and exams questions are derived from the learning objectives
Presentations	Power point presentations of assignments or homework	Same	Team or individual work was presented to the whole group in both cases
Course text	Two text books as support	Same	There were the same books as support for the classes
Evidences portfolio	Created at the end of every module. This folder was digital	Same but this folder was created with all the homework, exercises and all other reports	The creation and updating of the folder have significant value of the final grade
Lecture material	All prepared beforehand by the designing team with the help of the expert instructional designers of the ITESM	Same	The FIT course students had access to every recorded classes. The classroom session were not recorded
Enrollment number	23 (TA) 26 (TB)	23 (TA)	The three courses were offered during the 2018 spring semester
Team work, case assignment	Work in teams was carried out in and outside the class	Same	Collaborative work is a transversal competence that is developed in all engineering careers at ITESM

subject. Once the course is finished, a final review of it on the CANVAS platform with expert instructional designers from the area of digital programs of ITESM is performed and training sessions are set up.

2.5   Indicators

An anonymous student opinion survey (ECOA) was made to detect potential differences between the two groups of students. These ECOAs are applied in two moments, after the fourth week of instruction and at the end of the course (after week 15). In addition to this, an additional survey was held on what most pleased the students and what they liked least.

3   Results

3.1   Classroom vs. FIT

During the spring 2018 semester, TA was assigned with a classroom and a FIT groups; on the other hand, TB was assigned with a classroom group. In both types of class, the contents, quizzes, exams, schedule of each subject and didactic strategy were determined prior to the beginning of the courses. Table 2 shows the objective learning outcomes. The average grades in the first partial exam that included the subjects of a historical reflection of the microbiology, microscopy, techniques of identification of microorganisms and microbial metabolism was of 89.68 and 91.00 in the classroom classes, in comparison with the FIT course that was 96.38. Similarly, in the case of the second partial, which included the topics of classification of the microorganisms with emphasis in the environmental microbiology, the average grades were 93.00 and 84.19 in face-to-face courses and 91.38 in the FIT course. It is important to note that in this part of the course a challenge on bioremediation is developed. In the final exam, whose subjects include industrial, food and biotechnological microbiology, the students of the classroom obtained the highest score of 96.15 compared to the students of the FIT course who reached 82.04. Interestingly, 96.5% of the students in the face-to-face course passed (higher than 70/100), in contrast to 100% of the FIT course. One of the classroom class had a 3.85% withdrawal rate, compared to 0% of the FIT course (Table 2).

Table 2. Objective measurements of learning: course grades (maximum 100) and withdrawal rates (SD was less than 20% in all cases and it is omitted for clarity)

	Classroom Teacher A	Classroom Teacher B	FIT Teacher A	Historical Classroom AVG[a]	Historical FIT AVG[a]
First partial	91.50	89.68	96.38	72.32	93.42
Second partial	93.45	84.19	91.38	86.91	89.24
Final exam	93.35	93.50	82.04	ND	84.97
Percentage of students passing the course (above 70%)	100%	96.15%	100%	93.75%	100%
Withdrawal rate	0%	3.85%	0%	ND	0%

[a]It corresponds to all data available for the same course for at least 8 groups of at least 19 students

Our data showed that the students who took the FIT course had better grades than those who took the classroom course. To further explore this, we check the available historical record of grades from classroom groups of the Microbiology subject at the Bioengineering Department of the ITESM Mexico City Campus and the Microbiology FIT groups of the ITESM Monterrey Campus (to compare unrelated students). As shown in Table 2, the students of the classroom groups had consistently lower grades than the FIT groups, which supports the results in our experimental design. In the same line, the levels of withdrawal and approval of the course in the historical records were also similar to those obtained in our experimental design strongly suggesting that regardless the teacher, a well-designed course like the FIT course has a higher level of achievement as judged by the grades reached by the students.

Our data showed that the students who took the FIT course had better grades than those who took the classroom course. To further explore this, we check the available historical record of grades from classroom groups of the Microbiology subject at the Bioengineering Department of the ITESM Mexico City Campus and the Microbiology FIT groups of the ITESM Monterrey Campus (to compare unrelated students). As shown in Table 2, the students of the classroom groups had consistently lower grades than the FIT groups, which supports the results in our experimental design. In the same line, the levels of withdrawal and approval of the course in the historical records were also similar to those obtained in our experimental design strongly suggesting that regardless the teacher, a well-designed course like the FIT course has a higher level of achievement as judged by the grades reached by the students.

3.2 Satisfaction Surveys

An anonymous student opinion survey (ECOA) was made to detect potential differences between the two groups of students. The results are shown in Table 3. Students in the classroom felt that they had clearer explanations than those taking the FIT course (9.12/9.28 compared to 8.79; Q1), in addition classroom students felt that the interaction with the teacher and the feedback sessions were fundamental during the course (9.54/9.94 vs. 8.89, Q3). They also commented positively on the follow-up to the specific cases of study where they were challenged in their acquired knowledge (9.52/9.83 vs. 8.63; Q2), in the same way, the students of the classroom felt intellectually challenged by the course although at a lower level (9.20/9.67 vs. 8.58; Q5). Finally, the role as a learning guide, the teacher of a face to face course was better evaluated than the FIT course (9.52/9.39 vs 8.84; Q6).

4 Discussion

The availability of online courses has increased dramatically in the last few years, however, the effectiveness of online teaching remains uncertain. In the case of science education and especially engineering, online courses are complex since considerably interaction is required between the teacher and the students. Nowadays, there is an increasing demand for flexibility in time and space to take classes. In this work, the effectiveness of the FIT courses was analyzed comparing them with the same course

Table 3. Comparison of the anonymous ECOA surveys applied to both, FIT and classroom Microbiology courses students in the semester Jan-May 2018 at the Tecnologico de Monterrey. Results from Teacher A (TA) and Teacher B (TB) are indicated (SD was less than 20% in all cases and it is omitted for clarity)

		N	Mean (Max 10.0)
Q1. Regarding the methodology and learning activities (the teacher gave me clear and precise explanations, innovative means and techniques or technological tools that facilitated and supported my learning)	Classroom	25	9.12 (TB)
		26	9.28 (TA)
	FIT	23	8.79 (TA)
Q2. Regarding the understanding of concepts in terms of their application in practice (I solved cases, projects or real problems, I did internships in laboratories or workshops, visits to companies or organizations, or interacted with people who work applying the topics of the class)	Classroom	25	9.52(TB)
		26	9.83 (TA)
	FIT	23	8.63 (TA)
Q3. Regarding the interaction with the teacher and the advice received during the learning process (he supported me to answer questions, the teacher was available in previously agreed sessions and schedules, there was a respectful and open learning environment)	Classroom	25	9.54 (TB)
		26	9.94 (TA)
	FIT	23	8.89
Q4. Regarding the evaluation system (a set of tools was used that gave me feedback on my strengths and weaknesses in the course based on policies and criteria established in a timely manner)	Classroom	25	9.52(TB)
		26	9.50 (TA)
	FIT	23	8.89 (TA)
Q5. Regarding the level of intellectual challenge (motivated me and demanded that I give my best effort and fulfill quality in benefit of my learning and my personal growth)	Classroom	25	9.20 (TB)
		26	9.67 (TA)
	FIT	23	8.58
Q6. Regarding its role as a learning guide (it inspired me and showed commitment to my learning, development and integral growth)	Classroom	25	9.52 (TB)
		26	9.39 (TA)
	FIT	23	8.84 (TA)
Q7. Would you recommend a friend to take classes with this teacher?	Classroom	25	8.92 (TB)
		26	9.67 (TA)
	FIT	23	8.63 (TA)

taken in a classroom. The online courses have traditionally been sessions of consultation, review and monitoring of tasks on contents that are stipulated previously. This, in many cases, forces students to do research work or self-learning. Universities increasingly design specific online courses with modern tools that facilitate learning. The case of FIT courses is a special one, each FIT course is a product of ITESM's Digital Program Department. FIT courses belong to the exploration stage of an undergraduate engineering program, based on a pedagogical model, placing the student in challenging experiences where he builds his learning assuring academic quality.

They also standardize knowledge for all nationwide campuses of the Tecnologico de Monterrey. The FIT courses are Flexible to be able to take classes from any point that has a stable connection to the internet, Interactive, because students take their 2 classes (90 min each) per week live, have advice, as well as multimodal feedback, with use of the latest Technology for interaction. Although the contents of a classroom course are identical to a FIT course, the teaching technique is not. To design a FIT course, a team of experts from the direction of Digital Education Department of the ITESM is brought together to a series of face-to-face (ZOOM) work sessions, this includes the design teachers, pedagogical advisors, academic coordinators and administrators of the production of the courses. All the contents of the course are programmed on a WEB platform under a strict schedule for sessions. The peculiarity of this course is that it has a backbone (core) activity that guides the students during the course and goes, through the challenges that it presents, developing the contents of the course. In the case of Microbiology, the core activity consisted in creating student teams as part of a fictitious MicroSquad Inc Consultancy that solved three interesting microbiology challenges, one for bioremediation, one for public health and another for food consumption. The main technological tools used are the platform of the course: CANVAS Instructure (https://canvas.instructure.com/) and the live interaction tool for the classes: ZOOM (https://zoom.us/).

Our didactic experiment was designed to examine the effectiveness of the FIT courses with respect to the face-to-face courses, we compared the results of two groups, one with 23 students who took the FIT course and another group of 26 students that took the classroom course, both instructed by the same teacher (TA) in separate sessions during the 2018 spring semester. Clearly the results indicated that the students of the FIT course had better grades than those of the classroom courses. For the comparison to be validated, another face-to-face group with a different teacher (TB) but strictly following the times and strategies of the original FIT course was included, interestingly, same results were observed, the students of the FIT course achieved better grades than the face-to-face courses. This was further reinforced by the revision of the historical archives of achievement of the FIT courses (only two previous semesters) and of the classroom courses (four previous semesters). Clearly, the knowledge acquired by the students of the FIT course was significantly higher than the face-to-face courses.

It is noteworthy that in the student satisfaction surveys, we observed the opposite, the students feel more comfortable with the face-to-face classes and appreciate the present teacher more than the FIT on-line courses. This maybe because the FIT courses are new and we are in early times to establish the potential of these well- structured and well-designed courses. When the students were asked to indicate the best and the worst of a FIT course, most of the opinions (90%) identified the flexibility as the best feature, on the other hand, what they liked the least was the instability of internet connection and the difficulty of doing teamwork with colleagues who are in different campuses.

Our educational research on the efficiency of the FIT courses indicates that although students obtain better grades, we are still ignorant of the total potential of these courses, we must work more on the design with new tools, to make the FIT courses more attractive to students and integrate spaces for recent technological advances such as augmented reality, hologram teachers or even remote blackboard. In any case, this type of courses helps to reduce the gap between professors (digital migrants) and students (digital natives).

Acknowledgements. We deeply acknowledge all the personnel of the School of Engineering and Sciences and the Department of Digital Programs of the ITESM for their support throughout all the steps of this experimental set up.

References

1. Hansen, J.D., Reich, J.: Democratizing education? Examining access and usage patterns in massive open online courses. Science **350**, 1245–1248 (2015)
2. Means, B., Toyama, Y., Murphy, R., Bakia, M., Jones, K.: Evaluation of evidence-based practices in online learning: a meta-analysis and review of online learning studies, p. xvi, xvii. US Department of Education. Office of Planning Evaluation, and Policy Development. Policy and Program Studies Service (2010). ED-04-CO-0040
3. Membrillo-Hernández, J., et al.: Challenge based learning: the case of sustainable development engineering at the Tecnologico de Monterrey, Mexico City campus. In: Auer, M., Guralnick, D., Simonics, I. (eds) Teaching and Learning in a Digital World, ICL 2017. Advances in Intelligent Systems and Computing, vol. 715. Springer (2018). https://doi.org/10.1007/978-3-319-73210-7_103
4. Schoenfeld-Tacher, R., McConnell, S., Graham, M.: Do no harm: a comparison of the effects of on-line vs. traditional delivery media on a science course. J. Sci. Educ. Technol. **10**, 257–265 (2001)
5. Biel, R., Brame, C.J.: Traditional versus online biology courses: connecting course design and student learning in an online setting. J. Microbiol. Biology Educ. **17**, 417–422 (2016)
6. Allen, I.E., Seaman, J.: On-line report card: tracking online education in the United States (Rep.) Sloan Consortium (Sloan-C) (2016). http://onlinelearningconsortium.org/read/online-education-united-states-2015
7. Barbeau, M.L., Johnson, M., Gibson, C., Rogers, K.A.: The development and assessment of an online microscopic anatomy laboratory course. Anat. Sci. Educ. **6**, 246–256 (2013)
8. Collins, M.: Comparing web, correspondence, and lecture versions of a second-year, non-major biology course. Br. J. Educ. Technol. **31**, 21–27 (2000)
9. Johnson, M.: Introductory biology online: assessing outcomes of two student populations. J. College Sci. Teach. **31**, 312–317 (2002)
10. King, P., Hildreth, D.: Internet courses: are they worth the effort? J. College Sci. Teach. **31**, 112–115 (2001)
11. Lunsford, E., Bolton, K.: Coming to terms with the online instructional revolution: a success story revelaed through action research. Bioscene **32**, 12–16 (2006)
12. Siemens, G., Gasevic, D., Dawson, S.: Preparing for the digital university: a review of the history and current state of distance, blended and online learning 97, 120 (2015). http://linkresearchlab.org/PreparingDigitalUniversity.pdf
13. Somenarain, L., Akkaraju, S., Gharbaran, R.: Student perceptions and learning outcomes in asynchronous and synchronous online learning environments in a biology course. MERLOT J. Online Teach. Learn. **6**, 353–356 (2010)
14. Reuter, R.: Online versus in the classroom: student success in a hands-on lab class. Am. J. Distance Educ. **23**, 151–162 (2009)

Educational Model for Improving Programming Skills Based on Conceptual Microlearning Framework

Ján Skalka and Martin Drlík[✉]

Constantine the Philosopher University in Nitra, Tr. A. Hlinku 1,
949 74 Nitra, Slovakia
{jskalka,mdrlik}@ukf.sk

Abstract. The teachers of programming ask often students to develop complete programs in the early stages of the course. This strategy is inadequate for many students because learning programming is a complicated process. Taxonomies of educational objectives, such as Bloom's and its derivatives can be an excellent source to define and validate proposed educational models developed for teaching programming not only at the introductory programming level at the universities but also for teaching quite complex programming tasks, which require specialized skills and technologies. Several learning approaches and taxonomies from the teaching programming point of view are analyzed in the paper. Subsequently, individual phases of the selected taxonomies are mapped to the interrelated parts of the proposed conceptual model of microlearning framework prepared in the university environment. Finally, their mutual consistency and contribution to the teaching programming theory are discussed.

Keywords: Educational taxonomies · Micro-learning · Programming languages · Teaching · Bloom's taxonomy

1 Introduction

The digitalisation of the society and the automation of many processes bring new opportunities and types of jobs. The number of employees employed in IT sector is continually growing. The employees urge that there is an increasing lack of IT specialists, mainly in the field of software development, data analysis, and data science.

This problem can be solved by increasing the interest of students in IT study programs or by decreasing the number of students, who leave the study too early due to the learning failure in some courses. While the first solution requires the engagement of all stakeholders, who influence the students (parents, teachers, schools, friends), in the second case, the students have already decided to study IT study programs. Therefore, it is "easier" to invest time and resources to designing a suitable educational concept, which will be able to increase the probability, the student will successfully graduate and will have an appropriate knowledge and skills required by the job market.

However, young people are not nowadays interested in spending time by listening to the lectures or watching long-lasting educational videos. They prefer immediate use

© Springer Nature Switzerland AG 2020
M. E. Auer and T. Tsiatsos (Eds.): ICL 2018, AISC 916, pp. 923–934, 2020.
https://doi.org/10.1007/978-3-030-11932-4_85

or application of the obtained knowledge and skills. They require the options to learn anytime and anywhere, not only at the university. Moreover, they prefer selecting the knowledge and skills, which usefulness they can imagine or prove in a short time.

Innovative educational models and frameworks, which try to minimise the obstacles and fit the requirements and expectations of a new generation of students should utilize the well-known learning approaches and develop required skills and knowledge with respect to the known taxonomies. Therefore, the main aim of the paper is to map the individual parts of the selected taxonomies to the proposed educational model of microlearning framework developed for teaching programming at the university [1].

The paper has the following structure. The next section analyses several learning approaches and pedagogical theories related to the topic of teaching programming. Simultaneously, the related taxonomies are shortly discussed in the third section. Considering their strong features, the fourth section deals with mapping the individual parts of the selected taxonomies to the proposed educational model of microlearning framework developed for teaching programming at the university in detail. Finally, the last section discusses the proposed educational model from different points of view.

2   Educational Models Behind the Teaching Programming

The teaching of programming is a very complicated educational process. Educational theories and taxonomies are considered useful tools in developing learning objectives and assessing student's attainment, but they are not applicable in this area directly. According to Looi and Seyl [2] programming requires students not only to understand the relevant theory but also they should be able to apply it to solving real problems.

Different researchers use different approaches to teaching programming based on different pedagogical theories, didactical frameworks or educational approach. Some works tried to simplify the complexity of knowledge and skills acquisition process. On the other hand, other described and mapped it very precisely.

The ADRI (Approach, Deployment, Result, and Improvement) model was first known internationally as PDCA (Plan Do Check Act) developed in the 1920s by Walter Shewhart. It is originally a quality assurance model used extensively in the education and business sectors [3]. The four stages to enhance students learning outcomes in the introduction programming (IP) course was defined in [4]:

1. **A**pproach – covers problem-solving skills (pseudocode and flowchart).
2. **D**eployment – emphasises programming knowledge (syntax and semantics).
3. **R**esult – explains input, output, and process used to solve a problem statement.
4. **I**mprovement – emphasises different programming constructs.

The authors used the development environment, which supported all presented stages, and concluded that ADRI approach in the teaching and learning process of the IP course is beneficial because of providing all the necessary skills (problem-solving strategies and programming knowledge) required by the novices in comprehending the programming concepts.

The ARCS model of motivation proposed in [5] comprises four elements: **A**ttention, **R**elevance, **C**onfidence, and **S**atisfaction. Model is based upon the macro theory of motivation and instructional design developed by Keller and expectancy-value theory (by Tholman and Lewin) that assumes that people are motivated to engage in the activity if it is perceived in accordance with the satisfaction of personal needs (the value aspect), and if there is a positive expectancy for success (the expectancy aspect).

The objective of the model is to assist curriculum design or improve teaching, emphasising the triggering of learners' motivation through elements mentioned above to stimulate students' learning. From the perspective of teaching designers, inspiring learners' curiosity, enhancing students' confidence, and improving learners' satisfaction in their learning outcomes will cause learners to work harder and lead to a virtuous cycle [6].

In recent years, numerous studies have introduced the ARCS model of motivation (not only) into programming. The main methods employed have been visual graphics or questions to trigger learners' learning motivation [6–9].

Problem-based learning (PBL) derives from McMaster University in the 1960s and expands via medical education to other educational areas [10]. It is based on the theory that learning is a process in which the learner actively constructs knowledge. The theory underlying PBL is the constructivism – a cognitive approach to learning with an emphasis on the importance of the learners' previous knowledge. Kolmos's [11] characteristics of PBL can be summarised as follows:

- Ill-structured, complex problems that are often drawn from the real-world provide the focal points and act as stimuli for the course, curriculum or program.
- Learning is student-centered.
- The instructor takes the role of a supervisor, as a coach or facilitator.
- Learning is realised in small groups of students who analyse, study, discuss and propose solutions to (possibly) open-ended problems.
- Self and peer assessment enhance learner assessment.

Bransford et al. [12] defined the IDEAL model, which described the educational process in five steps:

- **I**dentification of the problem.
- **D**efinition of the problem with precision.
- **E**xploration of strategies to reach the problem solution (based on previous knowledge and experiences).
- **A**ction, in the sense of the execution of the previously planned.
- **L**earn (or Looking back) relative to the observation of the effect of the carried through actions and learning according to the evaluation of the results of these actions.

PBL is popular as an effective approach constructed computational and critical thinking, abstraction, generalisation and transferability, and problem-solving skills [13]. Many researchers [2, 14, 15] publish positive results, but some applications of PBL did not reach strong effects [16] because of insufficient entry level of educates. Using PBL is an appropriate tool in the later stages of teaching programming.

Game-based learning (GBL) is usually associated with some motivation theory. ARCS [9] presents game-based learning environment using the ARCS model in Table 1.

Table 1. The formula for the study's educational material [9]

Attention	A game-based design that is attractive to students; Cute characters that resonate with students; A fieldwork-style educational design that concentrates on student learning
Relevance	Useful content that assists students in tackling their academic tasks
Confidence	Varying difficulty levels that enable students to play both easy and difficult games; Several hints to assist students in reaching the correct answer; A practical experience that leads students to actually using a technique in a library context
Satisfaction	Students feel a sense of accomplishment when a task is completed; Students experience slight difficulty in completing the game in order to inspire learning; Students gain knowledge of the library's services and come to appreciate its usefulness, even without borrowing a book

Miljanovic and Bradbury [17] used GBL in teaching debugging by the ROboBUG (a serious game intended to be played by first-year computer science students who are learning to debug for the first time). The results showed a positive impact on the learning process of novices.

Virtual Worlds bring virtual environments that are able to engaged students in programming by controlling objects similar to real ones. The benefits are availability from anywhere and no need for any other devices. Popular and verified environments are Greenfoot, Alice, Scratch, Field and Meadow, etc. These tools usually remove the syntax of a programming language and present a simple interface through drag and drop interactions [18].

An interesting approach is presented in [19]. The authors used 3D environment for teaching programming in Prolog. Students and instructors are able not only to explore virtual word but can construct new representations in it, modify the existing settings and also participate in the problem solving through their avatars.

Game-themed programming is a little bit different approach where abstract programming concepts are taught by exploring small game applications. The main aim is not to teach game programming, but help students to understand programming concepts through simple game assignments [20].

3   Taxonomies

Taxonomies of educational objectives are used worldwide to describe learning outcomes and assessment results, reflecting a student learning stage. Usually, they divide educational objectives into three domains: cognitive, affective and psychomotor [21]. In [22] used Bloom's taxonomy and implemented a cognitive-based approach to the first two years of a computing degree. They used course micro-objectives mapped to specific levels in Bloom's taxonomy.

An additional way for taxonomies application is to use them in designing teaching materials and assessments. Structuring materials help students to move through a taxonomy, structuring assessments divided to a wide range of levels, help them to gain more in-depth knowledge and closer engagement with this material. Some universities have also gone further and developed frameworks to successfully leverage the learning outcomes and competencies in a systematic way when designing, delivering or revising a course within the program [23–25].

Bloom's taxonomy is the classification system used to define and distinguish different levels of human cognition [26] based on six main categories: Knowledge, Comprehension, Application, Analysis, Synthesis, and Evaluation.

This one-dimensional continuum has been revised by Anderson [27] and extended to two-dimensional model: the fact that any objection would be represented in two dimensions immediately suggested the possibility of constructing a two-dimensional taxonomy table (Table 2).

Table 2. Revised Bloom's taxonomy [27].

Knowledge	Cognitive processes
1. Remember	Recognizing, Recalling
2. Understand	Interpreting, Exemplifying, Classifying, Summarizing, Inferring, Comparing, Explaining
3. Apply	Executing, Implementing
4. Analyze	Differentiating, organizing, Attributing
5. Evaluate	Checking, Critiquing
6. Create	Generating, Planning, Producing

The Knowledge dimension would form the vertical axis of the table, whereas the Cognitive Process dimension would form the horizontal axis. The intersections of the knowledge and cognitive process categories would form the cells [28].

Some researchers applied Bloom's taxonomy or Revised Bloom's taxonomy in the design of computer science or introduction programming [29–31].

Solo taxonomy (Structure of the Observed Learning Outcomes) is a general theory which was not originally designed to be used in a programming context defined in [32]. This taxonomy helps teachers or content producers create and optimise the content for the students. It endeavours to identify the nature of that content and the structural relationships within that content. The content must be designed to assess knowledge and cognitive skills [23].

Lister [33] says that traditional approach in teaching programming jumps to the fifth and sixth levels of Bloom's Taxonomy of Educational Objectives when these last two levels depend upon competence in the first four levels. Other authors have used the taxonomy to describe how coding is understood by novice programmers and mapped levels to levels in SOLO taxonomy [34, 35]. Thompson [36] mapped the SOLO taxonomy to a programming course, and in particular, using this taxonomy to help students understand the grade that they have been assigned.

A two-dimensional adaptation of Bloom's taxonomy – The Matrix Taxonomy is presented in [23]. It represents a more practical framework for assessing learner capabilities in computer science and engineering. Though many benefits of Bloom's taxonomies, their principal weakness is that the levels do not appear to be well ordered when used to assess practical subjects such as programming. This taxonomy is based on the complexity of intrinsic characteristics of computer science: problem-solving, domain modelling, knowledge representation, efficiency in problem-solving, abstraction/modularity, novelty/creativity, categorisation, and communication skills with experts in other domains, adoption of good practice in software engineering.

The model reflects the fact that comprehension of program code and the ability to produce program code are two semi-independent capabilities. Students who can read programs may not necessarily be able to write programs of their own. Also, the ability to write program code does not imply the ability to debug it.

Visualization of this distinction and the semi-independent skills of reading and writing program code is projected into two-dimensional matrix with an adaptation of Bloom's taxonomy. The dimensions of the matrix represent the two separate ranges of competencies: the ability to understand and interpret an existing product (program code), and the ability to design and build a new program. Different students take different "learning paths" in the matrix. Some students are able to read and debug code at first, and some students are able to write their own code before they are able to read or debug unfamiliar ideas.

Figure 1 presents different learning paths and mapped programming activities. Mapping is feasible and results in fairly complete coverage of the grid. Furthermore, most of the activities are general enough to be immediately applicable to other fields of engineering [23].

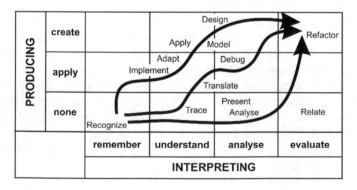

Fig. 1. Learning paths and localization of programming activities [23]

4 Mapping Levels of Taxonomies to the Proposed Microlearning-Based Framework

The long-term aim of programming teaching is to acquire the knowledge and skills needed to solve real-world problems through the appropriate programming language. Problem-solving is always a part of technology courses and sometimes a part of IP where its placement may not always be successful [16]. Gomes and Mendes [13] reported that novices try to solve the problem immediately, spending little time in its interpretation while experts come close to the solution through a process of successive refinements.

Therefore, we decided to divide a proposed educational model of teaching programming into two levels:

- IP courses are dedicated to mastering computational thinking and create a basis of programming language. Their content is widely used for different computer science programs.
- Technology courses are focused on specific technologies (web, server, mobile, database, IoT, etc.). This type of courses teaches students to use technologies and solve real problems.

IP courses must be prepared with respect to different students' learning paths presented in Fig. 1, and in accordance with ARCS model – motivation, engagement and satisfaction are the critical aspects of prevention the learning failure. These characteristics are involved via interactive assignments, immediately feedback and gamification elements that create an evitable part of the model. They support achieving the objectives, gaining badges and placing in the rankings.

We described conceptual framework of microlearning-based training for improving programming skills (Fig. 2) covering needs of IP courses with interactive microlearning lessons and using automated assessment of source code [1].

The central part of the framework is presentation module used for communication with the user via web or mobile application. This module reads content activities, interactive activities and activities for programming source code grouped into lessons in a database and present them in the attractive form. Students can discuss every activity and comments content, questions, and tasks. Students' success is expressed by the number of points earned by solving assignments. Students are ordered in the leaderboard and in addition to points, they can also earn badges associated with special types of activities.

Competitions allow students to compete with each other (answer to set of questions, write source code, etc.) and students in the role of experts – can define the new content of interactive activities or new competition tasks. Creation activities are rated, and students can earn points this way too.

Learning analytics module is intended to analyse students' behaviour – identify students with learning embarrassment, identify unfit content and activities and measures effectiveness of the educational process. Presented modules of framework cover elements of Bloom's taxonomy as well as programming activities in the matrix mentioned by Fuller [23].

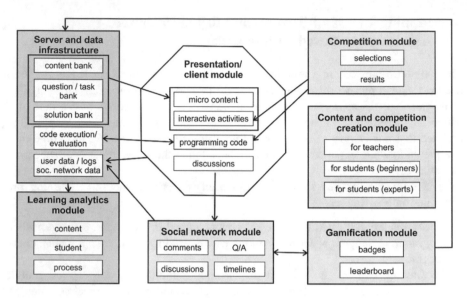

Fig. 2. Simplified framework structure

Moving between levels of taxonomies is provided by the content structure, and personalisation of the content selecting and ordering. The student with better reading skills starts lessons with tasks focused on reading and filling source code and the questions of learning content, while students who preferred code-writing can solve programming problems at first.

Solving simple and artificial problems corresponds to the content of IP and usually represents a syllabus of the first two semesters of computer science study programs (e.g. in Java it is: fundamental + objects, GUI, data structure, programming methods, threads, etc.).

Framework activities can be mapped to the first four phases of Bloom's taxonomy as follows:

- Remember – student obtains basic knowledge using suitable closed micro-learning units, which are connected into logical units. Their understanding by the student, as well as the duration of the knowledge, is evaluated on a sporadic basis;
- Understand – student completes source code, answer questions of programming theory, chooses the right form of the proposed algorithm, statements accuracy, syntax analysis, manually executes source code and writes programming codes, which are evaluated using methods like automated tests, manual statements executing, etc.;
- Apply – student solves school-based problems by means of writing own software programmes. The evaluation of their accuracy is automated. The user receives personalised feedback. Moreover, he/she can ask for additional automated help or guidance;
- Analyse – student solves predefined problems and design of programs and data structures for more complex tasks;

After the transition through four phases, students are ready for solving more difficult and complex problems. The next ideas can be started at the end of IP or at the beginning of the technology courses. Students add new content to the system and suggest new activities enjoying the acquired knowledge and skills - continuous system development is ensured.

Therefore, the following courses are focused on the technologies. The idea of course realisation is based on preparing students to long-life learning. Students receive assignments and some knowledge base. Subsequently, they have to develop the application or set of applications to solve assignment. This approach copies situations of real life in employment. Students can discuss, communicate in teams and with the tutor. They are continually encouraged to use given technologies. The recommended approach is to use principles of PBL. Mapping to the last two phases of Bloom's taxonomy is as follows:

- Evaluate - the student is involved in the learning process of his/her peers, creates assignments, writes automated tests, discusses in the community, evaluates the ideas and solutions of others;
- Create - the student is involved in the process of educational environment development, creates and modifies modules in the courses of his/her study and work on problems of the real world using approach and procedures of the corporate environment.

Student gradually passes through all phases of Bloom's taxonomy and all roles in microlearning framework.

5   Conclusions

The presented model of the IT specialists' education uses an educational environment for improving programming skills based on the microlearning, automated programming code evaluation, and modern learning environment. The model closely relates to the individual phases of the Bloom's taxonomy for the cognitive domain, which is based on active using of the specialised educational environment, as well as active participating of the students in educational content development. This model is considered suitable because it can cover almost all contemporary requirements for teaching future IT specialists in the fields of programming, software development, and database and information systems.

Application of microlearning-based training in higher education raises several new questions about the didactical design of learning activities, effective methodologies as well as technical implementation issues. On the other hand, teaching programming languages provide a wide range of educational resources, strategies, and methods suitable for the development of critical and innovative thinking. Some of these activities are possible to design as microlearning units, but many involved problems are out of microlearning time interval. It is crucial to complement micro-learning activities with program writing tasks, for improving the successful result.

A current trend requires that the IT graduates should not have specific highly specialised skills and should not know particular technology. On the contrary, the job market requires that the graduates are able to self-motivate and continually study new approaches and technologies.

The universities should be therefore a place, where the students obtain enough knowledge for life-long learning, develop students' learning habits, and transfer the responsibility for their further education on themselves. The weighted combination of the features of the work-based approach, involving students in the real-world projects and situations, which require appropriate utilising of provided study resources, should be considered as the suitable starting point for this process.

The proposed model of education of IT specialists is in its pilot phase. The prototype is available, as well as the initial set of educational content. The realisation of the proposed solution should have several impacts. In the short term horizon, a developed educational platform will provide a set of innovative educational tools. As a result, the overall quality of training and education of IT students will increase. The portion of students at risk of learning failure will decrease. In a long-term horizon, a higher level of IT knowledge and skills obtained by the target group will improve its readiness for the job market.

Acknowledgement. The research for this paper was financially supported by grant KEGA - 029UKF-4/2018 Innovative Methods in Programming Education in the University Education of Teachers and IT Professionals.

References

1. Skalka, J., Drlik, M.: Conceptual framework of microlearning-based training mobile application for improving programming skills. In: Auer M., Tsiatsos T. (eds.) Interactive Mobile Communication Technologies and Learning, IMCL 2017. Advances in Intelligent Systems and Computing, vol. 725 (2018)
2. Looi, H.C., Seyal, A.H.: Problem-based learning: an analysis of its application to the teaching of programming. In: Conference Problem-Based Learning: An Analysis of its Application to the Teaching of Programming, pp. 68–75 (2014)
3. Razvi, S., Trevor-Roper, S., Goodliffe, T., Al-Habsi, F., Al-Rawahi, A.: Evolution of OAAA strategic planning: using ADRI as an analytical tool to review its activities and strategic planning. In: Proceedings of 7th Annual International Conference on Strategic Planning for Quality Assurance and Accreditation of Universities and Educational Arab Institutions (2012)
4. Sohail, M., Coldwell-Neilson, J.: Comparison of traditional and ADRI based teaching approaches in an introductory programming course. J. Inf. Technol. Educ. Res. **16**, 267–283 (2016)
5. Keller, J.M.: The use of the ARCS model of motivation in teacher training. In: Aspects of Educational Technology, vol. 17, pp. 140–145 (1984)
6. Chang, Y.-H., Song, A.-C., Fang, R.-J.: The Study of Programming Language Learning by Applying Flipped Classroom (2018)
7. Alhazbi, S.: ARCS-based tactics to improve students' motivation in computer programming course. In: 10th International Conference on Computer Science & Education (ICCSE), pp. 317–321 (2015)

8. Alhassan, R.: The effect of project-based learning and the ARCS motivational model on students' achievement and motivation to acquire database program skills. J. Educ. Pract. **5**(21), 158–164 (2014)

9. Saito, Y., Kaneko, K., Nohara, Y., Kudo, E., Yamada, M.: A game-based learning environment using the ARCS model at a university library. In: 2015 IIAI 4th International Congress on Advanced Applied Informatics (IIAI-AAI), pp. 403–408. IEEE (2015)

10. Albanese, M.A., Mitchell, S.: Problem-based learning: a review of literature on its outcomes and implementation issues. Acad. Med. 52–81 (1993)

11. Kolmos, A.e.a.: Problem Based Learning (2007)

12. Bransford, J.D., Stein, B.S.: The IDEAL Problem Solver: A Guide for Improving. WH Freeman and Company, New York (1984)

13. Gomes, A., Mendes, A.C.: An environment to improve programming education. Rousse. In: Proceedings of the 2007 International Conference on Computer systems and Technologies. ACM, p. 88 (2007)

14. Yip, W.: Generic skills development through the problem-based learning and information technology. In: Hamza, M.H., Potaturkin, 0.1., Shokin, Yu.1. (eds.) Proceeding of Automation, Control, and Information, pp. 72–80 (2002)

15. Huang, Y.-R., Cheng, Z., Feng, Y., Meng-Xiao, Y.: Research on teaching operating systems course using problem-based learning. In: 2010 5th International Conference on Computer Science and Education (ICCSE), pp. 691–694. IEEE (2010)

16. Ambrosio, A.P., Costa, F.M., Almeida, L., Franco, A., Macedo, J.: Identifying cognitive abilities to improve CS1 outcome. In: Frontiers in Education Conference (FIE), 2011, pp. F3G1–F3G7. IEEE (2011)

17. Miljanovic, M.A., Bradbury, J.S.: RoboBUG: a serious game for learning debugging techniques. In: Proceedings of the 2017 ACM Conference on International Computing Education Research, pp. 93–100 (2017)

18. Kazimoglu, C., Kiernan, M., Bacon, L., Mackinnon, L.: A serious game for developing computational thinking and learning introductory computer programming. Procedia-Soc. Behav. Sci. **47**(2012), 1991–1999 (2012)

19. Vosinakis, S., Anastassakis, G., Koutsabasis, P.: Teaching and learning logic programming in virtual worlds using interactive microworld representations. Br. J. Edu. Technol. **49**(1), 30–44 (2018)

20. Sung, K., Hillyard, C., Angotti, R., Panitz, M., Goldstein, D., Nordlinger, J.: Game-themed programming assignment modules: a pathway for gradual integration of gaming context into existing in-troductory programming courses. IEEE Trans. Educ. **54**(3), 416–427 (2011)

21. Gomes, A., Mendes, A.J.: Bloom's taxonomy based approach to learn basic programming. In: EdMedia: World Conference on Educational Media and Technology. Association for the Advancement of Computing in Education (AACE), pp. 2547–2554 (2009)

22. Doran, M.V., Langan, D.D.: A cognitive based approach to introductory computer science courses: lesson learned. In: Proceeding SIGCSE 1995 Proceedings of the 26th SIGCSE Technical Symposium on Computer Science Education, pp. 218–222. ACM (1995)

23. Fuller, U., Johnson, C.G., Ahoniemi, T., Cukierman, D., Hernán-Losada, I., Jackova, J., Lahtinen, E., Lewis, T.L., Thompson, D.M., Riedesel, C., Thompson, E.: Developing a computer science-specific learning taxonomy. ACM SIGCSE Bull. **37**(4), 152–170 (2007)

24. Tovar, E., Soto, Ó.: Are new coming computer engineering students well prepared to begin future studies programs based on competences in the European Higher Education Area? In: 39th Frontiers in Education Conference, 2009, FIE 2009, pp. 1–6. IEEE (2009)

25. Bekki, J.M., Dalrymple, O., Butler, C.S.: A mastery-based learning approach for undergraduate engineering programs. In: Frontiers in Education Conference (FIE), pp. 1–6. IEEE (2012)

26. Bloom, B.S.: Taxonomy of educational objectives. Cogn. Domain **1**, 20–24 (1956)
27. Anderson, L.W., Krathwohl, D.R., Airasian, P.W., Cruikshank, K.A., Mayer, R.E., Pintrich, P.R., Wittrock, M.C.: A Taxonomy for Learning, Teaching, and Assessing: A Revision of Bloom's Taxonomy of Educational Objectives, abridged edn. Addison Wesley Longman Inc, New York (2001)
28. Krathwohl, D.R.: A revision of Bloom's taxonomy: an overview. Theory Pract. **41**(4), 212–218 (2002)
29. Machanick, P.: Experience of applying Bloom's taxonomy. In: Proceeding Southern African Computer Lecturers' Association Conference, pp. 135–144 (2000)
30. Azuma, M., Coallier, F., Garbajosa, J.: How to apply the Bloom taxonomy to software engineering. In: 11th Annual International Workshop on Software Technology and Engineering Practice, 2003, pp. 117–122. IEEE (2003)
31. Johnson, C.G., Fuller, U.: Is Bloom's taxonomy appropriate for computer science? Proceedings of the 6th Baltic Sea conference on Computing Education Research: Koli Calling. ACM, pp. 120–123 (2006)
32. Biggs, J.B., Collis, K.F.: Evaluating the Quality of Learning. Academic Press, New York (1982)
33. Lister, R.: On blooming first year programming, and its blooming assessment. In: Australasian Conference on Computing Education, pp. 158–162. ACM (2000)
34. Lister, R., Simon, B., Thompson, E., Whalley, J.L., Prasad, C.: Not seeing the forest for the trees: novice programmers and the SOLO taxonomy. ACM SIGCSE Bull. **38**(3) (2006)
35. Jimoyiannis, A.: Using SOLO taxonomy to explore students' mental models of the programming variable and the assignment statement. Themes Sci. Technol. Educ. **4**(2) (2013)
36. Thompson, E.: Holistic assessment criteria: applying SOLO to programming projects. In: Proceedings of the 9th Australasian Conference on Computing Education, vol. 66. Australian Computer Society (2007)

IT and Educational Environment
of an Engineering University

Galina Ivshina[1], Yakov Ivshin[2], Tatiana Polyakova[3(✉)],
and Viacheslav Prikhodko[3]

[1] Kazan National Research Technical University A.N. Tupolev – (KAI), Kazan,
Russia
gvivshina@kai.ru
[2] Kazan National Research Technological University, Kazan, Russia
Ivshin@kstu.ru
[3] Moscow Automobile and Road Construction State Technical University
(MADI), Moscow, Russia
kafedral01@mail.ru, prikhodko@madi.ru

Abstract. The paper is devoted to the description of the experience of creating electronic information and educational environment (hereinafter - EIOS) in an engineering university, the application of distance educational technologies in engineering education abroad and in Russia. It also describes the main problems of creation and introduction of the system of e-learning in engineering education in preparation of engineers of the XXI century. In Russia, the use of EIOS in the educational process is provided for by Federal state educational standards (FSES) and Secondary vocational education (SVE) under all programs of bachelor, specialist, master, and postgraduate training. Also describes first results of the Project "National Platform for Open Education".

Keywords: Open educational resources · Engineering education ·
CDIO-approach · E-learning system · Distance educational technologies ·
Massive open online courses -MOOC

1 Context

The work describes the experience obtained in the process of introduction of an IT and educational environment (hereinafter - ITEE) in an engineering education. It also covers an application of distance educational technologies in engineering education abroad and in Russia. In Russia the use of ITEE in the educational process is provided f by Federal state educational standard (FSES) and Secondary vocational education (SVE) under all programs of bachelor, specialist, master and postgraduate training. In 2016, the Government of the Russian Federation approved the Concept of the priority project "Modern IT educational environment in the Russian Federation". The goal of the project was "to create by 2018 conditions for sustainable improvement of the quality and to expand opportunities for life-long education for all the citizens through the development of the Russian IT educational environment. The number of on-line students in educational institutions should rise up to 11 million by the end of 2025.

© Springer Nature Switzerland AG 2020
M. E. Auer and T. Tsiatsos (Eds.): ICL 2018, AISC 916, pp. 935–945, 2020.
https://doi.org/10.1007/978-3-030-11932-4_86

The authoring session "The experience of development and introduction of the online courses (MOOC) in the educational programs of universities" was held on February 21, 2018. More than 300 representatives of universities from 52 regions of Russia and CIS took part in the event. The conference speakers emphasized that online education is to make a regional educational system more competitive, to integrate the higher education institutions of the country into the general educational system. This would help to resolve the problem of brain drain from the regions in the long run [1].

We believe that the wide dissemination of the Massive Open Online Courses (MOOC) - Internet courses with mass interactive participation and open access – is a kind of distance education pressing to rethink the principles of IT education of future engineers [2].

2 Goal

In-depth study of MOOCS application in supplementary education abroad using the well-known Coursera, EdX and other Russian sites has shown that the introduction of on-line learning in engineering education has many problems and issues that are to be addressed together. On the one hand, the problems, such as technologies for the creation and application of the MEP in training, are sufficiently thought through by both theorists and e-learning practitioners; on the other hand, the question how to ensure the quality of on-line course content and learning outcomes has not been explored well enough. The purpose of the study was to identify the difficulties of using the MEP, in particular of e-courses, and the ways to train future engineers. We believe that the gap in preparedness of both the teachers and the trainees to a new education paradigm - "lifelong learning" (when everyone sets himself/herself an educational goal and builds his/her individual educational trajectory) is the main challenge. How can I increase the? Let's consider the history of motivation of both teachers and trainees in a new open educational environment by the two Kazan engineering universities - KNITU-KAI and KNITU.

3 Approach

At the end of July 2017 the Russian government approved the program "Digital Economy" (DE), which aims to systemic development and introduction of digital technologies in all areas of life until 2024.

In Tatarstan it is, first of all, the site "Electronic Tatarstan". The Service is a vivid example of the relationship between citizens and the state in the era of the digital economy. Thus, the creation of advanced training and retraining programs for the digital economy is a pressing task for the universities now. There is a good reason that every university should now obligatory provide an electronic information and educational environment. According to the current standards, such environment is to work to the required competencies of engineers for the digital economy. In accordance with Federal Law No. 273-FZ of December 29, 2012, "On Education in the Russian Federation" and in accordance with the requirements of the Federal State Educational

Standards for the New Generation (in particular " + " generations), educational institutions must create and maintain electronic information educational environment (EIEE).

According to the russian statutory requirements of the EIEE, the university should provide:

1. Access to curricula, work programs of disciplines (modules), practices, electronic library systems and electronic educational resources specified in work programs;
2. Monitoring of the course of the educational process, the results of interim certification and the results of mastering the basic educational program;
3. Provision of various teaching sessions, procedures for evaluating learning outcomes with the use of e-learning, distance educational technologies;
4. Adjusting of the student's electronic portfolio with archiving routing of his/her works, reviews and evaluations on these works by any participant in the educational process;
5. Interaction between participants of the educational process, including in-sync or non-sync communications through the Internet.

The functioning of the electronic information and educational environment is provided by appropriate means of information and communication technologies and by the qualification of employees who use and support it.

The requirements for EIEE determine the availability in the educational organization of the appropriate information and communication tools to generate and to manage the necessary databases.

For example, in EIEE of KNITU-KAI (https://kai.ru/e-learning-environment)) meet the above mentioned requirements are met by the following:

- The official site of KNITU-KAI (https://kai.ru) - a single point of entry into all EIEE resources for students and employees of the University;
- "personal accounts" for the trainees with an access to the components of the EIEE, including data services: "Educational program", "Schedule", "Assessment", "Electronic Portfolio", "e-learning system", "Electronic Library System", "Announcements", "Internet waiting-room" – provides personal access to the University authorities with monitoring the process of review. In addition, the service allows you to make any enquiry. The service is available in the public section of the KNITU-KAI website and in private offices, "E-mail", etc.;
- section "Information on the educational organization";
- basic educational programs, curricula, working programs of disciplines and other regulatory and methodological materials;
- computer assisted management of educational process (CAM "Dean office");
- a system for recording tuition fees and calculation of scholarships;
- an entrance management system; admission information is available in the "Applicant" section of the official site of KNITU-KAI and in an applicant "personal account";
- e-learning system powered by LMS Blackboard; - conduct all types of classes, procedures for evaluating e-learning outcomes, distance learning educational programs for students of KNITU-KAI;

- e-learning system for supplementary education programs for KNITU-KAI powered by LMS MOODLE;
- electronic information and educational resources of the Chetaev Sci-Tech Library: electronic catalogue, electronic library system KNITU-KAI, external electronic library systems and databases;
- catalogue of electronic and educational resources of KNITU-KAI;
- system of testing school leavers for entrance examinations;
- service for conducting webinars and videoconferences;
- SED KNITU-KAI provides automation of processes of receiving, processing and distribution of incoming correspondence; signing, registration and dispatch of outgoing correspondence; signing, registration and sending of organizational and administrative documents;
- electronic system for checking the linguistic borrowings;
- corporate e-mail;
- Lync electronic instant messenger for synchronous interaction between participants of the educational process;
- Cloud service "Office 365" for cooperation of teachers and students of KNITU-KAI. Provision of the web tools that provide access to e-mail, documents, contacts and calendar from any place where there is an access to the Internet;
- reference and legal system;
- application software used in the educational process.

Hereinafter contains the analysis of the "challenges" and "how-to meet-them" in engineering universities by means of e-learning and digital educational environment and CDIO philosophy. CDIO approach –"Conceive - Design - Implement – Operate" is based on the students' mastering of engineering skills in accordance with the "Plan - Design - Produce - Apply" model. CDIO reflects "the main features of modern engineering education - enthusiasm for engineering, deep mastering of basic skills and understanding of the contribution of engineers in the development of society. The CDIO approach allows to inflame a passion for our profession in our students " [3].

For example KNITU-KAI is implementing the concept "Think-Project-Implement-Operate" in the Engineering Lyceum of KNITU-KAI for gifted children (hereinafter - the Lyceum), which has been working since September 1, 2015. The main objective of the Lyceum is to create conditions for strengthening the region's human resources potential in the field of engineering. The first steps in engineering are done by children, combining school subjects with the study of specialized engineering disciplines and engaging in technical creativity. In 2017, five teams of the Kazan engineering Lyceum KNITU-KAI, who were included in the national team of the Republic of Tatarstan to take part in the finals of the V national championship "Young Professionals" WorldSkills Russia-2017, won the prizes in their competencies (http://www.tatar-inform.ru). In February 2018, students of the Lyceum won seven gold and two bronze medals (https://kai.ru/news/new?id=9540475) at the JuniorSkills Regional Championship of the Republic of Tatarstan held according to the following backgrounds: Electronics 10+ and 14+, Electrical work 10+ and 14+, Engineering design CAD 10 + and 14+, Turning work on CNC machine tools 14+, Milling work on machine tools with CNC 14+.

The way to organize an education process on the basis of the CDIO approach will be considered from the standpoint of open education, the application of distance education technologies.

Open education involves the development and implementation of open educational resources. The term "Open Educational Resources" (OER) was first introduced at the Open Educational Systems Forum for Developing Countries organized by UNESCO in July 2002. Open Educational Resources (OER) refers to any type of publicly available educational material, which are placed in accordance with "open licenses" foe free use by any individual - copy, modify, create on their basis new resources [4].

The development of the open digital educational environment in KNITU-KAI will be considered as an example of LMS Black Board and LMS MOODLE application in intramural and supplementary education respectively.

In 2015–2017, four questionnaires were developed to interview students and teachers of the KNITU-KAI on the advisability of restructuring engineering education on the basis of foreign and domestic experience and using distance education technologies and MOOCS.

To support a full-time BEP training in KNITU-KAI, teachers create electronic courses (EC) in their LMS Black Board disciplines. In 2015 the questionnaire revealed that 40% of the teachers-authors of the EC- independently mastered the work on content creation and its placement in the LMS Black Board, and 60% - were pre-trained. The system has turned to have a user-friendly interface requiring no special knowledge and skills. Furthermore, 52% of teachers indicated that the introduction of e-learning and distance education technologies (DOT) as a support for intramural education led to a decrease in the quality of instruction in the discipline taught, while 44% noted improvement in quality and 4% did not feel any difference. We have tried to found out the reasons for such answers, and suggested that, firstly, it is necessary to realize what level of the quality of education should be to motivate the teachers' conclusions. Secondly, what is the quality of the EC, how much they meet the requirements of pedagogical design, etc. Therefore, it was necessary to specify the criteria of quality assessment. The Regulations on the examination of electronic courses in the KNITU-KAI were developed, materials were prepared to help teachers.

We studied out the teachers' evaluation of the labor intensity (development and use of EC): 44% answered - increased, 40% - decreased, 16% - remained unchanged. An analysis of these answers led to the conclusion that the difficulties of using EC could only increase due to the lack of an effective educational trajectory and poor organization of the content.

In 2015–2018 the Department of Information Technologies and the Department of Electronic Technologies in Education of KNITU-KAI prepared and conducted a number of activities to eliminate technical and methodological barriers to the introduction of EC and DET (distant educational technologies) in the training of engineers:

- a distance-learning program for teachers of the KNITU-KAI "Methodology for the creation and use of e-learning courses" was developed and introduced, where the author-teachers themselves study remotely, and after their studies either create a new EC or modernize the existing EC ready for examination and implementation in training;

- seminars, round tables, consultations on technologies for the creation and application of EC and DET in the training of engineers;
- regular sections for teachers are created on the site: to help the teacher, frequently asked questions, consultations, useful materials, etc. These materials can be used in the process of professional development in the Engineering Education Centers of the IGIP (International Society for Engineering Pedagogy). These materials can be used in the process of teachers' professional development in the IGIP Centers of Engineering Pedagogy (International Society for Engineering Pedagogy) [5, 6].

The first questionnaire showed the main problems of that period: technical ones related to the use of LMS Black Board (malfunctions, dead halts, lack of examples for creating EC, complex registration of EC); methodological: lack of knowledge on EC content development, electronic didactics; Unwillingness of both teachers and students to work with EC.

The questioning of students showed that 69.4% of students use EC, developed by the teachers of KNITU-KAI; three ECs are used by 23.9%, two ECs – 15.3%, one EC - 21.6%, more than 6 EC are used by 10.2%. When assessing the quality of the EC used on a 5-point scale, students chose: 1 point - 13.8%; 2 points - 9,9%; 3 points - 25,4%; 4 points - 28.7%; 5 points - 17.7%. Thus, 46.4% of the students appreciated the quality of the ECs used, but 49.1% of the respondents gave low marks. We have analyzed the reasons for the non-availability and low estimation of EC and received information that in the KNITU-KAI students (41.2%) had technical difficulties when working with the BlackBoard system. Further, 38.2% of students answered that they prefer to work with "paper" books and notes and "distributed" teachers electronic materials (28.7%). On poor filling of EC students reported: "they are uninteresting" - 25% and "not informative" - 27.9%.

Used the results was developed a plan of measures to increase the "attractiveness" of EC, improve the quality of information materials displayed in the EC. To enhance the competencies of teachers in the creation and application of EC in training, two advanced training programs were developed and implemented: "Methodology for the Creation and Use of Electronic Courses" and "Development of the Digital Educational Environment of an Engineering University." Both programs are remote, with forums, polls and chats, mentors and tutors are used to accompany the training. For each new group on the input poll, the current EC content and the form of the final control are determined.

Thus, the experience of KNITU-KAI showed that the introduction of e-learning in an engineering university should be carried out in stages: first, ECs are created and used to support full-time education, to organize independent work of students under the supervision of the teacher. Next, several on-line courses from the informational platforms of "Open Education" (https://openedu.ru/) or "MDEE (modern digital educational environment) in the Russian Federation" (https://online.edu.) are selected for information support of the work program of the academic course areas), for which trainees are enrolled on their own. The teacher must be also enrolled to this course to determine whether the content can be matched to its work program and the trainee's work in his discipline can be accredited. The ultimate goal is to create an online course in discipline and participate in the "single window" project as an author.

Education, organized with the CDIO approach, is based on the formation of basic technical knowledge in planning, design, production and application of objects, processes and systems. This approach works to create an "educational context", in other words an environment conducive to understanding and acquiring knowledge and skills. Further, an integrated approach has been worked out to determine the educational needs of students and the sequence of training activities to meet these needs. At the same time, an educational context has a double impact on students as it contributes to a deep understanding of the theoretical foundations of engineering and the acquisition of practical skills. Therefore, we plan with the help of our teachers to introduce a CDIO approach in the creation and application of original EC and external online courses in the training of engineers.

Experience of KNITU has shown, that for the electronic training of the industry experts is the most promising form to advance professional skill and retraining. Therewith they acquire actual knowledge, skills and competencies with on-the-job training taking part in day-to-day production activities, IT fora and projects, applying the latest achievements of science and practice, exchanging their experience.

Here are the results of the questionnaire survey of "Gazprom" experts, trained with the original EC "Corrosion and protection of pipelines" and DOT in KNITU.

Experience KNITU has shown, that electronic training of the experts working on manufacture, the most perspective form of improvement of professional skill and retraining. At the same time, specialists without disruption from the main work acquire actual, skills and competencies, discuss and solve real production tasks through forums and projects, applying the latest achievements of science and practice, exchanging their own experience.

Here are the results of the questionnaire survey of specialists of PJSC Gazprom, trained with the help of the author's EC "Corrosion and protection of pipelines" and DET in KNITU.

First, to determine the quality of EC, the students were asked if the proposed content corresponded to the expectations: 87.5% answered t "yes", and 100% pointed out the relevance of topics for their practical work. Secondly, to improve the content and amount of time allocated for certain topics in the future, the students proposed the most important topic (ways to protect metals from corrosion - 87.5%). For this group, the least important was the topic - technical and economic aspects of protection of the main pipelines (25%). Almost all participants of this program upgraded their skills using DET for the first time. So it was very interesting to learn about the organization of training and learning environment: 88.89% were satisfied with the training, and 100% - with work in LMS MOODLE and EC. It is clear that the training of practitioners requires of the EC author special professional competencies, and the content should be of immediate interest. In assessing the content and completeness of the stated EC material, 55.56% of participants rated it as "5 -very good", 33.33% - "4-good" and 11.11% - "satisfactory", although the relevance and practical value of the material were rated as "very good" and "good" (55.56% -5, 44.44-4), and 88.89% of trainees assessed the effectiveness of studying theoretical issues as "good' and "very good". And 100% appreciated the availability of intermediate control tasks. Some contradiction in the estimates was obtained with the answers to the question about the novelty of the information presented: 44.44%- "very good" and 55.56% - "good". The traditional system of education ("façe-to-façe") envisages the teacher's online answer to the questions of the students in the classroom. And in EC different ways and means of feedback can be implemented. In the proposed EC, consultations of the tutor were offered, the degree of efficiency of which was evaluated as "very good" by 44.44% of the students; and by 55.56% - as "good". 100% of the listeners reported "good" and "very good" on EC activity in the personal profile. For the self-work, a well-designed calendar of upcoming events in the EC is very important. Here 88.89% of the students assigned "good" and "very good" to the calendar of events proposed in the EC. After completion of the proposed EC, all the answers to the questions were reviewed, and the author of the EC received recommendations for finalizing and updating the content for the next group.

Analysis of all the responses of specialists showed that

- practical and laboratory work, which in engineering education play a leading role in training, require the development and use of special software (virtual worlds, remote laboratories, automated control systems, simulators, etc.);
- work programs on special disciplines for training engineers should contain both practice-oriented modules and design research assignments;
- the main directions of engineering education require an integrated approach to the development of both e-courses in basic and special disciplines of the PLO, as well as online courses, involving leading teachers and practitioners from the production, using network interaction;
- There is a need for a single regulatory and legal framework for the implementation of EC, BT, and online courses (LHS) in engineering education.

Thus, at currently the KNITU-KAI and KNITU in full-time education, EC and DOT are used to support and organize independent work of students on PLO. We can say that these universities are ready for "blended learning":

- in the training of engineers for the PLO under a bachelor degree program - support for classroom activities, student self-study (SSS);
- when training engineers for the master's degree- "blended learning", organization and support of the SSS.

As for the supplementary education, both the KNITU-KAI and KNITU actively develop and implement the EC and DET in all programs, but for the time being they do not use MOOCS due to the unavailability of both teachers and students, as well as the lack of a local regulatory framework.

The results of the questionnaire survey of teachers of KNITU-KAI in 2017 showed:

- There is no clear understanding of the terms of e-learning;
- 55% acknowledged that MOOCS is a fact of the world education, 25% deny their need for engineering education, 20% consider it a "fashion";
- answers to the question on the need for information about MOOCS: 68.6% consider it useful, 23.5% see it as a threat, 7.8% - just do not care.

The foreign authors consider that such answers are neither more nor less than a syndrome of "rejection of someone else's development", "imaginary risks". We believe that the psychological, social and other barriers in the KNITU-KAI can be overcome by increasing the awareness and motivation of both teachers and students. To do this, it is necessary to unite the efforts of the administration, the Department of Information Technologies, the Department of Electronic Technologies in Education and the teachers of KNITU-KAI. As the creators of MOOCS note, this process is very laborious and can be substantiated by the teamwork only.

- As the results of the survey showed the motivation of teachers to participate in the creation of MOOCS, first of all, depends on the guaranteed high level of income (47.1%), while 17.6% of the professors consider this not to be serious, and 11.8% are not yet ready, finally, 7.8% -against.

According to the results of the study by American universities, about 66% consider online education a strategic priority. The general trend is: the more students in the university and the higher the level of programs offered by the university (master's, doctoral, retraining), the more EE is used. But, perhaps, colleges and universities with up to 1,500 students do not really need distance education.

The new needs of students are the real challenges for the universities: "Institutions are losing their social role," stated András Syuch, Secretary-General of the European Network for Distance and Electronic Learning. "Knowledge ceases to be born in traditional higher education institutions, and training is provided by the independent centers."

For engineering universities, the following principles of using open online courses in PLO can be proposed:

- access to a getting-to-know mode of learning should be open and free for all comers, the course should be easily scaled by the number of students;
- should ensure the possibility of a reliable evaluation of the learning outcomes in the context of identification of the trainee's personality (the evaluation procedure can be paid for);
- The University independently decides whether a student can get a credit for the learning outcomes of the online course in the context of compliance with the content of PLO.

Today, the number of domestic and foreign training platforms is very high, and there are thousands of on-line courses. But in order to find the right online course, you need to spend a considerable amount of time, and it could be of inadequate quality and ineffective. Obviously, one of the main goals of the priority project "Modern digital educational environment in the Russian Federation" is the integration of online platforms and individual online courses under the auspices of an information resource that provides access to them on a "one-stop" basis [7].

Anyone can get access to a "one-stop" resource, regardless of place of living and level of education and this could definitely forge the practice of a genuine academic mobility for the students, giving an access to quality educational content from the leading universities of the country. In this case, the results of the online course will be credited on a par with the results of full-time study.

Teachers would be able to learn the resource of the best domestic pedagogical experience and could allocate more time for practical classes with students and upgrade their qualifications. Those who are ready to get additional knowledge or keep their skills up to date will get a convenient and high-quality service. Employers will be able to directly address to a teaching content, in order to bring it in line with the demands of the labor market. Educational platforms and creators of on-line courses "one-stop" resources will be provided with a unique opportunity to expand the audience, to improve the quality of the product, to offer a flexible and convenient analytical tool.

Thus, the "one-stop" resource will have a number of key features that make online learning convenient, affordable and of high quality:

- work on the principle of "one-stop", when one resource is used to receive anaccess to the materials of dozens of educational platforms;
- provide a unified user authentication;
- monitor the learning process and record educational successes;
- implement the mechanism of multistage evaluation of the quality of online courses.

Currently, the "one-stop" resource of is being tested, the access procedure to the resource for educational organizations and integrating with the online learning platforms are being implemented. Our university (KNITU-KAI) has an access to this resource.

On March 13, 2018, the results of an experiment conducted at the Center for Sociology of Higher Education of the Higher School of Economics at the Higher School of Sociology (I. Chirikov, T. Semenova, N. Maloshonok) were reported at the Institute of Education of the Higher School of Economics (HSE) (Moscow). The researches were carried out in accordance with the project of FSES of HE (Federal State Educational Standards of Higher Education) "Research of new forms of organization of the educational process using open online courses» commissioned by the Ministry of Education and Science of the Russian Federation. 3 universities were selected as a testing site: 2 technical and 1 classical university. The course units were selected according to 5 criteria: mandatory course within the framework of the PLO; students must complete the course at the beginning of the second year of study; a similar course should be posted at the National Platform of Open Education (NPOE); the online course must be consistent with the FSES; the online course should be designed for the selected training areas. Engineering mechanics and technology of structural materials were chosen for the test. Training was conducted in three formats: traditional, mixed and online. Comparison of groups was carried out in line with the levels of content uptake, satisfaction with the format of the course and involvement in the knowledge deepening.

4 Conclusions

Brief review of the results obtained:

- The format of teaching does not affect the results of students;
- Students, who studied online, are less satisfied with their course;
- Students prefer a traditional or mixed format of education;

- One of the courses was found more difficult to get online;
- Students spend less time on some courses to learn online.

   Generalizing conclusions:

- The training format (online, mixed, traditional) does not affect the educational outcomes of students.
- Students are still wary of the online format of education, preferring mixed or traditional ones.
- Some MEPs may be more difficult to complete in a fully online format.
- Within the framework of some MOOCs less time required to get through by the students.
- It was also noted that the low level of student self-efficacy adversely affects online training.

   Could these results be illustrative for a wide range of engineering universities? Of course, not yet, as it was a "field" experiment with certain limitations, but some conclusions agree with the results obtained by the employees of the Department of Electronic Technologies in Education KNITU-KAI (see https://kai.ru/for-staff/news/new?id=9555094).

References

1. Davis, D., Kizilcec, R. F., Hauff, C., Houben, G.-J.: The half-life of MOOC knowledge: a randomized trial evaluating the testing effect in MOOCs. In: Proceedings of the 8th International Conference on Learning Analytics and Knowledge (LAK) (2018)
2. http://web-in-learning.blogspot.ru/2014/05/mooc.html
3. Edward, F., Malmquist, Johan, Ostlund, Soren, Broder, Doris R., Edström, Christina: Rethinking Engineering Education, p. 503. The Publishing House of the Higher School of Economics, CDIO-M Approach (2015)
4. Ivshyna, G.V.: The paradigm of open education in the framework of reforming the educational space of the university. Uchenye zapiski ISGZ 1–1, 144–150 (2013)
5. Prikhodko, V., Polyakova, T.: IGIP International Society for Engineering Pedagogy. The Present, Past and Future, p. 143. Technopoligraphcenter (2015)
6. Solovyev, A.N., Prikhodko, V.M., Polyakova, T.Yu, Sazonova, Z.S.: Russian engineering teachers as an important part of IGIP. High. Educ. Russia Sci. Pedag. J. Typogr. Sci. №1659, 38–45 (2018)
7. https://vo.hse.ru/data/2014/08/04/1314334660/2013-3_Barber%20et%20al.pdf

Mobile Technologies in Education: Student Expectations – Teaching Reality Gap

Ekaterina Dvoretskaya[✉], Elena S. Mishchenko,
and Dmitry Dvoretsky

Tambov State Technical University, Tambov, Russia
dvkaterina@yandex.ru, dvoretsky@tambov.ru

Abstract. In the world where mobile technologies are generally available to the majority of population, there is an obvious demand to make use of them in education. But, despite the fact that there is no lack in creative ideas on how to incorporate them in the learning process, most formal educational systems do not specifically support nor encourage teachers to use mobile devices in their teaching. As a result, there is a discrepancy between the expectations of students and working practices of teachers. The paper intends to discuss this discrepancy using the data of the surveys on the implementation of mobile learning conducted among the students and teachers of Tambov State Technical University (Tambov, Russia).

Keywords: Mobile learning · Mobile devices · Survey

1 Introduction

The emergence of mobile technologies has changed the way we live and they are beginning to change the way we learn. According to UNESCO statistics, today over 6 billion people have access to a connected mobile device, and in 2017, 50.3 percent of all website traffic worldwide was generated through mobile phones [1]. The ubiquity of mobile devices is reflected in the fact that more and more mobile applications are being developed – during just one year, over the period from December 2016 to December 2017, the number of mobile applications offered by Google Play, the largest application store as so far, has grown from 2.6 million to 3.5 million [2]. The majority of service providers on the Internet choose to offer a mobile version along with a regular website interface to their clients.

And it is also worth mentioning that education is the category with the one of the largest number of apps available at the moment – as of the beginning of 2018, more than 300,000 educational apps were available on Google Play [3], and in Apple's App Store it is the third largest category with 8.5% share of the total amount, outranked only by gaming and business apps [4].

The implications of such wide spread of mobile technologies for the education include the possibility to engage in learning activities at any time and any place, with students being mobile both inside and outside the classroom, as well as to implement a variety of media and resources previously not available to teachers: audio, video, hypertext, geographical positioning data, augmented reality, real time broadcasting, community reviews, etc.

© Springer Nature Switzerland AG 2020
M. E. Auer and T. Tsiatsos (Eds.): ICL 2018, AISC 916, pp. 946–957, 2020.
https://doi.org/10.1007/978-3-030-11932-4_87

Students are generally eager to embrace new educational opportunities of the digital world. For instance, among more than 3000 US college students surveyed in 2016, 63% agreed that their usage of digital learning technology resulted in them being better prepared for classes and having improved studying efficiency. Other benefits of the digital technologies mentioned by respondents were more confidence in the knowledge of course materials and less stress related to studying [5]. Despite these figures presenting a somewhat subjective evaluation, it is clear that more students rely on digital resources and materials in their learning, and, given the amount of mobile device users among young population, it can be safe to assume that many of them do that with the help of mobile technology.

However, despite the availability of mobile technologies and the fact that there is no lack in creative ideas on how to incorporate them in the learning process, most formal educational systems do not specifically support nor encourage teachers to use mobile devices in their teaching. As a result, there is a discrepancy between the expectations of students and working practices of teachers. The paper intends to discuss this discrepancy using the data of the surveys on the implementation of mobile learning conducted among the students and teachers of Tambov State Technical University (Tambov, Russia).

2 Mobile Devices and Mobile Learning

There is probably little misunderstanding today regarding what a mobile device is, however, for clarification purposes, we shall consider that any device that "can be used at Point A, Point B and everywhere in between without stopping" is mobile, as opposed to a portable device that is "used at Point A, closed down and transported, then opened up again at Point B" [6]. Thus, the most common mobile devices include mobile phones, tablets, and media players (which are being used less and less as smartphones and tablets are able to perform their functions), and newer devices such as fitness trackers, smartwatches and smart glasses (or augmented reality glasses).

With the ubiquity of these mobile devices in our lives, it is most natural that they began to be used in education. At first, mobile learning was defined simply as "learning facilitated by mobile devices" [7], but as the focus shifted toward new opportunities mobile devices have created for student-teacher interactions, mobile learning is being described as "learning across multiple contexts, through social and content interactions, using personal electronic devices" [8].

Mobile resources available to students and teachers can be categorized in two groups: mobile materials and mobile activities. The first one implies the use of web services either through a regular browser or by downloading a specialized piece of software adapted for mobile use – an app. Apps provide a more streamlined experience, with many educational services offering app (and sometimes app-only) versions. By mobile activities we mean educational activities designed around the use of mobile devices, websites or apps. These activities may be fully digital or delivered in the context of blended learning, but it is essential that their "pedagogical use needs to be carefully considered in the overall design of a learning activity" [6].

Among educational apps ranked most popular by users (according to App Store and Google Play statistics), there are both targeted single-discipline apps, such as foreign language learning apps or standardized test preparation apps, and a variety of apps representing large online learning platforms like Coursera, Udacity, etc. which give access to lectures, courses or even degree programmes via mobile. An interesting category is apps that develop academic skills and general learning abilities, for example, GradeProof – a 'personal editor' apps which improves style, grammatical structure and checks originality of student writing, or Mendeley, a mobile version of a popular desktop service for annotating academic texts and managing research citations. A growing number of apps are designed to assist teachers in developing and managing lessons.

At the same time, researchers and practitioners try to evaluate and balance the advantages and challenges of implementing mobile technologies for learning. On the one hand, mobile devices with Internet access allow us to have knowledge literally at our fingertips, anytime and anywhere. They support situated learning, when knowledge acquisition is not confined to the walls of a classroom, first of all, and when learning occurs in a variety of contexts. Mobile technologies also give us an opportunity to learn with the others, sharing our educational experiences and results with wider audiences, be it classmates or the whole world (via social media platforms, for instance). For this reason, mobile learning has been endorsed by supporters of social constructivism theory of learning. As Cochrane writes, "The ubiquitous connection to web 2.0 tools and collaborative communication and user generated content creation capabilities of these devices make them ideal tools for facilitating social constructivist learning environments across multiple learning contexts" [9]. Pegrum [10] stresses the inter-cultural dimension of m-learning, focusing on the importance of digital technologies in the development of intercultural literacy across different cultural contexts by merging global and local contexts in a mobile classroom. This aspect of mobile learning is especially promising for foreign language teaching, which fact has already led to the emergence of numerous studies into mobile-assisted language learning (MALL) [10].

However, the critics of mobile learning point out that there is still a need of more extensive research on the interference of technology in the classroom. Pedro et al. [11], for example, highlight that a teacher in a mobile classroom must have excellent mul-titasking abilities since their role in-class changes to that of a media orchestrator and learning facilitator. Not many teachers nowadays are trained to perform such chal-lenging tasks, nor to properly utilize mobile resources.

3   Student and Teacher Surveys

3.1   Approach

The aim of the research was to identify and analyse the role of mobile technologies in educational activities as implemented at Tambov State Technical University (TSTU). For that purpose, a sample array of students and teachers were asked to complete a questionnaire outlining their use of mobile devices in personal life and education.

TSTU is located in the Central European part of Russia in the city of Tambov with a population of around 300,000 people. With about 9,000 students and 600 members of the teaching body, it is an average-sized university offering 42 undergraduate programmes of study, 62 master programmes and 20 Ph.D. programmes in a variety of fields. The main areas of the university's expertise are chemical and biochemical engineering, nanotechnologies and material science, information systems, as well as architecture and civil engineering.

The survey on mobile learning was conducted among 64 students and 20 teachers of TSTU. The majority of students surveyed were bachelor students of various engineering programmes, and the teachers represented different subject areas, from mathematics to engineering to humanities. One common denominator was that all participants either studied or taught at bilingual (Russian and English) programmes at TSTU Center for International Professional Training, and the survey was intended to assess the perspectives of introducing elements of mobile learning into its courses.

3.2 Students' Responses

Sixty four students aged 18 to 23 participated in the survey, 26 females and 38 males. The survey included questions about which mobile devices students own, how often they use them, which of them they use for the purpose of study. The examples of mobile devices included smartphones, tablets, media players, smartwatches, fitness trackers, virtual and augmented reality sets, robots, etc. The full list of survey questions is given in the Appendix.

100% of the participants had access to the Internet from a computer at home and all owned at least one mobile device – a smartphone with Internet connection. Other mobile devices mentioned were tablets (38 people (59% of the group) owned them), media players (iPod, MP3 players) – 32 (50%), smartwatches and fitness trackers – 4 people (6%) mentioned owning each of the devices. The most 'exotic' devices were drones (2 students had them) and robots (marked by 1 person). Other types of mobile devices were not mentioned. There were no specific differences in the number and type of a mobile device owned in males and females. On average, every student owned 2.37 mobile devices, with this figure being insignificantly higher for male students (2.52) as opposed to female ones (2.15). Overall, we could conclude that affordability of mobile technologies was not an issue for the surveyed group of students.

The dominant majority of the students used their mobile devices to access the Internet several times a day (58 students – 90%), with only two students using them just once a day and four students admitting to use them occasionally 3–5 days a week. As for the purpose of using these devices, all students mentioned personal communication (making calls, messaging, using social media, etc.), accessing information resources on the Internet and in apps, and all but one student said they used mobile technologies for learning purposes. Sixty students (93%) used smartphones, 36 students (56%) additionally made use of tablet computers for the purposes of learning or finding educational information and six students (9%) wrote about using media players, to make and then listen to recordings of lectures, for instance. Again, it became clear that the participants were accustomed and fully ready to implement their mobile devices for educational needs.

Moreover, when asked whether they wanted their teachers to use mobile devices during courses, the majority of students welcomed that: 46 (72%) said 'yes', 12 (19%) – 'no' and 6 (9%) were not sure.

Students were asked to identify which learning resources they usually access on mobile devices and briefly describe devices/resources/applications which they find most useful. Their answers are presented in Fig. 1.

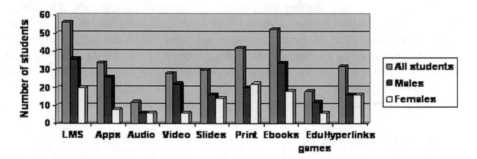

Fig. 1. Learning resources accessed via mobile devices

The most popular resources were university LMS and ebooks. Students also used mobile devices to access print content (PDFs, Office documents) and lecture slides to use in class, they watched educational videos, had some educational apps installed and kept record of hyperlinks and web addresses of course-related materials. The least accessed resources were educational games, such as flashcards (surprisingly!), and audio files. Also, for some reason educational apps were not very much used by female students (only 8 out of 26 used them), while for male students it was the third most popular resource after LMS and ebooks (used by 26 out of 38 participants).

A word must be said about the types of educational apps used by the students. Many respondents mentioned various language learning and translation apps, which is understandable given they were all enrolled in English-medium courses. The second most popular type of app were apps aimed at learning math.

Question 13 of the survey asked students to evaluate whether and how mobile technologies affect their learning. Their responses can be found in Table 1.

The results suggest that generally the majority of students embrace the idea of using mobile devices for studying (82% of respondents agree and strongly agree that they may be useful) and think that they significantly increase the speed of task completion (81% agree and strongly agree). However, when it comes to assessing the productivity of learning with mobile technologies, there is no unanimity. In fact, a significant number of students were not sure whether mobile learning made them more productive (37.5%) and whether mobile devices helped them to achieve better grades (28%). There was little statistical variation in male and female groups of respondents.

Despite the seemingly positive findings of the research, we found it alarming that the students only used their mobile devices 'in receptive mode' – to take notes in classes, to access class material available in digital form – and basically did not engage with mobile technologies in their full potential. The unstructured descriptions of their

Table 1. Students' self-evaluation of the effect of mobile technology implementation in their studies.

	Strongly disagree	Disagree	Neither agree nor disagree	Agree	Strongly agree
I would find mobile learning useful in my university studies		6 (9%)	6 (9%)	30 (47%)	22 (35%)
Using mobile learning enables me to accomplish learning activities more quickly		2 (3%)	10 (16%)	38 (59%)	14 (22%)
Using mobile learning increases my learning productivity		2 (3%)	24 (37.5%)	24 (37.5%)	14 (22%)
If I use mobile learning, I will increase my chances of getting a better grade		10 (16%)	18 (28%)	18 (28%)	18 (28%)

mobile learning experiences rarely referred to sharing knowledge with their classmates via mobile platforms or contributing to a collective resource. In our opinion, this may be due to the fact that often teachers themselves do not realise these opportunities of mobile technologies and do not encourage students to use them. This is suggested by the results of the teacher survey discussed below.

3.3 Teachers' Responses

For the teacher part of the research we have asked twenty teachers of various subjects to complete the questionnaire. The teacher questionnaire contained a section on individual's teaching background (subject area, years of experience) and implementation of mobile technologies for personal aims (questions 1–5). The main part of the survey focused on whether and which mobile technologies are used for educational purposes. The aim was to establish whether teachers perceive the benefits of mobile learning in terms of personalisation of learning (learn what, where and when a student prefers), improved collaboration between students outside class (virtual meetings, social media discussions, etc.), increased possibilities for authenticity in learning (i.e. using real-time information from web cameras, conducting virtual visits to cultural or industrial sites, etc.), and how willing the teachers are to design special learning activities with mobile technologies in mind. Questions 6–9 concerned learning context, questions 10–15 were about personalization of learning paths, questions 16–19 focused on opportunities for collaboration and questions 20–24 – on sharing students' learning results. Teachers were also asked to describe their activities and practices with the use of mobile devices.

The general structure of the questionnaire for teachers was based on another survey developed by an international team of academics from the University of Hull (UK), the University of Western Australia (Australia), the University of Auckland (New Zealand) and Guangxi Normal University (China). The original MTech Mobile Learning Survey for Teacher Educators was intended for teacher trainers in pre-service and professional development courses and aimed to assess what kind of training on mobile learning is given to future educators in different countries. It can be accessed at https://hull. onlinesurveys.ac.uk/mtech-mobile-learning-survey [12].

The participating teachers had from 5 to 22 years of teaching experience, ten of them taught Humanities, 6 teachers were from STEM departments and 4 teachers represented Social Science and Economics. In preliminary conversations it was established that neither of them had any specialised training or coaching in the implementation of mobile learning.

Probably the most significant yet not totally unexpected finding of the survey was that half of the participating teachers (10 out of 20) had not planned or designed any educational activities with mobile devices. Among other participants, six said they had been using elements of mobile learning for a period of 2–5 years already and four teachers had used them for less than two years. Only smartphones (10 teachers) and tablets (5 teachers) were used. Thus, the research basically focused on the experiences of those ten teachers.

All these teachers admitted to use mobile devices sparingly in the classroom, 'rarely' and 'sometimes' being the most often chosen answers. Seven teachers said they used mobile devices only in conjunction with non-digital materials (coursebooks, pens and paper) and they rarely designed any activities where students could be mobile themselves and interact with their surroundings.

Another important result was that most teachers (80%) preferred to prescribe students when, where and how exactly a mobile learning activity would occur: students either used a specific app, or were told to go to a particular resource, the activity could be in-class or given as a home assignment and its results had to be presented to a group. Overall, students had very little control over the context of the activity. Slightly more freedom was given to students in choosing what they wanted to learn (e.g., choosing their own question, problem or project to explore) and how to express their thinking (e.g., through a text, diagram, annotated image, or narrated animation), with 50% of respondents saying they allowed it. Only two teachers admitted to use adaptive apps (platforms that adapt automatically to students' preferences) and just three – to design mobile learning activities where the students could receive instant feedback adapted to their context (e.g., real-time weather data, exercise or fitness data, or information triggered by QR codes). In any case, it was obvious that personalisation of learning experiences through mobile technologies was not a top priority for many teachers.

Another extremely underused feature of mobile learning was opportunity to network and share the results of learning. Only three teachers 'sometimes' designed mobile learning activities where the students could participate in peer online discussions (e.g., via email, SMS, social messaging platforms, Skype, or social media platforms), two teachers created activities where the students could interact with experts they would not normally work with (e.g., through public discussion boards, blogs, or social media platforms), and no one admitted to design activities where students could interact with peers at other institutions, thus missing out on one of the most valuable opportunities of the modern technologies – collaborative learning.

At the same time, teachers were more eager to encourage students to share opinions and give feedback on each other's work within a group, after digital content has been emailed or shared in an LMS or on another class platform: 1 respondent answered 'often', 5 - 'sometimes', 3 -'rarely', and 1 – 'never'. However, that was the most overwhelming response to the survey.

3.4   Outcomes

The surveys have proven that, as expected, all participants have access to at least one mobile device (usually smartphone) and actively use them in their personal activities – sending and receiving messages, e-mails, interacting on social networks, making appointments, etc. At the same time, both students and teachers implement mobile technologies for information search, such as browsing the web, using online maps, checking the news, watching educational videos/tutorials, reading books and other activities. A sufficient number of students admitted to having installed educational apps on their smartphones or tablets, mostly foreign language learning apps. The majority of respondents agreed that they find mobile devices useful both for in-class and outside the classroom learning: they read electronic books, access lecture slides, make recordings and take notes during classes. When asked if they wanted their teachers to use mobile devices in educational process, many students answered 'yes'.

As for teachers, the survey has shown that, despite being users of mobile devices in their personal life, they were reluctant to implement them in their professional activities. Out of an array of options, the only devices mentioned in teacher surveys were smartphones and tablets, with half of the respondents answering that they did not use any mobile devices. Among those who did design special activities with the use of mobile technologies, the majority preferred a rather rigid approach: students were assigned a specific app or a platform to use, a prescribed form of material presentation, fixed learning schedule, annihilating in some cases the very flexible nature of mobile learning. Another interesting outcome of the surveys was that teachers generally were less aware than students of the apps available on mobile platforms for their subjects. For instance, several students mentioned specific educational apps that helped them in learning math, but none of the mathematics teachers who participated knew about them.

On the whole, it became obvious that to properly implement mobile devices in the educational process both teachers and students need preparation and special training. The teachers especially required such training because for many of them mobile learning constitutes a new, unknown field and the pedagogy behind mobile learning is often different from traditionally used approaches.

4   Conclusion

As can be seen from the conducted surveys, implementation of mobile devices for learning purposes has long become a reality for a dominant number of students. However, it is not so for many teachers. Even when creating student activities or tasks with mobile technologies in mind, teachers often fail to incorporate elements of personalisation, collaboration, and instant knowledge sharing that are perceived as strengths of mobile learning. Thus, it becomes clear that in order to meet the expectations of current day students, teachers of all subjects need specialised support in, first of all, informing them about the opportunities of mobile educational technologies and in design of educational activities for the age of mobile internet.

Appendix

Student Survey

1. Gender: Male/Female	2. Age:

3. Do you have access to a computer with Internet access in your home? Yes/NoWhich mobile devices do you own? (You can select multiple devices.)

• Basic feature phones	• Virtual reality headsets (e.g., Google Cardboard)
• Smartphones	
• Tablets	• Augmented reality headsets (e.g., Google Glass)
• Media players (e.g., iPods & iPod Touches)	• Drones
• Smartwatches	• Robots
• Fitness trackers	• Other
	• I don't own mobile devices

4. How often do you access the Internet from your mobile device?

5. Several times a day/about once a day/3–5 days a week/1–2 days a week/every few weeks/less often/never

6. What do you use mobile devices for?

7. Which mobile devices do you use for your personal activities? (You can name multiple devices.)

8. Which of the following *personal activities* do you currently engage in on your mobile device? (Select all that apply.)

• Phone communication	• Send and receive email
• Send and receive text messages	• Read and/or edit documents such as PDF, Word and Excel documents
• Schedule appointments or tasks	
• Banking	• I do not engage in personal activities on a mobile device
• Play non-academic interactive games	

9. Which of the following *information resources* do you currently access in on your mobile device? (Select all that apply.)

• Ebooks or print content	• Social Networks (such as Facebook, LinkedIn)
• Internet	
• Library	• Video, audio
• Movie Times/Reviews	• Weather
• Online Maps	• Other mobile information gathering applications
• Sports/News/	

10. Which mobile devices do you use for learning purposes? (You can name multiple devices.)

11. Which of the following *learning resources* do you access on a mobile device?

• School/university LMS	• Print content
• Educational apps	• Ebooks

• Lecture PPT slides	• Flashcards and other interactive educational games
• Audio recordings (e.g., recordings of lectures, school information)	• Hyperlinks to course related reference material
• Videos (e.g., course related, recordings of lectures, school information)	• I don't access learning resources on mobile devices

12. Would you want your teachers to use mobile devices in educational process? Yes/No/I don't know

13. How far do you agree or disagree with the following statements? (You can choose between Strongly disagree/Disagree/Neither agree nor disagree/Agree/Strongly agree)
• I would find mobile learning useful in my university studies
• Using mobile learning enables me to accomplish learning activities more quickly
• Using mobile learning increases my learning productivity
• If I use mobile learning, I will increase my chances of getting a better grade

14. If you use mobile devices for learning purposes, please give a brief example of one activity where you do this.

Teacher Survey

1. How many years in total have you been a teacher/lecturer? _____
2. Which subjects(s) do you teach? _____
3. How long have you been using mobile devices in your teaching? Less than 2 years/2–5 years/More than 5 years/I don't use mobile devices in my teaching
4. Which mobile devices do you use in your personal life?
5. Which mobile devices do you use in your teaching?

For questions 5–24 choose between Never/Rarely/Sometimes/Often/Always.

6. I design mobile learning activities where the students use devices which are mobile (e.g., feature phones, smartphones or tablets).
7. I design mobile learning activities where the students are mobile themselves (e.g., the students are moving around a classroom to interact, or they are using their devices while on the move on a bus or train).
8. I design mobile learning activities where the students take part in learning experiences which are mobile (i.e., the learning experiences are directly affected by the contexts and environments through which students are moving, such as when students are recording or interacting with their surroundings).
9. I design mobile learning activities where the students use the mobile devices in conjunction with non-digital materials (e.g., books, paper or pens).
10. I design mobile learning activities where the students can choose and control the context (e.g., where, when and/or how the activity occurs).
11. I design mobile learning activities where the students can choose the pace at which they progress through the activity.
12. I design mobile learning activities where the students can choose what they want to learn (e.g., choosing their own question, problem or project to explore).

13. I design mobile learning activities where the students can choose how to express their thinking (e.g., through a text, diagram, annotated image, or narrated animation).
14. I design mobile learning activities where the students can choose their own apps and platforms to support their learning.
15. I design mobile learning activities where the students can use apps or platforms that adapt automatically to their preferences and learning (e.g., adaptive apps or learning management systems).
16. I design mobile learning activities where the students can participate in peer face-to-face discussions (e.g., around an iPad screen).
17. I design mobile learning activities where the students can participate in peer online discussions (e.g., via email, SMS, messengers, Skype, or social media platforms).
18. I design mobile learning activities where the students can network with peers they would not normally work with (e.g., interacting with peers at other institutions through discussion boards, blogs, or social media platforms or multiplayer games).
19. I design mobile learning activities where the students can network with experts they would not normally work with (e.g., interacting with experts through discussion boards, blogs, or social media platforms).
20. I design mobile learning activities where the students can work together and create their own digital products (e.g., photos, audio podcasts, videos, digital posters, digital stories, or multimedia artefacts).
21. I design mobile learning activities where the students can share their own digital content with peers (e.g., emailing or sharing content in an LMS or on another class platform).
22. I design mobile learning activities where the students can receive feedback on their own digital content from peers (e.g., after emailing or sharing content in an LMS or on another class platform).
23. I design mobile learning activities where the students can offer feedback on their peers' digital content (e.g., after it has been emailed or shared in an LMS or on another class platform).
24. I design mobile learning activities where the students can publish their own digital content on the internet (e.g., posting it on a website, blog or wiki).
25. If you selected "Always" or "Often" for any of these statements, please give a brief example of one activity where you do this.

References

1. Digital in 2018. Share of mobile phone website traffic worldwide 2018, Statista. https://www.statista.com/statistics/241462/global-mobile-phone-website-traffic-share/ (2018)
2. Number of Available Applications in the Google Play Store. 2018, Statista. https://www.statista.com/statistics/266210/number-of-available-applications-in-the-google-play-store/ (2018)
3. Android Apps by Category. http://www.appbrain.com/stats/android-market-app-categories
4. Popular Apps in AppStore, 2018, Statista. https://www.statista.com/statistics/270291/popular-categories-in-the-app-store/ (2018)

5. McGraw-Hill Education 2016 Digital Study Trends Survey, Prepared by Hanover Research October, 2016, p. 40. http://www.infodocket.com/wp-content/uploads/2016/10/2016-Digital-Trends-Survey-Results1.pdf (2016)
6. Reinders, H., Pegrum, M.: Supporting language learning on the move. An evaluative framework for mobile language learning resources. In: Tomlinson, B. (ed.) Second Language Acquisition Research and Materials Development for Language Learning, pp. 116–141. Taylor & Francis, London (2015)
7. Gikas, J., Grant, M.M.: Mobile computing devices in higher education: student perspectives on learning with cellphones, smartphones & social media. Internet High. Educ. **19**, 18–26 (2013)
8. Crompton, H.: A historical overview of mobile learning: toward learner-centered education. In: Berge, Z.L., Muilenburg, L.Y. (eds.) Handbook of Mobile Learning, pp. 3–14. Routledge, Florence, KY (2013)
9. Cochrane, T.: Mobile web 2.0: the new frontier. Paper presented at the ASCILITE 2008 - The Australasian society for computers in learning in tertiary education, pp. 177–186 (2008)
10. Pegrum, M.: Mobile Learning: Languages, Literacies and Culture. Palgrave Macmillan, London (2014)
11. Pedro, L.F.M.G., Barbosa, C.M.M.O., & Santos, C.M.N.: A critical review of mobile learning integration in formal educational contexts. Int. J. Educ. Technol. High. Educ. **15**(1) (2018). https://doi.org/10.1186/s41239-018-0091-4
12. MTech Mobile Learning Survey for Teacher Educators. https://hull.onlinesurveys.ac.uk/mtech-mobile-learning-survey (2017)

School Without Walls, Expanding School Curricula Outside the School Walls with an Innovative Education Tool

Carole Salis(✉), Marie Florence Wilson, Stefano Leone Monni,
Franco Atzori, Giuliana Brunetti, and Fabrizio Murgia

CRS4, Loc. Piscina Manna Ed. 1, 09010 Pula, CA, Italy
{calis,marieflorence.wilson,stefano.monni,fatzori,
brunetti,fmurgia}@crs4.it

Abstract. School Without Walls - SWW is a tool conceived to address some of the challenges presented by today's Education in a digital world. Its educational features should promote students active role in their own learning experience, and improve their school engagement. Through the use of SWW, teachers and educators will create, access, use and reuse free educational material. Learning will no longer be confined to the school walls but will include any spot of natural surroundings or urban environment that has previously been geotagged with an educational content. Learning in and outside school walls is one of the 11 promising Educational Technology Trends for 2017, as students will develop a greater interest for their environment and wider surroundings [1]. Using the tool should help students make the connection between what is being taught at school and its relationship with everyday life, situations, problems. The tool is presently used in extracurricular educational context, but will evolve to be included in the classrooms.

Keywords: Mobile learning · Education · Augmented Reality · Geotagging · Flipped Classroom

1 Introduction

School Without Walls – SWW is a Web/Mobile platform that extends the school learning area to beyond the school walls, connecting it to life around us [2]. The tool is developed by the Educational Technologies group of CRS4, a multidisciplinary research centre, to be used in an Italian Regional Program that addresses school disaffection through extracurricular activities based on the use of technology (Iscol@, Line B2) [3].

Following the principles followed in the Regional project, SWW activities involve three actors: Schools (teachers and students), Institutions (The Regional Authorities, the Regional Research and Technology Development Agency and the multidisciplinary Research Centre); and the Economic Actors (technological start-ups, cultural associations and SMEs) that have the responsibility to transfer technological know-how to participating teachers, so that they will be able to use the technologies involved in the activities to innovate their teaching methods.

© Springer Nature Switzerland AG 2020
M. E. Auer and T. Tsiatsos (Eds.): ICL 2018, AISC 916, pp. 958–965, 2020.
https://doi.org/10.1007/978-3-030-11932-4_88

In this paper, we shall present the philosophy behind SWW platform, born to help students understand that school subjects are linked to real life. Therefore, the tool is conceived to promote ubiquitous learning and encourage teachers to use technology in their lessons. In point 2 of this paper, we shall describe the approach followed and explain how SWW, through geotagged contents and the possibility to share and reuse scenarios, can contribute to complement the school educational contents, in line with the Open Educational Resource OER movement [4]. Points 3 and 4 will give a description of the tool architecture and the workflow, followed by the presentation of future developments and conclusions.

2 Approach

The choice of ubiquitous learning was made because, as mentioned by Virtanen et al. [5], ubiquitous learning environments announce a new era in higher education, in which learning experiences become available any time, and anywhere. In the same paper, the criteria for ubiquitous learning environments are defined and include: context-awareness, interactivity, personalization and flexibility; supported by learning management systems, functional objects, wireless networks and mobile devices. Although the SWW platform does not yet have a learning management system, we are heading in this direction.

SWW based laboratories use mainly two technologies: Augmented Reality (AR), to create a dynamic learning environment and Geographic Information Systems technology to highlight the existing connections between the topics covered by school curriculum and what surrounds us.

The educational potential of Augmented Reality is gaining educators' attention in formal education, as shown in the following two review studies.

In their paper [6] Phil Diegmann et al. identify the following positive effects of using Augmented Reality in education: a positive impact on the student's state of mind (motivation, attention, concentration, satisfaction); on the teaching concepts (increased student-centered learning, improved collaborative learning,); on the presentation of educational content (increased details, information accessibility, interactivity); on the students learning (improved learning curve and creativity); on the content understanding (development of spatial abilities, improved memory). They also identify as a positive aspect of AR the Reduction of costs.

Saltan et al. [7], in their scoping Review of the use of Augmented Reality identify situated learning, inquiry-based learning, collaborative learning, game-based learning as being the most common pedagogical approaches integrated to the use of AR in educational contexts. They also identify knowledge comprehension and acquisition, concept development and knowledge retention as the most important affordances of AR applications.

Most recent smartphones have a GPS chip that uses satellite data to calculate the owner's position. This opportunity is used in SWW to georeference scenarios to achieve educational content customization and delivery. Transposing one specific scenario in a different geographical context, should help students improve the awareness of the connections existing between a topic of their school subjects and the world around them. For example, a teacher living in Paris might upload a physics scenario, on gravity, setting an experiment with objects falling from the Eiffel tower, and an Italian teacher might transpose the terms of the problem to the Pisa tower reality.

Two figures shall enter information on the platform: teachers and students. Teachers create and upload their own educational scenarios, supervise and validate that of students. Scenarios consists in data, quizzes and possible answers. They are organized and can be selected by school subject, by school grade, etc. (See Fig. 1). Once the scenarios are geotagged, they become Points of Interest (POIs) that can be organized by teachers into a map, i.e. a set of scenarios belonging to the same learning unit.

Fig. 1. SWW Home page showing georeferenced scenarios selected per school grade

In line with the Open Educational Resource OER movement, an important aspect of the project is that all scenarios are meant to be shared and reused by registered users who will be able to adapt existing contents to different geographical sites by simply setting the new coordinates, edit text and data to match the new location.

The expression "Open Educational Resource" was used for the first time by UNESCO in the Forum on the Impact of Open Courseware for Higher Education in Developing countries in 2002 [8]. Their definition of an Open Educational resource is:

"The open provision of educational resources, enabled by information and communication technologies, for consultation, use and adaptation by a community of users for non-commercial purposes".

Other definitions can be that of the Commonwealth of Learning "materials offered freely and openly to use and adapt for teaching, learning, development and research"; or that stated in the Report to The William and Flora Hewlett Foundation, dated 2017,"A Review of the Open Educational Resources (OER) Movement: Achievements, Challenges, and New Opportunities" [9] where OER are defined as "teaching, learning, and research resources that reside in the public domain or have been released under an intellectual property license that permits their free use or re-purposing by others.

This sharing philosophy provides support for learner-centred, informal learning approaches, with an expected impact on students interest in the subjects taught, on their engagement as a consequence of experimenting and gaining familiarity with new ways of learning, and last an increased level of self-confidence and self-reliance.

Teachers and educators will be able to compare one's teaching strategy, methods and ideas with that of other teachers, enrich their educational material, use a student centred approach to teaching and make their didactics more appealing through the use of technology.

This collaborative strategy combined with the use of ICT technologies should contribute to bridge the gap between students and educational contents, foster school engagement and promote the reuse of learning contents.

3 Description of the Platform Architecture

The platform architecture (see Fig. 2) consists in a client-server architecture. The client side is presently a Layar mobile application [10, 11], whereas the server side is a Web platform developed at CRS4. For back-end the Python language and the Tornado framework based on MVC model were used. In view of future developments of our tools, especially learning analytics capabilities, to store the students' answers we used Neo4J, a Graph-oriented database [12]. This technical solution should help us better investigate learning/teaching patterns, as meaningful patterns can easily emerge from the nodes, edges and properties used. On the client side it is possible to use either the Layar mobile application (available for iOS and Android devices) or, alternatively the ad hoc SWW mobile application which is being developed, (still in beta version for Android devices only) that offers more features and a better interaction with the server side application.

4 How Does SWW Work Now

The SWW platform allows Teacher to create educational material in the form of scenarios that will be published on an Augmented Reality application LAYAR, that can be downloaded from the App Store (for devices running iOS) and from the Play Store (for Android devices). After registering to SWW, the Teacher will be able to create his

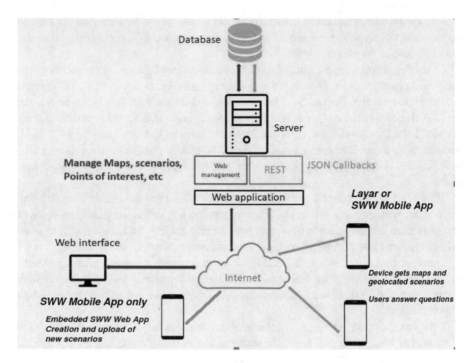

Fig. 2. SWW architecture

own profile from which to upload his original scenarios. He will also be able to visualise and share those created by other registered users. Presently, all uploaded scenarios will constitute a library of scenarios to be shared, modified and reused. Each scenario presents a georeferenced multiple choice question, i.e. a question that is associated through geographical coordinates with a physical space. The Teacher will be able to publish his scenarios on the application, make them visible to his students (through their mobile devices) and visualise their answers. Moreover, he will have the possibility to accept and publish scenarios created by his students.

Students role is twofold: they can be either scenario creators or end-users. As scenario creators, they will have the possibility to suggest scenarios that, if accepted by teachers, will be published on the platform. To that purpose, Teachers will check that students' scenarios respect the following criteria: quality of the query, the fact that it is georeferenced and in the form of multiple choice question and its link to the school program. As end-users, out of school, students will scan their neighbourhood with their smart device to look for the overlaid virtual elements that identify an existing Point of Interest and related educational scenario (see Fig. 3).

By selecting the overlaid elements (text, images, etc.), end-users will access a Point Of Interest (POI) and its related multimedia, queries and possible answers.

Scenarios not only draw attention to the existing links between real life and school programs, they will also encourage thinking and reasoning on the meaning and the

Fig. 3. SWW mobile App in action - Screenshot - In a range of 0.84 km, in the direction in which the mobile device is aiming at 2 Points of Interest were found. One 392 m away from where the student is, deals with "punch cards". The other, at a distance of 37 m deals with "Polaris Research Center". Clicking on the overlaid images students will access the questions

impact of objects and phenomena that surround us, thus favouring an active learning process. Students will go further on the concepts and the experiences they go through at school in line with the Flipped Classroom philosophy.

5   Present and Future Developments

We are about to release the ad-hoc SWW mobile version, for Android Operating System and to start developing the IOS version. Our App is to be used as an alternative to the Layar mobile augmented reality application. There are several reasons for this decision. Layar, from our point of view, presents some drawbacks: it is not open source and it is possible that the application will cease to be supported. Our mobile too will guarantee a long-term stability of the layer access system. Furthermore, being general purpose, the Layar application does not allow an easy interaction with our web application nor an easy upload of new scenarios from mobile devices. The App we are developing is open source and compatible with Android tablets and/or smartphones having Wifi connection, gyroscope and accelerometer. Geotagged educational material can easily be uploaded on the web platform directly from a mobile device. The code will be released under GPL license (version 3 or more recent).

SWW presents interesting educational potential that we plan to extend. For example:

- We are planning to experiment the platform and its content both in Italy and in other countries.
- We are working on making it at least tri-lingual (Italian, French and English). With no limitation to the sole Italian language, the tool will enable students and teachers from different countries to compare the educational paths, share their experiences and scenarios.
- We also intend to develop Moodle plug-in to facilitate interactions between SWW platform and that of Learning Management System (LMS) Moodle.
- So far SWW was developed for playful use in extracurricular activities, but we plan to expand it to become a tool able to help teachers to correlate students answers with learning data (Learning Analytics capabilities).

6 Conclusions

SWW in its complete version will prove to be a tool that can contribute to addressing the problems of Education in a digital world, but not only. One of the problems of the Italian school system is the ageing teaching staff [13] and the common use of very traditional teaching approaches in most of the teaching phases that are far from being attractive to students. By using our tool, on one hand, teachers will be able to experience innovative approach in their didactics and make it more attractive to their students; on the other, the use of our tool should promote students active role in their own learning experience and improve their school engagement.

SWW may also be used by teachers as an educational assessment tool. Indeed, the way students create new scenarios gives an indication of their understanding of the subject they write about. The geotagging of an object, a place or a specific phenomenon are constructive answers to a learning process that teachers can evaluate. The way students elaborate a scenario, their favourite ways to elaborate and conceive questions and possible answers are all important clues to be taken into consideration during the learning assessment. The way students transfer newly acquired concepts into educational scenarios gives an indication of how well they were understood and assimilated. Teachers are invited to pay attention to the students' active role when evaluating them.

Acknowledgements. The authors gratefully acknowledge the "Servizio Istruzione of Direzione Generale della Pubblica Istruzione of Assessorato della Pubblica Istruzione, Beni Culturali, Informazione, Spettacolo e Sport of RAS" and "Sardegna Ricerche".

References

1. Kelly, R.: 11 Ed Tech Trends to Watch in 2017, Virtual Roundtable, CampusTechnology. https://campustechnology.com/articles/2017/01/18/11-ed-tech-trends-to-watch-in-2017.aspx
2. Salis, C., Wilson, M.F., Atzori, F., Monni, S.L., Murgia, F., Brunetti,G.: School without walls, an open environment for the achievement of innovative learning loop. In: REV2018 Proceedings of Lecture Notes in Networks and Systems Proceedings, Springer

3. Salis, C., Wilson, M.F.H., Murgia, F., Monni, S.L., Atzori F.: Public investment on education in sardinia the "Tutti a Iscol@"; project. Introducing Innovative Technology in Didactics. In: Proceedings of 19th International Conference on Interactive Collaborative Learning (ICL-2016), vol. 1, pp. 558–565, September 2016
4. UNESCO: Forum on the Impact of Open Courseware for Higher Education in Developing countries in 2002, UNESCO doc. (CI-2002/CONF.803/CLD.1, Paris 1–3 July 2002. Final Report, UNESCO (2002)
5. Virtanen, M.A., Haavisto, E., Liikanen, E., et al.: Educ. Inf. Technol. **23**, p. 985 (2018). https://doi.org/10.1007/s10639-017-9646-6
6. Diegmann, P., Schmidt-Kraepelin, M., van den Eynden, S.: Dirk Basten benefits of augmented reality in educational environments. In: Wirtschaftsinformatik Proceedings 2015, vol. 103 (2015). http://aisel.aisnet.org/wi2015/103
7. Saltan, Faith, Arslan, Omer: The use of augmented reality in formal education: a scoping review. Eurasia J. Math. Sci. Technol. Educ. **13**(2), 503–520 (2017)
8. UNESCO: Forum on the Impact of Open Courseware for Higher Education in Developing countries in 2002, UNESCO doc. (CI-2002/CONF.803/CLD.1, Paris 1–3 July 2002. Final Report, UNESCO
9. Atkins, D., Brown, J., Hammond, A.: A Review of the Open Educational Resources (OER) Movement: Achievements, Challenges, and New Opportunities. (2007). https://www.hewlett.org/wp-content/uploads/2016/08/ReviewoftheOERMovement.pdf
10. https://www.layar.com
11. Layar, from Wikipedia. (https://en.wikipedia.org/wiki/Layar)
12. Graph oriented DB. https://neo4j.com/
13. Education and Training Monitor 2016–Italy Annual Report Released by European Union. (2016). https://ec.europa.eu/education/sites/education/files/monitor2016-it_en.pdf

Academic Integrity Matters: Successful Learning with Mobile Technology

Alice Schmidt Hanbidge[1][✉], Tony Tin[1][✉],
and Herbert H. Tsang[2][✉]

[1] Renison University College, University of Waterloo, Waterloo, ON, Canada
{ashanbidge, tony.tin}@uwaterloo.ca
[2] Trinity Western University, Langley, BC, Canada
herbert.tsang@twu.ca

Abstract. Educational institutions often struggle to identify what is the best pedagogical approach to engaging students with academic integrity content. As student plagiarism and cheating frequently occur, new strategies are needed to address this challenge in educational settings. The Canadian academic integrity mobile technology project developed one digital strategy, Integrity Matters, to enhance student academic integrity knowledge and understanding using an innovative pedagogical approach. Best strategies, from 774 undergraduate student users perspective, for accessing, delivering, assessing and learning this information with mobile technology (m-learning) are explored. Following completion of the six lessons and subsequent quizzes, academic integrity knowledge increased for the research study participants in engineering, math, computer science, and arts faculties. The Integrity Matters open access trilingual mobile application (available in Android and iOS) can be readily adopted across post-secondary colleges and universities and adapted to meet institutional priority needs.

Keywords: Academic integrity · Academic misconduct · Core values · Mobile learning · Digital badge

1 Academic Integrity Dilemma

Student cheating and plagiarism has plagued higher education and has continued with dogged determination. It is time for educators to determine more effective ways to deter student dishonesty! Significant efforts undertaken by institutions encourage students to participate in learning about academic integrity [5]. Typically, universities rely on teachers to cover this information; however, variation exists in the consistency and quality in the accessibility and delivery of this material. Discussions at orientation, instructor lectures in the first class, and notices in the course syllabus have been used to support first year students adjustment to university, but these have not been very effective at increasing student's level of academic integrity. Academic integrity material posted in schools tends to be difficult to locate or ignored by students. Globally, academic misconduct is on the rise despite well-intentioned efforts to effectively address cheating [7].

© Springer Nature Switzerland AG 2020
M. E. Auer and T. Tsiatsos (Eds.): ICL 2018, AISC 916, pp. 966–977, 2020.
https://doi.org/10.1007/978-3-030-11932-4_89

Academic institutions lose their credibility when students cheat on tests or plagiarize their work and academic integrity dissipates. Unless the importance of academic values is learned, students could graduate without proper training. Hence, it is imperative that students understand academic rules and demonstrate them in their scholarship. Academic Integrity (AI) "refers to a set of conventions that scholars follow in their work, and which generates credibility, trust, and respect within the academic community" [1].

This dilemma formed the basis for our Canadian team to develop a mobile technology solution that benefits and supports students for any time anywhere AI learning. Given that technology is rapidly changing the way that information is delivered and processed, the transfer of knowledge about academic integrity needs to keep pace with the best and most innovative ways to educate students. Digitizing AI provides many opportunities for access to material beyond what is possible with physical artifacts. Digitization makes it possible to enable global access to everyone in education who might have an interest in and could benefit from being able to view and study those materials. Although digitizing materials enables the potential for global distribution, legal or practical barriers may stand in the way. By making our mobile technology solution open access with a Creative Commons license; lessons, quizzes, modules and videos are freely accessible. Best strategies are explored, from a student user perspective, from 700 + university students who provided feedback about accessing and learning information with this open access mobile learning tool.

2 Academic Integrity in Education

According to the International Center for Academic Integrity (ICAI), academic integrity is defined "as a commitment, even in the face of adversity, to six fundamental values: honesty, trust, fairness, respect, responsibility, and courage ("Fundamental Values Project 2014). Over the past 20 years in North America, institutional approaches to academic integrity have shifted from being punitive or rules-focused to being educative (e.g. teachable moments) and values-based [6, 8, 9], as can be observed in the six fundamental values outlined above. Bertram Gallant (2011) emphasizes that "schools should aim to infuse the value of integrity into structures, processes and cultures of the organization" (p. 13). Therefore, to embrace the concept of academic integrity, students need to have scholarship and integrity role modeled and nurtured within an educational institution [3, 4, 10, 15].

Traditionally, scholarly rules to maintain academic integrity were passed onto students directly from their instructors. While a focused and interactive discussion about AI remains the most effective way to encourage students to embrace academic integrity, this practice not only varies between instructors, but there is vast inconsistency in both the AI content shared and the depth of the discussion. MacLeod, studying faculty attitudes on students academic integrity at 17 Canadian universities, concluded that [every university] mentions the importance of academic integrity and affirms that they expect students to act ethically. Regrettably, there are often no follow-up provisions for teaching students to do so (2014, p. 11). It is therefore imperative that students receive this foundational knowledge in a consistent manner that will augment any other instruction (or lack thereof) they have received about AI.

Boehm, Justice and Weeks [5] provide numerous best practices to adhere to when inspiring academic integrity in higher education. After surveying instructors in three higher education institutions in the United States to determine what were considered the best initiatives to promote academic integrity, it was determined that providing training for instructors, adhering to class-room management strategies to reduce the chance of cheating, and providing clear examples of what academic integrity constitutes, were all perceived by the instructors to be the most beneficial means to support academic integrity initiatives. Aside from the last strategy, however, these are all firmly situated as instructor-led initiatives, and the student is still considered a passive recipient of knowledge pertaining to academic integrity, rather than taking ownership over the understanding and application of this pertinent information. To inspire this, the ultimate goal of educating students about the fundamentals of academic integrity should be to recognize the importance of these values, and how they transcend their academic life and may apply to their personal lives and future careers (cf. Pfeiffer and Goodstein 1983).

Furthermore, East states that the challenge is not only to inform students about academic integrity, but also to engage students in this education and to provide them with opportunities to develop their scholarship capabilities (2016, p. 482). East and Scanlan [14], suggests that any AI digital module should include the following key elements: be engaging (e.g., use a progression of challenges and decision-making activities); use multi-languages to facilitate foreign language speakers, utilize multi-media rather than text to convey meaning; provide learners with immediate feedback; and incorporate interactive games to immerse students in the content.

Educational institutions need to go beyond detection, deterrence, and punishment, and take an innovative approach to promote a culture of academic integrity. Boehm, Justice and Weeks [5] and Weimer suggested different ways to promote and support academic integrity initiatives in higher education including best practices and research on academic integrity, training for instructors and learners, and class-room examples of what academic integrity constitutes, restructuring the course and assignments to discourage academic dishonesty Aside from the last strategy, however, these are all firmly situated as instructor-led initiatives, and the student is still considered a passive recipient of knowledge pertaining to academic integrity, rather than taking ownership over the understanding and application of this pertinent information. To inspire this, the ultimate goal of educating students about the fundamentals of academic integrity should be to recognize the importance of these values, and how they transcend their academic life and may apply to their personal lives and future careers (cf. Pfeiffer and Goodstein 1983).

Mobile learning is one effective mean to empowering students to better learn and understand the values of academic integrity, but as of now, there is currently little research on mobile AI training [13] and only one mobile application (uomfair.info) that offers AI content and this information is not contextualized to the Canadian higher education landscape.

2.1   Mobile Learning (M-Learning)

Mobile devices such as phones and tablets have become ubiquitous in our life. Mobile learning (m-learning) allows access to knowledge from anywhere at any time [2]. M-Learning technologies are an ideal complement to online learning while providing direct communication beyond traditional places of learning. As this is an emerging field, its full potential is untapped and best-practice guidelines for m-learning still require formulation [11]. Few academic institutions explicitly support the use of mobile learning to address specific learning outcomes. Indeed, as Herrington and Herrington state, few universities have adopted widespread [mobile] learning technologies, and in those that have, it is not clear that they are being used in pedagogically appropriate ways [12] (2007, p. 3). The Horizon Report: 2016 Higher Education Edition describes the pervasiveness of mobile devices and suggest integrating mobile learning applications on students personal devices as one of the drivers of innovation and change to be used by higher educational institutions (p. 1). As East states, all students take time and practice to become versed in academic codes and to understand academic culture (2016, p. 485), and this academic integrity application provides students with the opportunity to practice and learn from their mistakes in a simulated digital environment. Mobile technologies could be an effective mean to facilitate anytime, anywhere, academic integrity training for students.

More research is needed to discover the best strategies for maximizing m-learning in order to contribute to effective adult learning [11]. As advocates of mobile learning, our intention is to enhance learning experiences rather than to replace educational interactions. Prior to students arriving on our university campus, we were previously unable to offer AI modules in an online format or as a mobile application due to the limitations of our learning management system. Having a mobile application in place allows us to provide this information to students well before they arrive on campus, hence giving them an acculturation to the values at University of Waterloo, Ontario, Canada (UW) and how they can succeed in their academics before they begin classes. The *IntegrityMatters* app tracks which students have completed the module and has the capacity to immediately recognize their efforts by issuing a certificate and badge that signifies completion. We envisioned a mobile learning experience that embraces the affordances offered by mobile learning, while ensuring that the *IntegrityMatters* platform we designed met the needs of the emergent trends in higher education.

3   The Innovative Solution: UWAI

Based on an innovative mobile learning strategy, the main objective of the University of Waterloo Academic Integrity (UWAI) project was to build an application to enhance academic integrity at the University of Waterloo. UW caters to science, technology, engineering and math (STEM) disciplines while hosting a large population of international students.

Fig. 1. Home screen of the IntegrityMatters app.

With funding and a research grant from eCampusOntario[1] under the UWAI initiative, we created the "IntegrityMatters" application (available for both iOS and Android operating systems). To access the Android version of *IntegrityMatters* app in the Google Play Store, click on https://play.google.com/store/apps/details?id=uwai.dev.uwai&hl=en), while the iOS app version at the iTunes Apple Store can be accessed through this link https://itunes.apple.com/us/app/integritymatters/id1355112345?mt=8. The UWAI project was structured as an inter-university research project. In this project, UW collaborated with Trinity Western University researchers and created a system that presented mobile learning modules to undergraduate students. The system includes an evaluation module for students to gauge their understanding and learning about academic integrity concepts. Evaluation of this application with over 700 students was assessed in terms of both the efficacy and the effectiveness.

3.1 System Architecture

IntegrityMatters is a tri-lingual (English, French, Chinese) application designed primarily for higher education institutions. It encourages integrity through value-based active learning to inspire positive change that supports student academic success. Figure 1 shows the home screen of the *IntegrityMatters* app.

Figure 2 shows the system architecture. This is a modular design, which allows separate components to be upgraded over time. Users can access different mobile devices (e.g. smart phone or tablet) to download and install the app. After passing all the quizzes in each module, users are able to obtain a certificate of completion via email (see Fig. 3).

---

[1] https://www.ecampusontario.ca/research_funding/foundations-academic-success-innovative-mobile-learning-enhance-academic-integrity/.

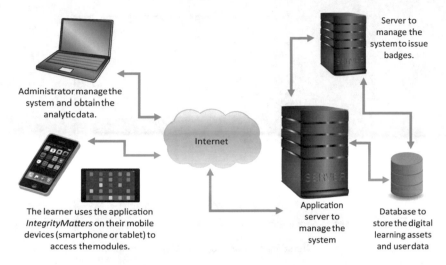

Fig. 2. The overall system architecture of the *IntegrityMatters* app.

Fig. 3. Example of the e-Certificate of Completion.

Digital Badge - In addition to obtaining the certificate of completion, users can demonstrate their achievement via various social media platforms using the digital badge (see Fig. 4).

A digital badge (e-badge) is awarded for successful completion of all the *IntegrityMatters* six modules at a passing grade of 75% or higher (see Fig. 4). Learners who successfully complete all modules are awarded a badge, thus validating their competency in this functional area of academic integrity knowledge. Additionally, learners may choose to make their badges publicly visible. When a badge is earned, it is claimed at the CanCred Passport site (https://passport.cancred.ca/). Badges are an alternative form of credentialing popular in informal learning contexts and somewhat

Fig. 4. Digital badge sample issued to a participant.

similar to "achievements" in video and computer games. Some academic institutions have begun to deploy badges within courses/programs (www.cancred.ca). Demonstration of specified learning outcomes is represented visually by a "badge": a digital image displayed on a website accompanied by written information detailing the accomplishment and criteria for earning the badge.

Lessons - Six academic integrity (AI) lessons educate students about the values of honesty, trust, respect, responsibility, fairness, and courage, which form the basis of academic integrity (ICAI 1999; https://academicintegrity.org/fundamental-values/). Each lesson is comprised of multiple components: a succinct definition of the core value related to AI and how it can be demonstrated and real-life scenarios with animated videos that encourage learners to apply the value to a problem and determine how they would react. The scenarios presented are rooted in typical contexts, such as collaboration on classroom projects, exams, and co-op placements, which are applicable to what a student at university or college would experience. Each module focuses on scenarios involving diverse aspects of student academic life (cultural difference and expectations, physical stress, peer pressure, time constraints, etc.). Learners are given a short, low-stakes quiz at the end of each module, and are provided with immediate feedback on how accurate their response was, as well as to help understand the complexities of value-based decisions.

Game - In addition, app functions engage users to learn about the six academic integrity values through an interactive skill-testing game that reinforces learning from the lessons. Figure 5 shows the game for users to learn by gaming.

Fig. 5. The "Values" game interface explores six academic integrity core values.

4   Research Findings

4.1   User Testing

This research was structured as a mixed-method, non-experimental research study ($n = 774$) with a comparison group ($n = 30$). Quantitative data collection compared test group (with intervention of AI lessons) to the comparison group (no AI lessons). While over 1,300 students attempted the lessons; due to missing data, 774 participants completed all components of the research study. Statistical analysis of the data collection was completed using the program Stata, SurveyMonkey and the app learning analytics.

4.2   Study Results

Preliminary findings suggest the *IntegrityMatters* module lessons significantly enhanced students AI knowledge. Statistical analysis revealed significant differences among groups regarding academic integrity testing results. Data analysis as seen in Fig. 6 indicates that students improved their test scores by a mean of 0.34; 95% confidence interval [CI] = 0.1753, 0.5047. The Pre-test mean = 7.525; while the Post-test mean = 7.86 and the difference was 4.45%. Figure 6 shows the pre and post outcomes for our study participants.

As shown in Figs. 7 and 8, citizenship and faculty were two major factors that significantly influenced test results. Canadian students and permanent residents outperformed International students in the Pre-test, shown in Fig. 7. However,

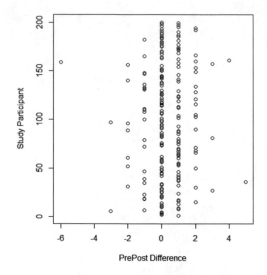

Fig. 6. The Pre and Post Outcomes for Studying repent.

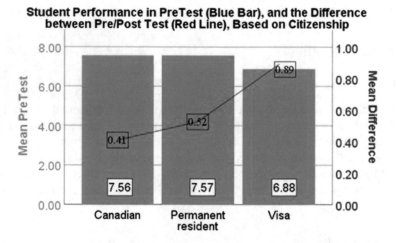

Fig. 7. The citizenship of the users.

international students improved their performance by 0.89 points, which is a 12.9% improvement. As demonstrated in Fig. 8, Students in the Math Faculty performed poorly in the Pre-test, but they improved the most after completing the lessons (0.73). One hundred and eighty-three Engineering students participated in the research study and they improved their academic integrity knowledge on an average of 0.42. Interestingly, there were no significant differences between males and females in the pre and post tests (differences were 0.06 and 0.092, respectively).

Figure 8 shows the participants and their faculties. Undergraduate students from six diverse academic faculties voluntarily participated in the study: Applied Heath

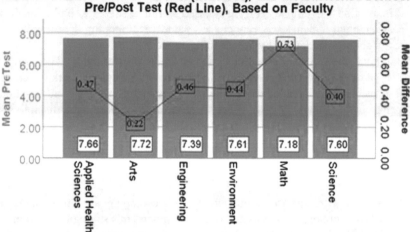

Fig. 8. The faculties of the participants.

Sciences, Arts, Engineering, Environment, Math, and Science. As indicated in Fig. 8, Math students performed poorly in the pre-test but they made the most significant gains at post-test. After Math students, Engineering faculty students most enhanced their academic integrity knowledge following completion of the lessons.

5 Discussion

Mobile technology seems to have increasing ubiquity with significant student engagement to fill a prominent role in the higher education landscape. Mobile technology can support enhancement of educational outcomes and can boost knowledge and understanding of competencies to meet current academic integrity challenges.

Development of the mobile application, *Integrity Matters* based on a foundation of six core values, was found to enhance academic integrity training for undergraduate students and the app aims to enhance the level of scholastic integrity across campuses. Findings from the current study will be used as basic evaluation research to develop best practices in the emerging field of digitization and mobile learning in AI instruction.

Our collaboration with the American Councils for International Education (https://www.americancouncils.org) brings the *IntegrityMatters* application to its broad networks in global educational institutions at all levels. *IntegrityMatters* is an open access application available to freely download. Also, the various assets that were created are available under a Creative Commons Attribution 4.0 International Public License (CC-BY 4.0). This license allows for wide adoption and integration across post-secondary colleges and universities.

The creation of this digital mobile platform and student-focused praxis has created both a curriculum and learning approach that can facilitate academic integrity understanding and knowledge construction in more dynamic ways with online participatory experiences that epitomize a new culture of learning. Both the technical and pedagogical innovation that emphasize values-based learning, strives to be hallmark of the best learning environments we can create, while incorporating learning tools and practices to support students' diverse learning modes.

We did not overlook the limitations inherent in the present study. The conclusions cannot be generalized, as this research was focused on a single university setting and undergraduate student population. While the participants in the study voluntarily participated in the study, they were drawn from a cross-section of disciplines in a university.

6   Conclusion

Institutions of higher education must focus on developing a culture of academic integrity across entire organizations and our app supports this strategic direction. Study findings conclude that completion of the *IntegrityMatters* module lessons and summative quizzes significantly increased students academic integrity knowl edge. By informing students of their institutional academic expectations well before learners set foot on campus and attend their first day of classes, this innovation supports students to be well-prepared for academic success.

Academic integrity content in this app extends beyond its utility to University of Waterloo students as the core values it promotes apply provincially, nationally and internationally. Its contents can be customized, through the Creative Commons license, to align with institutional strategic policies.

Students are more likely to act with integrity when mechanisms are in place to support their work. *IntegrityMatters* application helps to encourage value-based education that translates into lifelong skills. By giving voice to key values, students are armed with reasons to make informed ethical choices.

Acknowledgements. The authors wish to acknowledge research funding support from eCampusOntario. Also, the authors acknowledge support from the University of Waterloo, Renison University College and Trinity Western University.

References

1. Definition of Academic Integrity. http://www.yorku.ca/spark/academic_integrity. Accessed 01 Aug 2018
2. Ally, M.: Designing effective learning objects for distance education. In: McGreal, R. (ed.) Online Education Using Learning Objects, chap. 6, pp. 87–97. Routledge Falmer, London (2004)
3. Batane, T.: Turning to turnitin to fight plagiarism among university students. Educ. Technol. Sci. **12**(2), 1–12 (2010)
4. Batane, T.: Can turnitin be used to provide instant formative feedback? Br. J. Educ. Technol. **42**(4), 701–710 (2011)
5. Boehm, P.J., Justice, M., Weeks, S.: Promoting academic integrity in higher education. Community Coll. Enterp. **15**(1), 45–61 (2009)

6. Cole, S., Kiss, E.: What can we do about student cheating. About Campus **5**(2), 5–12 (2000)
7. Dee, T.S., Jacob, B.A.: Rational ignorance in education. A field experiment in student plagiarism. J. Hum. Resour. **47**(2), 397–434 (2012)
8. Gallant, T.B.: Moral panic: the contemporary context of academic integrity. ASHE High. Educ. Rep. **33**(5), 1551–6970 (2008)
9. Gallant, T.B.: Building a culture of academic integrity. Technical Report, A Magna Publications White Paper, Madison, WI (2011). https://www.depts.ttu.edu/tlpdc/Resources/Academic_Integrity/files/academicintegrity-magnawhitepaper.pdf
10. Glendinning, I.: Responses to student plagiarism in higher education across Europe. Int. J. Educ. Integr. **10**(1), 4–20 (2014)
11. Hanbidge, A.S., Sanderson, N., Tin, T.: Using mobile technology to enhance undergraduate student information literacy skills: a Canadian case study. IAFOR J. Educ. Technol. Educ. Spec. Ed. **3**(2), 108–118 (2015)
12. Herrington, A.J., Herrington, J.A.: Authentic mobile learning in higher education. In: Proceedings of the Australian Association for Research in Education (AARE) International Educational Research Conference, pp. 1–9 (2008)
13. Macfarlane, B., Zhang, J., Pun, A.: Academic integrity: a review of the literature. Stud. High. Educ. **39**(2), 339–358 (2014)
14. Scanlan, C.L.: Strategies to promote a climate of academic integrity and minimize student cheating and plagiarism. J. Allied Health **35**(3), 179–185 (2006)
15. Stappenbelt, B., Rowles, C.: The effectiveness of plagiarism detection software as a learning tool in academic writing education. In: The 4th Asia Pacific Conference on Educational Integrity (4APCEI (01 2010))

Author Index

© Springer Nature Switzerland AG 2020
M. E. Auer and T. Tsiatsos (Eds.): ICL 2018, AISC 916, pp. 979–982, 2020.
https://doi.org/10.1007/978-3-030-11932-4